ORGANIC CHEMISTRY

Volume I

ORGANIC CHEMISTRY

(As per UGC Syllabus)

Volume I

SECOND REVISED EDITION

S M Mukherji
Ex-Professor Emeritus
Department of Chemistry
MD University
Rohtak, Haryana

S P Singh
Professor Emeritus
Department of Chemistry
Kurukshetra University
Kurukshetra, Haryana

R P Kapoor
Professor Emeritus and Coordinator
Guru Gobind Singh College
of Pharmacy, Yamuna Nagar, Haryana
(Former Chairman, Deptt. of Chemistry
Kurukshetra University)

R Dass
Associate Professor and Head
Post Graduate Department of
Industrial Chemistry
Guru Nanak Khalsa College
Yamuna Nagar, Haryana

Committed to Educate the Nation

Publishing Globally
IN INDIA

NEW AGE INTERNATIONAL (P) LIMITED, PUBLISHERS
LONDON • NEW DELHI
Bangalore • Chennai • Cochin • Guwahati • Hyderabad • Kolkata • Mumbai
Visit us at **www.newagepublishers.com**

Published by New Age International (P) Ltd., Publishers
First Edition: 1985
Second Revised Edition: 2010
Reprint: 2021

GLOBAL OFFICES

- **New Delhi** **NEW AGE INTERNATIONAL (P) LIMITED, PUBLISHERS**
  7/30 A, Daryaganj, New Delhi-110002, INDIA
  Tel.: (011) 23253771, 23253472
  E-mail: contactus@newagepublishers.com • Visit us at www.newagepublishers.com
- **London** **NEW AGE INTERNATIONAL (UK) LTD.**
  27 Old Gloucester Street, London, WC1N 3AX, UK
  E-mail: info@newacademicscience.co.uk • Visit us at www.newacademicscience.co.uk

BRANCHES

- **Bangalore** 37/10, 8th Cross (Near Hanuman Temple), Azad Nagar, Chamarajpet, Bangalore- 560 018
  Tel.: (080) 26756823, **E-mail: bangalore@newagepublishers.com**
- **Chennai** 26, Damodaran Street, T. Nagar, Chennai-600 017, **Tel.:** (044) 24353401
  E-mail: chennai@newagepublishers.com
- **Cochin** CC-39/1016, Carrier Station Road, Ernakulam South, Cochin-682 016, **Tel.:** (0484) 4051304
  E-mail: cochin@newagepublishers.com
- **Guwahati** Hemsen Complex, Mohd. Shah Road, Paltan Bazar, Near Starline Hotel, Guwahati-781 008,
  Tel.: (0361) 2513881, **E-mail: guwahati@newagepublishers.com**
- **Hyderabad** 105, 1st Floor, Madhiray Kaveri Tower, 3-2-19, Azam Jahi Road, Near Kumar Theater, Nimboliadda Kachiguda, Hyderabad-500 027, **Tel.:** (040) 24652456
  E-mail: hyderabad@newagepublishers.com
- **Kolkata** RDB Chambers (Formerly Lotus Cinema) 106A, 1st Floor, S N Banerjee Road, Kolkata-700 014
  Tel.: (033) 22273773, **E-mail: kolkata@newagepublishers.com**
- **Mumbai** 142C, Victor House, Ground Floor, N.M. Joshi Marg, Lower Parel, Mumbai-400 013
  Tel.: (022) 24927869, 24915415, **E-mail: mumbai@newagepublishers.com**
- **New Delhi** 22, Golden House, Daryaganj, New Delhi-110 002, **Tel.:** (011) 23262368, 23262370
  E-mail: sales@newagepublishers.com

ISBN: 978-81-224-2738-7
C-21-03-12659

Printed in India at Glorious Printers, Delhi.
Typeset at Aakriti Graphics, Delhi.

NEW AGE INTERNATIONAL (P) LIMITED, PUBLISHERS
7/30 A, Daryaganj, New Delhi-110002
Visit us at **www.newagepublishers.com**
(CIN: U74899DL1966PTC004618)

Preface to the Second Edition

The authors are thankful to the academic community for its generous response to the first edition of the book and its subsequent reprints. Instead of having another reprint, we now thought it worthwhile to completely revise and update the material. We have rearranged the contents in such a manner so as to be in consonance with the syllabus proposed by the U.G.C. Curriculum Development Committee for Chemistry headed by one of us (SPS).

Thus, the course material of B.Sc. I year is reflected in eleven chapters of the present volume. A few salient features of this volume are: some typical problems along with the solutions are given at the end of each chapter, review of the important points is presented after every chapter, mechanism of all the reactions described in the text are illustrated and finally the subject matter on chemical bonding, stereochemistry and principles of reaction mechanism has been reoriented.

Volumes II and III of the book are also under the process of revision. The contents of these volumes will meet the requirements of the students of B.Sc. II year and III year, respectively. We are confident that presentation of these volumes in such a focussed manner will serve the students in a more effective way.

We look forward for constructive suggestions from the teaching community for further improvements.

AUTHORS

Preface to the First Edition

This book has primarily been written with the aim of meeting the needs and interests of B.Sc. (Pass) and B.Sc. (Hons) and M.Sc. (Part 1) students of Indian Universities. It is broadly divided into three parts: Fundamentals, Core Course and Special Topics. The chapters in each part are further divided into sections and sub-sections to facilitate the organization of the topics. We feel that some major advantages are associated with our arrangement of topics in the present text. New concepts are introduced at an early stage before they are to be applied in the chapters that follow. Thus the subject-matter of the first four chapters (Fundamentals) finds applications in the subsequent chapters.

The chapter sequence in the Core Course (Chapters 5-22) has been planned on the principle that the reactivities of organic functional groups follow certain general rules. The compounds are therefore arranged according to the functional groups without the traditional division into aliphatic and aromatic compounds.

The third part (Special Topics, Chapters 23-34) of the book contains detailed account of the topics of current interest in organic chemistry. As these topics are not required for an understanding of the material in the Core Course, they are placed in the end.

We believe that the most effective way to learn organic chemistry is by a problem-solving approach. This book thus contains problems at the end of each chapter which are designed to develop proficiency that can be acquired only by practice. A total of about 1000 drill questions and problems provide sufficient breadth to be adequately comprehensive. Solutions to selected problems are given at the end of the book.

Some highlights of the book are as follows: The Chapter on *Chemical Bond* includes discussion on localized as well as delocalized covalent bond. It demonstrates how the concept of molecular orbital is extended to explain the structures of carbon compounds and how the concepts of resonance and aromaticity lead to understanding of their properties. A new outlook on resonance energy has been introduced. This Chapter coupled with the second and fourth Chapters–which deal with *Stereoismerism* and *General Principles of Organic Reaction Mechanisms* from the background material for the subsequent chapters listed in the Core Course. An early description of the stereochemistry of organic compounds would greatly help the students in understanding the stereochemical implications of the chemical reactions of different families of compounds. Chapter 4 is designed to help the student appreciate how the reagents carrying a positive charge tend to attach a molecule at the point of high electron density, while reagents carrying a negative charge tend to attack a molecule at a point where the electron density is low. It also demonstrates that transformation of one compound into another depends in part on thermodynamics and in part upon kinetics. Types of organic reagents are also classified to simplify further discussion. The section on the methods of determination of reaction mechanism has been added with a view to familiarizing the student with the various techniques so that he may be able to appreciate these when applied in delineating the mechanistic path-ways of the various reactions in subsequent chapters. We believed that our treatment of Organic Spectroscopy (Chapter 3) is the most accessible and interesting of any that are available in the contemporary text books. The basic principles of the various spectroscopic techniques (including high resolution mass spectrometry) are adequately discussed. These techniques are subsequently used in almost all the chapters of the second and third parts for the characterization of compounds belonging to a particular class.

In the Core Course (Chapters 5-22), we have taken special care to discuss the spectral properties of various classes of compounds. The mechanisms involved in their formation and chemical reactions are critically examined. Special emphasis is given on individual compounds of industrial/biological importance. The organization of this section is somewhat different from what obtains in other books. For instance, all hydrocarbons have been treated in the first five chapters (Chapters 5-9) highlighting the fundamental principles underlying the free radical substitutions, eliminations, electrophilic, nucleophilic and free radical addition to carbon-carbon multiple bond. In Chapter 6, only β-eliminations have been discussed in detail. A cursory description of α-elimination has been given here but this topic is dealt with in more detail in Chapter 19, under diazoalkanes. Attempts have been made in the relevant chapters to include detailed discussion of the mechanisms of: nucleophilic displacement including benzyne mechanism (Chapter 10), nucleophilic addition to carbonyl group (Chapter 15) and esterification and ester hydrolysis (Chapter 18). Besides these, the mechanistic details of a large number of the name reactions have been discussed. Another salient feature of this section is the analysis of the mass spectral fragmentation of compounds having different functional groups.

We have identified twelve different topics ranging from *Synthetic Polymers to Organic Chemistry of Life* to be included in the last part (Chapters 23-34, Special Topics). In addition to the topics such as *Photochemistry* and *Pericylic Reactions*, which are invariably included in most of the undergraduate texts, we have included a chapter on *Petroleum and Petrochemical*. This has been done in the conviction that in this age of energy crisis, student should not only know as to what the 'Crude' is, but he should also have some feeling for the chemical processes associated with its refining and utilization on the by-products. A chapter on *Synthetic Polymers* reflects the growing importance of such materials in our daily life. In recognition of the growing importance of three-and four-membered heterocyclic compounds which normally escape attention in most of the courses, we have discussed their chemistry in reasonable detail (Chapter 29). The arrangement of Chapters on Special Topics is not without significance. Thus, *Polynuclear Hydrocarbons* appear before *Steroids* and *Heterocyclic Compounds* before *Alkaloids. Similarly, Carbohydrates, Amino Acids, Peptides and Proteins* have been discussed before the concluding chapter on the *Organic Chemistry of Life*. The salient features of the last chapter include a brief description of enzymes and co-enzymes along with their role in biological processes. The concept of 'active site' has been developed by discussing the mode of action of a proteolytic enzyme-chymotrypsin. Some of the complex processes occurring in the living systems such as energy-releasing cycle, biosynthesis of fatty acids and mutations are presented in a simple way.

Many scholars have been involved at different stages of the development of this text. We particularly thank Professors M.V. Bhatt (I. I. Sc., Bangalore), Krishna C. Joshi (University of Rajasthan, Jaipur),

G. Mehta (University of Hyderabad), J.R. Mechant (University of Bombay) K. S. Sharma (M. D. University, Rohtak) and P. R. Singh (I. I. T., Kanpur), who read with critical appraisal several portions of the manuscript.

We are grateful to Mr. Ram Gopal Gupta for typing the entire manuscript with great care and to M/s Ashu Applied Arts, Kurushetra for their valuable assistance and insight in the production of the illustrations and diagrams.

Finally, we thank the University Grants Commission, New Delhi for sponsoring this Project and to National Book Trust, India for financial subsidy.

Kurukshetra
May 1985

AUTHORS

Contents

Section I

Section II

Section III

Section IV

Section V

SECTION I

CHAPTER 1

Structure and Bonding

A molecule is formed by the combination of atoms and in the process they attain the electronic configuration of an inert gas. A *chemical bond* is formed as a result of attraction between the combining atoms. To explain the formation of bonds in diatomic molecules, such as H_2, O_2, N_2, Cl_2 etc., it was first suggested by G.N. Lewis that atoms of such molecules attain an inert gas electronic configuration by sharing one or more pairs of electrons, each atom contributing one or more electrons. The atoms are then held together by the shared pair(s) of electrons forming what is called a *covalent bond*. The structures in which the shared electrons are shown by *dots* were first used by Lewis and they are known as Lewis structures, and the system of using *dashes* to represent the shared electrons is due to Couper.

The structure of a compound means how different atoms in a molecule are joined to each other and arranged in the three-dimensional space. The properties of different compounds depend on their structures. Structural concepts of organic compounds can be understood by the nature and strength of bonds present in a molecule.

Now the question arises as to why two or more atoms combine together to form molecules? This is because, as a result of bond formation, the atoms reach a state of lower energy. In other words, a certain amount of energy is released during the bond formation. The same amount of energy called '*bond dissociation energy*', is needed to break the molecule into its component atoms. In this Chapter, we shall discuss hybridization and its types, the parameters of molecular structure and various types of bonds and their characteristics.

1.1 HYBRIDIZATION AND ITS TYPES (Geometry and Shapes of Molecules)

Generally, the number of covalent bonds formed by an atom is equal to the number of unpaired electrons in it. However, the atoms of beryllium, boron and carbon are exceptions to this generalization. The electronic configurations of these atoms are shown in Fig. 1.1.

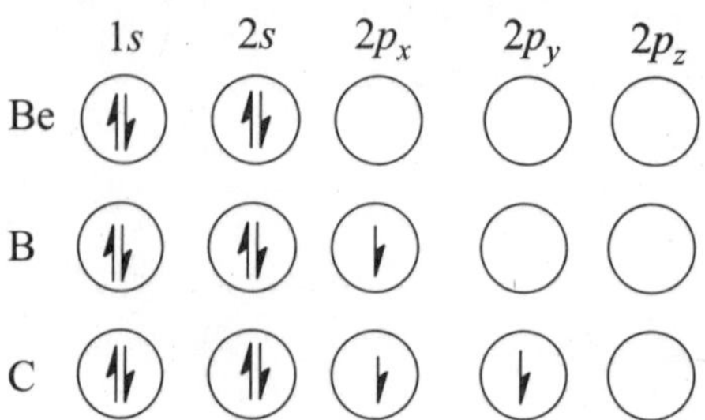

Fig. 1.1: Atomic orbital formulation of beryllium, boron and carbon

An inspection of the electronic configurations of these atoms reveals that beryllium should be an inert element, since it has no unpaired electron. Similarly, a boron atom would form only one bond and a carbon atom would form two bonds with other atoms. Beryllium, on the contrary, forms beryllium fluoride (BeF_2) and boron and carbon form compounds such as boron trifluoride (BF_3) and methane, (CH_4) respectively. Thus, Be, B and C are di-, tri- and tetravalent atoms, respectively. The above electronic formulations are, therefore, to be modified to explain the 'anomalous' behaviour of these atoms. Thus, we can assume that one of the electrons of 2*s* orbital in these atoms is promoted to a vacant *p* orbital before the bond formation takes place as shown in Fig. 1.2. Some energy must be supplied to the system in order to effect this promotion and this energy is more than compensated by the energy released during the covalent bond formation.

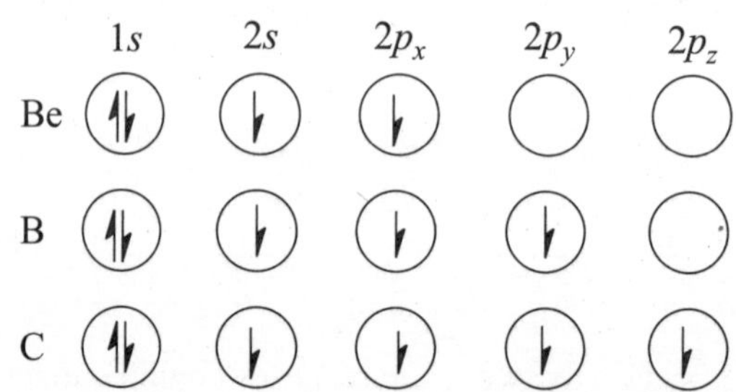

Fig. 1.2: Promotion of an electron from a 2*s* atomic orbital to a 2*p* atomic orbital

The promotion of a 2*s* electron to one of the vacant 2*p* orbitals explains the observed valencies of these elements. But we still have to account for the observed geometry of the compounds formed by these elements. For instance, we know that, the *s* orbital of carbon is spherical in shape whereas the *p* orbitals are dumbbell shaped and directed at right angles to each other. We should, therefore, expect that the three of the four bonds of carbon would be directed at right angles to each other whereas the fourth bond (formed by the overlap of *s* orbital) would have no definite orientation. However, it is well known that the four bonds formed by carbon in a molecule, such as, methane are equivalent and directed towards the four corners of a regular tetrahedron with an H—C—H angle of 109.5°.

In order to solve the problem of observed valency of atoms, and shape and geometry of molecules formed from these atoms, the concept of hybridization or a special type of mixing of different orbitals was evolved. Hybridization may thus be defined as *the mixing of two or more than two atomic orbitals of an atom having comparable energy to give an equal number of identical orbitals having same energy and shape. All hybrid orbitals are oriented symmetrically to have maximum distance from each other.*

1.1.1 Types of Hybridization

Carbon atom undergoes three types of hybridization depending upon the number and type of orbitals mixing together. These are discussed below:

1.1.1.1 sp^3 or Tetrahedral Hybridization

One *s* and three *p* orbitals mix to form a set of four identical orbitals called sp^3 hybrid orbitals which are directed towards the four corners of a regular tetrahedron. All saturated compounds involve sp^3 hybridization where carbon atom is bonded to four atoms, viz., alkanes, cycloalkanes, etc.

Each sp^3 hybrid orbital has 25% *s*-character and 75% *p*-character. A schematic representation of sp^3-hybridization is shown in Figs. 1.3 and 1.4.

On the basis of the concept of hybridization developed above, we shall discuss here geometry and shapes of some common molecules such as methane, ethane, ammonia and water.

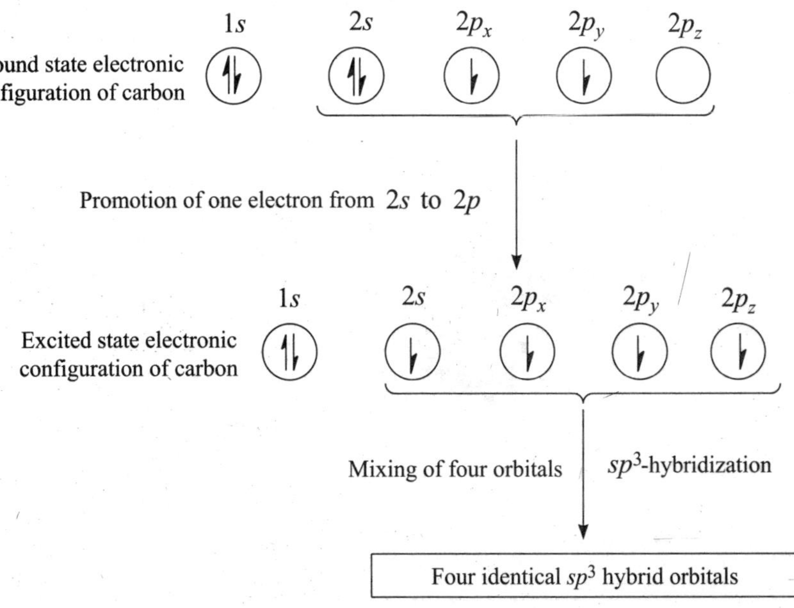

Fig. 1.3: sp^3-hybridization of carbon

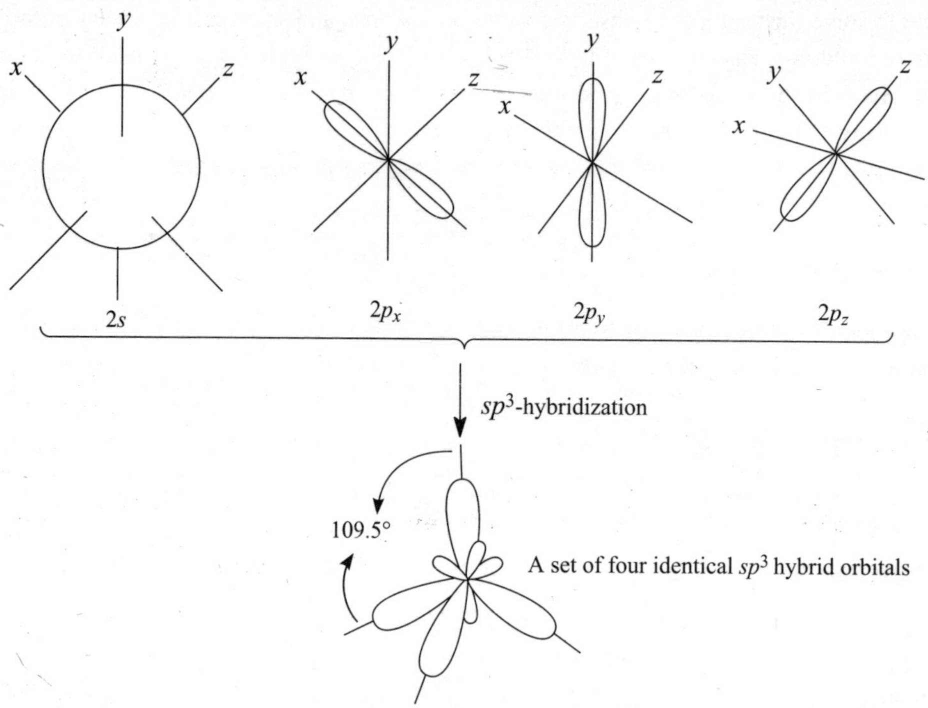

Fig. 1.4: sp^3-hybridization of carbon

(*i*) *Shape of Methane Molecule*, CH_4

As we have seen (Fig. 1.3), one of the electron from a 2*s* orbital is promoted to a vacant 2*p* orbital. The one 2*s* and the three 2*p* orbitals are then mixed (hybridized) to form four identical sp^3 orbitals. The four new orbitals are called sp^3 hybrid orbitals and the process is known as sp^3-hybridization each sp^3 orbital containing one electron.

The sp^3 orbitals are directed towards four corners of a regular tetrahedron with the carbon atom located at the centre. The orbital picture of sp^3-hybridization is depicted in Fig. 1.4. In methane molecule each of the four sp^3 hybrid orbitals then overlap with 1*s* orbitals of four hydrogen atoms to form four sigma C—H bonds (Fig. 1.5).

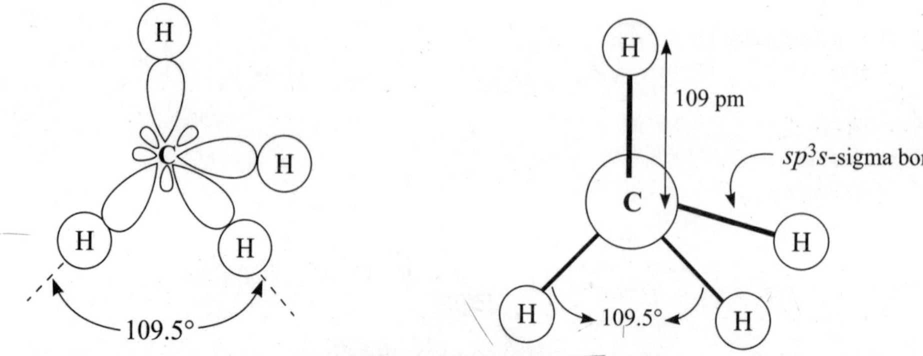

Fig. 1.5: Orbital picture of methane

Fig. 1.6: Ball and stick model of methane

These four bonds in methane are tetrahedrally arranged in space and the H—C—H bond angle is 109.5°, the C—H bond length is 109 pm. A ball and stick model (Fig. 1.6) of the molecule of methane helps to visualize its structure.

(ii) Shape of Ethane Molecule, C_2H_6

Both the carbon atoms in ethane molecule are sp^3-hybridized. One sp^3 hybrid orbital of each carbon atom overlaps with the sp^3-hybrid orbital of the other carbon along its internuclear axis forming sp^3-sp^3, C—C sigma bond. The remaining three sp^3 orbitals on each carbon overlap with 1*s*-orbital of three H-atoms forming six sp^3-*s*, C—H sigma bonds (Figs. 1.7 and 1.8).

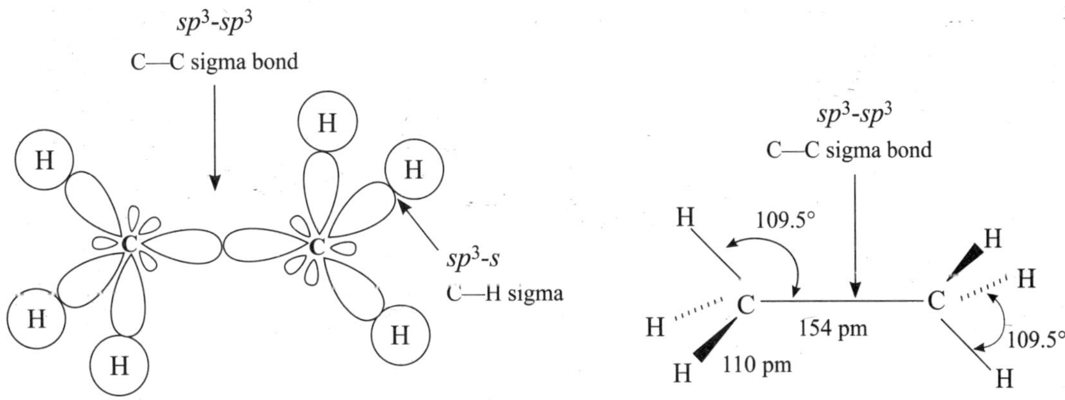

Fig. 1.7: Orbital picture of ethane

Fig. 1.8: Structure of ethane

(iii) Shape of Ammonia, NH_3

The ground state electronic configuration of nitrogen is shown in Fig. 1.9.

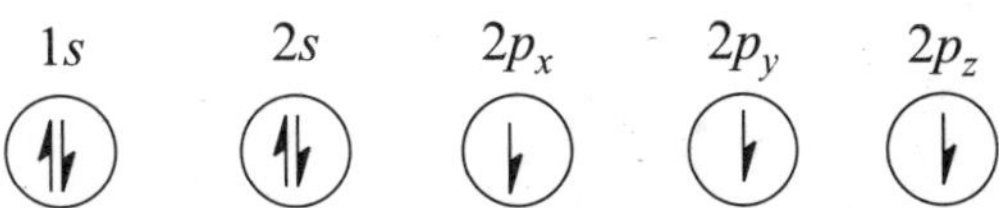

Fig. 1.9: Electronic configuration of nitrogen

In the molecule of ammonia, the nitrogen forms three bonds with hydrogen. Now question arises whether in the formation of a molecule of ammonia, there is an overlapping of three 2*p* orbitals of nitrogen with three 1*s*-orbitals of three hydrogen atoms. If this had been the case, the expected H—N—H bond angles in ammonia would have been 90°. However, the experimental value for this angle has been found to be 107° which is not far from the normal tetrahedral angle of 109.5° (in methane). How to account for this 'anomally'? It can be assumed that there is hybridization of the $2s^2$, $2p_x^1, 2p_y^1$ and $2p_z^1$ atomic orbitals to form four sp^3 hybrid orbitals (Fig. 1.10). Three of these hybrid orbitals form 3σ bonds by overlapping 1*s* orbitals of the three hydrogen atoms. The remaining (fourth) sp^3 hybrid orbital retains a pair of electrons commonly referred to as a *lone pair*.

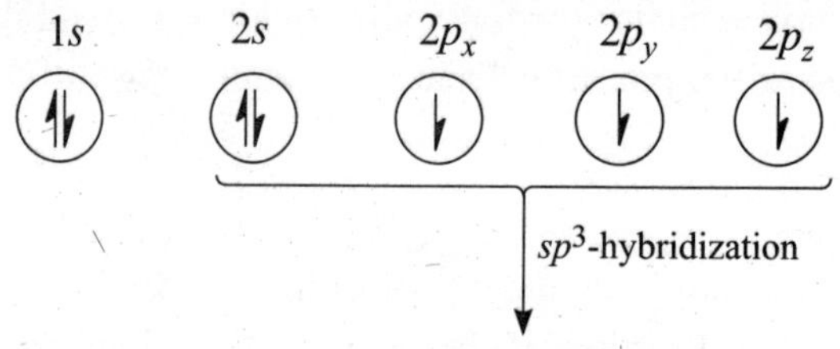

Fig. 1.10: sp^3-hybridization of nitrogen

Thus to achieve maximum overlapping, the three hydrogens should be situated at the three corners of a regular tetrahedron, the 'lone pair' occupying the fourth corner. In fact, ammonia is a pyramidal molecule with nitrogen at the apex and three hydrogens located at the corners of the triangular base. (Fig. 1.11).

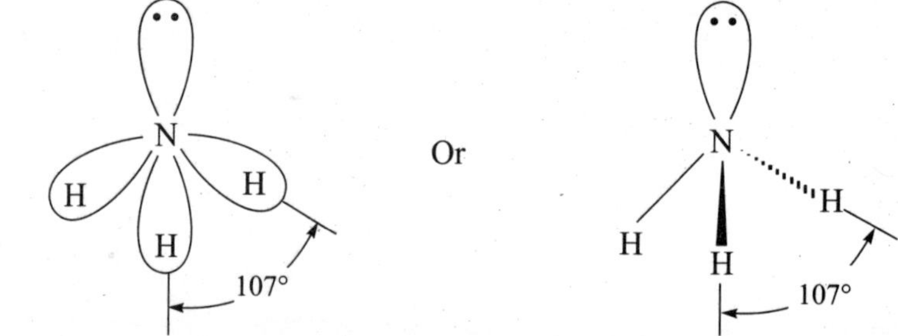

Fig. 1.11: Structure of ammonia molecule

Since, the four pairs of electrons in ammonia are not equivalent (three sp^3 bonding and one sp^3 non-bonding), the bond angles are slightly deviated from the ideal value of 109.5°.

(iv) *Shape of Water,* H_2O

The ground state electronic configuration of oxygen is shown in Fig. 1.12.

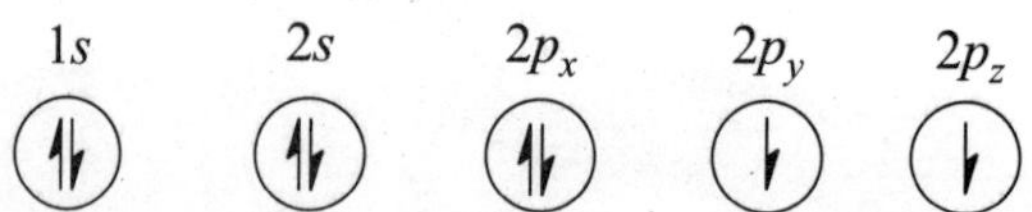

Fig. 1.12: Electronic configuration of oxygen

Because oxygen combines with two atoms of hydrogen, the H—O—H bond angle should have been 90°. But the experimental value for this angle is 104.5°.

This is better explained by assuming sp^3-hybridization of $2s^2$, $2p_x^2$, $2p_y^1$ and $2p_z^1$ orbitals Fig. 1.13.

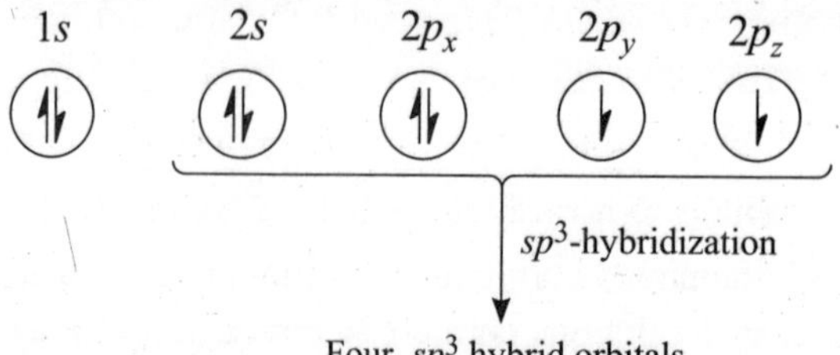

Fig. 1.13: sp^3-hybridization of oxygen

Out of the four sp^3 hybrid orbitals only two form 2σ bonds by overlapping 1*s* orbitals of the two hydrogen atoms. The remaining two sp^3 orbitals have a pair of electrons each (the lone pair) as shown in Fig. 1.14. Although the four sp^3 hybrid orbitals are directed towards the four corners of a tetrahedron, the deviation in H—O—H bond angle is due to non-equivalence of the hybrid orbitals (two of the four sp^3 hybrid orbitals contain a pair of electrons each).

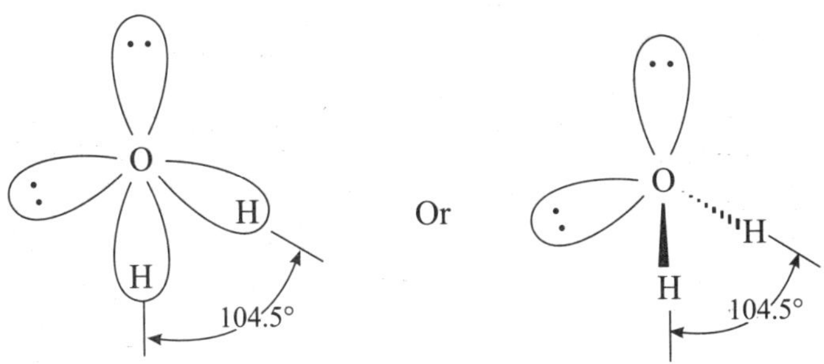

Fig. 1.14: Structure of water molecule

Cause of deviation in bond angles in ammonia and water

Although oxygen and nitrogen atoms in water and ammonia are sp^3-hybridized, their bond angles are not equal to the expected value of 109.5°. In water, the bond angles are 104.5° and in ammonia, they are 107°. These 'anomalies' can be explained on the basis of Valence Shell Electron Pair Repulsion Theory (VSEPRT). According to this theory, for the maximum stability of a molecule, the valence electrons should be at a maximum distance from each other because of mutual repulsion. It may be noted here that so long as the central atom is surrounded by σ bonded electrons, it will have a fixed geometry, *i.e.*, a regular tetrahedral arrangement as in methane. But if the central atom has a non-bonding pair (lone pair) of electrons, the molecule will have a slightly distorted shape. This is due to the fact that the non-bonding-non-bonding electron repulsion is more than the non-bonding-bonding electron repulsion, which, in turn, is more than the bonding-bonding electron repulsion *i.e.*, order of repulsion is: ***Lone pair-lone pair > lone pair-bond pair > bond pair-bond pair***. In a molecule of water, for instance, there are two pairs of non-bonding electrons in two of the sp^3 hybrid orbitals. Since the non-bonding-non-bonding electron repulsion is the greatest, it will result in the shortening of H—O—H bond angle to a greater extent. In the molecule of ammonia, on the other hand, there is only one pair of non-bonding electrons occupying one of the four sp^3 hybrid orbitals. So in this case there is only one non-bonding-bonding repulsion which is less than non-bonding-non-bonding repulsion in water. Hence the H—N—H bond angle in NH_3 would be distorted to a lesser extent as compared to H—O—H bond angle in water molecule. This explains why H—N—H bond angle in ammonia is 107° while the H—O—H bond angle in water is 104.5°.

1.1.1.2 *sp*² or Trigonal Hybridization

One *s* and two *p* orbitals of carbon atom mix to give a set of three sp^2 hybrid orbitals which are pointing at angles of 120° and each of them contains one electron. The lobes of sp^2-hybridized orbitals are directed towards the corners of an equilateral triangle. That is why sp^2-hybridization is called trigonal hybridization. The remaining unhybridized orbital is always perpendicular to the plane containing three sp^2 hybrid orbitals. All compounds with C$=$C bonds have sp^2-hybridized carbon atoms. Each sp^2 hybrid orbital has 33% s-character and 66% p-character.

A schematic representation of sp^2-hybridization is shown in Fig.1.15.

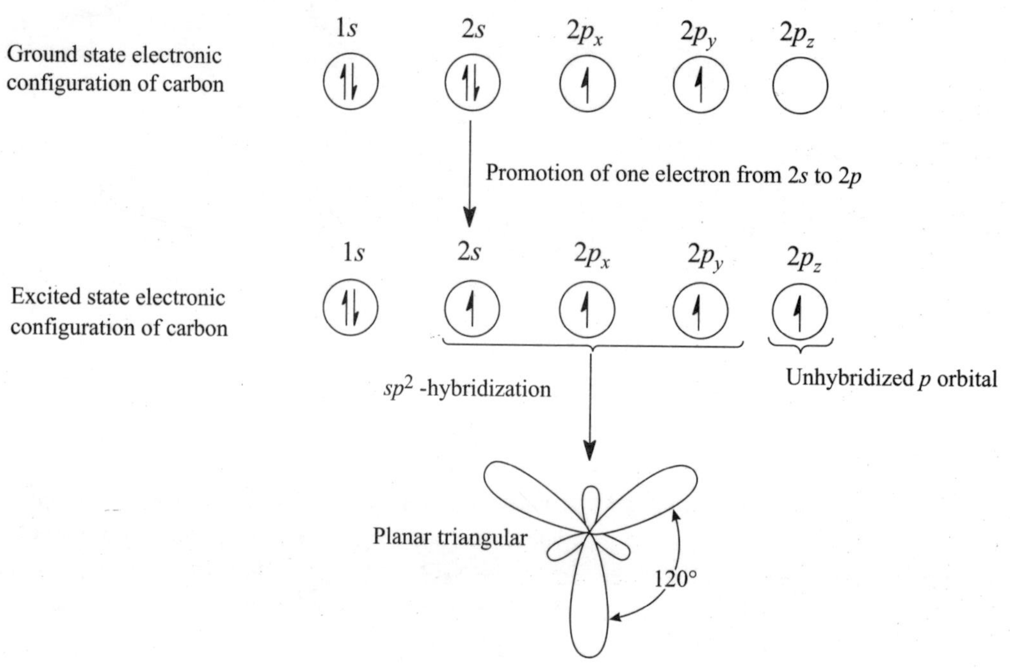

Fig. 1.15: sp^2-hybridization of carbon

Shape of Ethylene Molecule, C_2H_4

A molecule of ethylene has two electron pairs shared between two carbon atoms. A bond of this type is known as *carbon-carbon double bond*. Each carbon atom in the molecule of ethylene is attached to three other atoms by the overlapping of its three hybrid orbitals. These hybrid orbitals are formed by mixing of one 2s and two 2p orbitals. The three sp^2 orbitals are equivalent and each one of them has one electron.

Two sp^2-hybridized carbon atoms use one sp^2 orbital each to form a σ (sigma) bond between them. The rest of the molecule is formed by the overlapping of the remaining two sp^2 orbitals on each carbon atom with 1s orbitals of two hydrogen atoms. The structure of ethylene is still not complete as there are two unhybridized p orbitals, one on each carbon. These two orbitals overlap each other sideways producing a new type of bond, called π bond and the electrons involved in the formation of this bond are known as π electrons. Molecular orbital formulation of ethylene thus suggests that all the six atoms (two carbons and four hydrogens) should lie in one plane with a bond angle of 120°. The π electrons are distributed above and below the plane of the sigma bond. The molecular structure of ethylene is given in Fig. 1.16.

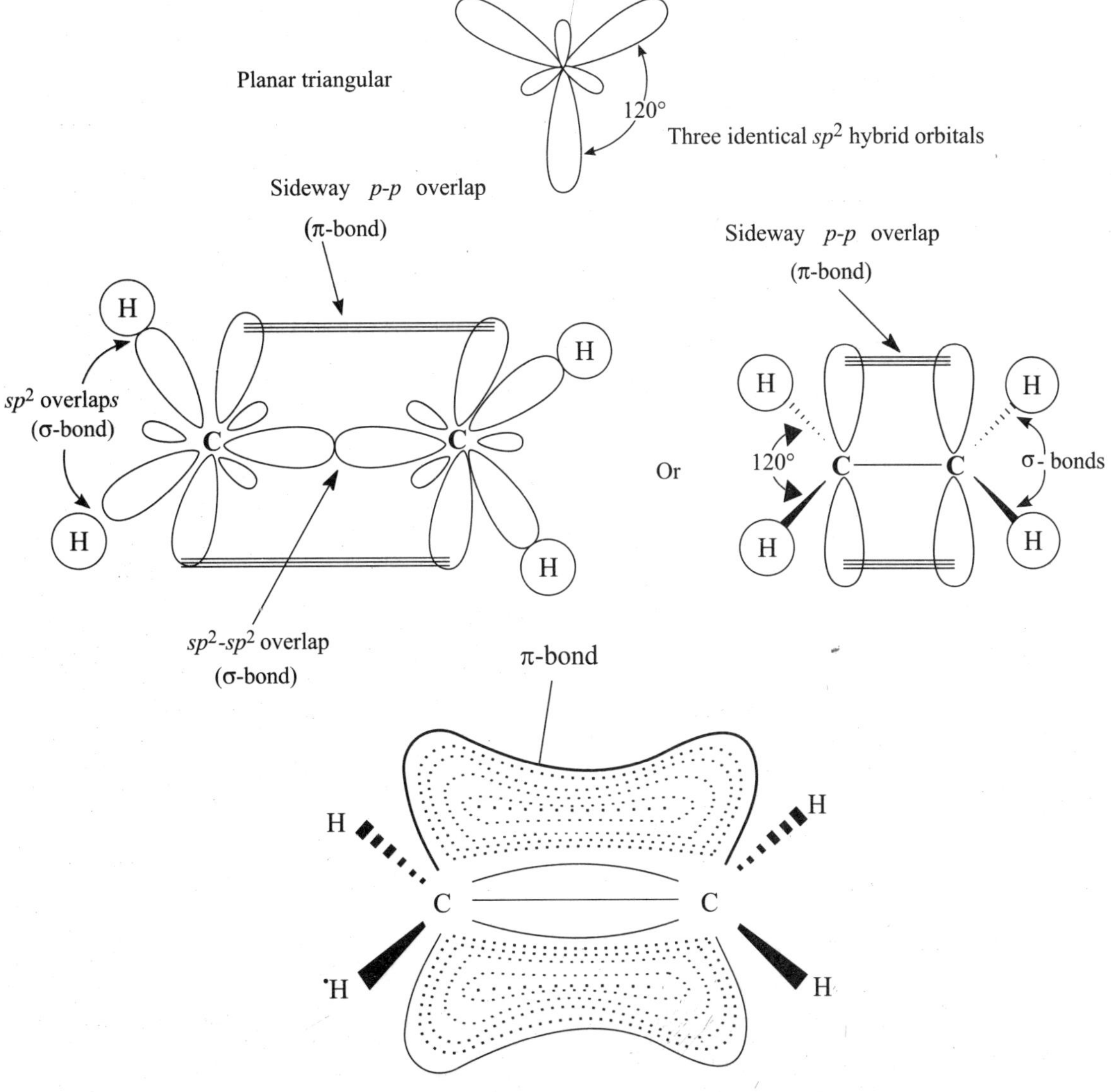

Fig. 1.16: Orbital model of ethylene

We conclude on the basis of the above discussion that a carbon-carbon double bond is made up of a σ-bond and a π-bond. The bond energy of a carbon-carbon π-bond is about 251 kJ mol^{-1} and, therefore, it is weaker than a C—C σ-bond which possesses 347 kJ mol^{-1} of energy. As the carbon atoms are held more tightly, the C=C bond length in ethylene is shorter (134 pm) than the C—C bond length in ethane (154 pm). The bond energy for C=C bond (σ + π) in ethylene is 598 kJ mol^{-1} and that for C—C bond in ethane is 347 kJ mol^{-1}.

1.1.1.3 *sp* or Diagonal or Linear Hybridization

These hybrid orbitals are formed by the mixing of a carbon 2*s* orbital with one of the 2*p* orbitals, and hence, they are called *sp* hybrid orbitals. This process of hybridization is called *sp-hybridization*. The two *sp* hybrid orbitals are equivalent and each one of them has one electron. The two *sp-hybridized* orbitals are directed in opposite directions at an angle of 180°. The remaining two unhybridized *p orbitals* (*i.e.*, $2p_y$ and $2p_z$) are always perpendicular to each other and also to the two *sp-hybridized* orbitals. Each *sp hybrid* orbital has 50% *s*-character and 50% *p*-character.

A schematic representation of *sp-hybridization* is given in Fig. 1.17.

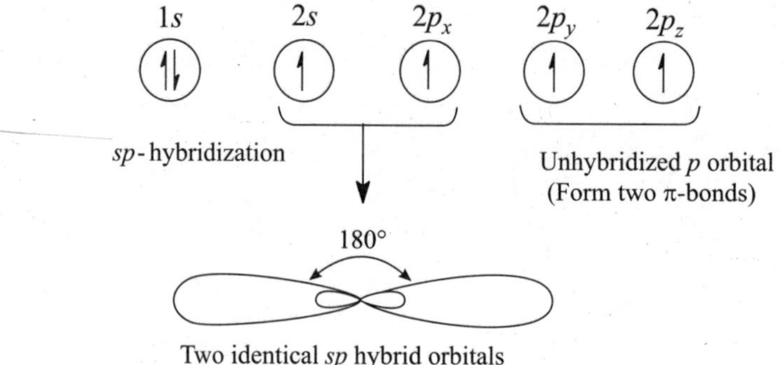

Fig. 1.17: *sp*-hybridization of carbon

Shape of Acetylene Molecule, C_2H_2

A molecule of acetylene has three electron pairs shared between two carbon atoms. A bond of this type is known as *carbon-carbon triple bond*. Thus each carbon atom is bonded to only two other atoms, a carbon and a hydrogen, by the overlapping of two hybrid orbitals. These hybrid orbitals are formed by the mixing of a carbon 2*s* orbital with one of the 2*p* orbitals, and hence they are called *sp hybrid* orbitals. The two *sp* hybrid orbitals are equivalent and each one of them has one electron. Carbon-hydrogen σ-bond is formed by the overlapping of *sp* orbitals on each carbon atom with a

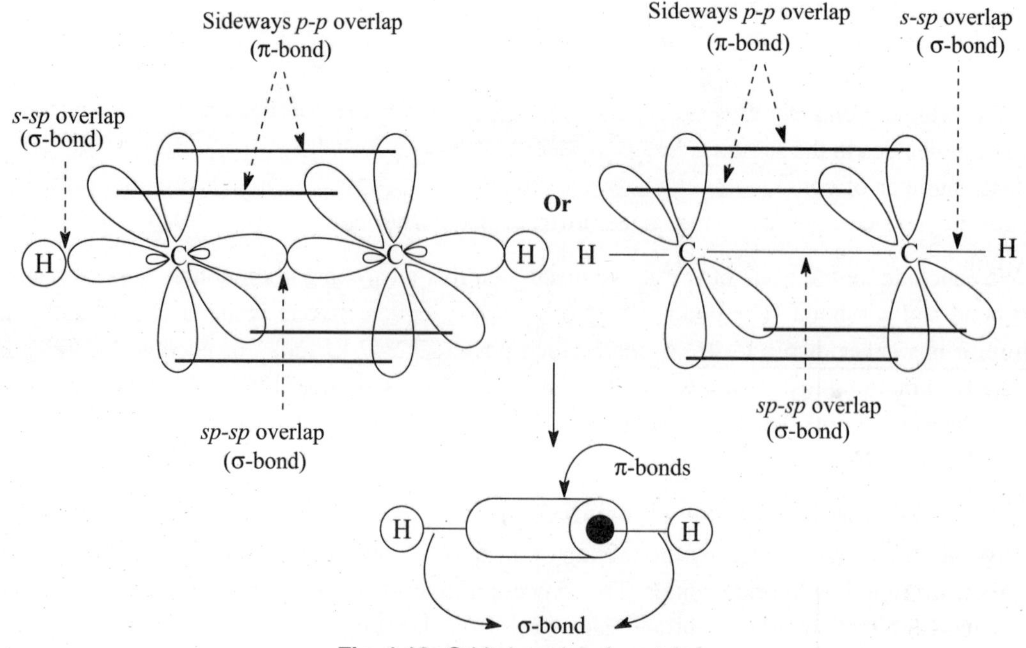

Fig. 1.18: Orbital model of acetylene

hydrogen 1*s* orbital, whereas *carbon-carbon sigma* (σ) *bond* is formed by the overlap of two *sp* orbitals of the two carbon atoms. The structure of acetylene is still not complete as two unhybridized *p* orbitals are left on each carbon atom. These orbitals are perpendicular to each other as also to the *sp* hybrid orbitals. The sideways overlap of two parallel pairs of *p* orbitals leads to the formation of two carbon-carbon π-bonds, which merge into a cylindrical π electron cloud. The molecular structure of acetylene is given in Fig. 1.18.

From Fig. 1.18, it is clear that a molecule of acetylene should be linear. This is supported by physical measurements, *i.e.*, H—C—C bond angle is 180°. The C≡C and C—H bond lengths in acetylene are 120 and 106 pm, respectively.

120 pm 106 pm

H—C≡C—H

180°

It is clear that C≡C triple bond is formed by one strong σ bond and two weak π bonds. The bond energy of a C≡C bond is 803.3 kJ mol^{-1} as compared to 598 kJ mol^{-1} for C=C bond in ethene and 347 kJ mol^{-1} for C—C bond in ethane.

1.2 PARAMETERS OF MOLECULAR STRUCTURE

It is very useful to have a knowledge of the various parameters of molecular structure *i.e.*, bond lengths, bond angles and bond energies for a proper understanding of the nature of chemical bonds.

1.2.1 Bond Length

As we will see later, the atoms forming a bond cannot come closer to each other than a certain distance. The minimum distance between the nuclei of two bonded atoms is known as *bond length.* It should be noted that because of the perpetual atomic vibrations, this distance does not remain constant. Thus the *bond length is actually the average distance between the centres of nuclei of the two bonded atoms where the attractive and repulsive forces just balance each other and the potential energy is minimum.* It is expressed either in Angstrom (Å) or Picometer (pm) units [1Å = 100 pm = 10^{-10} m].

The value of bond length between two atoms X and Y remains constant and is independent of the nature of molecules in the same class of compounds. The C—H distance, for instance, remains 100 pm in a large number of compounds like methane, ethane and propane. Various bond lengths have been measured by physical methods such as X-ray diffraction, electron diffraction and spectroscopic methods. Some typical bond lengths are listed in Table 1.1.

Table 1.1: Bond lengths

Bond	Bond length (pm)	Bond	Bond length (pm)
H—H	74	C—Br	191
N—N	109.4	C—I	213
Cl—Cl	199	C—C	154
C—Cl	176	C=C	134
C—H	110	C≡C	119
C—N	147	H—F	92
C—O	143	H—Cl	127
C—S	182	H—Br	141
C—F	213	H—I	161

Factors Affecting Bond Length

The bond length between any two atoms depends on the following factors:

(*i*) ***Size of the Atom*:** Bond length increases with an increase in size of the bonded atoms, *e.g.*, C—F (142 pm) < C—Cl (177 pm) < C—Br (191 pm) < C—I (213 pm).

(*ii*) ***Multiplicity of bonds*:** Bond length decreases with an increase in the multiplicity of the bond (or bond order), *e.g.*, C—C (154 pm) > C=C (134 pm) > C≡C (120 pm).

(*iii*) ***Type of hybridization*:** The size of hybrid orbitals decreases in the order: $sp^3 > sp^2 > sp$. As a larger orbital forms a longer bond, therefore, carbon-carbon bond length decreases in the order: C—C (sp^3-sp^3) > C=C (sp^2-sp^2) > C≡C (sp-sp).

Further, the order of electronegativity of hybrid orbitals is: $sp > sp^2 > sp^3$ *i.e.*, the electronegativity of carbon is maximum in the *sp*-hybridized state and minimum in sp^3-hybridized state and we also know that a bond formed with a more electronegative atom will be *shorter* than that formed with a less electronegative atom. Consequently, a C—H bond formed with a carbon orbital of high *s*-character will be shorter than the one formed with a carbon orbital of high *p*-character. The change in hybridization of the atomic orbitals in carbon thus produces a change in the covalent atomic radius which decreases in passing from the tetrahedral (sp^3) to diagonal (*sp*) type of hybridization. In fact the state of hybridization in which the bonded atoms exist is the most important factor in determining bond length. In Table 1.2 are listed the average bond lengths depending upon the state of hybridization of the bonded atoms.

Table 1.2: Bond lengths and hybridization

Bond type	Bond length (pm)	Typical compound
C—C		
sp^3-sp^3	154	$CH_3—CH_3$
sp^3-sp^2	150	$CH_3—CH=CH_2$
sp^3-sp	146	$CH_3—C\equiv CH$
sp^2-sp^2	148	$CH_2=CH—CH=CH_2$
sp^2-sp	143	$CH_2=CH—C\equiv CH$
sp-sp	138	$HC\equiv C—C\equiv CH$
C=C		
sp^2-sp^2	134	$CH_2=CH_2$
sp^2-sp	131	$CH_2=C=O$
sp-sp	128	$O=C=C=C=O$
C≡C		
sp-sp	120	$HC\equiv CH$
C—H		
sp^3-H	111	CH_4
sp^2-H	110	$CH_2=CH_2$
sp-H	108	$H—C\equiv C—H$

1.2.2 Bond Angles

A polyatomic molecule has more than one bonds which are formed by overlap of atomic or hybrid orbitals and due to directional nature of the hybrid or atomic orbitals these bonds make angles between them. *The angles between the lines representing the two bonds are known as bond angles* and are measured by X-ray diffraction and spectroscopic methods. Because of the constant atomic vibrations, the bond angles thus measured are really average bond angles. The shapes of the molecules are dependent on the bond angles. The molecule of methane, therefore, is tetrahedral because of the H—C—H bond angles of 109.5°, the molecule of water is V shaped because of the H—O—H bond angles of 104.5° and the molecule of NH_3 is pyramidal because of H—N—H bond angle of 107°.

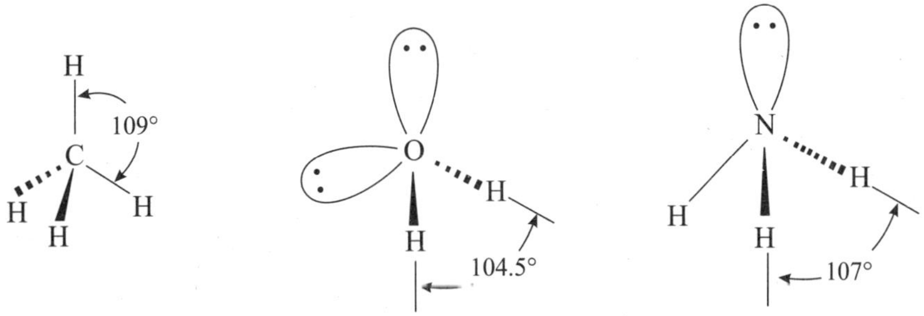

Factors Affecting Bond Angles

The bond angles between any two bonds depend on the following factors:

(*i*) ***Type of hybridization*:** As we know, sp^3 hybrid atom has bond angle of 109.5°, sp^2 has 120° and *sp* has 180°. But these angles may deviate from their regular geometry values due to some other electronic effects.

(*ii*) ***The number of lone pairs and Bond pairs*:** According to VSEPR (Valence Shell Electron Pair Repulsion) theory the magnitude of repulsions between lone pairs and bond pairs decreases in the order: Lone pair-lone pair > bond pair-lone pair > bond pair-bond pair.

Therefore, the regular geometry gets distorted with the presence of lone pairs and bond angles decrease from their expected values. We have already seen this in the case of CH_4, NH_3 and H_2O, where the central atom is sp^3-hybridized in each case but they have bond angles 109° 28′, 107° and 104°, respectively. This is due to the presence of one lone pair in NH_3 and two lone pairs in H_2O.

Similarly, variations are found from ideal values of 120° for sp^2 carbon. *e.g.*, in ethylene H—C—C and H—C—H bond angles are 121.7° and 116.6° respectively, due to the presence of π electrons which repel C—H bonds away.

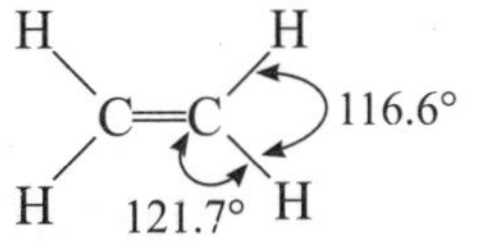

1.2.3 Bond Energy

During the formation of a bond, certain amount of energy is released. The same amount of energy will be needed to break this bond. Bond energy may thus be defined as the *energy required to break a bond between two atoms*. It is expressed in units of kJ mol^{-1}. For example, 435 kJ mol^{-1} of heat is needed to break a mole of hydrogen molecules into individual atoms. Therefore, the bond energy of hydrogen is 435 kJ mol^{-1}.

$$H—H \longrightarrow 2H \qquad \Delta H = 435 \text{ kJ mol}^{-1}.$$

The values of bond energies for certain diatomic molecules are given in Table 1.3.

Table 1.3: Bond energies of certain diatomic molecules

Bond	Energy (kJ mol^{-1})
H—H	435
N—N	297
Cl—Cl	247
Br—Br	192
F—F	155
I—I	150
O—O	146

These values are called *bond dissociation energies* and are symbolized by **D**. However, for a polyatomic molecule containing more than one covalent bond, the term bond energy may have two different meanings: (*i*) Bond dissociation energy (D) and (*ii*) Bond energy (E).

(*i*) *Bond dissociation energy (D) is the energy required to break one mole of a particular bond in the molecule.*

(*ii*) *Bond energy (E) is the average energy per bond, which is the average of the different bond dissociation energies for such bonds present in the molecule.*

Let us illustrate the difference between these two values by considering the example of methane. If we begin to remove the four hydrogen atoms one by one by splitting of carbon-hydrogen bonds, we get four different D values, Thus,

$$CH_4 \longrightarrow CH_3 + H \qquad D(CH_3—H) = 427 \text{ kJ mol}^{-1}$$

$$CH_3 \longrightarrow CH_2 + H \qquad D(CH_2—H) = 439 \text{ kJ mol}^{-1}$$

$$H—C—H \longrightarrow CH + H \qquad D(CH—H) = 452 \text{ kJ mol}^{-1}$$

$$C—H \longrightarrow C + H \qquad D(C—H) = 347 \text{ kJ mol}^{-1}$$

Why do we get four different values when we are breaking a C—H bond in each step? The reason is that the C — H bond dissociation energy not only includes the energy needed for rupturing the bond, but it also includes the energy changes accompanying the rehybridization at the carbon atom in each step. Direct atomization of methane requires 1665 kJ mol^{-1}, which is the sum of all the four D values listed above.

$$\begin{array}{c} H \\ | \\ H-C-H \\ | \\ H \end{array} \longrightarrow C + 4\,H \qquad \Delta H = 1665 \text{ kJ mol}^{-1}$$

The bond energy per C—H bond in methane is thus one fourth of this value, *i.e.*, 1665/4 = 416.25 kJ mol^{-1}.

Since, a diatomic molecule has only one bond, D is equal to E in such cases. Some bond energies obtained by taking the average from measurements on several polyatomic molecules are listed in Table 1.4.

Table 1.4: Average bond energies (E values)

Bond	Average bond energy (kJ mol^{-1})
O—H	448
C—II	414
C—F	448
C—Cl	326
C—Br	284
C—I	213
C—C	347
C=C	609
C≡C	804
C—O	360
C=O	740

Factors Effecting Bond Energies

The bond energy gives an approximate idea about the strength of a particular bond and depends upon the following factors:

***(i) Size of atom*:** Larger the size of atom forming covalent bond smaller will be the bond energy, *e.g.*, as the size of halogen atom (X) increases from F to I, the bond energy of C—X bond decreases *i.e.*, C—F (448) > C—Cl (326) > C—Br (284) > C—I (213) kJ mol^{-1}.

***(ii) Bond length*:** Bond energy increases with a decrease in bond length. *e.g.*, C—C (347 kJ mol^{-1}) < C = C (609 kJ mol^{-1}) < C ≡ C (804) kJ mol^{-1} for carbon-carbon bond where bond length decreases due to increase in multiplicity of the bond from single to triple bond.

***(iii) Type of Hybridization*:** We know that, the size of hybrid orbitals decreases in the order: $sp^3 > sp^2 > sp$ and their electronegativity increases in the order: $sp^3 < sp^2 < sp$. Both these factors cause a decrease in bond length (from $sp^3 \rightarrow sp^2 \rightarrow sp$) and hence bond energy increases, *e.g.*, for C — H bond.

C (sp^3) — H (416 kJ mol^{-1}) < C (sp^2) — H (443 kJ mol^{-1}) < C (sp) — H (50 kJ mol^{-1}) and for C — C bond C (sp^3) — C (sp^3) (347 kJ mol^{-1}) < C (sp^2) — C (sp^3) (383 kJ mol^{-1}) < C (sp) — C (sp^3) (433 kJ mol^{-1}).

1.3 LOCALIZED AND DELOCALIZED BONDS

Atomic or hybrid orbitals overlap to form covalent bonds where the electrons are either localized or delocalized.

When the electrons forming a bond spend most of their time in the space between the two bonded atoms, they are called localized electrons and such a bond is called localized bond, e.g., all σ -bonds are localized bonds.

On the other hand, when the electrons are moving in and out of the space between the two bonded atoms, they are called delocalized electrons and the bond formed by them is called delocalized bond.

Electrons forming π-bonds may be either localized or delocalized, *e.g.*, π-bond of ethylene is localized because the electrons forming the π-bond in ethylene are confined to the space between two carbon atoms in such a way that these electrons are distributed equally in the space above and below the plane of C—C, σ-bond (Fig. 1.19 A). However, the two π-bonds of acetylene are delocalized because the electrons forming the π-bond in the plane of paper do not remain confined to the space above and below the plane of C—C, σ-bond and similarly the electrons of other π-bond formed in perpendicular direction, do not remain confined to that space. Actually, all these electrons merge together to form a cylindrical electron cloud around C—C σ-bond (Fig. 1.19 B).

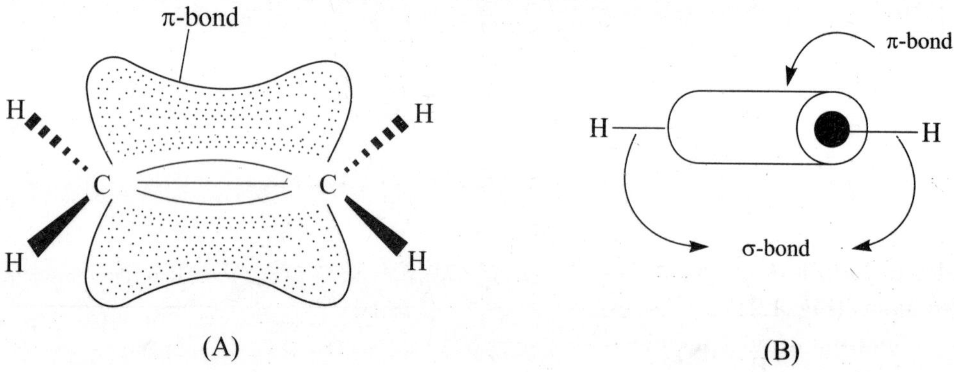

Fig. 1.19: (A) Localized bond in ethylene, (B) Delocalized bond in acetylene

Other examples of delocalized bonds are 1, 3-butadiene and benzene where two or more than two π-bonds are in conjugation. The electrons of one π-bond are delocalized into the space of other π-bond and *vice-versa*. This delocalization occurs through the overlap of unhybrid *p* orbitals present on each sp^2 carbon. Thus;

(*a*) *1, 3-Butadiene*: All the four carbon atoms of 1, 3-butadiene lie in the same plane due to which all the *p*-orbitals at four carbon atoms overlap with each other and the π-electrons can move to a limited extent over all the four carbon atoms *i.e.*, π-electrons of 1, 3-butadiene are delocalized as shown in Fig. 1.20(B). Hence, the π-bonds of butadiene are delocalized bonds.

In contrast, the π-*electrons and hence the* π *-bonds are localised in isolated dienes such as* 1,4-

pentadiene (Fig. 1.20 A), where each pair of π-electrons is confined to the space between two carbon atoms.

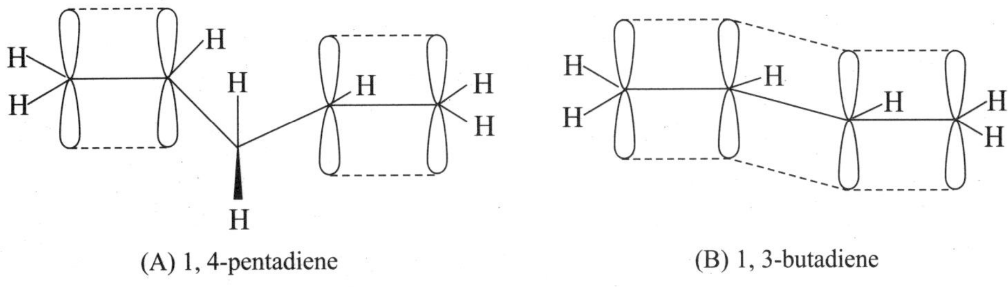

Fig. 1.20: (A) Localized π-bonds, (B) Delocalized π-bonds

(*b*) *Benzene*: In case of benzene there are six sp^2-hybridized carbon atoms and each sp^2-carbon has one unhybridized *p*-orbital containing one electron. These *p* orbitals are so close to each other that they can overlap sideways to form a π-bond. There are two modes for overlap of adjacent *p*-orbitals as shown in Fig. (1.21 A and B).

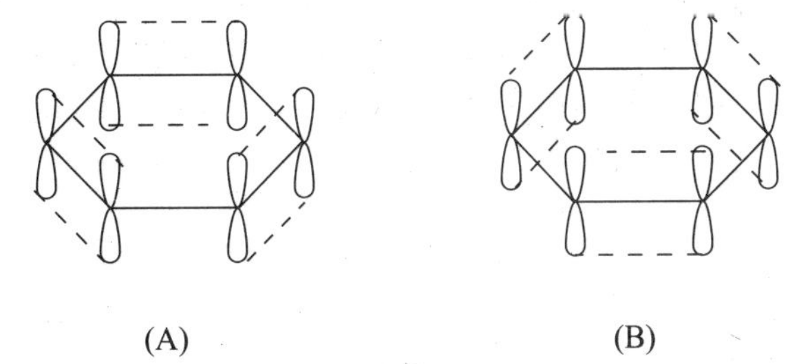

Fig. 1.21: (A) and (B) Two possible sideways overlap of six unhybridized *p*-orbitals to form three π-bonds in benzene

Actually, each *p* orbital overlaps equally well with the *p* orbitals on adjacent two carbon atoms on both sides to form a doughnut shaped π-electron cloud above and below the plane of carbon and hydrogen atoms (Fig. 1.22) *i.e.*, these three π-bonds of benzene are delocalized.

Fig. 1.22: π-Electron clouds lying above and below the plane of benzene ring (Delocalized π-bonds)

Armit and Robinson Structure

As the three π-bonds of benzene are completely delocalized, it is not proper to represent benzene with a hexagonal ring with three double bonds at alternate positions since the position of π-bonds is not fixed. Therefore, benzene is written as shown in Fig. 1.23. This representation was given by **Armit** and **Robinson**.

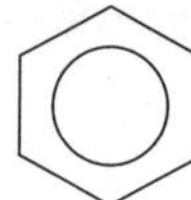

Fig. 1.23: Representation of benzene

1.4 VAN DER WAALS INTERACTIONS

It is well known that:

(*i*) *Gases do not obey the ideal gas equation, PV = nRT, and*

(*ii*) *A non-polar gaseous compound can be condensed to the liquid state which in turn can be made to solidify.*

In order to account for these observations *van der Waals*, a Dutch chemist, proposed the existence of some sort of attractive forces between the molecules of non-polar compounds. *These attractive forces between the molecules of non-polar compounds are called van der Waals forces. These are short range forces.*

1.4.1 Origin of van der Waals Attractive Forces

Normally in a non-polar neutral molecule, the centres of positive and negative charge densities coincide with each other and consequently these molecules should not have any dipole moment. But because of the constant random movement of electron clouds around the nucleus, it is expected that at some instant of time these clouds may get slightly distorted in one or some of the molecules due to mutual collisions or some other factors. In such a situation, the centres of positive and negative charge densities no longer coincide at that instant of time and this results in the creation of a small instantaneous local dipole. This instantaneous dipole moment induces oppositely oriented *dipole moment* known as induced dipole moment in the neighbouring molecules and when they are close enough these dipoles attract each other and cause the molecules to cling together. Because of this, such attractions are also referred to as *instantaneous dipole-induced dipole attractions*. These instantaneous and induced dipoles are constantly changing but the overall result is the attraction between such molecules. Such attractions are also called **london forces** or **dispersive forces**.

These forces are very weak and are at maximum in the solid and minimum in the gases because of difference in the intermolecular distances in the two cases. Moreover, these short range forces are appreciable only between those portions of the molecules over which they can *touch* each other, *i.e.*, between the surfaces of the molecules. Thus, *it is reasonable to believe that greater the surface over which the molecules can touch each other, the greater is the overall van der Waals attraction.*

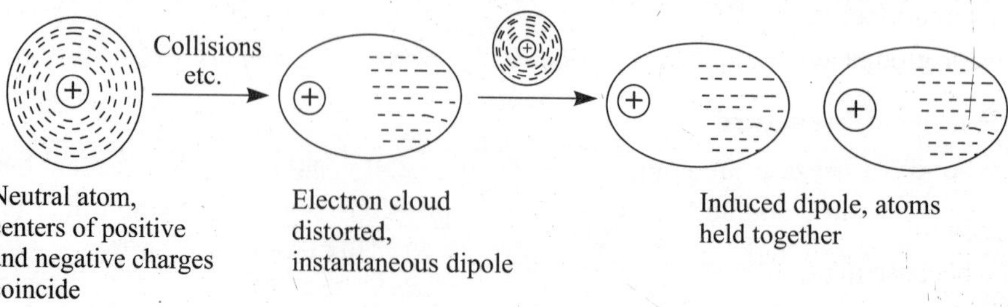

This helps us in understanding the influence of molecular size and shape on physical properties of many substances. For instance,

(*i*) *The boiling points in the homologous series of hydrocarbons show a regular increase per* CH_2 *unit.* This is because the increase in the molecular weight increases the molecular size and hence the van der Waals forces of attraction.

(*ii*) *Among the isomers a straight chain isomer has higher boiling point than a branched chain isomer. In other words, the boiling point decreases with the branching of the chain.* This is observed practically in all the families of organic compounds. It can be explained on the basis of the fact that with branching the shape of the molecule tends to approach that of a sphere, thereby decreasing its surface area. This results in decrease in van der Waals forces of attraction which can be overcome at a lower temperature for boiling.

1.4.2 van der Waals Repulsions

As mentioned above, the van der Waals forces of attraction increase in magnitude with decrease in the distance between non-bonded atoms or groups of atoms (molecules). But this attraction works up to a certain point when such moieties are at a minimum distance so that they can touch each other, and *if they are forced closer than this minimum distance, the attractive forces turn into great repulsive forces, called van der Waals* repulsions. Thus it can be said that every atom or a group of atoms has an effective size referred to as *van der Waals radius* and the minimum distance beyond which the forces of attraction turn into forces of repulsion is equal to the sum of the van der Waals radii of two atoms or group of atoms. In Table 1.5 are listed van der Waals radii of some atoms and groups of atoms. It may be mentioned that the van der Waals radii of various atoms are greater than their covalent radii by about 80 pm.

Table 1.5: van der Waals radii (pm) of some atoms and groups

Atoms and Groups	Van der Waals Radii (pm)	Atoms and Groups	Van der Waals Radii (pm)
H	120	Cl	180
N	150	Br	195
O	140	I	220
S	190	CH_3	200
F	135	Half the thickness of benzene nucleus	170

Van der Waals repulsive forces greatly influence the shapes and geometries of the molecules. Because of the directional nature of covalent bonds the molecules have certain spatial requirements, and if the atoms or groups are situated at such positions that the distance between them is smaller than the sum of their van der Waals radii, they are said to be *crowding* together. In order to achieve maximum stability—a situation where such repulsive forces could be avoided—the molecule orients itself in such a way that the distance between these atoms increases.

For instance, *cis*-1, 3-dimethylcyclohexane is more stable when the substituents are equatorial rather than axial because in the latter case, 1, 3-diaxial interactions (because of van der Waals repulsions) make the molecule less stable.

The reason for this is not for to seek. In 1, 3-diaxial conformation the distance between the two axial methyl groups and one axial hydrogen is much less than the sum of their van der Waals radii (Fig. 1.24) which results is setting up of repulsive forces which are called 1, 3-*diaxial* interactions in these cases. However, in the diequatorial conformation, the two methyl groups are held at a distance which is more than the sum of their van der Waals radii and hence such repulsive forces do not exist. Therefore, *cis*-1, 3-dimethylcyclohexane is more stable in its diequatorial conformation than in diaxial.

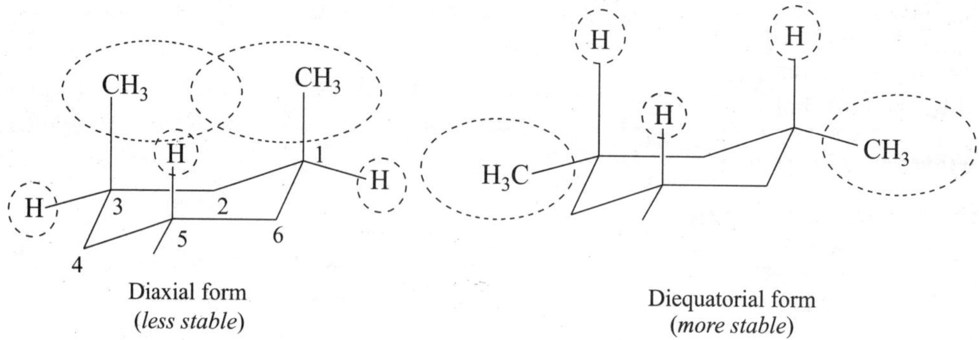

Fig. 1.24: *Cis*-1, 3-dimethylcyclohexane conformations

1.4.3 van der Waals Attraction in Polar Molecules

Apart from non-polar molecules the van der Waals forces of attraction also occur in polar molecules. They may originate from *dipole-dipole* and *dipole-induced dipole* interactions.

***(i) Dipole - dipole interaction*:** Polar molecules having permanent dipole moment are held together by dipole-dipole interactions (Fig. 1.25 A). For instance, in gases such as NH_3, HCl, HF, SO_2, etc., there are significant dipole-dipole attractions between the molecules of these gases due to the presence of permanent dipole moment in the molecules. The extent of this type of interaction depends upon the magnitude of the dipole-moment of the molecule. Thus, greater the dipole moment, stronger is the dipole-dipole interaction.

***(ii) Dipole-induced dipole interaction*:** They are attractive interactions between polar and non-polar molecules. A polar molecule may induce polarization in a non-polar molecule present in its vicinity. This induced dipole then interacts with the dipole moment of the first molecule and thereby the two molecules are attracted to each other [Fig. 1.25 B]. The extent of such interaction depends upon the magnitude of the dipole moment of the polar molecule and the polarizability of the non-polar molecule. For instance, the increasing order of water solubility of noble gases from He to Rn is attributed to increase in the magnitude of dipole induced dipole intraction due to increase in their polarizability in the same order.

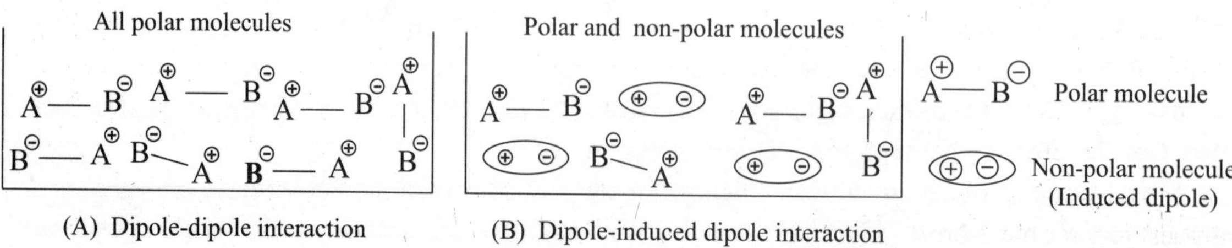

Fig. 1.25: van der Waals interactions in polar molecules

1.5 INCLUSION COMPOUNDS AND CLATHERATES

Certain organic solids such as urea, thiourea, hydroquinone, etc. have crystalline shapes. However, in some cases, the crystalline shapes of these compounds undergo a change in presence of certain other compounds. *The former compounds whose crystalline structures change are called* **hosts** *while the latter ones in whose presence the crystalline structures of the former change are called* **guests**. The crystal lattice of host compounds forming long channels or cage like structures has enough space to trap the guest molecules. In fact there is no chemical bonding between the host and the guest molecules. There are only weak van der Waals forces of attraction which hold the guest molecule in the space provided by the host molecules.

1.5.1 Types of Host-Guest Addition Compounds

Depending upon the type of the space available within the crystal lattice of the host molecule the host-guest addition compounds are divided into following two categories:

1. Inclusion compounds: The host-guest addition compounds are known as inclusion compounds when the space available within the crystal lattice of host molecule is in the form of long channels.

2. Clatherates: The host-guest addition compounds are known as clatherates, when the space available within the crystal lattice of the host molecule is in the form of a cage.

Both these types of addition compounds are well defined crystalline solids but they are not useful for derivatization since they decompose at the melting point of the host compounds. Although the two types of addition compounds differ from each other yet they have one thing in common that is the guest molecules of only right size can be trapped into the crystal lattice of the host molecules. We will now discuss both types of addition compounds in more details.

1. Inclusion Compounds

***(i) Urea as host*:** The most common host molecule for these compounds is urea. Ordinarily urea crystallizes in tetragonal shape. However, when a guest is present, urea crystallizes in the hexagonal lattice trapping the guest molecules (*n*-alkanes) in the space (Fig 1.26). Since hexagonal type of lattice is formed only in presence of some guest molecule, it may be reasonable to believe that the van der Waals forces of attraction between the host and the guest molecule are essential for the stability of inclusion compounds. The diameter of the urea channel is about 500 pm and the type of guest molecules will depend upon their shape and size only. Some examples are illustrative.

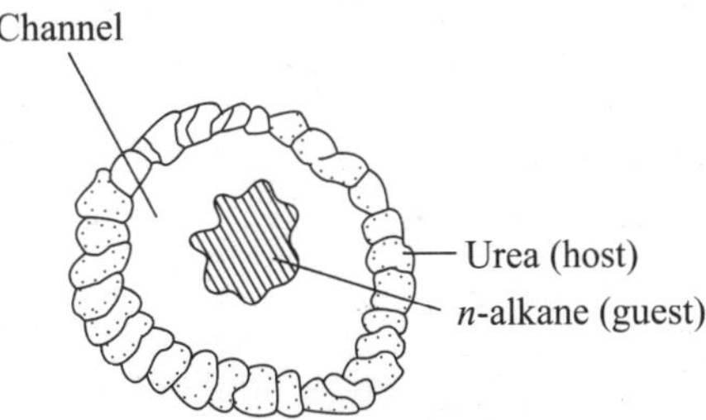

Fig. 1.26: Inclusion compounds

(*a*) *n*-Octane, 2-chlorooctane and 1-bromooctane are of appropriate size and shape and thus act as guest for urea but 2-bromooctane, 2-methylheptane and 2-methyloctane do not fit the size of channels. Hence, these do not act as guests for urea lattice.

(*b*) Both dibutyl maleate and dibutyl fumarate are guests but neither diethyl maleate nor diethyl fumarate is a guest.

(*c*) Dipropyl fumarate is a guest but dipropyl maleate is not.

In these complexes there is no fixed integral stochiometry (*i.e.*, integral molar ratio) of host and guest though by chance it may be, for instance, the octane-urea ratio is just 1 : 6.73. Further in the inclusion compounds of urea, the urea molecules are not rigid but undergo 180° flip about the C═O axis at a rate of more than 10^6 per second at 303 K which has been shown by spectroscopic studies of these compounds.

***(ii) Thiourea as host*:** Thiourea also forms inclusion compounds with channels of larger diameter. Consequently, *n*-alkanes cannot act as guests for thiourea. However, 2-bromooctane, cyclohexane, chloroform etc., have the right size and shape to be trapped in the channels of thiourea.

***(iii) Amylose as host*:** Starch gives a deep blue colour with iodine. In fact it is a mixture of two components *i.e.*, amylose which is water soluble and amylopectin which is water insoluble. Amylose having helical structure, has the inside of the helix of the appropriate size and polarity to accommodate an iodine molecule. Thus the blue colour is due to the formation of an inclusion complex between amylose and iodine (Fig. 1.27).

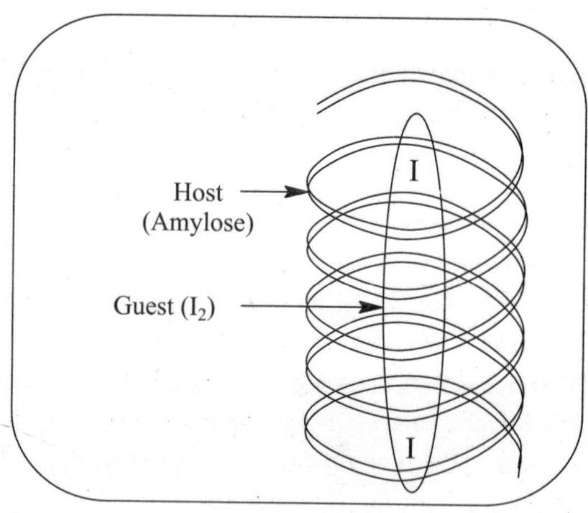

Fig. 1.27: Inclusion complex of amylose and iodine

Uses of inclusion compounds: Some of the more important uses of inclusion compounds are as under:

(*i*) These complexes are quite useful in separating certain isomers that would be otherwise difficult to separate. For example, *n*-octane can be separated from its branched chain isomers, because only *n*-octane can form inclusion compound with urea.

(*ii*) Urea inclusion compounds have also been used for resolving racemic mixtures. For example, (±) racemic mixture of 2-chlorooctane forms two different inclusion compounds (diastereomeric) which can be separated by fractional crystallization.

2. Clatherates

The adducts in which the host molecule has cage like cavities are called **clatherates** or **cage compounds**. Unlike inclusion compounds, the crystal lattice of clatherates is not fully occupied by the guest.

Quinol or hydroquinone, water, sodium chloride and some alkali metal halides are the most important hosts for clatherates. On the basis of the nature of host, clatherates are divided in two categories as under:

***(i) Quinol clatherates*:** Quinol clatherates have hydroquinone as host. In these complexes, three molecules of hydroquinone which are held together by hydrogen bonding form a cage containing several cavities of 400 pm diameter. The guest molecules of appropriate size fit into these cavities. The best typical guests for quinol clatherates are methanol (but not ethanol), SO_2, CO_2, H_2S, CH_3CN and noble gases such as Ar, Kr, Xe, etc. It may be mentioned here that normally all the cavities are not filled and thus clatherates have some vacant space in the cage structures.

***(ii) Water clatherates*.** Water is another important host. Generally six molecules of water form the cage in which guest molecules of appropriate shape and size, *i.e.*, Cl_2, propane, CH_3I and noble gases such as Ar, Kr, and Xe can get trapped. When water is frozen under a high pressure of any one of the above noble gases, solid crystalline addition compounds of the formula H_2O : gas (5.75 : 1) are obtained. These clatherates are more commonly called as **noble gas hydrates**. These are normally preserved at low temperature because at room temperature they decompose whereby trapped gases get released. He and Ne, however, do not form clatherates since their atoms are too small to be got entrapped.

Uses of Clatherates: Some of the more important uses of clatherates of noble gases are as under:

(*i*) He and Ne can be separated from other noble gases present in a mixture because both He and Ne do not form clatherates.

(*ii*) Radioactive isotopes of Kr and Xe which are produced in nuclear reactors can be stored as clatherates.

1.6 CHARGE-TRANSFER COMPLEXES

When an electron rich compound (called the donor molecule and designated as D) is mixed with an electron - poor or deficient compound (called the acceptor molecule and designated as A) in 1 : 1 ratio, it forms a complex called **charge transfer complex**. In the formation of charge transfer complexes, the donor molecule may donate either a non-bonding (or an unshared) pair of electrons or a pair of electrons of π-system (an olefinic double bond or the aromatic ring system) to the acceptor molecule. The former donor is called an *n*-donor and the latter as a π-donor. Such complexes are also known as *electron-donor-acceptor complexes or EDA complexes*. The formation of a charge transfer complex may be represented as follows:

$$\underset{\text{(Donor)}}{D} + \underset{\text{(Acceptor)}}{A} \longrightarrow \underset{\text{Charge-transfer complex}}{[D \text{------} A]} \longleftrightarrow D^+ - A^-$$

In these complexes, the bonding forces between the donor and the acceptor molecules are the weak coloumbic forces. A bonding of such type which results from attractive forces between donor and acceptor molecules is called *charge transfer interaction*.

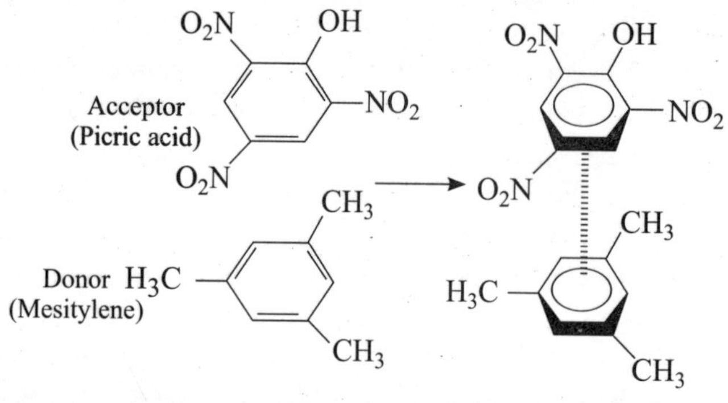

Acceptor (Picric acid) O_2N, OH, NO_2, O_2N

Donor (Mesitylene) H_3C, CH_3, CH_3

(Picrate)
(π-complex)

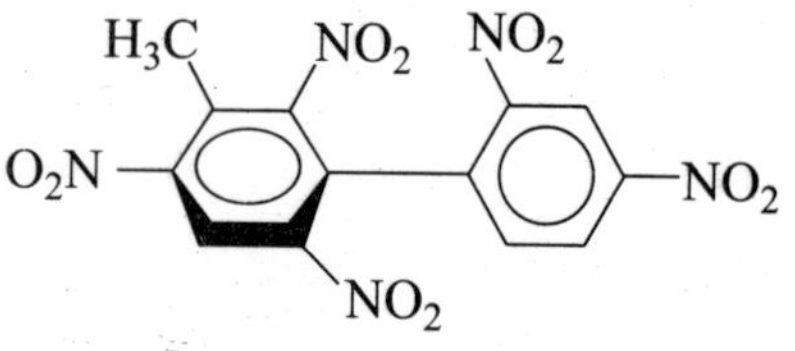

(Z)

1.6.1 Structure of Charge-Transfer Complexes

Picrates: Formation of the so called "picrates" by the action of 2, 4, 6-trinitrophenol (picric acid) and an aromatic hydrocarbon such as naphthalene, anthracene, mesitylene, etc., represents the most common examples of charge transfer complexes. The picrates are yellow, orange or red coloured solids having definite melting point and hence are oftenly used as derivatives for characterization of different compounds. The nitro groups rather than a hydroxyl group are essential for the success of the reaction, as not only picric acid but other polynitro aromatic compounds such 1, 3, 5-trinitrobenzene, form similar addition compounds. The formation of charge-transfer complexes is facilitated by the presence of electron-donating substituents in D and electron-withdrawing substituents in A.

Donor has a low ionization potential and acceptor has a high electron affinity with a low energy vacant orbital. Due to parallel arrangement of the two rings, overlap occurs between molecular orbitals of D and A (HOMO-LUMO) causing electron transfer (D → A).

The distance between the donor and the acceptor compounds is 300–350 pm, which is much larger than the typical covalent bond length and shows that though there is a slight overlap of π-orbitals of both the rings, it is not so strong as to form a pure covalent bond. Determination of crystal structure has shown that in such a complex the aromatic rings are held in *parallel planes*. This is further supported by the observation that compounds such as (Z) in which the two rings cannot attain *coplanarity* because of steric interactions (due to the presence of nitro groups at ortho positions) fail to form a charge transfer complexes.

The extent of charge separation in a charge transfer complex is often reflected in ground state properties such as dipole moment. The dipole moment of pyridine (2.28 D), for instance, is raised significantly (4.90 D) in the pyridine-iodine complex. Similarly, 1, 3, 5-trinitrobenzene does not display dipole moment in carbon tetrachloride but does show a significant dipole moment in benzene due to formation of charge transfer complex.

1.6.2 Charge-Transfer Spectra

The spectroscopic properties of charge transfer complexes are changed due to the weak ground-state interactions and this results in the appearance of a new type of electronic transition called a *charge transfer transition*. The absorption spectra of these complexes are thus quite characteristic as there is a new absorption band in addition to the original bands of the *donor* and *acceptor* components. In the case of 1-alkylpyridinium iodides where the ground state is an ion-pair, the light absorption process has been shown to involve the loss of an electron from the iodide ion and the gain of the electron by the pyridinium ring.

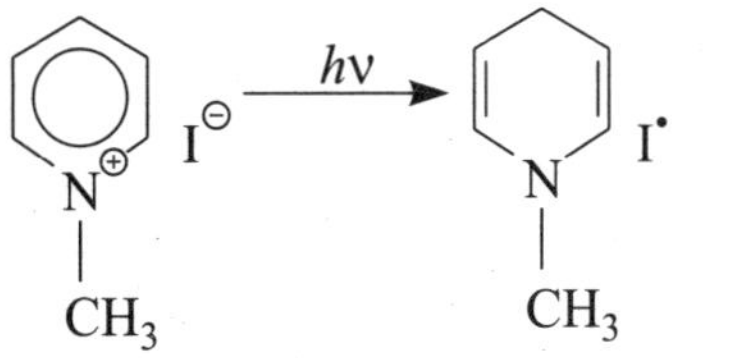

$$\text{C}_5\text{H}_5\overset{\oplus}{\text{N}}\text{-CH}_3\ \text{I}^{\ominus} \xrightarrow{h\nu} \text{C}_5\text{H}_5\text{N-CH}_3\ \text{I}^{\bullet}$$

Since, the appearance of this new absorption band is due to the change in the position of an electron from the donor to the acceptor component, the nature of the substituents in both the components should affect its position. Indeed, its position has been found to vary with the ionization potential of the donor and with the electron affinity of the acceptor.

1.7 INDUCTIVE AND FIELD EFFECTS

1.7.1 Inductive Effect

When a covalent bond is formed between two atoms having different electronegativities, there is induced a permanent dipole in such a bond. This induction of dipole due to polarization is referred to as the **inductive effect**. Actually this polarization *is due to drift of the shared pair of electrons towards the more electronegative atom resulting in the origin of small fractional charges on the constituent atoms* (δ^+ *and* δ^-). When a carbon is bonded to a hydrogen (C—H) or another carbon (C—C) by a covalent bond such as in alkanes, the sharing of electron pair is symmetrical between them, because there is no significant difference in the electronegativity values of the two bonded atoms (C and H).

$$-\overset{|}{\underset{|}{\text{C}}}-\text{H} \quad \text{or} \quad -\overset{|}{\underset{|}{\text{C}}}:\text{H}$$

(Symmetrical sharing of bonded electron pair)

$$-\overset{|}{\underset{|}{\text{C}}}-\text{C} \quad \text{or} \quad -\overset{|}{\underset{|}{\text{C}}}:\text{C}$$

(Symmetrical sharing of bonded electron pair)

However, when the carbon atom of the alkane has a more electronegative atom or group (X) attached to it, the electron pair of covalent bond shifts slightly towards X resulting in the distortion of the molecular orbital. Consequently, the atom or group (X) acquires a partial negative charge (δ^-) and the carbon a partial positive charge (δ^+). Thus,

$$-\overset{|}{\underset{|}{\text{C}}}\ :\text{X} \quad \text{or} \quad -\overset{|}{\underset{|}{\text{C}}}{}^{\delta^+}:\text{X}^{\delta^-}$$

(Unsymmetrical sharing of bonded electron pair — *origin of fractional charges due to higher electronegativity of X*)

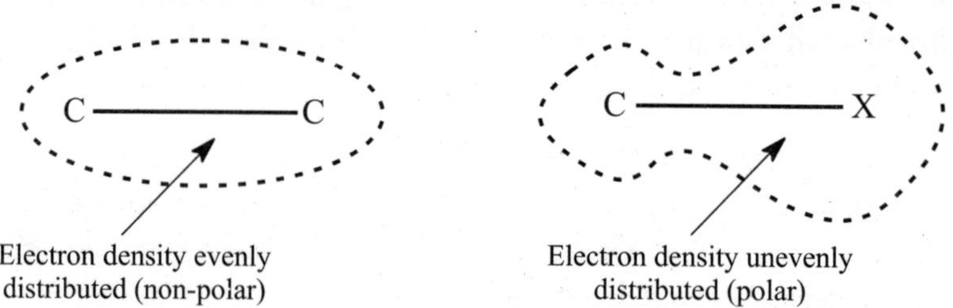

Electron density evenly distributed (non-polar)

Electron density unevenly distributed (polar)

The inductive effect is represented by an arrow with its head in the middle of the covalent bond pointing in the direction of electron displacement.

$$\overset{\delta^+}{C_3} \rightarrow \overset{\delta^+}{C_2} \rightarrow\!\!\rightarrow \overset{\delta^+}{C_1} \rightarrow\!\!\rightarrow\!\!\rightarrow \overset{\delta^-}{C_1}$$

This induction travels along the chain through the bonds. The extent of this effect, however, diminishes with increase in distance.

This type of σ-electron shift along a saturated chain of atoms due to presence of a polar covalent bond is called **inductive effect or I-effect**. *Sometimes it is also called* **transmission effect or T-effect**.

1.7.1.1 Types of Inductive Effect

The inductive effect is designated as +I or –I-effect depending upon whether the atom or group is electron repelling (donor group) or electron-withdrawing (acceptor group), respectively, for example, chlorine shows –I-effect whereas methyl group shows +I-effect.

It may be mentioned here that inductive effect is permanent effect involving polarization of a covalent bond.

$$C \rightarrow Cl \qquad\qquad C \leftarrow CH_3$$

(– I-effect) (+ I-effect)

Some of the more common groups exerting negative or positive inductive effect are listed below:

Groups Showing – I-Effect

–H, $-C_6H_5$, –OH, $-OCH_3$, –I, –Br, –F, –COOH, –CN, $-NO_2$

The –I-effect of the groups having hetero atoms such as oxygen, nitrogen, halogen, etc., is due to the greater electronegativity of these groups as compared to carbon. These groups thus cause –I-effect due to their electron attracting nature.

Groups Showing + I-Effect

–H, –D, $-CH_3$, $-CH_2CH_3$, $-CH(CH_3)_2$, $-COO^-$.

The +I-effect of these groups is due to electron-donating nature of alkyl groups. Deuterium causes +I-effect greater than hydrogen probably, because of larger atomic radius which makes it more electropositive than hydrogen. The *more highly substituted the alkyl group, stronger is its +I-effect*. Thus, the decreasing order of +I-effect of some alkyl groups relative to hydrogen when attached to a saturated carbon chain follows the order:

Tert-butyl > Isopropyl > *n*-Propyl > Ethyl > Methyl > Hydrogen.

This order of +I-effect can also be easily explained on the basis of hyperconjugation which will be discussed later in this unit.

1.7.1.2 Inductive Effect and Hybridization

Inductive effect also depends on the state of hybridization of the atom linked by a covalent bond. For example, acrylic acid ($CH_2=CHCOOH$) is stronger acid than propionic acid (CH_3CH_2COOH). It may be attributed to the electron-withdrawing inductive effect (–I) of the sp^2-hybridized carbon atom next to carbonyl group although the resonance effect of α, β-unsaturated carbonyl system would tend to decrease the acid strength.

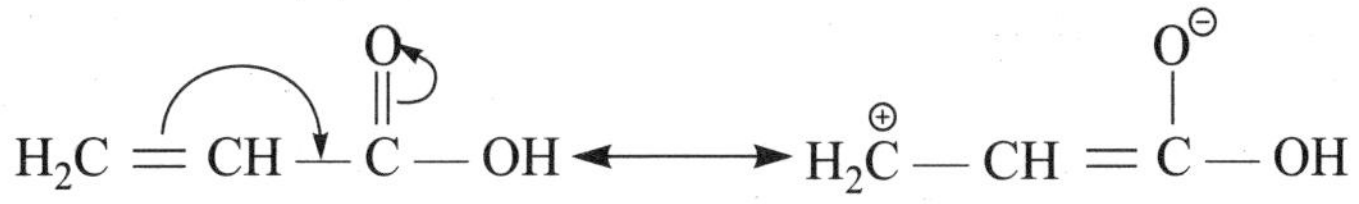

$$H_2C=CH-\overset{\overset{\displaystyle O}{\|}}{C}-OH \longleftrightarrow H_2\overset{\oplus}{C}-CH=\overset{\overset{\displaystyle O^{\ominus}}{|}}{C}-OH$$

Further, the acid strength of propiolic acid ($CH\equiv C-COOH$) is many times as much as that of propionic acid because of strong electron-withdrawing inductive effect (–I) of *sp*-hybridized carbon atom next to carbonyl group.

1.7.1.3 Characteristics of Inductive Effect

(*i*) I-effect involves only the tendency of displacement of σ-electrons and it occurs in polar single covalent bonds.

(*ii*) It is a permanent effect which cannot be reversed.

(*iii*) The inductive effect decreases progressively with an increasing distance between the carbon atom and the other atom or group causing this effect.

(*iv*) The electrons are shifted slightly without leaving their orbitals, *i.e.*, only little distortion of the orbital occurs which leads to polarization.

(*v*) Different atoms or groups cause polarization of the covalent bond to different extent depending upon their power of accepting or donating the electrons.

1.7.1.4 Uses of Concept of Inductive Effect in Organic Chemistry

1. *Stability of Reactive Intermediates*

As we shall see later (Chapter 2, Sec. 2.5) the relative stability of free radicals, carbocations, and carbanions can be explained on the basis of inductive effect.

2. *Dipole Moment of Organic Compounds*

Due to inductive effect covalent bonds between two dissimilar atoms get polarized. Due to this polarization such covalent bonds have a definite dipole moment. The magnitude of this dipole moment, however, depends upon:

(*i*) Difference in electronegativity.

(*ii*) Orientation of various polar covalent bonds.

(*i*) *Difference in electronegativity*: Greater the difference in electronegativity between the two atoms joined by covalent bond, more polar is the covalent bond and hence greater is the dipole moment. The data given in table 1.6 are in good agreement with this.

Table 1.6: Increase in dipole moment with increase in difference in electronegativity (E.N.) of bonded atoms

Compounds	Difference in E.N.	Dipole moment (D)
CH_3—CH_3	0	0
CH_3—NH_2	3.0 – 2.5 = 0.5	1.32
CH_3—OH	3.5 – 2.5 = 1.0	1.71
CH_3—F	4.0 – 2.5 = 1.5	1.82

Since, the difference in electronegativity of C and H is very small, the dipole moment of alkanes are so small that they are difficult to measure. For example, dipole moment of propane is 0.083 D and that of isobutane is 0.132 D.

***(ii) Orientation of various polar covalent bonds*:** For a molecule having two or more covalent bonds, the actual dipole moment depends upon the orientation of these bonds in space. In other words, the actual dipole moment of the molecule is a vector sum of all the individual bond moments (Fig. 1.28). For instance, the dipole moment of methylene chloride (CH_2Cl_2), chloroform ($CHCl_3$) and carbon tetrachloride (CCl_4) are 1.62, 1.01 and 0D, respectively.

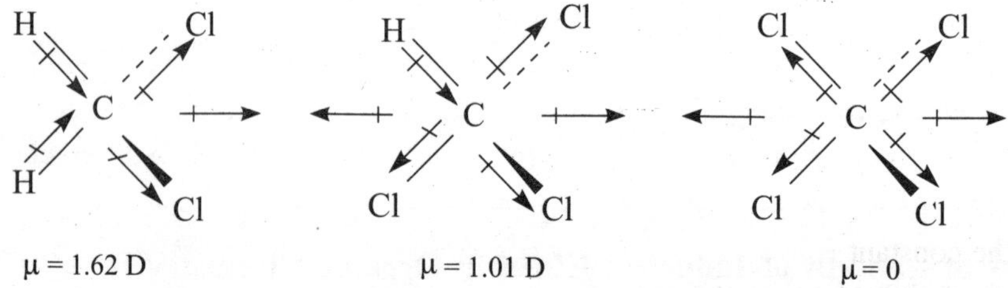

Fig. 1.28: Dipole moments of CH_2Cl_2, $CHCl_3$ and CCl_4

The molecule of carbon tetrachloride has four polar C—Cl bonds. As the molecule has tetrahedral geometry, the resultant dipole moment of any two adjacent (C—Cl) bonds gets cancelled by an equal and opposite resultant dipole moment of the remaining two C—Cl bonds. As a result, the net dipole moment of the molecule is zero.

In methylene chloride, the resultant of the two C—H dipoles reinforces the resultant dipole moment of the two C—Cl bonds and hence the molecule has a net dipole moment of 1.62 D. However, in chloroform, the resultant of one small C—H bond dipole and one large C—Cl dipole opposes the resultant of the remaining two C—Cl bonds dipoles. As a result, the net dipole moment of chloroform is lower (1.01 D) than that of methylene chloride (1.62 D).

Similarly, we can explain difference in dipole moments of three isomeric dichlorobenzenes (Fig. 1.29).

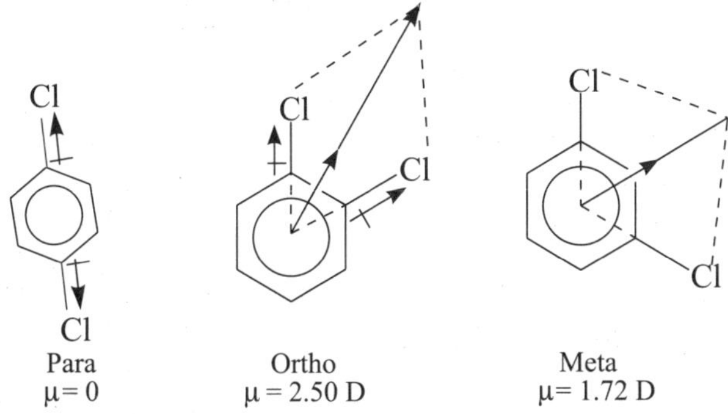

Fig. 1.29: Dipole moments of *o, m* and *p*-dichlorobenzenes

In *p*-dichlorobenzene, the two C—Cl dipoles are equal and opposite in direction and hence cancel out each other. Thus its net dipole moment is zero.

In *o*-and *m*-dichlorobenzenes, the angles between the two dipoles are 60° and 120°, respectively. According to law of parallelogram of forces*, the magnitude of the resultant force increases with a decrease in the angle between the two forces. Thus, dipole moment of *o*-isomer is higher (2.50 D) than that of *m*-isomer (1.72 D).

3. *Strength of Carboxylic Acids*

In aqueous solution, a carboxylic acid ionises and the following equilibrium is established:

$$RCOOH + H_2O \rightleftharpoons RCOO^- + H_3O^+$$

The strength of an acid is usually expressed in terms of its dissociation or acidity constant (K_a) or pK_a, which is calculated as follows:

$$K_a = \frac{[RCOO^-][H_3O^+]}{[RCOOH]}$$

and $pK_a = -\log K_a$

Since, the constant K_a is directly proportional to the concentration of H_3O^+, higher the value of K_a or lower the value of pK_a, stronger is the acid. Further, the value of K_a also depends upon the concentration of the carboxylate anion ($RCOO^-$) which, in turn, will depend upon its stability. Thus any factor which increases the stability of the $RCOO^-$ anion relative to RCOOH, should increase the acid strength while any factor which decreases the stability of the $RCOO^-$ anion relative to RCOOH decreases the acid strength.

It has been observed that electron-withdrawing groups increase the acid strength while electron-donating groups decrease the acid strength of carboxylic acids.

Reason: As expected electron-donating groups (D), because of their +I-effect, would decrease the stability of the carboxylate anion relative to carboxylic acid by intensifying the electron density thereby decreasing the acid strength of the carboxylic acid. Conversely, electron-accepting groups (A), because of their –I-effect, would increase the stability of the carboxylate anion by dispersing its negative charge, *i.e.*, electron density thereby increasing the acid strength of the acid.

* If A and B are two forces inclined at an angle θ, then their resultant $\sqrt{A^2 + B^2 + 2\,AB\cos\theta}$

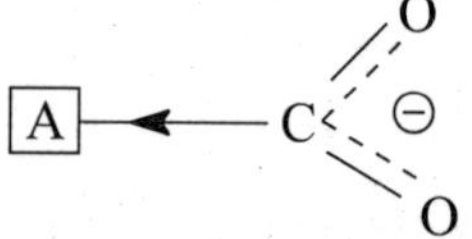

[D] =Donar group increases the electron density, destabilizes the carboxylate anion and hence decreases the acid strength.

[A] = Acceptor group decreases the electron density, stabilizes the carboxylate anion and hence increases the acid strength.

With the help of an example, we shall now discuss the effect of substituents on strength of acids.

(*i*) *Effect of electron-donating substituents* (*Effect of alkyl groups*): The pK_a values of formic acid and acetic acid are as follows:

Formic acid (pK_a = 3.75) Acetic acid (pK_a = 3.76)

Obviously, acetic acid is weaker than formic acid.

Reason: In acetic acid, the CH_3 group, due to its +I-effect increases the electron density in the O—H bond which makes the release of H^+ more difficult. Alternatively, the +I-effect of CH_3 group intensifies the negative charge on the acetate ion and thus destabilizes it relative to formate ion.

Formate ion Acetate ion

(*ii*) *Effect of electron-withdrawing groups* (*Effect of halogens*): The pK_a values of acetic acid and chloroacetic acid are as follows:

Acetic acid (pK_a = 4.76) Chloroacetic acid (pK_a = 2.86)

Obviously, chloroacetic acid is stronger than acetic acid.

Reason: In chloroacetic acid the chlorine atom due to its –I-effect, withdraws electrons from the O—H bond which weakens the O—H bond thereby facilitating the release of H^+ from this acid. Hence, chloroacetic acid is a stronger acid than acetic acid.

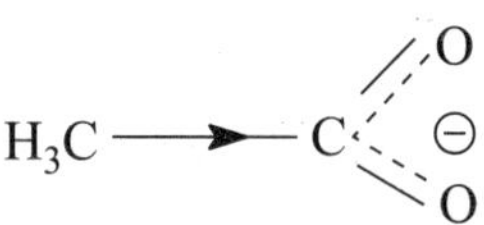

CH_3 group donates electrons and thus destabilize the acetate ion.

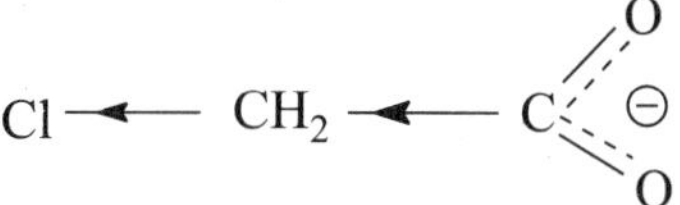

Cl-atom withdraws electrons and thus stabilizes the chloroacetate ion

Alternatively, the –I-effect of chlorine disperses the negative charge on chloroacetate ion while CH_3 group due to its +I-effect enhances the negative charge on acetate ion. Therefore, chloroacetate ion is more stable than acetate ion and hence chloroacetic acid is a stronger acid than acetic acid.

Some Important Generalizations

(*i*) *The greater is the magnitude of* +I-*effect of a substituent or larger is the number of electron-donating substituents, the weaker is the acid.* For example, strengths of some aliphatic acids are in the following order:

$$CH_3 \rightarrow COOH > CH_3 \rightarrow CH_2 \rightarrow COOH > (CH_3)_2HC \rightarrow COOH > (CH_3)_3C \rightarrow COOH$$

Ethanoic acid (no methyl group) > Propanoic acid (one methyl group) > 2-methylpropanoic acid (two methyl groups) > 2,2-dimethylpropanoic acid (three methyl groups)

(*ii*) *The greater is the magnitude of –I-effect of a substituent stronger is the acid.* The order of –I-effect of some substituents is as follows:

$$-NO_2 > -CN > -F > -Cl > -Br > -I$$

For example, the acid strength of some acids having these groups follows the same order:

$$O_2N \leftarrow CH_2 \leftarrow COOH > NC \leftarrow CH_2 \leftarrow COOH > F \leftarrow CH_2 \leftarrow COOH$$

Nitroacetic acid ($pK_a = 1.68$) > Cyanoacetic acid ($pK_a = 2.47$) > Fluoroacetic acid ($pK_a = 2.57$)

$$Cl \leftarrow CH_2 \leftarrow COOH > Br \leftarrow CH_2 \leftarrow COOH > I \leftarrow CH_2 \leftarrow COOH$$

Chloroacetic acid ($pK_a = 2.86$) > Bromoacetic acid ($pK_a = 2.90$) > Iodoacetic acid ($pK_a = 3.16$)

(*iii*) *Increase in number of groups showing –I-effect makes the acid stronger.* For example,

$$\mathrm{Cl} \leftarrow \underset{\underset{\mathrm{Cl}}{\downarrow}}{\overset{\overset{\mathrm{Cl}}{\uparrow}}{\mathrm{C}}} \leftarrow \mathrm{COOH} \;>\; \underset{\underset{\mathrm{Cl}}{\downarrow}}{\overset{\overset{\mathrm{Cl}}{\uparrow}}{\mathrm{HC}}} \leftarrow \mathrm{COOH} \;>\; \mathrm{Cl} \leftarrow \mathrm{CH_2} \leftarrow \mathrm{COOH}$$

Trichloroacetic acid	Dichloroacetic acid	Chloroacetic acid
($pK_a = 0.65$)	($pK_a = 1.25$)	($pK_a = 2.86$)
(three Cl atoms)	(two Cl atoms)	(one Cl atom)
Strongest acid		***Weakest acid***

(*iv*) *The strength of an acid decreases with increase in distance between the –I-group and COOH group.* This is due to the reason that inductive effect decreases rapidly with an increase in number of intervening covalent bonds. Some examples are:

$$H_3C—CH_2—\overset{\overset{\mathrm{Cl}}{|}}{\mathrm{CH}}—COOH > H_2C—\overset{\overset{\mathrm{Cl}}{|}}{\mathrm{CH}}—CH_2—COOH > Cl—CH_2CH_2CH_2—COOH$$

α-Chlorobutyric acid	β-Chlorobutyric acid	γ-Chlorobutyric acid
($pK_a = 2.84$)	($pK_a = 4.06$)	($pK_a = 4.52$)
Strongest acid		***Weakest acid***

4. *Acid Strength of Alkynes*

The acid strength of alkynes decreases in the order:

$$HC{\equiv}CH > CH_3 \rightarrow C{\equiv}CH > CH_3CH_2 \rightarrow C{\equiv}CH$$

Reason: The magnitude of +I-effect of the alkyl groups follows the order $CH_3CH_2 > CH_3 > H$. Now as we replace H by CH_3 or CH_3CH_2, the electron density around the $\equiv C—H$ bond increases in the same order. Thus, the $\equiv C—H$ bond becomes stronger thereby making the release of the proton more and more difficult and hence the acid strength decreases.

5. *Base Strength of Amines*

Like ammonia, all the aliphatic amines have a pair of non-bonding electrons on the nitrogen atom which can readily capture a proton. Thus all the amines (1°, 2° and 3°) are basic in nature. Therefore, we can say that *the greater is the tendency of an amine to donate its lone-pair of electrons greater is its base strength.*

The strength of an amine in aqueous solution is generally expressed in terms of dissociation constant K_b or pK_b which is calculated as follows:

An amine in aqueous solution exists in the following equilibrium:

$$RNH_2 + H_2O \longrightarrow R\overset{\oplus}{N}H_3 + \overset{\ominus}{O}H$$

$$\therefore\; K_b = \frac{[R\overset{\oplus}{N}H_3] + [\overset{\ominus}{O}H]}{[RNH_2]}$$

and $$pK_b = -\log K_b$$

Since the constant K_b is directly proportional to the concentration of $\overset{\ominus}{O}H$, the higher is the value of K_b or lower is the value of pK_b stronger is the base. Similar to carboxylic acids, the strength of amines depends upon the inductive effect of the substituents. We shall discuss these effects below:

(i) Effect of electron-donating substituents (Alkyl group): In aliphatic 1°, 2°, and 3° amines, one, two and three alkyl groups having +I-effect are attached to the nitrogen atom, respectively. These groups increase the electron density on the nitrogen atom. Thus, *all the amines are capable of donating a pair of electrons more easily than ammonia and hence are stronger bases than ammonia.*

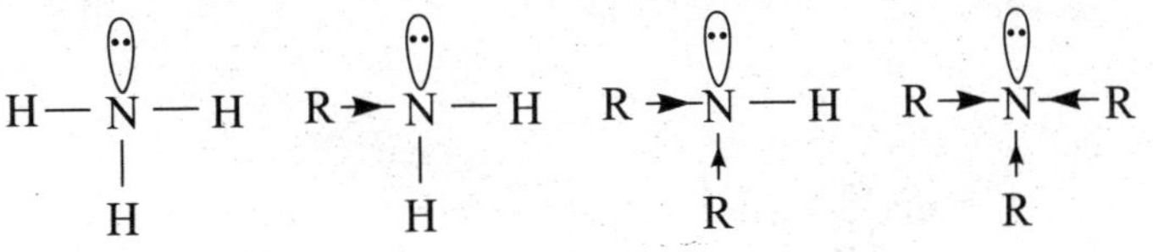

H—N—H (with H below) — Ammonia; R→N—H (with H below) — 1° Amine; R→N—H (with R↑ below) — 2° Amine; R→N←R (with R↑ below) — 3° Amine

Ammonia 1° Amine 2° Amine 3° Amine

The expected order of base strength on the basis of +I-effect of alkyl groups alone should be 3° > 2° > 1° > NH_3.

However, the observed order has been found to be 2° > 1° > 3° > NH_3 in aqueous solution as evident from the following pK_b values in water:

$(CH_3)_2NH$	>	CH_3NH_2	>	$(CH_3)_3N$	>	NH_3
(pK_b = 3.23)		(pK_b = 3.36)		(pK_b = 4.20)		(pK_b = 4.65)
Dimethylamine		Methylamine		Trimethylamine		Ammonia

Reason: The tertiary amines are weaker than secondary amines. This ***anomaly*** has been explained on the basis of solvation and subsequent stabilization of the conjugate acids of amines by water molecules through the formation of hydrogen bonds. The conjugate acid of tertiary amine has less capacity to enter into hydrogen bonding as it has only one hydrogen as compared to two and three hydrogens in the conjugate acids of secondary and primary amines, respectively.

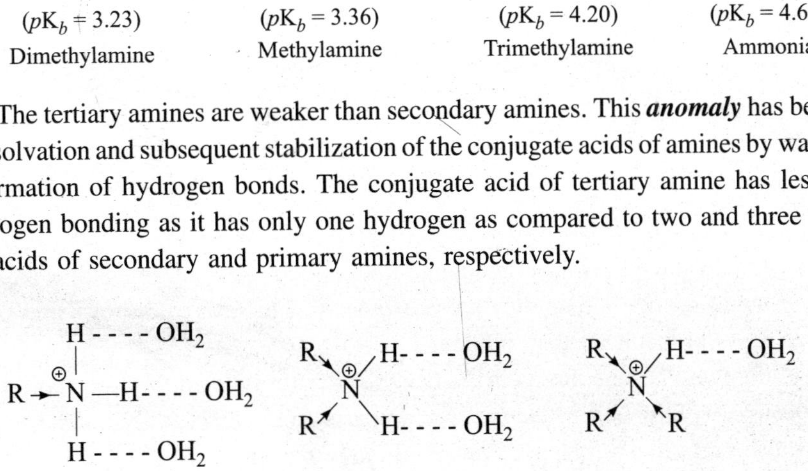

$R\rightarrow\overset{\oplus}{N}$(—H----$OH_2$)$_3$ — Most stable; $R_2\overset{\oplus}{N}$(—H----OH_2)$_2$ — Less stable; $R_3\overset{\oplus}{N}$—H----OH_2 — Least stable

Most stable Less stable Least stable

Actually, these two opposing factors (+I-effect of alkyl groups and stability of conjugate acids) balance each other in case of 2° amines, which makes 2° amines to be the strongest.

3° Amines are weaker than 2° amines due to less stability of conjugate acids of 3° amines while 1° amines are weaker than 2° amines due to less +I-effect of one alkyl group as compared to that of two alkyl groups in 2° amines.

***(ii) Effect of electron-withdrawing substituents*:** Electron-withdrawing groups (–I-effect) tend to decrease the electron density on the N-atom due to which the lone pair of electrons is not easily available for protonation. Therefore, the presence of such groups on the N-atom decreases the base strength of amines. For example, tris–(trifluoromethyl) amine is virtually non basic due to powerful –I-effect of the three-CF_3 groups.

$$F_3C \leftarrow \ddot{N} \rightarrow CF_3$$
$$\downarrow$$
$$CF_3$$

1.7.2 Field Effect

We have learnt that inductive effect operates through bonds. However, there is another effect operating not through the bonds but through space or solvent molecules. This effect is called **field effect**. As such it is not very easy to distinguish inductive effect from field effect. However, it has been made possible to recognise the operation of field effect by taking advantage of the fact that the field effect depends upon the geometry of the molecule whereas inductive effect depends upon the nature of bonds. For example, in isomers A and B, the inductive effect of the chlorine atoms on the COOH group must be the same since the same number of bonds intervene. Hence, acid strength of the two acids (A and B) should be the same. Actually, *the acid strength of the two acids has been found to be different.*

(A) pK_a : 5.67 (Stronger acid)

(B) pK_a : 6.07 (Weaker acid)

Reason: The acid strength is different due to field effects. The chlorine atoms are closer in space to the COOH group in isomer A than they are in isomer B. This example shows that sometimes field effects are even more pronounced than inductive effects.

Therefore, the two chlorine atoms in A, being closer to the acid group as compared to B, will exert greater electron-withdrawing effect in A as compared to B. Hence, the isomer A is a stronger acid than B. This is also supported by the pK_a values of the two acids.

The best example of field effects occurring through solvent molecules is provided by comparison of first and second ionisation constants of dicarboxylic acids (Table 1.7).

Table 1.7: Dissociation or ionisation constants of dicarboxylic acids

Acid	pK_a^1	pK_a^2	K_a^1 / K_a^2
Oxalic acid	1.27	4.27	989
Malonic acid	2.86	5.70	692
Succinic acid	4.21	5.64	26.9
Glutaric acid	4.34	5.27	8.51

All these acids have two carboxyl groups which undergo ionization in two stages as shown below:

First ionization stage

$$(CH_2)_n\begin{matrix} COOH \\ COOH \end{matrix} \quad H_2O \xrightleftharpoons{K^1} (CH_2)_n\begin{matrix} COOH \\ COO^{\ominus} \end{matrix} + H_3\overset{\oplus}{O}$$

First conjugate base (CB 1)

Second ionization stage

$$(CH_2)_n\begin{matrix} COOH \\ COO^{\ominus} \end{matrix} \quad H_2O \xrightleftharpoons{K^2} (CH_2)_n\begin{matrix} COO^{\ominus} \\ COO^{\ominus} \end{matrix} + H_3\overset{\oplus}{O}$$

Second conjugate base (CB 2)

During the first ionization stage, the acid can lose a proton from any one of the two COOH groups but the conjugate base (CB 1) can add a proton only at one —COO^- site. Therefore, it can be assumed that the acid should be twice as strong as a monocarboxylic acid for first ionization stage, *i.e.*,

$$K_a^1 = 2 \cdot K_a \qquad ...(i)$$

where K_a^1 is first dissociation constant of dicarboxylic acid, and

K_a is dissociation constant of any monocarboxylic acid (RCOOH).

During the second stage of ionization, the half ion can lose a proton from only one position but the dianion *i.e.*, second conjugate base (CB 2) can add a proton to two positions. Therefore, the acid is half as strong as a monocarboxylic acid for second ionization stage, *i.e.*,

$$K_a^2 = \frac{1}{2} \cdot K_a \qquad ...(ii)$$

From equations (*i*) and (*ii*) it is clear that $K_a^1 = 4 \cdot K_a^2$ *i.e.*, K_a^1 is expected to be four times K_a^2. But in actual practice, the two constants differ by a factor larger than 4 (Table 1.7).

Reason: The above experimental observation can be explained in terms of field effects occuring through the solvent molecule which increase the ratio K_a^1 / K_a^2 above 4. Further, as the chain length increases *i.e.*, as *n*-increases, both the field effect and the +I-effect would decrease and so, the ratio would approach 4.

1.7.3 Electromeric Effect

Contrary to inductive effect, electromeric effect is a temporary effect which involves a complete transfer of π-electrons of a multiple bond to one of the bonded atoms. Further, this effect operates only on the demand of an attacking reagent. Let us consider, for example, the nucleophilic addition of cyanide ion to a carbonyl group. The π-electrons of the C=O bond are completely transferred to more electronegative oxygen atom. However, the molecule will revert back to its original electronic state when the attacking reagent is removed. Thus this effect is temporary.

$$\rangle C{=}O \underset{\text{No nucleophile}}{\overset{\text{Nucleophile}}{\rightleftharpoons}} \rangle \overset{\oplus}{C}-\overset{\ominus}{C}$$

The net result of the above electron displacement is generation of positive charge on the carbon atom which facilitates the attack of the nucleophile on it. This type of transfer of π-electrons of a multiple bond to one of its bonded atoms in presence of an attacking reagent is called the **electromeric effect** and is designated as **E-effect**.

1.7.3.1 Types of Electromeric Effect

Like inductive effect, the electromeric effect is also of two types:

(*i*) Electron-donating electromeric effect or +E-effect.

(*ii*) Electron-withdrawing electrometric effect or –E-effect.

***(i) Electron-donating electromeric effect*:** When π-electrons of the double bond are transferred to the atom to which attacking reagent gets finally attached , the effect is called **+E-effect.** For example, the addition of *electrophiles* (say $H^{\oplus}$) to alkenes.

$$\overset{\oplus}{H} + \rangle C{=}C\langle \longrightarrow \rangle \underset{H}{\underset{|}{C}}-\overset{\oplus}{C}\langle \qquad \text{(+ E-effect)}$$

***(ii) Electron-withdrawing electromeric effect*:** When π-electrons of the double bond are transferred to the atom to which the attacking reagent does not get attached, the effect is called **–E-effect.** For example, the addition of *nucleophile* (say $\overset{\ominus}{C}N$) to the carbonyl group.

$$\overset{\ominus}{C}N + \rangle C{=}O \longrightarrow NC-\underset{|}{\overset{|}{C}}-\overset{\ominus}{O} \qquad \text{(– E-effect)}$$

Therefore, *electromeric effect is said to be* **–E-effect** *when displacement of electrons is away from the atom or group and* **+E-effect** *when it is towards the atom or group on which attacking reagent is added.*

Table 1.8: Comparison of I-effect and E-effect

I-Effect	E-Effect
1. It is a **permanent effect** which involves slight drifting of shared σ-electrons towards more electronegative atom.	1. It is a **temporary effect** which involves complete transfer of π-electrons to one of the bonded atoms.
2. It operates only in saturated compounds which contain at least one polar bond.	2. It operates in unsaturated compounds which contain at least one multiple bond which may be polar or non-polar.
3. It does not need any outside reagent for its operation.	3. It operates only in the presence of an outside reagent.
4. There is only a slight distortion in the shape of molecular orbital of the σ-bond because electron pair is slightly drifted without leaving its molecualr orbital.	4. Here the π-electrons are transferred to a new molecular orbital because the electron pair is completely transferred to one of the bonded atoms.
5. Inductive effect involves a partial separation of charges without formation of ions.	5. Electromeric effect involves complete transfer of electrons from the reagent to substrate and *vice-versa*. Therefore, ions are always formed.

1.7.3.2 Importance of Electromeric Effect

It may be worthwhile to mention here that a distinction between E- and I-effects has not been made so far. However, close approach of the reagent significantly enhances the reactivity of the molecule which suggests that some electronic effect such as electromeric effect is involved. In certain molecules both I- and E-effects operate together. In some cases they may reinforce each other whereas in others they may oppose each other. In case they oppose each other, the E-effect always dominates over the I-effect.

1.8 THE CONCEPT OF RESONANCE (MESOMERISM)

There are a number of compounds for which it is possible to write more than one Lewis structures. However, none of these structures conforms to all the observed properties of these substances. For instance, following three Lewis structures can be written for the carbonate anion.

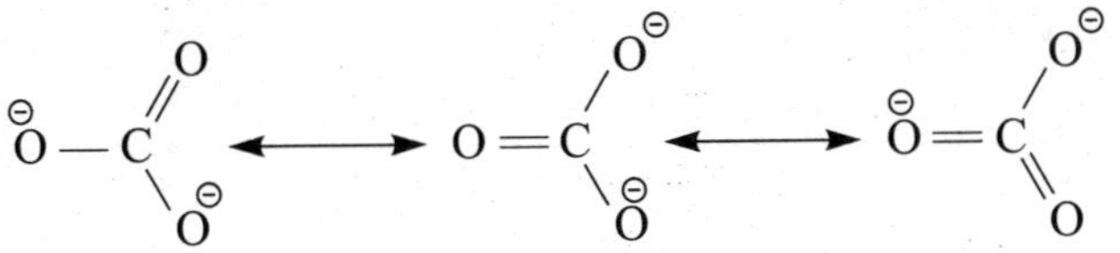

$$\overset{\ominus}{O}-C(=O)-O^{\ominus} \longleftrightarrow O=C(-O^{\ominus})-O^{\ominus} \longleftrightarrow \overset{\ominus}{O}=C(-O^{\ominus})=O$$

A look at these structures reveals that this anion has two carbon-oxygen single bonds and one carbon-oxygen double bond but neither of these predictions is consistent with the observation that all the C—O bonds in carbonate anion are of equal length (130 pm). The concept of *resonance* was introduced to explain such limitations of the classical Lewis structures.

According to the theory of **resonance** *(or mesomerism) for a molecule two or more Lewis structures (called resonance structures) can be written which differ only in the position of electrons but not in relative position of atoms.*

The actual structure of the molecule is not represented by anyone of the Resonance structure but is considered to be hybrid of all these structures. In other words, the actual molecule is considered to be a **hybrid** or **mesomer** (G. between the parts) of all the possible Lewis structures which are referred to as *contributing* or *canonical forms*. One of the important assumptions of this theory is that the contributing or canonical forms have no real existence.

As shown above, the situation of resonance is depicted using a double headed arrow (↔) which is placed between each pair of the contributing forms. It is important to note that the double headed arrow does not signify the oscillation of molecule from one Lewis structure to another. In fact, the word resonance does not mean the mixing of formal structures and, therefore, a less confusing term for this phenomenon *π-electron delocalization* has been coined.

Several analogies have been drawn to understand the meaning of the term resonance hybrid. According to one analogy advanced by **Wheland**, a mule is a hybrid of a horse and a donkey but it does not mean that some mules are horses and some are donkeys nor does it mean that a mule is a donkey part of the time and horse part of the time. Roberts questioned the validity of this analogy as it had a serious drawback, *i.e.*, horse and donkey are real species which wheland compared with the contributing forms having no real existence, and are merely imaginary structures. So, he proposed that an ideal analogy would be the description of a real animal in terms of some mythological creatures. Thus a closer analogy is to describe a rhinoceros as a sort of hybrid of a dragon and a unicorn which are purely mythological animals having no real existence. A person who has never seen a rhinoceros (as is really the case with a molecule capable of having resonance forms) but has some idea of the above mentioned imaginary animals (resonance forms) gets a good mental picture of the former.

An important example of resonance is that of benzene, which can be considered to be hybrid of two Kekulé forms, and three Dewar forms with the latter contributing only little. The real structure of benzene molecule cannot be represented by any of these formulations, as all the carbon-carbon lengths are equal (139 pm). This value is somewhat intermediate between carbon-carbon single bond (154 pm) and carbon-carbon double bond (134 pm). Further,

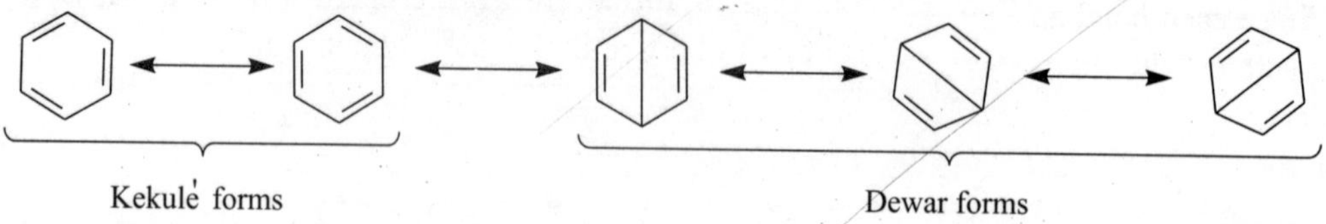

Kekulé forms Dewar forms

benzene resists addition reactions and is a highly stable molecule. The above structures on the other hand show it to be a highly unsaturated compound.

Similarly, carbon dioxide can be written as a resonance hybrid of the following three structures:

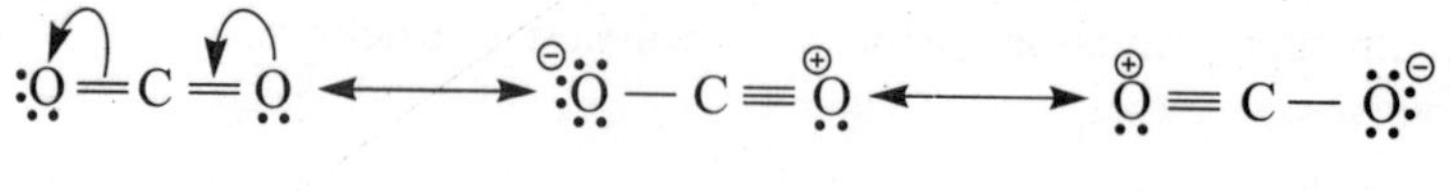

$$:\ddot{O}=C=\ddot{O} \longleftrightarrow \ominus:\ddot{O}-C\equiv\overset{\oplus}{O} \longleftrightarrow \overset{\oplus}{O}\equiv C-\ddot{O}:\ominus$$

According to valence bond method the concept of real structure as the *weighted average* of various possible contributing forms, with localized bonds, is called *resonance*.

1.8.1 Rules for the Resonance Method

The following rules for the use of resonance method enable us to predict when and to what extent resonance is important in determining the properties of various molecules.

(*i*) *Structures contributing towards the hybrid must conform to real Lewis structures, i.e.*, there can be no structure with penta-covalent carbon and bi-covalent hydrogen.

(*ii*) *Resonance involves movement of only π or n-electrons and a resonance structure can be derived from another by a series of one or more short electron shifts without changing the relative position of atoms*. In other words, if two or more double bonds or a double bond and an unshared pair of electrons are in conjugation, there is delocalization of π or *n*-electrons. For examples,

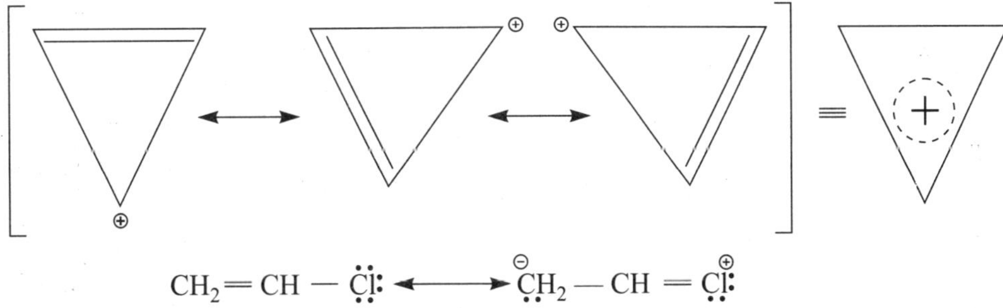

It may be mentioned that in these canonical forms the relative positions of atomic nuclei remain unchanged, otherwise resonance would imply tautomerism*.

(*iii*) *All resonance structures must have same number of paired electrons*. Thus $\dot{C}H_2—\dot{C}H_2$ does not contribute towards resonance of ethylene.

(*iv*) *All resonance structures do not contribute equally to the hybrid*. The relative energies of the various contributing structures are estimated by considering bond energies.

Since each bond adds about 210–420 kJ mol^{-1} to the stability of the system, it is to be expected that structures with more covalent bonds are more stable than those with fewer. The non-polar form of butadiene, for instance, is the most stable of all the resonance structures and the other forms make only minor contribution, if at all, to the resonance hybrid.

$$CH_2=CH-CH=CH_2 \longleftrightarrow \overset{\oplus}{C}H_2-CH=CH-\overset{\ominus}{C}H_2 \longleftrightarrow \overset{\ominus}{C}H_2-CH=CH-\overset{\oplus}{C}H_2$$

(*v*) *When atoms of different electronegativities are involved, the structure which places a negative charge on the most electronegative atom would be more important.* For instance,

---

* **Tautomerism** involves reversible interconversion of structural isomers and is a chemical change symbolized by the reversible arrow (⇌) as against double headed arrow (↔) used to indicate resonance.

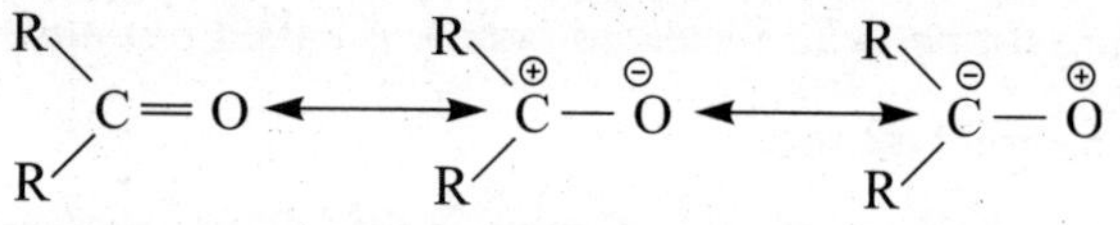

$$R_2C{=}O \longleftrightarrow R_2\overset{\oplus}{C}{-}\overset{\ominus}{O} \longleftrightarrow R_2\overset{\ominus}{C}{-}\overset{\oplus}{O}$$

Major contributor — Minor contributor

(vi) *Resonance contribution is the greatest when there are two or more equivalent contributing forms of lowest energy.* The species involved may be neutral or ionic. For example, resonance in allyl cation, cyclopentadienyl anion and benzene involves contributing forms of comparable energy and renders them quite stable.

$$[\overset{\oplus}{C}H_2{-}CH{=}CH_2 \longleftrightarrow CH_2{=}CH{-}\overset{\oplus}{C}H_2] \equiv [CH_2 \overset{---}{} CH \overset{---}{} \overset{\oplus}{C}H_2]$$

Allyl cation (a stable carbocation)

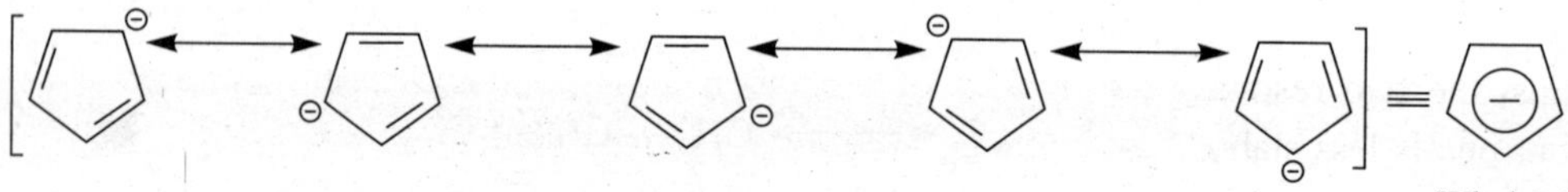

Cyclopentadienyl anion — Wheland structure

It may be noted that ionic species are especially stabilized if resonance is possible in them. This is because of the fact that due to the movement of electrons there is a dispersal of charge over the whole system rather than being localized at a particular atom.

(vii) *If there is only a single contributing structure of lowest energy, the resonance hybrid possesses properties expected for that structure.* Ethylene and butadiene are thus well represented by their conventional structures, but benzene and carbonate anion are not.

(viii) In order to *achieve maximum overlap of p-orbitals, which is necessary for delocalization, it is essential that the skeleton should be planar in conjugated system.* In other words, any structural feature that destroys this coplanarity of the conjugated system inhibits resonance. This inhibition is referred to as *steric inhibition of resonance*. Picryl iodide provides a good example. A comparison of C—N bond lengths of nitro group in the *para* position (135 pm) with that of the nitro groups in the *ortho* position (145 pm) reveals that the former has some double bond character. This is explained on the basis of involvement of the *para* nitro group in resonance because of coplanarity of benzene ring and nitro group. On the other hand, the *ortho* nitro groups do not participate in resonance as they are pushed out of the plane of the ring by bulky iodine atom. Consequently, the C — N bond length of *ortho* nitro groups is very close to that of aliphatic nitro compounds.

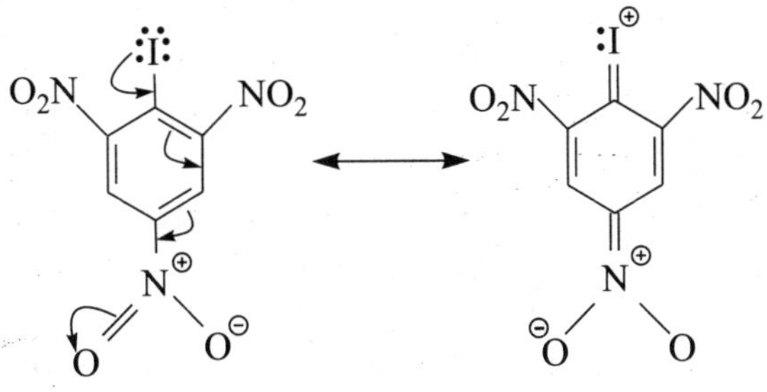

(*ix*) *Resonance structures involving charge separation are less stable than non-polar structures.*

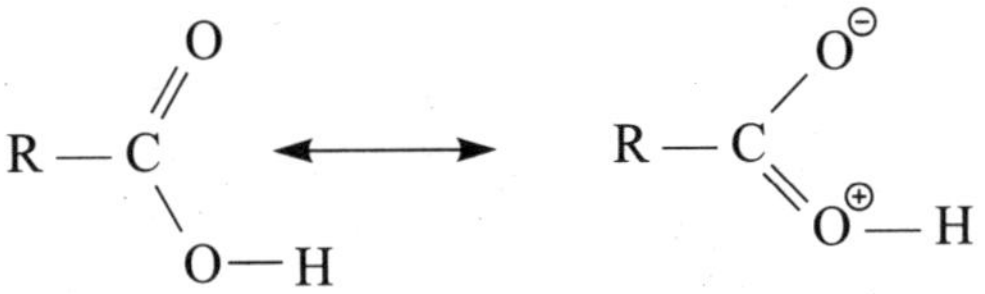

More stable
(*Contributes more towards the resonance hybrid*)

Less stable
(*Contributes less towards the resonance hybrid*)

Thus, the two resonance structures of a carboxylic acid have the same number of covalent bonds, but one is less stable than the other because of charge separation.

(*x*) *Resonance structures having like charges on adjacent atoms are of high energy and hence unimportant.*

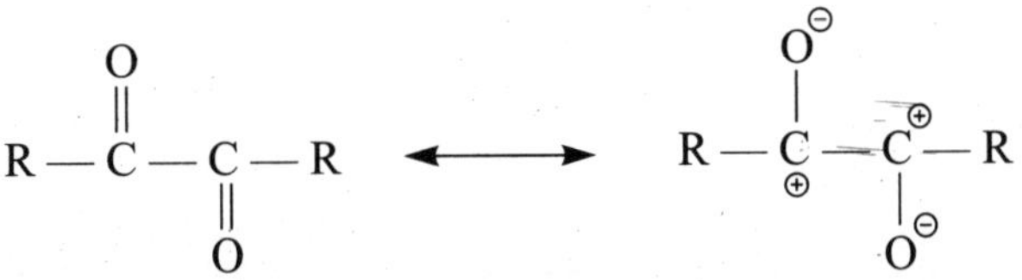

Less stable (*Unimportant*)

(*xi*) *Structures with greater number of covalent bonds contribute more towards the resonance hybrid.*

$$H_2C{=}CH{-}CH{=}CH_2 \longleftrightarrow H_2\overset{\oplus}{C}{-}CH{=}CH{-}\overset{\ominus}{C}H_2 \longleftrightarrow H_2\overset{\ominus}{C}{-}CH{=}CH{-}\overset{\oplus}{C}H_2$$

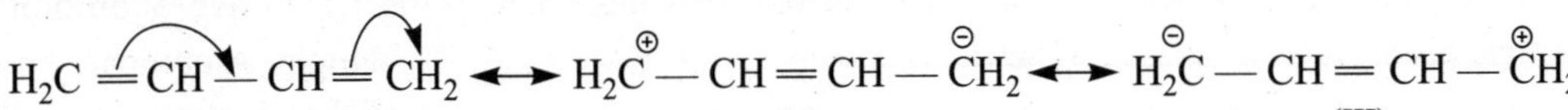

(I) (II) (III)

Since, formation of a bond leads to a decrease in energy, therefore, structure (I) with two π-bonds is more stable than structures (II and III). Hence, structure (I) makes more contribution towards resonance hybrid.

1.8.2 Resonance Energy

One of the important consequences of resonance is that the hybrid (actual molecule) has much lower energy than the energy calculated for any of the contributing structures. This increased stabilization in such systems is expressed in terms of *resonance energy*. It may be defined as the difference between the actual energy of a molecule, capable of resonance, and the energy calculated for the most stable of the resonating structures, *i.e.*, having the minimum energy among the resonanting structures. An idea of the concept of resonance energy is conveyed in Fig 1.30.

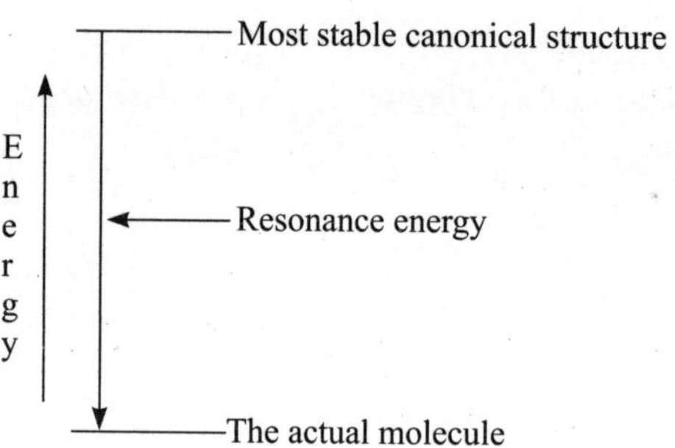

Fig. 1.30: Resonance energy

The resonance energy has been computed from the heats of hydrogenation and heats of formation. Both the methods are given below with bnezene as an example.

***(i) From heats of hydrogenation*:** The resonance energy of benzene has been calculated by comparing heats of hydrogenation of benzene with those of cyclohexene and 1, 3-cyclohexadiene. The heat of hydrogenation of cyclohexene is –119.6 kJ mol^{-1} while that of cyclohexadiene is –232.5 kJ mol^{-1} (nearly double).

Cyclohexene $+ H_2 \xrightarrow{\text{Catalyst}}$ Cyclohexane $\Delta H = -119.6 \text{ kJ mol}^{-1}$

1,3-Cyclohexadiene $+ 2\,H_2 \longrightarrow$ Cyclohexane $\Delta H = -232.5 \text{ kJ mol}^{-1}$

On the basis of these values one would expect the heat of hydrogenation for Kekulé structure of benzene to be $-119.6 \times 3 = -358.8$ kJ mol^{-1}. But the experimental value of the heat of hydrogenation of benzene has been found to be –208.4 kJ mol^{-1}.

Benzene $+ 3\,H_2 \longrightarrow$ Cyclohexane $\Delta H = -208.36 \text{ kJ mol}^{-1}$

Thus, the molecule of benzene is stabilized by $358.98 - 208.36 = 150.62$ kJ mol^{-1}. This difference is referred to as **resonance energy**.

A diagrammatic representation of the concept of resonance energy is given in Fig. 1.31.

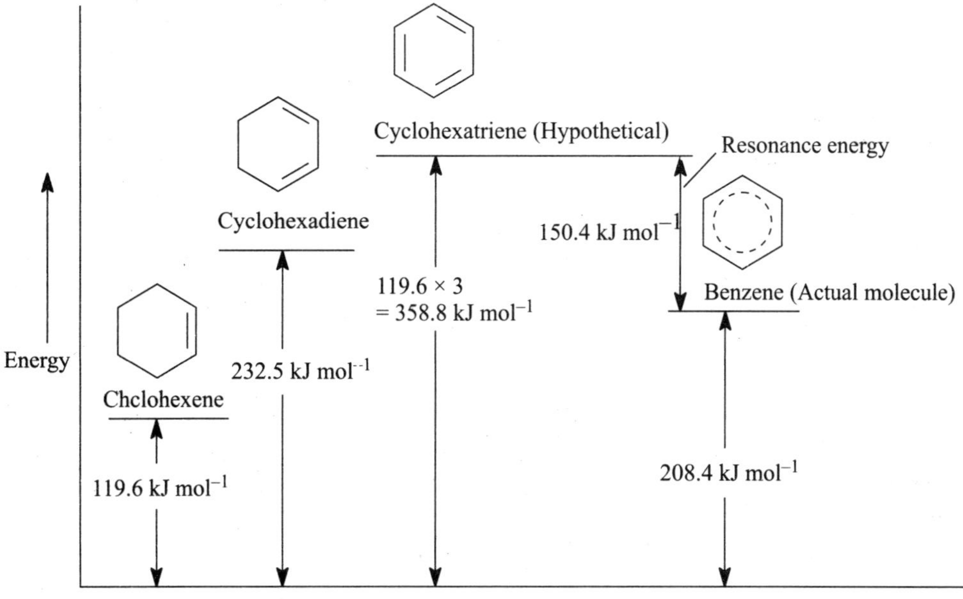

Fig. 1.31: Resonance energy of benzene

***(ii) From heats of formation*:** Heat of formation of Kekulé structure of benzene (ΔH_J) can be calculated from various bond dissociation energies. Thus,

$$\Delta H_J = 6\,(\text{Bond energy of C—H bond}) + 3\,(\text{Bond energy of C—C bond}) + 3\,(\text{Bond energy of C=C bond})$$
$$= 6 \times 414 + 3 \times 347 + 3 \times 609$$
$$= 5352 \text{ kJ mol}^{-1}.$$

The actual value (experimental) for the heat of formation of benzene comes to be 5503.8 kJ mol^{-1}. Hence, the resonance energy of benzene is 5503.8 – 5352 = 151.8 kJ mol^{-1}.

1.8.3 Characteristics of a Resonance Hybrid

A molecule which exhibits the phenomenon of resonance has the following properties:

***(i)* Stability:** A resonance hybrid (or the actual molecule) is always more stable than any of its contributing forms. This stability is attributed to delocalization of electrons and is measured in terms of **resonance energy**. It has been observed that greater is the number of equivalent resonance structures, greater is the delocalization of electrons, larger is the resonance energy and hence stabler is the actual molecule. In other words, larger is the resonance energy, more stable is the compound. For instance, both benzene and 1, 3-butadiene have conjugated double bonds but the resonance energy of benzene (151 kJ mol^{-1}) is larger than that of 1, 3-butadiene (12.5 kJ mol^{-1}). Consequently, benzene is more stable than 1, 3-butadiene. The former is not oxidised with $KMnO_4$ whereas the latter is oxidised readily with $KMnO_4$.

***(ii)* Bond length:** The bond lengths in the resonance hybrid (or actual molecule) differ from those in the respective contributing structures. For instance, all the C — C bond lengths in benzene are 139 pm while each of the two contributing Kekules′ structures has two types of carbon-carbon bond lengths (C — C equal to 154 pm and C = C equal to 134 pm).

1.8.4 Tautomerism *Vs.* Resonance

Tautomerism is a dynamic equilibrium between two functional isomers. It arises due to 1, 3-shift of an ion usually proton from one polyvalent atom to the other within the same molecule. The two isomers thus obtained which exist in dynamic equilibrium with each other are called **tautomers**. For example, propanone and prop-1-en-2-ol are tautomers.

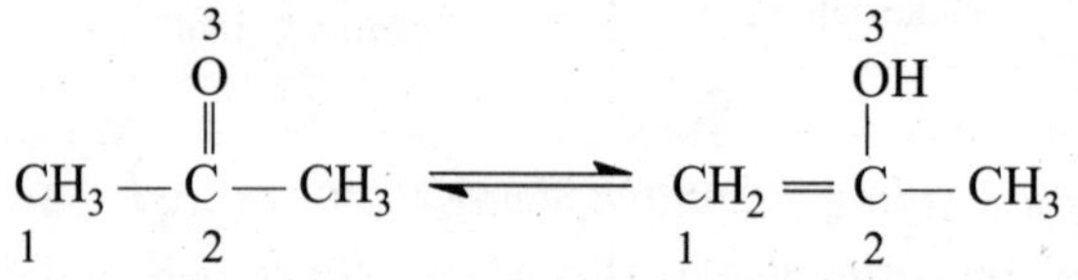

$$\underset{1}{CH_3}-\underset{2}{\overset{\overset{3}{O}}{\overset{\|}{C}}}-CH_3 \rightleftharpoons \underset{1}{CH_2}=\underset{2}{\overset{\overset{3}{OH}}{\overset{|}{C}}}-CH_3$$

(Keto-form) (enol-form)

This type of tautomerism is generally referred to as *keto-enol tautomerism*, since one isomer, *i.e.*, propanone contains the keto group ($>C=O$) and the other isomer, *i.e.*, prop-1-en-2-ol contains the enolic group.

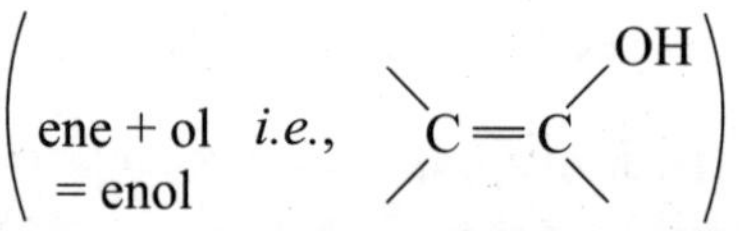

$$\left(\begin{array}{l}\text{ene + ol} \\ \text{= enol}\end{array} \quad i.e., \quad >C=C\begin{array}{l}OH \\ \end{array}\right)$$

The main points of difference between resonance and tautomerism are listed in Table 1.9.

Table 1.9: Difference between resonance and tautomerism

Resonance	Tautomerism
1. Various resonance structures are separated by a double head arrow ($\longleftrightarrow$).	1. Tautomers are separated by an equilibrium arrow ($\rightleftharpoons$).
2. Resonating structures (contributing forms) differ only in the position of electrons. The position of atoms, however, remains fixed.	2. Tautomers differ both in position of electrons as well as of atoms.
3. Various contributing forms (resonating structures) are only imaginary and do not exist at all. Hence, the question of their isolation does not arise.	3. Tautomers are real molecules and actually exist under suitable conditions. They can even be isolated.
4. The contributing forms (resonance structures) are not in dynamic equilibrium.	4. Tautomers exist in a dynamic equilibrium with each other.
5. Resonance leads to stabilization of the molecule.	5. Tautomerism does not lead to stabilization of the molecule.

1.8.5 Resonance Effect *or* Mesomeric Effect

In molecules exhibiting resonance as a result of the displacement of *n*- and π-electrons in a conjugated system, centres of low and high electron densities are created which, in turn, effect the reactivity. *The generation of centres of low and high electron density in conjugated systems as a result of resonance (or mesomerism)* is called **resonance effect (R-effect) or mesomeric effect (M-effect).**

Types of R (or M) Effect

Resonance or mesomeric effects can be divided into two categories depending on the nature of the functional group present adjacent to the multiple bond (double or triple bond).

1. Electron-withdrawing resonance or mesomeric effect (–R or –M-effect): When electron-withdrawing groups such as >C = O, —COOR, —C ≡ N, —NO_2 etc., are present next to a double bond, they withdraw the electrons of the double bond towards themselves. Such groups which withdraw electrons from the double bond as a result of resonance are said to have –R (or –M)-effect. For example,

$$CH_3-CH=CH-CH=O \longleftrightarrow CH_3-\overset{\oplus}{C}H-CH=CH-\overset{\ominus}{O}:$$

Crotonaldehyde

$$CH_2=CH-C\equiv N: \longleftrightarrow \overset{\oplus}{C}H_2-CH=C=\overset{\ominus}{N}:$$

Acrylonitrile

Nitrobenzene

2. Electron-donating resonance or mesomeric Effect (+R or +M-effect): When electron-donating groups having one or more non-bonding electron pairs such as

$$-\ddot{O}-H,\ -\ddot{O}-R,\ -\ddot{S}-R,\ -\ddot{N}H_2,\ -\ddot{N}HR,\ -\ddot{Cl}:,\ -\ddot{Br}:,\ -\ddot{I}: \quad \text{etc.,}$$

are present adjacent to a double bond, they donate (or release) electrons to the double bond. Such groups which donate electrons to the double bond through resonance are said to exhibit +R (or +M)-effect. For example,

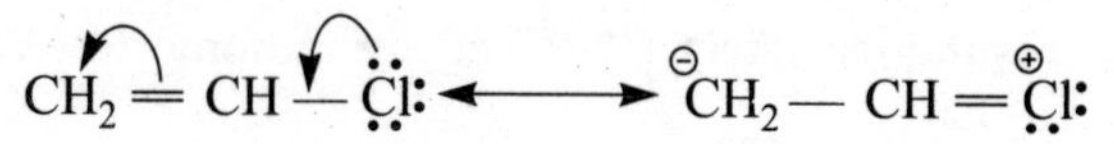

$$CH_2{=}CH{-}\ddot{Cl}: \longleftrightarrow \overset{\ominus}{C}H_2{-}CH{=}\overset{\oplus}{Cl}:$$

Vinyl chloride

$$CH_2{=}CH{-}\ddot{O}{-}CH_3 \longleftrightarrow \overset{\ominus}{C}H_2{-}CH{=}\overset{\oplus}{O}{-}CH_3$$

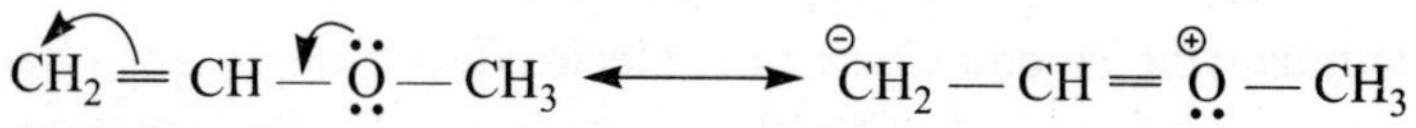

Vinyl methyl ether

Aniline ($C_6H_5\ddot{N}H_2$) $\longleftrightarrow$ resonance structures with $\overset{\oplus}{N}H_2$ and negative charge at ortho and para positions

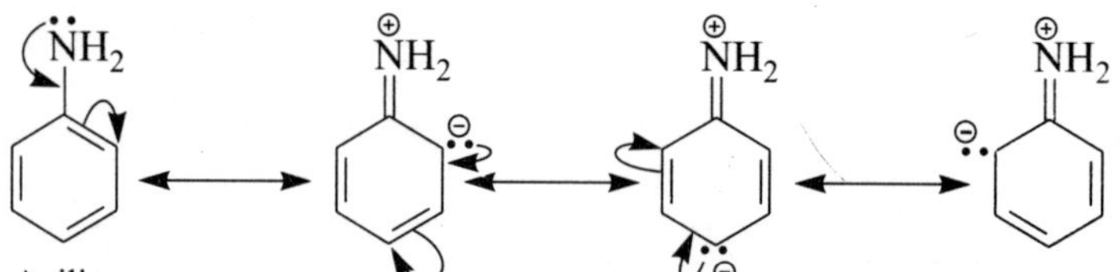

Aniline

Invariably both inductive and mesomeric effects occur together. In some cases they reinforce each other whereas in other cases they oppose each other. Whenever they oppose each other, it is the resonance effect which dominates over the inductive effect. For example, in vinyl chloride if Cl exhibits –I-effect it should attain δ^- charge and if it exhibits +R-effect it should attain full positive charge. So, **+R-effect** of chlorine dominates over its **–I-effect**.

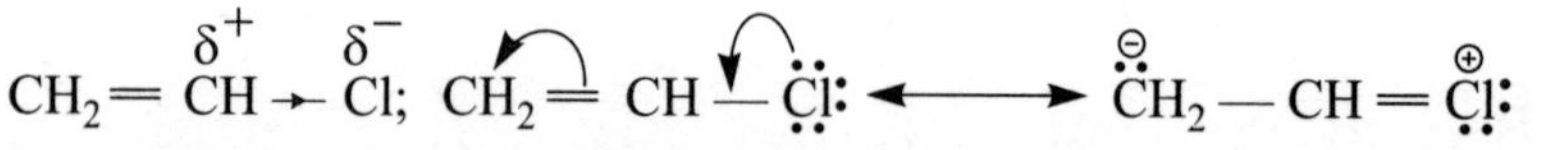

$$CH_2{=}\overset{\delta+}{C}H \rightarrow \overset{\delta-}{Cl}; \quad CH_2{=}CH{-}\ddot{Cl}: \longleftrightarrow \overset{\ominus}{C}H_2{-}CH{=}\overset{\oplus}{Cl}:$$

–I-effect +R-effect

The predominance of +R-effect of Cl is reflected in the addition of HCl to vinyl chloride where H^+ goes to CH_2 and Cl^- goes to that carbon on which Cl is already attached to give ethyledene dichloride (1,1-dichloroethane).

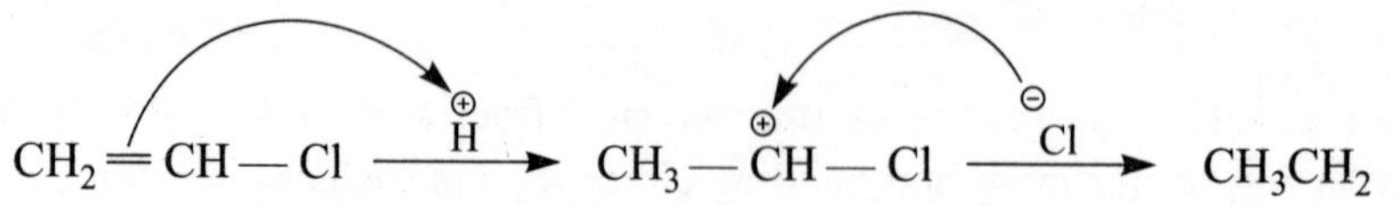

$$CH_2{=}CH{-}Cl \xrightarrow{\overset{\oplus}{H}} CH_3{-}\overset{\oplus}{C}H{-}Cl \xrightarrow{\overset{\ominus}{Cl}} CH_3CH_2$$

Vinyl chloride Ethyledene dichloride

1.8.6 Resonance *Vs.* Inductive Effect

The more important points of difference between R (or M) effect and I-effect are listed in Table 1.10.

Table 1.10: Difference between resonance and inductive effect

R (or M) — effect	Inductive effect
1. Resonance effect involves displacement of n - and π - electrons.	1. Inductive effect involves drifting of σ – electrons.
2. It operates only in conjugated unsaturated systems.	2. It operates in saturated compounds.
3. It involves delocalization of electrons.	3. It does not involve delocalization of electrons.
4. Resonance effect involves complete transfer of electron pair and hence full +ve and -ve charges are developed.	4. Inductive effect involves slight drift of σ - electrons towards the more electronegative atom and hence only partial charges (δ^+ and δ^-) are developed.
5. Resonance effects are transmitted over large distances with almost no change in their magnitude. For example, C_3 of crotonaldehyde is equally susceptible to addition of nucleophile as C_1. $\underset{4}{CH_3}-\underset{3}{CH}=\underset{2}{CH}-\underset{1}{CH}=\ddot{O}: \longleftrightarrow$ $\underset{4}{CH_3}-\underset{3}{CH}=\underset{2}{CH}-\underset{1}{\overset{\oplus}{CH}}-\overset{\ominus}{\ddot{O}}: \longleftrightarrow$ $\underset{4}{CH_3}-\underset{3}{\overset{\oplus}{CH}}-\underset{2}{CH}=\underset{1}{CH}-\overset{\ominus}{\ddot{O}}:$	5. Inductive effects are transmitted only over short distances in saturated chains and the magnitude of the effect decreases rapidly with increasing number of intervening σ - bonds.

1.8.7 Importance of Resonance Effects

Resonance effect plays key role in understanding the reactivity of organic compounds. Some important uses of resonance effect are as follows:

1. Low reactivity of vinyl and aryl halides towards nucleophilic substitutions: In contrast to alkyl halides, vinyl and aryl halides are much less reactive towards nucleophilic substitution reactions.

Reason: Both vinyl chloride and chlorobenzene can be regarded as hybrids of canonical forms I, II and III and VI, respectively.

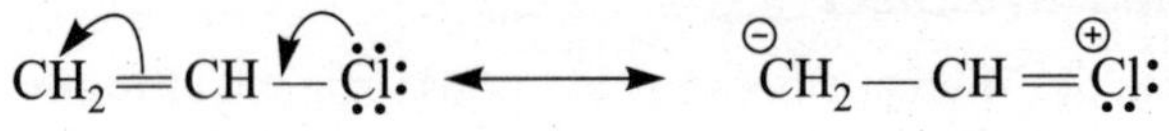

$CH_2{=}CH{-}\ddot{C}l: \longleftrightarrow \overset{\ominus}{C}H_2{-}CH{=}\overset{\oplus}{C}l:$

I Vinyl chloride — II

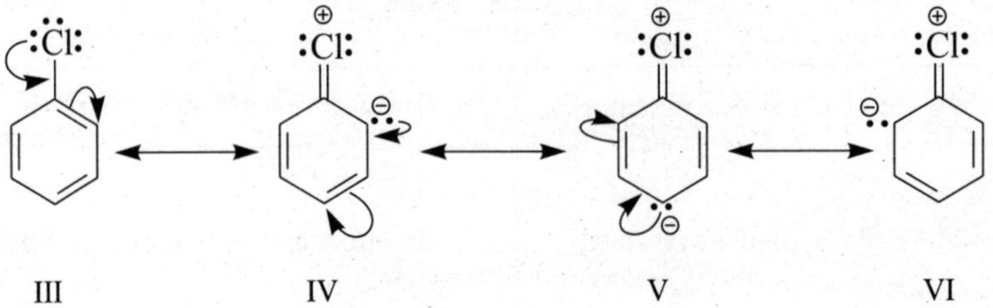

III ⟷ IV ⟷ V ⟷ VI

Chlorobenzene

Due to the above resonance the carbon-chlorine bond acquires some double bond character, thus binding halogen to carbon atom more tightly. Hence, the chlorine atom cannot be easily displaced in a substitution reaction.

2. High reactivity of allyl and benzyl halides: Allyl and benzyl halides undergo nucleophilic substitution reactions more readily than alkyl halides.

Reason: Both allyl and benzyl halides undergo ionization to give resonance stabilized allyl and benzyl cations.

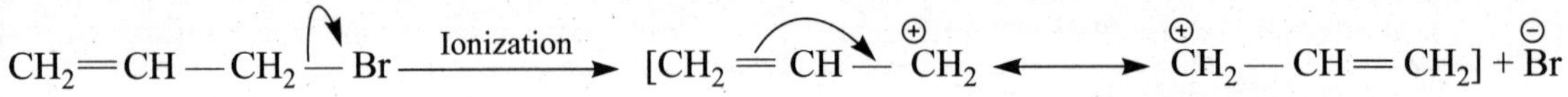

$$CH_2{=}CH{-}CH_2{-}Br \xrightarrow{\text{Ionization}} [CH_2{=}CH{-}\overset{\oplus}{C}H_2 \longleftrightarrow \overset{\oplus}{C}H_2{-}CH{=}CH_2] + \overset{\ominus}{Br}$$

Allyl bromide — Allyl cation (Resonance stabilized)

$C_6H_5CH_2{-}Br \xrightarrow{\text{Ionization}}$ [resonance structures of benzyl cation] $+ \overset{\ominus}{Br}$

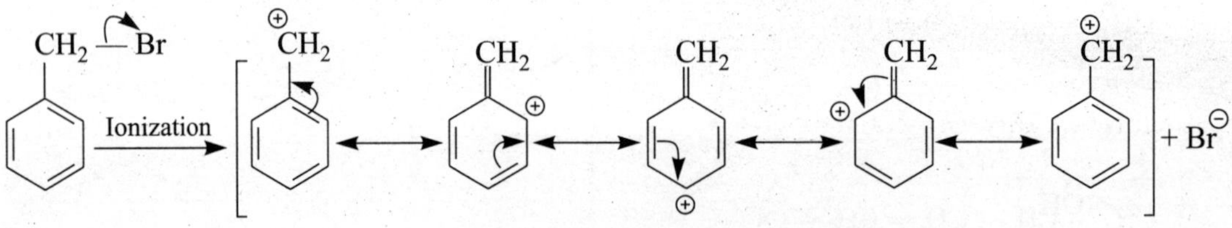

Benzyl bromide — Benzyl cation (Resonance stabilized)

In contrast, the carbocations obtained from primary alkyl halides are not stabilized by resonance and hence are not formed easily.

3. Acid strength of carboxylic acids: The acid strength of the carboxylic acids depends upon the ease of release of proton in aqueous solution to give carboxylate anion. Both carboxylic acids and carboxylate anions are stabilized by resonance. Whereas canonical forms (III and IV) of carboxylate anion are **equivalent** involving dispersal of negative charge, the canonical forms (I and II) of the undissociated carboxylic acid are not only **non-equivalent** but also involve charge separation. We know that resonance involving dispersal of negative charge makes greater contribution than resonance involving separation of charge. Hence, the carboxylate anion will be more stable than carboxylic acid. Hence, the acidic character of carboxylic acids.

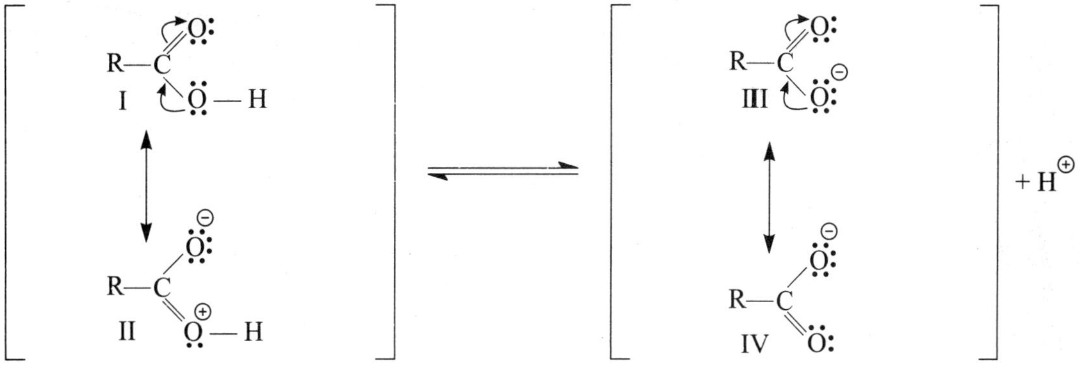

R—C(=O:)—Ö—H (I) ⟷ R—C(—Ö:⊖)=O⊕—H (II) ⇌ R—C(=O:)—Ö:⊖ (III) ⟷ R—C(—Ö:⊖)=Ö: (IV) + $H^{\oplus}$

Carboxylic acid
(*separation of charge*)

Carboxylic anion
(*dispersal of charge*)

4. Greater acid strength of phenol compared to alcohols: In aqueous solution phenol releases a proton to form phenoxide ion due to which it behaves as an acid. Any factor which renders phenoxide ion more stable than phenol will make phenol acidic.

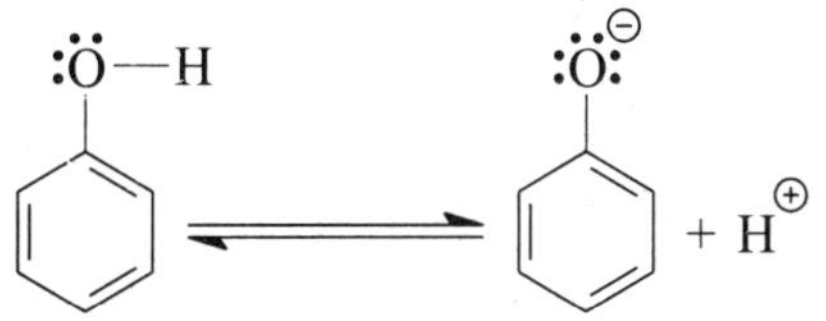

:Ö—H ⇌ :Ö:⊖ + $H^{\oplus}$

Phenol

Phenoxide ion

Both phenol and phenoxide ion are stabilized by resonance. Whereas resonance of *phenol involves separation of charge*, that of *phenoxide involves dispersal of charge*. Since, resonance involving dispersal of charge makes greater contribution than separation of charge, phenoxide ion is more stable than phenol. Hence, *phenol is acidic in nature.*

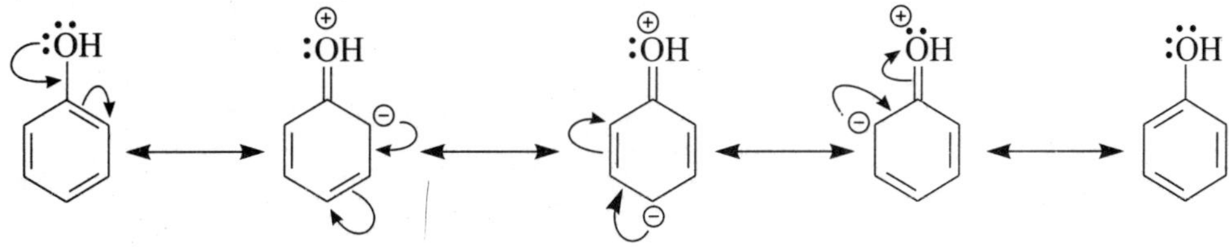

:ÖH ⟷ :OH⊕ ⟷ :OH⊕ ⟷ ÖH⊕ ⟷ :ÖH

Phenol (*separation of charge*) less stable

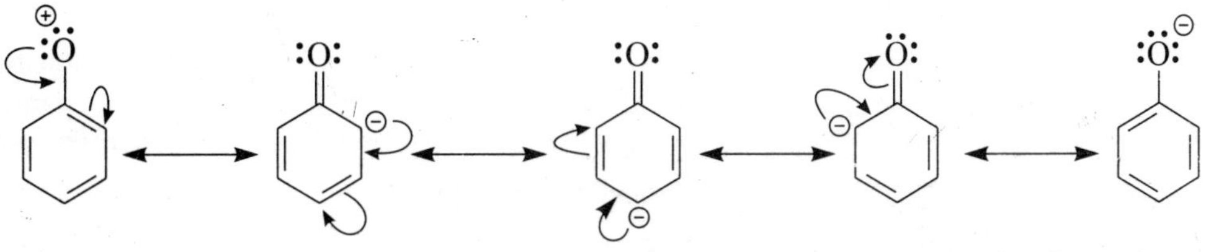

:Ö⊖ ⟷ :O: ⟷ :O: ⟷ Ö: ⟷ :Ö:⊖

Phenoxide ion (*dispersal of negative change*) more stable

Contrary to this an alcohol molecule and the alkoxide ion both are not stabilized by resonance. Hence, alcohols are weak acids as compared to phenols.

$$R-\ddot{\underset{..}{O}}-H \rightleftharpoons R-\ddot{\underset{..}{O}}:^{\ominus} + H^{\oplus}$$

Alcohol (*more stable*) Alkoxide (*less stable*)

Moreover, the alkoxide ion having a formal negative charge on oxygen atom has higher energy than alcohol molecule. Therefore, alcohols have very low tendency to ionize and form less stable alkoxide anion.

Thus, it may be concluded that phenols are stronger acids ($K_a = 10^{-8}$ to 10^{-10}) than alcohols ($K_a = 10^{-6}$ to 10^{-8}). It may be due to the fact that the phenoxide anion formed by the release of proton by phenol is stabilized by resonance but alkoxide anion is not.

5. Comparison of relative base strength of aliphatic and aromatic primary amines: Aliphatic 1° amines *e.g.*, ethylamine ($K_b = 5.6 \times 10^{-5}$) are much stronger bases than aromatic 1° amines *e.g.*, aniline ($K_b = 4.2 \times 10^{-10}$). The +I-effect of ethyl group in ethylamine increases the electron density on the nitrogen atom and hence increases its base strength in comparison to ammonia ($K_b = 1.8 \times 10^{-5}$).

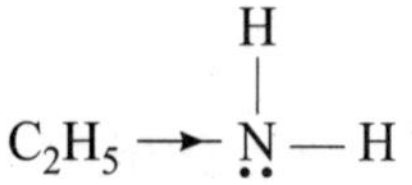

$$C_2H_5 \rightarrow \underset{..}{\overset{H}{\overset{|}{N}}}-H$$

+ I effect of C_2H_5 group

Stronger base *Increased electron density at nitrogen atom*

In contrast aniline is a hybrid of the following contributing forms. Due to resonance the lone pair of electrons on the nitrogen atom gets delocalized over the benzene ring and is thus less readily available to capture a proton as compared to ammonia. Hence, it is weaker base.

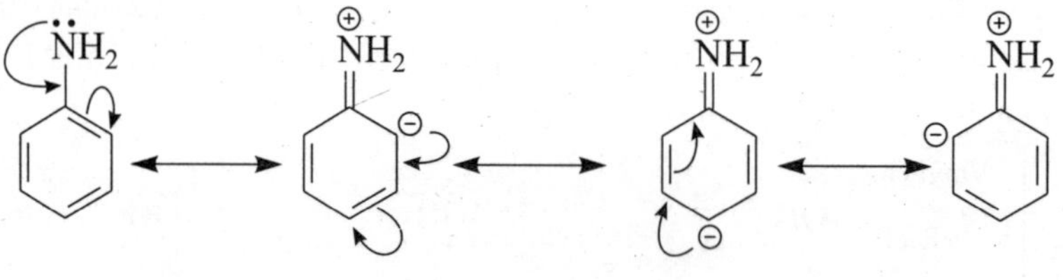

$\ddot{N}H_2$ ⟷ $\overset{\oplus}{N}H_2$ ⟷ $\overset{\oplus}{N}H_2$ ⟷ $\overset{\oplus}{N}H_2$

+ R-effect of NH_2 group
(*Decreased electron density at nitrogen atom*)

6. Relative base strength of ethylamine and acetamide: Ethylamine ($K_b = 5.6 \times 10^{-5}$) is many times stronger base than acetamide ($K_b = 4.3 \times 10^{-13}$). It may be due to following reason:

In ethylamine, the +I-effect of ethyl group increases the electron density on the nitrogen atom due to which its base strength increases.

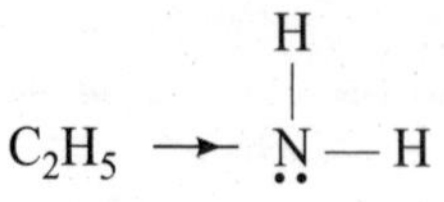

$$C_2H_5 \rightarrow \underset{..}{\overset{H}{\overset{|}{N}}}-H$$

+ I-effect of C_2H_5 group

Ethylamine
(Stronger base)

(*Increased electron density at nitrogen atom*)

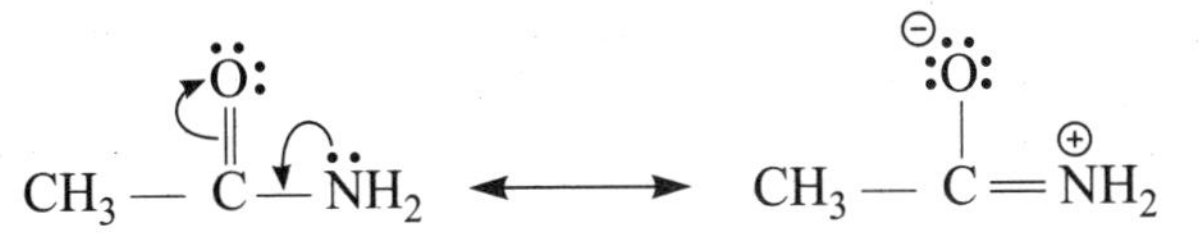

$$CH_3-\overset{\overset{\ddot{O}:}{\|}}{C}-\ddot{N}H_2 \longleftrightarrow CH_3-\overset{\overset{\ominus}{:\ddot{O}:}}{\overset{|}{C}}=\overset{\oplus}{N}H_2$$

I
Acetamide
Weaker base

II
+ R-effect of amino group
(Decreased electron desnity at N atom)

However, acetamide can be regarded as a resonance hybrid of the contributing forms (I and II) where + R-effect of amino group decreases the electron density at the nitrogen atom. Due to this the lone pair of electrons on nitrogen is much less available to capture the proton. Hence, it is a weaker base than ethylamine.

7. Effect on the dipole moment: It is well known that the overall dipole moment of a molecule is the resultant of dipole moments of all the polar bonds. However, resonance may increase or decrease the magnitude of this resultant dipole moment depending upon the nature of the polar bonds present in the molecule. Following examples are illustrative:

(*a*) *Lowering of dipole moment*: The dipole moment of vinyl chloride (1.45 D) is much lower than that of ethyl chloride (2.05 D). This is due to the reason that in vinyl chloride, the –I-effect and +R-effect of chlorine operate in opposite directions thereby decreasing its dipole moment as compared to that of ethyl chloride where only –I-effect of chlorine operates.

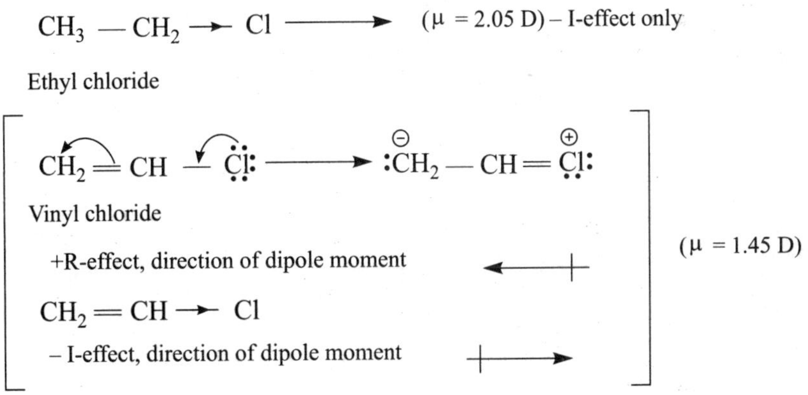

$$CH_3-CH_2 \rightarrow Cl \longrightarrow \quad (\mu = 2.05\ D) - \text{I-effect only}$$

Ethyl chloride

$$CH_2=CH-\ddot{\underset{..}{Cl}}: \longrightarrow :\overset{\ominus}{C}H_2-CH=\overset{\oplus}{\underset{..}{Cl}}:$$

Vinyl chloride

+R-effect, direction of dipole moment $\longleftarrow\!\!+$

$$CH_2=CH \rightarrow Cl$$

– I-effect, direction of dipole moment $+\!\!\longrightarrow$

$(\mu = 1.45\ D)$

In a similar manner, it can be explained why the dipole moment of chlorobenzene (1.69 D) is lower than that of ethyl chloride (2.05 D).

(*b*) *Increasing the dipole moment*: The dipole moment of *p*-nitroaniline (4.92 D) is higher than that of nitrobenzene (4.22 D). In *p*-nitroaniline, the –I-effect of nitro group and +R-effect of amino group reinforce each other (act in the same direction). Hence, the observed higher dipole moment of *p*-nitroaniline as compared to nitrobenzene.

NO_2 – I-effect of NO_2 group
Direction of dipole moment due to NO_2 group

Direction of dipole moment due to –I-effect of NO_2
Direction of dipole moment due to +I-effect of NH_2

8. Difference in C—N bond length for *o* and *p*-nitro groups in picryl iodide: The large iodide atom forces *o*-nitro groups out of the plane of benzene ring and has no effect on *p*-nitro group which remains in the plane of the ring. Therefore, resonance operates in benzene ring involving *p*-nitro group only. As a result, carbon-nitrogen bond acquires some double bond character in *p*-nitro group and hence, the C—N bond length '*a*' is longer (145 pm) than '*b*' which is only 135 pm.

Picryl iodide

1.9 HYPERCONJUGATION

There is another kind of delocalization, called *hyperconjugation*, which is a special type of resonance. It is attributed to σ-π orbital overlap against π-π overlap in butadiene or benzene (Fig.1.32).

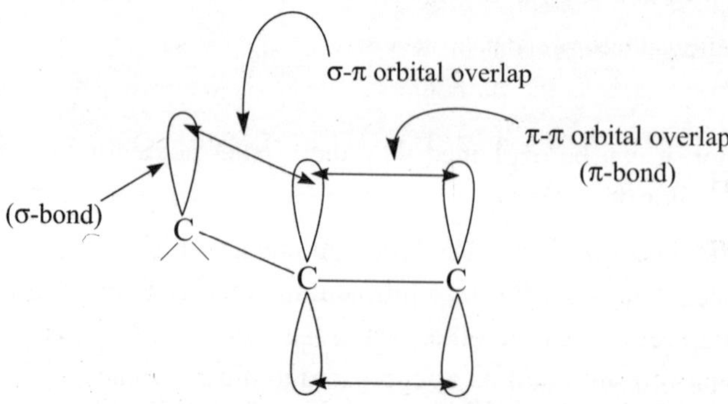

Fig. 1.32: Hyperconjugation (σ-π and π-π orbital overlap)

Baker and Nathan discovered that the rate of the reaction of *p*-alkylbenzyl bromides with pyridine was contrary to what would be expected from the order of the *inductive effect* of the substituent alkyl groups. The order of inductive effect is: R = *tert*-butyl > isopropyl > ethyl > methyl, whereas the rate of the above reaction followed the order: methyl > ethyl > isopropyl > *tert*-butyl *i.e.*, when these alkyl groups are attached to an unsaturated system, the order of +I effect is reversed.

$$R{-}C_6H_4{-}CH_2Br + C_5H_5N \longrightarrow R{-}C_6H_4{-}CH_2\overset{\oplus}{N}C_5H_5 + \overset{\ominus}{Br}$$

p-Alkylbenzyl bromide   Pyridine

This effect is known as the **Baker-Nathan effect** or **no bond resonance** or **hyperconjugation effect**.

It was explained on the basis of hyperconjugative forms, as shown below, contributing to the overall electron donating effect for the substituent having hydrogen attached to it, methyl group showing maximum effect because of the maximum number of hydrogens at the α-position to an unsaturated system. Since, there is no bond between the carbon and hydrogen in the canonical forms, this phenomenon is sometimes also called **no bond resonance.**

Hyperconjugation or no-bond resonance may be shown as below:

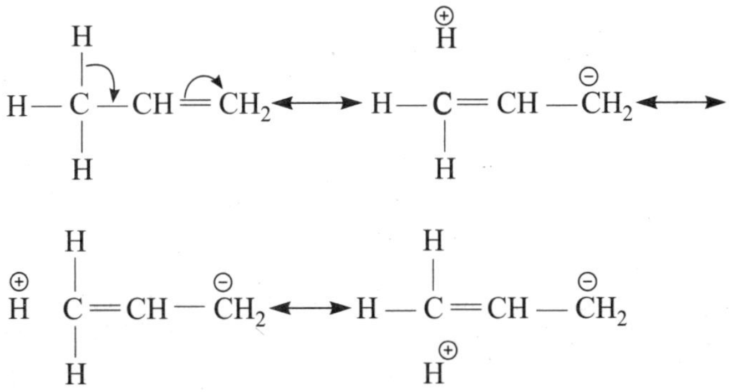

$$H_3C{-}CH{=}CH_2 \longleftrightarrow \overset{\oplus}{H}\;H_2C{=}CH{-}\overset{\ominus}{C}H_2 \longleftrightarrow \overset{\oplus}{H}\;H_2C{=}CH{-}\overset{\ominus}{C}H_2 \longleftrightarrow \overset{\oplus}{H}\;H_2C{=}CH{-}\overset{\ominus}{C}H_2$$

These hyperconjugative contributing forms of propene clearly show that hyperconjugation occurs through σ bonded hydrogen atoms present on the carbon adjacent to the doubly bonded carbon *i.e.*, α-hydrogen atoms. Therefore, larger the number of such α-hydrogen atoms, more are the number of such hyperconjugative structures. Thus, the order of hyperconjugation effect decreases in the order:

$$—CH_3 > —CH_2CH_3 > —CH(CH_3)_2 > —C(CH_3)_3$$

1.9.1 Importance of Hyperconjugation Effect

Both resonance and hyperconjugation are the results of delocalizaiton of electrons involving π–π conjugation and σ–π conjugation, respectively. The conjugation effect due to hyperconjugation is weaker than resonance but even then it is very useful in explaining some physical and chemical properties of organic molecules as discussed below:

1.9.1.1 Stability of Carbocations and Free Radicals

The order of stability of carbocations and free radicals can be explained on the basis of hyperconjugation. The decreasing order of stability of carbocations and free radicals is : 3° > 2° > 1° (For details refer to Chapter 2, *pp*. 25 and 43).

1.9.1.2 Stability of Alkenes

The alkene which has more alkyl groups attached to the doubly bonded carbon atoms is the most stable. This can be explained on the basis of hyperconjugation. Larger the

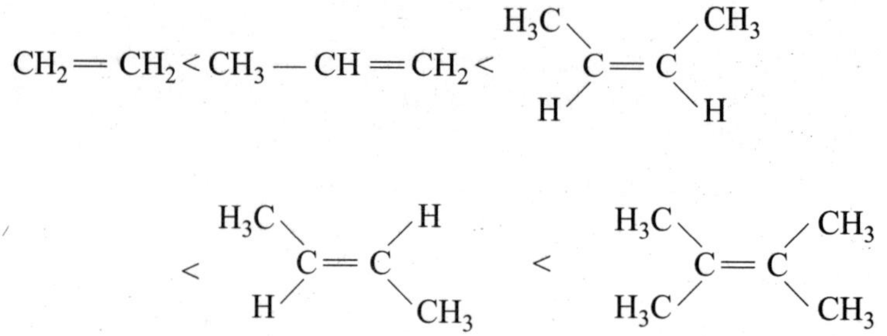

$$CH_2{=}CH_2 < CH_3{-}CH{=}CH_2 < (H_3C)(H)C{=}C(CH_3)(H) < (H_3C)(H)C{=}C(H)(CH_3) < (H_3C)_2C{=}C(CH_3)_2$$

number of alkyl groups more are the number of hyperconjugation structures that can be written for each one of the alkene. Therefore, greater the number of such structures, more stable is the alkene.

For but-2-ene, *trans*-form is more stable than *cis* due to electronic repulsions present in *cis* arrangement where both methyl groups are on the same side *i.e.*, close to each other.

1.9.1.3 Directive Influence of Alkyl Groups

The *o, p*-directing influence of alkyl groups can be explained on the basis of hyperconjugation. For example, toluene can be depicted as a hybrid of the following contributing forms:

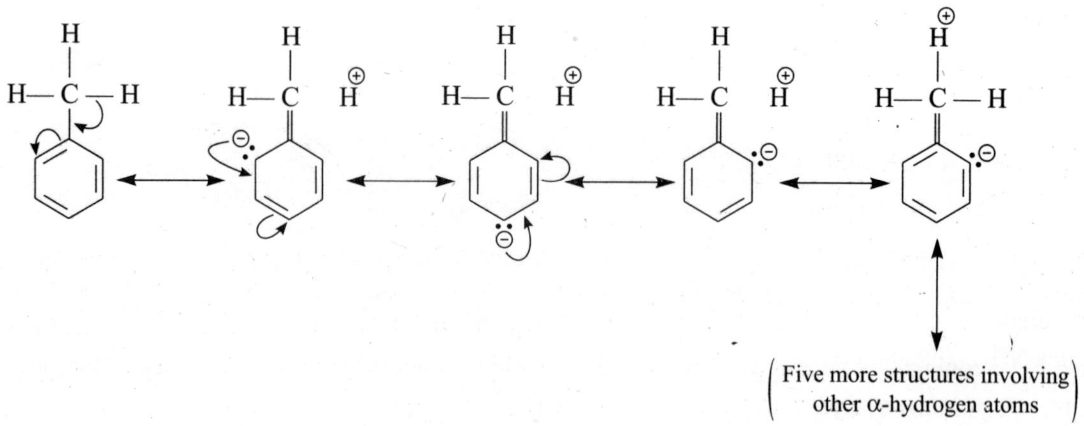

(Five more structures involving other α-hydrogen atoms)

It is clear from these structures that the electron density at *o*- and *p*-positions with respect to CH_3 group increases and hence the electrophilic substitution reactions in toluene prefer attack at *o*-and *p*-positions.

1.9.1.4 Shortening of Carbon-Carbon Bonds Adjacent to Unsaturated Bond

In propene $C_2 — C_3$ single bond acquires some double bond character *i.e.*, it is a little shorter (149 pm) as compared to normal single C — C bond length (154 pm) in propane. This can be explained on the basis of hyperconjugation.

$$\underset{\substack{|3\\H}}{\overset{H}{\underset{|}{\overset{|}{H-C}}}}-\underset{\substack{|2\\H}}{C}=\underset{1}{CH_2} \longleftrightarrow H-\underset{\substack{3|\\H}}{C}=\underset{\substack{|2\\H}}{C}-\underset{1}{\overset{\ominus}{C}H_2} \longleftrightarrow \overset{\oplus}{H} \quad \overset{H}{\underset{H}{C}}=CH-\overset{\ominus}{C}H_2$$

(the middle structure bears $\overset{\oplus}{H}$ on carbon 3)

and so on

1.10 AROMATICITY

During the nineteenth century, a large group of organic compounds, isolated from balsams, resins, etc., were termed as 'aromatic' since many of them had *fragrant smell*. These compounds differ widely from the alphatics in their chemical behaviour and all of them have been shown to be closely related to the parent cyclic hydrocarbon, *benzene* (C_6H_6). This hydrocarbon, though expected to be highly unsaturated, shows a high degree of stability. Benzene and all aromatic compounds show a tendency to undergo *substitution* rather than *addition* reactions. In all such reactions the parent structure, called *benzene nucleus*, is retained. Thus aromatic compounds are also termed as *benzenoid* compounds. The term aromatic, however, also includes another class of compounds possessing aromatic character but having structures that differ considerably from that of benzene. Such compounds are termed as *non-benzenoid aromatics*, and include compounds such as azulene, tropolone, ferrocene, etc.

The distinct behaviour of all the aromtic compounds suggests that they must have some common structural features which account for their aromatic character.

There are many theories which explain aromaticity but Hückle rule is more important.

Hückle rule: According to this rule *any monocyclic planar completely conjugated polyene containing* ($4n + 2$) *π-electrons* (*where* $n = 0, 1, 2, ...$ *etc.*) *will be aromatic in nature.* (For details refer to unit **arenes and aromaticity**).

1.11 THE HYDROGEN BOND

It is well known that in a family of elements in the periodic table, the physical properties of the compounds show a regular gradation. For example, the melting and boiling points of the hydrides of group IV elements increase progressively with molecular mass. However, hydrides of groups V, VI and VII elements do not follow this generalization as in each of the groups, the first members of the series NH_3, H_2O and HF show abnormally high melting and boiling points (Table 1.10).

1.11.1 Formation of Hydrogen Bond and its Types

The above deviations have been traced to the special ability of the hydrogen atoms in the molecules of ammonia, water and hydrogen fluoride to form a link between two highly electronegative atoms, one of which is joined to it by a polar covalent bond and the other by electrostatic attraction. This electrostatic attraction comes into play due to the uneven sharing of the pair of bonding electrons between hydrogen and a highly electronegative atom (F, N or O). Consequently, there is polarizaiton of the molecule with

Table 1.11: Melting and boiling points of hydrides of group VI and VII elements

Group	Compound	Molecular weight	Melting point(°C)	Boiling point(°C)
VI	H_2O	18	0	100
	H_2S	34	–83	–59.6
	H_2Se	81	–64	–42.1
	H_2Te	130	–54	–1.8
VII	HF	20	–92	19.4
	HCl	36	–112	–85.1
	HBr	81	–88	–67.2
	HI	128	–51	–35.4

hydrogen developing a partial positive charge (δ^+) and the electronegative atom developing a partial negative charge (δ^-).

$$\overset{\delta^+}{H} : \overset{\delta^-}{X}$$

(X = electronegative atom, N, O or F)

In such a situation the *hydrogen, with a partial positive charge, is attracted towards the electron cloud of the electronegative atom of the same or other molecules forming a relatively weak bond. This type of bond, called the* ***hydrogen bond****, is the most powerful kind of dipole-dipole attraction.* Being electrostatic in nature, it is much weaker than a covalent bond; the hydrogen bond energy is only 12.5–42 kJ mol^{-1} as compared to the normal covalent bond energy of 210–420 kJ mol^{-1}.

While writing the structure of hydrogen bonded molecules, the hydrogen bonds are usually represented by *dotted lines* in contrast to a normal covalent bond which is shown by a dash. For instance, the structure of $HF_2^{\ominus}$ ion in KHF_2 can be written as:

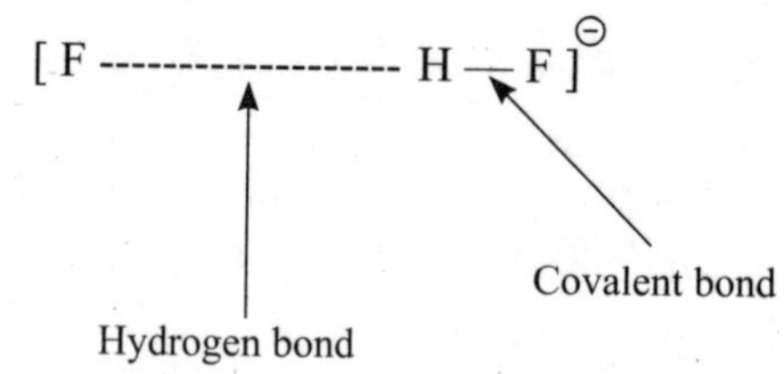

There are two types of hydrogen bonds:

***(i)* Intermolecular hydrogen bond** exists between atoms of two or more molecules, resulting in their association. For instance, water and alcohols are associated as polymeric aggregates in the solid and liquid state whereas carboxylic acids and amides exist as dimers due to intermolecular hydrogen bonding (Fig. 1.33).

X = H, water
X = R, alcohol

Amide dimer

Acid dimer

Fig. 1.33: Some examples of intermolecular hydrogen bonding

***(ii)* Intramolecular hydrogen bond** is formed between two atoms of the same molecule. As a consequence of this, a five or six membered ring called a *chelate ring* is formed. Examples of compounds exhibiting this types of hydrogen bond are *o*-nitrophenol, salicylaldehyde and *o*-chlorophenol (Fig. 1.34).

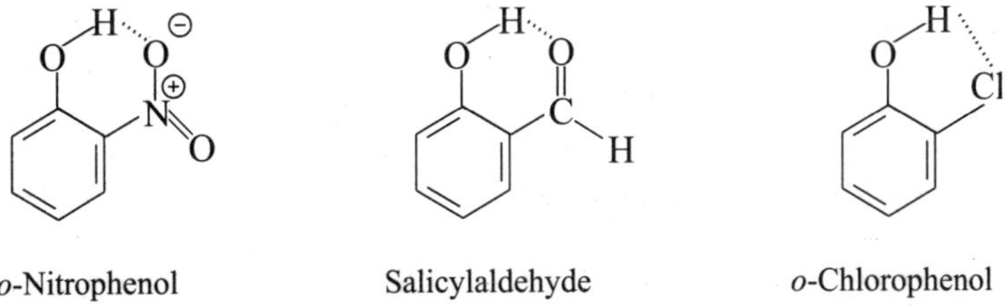

o-Nitrophenol Salicylaldehyde *o*-Chlorophenol

Fig. 1.34: Some examples of intramolecular hydrogen bonding

Conditions for Intramolecular Hydrogen Bonding

For exhibiting intramolecular hydrogen bonding, the molecule must satisfy the following conditions:

1. *The molecule must have two groups in such a way that one group contains H-atom linked to a highly electronegative atom and the other group also contains an electronegative atom.*
2. *The molecule must be planar.*
3. *The H-bond must lead to the formation of five or six membered ring.*

1.11.2 Effect of Hydrogen Bonding

In fact hydrogen bond affects all the properties of compounds in which it exists but we shall limit our discussion to only a few important ones.

1. Effect of intermolecular hydrogen bonding on boiling and melting points: These are the temperatures at which thermal energy supplied is sufficient to overcome the intermolecular forces of attraction. As some extra energy is needed to break the *intermolecular hydrogen* bond, it is responsible for *higher boiling and melting points of the hydrogen bonded compounds* in comparison to those compounds, of comparable molecular weight, in which this type of bond does not exist. Table 1.12 lists the boiling points and molecular weights of a few substances to illustrate this point.

Table 1.12: Boiling points of alkanes, ethers and alcohols of comparable molecular weight

Compound	Molecular weight	Boiling point	Compound	Molecular weight	Boiling point
$CH_3CH_2CH_3$	44	–44	$CH_3CH_2CH_2CH_2CH_3$	72	36
CH_3OCH_3	46	–25	$CH_3CH_2OCH_2CH_3$	74	35
CH_3CH_2OH	46	78	$CH_3CH_2CH_2CH_2OH$	74	118

The effect of intermolecular hydrogen bonding on melting and boiling points of different compounds are discussed below:

***(a) Water*:** Water has the lowest molecular mass but still it has the highest boiling point and melting point (Fig. 1.35). Furthermore, water is a liquid at room temperature, while all other hydrides (H_2S, H_2Se, and H_2Te) of this group are gases. This may be explained as follows:

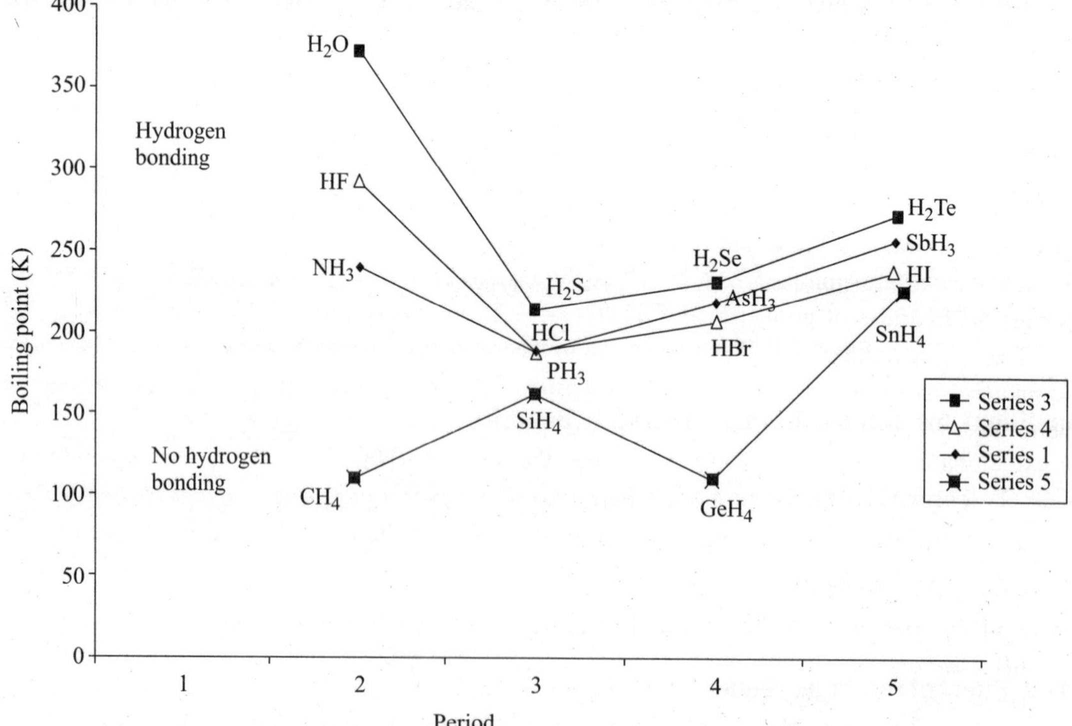

Fig. 1.35: Boiling points of hydrides of second, third, fourth and fifth period elements

In case of water, *the H-atoms are bonded to highly electronegative oxygen atom and hence undergoes intermolecular H-bonding.* As a result water exists as an associated liquid.

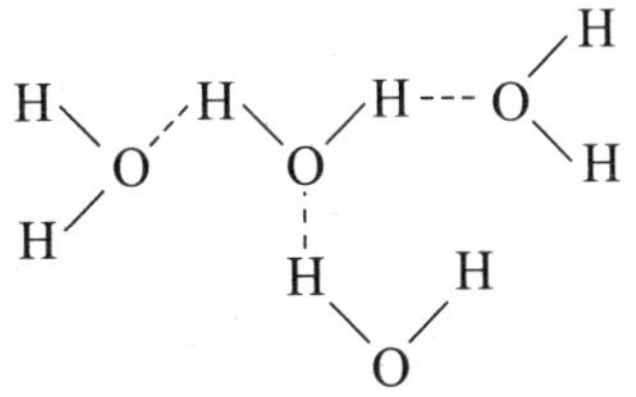

```
                            H
                           /
  H\     ,H\     /H---O
    O'      O        \H
  H/        :
            H    H
             \  /
              O
```

However, H_2S, H_2Se and H_2Te do not form H-bonds and hence exist as discrete molecules because S, Se and Te are not only less electronegative than oxygen but are also bigger in size than it. For boiling water in addition to energy required for vapourization, *a large amount of energy is needed to break these hydrogen bonds.* On the other hand, since H_2S does not form hydrogen bonds, no additional amount of energy is required. In other words, the energy required to boil water is greater than that of H_2S. Actually, the energy required to boil H_2S is so small that it is easily availble by the collisions of the molecules at room temperature. Hence, H_2S exists as a gas. On the other hand, energy required to boil water is so large that it cannot be provided by the collisions of the molecules at room temperature. Hence, water exists as a liquid.

***(b) Hydrogen fluoride*:** Hydrogen fluoride has the lowest molecular mass but still it is a liquid and has the highest boiling point amongst the hydrides of group 17 (Fig. 1.35).

The abnormally high boiling point of HF is attributed to the fact that *fluorine having highest electronegativity and smallest size from amongst halogens forms much stronger intermolecular H-bonds.* Consequently, HF exists as associated molecules as shown below:

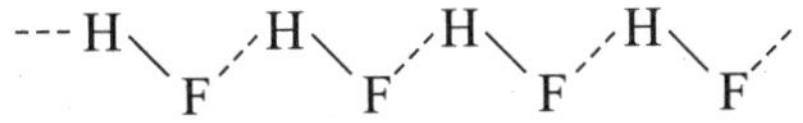

```
---H\    ,H\    ,H\    ,H\    ,
     F'     F'     F'     F'
```

On the other hand, chlorine is not only less electronegative but also larger in size than fluorine and hence it forms very weak H-bonds. However, other elements of this group do not form H-bonds at all.

***(c) Ammonia*:** Ammonia has the lowest molecular mass but still has the highest boiling point amongst the hydrides of group 15 (Fig. 1.35).

The boiling point of NH_3 is very high due to the fact that *nitrogen having highest electronegativity and smallest size from amongst the elements of group 15 forms much stronger intermolecular H-bonds.* Due to this hydrogen bonding NH_3 exists in associated form as shown below:

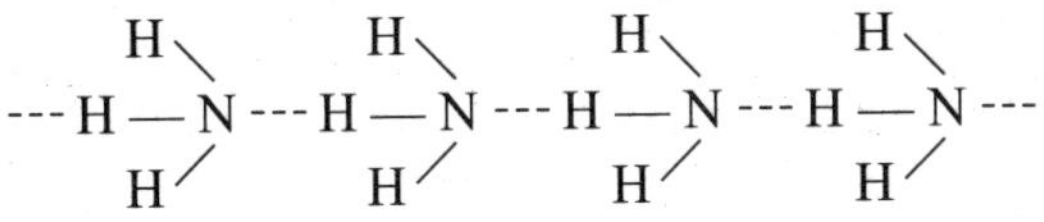

```
      H\         H\         H\         H\
---H—N---H—N---H—N---H—N---
      H/         H/         H/         H/
```

On the other hand, P, As and Sb are less electronegative and larger in size than N atom. Their hydrides PH_3, AsH_3 and SbH_3 do not form H-bonds and have lower boiling points than NH_3.

***(d) Alcohols*:**

(*i*) *Boiling points of alcohols, ethers and thioalcohols:* The boiling points of alcohols are much higher than those of the corresponding thioalcohols (mercaptans) and ethers (Table 1.13).

Table 1.13: Boiling points of alcohols, thioalcohols and ethers

Alcohols	b. pt. (K)	Thioalcohols	b. pt. (K)	Ethers	b. pt. (K)
CH_3OH	337.5	CH_3SH	279.8		
C_2H_5OH	351	C_2H_5SH	310	CH_3OCH_3	249
n-C_3H_7OH	370	n-C_3H_7SH	340	$CH_3OC_2H_5$	282.9
n-C_4H_9OH	390	n-C_4H_9SH	370	$C_2H_5OC_2H_5$	308

The reason for higher boiling points of alcohols compared to thioalcohols is that former exhibit intermolecular hydrogen bonding leading to association of these molecules.

$$\cdots H-O(R)\cdots H-O(R)\cdots H-O(R)\cdots H-O(R)\cdots$$

Due to low electronegativity and larger size of sulphur atom thioalcohols do not form H-bonds and thus exist as discrete molecules. Therefore, they have lower boiling points as compared to those of alcohols.

Ethers which are isomeric with alcohols have boiling points much lower than the latter. This is due to the fact that unlike alcohols, ethers which cannot form H-bonds do not exist as molecular aggregates. On similar lines we can explain why the boiling points of dihydric, trihydric and polyhydric alcohols are much higher than those of corresponding ethers. The replacement of each H-atom of the OH group of a polyhydric alcohol by an alkyl group decreases the boiling point of the corresponding ether although its molecular mass is higher.

CH_2OH-CH_2OH	$CH_2OCH_3-CH_2OH$	$CH_2OCH_3-CH_2OCH_3$
Ethylene glycol b.pt. 470 K	Ethylene glycol monomethyl ether b.pt. 398 K	Ethylene glycol dimethyl ether b.pt. 357 K

It may, however, be mentioned that polyhydric alcohols boil at much higher temperature than corresponding monohydric alcohols. Reason for this is that *polyhydric alcohols contain two or more OH groups due to which they undergo extensive intermolecular hydrogen bonding*. Thus, polyhydric alcohol molecules are highly associated and hence have much higher boiling points than corresponding monohydric alcohols. For examples:

CH_3CH_2OH	$HOCH_2CH_2OH$	$HOCH_2-CHOH-CH_2OH$
Ethyl alcohol b.pt. 351 K	Ethylene glycol b.pt. 470 K	Glycerol b.pt. 563 K

(*ii*) *Boiling points of water and ethyl alcohol*: Water and ethyl alcohol both exhibit intermolecular hydrogen bonding. But water, having lower molecular mass (18 amu), boils at a higher temperature (373 K) than ethyl alcohol (351 K) which has higher molecular mass (46 amu). Clearly water undergoes a higher degree of hydrogen bonding than ethyl alcohol. *i.e.*, the number and the strength of H-bonds in water molecules is higher than that in alcohol molecules. The reason for this is that replacement of one single atom of H in water by an alkyl group reduces the chances of hydrogen bonding in alcohols and hence water has higher boiling point than alcohol.

(*iii*) *Boiling points of* 1°, 2° *and* 3° *alcohols*: We have discussed above that the replacement of a smaller hydrogen atom of H_2O molecule by a bigger alkyl group reduces the tendency of the resulting alcohol to form H-bonds. Hence it may be said that an increase in size or number of alkyl groups in an alcohol decreases the tendency of H-bond formation thereby lowering the boiling point. Thus, 1° alcohols with one alkyl group should have higher boiling points than 2° alcohols with two alkyl groups which in turn, should have higher boiling points than 3° alcohols with three alkyl groups. In agreement with this generalization it has been found that the boiling points of isomeric alcohols decrease in the order: 1° > 2° > 3°. For examples,

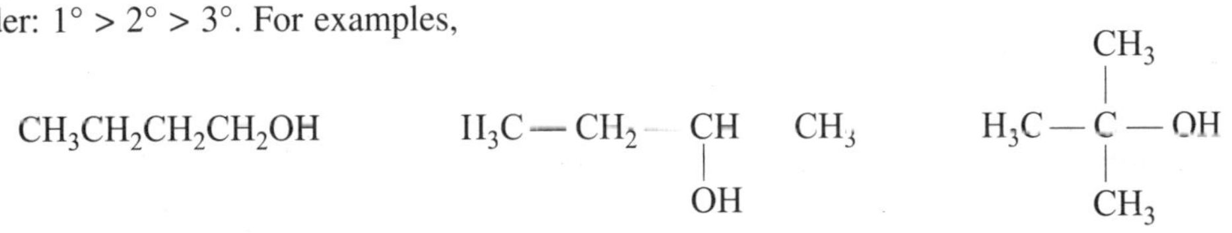

$CH_3CH_2CH_2CH_2OH$	$H_3C—CH_2—CH(OH)—CH_3$	$(CH_3)_2C(OH)—CH_3$ i.e. $H_3C—C(CH_3)_2—OH$
Butan-1-ol (1°) b.pt. 391 K	Butan-2-ol (2°) b.pt. 372 K	2-Methylpropan-2-ol (3°) b.pt. 356 K

(*e*) ***Boiling point of 1°, 2° and 3° amines***: Primary and secondary amines having two and one hydrogen atoms attached to nitrogen exhibit intermolecular hydrogen bonding and hence have higher boiling points than the isomeric tertiary amines (with no hydrogen at nitrogen).

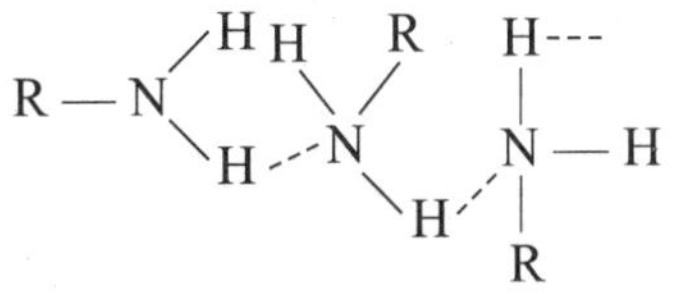

```
       H  H   R   H---
      /    \ /    |
R — N       N     N — H
      \    ·  \  ·|
       H···    H  |
                  R
```

where R = alkyl or aryl group

Further, the boiling points of amines are lower than those of alcohols and carboxylic acids of comparable molecular masses. For example,

$CH_3CH_2NH_2$	CH_3CH_2OH	HCOOH
Ethylamine Mol. mass 45 a.m.u. b.pt. 292 K	Ethyl alcohol Mol. mass 46 a.m.u. b.pt. 351 K	Formic acid Mol. mass 46 a.m.u. b.pt. 374 K

This is due to the fact that nitrogen has lower electronegativity than oxygen. Thus amines form weaker H-bonds as compared to alcohols and carboxylic acids. As a result, amines are not associated as extensively as alcohols and carboxylic acids and hence their boiling points.

2. Effect of intramolecular hydrogen bonding on melting and boiling points: It has been observed that *intramolecular hydrogen bonding usually lowers the melting and boiling points compared to intermolecular hydrogen bonding which increases the boiling and melting points.* It is due to the fact that intramolecular hydrogen bonding generally involves two *ortho* substituents which prevent association of the molecules. Therefore, such compounds exist as discrete molecules and thus require much less energy for their melting and boiling.

In contrast intramolecular hydrogen bonding is not possible in case of *m*- and *p*-isomers because it would involve a large ring which cannot be formed. So, these isomers undergo intermolecular H-bonding which causes some degree of association of their molecules and hence the melting and boiling points of *m*- and *p*-isomers are higher than those of the corresponding *o*-isomers.

The comparison of melting and boiling points of a few *o*- and *p*-hydroxy nitro and carbonyl compounds listed in Table 1.14 reveals that the melting points of *o*-isomers are lower than those of *p*-isomers.

Table 1.13: Melting points of some *o*-and *p*-hydroxy nitro and carbonyl compounds

ortho (intramolecular hydrogen bond)		*para* (intermolecular hydrogen bond)	
Compound	**m. pt. (K)**	**Compound**	**m. pt. (K)**
o-Nitrophenol	317	*p*-Nitrophenol	378
Salicylic acid	413	*p*-Hydroxybenzoic acid	487
Catechol	377	Hydroquinone	442
o-Chlorophenol	451	*p*-Chlorophenol	490

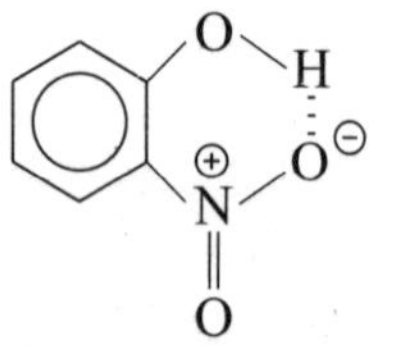

o-Nitrophenol (m.pt. 317 K)

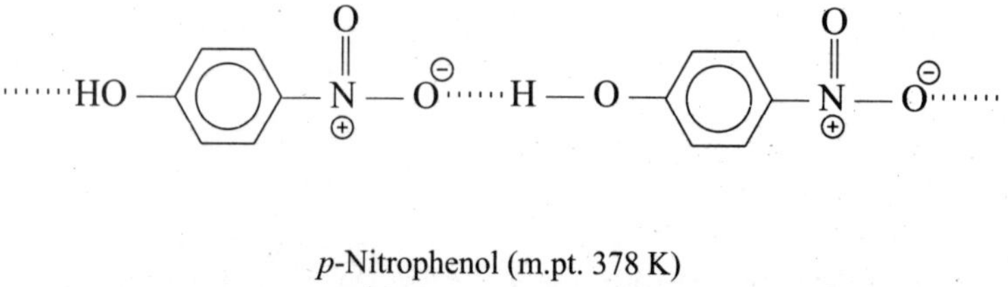

p-Nitrophenol (m.pt. 378 K)

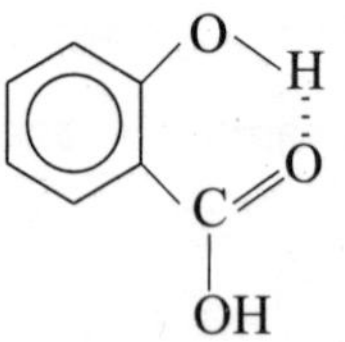

Salicylic acid (m.pt. 413 K)

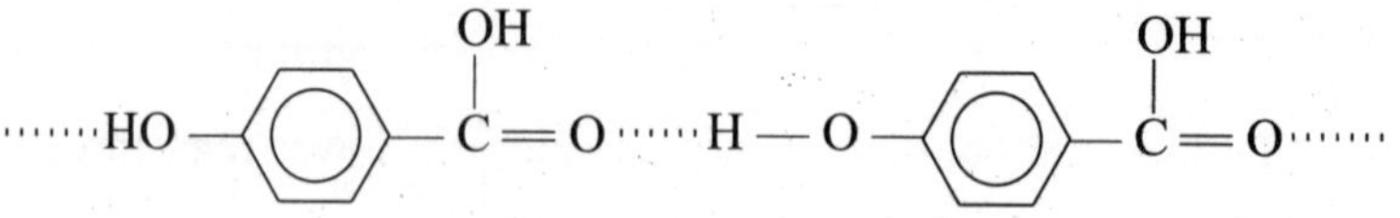

p-Hydroxybenzoic acid (m.pt. 487 K)

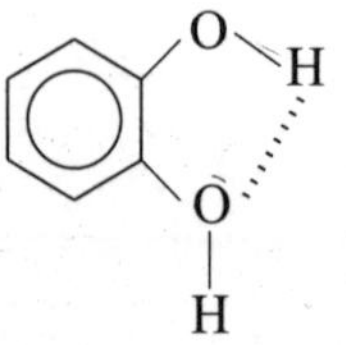

Catechol (m.pt. 377 K)

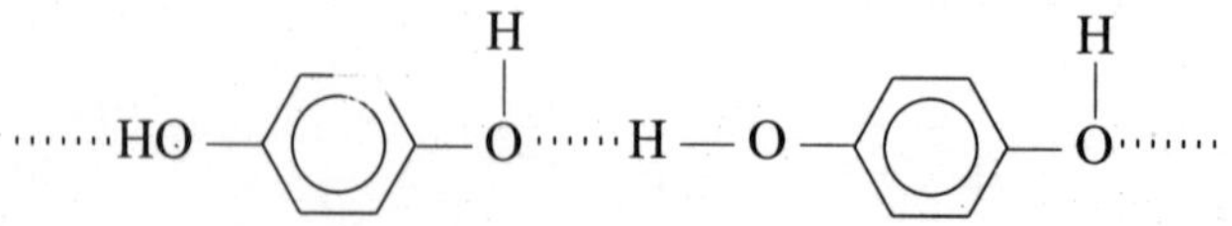

Hydroquinone (p-dihydroxybenzene) (m.pt. 442 K)

o-Chlorophenol (m.pt. 451 K)

······Cl—C6H4—O—H······Cl—C6H4—O—H······

p-Chlorophenol (m.pt. 490 K)

3. *Water solubility*: It is said that *like dissolves like.* Compounds which can form hydrogen bonds with themselves as well as with water (*cross hydrogen bonding*) are soluble in water. On the other hand, compounds in which this hydrogen bonding is either not possible or is prevented due to chelation (intramolecular hydrogen bonding) are either less soluble or completely insoluble in water. For example, alkanes, alkenes, alkynes, halogen derivatives, ethers etc., which are incapable of forming hydrogen bonds with water are insoluble in it while alcohols and carboxylic acids which are capable of forming cross hydrogen bonds with water are readily soluble in it.

---H—O---H—O---H—O---H—O---
(O atoms bearing, in order: H, R, H, R)

Further, it may be mentioned that all alcohols are not equally soluble in water. The lower alcohols such as CH_3OH, C_2H_5OH and C_3H_7OH are highly soluble in water and the solubility of higher alcohols (with four or more carbon atoms) decreases with increasing molecular mass.

Reason: The extent of solubility of any alcohol in water depends upon the capability of its molecules to form cross H-bonds with water. As the molecular mass increases, the hydrocarbon part (*i.e.*, alkyl group) becomes larger which resists the formation of hydrogen bonds with water molecules and hence the solubility goes on decreasing. In others words, the solubility of alcohols in water decreases with increase in the size of hydrocarbon moiety.

However, when the ratio of the OH groups to the hydrocarbon part is higher, the solubility of the organic compound in water increases correspondingly. For example, ethylene glycol, glycerol, sugars, dihydric phenols (catechol, resorcinol, hydroquinone) and trihydric phenols (pyrogallol, phlorogucinol etc.) are highly soluble in water. Similarly we can explain the solubility of 1° and 2° amines in water since these can also form cross H-bonds with water.

---H—N(H)(R)---H—O(H)---H—N(H)(R)---H—O(H)---

However, 3° amines are less soluble in water even though they can form H-bonds with water. This is probably due to steric hindrance of the three alkyl groups. But those 3° amines are water soluble in which the alkyl groups are tied back by ring formation. For example, pyridine is water soluble.

Another noteworthy point is that *intermolecular H-bonding (cross H-bonds) increases the solubility of a compound in water while the intramolecular hydrogen bonding (chelation) decreases it.* This is due to the fact that the already existing internal H-bonds in these compounds prevent the formation of cross H-bonds between the compounds and the water molecules and thus reduces the solubility of these compounds in water.

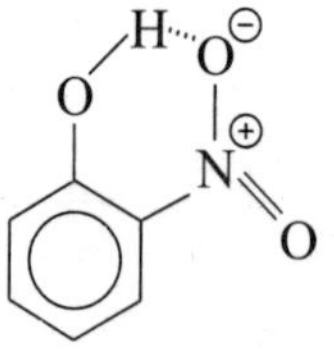

ortho-Nitrophenol

The OH group is not free due to intramolecular H-bonding and thus does not form cross H-bonding with H_2O. As a result, it is only sparingly soluble in water. Further it exists as discrete moleculre and hence has lower boiling point.

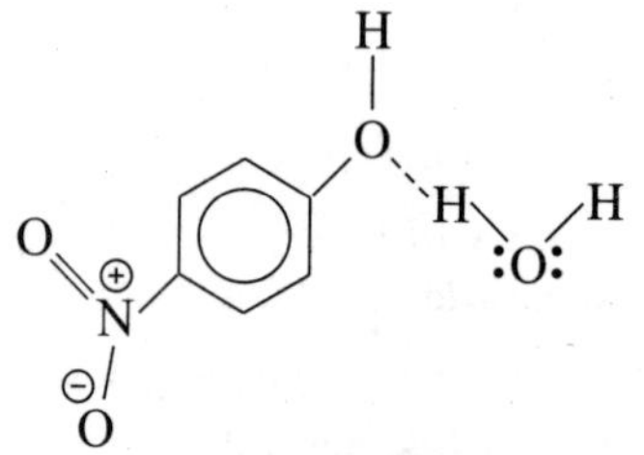

para-Nitrophenol

The OH group is available for cross H-bonding with H_2O and hence it is highly soluble in water. Further, due to intermolecular H-bonding it has higher boiling point.

The difference in physical properties of organic compounds due to inter and intramolecular H-bonding is quite useful in separation of *ortho* and *para*-hydroxy nitro and carbonyl compounds. Because of intramolecular H-bonding the *ortho*-isomers are only sparingly soluble in water besides possessing lower boiling points, while the *para*-isomers due to formation of cross H-bonding with water molecules are more water soluble and also possess higher boiling points due to intermolecular H-bonding. In other words, *ortho*-isomers are more volatile than *para*-isomers and hence can be easily separated from a mixture of the two by steam distillation.

4. Strength of acids: Any structural feature that is responsible for greater stability of the anion in comparison to free acid, will shift the ionization equilibrium to the right, thereby making the acid stronger.

$$RCOOH + H_2O \rightleftharpoons RCOO^{\ominus} + H_3O^{\oplus}$$

Anion
(More stable than free acid)

Thus if the anion is stabilized due to intramolecular H-bonding, there would be increase in strength of the acid from which it is derived. For instance, the ionization constants of benzoic acid and salicylic acid (*o*-hydroxybenzoic acid) and *o*-anisic acid (*o*-methoxybenzoic acid) as given below show that salicylic acid is the strongest acid. It can be explained on the basis of the fact that after ionization the salicylate ion is stabilized by intramolecular H-bonding which is not possible in the anions derived from the other two acids.

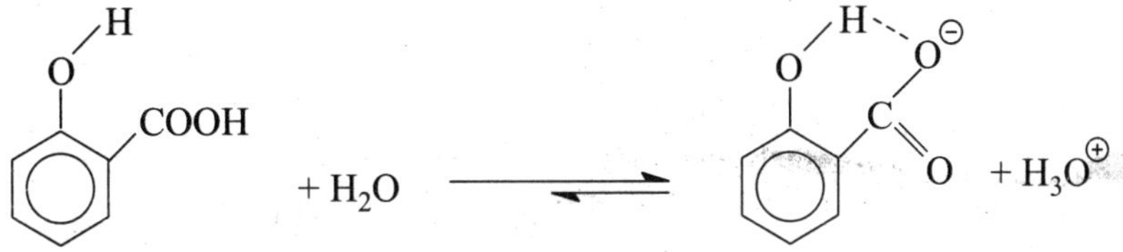

Salicylic acid + H_2O ⇌ Salicylate ion + $H_3O^{\oplus}$

Salicylic acid
$K_a = 10.5 \times 10^{-4}$
Strong acid

Salicylate ion
(Stabilized by chelation)

Benzoic acid + H_2O ⇌ Benzoate ion + $H_3O^{\oplus}$

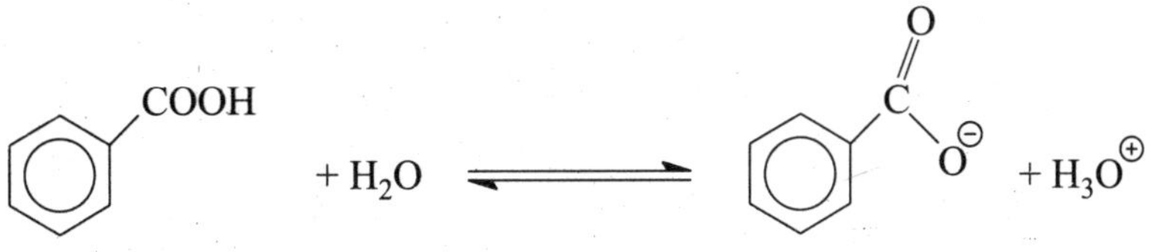

Benzoic acid
$K_a = 6.3 \times 10^{-5}$
Weak acid

Benzoate ion
(Not stabilized by chelation)

o-Anisic acid + H_2O ⇌ *o*-Anisate ion + $H_3O^{\oplus}$

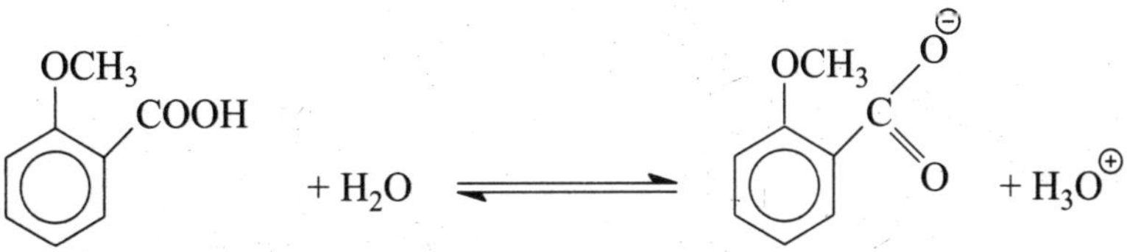

o-Anisic acid
$K_a = 8.3 \times 10^{-5}$
Weak acid

o-Anisate ion
(Not stabilized by chelation)

Similarly, 2, – 6-dihydroxybenzoic acid is 800 times stronger than benzoic acid. This is due to greater stabilization of the corresponding anion by chelation with two *ortho* OH groups as shown below:

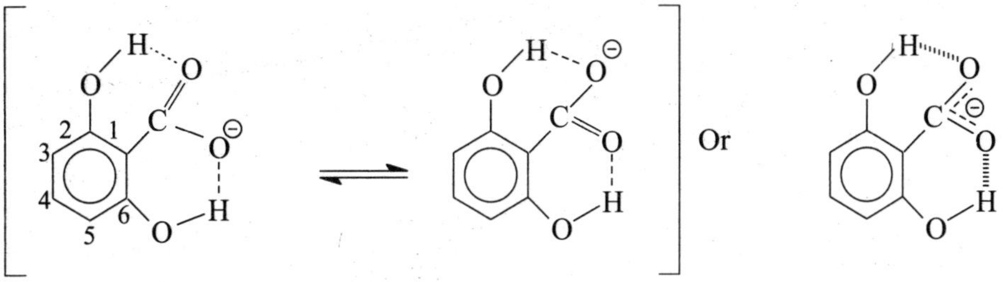

Or

Similarly, it has been observed that maleic acid ($K_a = 1.42 \times 10^{-2}$) is a much stronger acid than fumaric acid ($K_a = 9.3 \times 10^{-4}$). In this case maleate ion obtained by ionization of maleic acid is stabilized by intramolecular H-bonding whereas no such stabilization is possible in the case of fumarate ion.

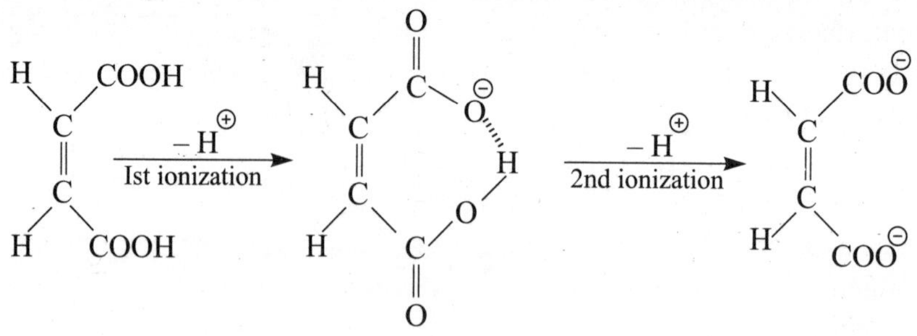

Maleic acid —($-\overset{\oplus}{H}$, Ist ionization)→ Maleate ion (Stabilized by chelation) —($-\overset{\oplus}{H}$, 2nd ionization)→

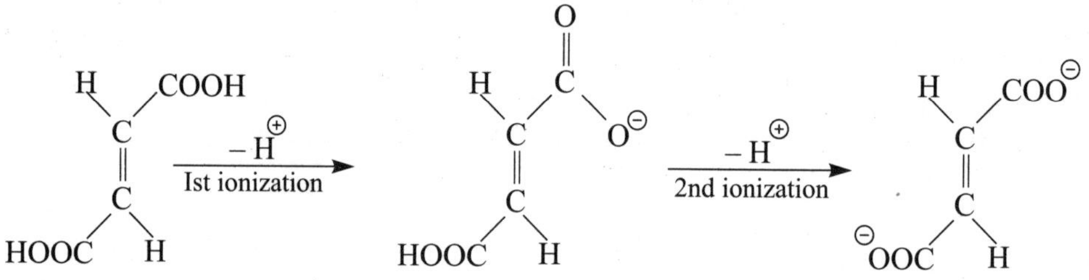

Fumaric acid —($-\overset{\oplus}{H}$, Ist ionization)→ Fumarate ion (Not stabilized by chelation) —($-\overset{\oplus}{H}$, 2nd ionization)→

However, it may be worthwhile to mention here that the *second ionization of fumaric acid is 24 times less than its first ionization whereas it is 24,000 times less in the case of maleic acid.* In other words, the second ionization constant of fumaric acid ($K_a = 3.62 \times 10^{-5}$) is much higher than that of maleic acid ($K_a = 8.5 \times 10^{-7}$). This is due to the fact that the removal of second proton from the chelated maleate ion is very difficult compared to the removal of second proton from the fumarate ion.

5. Stability: Exceptional stability of certain compounds is also due to H-bonding. We know that organic compounds containing two OH groups attached to the same carbon are usually unstable. But chloral hydrate which has two OH groups on the same carbon is a stable compound. This unexpected stability is partly due to intramolecular H-bonding as shown below:

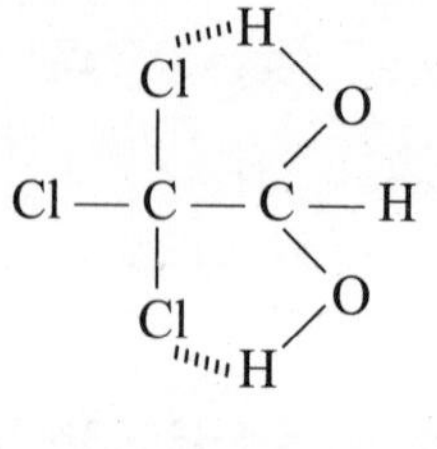

Chloral hydrate
(Stable due to chelation)

6. Viscosity and surface tension: As we have seen that the compounds exhibiting intermolecular H-bonding exist as associated molecules. Due to this association they have high viscosity and high surface tension as compared to those exhibiting intramolecular H-bonding which exist as discrete molecules and hence have low viscosity and surface tension. For example, glycerol having three OH groups per molecule shows extensive intermolecular H-bonding. As a result, its molecules are highly associated. This is precisely the reason why it has a high viscosity and much higher boiling point compared to ethylene glycol having two OH groups per molecule. Similarly, H_2SO_4 with two OH groups and H_3PO_4 with three OH groups are also highly viscous.

7. Importance of H-bonding in biological systems: Hydrogen bonding also plays an important role in biological systems. In plants and animals, most of the water is present as attached to protein molecules by H-bonding.

Furthermore, the secondary and tertiary structures of proteins are stabilized by H-bonding. For instance, the unique stability of α-helical structure of certain proteins such as keratin in wool, silk, nails, horns, etc., is due to the intermolecular H-bonding between the C = O of one amino acid and the NH of the other (fourth) amino acid residue in the chain.The specific double helical structure of DNA is also due to intermolecular hydrogen bonding which involves the purine and pyrimidine base.

A BRIEF REVIEW

- **Hybridization** means the mixing of two or more atomic orbitals of an atom having comparable energies to give a set of equal number of new orbitals which have equal energy, identical shape and size and are symmetrically oriented in three dimensional space to achieve maximum distance between their lobes. Such newly formed orbitals are called *hybrid orbitals.*

  – In carbon atom *promotion of electrons* occurs before hybridization.

  – Hybridization occurs before the bond formation.

  – The carbon atom can undergo *three types* of hybridizations *viz. sp, sp^2*, and *sp^3*. All saturated compounds such as alkanes and cycloalkanes involve *sp^3*-hybridization; alkenes involve *sp^2* and alkynes involve *sp*-hybridization.

- **Bond length** is the minimum distance between the centres of nuclei of two bonded atoms where the potential energy of the system becomes minimum.

  ♦ *Bond length decreases due to*

  (*a*) the increase in *s*-character of hybrid orbital.

  $$\underset{(110\ \text{pm})}{C\,(sp^3)-H} \quad > \quad \underset{(107.6\ \text{pm})}{C\,(sp^2)-H} \quad > \quad \underset{(106\ \text{pm})}{C\,(sp)-H}$$

  (*b*) the decrease in size of hybrid orbital.

  $$\underset{(154\ \text{pm})}{C\,(sp^3)-C\,(sp^3)} \quad > \quad \underset{(148\ \text{pm})}{C(sp^2)-C\,(sp^2)} \quad > \quad \underset{(138\ \text{pm})}{C\,(sp)-C\,(sp)}$$

  (*c*) the increase in the multiplicity of the bond

  $$\underset{(154\ \text{pm})}{C-C} \quad > \quad \underset{(134\ \text{pm})}{C=C} \quad > \quad \underset{(120\ \text{pm})}{C\equiv C}$$

(*d*) the decrease in size of the bonding atoms.

C — I	>	C — Br	>	C — Cl
(214 pm)		(194 pm)		(178 pm)

- **Bond angle** is the angle between two adjacent bonds in a polyatomic molecule.
  - *Bond angle depends upon*

    (*a*) Type of hybridization.

    (*b*) Number of shared and lone pair of electrons around the central atom.
  - The order of repulsions between lone pair and bond pair is -

    lp – lp > lp – bp > bp – bp
  - Larger the number of lone pairs of electrons on central atom, smaller will be the bond angle. For sp^3-hybridization, bond angle decrease as

CH_4	>	$\ddot{N}H_3$	>	$H_2\ddot{\underset{\cdot\cdot}{O}}$
(109°28′)		(107°)		(104.5°)

- **Bond dissociation energy** is the energy required to break one mole of a particular bond.
  - *Bond dissociation energy increases due to*

(*a*) the increase in *s*-character of hybrid orbitals

C (sp^3) — H	<	C (sp^2) — H	<	C (sp) — H
(435 kJ mol^{-1})		(443 kJ mol^{-1})		(507 kJ mol^{-1})

(*b*) the decrease in size of hybrid orbitals

C (sp^3) — C (sp^3)	<	C (sp^2) — C (sp^2)	<	C (sp) — C (sp)
(347 kJ mol^{-1})		(383 kJ mol^{-1})		(433 kJ mol^{-1})

(*c*) the decrease in size of the bonded atoms

C — I	<	C — Br	<	C — Cl	<	C — F
(215 kJ mol^{-1})		(284 kJ mol^{-1})		(320 kJ mol^{-1})		(448 kJ mol^{-1})

(*d*) the decrease in the multiplicity of bonds

C — C	<	C = C	<	C ≡ C
(347 kJ mol^{-1})		(609 kJ mol^{-1})		(804 kJ mol^{-1})

- **Localized electrons** spend their maximum time in the space between the two bonding atoms. Such electrons form a **localized** covalent bond. All *σ-bonds are localized bonds.*
- **Delocalized electrons** are not confined to the space between the two bonding atoms and thus form delocalized bond. The *π-bond may be localized or delocalized.* Isolated double bonds such as in ethylene are localized, while the conjugated double bond such as in 1, 3-butadiene are somewhat delocalized and the double bonds in benzene are completely delocalized.
- The bond length of C — C σ-bonds decreases due to delocalization while that of C = C (double bond) increases.

- The **van der Waals** forces of attraction are the *induced dipole-induced dipole* attractions. These are short range forces which are directly proportional to the surface area of the molecules.
- Boiling points of alkanes decrease as branching increases since the surface area of the molecule decreases with branching.
- **Boat-form** of cyclohexane is destabilized by flagpole-flagpole interactions and 1, 3-diaxial conformation of *cis*-1, 3-dimethylcyclohexane is less stable than the corresponding diequatorial conformation. This is due to *van der Waals repulsions* which occur when the distance between the two non-bonded atoms in less than the sum of their van der Waals radii.
- **Inclusion compounds** are the *guest-host addition compounds* in which the crystal lattice of host molecules provides a space in the form of *tunnels or channels* to accommodate the guest molecules. Only those guest molecules whose shape and size match these channels can form inclusion compounds. For instance, octane and 2-chlorooctane act as guest for urea lattice but 2-bromooctane and 2-methyloctane do not fit the size of channels in urea lattice and hence do not form inclusion compounds.
- **Clatherates or cage compounds** are the *guest-host addition compounds* in which the crystal lattice of the host molecules provides a space in the form of a *cage or cavity* to accomodate the guest molecules. The most important hosts for clatherate compounds are *quinol and water* which form clatherates with noble gases such as Ar, Kr and Xe but not with He and Ne. Radioactive isotopes of Kr and Xe are stored in the form of their clatherates.
- **Charge transfer complexes** are 1 : 1 molecular complexes obtained by mixing of electron donor compounds (D) with the electron acceptor compounds (A). **Picrates** are formed by mixing picric acid (A) with the donor molecules (D) such as hydrocarbons, aromatic amines and phenols.
- **Inductive effect** is the drift of σ-electrons along a saturated carbon chain due to the presence of electron-accepting or electron-donating group. Inductive effect is of two types: **–I effect** and **+I-effect.**
- **–I-effect** operates due to more electronegativity of the group, *i.e.*, in presence of electron withdrawing groups. It decreases in the order:

$$NO_2 > CN > COOH > F > Cl > Br > I$$

- **+I-effect** operates in presence of electron-donating groups. It decreases in the order:

$$O^- > COO^- > (CH_3)_3C > (CH_3)_2CH > CH_3CH_2 > CH_3 > D > H$$

- *Inductive effect in a permanent effect* which influences the melting point, boiling point and dipole moment of polar compounds.
- **Electromeric effect** involves complete shift of π-electrons of a multiple bond to the more electronegative atom in presence of an attacking reagent. This effect is ***purely temporary*** and is a reversible process where the electrons are transfered back to the original position when the attacking reagent is withdrawn.

- **Resonance** means that for a molecule two or more Lewis structures can be written which differ in the relative positions of electrons and not that of the atoms. None of such Lewis structures represents the actual structure of the molecule. However, the actual structure is a resonance hybrid of all these Lewis structures which are also called canonical forms (structures) or resonance forms.
- **Resonance** *leads to stability of the molecules.*
- **Aromaticity:** According to **Hückel's rule** the monocyclic, planar, conjugated polyenes containing $(4n + 2)$ π-electrons are **aromatic compounds**. They have π-electron cloud around all the atoms of the ring without any node in the electron density.
- **Aromatic compounds** resist oxidation but readily undergo electrophilic substitution reactions.
- **Hydrogen bond** is the electrostatic attraction between the slightly positively charged hydrogen atom and negatively charged heteroatom. Hydrogen bonding is of two types : *Intramolecular* and *Intermolecular.*
- **Intramolecular H-bonding** is formed between two atoms of the same molecule. It is also called *chelation*. Compounds exhibiting this type of H-bonding are *o*-nitrophenol, salicyldehyde, salicylic acids, etc.
- **Intermolecular H-bonding** is formed between the two atoms of different molecules.
- In case when the two molecules are of different compounds the intermolecular H-bonding is also called *cross H-bonding*.
- The compounds forming intermolecular H-bonding possess high melting point, boiling point, viscosity and surface tension.
- Hydrogen bonding also affects the acid strength of carboxylic acids, solubility and stability of different organic compounds.

PROBLEMS

1. What do you mean by hybridization? Give its significance.
2. What are different types of hybridization? Explain with help of orbital overlap diagrams.
3. Make orbital diagrams for the following compounds:

   (*i*) $CH_3—CH_2—CH_3$ (*ii*) $CH_3—CH=CH_2$

   (*iii*) $CH_3—C\equiv CH$ (*iv*) $CH_3—CH_2—CH_2Cl$
4. In terms of hybridization of carbon atom, discuss briefly the shapes of CH_4, C_2H_4 and C_2H_2.
5. Define and explain with suitable examples, the following:

   (*a*) Bond length, (*b*) Bond angle, (*c*) Bond energy and (*d*) σ- and π-bonds.
6. Why the bond angles of CH_4, H_2O and NH_3 are different?

7. Which of the following compounds has the largest dipole moment:

(*i*) CH_4 (*ii*) CH_3OH (*iii*) CBr_4

(*iv*) CF_4 (*v*) CO_2 (*vi*) CH_3F

8. Show by both partial charges and the crossed arrow ($\mapsto$) the polarity of the following compounds:

(*i*) CH_3Cl (*ii*) CH_3CN (*iii*) CH_3COOH

(*iv*) $HC{\equiv}CH$ (*v*) CO_2 (*vi*) CH_3OH

9. Arrange the bonds listed below in order of decreasing polarity (**most polar first**).

(*i*) H—C (*ii*) H—O (*iii*) H—B

(*iv*) H—F (*v*) H—N (*vi*) C—O

(**Hint:** *Polarity of a bond depends upon the difference in electronegativity of the two bonded atoms)*

10. Give the complete structural formula for each of the following compounds and tell which type of orbitals overlap to form each bond:

(*i*) $CH_3-CH{=}\ddot{N}-CH_2-\overset{\overset{\displaystyle O}{\|}}{C}-O-H$

(*ii*) $CH_3-CH_2-CH_2-C{\equiv}N$

(*iii*) $(CH_3)_2\ddot{N}CH_2NH_2$

(*iv*) $CH{\equiv}C-CH_2NH\underset{\underset{\displaystyle NH}{\|}}{C}-NH_2$

11. What is the predicted geometry of allenes? Explain your answer with suitable drawing of orbitals overlap.

12. Define and illustrate localized and delocalized bonds.

13. What are inclusion compounds and clatherates? Mention some of their important uses.

14. What are charge transfer complexes? Discuss the nature of bonding and structure of these complexes.

15. Define inductive effect. Discuss its various types. How does it affect the reactivity of a molecule?

16. How does inductive effect help in explaining?

(*a*) Relative strength of acids and bases.
(*b*) Polarity of organic compounds.

17. Explain why chloroacetic acid is stronger acid than acetic acid?

18. Define aromaticity. Mention some characteristics of aromatic compounds.

19. What is hyperconjugation? How does it explain the stability of alkenes ? Why is it called no bond resonance?

20. In the following structures some bonds are labelled *a*, *b*, *c*, etc. List these bonds in order of decreasing bond length:

(*i*) $CH_3 \overset{}{\underset{a}{-}} CH_2 \underset{b}{-} CH \underset{a}{=} CH_2$

(*ii*) $CH_3 \underset{a}{-} CH_2 - C \underset{c}{\equiv} C \underset{b}{-} CH_3$

(*iii*) $CH_3 \underset{a}{-} CH_2 - CH_2 - CH = CH \underset{b}{-} H$ (drawn as $CH_3-\overset{H}{\underset{H}{C}}-CH_2-CH=\overset{H}{C}-H$)

(*iv*) $CH_3 \underset{a}{-} C \equiv C - CH_2 - CH = \overset{H}{C} - \overset{H}{\underset{H}{C}} \underset{b}{-} H$ (bond *c*: the C—H bond on the sp^2 carbon)

(**Hint:** *Effect of hybridization on bond properties.*)

21. Write notes on:

(*i*) Resonance energy (*ii*) Hyperconjugation

(*iii*) Clatherates (*iv*) Picrates

22. (*i*) Which resonance form in each of the following set is the major contributor to the real structure:

(*a*) $\underset{A}{CH_2 = \underset{CH_3}{C} - CH = CH_2} \longleftrightarrow \underset{B}{\overset{\oplus}{C}H_2 - \underset{CH_3}{C} = CH - \overset{\ominus}{C}H_2} \longleftrightarrow \underset{C}{\overset{\ominus}{C}H_2 - \underset{CH_3}{C} = CH - \overset{\oplus}{C}H_2}$

(*b*) $\underset{A}{CH_3 - \overset{O}{\overset{\|}{C}} - CH = CH_2} \longleftrightarrow \underset{B}{CH_3 - \overset{\overset{\ominus}{O}}{\underset{\oplus}{C}} - CH = CH_2} \longleftrightarrow \underset{C}{CH_3 - \overset{\overset{\ominus}{O}}{C} = CH - \overset{\oplus}{C}H_2}$

(*c*) $\underset{A}{CH_3 - \overset{O}{\overset{\|}{C}} - \ddot{N}H_2} \longleftrightarrow \underset{B}{CH_3 - \overset{\overset{\ominus}{O}}{\underset{\oplus}{C}} - \ddot{N}H_2} \longleftrightarrow \underset{C}{CH_3 - \overset{\overset{\ominus}{O}}{\underset{\ominus}{C}} - \ddot{N}H_2} \longleftrightarrow \underset{D}{CH_3 - \overset{\overset{\ominus}{O}}{C} - NH_2}$

(*ii*) Draw the structures of a hydrocarbon having eight carbon atoms with three isolated double bonds, two conjugated and one isolated double bond, and three conjugated double bonds.

23. Classify the following pairs as resonance contributors or structural isomers:

(*i*) $CH_3-\overset{O}{\overset{\|}{C}}-CH_2-\overset{O}{\overset{\|}{C}}-OEt$, $CH_3-\overset{OH}{\overset{|}{C}}=CH-\overset{O}{\overset{\|}{C}}OEt$

(*ii*) $CH_3-\overset{O}{\overset{\|}{C}}-\overset{\ominus}{C}H-\overset{O}{\overset{\|}{C}}-OEt$, $CH_3-\overset{\overset{\ominus}{O}}{\overset{|}{C}}=CH-\overset{O}{\overset{\|}{C}}-OEt$

(*iii*) Ph–C(=O)–CH=C(–O–H)–CH₃ ; Ph–C(–O–H)=CH–C(=O)–CH₃

(*iv*)

(*v*) $\overset{\ominus}{C}H_2-C\equiv N$ $CH_2=C=\overset{\ominus}{N}$

(*vi*) $H_2N-\overset{\overset{\oplus}{N}H_2}{\overset{\|}{C}}-NH_2$, $N_2\overset{\oplus}{N}=\overset{NH_2}{\overset{|}{C}}-NH_2$, $N_2N-\overset{NH_2}{\overset{|}{C}}=\overset{\oplus}{N}H_2$

24. What is the significance of the concept of hyperconjugation in explaining the stability of olefins?

25. Comment upon the statement, "usual order of the inductive effects of the alkyl groups is often reversed when attached to a double bond or a benzene ring."

26. Name the various electronic effects and illustrate your answer with suitable examples.

27. Explain the effects of hydrogen bonding on the boiling points and solubilities of alcohols and carboxylic acids.

28. Explain why aryl halides are less reactive than alkyl halides?

29. Explain why:

(*a*) *o*-Nitrophenol is steam volatile but *p*-nitrophenol is not.

(*b*) Boiling point of ethylene glycol is much lower than that of glycerol.

(*c*) 2-Chlorooctane forms an inclusion compound but 2-bromooctane does not.

(*d*) He and Ne do not form clatherates with water.

(*e*) Acetamide is a weaker base than ethylamine.

(*f*) Methyl chloride has a higher dipole moment than chloroform.

(*g*) Allyl and benzyl halides are more reactive than alkyl halides towards nucleophilic substitution reactions.

(*h*) All the carbon-carbon bond lengths in benzene are equal.

(*i*) Diequatorial conformation is more stable for *cis*-1, 3-dimethylcyclohexane.

(*j*) *n*-Butyl alcohol has much higher boiling point than its isomers— isobutyl alcohols as well as diethyl ether.

(*k*) The boiling point of methanol (CH_3OH) is 66°C and that of methyl mercaptan is 6°C whereas the boiling points of diethyl ether and diethyl sulphides are 35°C and 92°C, respectively.

(*l*) 8-Hydroxyquinoline can be separated from 4-hydroxyquinoline by steam distillation.

(*m*) Ethanol (M.W. 45) boils at 78°C whereas ethylamine (M.W. 46) has b.pt. of 17°C.

(*n*) The dipole moment of CH_3F is 1.85 D and that of CD_3F is 1.86 D.

(*o*) Oxygen containing organic compounds dissolve in cold conc. H_2SO_4 to yield a solution from which the original compound can be liberated by dilution with water.

(*p*) The second ionization of fumaric acid is 24 times less than the first ionization whereas it is 24,000 times less in the case of maleic acid.

(*q*) Mesitylene forms a picrate with 1, 3, 5-trinitrobenzene but benzene does not.

(*r*) Salicylic acid is 17 times as strong as benzoic acid.

30. What conclusion could you draw from the given dipole moment regarding geometrical shapes of the following molecules?

$$NH_3 \text{ (1.46 D)}, \; BeF_2 \text{ (0.0 D)}, \; CH_4 \text{ (0.0 D)}$$

CHAPTER 2

General Principles of Organic Reaction Mechanism

A molecule may be converted into another by a chemical reaction. The new molecule (*product*) has different arrangement of atoms compared to the starting molecule (*reactant*). This change is accompanied by a redistribution of electrons. In most of the organic reactions, transformation of reactant to product takes place through some definite steps. *Organic reaction mechanism is a detailed description of these steps.* It not only tells us the number of steps involved in the reaction, but also gives us information regarding the sequence of breaking of old and making of new bonds. Study of reaction mechanisms is of paramount importance in organic chemistry as thousands of apparently diverse reactions proceed through a limited number of common steps.

2.1 FISSION OF COVALENT BONDS

As described above, organic reactions usually involve the fission and formation of covalent bonds. The covalent bond is often represented by a dash (—) and the movement of a pair of electrons is shown by a curved arrow (⤻), whereas the movement of a single electron is shown by *fish hook arrow* (⤻). There are two ways in which a covalent bond may be broken.

(*i*) Homolytic fission: Suppose the atoms X and Y are bonded through a covalent bond. This bond may break in such a way that each fragment carries one unpaired electron. The species formed by this fission are known as *free radical intermediates* and this manner of cleavage is known as *homolytic fission*. Homolytic fission is also called as **homolysis**. Generally non-polar bonds undergo homolytic fission which is favoured by high temperature, UV-irradiation or by the presence of radical initiators such as peroxides.

$$X{-}Y \xrightarrow{\text{Homolytic fission}} \dot{X} + \dot{Y}$$

Free radicals

(*ii*) Heterolytic fission: When the covalent bond joining the atoms X and Y breaks in such a way that the pair of electrons stays on any one of the atoms, such a fission is called as *heterolytic fission*. This type of fission results in the formation of *ionic intermediates*. Heterolytic fission is also called

heterolysis, which occurs in polar bonds and is favoured by polar solvents.

$$X - Y \xrightarrow{\text{Heterolytic fission}} \overset{\oplus}{X} + \overset{\ominus}{Y}$$

or

$$X - Y \xrightarrow{\text{Heterolytic fission}} \overset{\ominus}{X} + \overset{\oplus}{Y}$$

If the initially bonded atoms are of different electronegativities, the electron pair becomes attached to the more electronegative atom. Further, when as a result of heterolytic fission a carbon atom develops negative charge, it is known as *carbanion*. On the other hand, a positively charged carbon atom is called *carbocation*.

It is important to note that depending upon the reaction conditions the same covalent bond can undergo homolytic or heterolytic fission. Generally, the bond dissociation energy for homolytic fission is lower than that for heterolytic fission. For example,

$$CH_3 - Cl \xrightarrow{\text{Homolytic fission}} \dot{C}H_3 + \dot{C}l \quad \Delta H = +352 \text{ kJ mol}^{-1}$$

$$CH_3 - Cl \xrightarrow{\text{Heterolytic fission}} \overset{\oplus}{C}H_3 + \overset{\ominus}{Cl} \quad \Delta H = +952 \text{ kJ mol}^{-1}$$

SOME OTHER NOTATIONS

(*i*) Double Headed Arrow ($\longleftrightarrow$)

A double headed arrow ($\longleftrightarrow$) is used to separate two or more resonance structures (and not real structures). In other words it is written between two imaginary species. For example,

$$\text{(benzene Kekulé structure)} \longleftrightarrow \text{(benzene Kekulé structure)}$$

(*ii*) Double Arrow ($\rightleftharpoons$ Reversible Arrow)

Double arrow is used to represent an equilibrium between the reactants and the products of a reaction in a reversible reaction. For example,

1. $2\,H_2O \rightleftharpoons H_3\overset{\oplus}{O} + \overset{\ominus}{O}H$

2. $CH_3COCH_2\,COOEt \rightleftharpoons CH_3 - \underset{\displaystyle OH}{\underset{|}{C}} = CH - COOEt$

   Acetoacetic ester (Keto form) (Enol form)

($\rightleftharpoons$) is written between two or more real species including two or more tautomers in equilibrium.

(*iii*) Half Headed Arrow (→—)

A half headed arrow (→—) is used to show the inductive effect symbolized by I. It points in the direction of electron drift.

C —←X
+ I effect

C →— Y
– I effect

2.2 TYPES OF REAGENTS

For an organic reaction to take place some starting materials are required. These materials are called **reactants** or **reacting molecules**. The reactants are classified into two categories the **substrates** and the **reagents**. In fact no clearcut distinction between these two types of reactants is possible. However, by convention more reactive reactant is called the reagent and the less reactive one is called the substrate. A substrate can also be defined as a molecule in which old bonds are broken and new bonds are formed or in other words a molecule in which one functional group changes into another. For example, in the following reaction, ethyl alcohol is designated as substrate (*i.e.*, the compounds in which old C—O bond is broken and a new C—Br bond is formed or functional group OH is replaced by Br).

$$\underset{\text{Substrate}}{C_2H_5OH} + \underset{\text{Reagent}}{HBr} \longrightarrow C_2H_5Br + H_2O$$

With this background let us discuss various types of reagents. Broadly speaking the reagents can be divided into following two categories:

1. Nucleophilic reagents (or Nucleophiles).
2. Electrophilic reagents (or Electrophiles).

2.2.1 Nucleophilic Reagents (or Nucleophiles)

A nucleophilic reagent or nucleophile (L., **nucleus**, kernel; Gr, **philos**, loving; hence "nucleus loving") is an atom or a group of atoms that has available an unshared pair of electrons. Thus nucelophiles must be electron rich species. They are of two types:

(*i*) Charged nucleophiles: They are negatively charged species. They are also called as **anionoid reagents**.

Examples of charged nucleophiles or anionoid reagents are: Chloride ion (Cl^-), bromide ion (Br^-), iodide ion (I^-), hydroxide ion (HO^-), alkoxide ion (RO^-), amide ion (NH_2^-), cyanide ion (CN^-), carbanions (R^-), $CH^-(COO\ Et)_2$, hydride ion (H^-), carboxylate ion (COO^-), acetylide ion ($R-C\equiv C^-$), etc.

(*ii*) Neutral nucleophiles: Many neutral compounds containing oxygen, nitrogen and sulphur are also nucleophilic because of the presence of unshared pair(s) of electron(s) on these atoms. Examples of neutral nucleophiles are R_3N (tertiary amines), NH_3, H_2O, R—O—H (alcohols), R—S—R (alkyl sulphides), R— S—H (mercaptans).

As the nucleophiles (neutral and charged both) contain at least one unshared pair of electrons, they can easily donate this pair to electron deficient atoms or substrates and hence behave as Lewis bases.

$$\overset{\ominus}{X} + \overset{\oplus}{C}{<} \longrightarrow X-C{<}$$

Nucleophile

Nucleophilicity: The ability of a nucleophile to donate its unshared pair of electrons is called nucleophilicity. Fully charged ions such as hydroxide ($\overset{\ominus}{O}H$) and alkoxide ($R\overset{\ominus}{O}$) ions have more nucleophilicity than neutral nucleophiles such as water (H_2O) or alcohols (ROH). As expected nucleophilicity also increases in the presence of electron-donating groups such as alkyl groups which intensify negative charge on the atom having unshared pair of electrons. Thus, methoxide ion, ($CH_3\overset{\ominus}{O}$) has greater nucleophilicity than hydroxide ion ($H\overset{\ominus}{O}$), due to +I-effect of CH_3 group in ($CH_3\overset{\ominus}{O}$). Some common nucleophiles are arranged in increasing order of nucleophilicity as under:

$$H_2O < ROH < \overset{\ominus}{Cl} < \overset{\ominus}{Br} < \overset{\ominus}{O}H < \overset{\ominus}{O}R < \overset{\ominus}{I} < \overset{\ominus}{C}N$$

In general **nucleophilicity** of a reagent is related to its **basic property**. However, this is not always true. For example, methyl mercaptide ($CH_3\overset{\ominus}{S}$) is stronger nuclophile than methoxide ($CH_3\overset{\ominus}{O}$) although the former is much weaker base than the latter.

2.2.2 Electrophilic Reagents (or Electrophiles)

An electrophilic reagent or electrophile (**electron loving**) is a reagent which is deficient in electrons (usually a pair). The electrophiles must be electron poor species. They are of two types:

(*i*) **Charged electrophiles**: These reagents are deficient in a pair of electrons and are, therefore, positively charged. Due to positive charge they are also called as **cationoid reagents**. Examples of charged electrophiles are: Proton (H^+), hydronium ion ($H_3\overset{\oplus}{O}$), nitronium ion ($\overset{\oplus}{N}O_2$), bromonium ion ($\overset{\oplus}{Br}$), carbocations ($\overset{\oplus}{R}$), etc.

(*ii*) **Neutral electrophiles**: They are reagents in which the electron deficient atom does not carry any charge. Some examples of neutral electrophiles are: Boron trifluoride (BF_3), zinc chloride ($ZnCl_2$), aluminium chloride ($AlCl_3$), sulphur trioxide (SO_3), dichlorocarbene ($:CCl_2$). These reagents behave as Lewis acids.

Electrophiles react with substrates at a negative or partially negative site. The charged electrophiles are more reactive than neutral ones. The presence of electron-withdrawing substituents intesify the positive charge on the reactive site of the electrophile.

$$\overset{\oplus}{E} + \overset{\ominus}{:C}{<} \longrightarrow E-C{<}$$

Electrophile

Distinction between nucleophiles and electrophiles: Some of the important points of difference between nucleophiles and electrophiles are listed in Table 2.1.

Table 2.1: Distinction between electrophiles and nucleophiles

Electrophiles	Nucleophiles
1. They are deficient in electrons.	1. They are rich in electrons.
2. They accept a pair of electrons from the substrate.	2. They donate a pair of electrons to the substrate.
3. They behave as Lewis acids.	3. They behave as Lewis bases.
4. They have one empty orbital due to which they accept electrons from the substrate molecule.	4. They have at least one lone pair of electrons which can easily be donated to the substrate molecule.
5. They are either neutral or positively charged chemical species.	5. They are either neutral or negatively charged chemical species.

Nucleophilic and Electrophilic Reactions: A reaction is said to be nucleophilic if the reagent is a nucleophile and elctrophilic if the reagent is an electrophile.

Electron density on the carbon atoms of organic compounds may increase or decrease depending upon the electronegativities of the bonded atoms, and also on reasonance effects. If the electron density on a given carbon is reduced, a nucleophilic reagent (**nucleophile**) may bring a pair of electrons to the carbon. Such reaction is called a *nucleophilic reaction*. The carbon atom of methyl chloride, for instance, becomes partially positive due to the inductive effect of chlorine and a nucleophile such as hydroxide ion attacks this carbon. The whole sequence may be written as below:

$$\overset{\ominus}{HO} + H-\overset{\delta^+}{C}H_2 \longrightarrow \overset{\delta^-}{Cl} \longrightarrow \left[\overset{\delta^-}{HO} \cdots\cdots CH_3 \cdots\cdots \overset{\delta^-}{Cl} \right] \longrightarrow HO-CH_2-H + \overset{\ominus}{Cl}$$

In an *electrophilic reaction*, the reagent takes a pair of electrons from the substrate. A common example is the addition of hydrogen bromide to the double bond of an alkene through the initial electrophilic attack of a proton to form a carbocation which subsequently combines with a nucleophile, *e.g.*, bromide ion.

$$H-CH=CH-H + \overset{\oplus}{H} \longrightarrow H-\overset{\oplus}{C}H-CH_2-H \quad \text{(Electrophillic attack)}$$

$$\overset{\ominus}{Br} + H-\overset{\oplus}{C}H-CH_2-H \longrightarrow H-CHBr-CH_2-H \quad \text{(Nucleophilic attack)}$$

It may be noted that in a two step process as above, if one step is electrophilic, the other will be nucleophilic.

2.3 TYPES OF ORGANIC REACTIONS

Broadly speaking, organic compounds may undergo eight kinds of reactions, namely: (1) Substitution reactions, (2) Addition reactions, (3) Elimination reactions, (4) Condensation reactions, (5) Molecular rearrangements, (6) Isomerization reactions, (7) Molecular reactions/pericyclic reactions and (8) Disproportionation reactions.

2.3.1 Substitution Reactions

A substitution reaction is the one in which an atom or a group in a molecule is replaced by another. The incoming group becomes attached to the same carbon to which the outgoing group was attached. The substituting species may be either a nucleophile, an electrophile or a free radical.

(*i*) Nucleophilic substitution reaction: In *nucleophilic substitution* (substituting species a nucleophile) a nucleophile brings an electron pair to the substrate and the leaving group departs with an electron pair.

$$\overset{\ominus}{Nu}: + R - X \longrightarrow R - Nu + \overset{\ominus}{X}:$$

(*ii*) Electrophilic substitution reaction: In an electrophilic substitution, the attacking species is an electrophile and the leaving group departs without the pair of bonding electrons. This type of reaction occurs mostly in aromatic systems where the high electron density attracts electrophiles.

$$C_6H_6 + \overset{\oplus}{NO_2} \longrightarrow [C_6H_6NO_2]^{\oplus} \longrightarrow C_6H_5NO_2 + H^{\oplus}$$

(*iii*) Free radical substitution reaction: *The reaction in which substituting species is a free radical is called free radical substitution reaction.* Such reactions are common in alkanes and other saturated compounds. For example, chlorination of methane in presence of heat or light to give methyl chloride and hydrogen chloride.

$$\underset{\text{Methane}}{CH_3 - H} + Cl_2 \xrightarrow[\text{or light}]{\text{520-670 K}} \underset{\text{Methyl chloride}}{CH_3 - Cl} + HCl$$

It is a chain reaction which takes place through a free radical mechanism and involves the following steps:

Chain initiation

$$\underset{\text{Chlorine}}{Cl - Cl} \xrightarrow[\text{520-670 K or light}]{\text{Homolysis}} 2\,\dot{Cl} \text{ (Chlorine atoms)}$$

Chain propagation

$$CH_3-H+\dot{C}l \longrightarrow \underset{\text{Methyl free radical}}{\dot{C}H_3} + HCl$$

$$\dot{C}H_3 + H-Cl \longrightarrow CH_3Cl + \dot{C}l$$

Chain termination

$$\dot{C}l + \dot{C}l \longrightarrow Cl_2$$

$$\dot{C}H_3 + \dot{C}H_3 \longrightarrow CH_3-CH_3$$

It may be noted that the steps involved in chain propagation occur over and over again until the completion of the reaction.

2.3.2 Addition Reactions

A reaction in which two reacting species combine to give a single molecule of the product is called an addition reaction. Compounds containing multiple (double or triple) bonds undergo this type of reaction. As with substitution reactions in these reactions also the attacking species may be a nucleophile, an electrophile or a free radical.

(*i*) Nucleophilic addition reactions: *Addition reactions in which the attacking species is a nucleophile are called nucleophilic additions.* In such reactions, which are exhibited by aldehydes and ketones, a nucleophile brings a pair of electrons to the carbon atom linked to a hetero atom by a multiple bond. The π-electrons of the multiple bond are forced to shift to the adjacent atom. The reaction is then completed by combination of the anion so formed with a positive species. For example, base catalyzed addition of HCN to acetone.

$$\overset{\ominus}{O}H + H-CN \longrightarrow H_2O + \overset{\ominus}{C}N$$

$$\underset{\text{Acetone}}{(H_3C)_2C{=}O} + \overset{\ominus}{C}N \longrightarrow H_3C-\underset{CN}{\overset{CH_3}{C}}-\overset{\ominus}{O} \xrightarrow[-\overset{\ominus}{C}N]{H-CN} \underset{\text{Acetone cyanohydrin (addition product)}}{H_3C-\underset{CN}{\overset{CH_3}{C}}-OH}$$

(*ii*) Electrophilic addition reactions: *Addition reactions in which the attacking species is an electrophile are called electrophilic additions.* In these reactions a positive or electron deficient species (electrophile) adds to one of the atoms joined by the multiple bond and converts π-bond into σ-bond forming a carbocation which combines with a nucleophile to give the addition product. For example, addition of Br_2 or HBr to propylene.

$$\underset{\text{Propylene}}{CH_3CH{=}CH_2} + \overset{\oplus}{H} \xrightarrow{\text{Slow}} \underset{\text{Isopropyl carbocation (2°)}}{CH_3-\overset{\oplus}{C}H-CH_3}$$

$$\overset{\ominus}{Br} + CH_3-\overset{\oplus}{C}H-CH_3 \xrightarrow{\text{Fast}} \underset{\text{2-Bromopropane (addition product)}}{CH_3-\underset{|}{\overset{Br}{C}}H-CH_3}$$

(*iii*) Free radical addition reactions: *Addition reactions in which the attacking species is a free radical are called free radical addition reactions.* For example, addition of HBr to alkenes in presence of peroxides.

$$\underset{\text{Propylene}}{CH_3CH{=}CH_2} + HBr \xrightarrow{\text{Peroxides}} \underset{n\text{-Propyl bromide}}{CH_3-CH_2-CH_2Br}$$

It is a chain reaction which occurs in three stages as shown below:

Chain initiation

$$\underset{\text{Peroxide}}{RO \vdots OR} \xrightarrow[\text{Homolytic fission}]{\text{Heat or light}} \underset{\text{Alkoxy free radical}}{2\,\dot{O}R}$$

$$\dot{O}R + HBr \longrightarrow ROH + \dot{Br}$$

Chain propagation

$$CH_3-CH{=}CH_2 + \dot{Br} \xrightarrow{\text{Slow}} \underset{\text{Free radical (2°)}}{CH_3-\dot{C}H-CH_2Br}$$

$$CH_3-\dot{C}H-CH_2Br + HBr \xrightarrow{\text{Fast}} \underset{n\text{-Propyl bromide}}{CH_3-CH_2-CH_2Br} + \dot{Br}$$

Chain termination

$$\dot{Br} + \dot{Br} \longrightarrow Br_2$$

It may be noted that the steps involved in chain propagation stage are repeated again and again until the reaction is complete.

2.3.3 Elimination Reactions

A reaction which involves the loss of two atoms or groups of atoms from the same or adjacent atoms of a substrate molecule is called elimination reaction. Such reactions produce compounds containing double or triple bonds. These reactions are classified as α (alpha), β (beta) and γ (gama) elimination reactions depending upon the relative positions of atoms or groups being eliminated.

(*i*) **α-Elimination reactions:** *These reactions take place by the loss of two atoms or groups from the same atom of the substrate molecule.* For example, base-catalyzed dehydrochlorination of chloroform to form dichlorocarbene.

$$H\overset{\ominus}{O} + H-CCl_3 \longrightarrow H_2O + :\overset{\ominus}{C}Cl_3$$

$$:\overset{\ominus}{C}Cl_3 \longrightarrow :CCl_2 + \overset{\ominus}{Cl}$$

Dichlorocarbene

Dichlorocarbene is a very useful reactive intermediate which is involved in many reactions such as *carbylamine reaction* and *Reimer-Tieman reaction.*

(*ii*) **β-Elimination reactions:** *The reactions involving loss of two atoms or groups from adjacent atoms of the substrate molecule are called β-elimination reactions.* For example, acid-catalyzed dehydration of alcohols and base catalysed dehydrohalogenation of alkyl halides.

Acid catalyzed dehydration of alcohols

$$H-\overset{\alpha}{C}H_2-\overset{\beta}{C}H_2-\ddot{O}H + \overset{\oplus}{H} \longrightarrow H-CH_2-CH_2-\overset{\oplus}{O}H_2 \longrightarrow CH_2{=}CH_2 + H_2O + \overset{\oplus}{H}$$

Ethanol → Ethene

Base catalyzed dehydrohalogenation of alkyl halides

$$H\overset{\ominus}{O} + H-\underset{\alpha}{C}H_2-\underset{\beta}{C}H_2-Br \xrightarrow[\Delta]{\text{KOH (alc.)}} CH_2=CH_2 + HBr + H_2O$$

Ethyl bromide → Ethene

(*iii*) **γ-Elimination reactions:** *The reactions involving loss of two atoms or groups from α and γ positions (i.e., three bonds away) are called γ-eliminations.* Such reactions produce three membered rings. For example, dehalogenation of 1, 3-dibromopropane in presence of Zn dust.

$$BrH_2\overset{\alpha}{C}-\overset{\beta}{C}H_2-\overset{\gamma}{C}H_2Br \xrightarrow[\gamma\text{-elimination},\ -Br_2]{\text{Zn dust}} \text{Cyclopropane } (H_2C-CH_2 \text{ ring with } CH_2)$$

Cyclopropane

This reaction is called **Freund reaction** and is used for the synthesis of cyclopropanes.

2.3.4 Condensation Reactions

The reactions in which two or more molecules of the same or different compounds combine to form a product with or without the elimination of simple molecules such as H_2O, HCl, NH_3, ROH etc., are called condensation reactions. For example, two molecules of acetaldehyde condense in presence of dilute alkali to form 3-hydroxybutanal. Since β-hydroxyaldehydes or ketones are commonly known as aldols, therefore, this reaction is called **aldol condensation**. The reaction scheme is as under:

$$\underset{\text{Base}}{\overset{\ominus}{HO}} + \underset{\text{Acetaldehyde}}{H-CH_2-\overset{O}{\overset{\|}{C}}-H} \xrightarrow{\text{Dil. NaOH}} \underset{\text{Carbanion}}{:\overset{\ominus}{C}H_2-\overset{O}{\overset{\|}{C}}-H}$$

$$CH_3-\overset{O}{\overset{\|}{C}}-H + :\overset{\ominus}{C}H_2-\overset{O}{\overset{\|}{C}}-H \longrightarrow CH_3-\overset{\overset{\ominus}{O}}{\underset{H}{C}}-CH_2-\overset{O}{\overset{\|}{C}}-H \xrightarrow[-\overset{\ominus}{O}H]{H-OH} \underset{\text{3-Hydroxybutanal (aldol)}}{\overset{4}{C}H_3-\overset{OH}{\underset{H}{\overset{3}{C}}}-\overset{2}{C}H_2-\overset{O}{\overset{\|}{\overset{1}{C}}}-H}$$

Aldols on heating with dilute mineral acids readily undergo dehydration to form α, β-unsaturated aldehydes or ketones.

Thus,

$$\underset{\text{Aldol}}{CH_3-\overset{OH}{\underset{H}{C}}-CH_2-\overset{O}{\overset{\|}{C}}-H} \xrightarrow[\Delta]{\overset{\oplus}{H}} \underset{\text{Crotonaldehyde (But-2-enal)}}{CH_3-CH=CH-\overset{O}{\overset{\|}{C}}-H}$$

2.3.5 Molecular Rearrangement Reactions

These are the reactions which involve the shifting of an atom or a group from one atom to another within the same molecule. For example,

(*i*) Treatment of 3-methylbutan-2-ol with conc. H_2SO_4 to give 2-methylbut-2-ene which is a rearranged product.

$$\underset{\text{3-Methylbutan-2-ol}}{CH_3-\underset{H}{\overset{CH_3}{C}}-\underset{OH}{CH}-CH_3} \xrightarrow{\overset{\oplus}{H}(H_2SO_4)} CH_3-\underset{H}{\overset{CH_3}{C}}-\underset{\overset{\oplus}{H_2O}}{CH}-CH_3$$

$$\xrightarrow{-H_2O}$$

$$\underset{\text{2° Carbocation (\textit{less stable})}}{CH_3-\underset{H}{\overset{CH_3}{C}}-\overset{\oplus}{CH}-CH_3} \xrightarrow[\text{shift}]{\text{1, 2-Hydride}} \underset{\text{3° Carbocation (\textit{more stable})}}{CH_3-\underset{\oplus}{\overset{CH_3}{C}}-CH_2-CH_3}$$

$$\xrightarrow[\text{Saytzeff rule}]{-\overset{\oplus}{H}}$$

$$\underset{\text{2-Methylbut-2-ene (\textit{major product})}}{CH_3-\overset{CH_3}{C}-CH-CH_3}$$

(*ii*) Treatment of 2, 2-dimethylpropan-1-ol with conc. H_2SO_4 to give 2-methylbut-2-ene, which is again a rearranged product.

$$CH_3-\underset{CH_3}{\overset{CH_3}{C}}-CH_2-OH \xrightarrow{\overset{\oplus}{H}\text{(conc. }H_2SO_4)} CH_3-\underset{CH_3}{\overset{CH_3}{C}}-CH_2-\overset{\oplus}{O}H_2 \xrightarrow{-H_2O}$$

$$\underset{\text{1° Carbocation (\textit{less stable})}}{CH_3-\underset{CH_3}{\overset{CH_3}{C}}-\overset{\oplus}{CH_2}} \xrightarrow[\text{shift}]{\text{1, 2-Methyl}} \underset{\text{3° Carbocation (\textit{more stable})}}{CH_3-\underset{\oplus}{\overset{CH_3}{C}}-CH_2CH_3} \xrightarrow[\text{Saytzeff rule}]{-\overset{\oplus}{H}}$$

$$\underset{\text{2-Methylbut-2-ene (\textit{major product})}}{CH_3-\overset{CH_3}{C}=CHCH_3}$$

(*iii*) **Pinacol-pinacolone rearrangement**

$$H_3C-\underset{\underset{HO:}{|}}{\overset{\overset{CH_3}{|}}{C}}-\underset{\underset{:OH}{|}}{\overset{\overset{CH_3}{|}}{C}}-CH_3 \xrightarrow{H^{\oplus}} H_3C-\underset{\underset{H_2\overset{\oplus}{O}}{|}}{\overset{\overset{CH_3}{|}}{C}}-\underset{\underset{:OH}{|}}{\overset{\overset{CH_3}{|}}{C}}-CH_3 \xrightarrow{-H_2O}$$

Pinacol

$$H_3C-\underset{\oplus}{\overset{\overset{CH_3}{|}}{C}}-\underset{\underset{:OH}{|}}{\overset{\overset{CH_3}{|}}{C}}-CH_3 \xrightarrow[\text{shift}]{\text{1, 2-Methyl}} H_3C-\underset{\underset{H_3C}{|}}{\overset{\overset{CH_3}{|}}{C}}-\underset{\underset{\overset{\oplus}{O}H}{||}}{C}-CH_3 \xrightarrow{-H^{\oplus}} H_3C-\underset{\underset{CH_3}{|}}{\overset{\overset{CH_3}{|}}{C}}-\underset{\underset{O}{||}}{C}-CH_3$$

Pinacolone

(*iv*) Hoffmann rearrangement in which 1° amides are converted into 1° amines on treatment with Br_2/KOH.

$$R-\overset{\overset{O}{||}}{C}-NH_2 \xrightarrow[-KBr,\ -H_2O]{Br_2/KOH} R-\overset{\overset{O}{||}}{C}-N\begin{matrix}H\\Br\end{matrix} \xrightarrow[-H_2O]{KOH} \left[R-\overset{\overset{O}{||}}{C}-\overset{\ominus}{\ddot{N}}-Br\right]\overset{\oplus}{K}$$

1° Amide

$$\downarrow -KBr$$

$$R-NH_2 \xleftarrow[\substack{(\text{Hydrolysis})\\ -K_2CO_3}]{2\ KOH} R-N=C=O \xleftarrow[\text{Migration}]{\text{Alkyl}} R-\overset{\overset{O}{||}}{C}-\ddot{N}:$$

1° Amide | Alkyl isocyanate | Acylnitrene

Here migration of R(alkyl) from carbon to nitrogen gives an alkyl isocyanate which upon hydrolysis gives a 1° amine with one carbon atom less than the original amide.

2.3.6 Isomerization Reactions

Interconversion of one isomer into another keeping the molecular formula as well as skeletons of the reactants and the products intact is called isomerization reaction. For example, interconversion of *cis*-but-2-ene to *trans*-but-2-ene and *vice versa*.

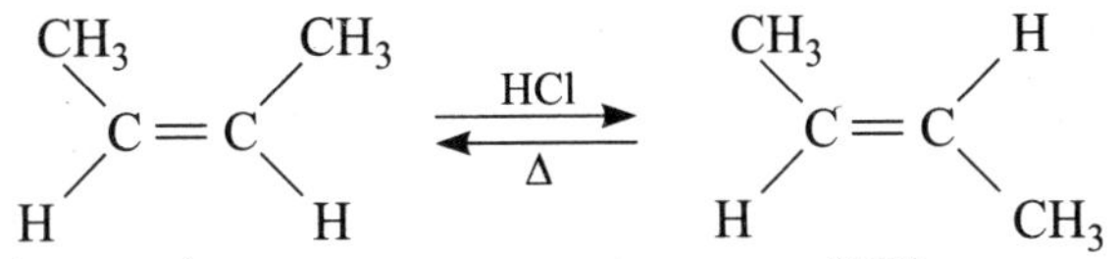

$$\underset{cis}{\begin{matrix} CH_3 & & CH_3 \\ & C=C & \\ H & & H \end{matrix}} \underset{\Delta}{\overset{HCl}{\rightleftharpoons}} \underset{trans}{\begin{matrix} CH_3 & & H \\ & C=C & \\ H & & CH_3 \end{matrix}}$$

Similarly, 1-bromobutane isomerises to 2-bromobutane, on heating with anhydrous $AlCl_3$ at 575 K.

$$\underset{\text{1-Bromobutane}}{CH_3CH_2CH_2CH_2Br} \xrightarrow[575\ K]{\text{Anhyd. } AlCl_3} \underset{\text{2-Bromobutane}}{CH_3CH_2-\overset{\overset{Br}{|}}{CH}-CH_3}$$

2.3.7 Molecular Reactions or Pericyclic Reactions

These are reactions in which the covalent bonds undergo reorganization either intramolecularly or intermolecularly through *cyclic transition states* and without the intervention of ionic or radical intermediates; acids and bases having no effect. These reactions take place under the influence of heat or light and are characterized by a high degree of stereospecificity. *Electrocylic* and *cycloaddition* reactions are typical examples.

Electrocyclic reaction

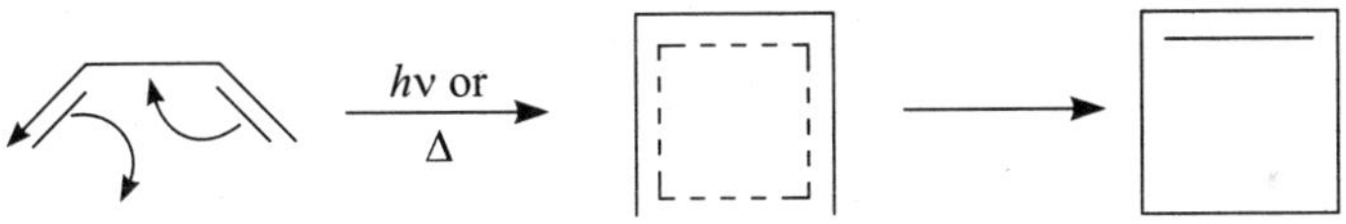

$h\nu$ or Δ

Transition state

Cycloaddition reaction

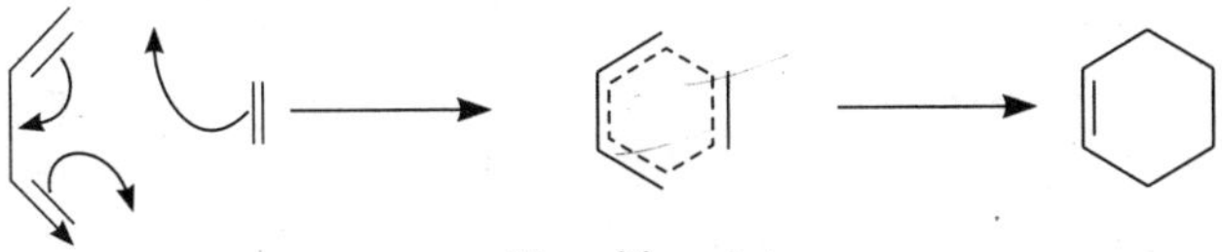

Transition state

2.3.8 Disproportionation Reaction (Decomposition Reactions)

Following reactions are the typical examples of disproportionation reactions.

$$CH_3COOH \xrightarrow{\Delta,\ MnO} CH_4 + CO_2$$

$$\dot{C}_2H_5 + \dot{C}_2H_5 \longrightarrow CH_3-CH_3 + CH_2=CH_2$$

2.4 ENERGY CONSIDERATIONS

Chemical reactions may be regarded as equilibrium processes. The point of equilibrium is decided by the relative energies of the reactants and the products. Reactions which do not proceed at all are also equilibrium processes for which the equilibrium constant has a value near zero. A chemical reaction proceeds when it has a favorable equilibrium or, in other words, when the products are more stable than the reactants.

2.4.1 Criterion of Feasibility of a Chemical Reaction

The equilibrium constant of a reaction depends on *the enthalpy change* (ΔH) *which is the quantity of heat released or absorbed when a mole of reactant is converted into a mole of product, and on the entropy change* (ΔS) *which is a measure of the energy consumed* (*or released*) *in the reaction to create more order (or disorder) in products relative to reactants.* When entropy change is taken with the absolute temperature (T) it gives the free energy change (ΔF) according to the relationship,

$$\Delta F = \Delta H - T\Delta S$$

It should be noted that *a reaction proceeds in the forward direction only when ΔF is negative that is there is overall release of energy. A positive ΔF, on the other hand, indicates that the reaction has a tendency to go backward.*

Since, **entropy** is a measure of the degree of freedom (or disorder) of a system, *a system is most stable when it has maximum entropy*. Reactions in which there is loss of freedom should, therefore, not be favourable processes and it has indeed been found that reactions involving a decrease in the number of molecular species on going from starting materials to products proceed only if ΔH is negative (exothermic reactions) and is large enough to give overall negative value for ΔF. On the other hand, when the number of molecular species increases in going from starting materials to products there is increase in entropy and the $-T\Delta S$ term which is a negative quantity then becomes large enough to make the reaction move in the forward direction even when it is not an exothermic reacticn. When all the substances are in their standard states (at 25°C and 1 atmosphere pressure), the above equation may be written as

$$\Delta F^0 = \Delta H^0 - T\Delta S^0$$

in which the superscript zero is used to denote *standard states*, and ΔF^0 *is the difference in standard free energies of products and starting materials*: $\Delta F^0 = F^0$ (products) – F^0 (starting materials). The relationship between the equilibrium constant (K) and ΔF^0 is given by the expression:

$$\Delta F^0 = -RT \ln K$$

where K = equilibrium constant, R = gas constant and T = absolute temperature.

2.4.2 Kinetics

Although a negative value of ΔF tells us that the thermodynamics is favourable for the reaction to proceed in the forward direction, it does not necessarily tell us that the reaction will proceed at a measurable rate. This is quite understandable as thermodynamics only predicts the position of equilibrium between the reactant and product without giving any information regarding the rate of the reaction.

Kinetic studies deal with the *rates of reactions and their dependence on the concentration* of various reacting species. In contrast to thermodynamics, time is the most important variable in kinetics. Kinetics is often used to gain an insight into reaction mechanisms. A typical organic reaction is a multistep process in which one of the steps, which is the slowest, is the ratin-determining step; kinetic studies give us the information that helps in identifying the species that participate in this step. Generally speaking, the rate of a reaction is proportional to the product of concentrations of the species that are involved in the rate determining step.

2.4.3 Order of Reaction

The rates of many organic reactions are found to vary with the concentrations of one or more of the reactants. This dependence is usually determined by measuring the rate of a reaction by performing several experiments in each of which all reactants except one are present in such a large excess that their concentrations may be regarded as constant. Obviously, the rate of reaction will vary only with the concentration of the species which are involved in the rate-determining step. Let us illustrate this by considering a reaction between A and B to give C.

$$A + B \rightleftharpoons C$$

When the concentration of either A or B is doubled, the rate is doubled and when the concentration of either of them in halved, the rate is halved. Thus, the rate of this reaction depends upon the concentration of both A and B. The reaction rate may be measured by observing either (*a*) the disappearance of reactants A or B with time, or (*b*) the appearance of product C with time. A rate equation can thus be written as follows:

$$\text{Rate} = \frac{-d\,[A]}{dt} = \frac{-d\,[B]}{dt} = \frac{d\,[C]}{dt} = k[A]\,[B]$$

where k is the *rate constant, dt is the incremental amount of time and the terms in square brackets denote concentrations.* The minus sign is used to signify that the concentration of the reactants is decreasing with time.

The above rate equation expresses the kinetic order of a reaction, and the reaction is said to be second order reaction as the right hand side of the equation is the product of two concentrations, [A] and [B]. In addition to the overall order of a reaction, it is often useful to indicate the order of a reaction, with respect to a specific reactant that reflects the power to which the concentration of that particular reagent must be raised to have a direct proportionality between concentration and reaction rates. Thus, the above reaction is first order with respect to A and first order with respect to B.

Order of reaction and molecularity: *Order of a reaction, n, is the sum of all the exponents of the concentrations of various reactants*, A, B, C,......

$$\text{Rate} = k[A]^{n_1}\,[B]^{n_2}\,[C]^{n_3}$$

$$\therefore \qquad n = n_1 + n_2 + n_3 + \ldots\ldots.$$

The reaction is said to be of total order n, and each individual exponent is the order with respect to that component. *e.g.*, n_1th order with respect to A, n_2th order with respect to B, and n_3th order with respect to C. The powers, n_1, n_2 and n_3 may not necessarily be whole numbers. The decomposition of acetaldehyde into methane and carbon monoxide, for instance, is of three-halves order, conforming to the relation.

$$\frac{-d[CH_3CHO]}{dt} = k\,[CH_3CHO]^{3/2}$$

The molecularity of a reaction, on the other hand, *may be defined as the number of species that are undergoing covalency changes in the rate-determining step.*

The main points of difference between order of reaction and molecularity are listed below:

Order of a reaction	Molecularity of a reaction
1. It is the sum of exponents of the concentrations of all the reactants involved in the rate-determining step.	1. It is the number of species undergoing covalency changes in the rate-determining step.
2. It is determined experimentally.	2. It is a theoretical concept.
3. It can be a fractional quantity (*e.g.*, the order of a reaction can be 3.5, etc.)	3. It is always a whole number.

Here it may be noted that order and molecularity of those reactions will be the same which take place in a single step (concerted reactions). However, for multistep reactions the order for each step is the same as molecularity of that step.

2.4.4 Reactivity and Rates of Reactions: Energy Profile Diagrams, Transition States and Hammond's Postulate

From the above discussion, it is clear that a reaction will proceed only when:

(*a*) The products are more stable than, or of the same stability as the reactants.

(*b*) There is a suitable pathway between reactants and products *i.e.*, the rate constant (*k*) is favourable. There is a characteristic rate constant associated with every reaction. The larger is the value of *k*, the faster is the reaction. Now the question arises as to what are the factors that make one rate constant larger than another? Following two theories have been proposed to account for this variation:

(A) Collision Theory.

(B) Transition State Theory.

A. Collision Theory

According to this theory a chemical reaction will take place only when the molecules, which are in the state of random motion, collide with each other. Thus the rate of the reaction will be proportional to the number of molecules per litre which collide with each other in unit time. Thus

Rate of the reaction $\propto$ *Number of colliding molecules/litre/sec.*

The number of collisions taking place per second per unit volume of a reaction mixture is called collision frequency. The collision frequency depends upon the following factors:

(*a*) *Concentration of reactants*: The greater is the number of molecules per unit volume, the greater will be the number of collisions per second. Hence, faster will be the rate of the reaction.

(*b*) *Temperature*: Rise in temperature of the system leads to increase in molecular motions (kinetic energy) of the component molecules which in turn lead to increase in collision frequency. Hence, faster will be the rate of the reaction.

All collisions do not form products: Under ordinary conditions of temperature and pressure the collision frequencies of gaseous reactants are very high ($\simeq 10^{25}$ to 10^{28} collision s^{-1}). Now if all collisions were effective (*i.e.*, they would have resulted in the formation of products), the reaction between gaseous reactants should have taken place in fraction of a second. However, in practice this does not happen. Most of the reactions are either very slow or do not take place at all at room temperature. In fact only one in 10^{15} collisions leads to the formation of products. Thus, it may be said that the feasibility and rate of a reaction does not depend only on number of collisions per second between reactant molecules but upon what are called as the number of effective collisions, *i.e.*, the number of collisions which actually lead to product formation. The number of effective collisions is governed by the following parameters:

(*a*) Energy Factor.

(*b*) Probability Factor.

(*a*) Energy factor: It has been shown that a collision can become "effective" if the colliding molecules possess a certain minimum amount of energy which is known as threshold energy. Under ordinary conditions of temperature and pressure most of molecules undergoing chemical reaction do not possess energy which is equal to threshold energy. So, they have to be supplied "extra" energy to make the collision "effective". This minimum amount of "extra" energy which has to be supplied to the reactant molecules in their ground state, to raise them to the level of threshold energy is called **activation energy**. Thus,

Activation energy = Threshold energy – Energy of the reactants in their ground state

From the above equation it is quite clear that lesser the activation energy faster will be the reaction and more the activation energy slower will be the reaction.

Activation energy is required even for a thermodynamically favoured reaction as energy is needed for breaking the bonds in the reactants before new bonds are formed in the products. Thus, there is a geometrical arrangement of atoms possessing higher energy than that of either reactants or products. Reactants have to pass through this highest energy state, called an *activated complex*, before their conversion to product (Fig. 2.1).

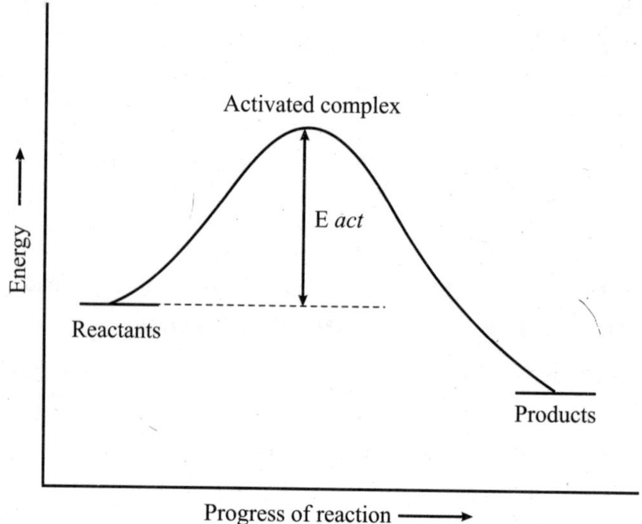

Fig. 2.1: Collision theory energy diagram of a typical reaction

(*b*) **Probability factor:** It is also called **orientation factor**. It has been shown than even if the energy of colliding molecules is higher than their threshold energy, still the reaction is either slow or does not take place. This means that apart from threshold energy, the rate of the reaction depends upon some other parameter. It has been proposed that apart from collision frequency and threshold energy, a reaction will take place only when the reactant molecules have proper **orientation or probability** factor. Thus

$$\text{Rate of the reaction} = PZe^{-E_a/RT}$$

where

P = probability factor
Z = collision frequency
E_a = energy of activation
R = gas constant
T = absolute temperature

Here $e^{-E_a/RT}$ is the fraction of collisions having energy greater than E_a.

B. Transition State Theory

Transition state theory has been found to be more useful in understanding the mechanism of organic reactions. This theory, envisages that collisions between the reactant molecules (say A and B—C) lead to the formation of highly energetic species called the **activated complex** or **transition state** (A------B------C) which lies at the top of the energy barrier between the reactants and the products (Fig. 2.2). The activated complex subsequently decomposes to give the products (A—B and C). In other words, *reactants on their way to products pass through a transition state*. The sequence of events is shown in energy profile diagram of a one-step reaction in Fig. 2.2.

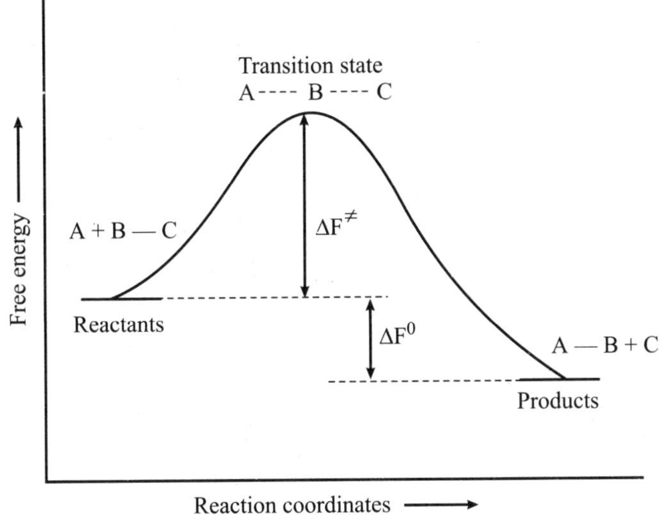

Fig. 2.2: Energy profile diagram for a single step reaction

A look at the Fig. 2.2 reveals that the rate of the reaction depends on the rate at which the *reactants reach the transition state* instead of *frequency of collision* as envisaged in the collision theory. This species is in equilibrium with reactants and hence its concentration is determined by an equilibrium constant denoted as $K^{\neq}$ which is related to the enthalpy of activation in the usual way.

$$\Delta F^{\neq} = \Delta H^{\neq} — T\Delta S^{\neq} = -2.303 \text{ RT} \log K^{\neq}$$

Where transition state is denoted by ≠ and F is the free energy of activation, $\Delta H^{\neq}$ is the heat of activation, $\Delta S^{\neq}$ is the entropy of activation.

Let us exemplify transition state theory by considering a general reaction

$$A + B—C \longrightarrow A—B + C$$

without involving an intermediate *i.e.*, a *single step reaction*. It has been shown by making certain approximations, that a minimum activation energy is required when A approaches B—C along the bonding line of B—C and on the side remote from C. The bond A—B begins to form whereas bond B—C begins to stretch. During the course of the reaction, each bond may be regarded as a partial one with both A—B and B—C bond distances greater than those in the product and the reactant. However, the amount of energy released during the partial formation of bond A—B is less than the energy absorbed by the corresponding partial separation of the bond B—C and as a result the energy of the system begins to *rise* with the process of bond transfer. The rise in energy finally reaches a maximum, (where the system is least stable), followed by a fall. At the end of the reaction the bond B—C has been completely replaced by the bond A—B. *The transition state is the state of highest potential energy acquired by the reactants during the transformation*. The system at this point is also called an *activated complex*. It possesses a definite geometry and charge distribution. However, due to the associated high energy it is highly unstable and cannot be isolated. We study it indirectly with the help of models.

The activation energy, among other factors, depends upon the strengths of B—C and A— B bonds. The stronger the B—C bond, the greater will be $\Delta F^{\neq}$ and conversely, the greater the strength of A—B bond, the lower is $\Delta F^{\neq}$. The energy changes that accompany a chemical reaction are conveniently represented in the form of an energy diagram with free energy (F) plotted against the pathway of the reaction called reaction coordinate (Fig. 2.2).

Since $\Delta F^{\neq}$ represents the energy barrier, the greater the height of this barrier, the slower will be the reaction. ΔF^0, which reflects the difference in standard free energy of the reactants and the products, is a thermodynamic quantity and does not affect the rate of the reaction which is governed solely by $\Delta F^{\neq}$.

In addition to one-step reaction discussed above, many organic reactions proceed through intermediates which may or may not possess enough stability to permit isolation. Multistep reactions will involve more than one transition states besides intermediates. For example, a two-step reaction will involve two transition states and one intermediate between reactants and products represented by a minimum in the energy profile diagram. The *deeper* the depression, *the stabler* will be the intermediate, and conversely a shallow depression in the energy curve indicates a less stable intermediate. Although there are two transition states each with an energy higher than that of the intermediate, the overall rate of the reaction is reflected in the energy difference between the reactants and the highest energy transition state ($\Delta F^{\neq}$ in Fig. 2.3). The intermediate is very unstable and is transformed to products more rapidly than the reactants generate the intermediate. *i.e.*, the intermediate cannot be isolated when the free energy of activation of the first step transition state is higher than that of second step.

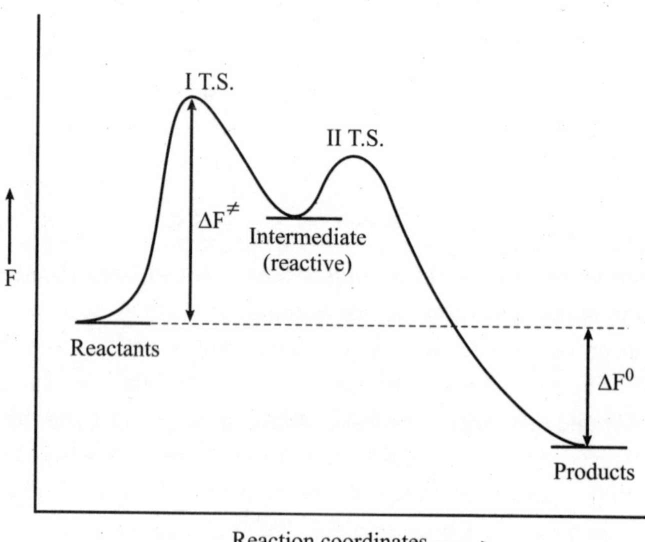

Fig. 2.3: Energy profile diagram for a two-step reaction with a less stable intermediate

Figure 2.4 represents the reverse situation in a two-step reaction. Here the free energy of activation of the second step is higher than that of the first. In this case the intermediate is very stable and can be isolated, and the second step is thus rate-determining.

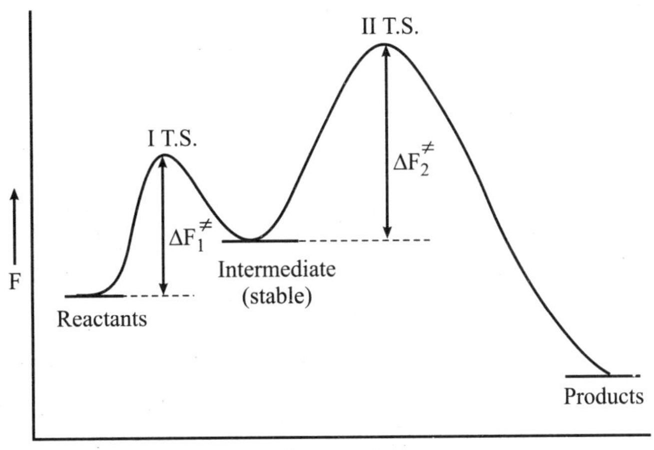

Fig. 2.4: Energy profile diagram for a two-step reaction with a stable intermediate

From the above discussion it may be concluded that a reaction may follow any one of the following routes:

Route I Reactants ⟶ Transition State ⟶ Products

Route II Reactants ⟶ Transition State I ⟶ Intermediate ⟶ Transition State II ⟶ Products

A detailed study of all steps of a chemical reaction showing as to how reactants are converted into products is called the **mechanism of the reaction**. A reaction involving simultaneous making and breaking of bonds in one step without the formation of any intermediate, is called a **concerted reaction**. Such a reaction involves a single transition state (Fig. 2.2). On the other hand, if a reaction occurs in two or more steps with the formation of a reactive or isolable intermediate, it is said to be a **stepwise** or **non-concerted reaction**. (Figs. 2.3 and 2.4).

The number of transition states formed during a stepwise reaction is equal to the number of steps in which the reaction occurs. Thus, a *two-step reaction involves two transition states of which the one having higher energy of activation constitutes the rate-determining step of the reaction.* The number of intermediates is always one less than the number of steps.

2.4.5 Difference Between Transition State and Intermediate: Hammond's Postulate

Intermediates are normal covalent molecules (or ions) possessing a definite geometry and can be isolated in many cases. Even when they are too unstable to be isolated, their properties can be studied directly by methods such as spectroscopy, or alternatively they can be trapped with substances, with which they are known to react. *The transition state, on the other hand, is not a normal covalent molecule (or ion) as some bonds are partially broken whereas others are only partially formed.* The properties of transition states are often studied by assuming that they resemble rather closely a real molecule (reactant, intermediate or product) that lies very close to it in energy content in a free energy diagram. This important generalization is a consequence of *Hammond's postulate which states that 'if two states, for example, a transition state and an unstable intermediate, occur consecutively during a reaction process and have nearly the same energy content, their inter-conversion will involve only a small reorganization of the molecular structure*'.

Thus the transition state may be described as *reactant-like*, *product-like* or *intermediate-like* depending upon the proximity of the former and any one of the latter. For example, for strongly exothermic reactions, the geometry of T.S. resembles that of the reactants *i.e.*, existing bonds are almost intact and new bonds are very little formed. Conversely, for strongly endothermic reactions, the geometry of the T.S. resembles that of the products.

This generalization also holds good when the reaction occurs through intermediates. Hammond's postulate is often invoked to explain the observed rate and course of reactions, for instance, it is well known that increase in branching at the α-carbon enhances the rate of S_N1 reactions. Generally, it is related to the stability of carbocations. But the rate enhancement does not depend upon the stability of a carbocation. Rather, it depends upon the ease with which the transition state leading to carbocation is reached, *i.e.*, upon $\Delta F^{\neq}$. As the transition state is not amneble we invoke Hammond's postulate assuming that in this reaction, the transition state is carbocation-like and any factor that lowers the energy of the carbocation will lower the energy of the transition state. Hence, such a transition state is reached more readily lowering the value of $\Delta F^{\neq}$.

$$R_3C{-}X \longrightarrow R_3\overset{\oplus}{C}\text{----}\overset{\ominus}{X}$$

Carbocation like transition state

2.4.6 Catalysis

A catalyst is a substance which usually increases the rate of a reaction without itself being consumed in the reaction. The mode of action of the catalyst may be described as under:

The catalyst first combines with one of the reactants to form an intermediate which subsequently decomposes or reacts with other reactants to give the product and is itself regenerated at the later stage. The rate enhancement in presence of the catalyst is mainly due to the fact that the catalyst provides an alternate low energy path (low energy transition state) from reactants to products. The catalytic effect of iodide ion on the hydrolysis of methyl chloride provides an illustrative example. Methyl chloride reacts with the hydroxide ion in a one-step process and, as expected, the rate depends on $\Delta F^{\neq}$.

$$CH_3Cl + H\overset{\ominus}{O} \longrightarrow CH_3OH + \overset{\ominus}{Cl}$$

Addition of iodide ion to the reaction mixture provides an alternate path consisting of a two-step mechanism in which methyl iodide is an intermediate.

$$CH_3Cl + \overset{\ominus}{I} \longrightarrow CH_3I + \overset{\ominus}{Cl}$$
(Intermediate)

$$CH_3I + H\overset{\ominus}{O} \longrightarrow CH_3OH + \overset{\ominus}{I}$$

Since, the iodide ion is both an excellent nucleophile and an excellent leaving group, the transition states of each step ($\Delta F_1^{\neq}$ and $\Delta F_2^{\neq}$) have lower energy than that for the one-step process. It should be noted that the $\Delta F°$ of the reaction remains unchanged by the addition of the catalyst; only the rate is increased. Figure 2.5 illustrates these points.

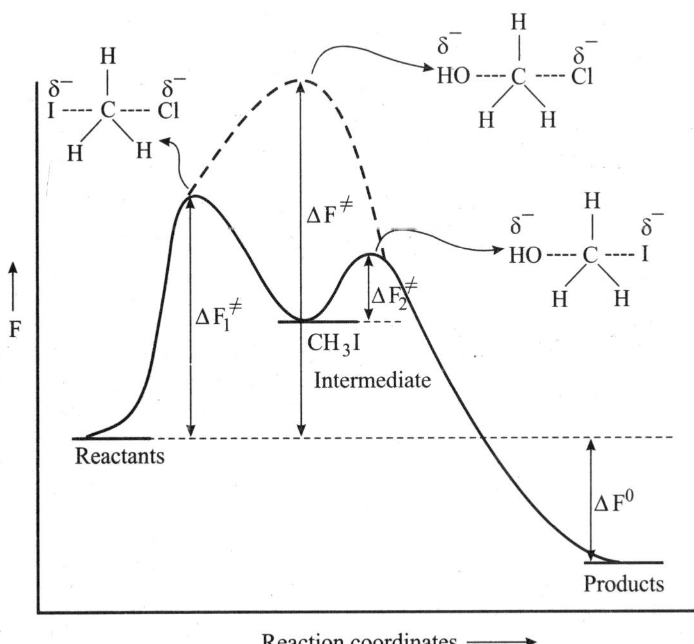

Fig. 2.5: Energy profile diagram for the uncatalyzed (broken line) and catalyzed (solid line) reaction

2.4.7 Competitive Reactions : Thermodynamic *Vs* Kinetic Control

Whenever a reaction yields more than one product, the relative amounts of these products depend on their ease of formation, which in turn is governed by the relative values of $\Delta F^{\neq}$ for each step. Thus, the rate of reaction will be slower when $\Delta F^{\neq}$ is relatively higher. Since $\Delta F^{\neq}$ is more important in deciding the rate of a reaction than ΔF^0 (which is a thermodynamic description of a reaction), these reactions are called *kinetically controlled reactions. A product which is formed faster because it has lower free energy of activation is called kinetically controlled product.*

Many reactions produce a mixture of products which are formed from a common intermediate. Very often the initially formed product (kinetically controlled) may possess higher energy than the other product, and under the conditions of the reaction may attain equilibrium with it which ultimately leads to the predominance of the most stable product. Such reactions are said to be *thermodynamically controlled.* Therefore, *when a product is formed slowly but due to its greater stability goes on forming (accumulating) in the reaction mixture it is called thermodynamically controlled product.* Addition of HBr to butadiene (I) elegantly exemplifies the importance of kinetic and thermodynamic control in product composition.

$$\underset{\text{I}}{CH_2{=}CH{-}CH{=}CH_2} \xrightarrow[\text{HBr}]{\overset{\oplus}{H}} \underset{\text{II}}{CH_3{-}\overset{\delta^+}{CH}{\overset{\cdots}{=}}CH{\overset{\cdots}{=}}\overset{\delta^+}{CH_2}}$$

$$\downarrow Br^{\ominus}$$

$$\underset{\substack{\text{III}\\ \text{1, 2 Addition}}}{CH_3{-}CHBr{-}CH{=}CH_2} \qquad \underset{\substack{\text{IV}\\ \text{1, 4 Addition}}}{CH_3{-}CH{=}CH{-}CH_2Br}$$

At low temperature III predominates, but when equilibrium sets in at higher temperatures in the presence of HBr and IV becomes the major product. Thus, at lower temperatures the reaction is *kinetically controlled* whereas it becomes thermodynamically controlled at elevated temperatures.The situation is best illustrated by the energy profile diagram (Fig. 2.6).

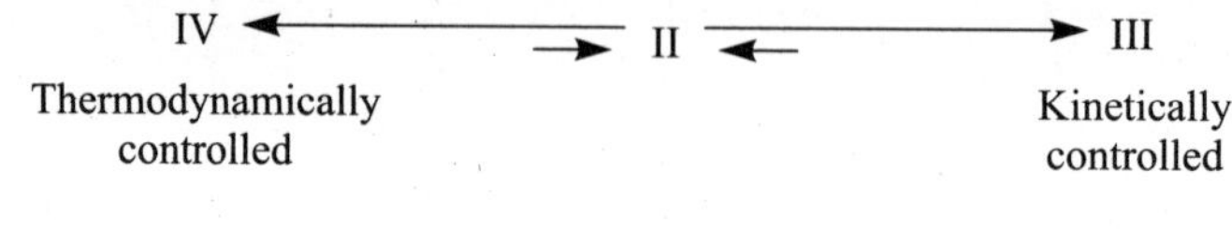

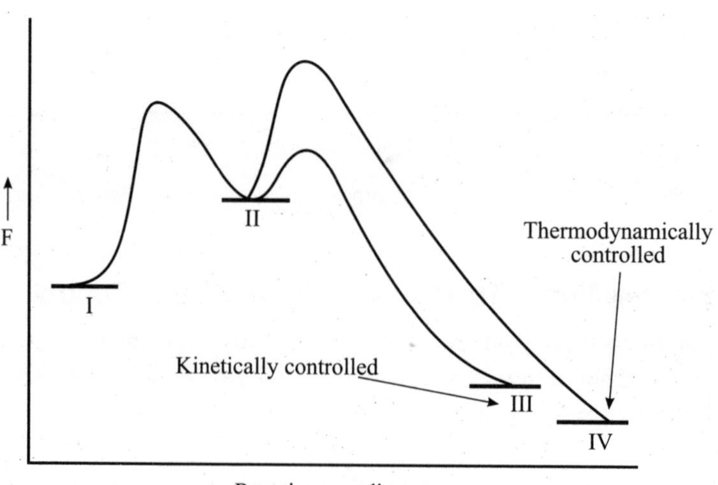

Fig. 2.6: Energy profile diagram for kinetically and thermodynamically controlled reactions

However, it must be remembered that if the conditions of the reaction do not permit interconversion, *i.e.*, the reaction is irreversible then the kinetically controlled product will invariably predominate.

2.5 REACTIVE INTERMEDIATES

A great majority of organic reactions are multistep processes involving intermediates whose lifetimes are usually very short. *These short lived highly reactive chemical species are called* **reactive intermediates or reaction intermediates.** An understanding of the structures and properties of these intermediates—which are normal covalent compounds, ions or radicals—is of paramount importance for understanding organic reaction mechanisms. As we have seen in the energy profile diagrams, intermediates possess sufficiently low energy to be *formed* under the reaction conditions, but in most of the cases are *not stable* enough to permit isolation. This is especially true of those intermediates in which carbon has fewer than four covalent bonds. These reactive intermediates are:

1. Carbocations, 2. Carbanions, 3. Free radicals, 4. Carbenes, 5. Nitrenes, 6. Arynes, 7. Enamines.

A brief description of these kinds of intermediates is given below:

2.5.1 Carbocations

Carbocations are cations which are obtained by the removal of one of the four groups attached to the carbon atom along with its bonding electrons. Thus, *they may be defined as organic species which have a positive charge on a carbon atom bearing only six electrons*. They are short lived and highly reactive species. Untill the early 1970s, the ions with positively charged carbon atoms were known as carbonium ions. As this name does not follow the usual pattern of nomenclature, the term *carbocation* was introduced to describe such ions. The term carbocation is consistent with the well-accepted name for the related anions, *i.e.*, *carbanions*. However, both these terms—carbocation and carbonium ion—are being used to describe an ion with a carbon atom having no unshared pair of electrons and only three covalent bonds. It has a positive charge.

In most of the cases carbocations cannot be isolated but this instability does not minimize, their importance as reactive intermediates in a great number of chemical reactions.

Classification of Carbocations

Alkyl carbocations are classified into three categories *i.e.*, primary (1°), secondary (2°) and tertiary (3°) depending upon whether the cationic carbon is primary, secondary or tertiary. Methyl carbocation with one carbon is a special case.

$\overset{\oplus}{C}H_3$
Methyl carbocation

$CH_3\overset{\oplus}{C}H_2$
Ethyl carbocation (1°)

$(CH_3)_2\overset{\oplus}{C}H$
iso-Propyl carbocation (2°)

$(CH_3)_3\overset{\oplus}{C}$
tert-Butyl carbocation (3°)

Structure

The carbon atom of the carbocation is *sp*²*-hybridized* and it uses the three hybrid orbitals for singly bonding to three substituents; the remaining unhybridized *p*-orbital is empty. The carobcation thus has a planar structure having all the three covalent bonds in one plane with the bond angles of 120° between them (Fig. 2.7). The empty *p*-orbital is at right angle to the plane of *sp*²-hybridized orbitals.

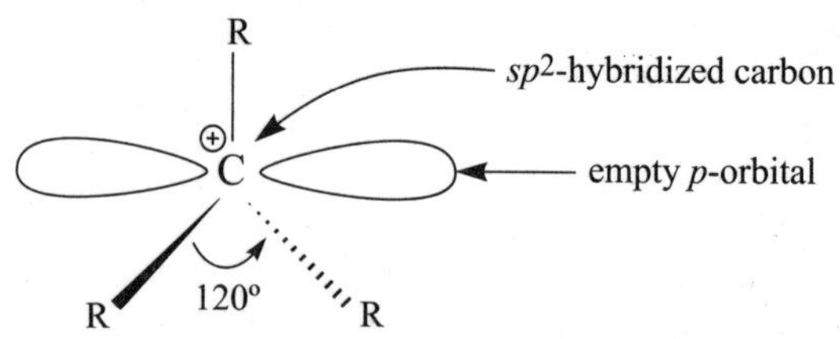

Fig. 2.7: Orbital picture of a carbocation

Carbocations have a strong tendency to complete the octet of carbon and consequently they are highly reactive species and participate in many chemical reactions, i.e., in unimolecular substitution and elimination reactions, in the decomposition of diazonium salts and in various rearrangements.

Stability of Carbocations

Any factor which tends to disperse the positive charge must increase the stability of carbocations while any factor which tends to intensify the positive charge must decrease the stability of the carbocation.

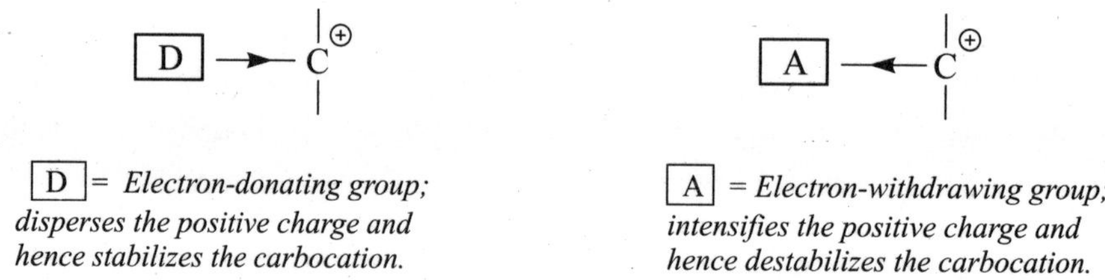

D = *Electron-donating group; disperses the positive charge and hence stabilizes the carbocation.*

A = *Electron-withdrawing group; intensifies the positive charge and hence destabilizes the carbocation.*

1. Stability of saturated carbocations (alkyl carbocations): The stability of alkyl carbocations decreases in the following order:

$$\underset{(R_3\overset{\oplus}{C})}{\text{Tertiary}} > \underset{(R_2\overset{\oplus}{C}H)}{\text{Secondary}} > \underset{(R\overset{\oplus}{C}H_2)}{\text{primary}}$$

This order can be explained on the basis of the following factors:

***(i) Inductive effect*:** There is strong evidence, both physical and chemical, that alkyl groups are electron-donating, *i.e.*, they exhibit +I-effect. Thus, when these groups are attached to the electron-deficient carbon of the carbocation, they release electrons towards this carbon. The positive charge thus gets dispersed over the alkyl groups and this dispersal of charge increases the stability of the carbocation. Obviously, the greater the number of such alkyl groups attached to the cationic carbon, the more is the dispersal of positive charge and hence stabler will be the carbocation. Thus a tertiary carbocation having three alkyl groups, should be more stable than secondary having two alkyl groups which in turn should be more stable than primary having one alkyl group.

$R_3C^{\oplus}$ (three R → C) > $R_2CH^{\oplus}$ (two R → C) > $RCH_2^{\oplus}$ (R → C)

Most stable (three +I-groups) … *Least stable* (one +I-groups)

***(ii) Hyperconjugation effect*:** The greater stability of alkyl substituted carbocations can also be explained on the basis of the phenomenon of **hyperconjugation**. According to this the σ-electrons of αC—H bonds are delocalized into vacant *p*-orbital of the positively charged carbon. This disperses the positive charge over all such bonds. Thus it is reasonable to say that greater is the number of α σC—H bonds stabler is the corresponding carbocation. In agreement with this the tertiary butyl carbocation having nine such σC—H bonds is most stable followed by isopropyl carbocation having six such σC—H bonds. The ethyl carbocation having only three such σC—H bonds is the least stable in this series. These cations are regarded as hybrids of various hyperconjugative resonance contributing forms and the number of these contributing forms depends upon the number of ασ C—H bonds. Thus greater is the number of ασ C—H bonds, the greater is the number of hyperconjugative resonance contributing forms and hence stabler is the carbocation.

$H_3C-\overset{\oplus}{C}(CH_3)-CH_3 \leftrightarrow \overset{\oplus}{H}\ H_2C{=}C(CH_3)-CH_3 \leftrightarrow$ … $\leftrightarrow$ …

tertiary-Butyl cation

Six more such equivalent structures

(Nine ασ C—H bonds, nine equivalent hyperconjugative structures, *most stable*)

$H_3C-\overset{\oplus}{C}H-CH_3 \leftrightarrow \overset{\oplus}{H}\ H_2C{=}CH-CH_3 \leftrightarrow$ … $\leftrightarrow$ …

iso-Propyl cation

Three more such equivalent structures

(Six ασ C—H bonds, six equivalent hyperconjugative structures, *less stable* than *tert*-butylcation).

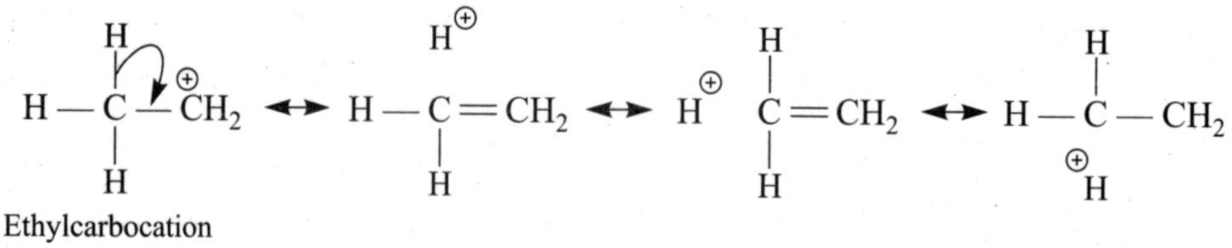

$$H-\underset{H}{\overset{H}{C}}-\overset{\oplus}{C}H_2 \leftrightarrow H-\underset{H}{\overset{H^{\oplus}}{C}}=CH_2 \leftrightarrow \overset{\oplus}{H}\ \underset{H}{\overset{H}{C}}=CH_2 \leftrightarrow H-\underset{\overset{\oplus}{H}}{\overset{H}{C}}-CH_2$$

Ethylcarbocation

(Three ασ C—H bonds, three equivalent hyperconjugative structures, *least stable* of the three)

***(iii) Steric effects (or better steric relief)*:** Steric effects play a key role in the ease of formation as well stability of carbocations. We know that carbocations are planar with bond angles of 120° (*i.e.*, having maximum distance between the substituents attached to the positively charged carbon) but the substrates from which they are formed have sp^3-hybridized carbon and have bond angles of 109.5° (*i.e.*, lesser distance between substituents, thus causing congestion or **steric crowding**). Now, if the four substitutents attached to an sp^3-hybridized carbon are bulky, the molecule will be highly strained* due to congestion. However, when such molecule ionises to produce a carbocation, the bond angles expand from 109.5° (sp^3) to 120° (sp^2) thereby resulting in the relief (removal of congestion) due to increase in space between the bulky substituents. A typical case of steric relief is the ionisation of tri-isopropylmethyl chloride.

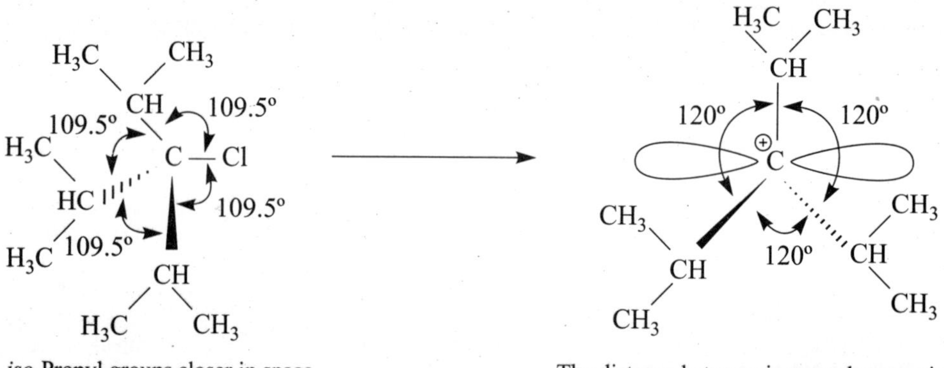

iso-Propyl groups closer in space
(*Highly strained*)

The distance between isopropyl groups increases
(*Steric relief*)

The order of stability of simpler 3°, 2°, 1° and methyl cations is also in tune with the concept of steric relief.

2. Stability of unsaturated and other types of carbocations: The carbocations having a double bond or a non-bonding pair of electrons next to the positively charged carbon are stabilized by **resonance** which involves the delocalization of π or *n*-electrons into vacant *p*-orbital of this carbon. For example,

***(i) Allyl and benzyl cations, diphenylmethyl and triphenyl methyl carbocations*:** In these cations the positively charged carbon atom is in conjugation with a double bond. This arrangement leads to increase in delocalisation due to resonance and because of this the positive charge is dispersed over the

* This type of strain is caused by pushing of the substituents together in the limited space and is called as **B strain** (or back strain)

carbon skeleton. Thus,

Allyl cation

$$H_2C{=}CH{-}\overset{\oplus}{C}H_2 \longleftrightarrow \overset{\oplus}{C}H_2{-}CH{=}CH_2$$

(Resonance stabilized allyl cation)

Benzyl cation

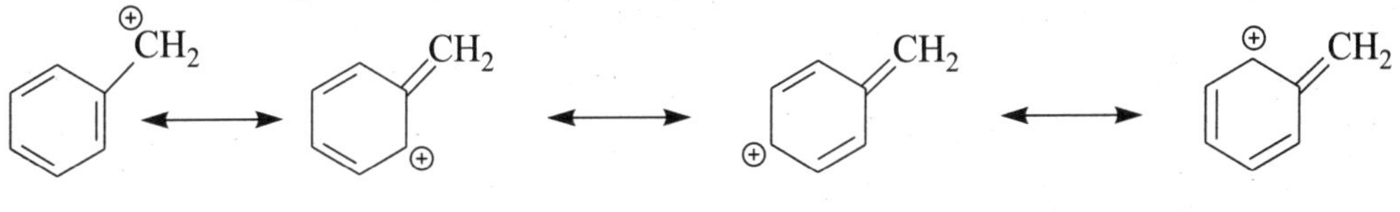

(Resonance stabilized benzyl cation)

Effect of substitutents on the stability of benzyl carbocation

The presence of various substituents at *o*-and *p*-positions of benzene ring in benzyl carbocation greatly influence the stability of substituted benzyl carbocations. **Electron-donating groups** such as alkyl groups, –OH, –OR, $–NH_2$, –NHR, $–NR_2$ etc., *stabilize* these cations. This may be attributed to the dispersal of positive charge by resonance (+R-effect) in case of groups having unshared pair of electrons and hyperconjugative effect of alkyl groups (CH_3 group having highest effect). In contrast, **electron-withdrawing groups** such as — CN, — NO_2, — CHO, — COR, — COOH, — COOR, etc., *destabilize* these cations. This may be due to the fact that electron-withdrawl by these substituents would intensify positive charge on the cationic carbon. Thus for example, *p*-methoxybenzyl cation is more stable than benzyl cation whereas *p*-nitrobenzyl cation is less stable than this.

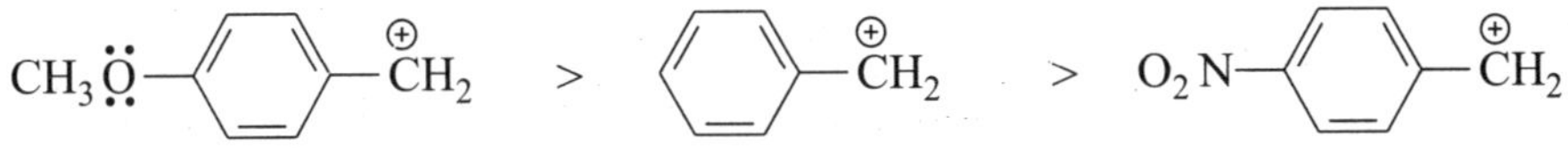

$$CH_3\ddot{O}{-}C_6H_4{-}\overset{\oplus}{C}H_2 > C_6H_5{-}\overset{\oplus}{C}H_2 > O_2N{-}C_6H_4{-}\overset{\oplus}{C}H_2$$

p-Methoxybenzyl cation (Dispersal of charge on methoxy group due to resonance, *more stable than benzyl cation*)

Benzyl cation

p-Nitrobenzyl cation (Intensification of charge due to electron-withdrawl by nitro group, *less stable than benzyl cation*)

The diphenyl and triphenyl carbocations are much more stable than benzyl cation. Obviously, the largerer the number of phenyl groups attached to cationic carbon, greater is the dispersal of positive charge. Hence, more stable are such carbocations. The decreasing order of stability of benzyl, diphenylmethyl and triphenylmethyl carbocations is as follows:

$$(C_6H_5)_3\overset{\oplus}{C} > (C_6H_5)_2\overset{\oplus}{C}H > C_6H_5\overset{\oplus}{C}H_2$$

Diphenylmethyl carbocation is more stable compared to benzyl carbocation due to greater number of resonating structures.

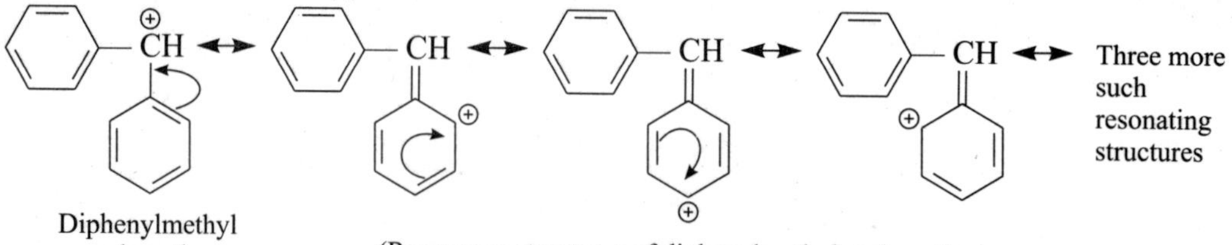

CH ↔ CH ↔ CH ↔ CH ↔ Three more such resonating structures

Diphenylmethyl carbocation

(Resonance structures of diphenylmethyl carbocation)

Similarly, triphenylmethyl carbocation is more stable than diphenylmethyl carbocation due to larger number of resonating structures which can be written for this as shown below:

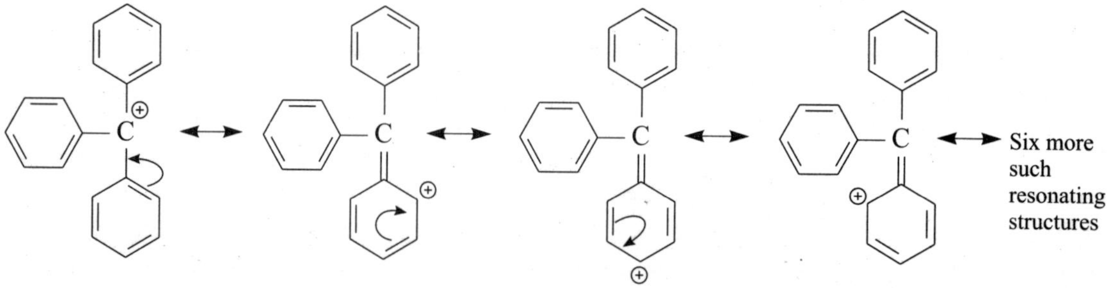

C ↔ C ↔ C ↔ C ↔ Six more such resonating structures

(Resonance structures of triphenylmethyl carbocation)

The high stability of triphenylmethyl carbocation is reflected in the fact that salts such as perchlorates, fluoroborates, etc., have been isolated and stored for a very long time.

Similary, triphenylcyclopropenyl cation is more stable than diphenylcyclopropenyl cation which in turn is more stable than phenylcyclopropenyl cation.

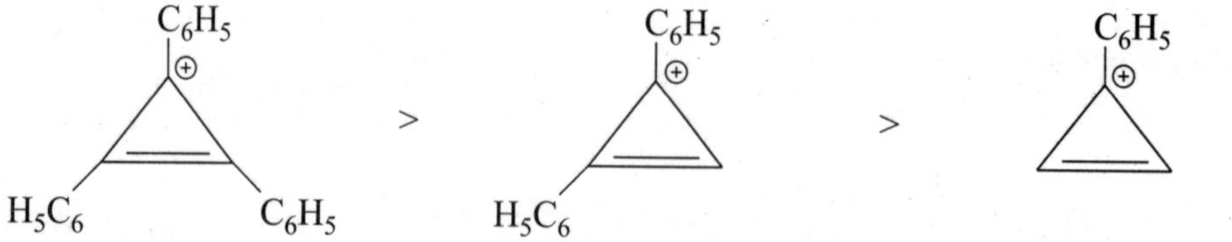

C_6H_5, H_5C_6, C_6H_5 > C_6H_5, H_5C_6 > C_6H_5

Triphenylcyclopropenyl carbocation

Diphenylcyclopropenyl carbocation

Phenylcyclopropenyl carbocation

***(ii) Stability of tropylium ion or cycloheptatrienyl cation*:** Tropylium ion is unusually stable. The reason for its unusual stability is that it is an aromatic cation containing 6π-electrons. As shown below the 6π-electrons are completely delocalized over all the seven sp^2-hybridized carbon atoms below just in the same manner as the 6π-electrons are delocalized in benzene ring. Various resonance contributing forms of tropylium ion are as under:

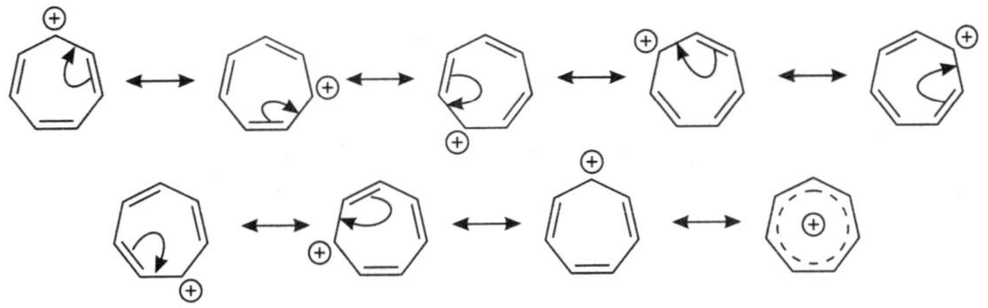

(Resonance structures of tropylium ion)

Actually the tropylium is so stable that its reaction with nucleophiles such as water, alcohols, etc., is very slow.

***(iii) Cyclopentadienyl carbocation*:** Like cycloheptatrienyl cation, it is expected that cyclopentadienyl cation should also be stabilized by resonance because the cationic carbon is conjugated with two double bonds. Actually this carbocation is highly unstable. The reason for this instability is that it has 4 π-electrons and hence is *anti-aromatic*. This is an example of *a species which is destabilized by resonance* (for details see Chapter 8).

***(iv) Stability of alkoxyalkyl carbocations*:** Carbocations in which the positively charged carbon atom is attached to a heteroatom (carrying non-bonding pair(s) of electrons) such as oxygen, nitrogen, halogen etc., are also stabilized by resonance. For example,

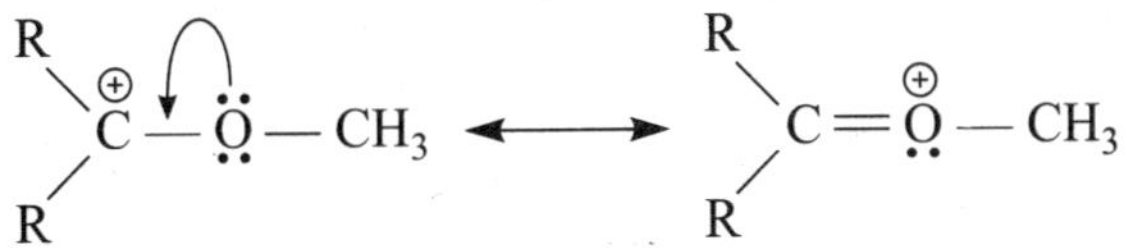

$$R_2\overset{\oplus}{C}-\ddot{O}-CH_3 \longleftrightarrow R_2C=\overset{\oplus}{\ddot{O}}-CH_3$$

The stability conferred by resonance on this carbocation is reflected in the fact that methoxymethyl cation can be isolated as a stable salt, $[CH_3O\overset{\oplus}{C}H_2\overset{\ominus}{Sb}F_6]$.

***(v) Stability of acyl cations*:** Like alkoxyalkyl cations, acyl cations can also be regarded as hybrids of the following contributing forms and hence are stabilized:

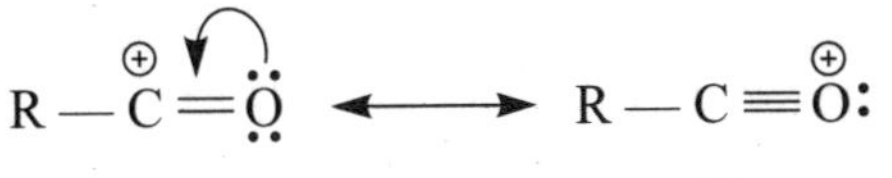

$$\underset{\text{I}}{R-\overset{\oplus}{C}=\ddot{O}} \longleftrightarrow \underset{\text{II}}{R-C\equiv\overset{\oplus}{O}:}$$

Although having positive charge on the more electronegative oxygen atom, the contribution of canonical structure (II) is less than that of structure (I) to the resonance hybrid, yet acetyl cation, CH_3CO^+ is almost as stable as *tert*-butyl carbocation.

***(vi) Vinyl and phenyl carbocations*:** Carbocation having positive charge on the doubly bonded carbon atom as in the case of vinyl ($CH_2 = CH^+$) and phenyl ($C_6H_5^+$) cations are not stabilized by resonance because the positively charged sp^2-orbital of the carbon atom is perpendicular to the p-orbitals of the π-bonds. Thus, these cations are very short lived. However, vinyl cations are more stable than methyl carbocation but less stable than other primary carbocations (Table 2.2).

Relative Stability of Carbocations

The relative stability of carbocations can be easily assessed by determining bond dissociation energy for heterolytic fission. *The lower the value of this energy, the greater is the stability of the carbocation.* The bond dissociation energies of some alkyl/aryl halides are given in Table 2.2.

Table 2.2: Bond dissociation energies for some alkyl/aryl chlorides for heterolytic fission (kJ mol^{-1})

Alkyl / aryl chloride	Bond dissociation energy
$CH_3—Cl$	952
$CH_3CH_2—Cl$	799
$CH_3CH_2CH_2—Cl$	744
$(CH_3)_2CH—Cl$	711
$(CH_3)_3C—Cl$	657
$CH_2{=}CH—Cl$	866
$CH_2{=}CHCH_2—Cl$	724
$C_6H_5—Cl$	916
$C_6H_5CH_2—Cl$	695

A look at the data listed in the Table 2.2 reveals that the order of stability of carbocations is: *Tert-butyl > Benzyl > Isopropyl > Allyl > n-Propyl > Ethyl > Vinyl > Methyl > Phenyl.*

Detection of Carbocations

Simple carbocations can be detected by making use of NMR spectroscopy. Their proton NMR spectra show extensive downfield shift of the hydrogen atoms (deshielding) attached to the cationic carbon.

Alkyl fluorides, react with SbF_5, a powerful Lewis acid, to yield the carbocations.

$$R—F + SbF_5 \xrightarrow{\text{Liq. } SO_2} \overset{\oplus}{R} + \overset{\ominus}{SbF_6}$$

The nine protons in *tert*-butyl fluroride in sulphur dioxide exhibit a singlet at δ 1.3. However, when SbF_5 is added to the solution, the protons show a large downfield shift appearing as a sharp singlet at δ 4.35. This large downfield shift of the signal due to methyl protons can be explained on the basis of the fact that positively charged carbon exerts large inductive and hyperconjugative effect on these protons.

$$H_3C—\underset{CH_3}{\overset{CH_3\ (\delta 1.3)}{C}}—F \xrightarrow[SbF_5]{\text{Liq. } SO_2} H_3C—\underset{CH_3}{\overset{CH_3\ (\delta 4.35)}{\overset{\oplus}{C}}} \quad \overset{\ominus}{SbF_6}$$

Similarly, the NMR spectrum of *iso*-propyl fluoride shows methyl protons at δ 1.23 while in SbF_5 — SO_2 medium they show up at δ 5.06. The secondary proton in *iso*-propyl fluoride gives a signal at δ 4.64 whereas in the carbocation, it is shifted to δ 13.50.

$$H_3C-\underset{\underset{H\ (\delta\ 4.64)}{|}}{\overset{\overset{CH_3\ (\delta\ 1.23)}{|}}{C}}-F \xrightarrow[\text{Liq. } SO_2]{SbF_5} H_3C-\underset{\underset{H\ (\delta\ 13.50)}{|}}{\overset{\overset{CH_3\ (\delta\ 5.06)}{|}}{C^{\oplus}}} \quad \overset{\ominus}{SbF_6}$$

The NMR spectroscopy has also provided evidence for the rearrangement of carbocations to yield more stable ions. Thus *n*-butyl, *iso*-butyl and *sec*-butyl fluorides all display the same NMR spectra in antimony pentafluoride and liquid SO_2 as *t*-butyl fluoride. Since, the NMR spectrum for *t*-butyl carbocation is unequivocal (one signal at δ 4.35), the other three primary and secondary carbocations (generated from *iso*-meric alkyl fluorides) must have rearranged to a more stable tertiary carbocation. Similarly, the NMR spectra of *n*-propyl and *iso*-propyl fluorides in SbF_5/liq. SO_2 are similar; presumably the less stable *n*-propyl carbocation rearranges to the more stable *iso*-propyl carbocation.

Formation of Carbocations

Carbocations can be generated in a variety of ways. Some of the reactions in which carbocations are formed are summarized below:

(i) Solvolysis of C—X bond (X = halogens, OBs, etc.) i.e., ionization of these compounds in polar solvents.

$$R-X \xrightarrow{\text{Polar solvent}} \overset{\oplus}{R} + :\overset{\ominus}{X}$$

(ii) Deamination of aliphatic amines by nitrous acid.

$$R-NH_2 \xrightarrow{HNO_2} R\overset{\oplus}{N_2} \longrightarrow \overset{\oplus}{R} + N_2$$

(iii) Reaction of halides with super acids (SbF_5, FSO_3H, etc.).

$$R-F + SbF_5 \longrightarrow \overset{\oplus}{R} + \overset{\ominus}{SbF_6}$$

(iv) Protonation of alcohols followed by dehydration.

$$R-\ddot{O}H \xrightarrow{\overset{\oplus}{H}} R-\overset{\oplus}{\ddot{O}H_2} \longrightarrow \overset{\oplus}{R} + H_2O$$

(v) Protonation of alkenes.

$$R-CH{=}CH_2 \xrightarrow{\overset{\oplus}{H}} R-\overset{\oplus}{C}H-CH_3$$

(*vi*) *Protonation of carbonyl compounds.*

$$>C=\ddot{O}: \xrightarrow{\overset{\oplus}{H}} \left[>C=\overset{\oplus}{\ddot{O}}H \longleftrightarrow >\overset{\oplus}{C}-\ddot{O}H \right]$$

(*vii*) *Protonation of nitriles.*

$$R-C\equiv N: \xrightarrow{\overset{\oplus}{H}} \left[R-C\equiv\overset{\oplus}{N}H \longleftrightarrow R-\overset{\oplus}{C}=NH \right]$$

Reactions of Carbocations

Broadly speaking the carbocations undergo four basic types of reactions which are listed below:

(*a*) Combination with a nucleophile.

(*b*) Elimination of a neighbouring proton.

(*c*) Addition to an unsaturated bond.

(*d*) Rearrangement.

(*a*) Combination with a nucleophile: A carbocation may combine with a species having unshared pair of electrons.

$$\ddot{N}u^{\ominus} + \overset{\oplus}{R} \longrightarrow Nu-R$$

Example: S_N1 reaction of *t*-butyl chloride to produce *t*-butyl alcohol (S_N1) takes place in the following two steps:

(*i*) Ionisation of *t*-butyl chloride to give *t*-butyl carbocation.

$$H_3C-\underset{CH_3}{\overset{CH_3}{C}}-Cl \xrightarrow[\text{Slow}]{H_2O} H_3C-\underset{CH_3}{\overset{CH_3}{\overset{\oplus}{C}}} + \overset{\ominus}{Cl}$$

t-Butyl cation

(*ii*) Combination with $H_2\ddot{O}$: (a weak nucleophile to produce *t*-butyl alcohol).

$$H_3C-\underset{H_3C}{\overset{H_3C}{\overset{\oplus}{C}}} + \underset{\text{Nucleophile}}{H_2\ddot{O}} \xrightarrow[\text{Slow}]{H_2O} H_3C-\underset{CH_3}{\overset{CH_3}{C}}-\underset{H}{\overset{\oplus}{O}}-H \xrightarrow{-H^{\oplus}} CH_3-\underset{CH_3}{\overset{CH_3}{C}}-OH$$

(*b*) Elimination of a neighbouring proton: A carbocation may eliminate a proton (or some other positively charged species) from the neighbouring atom.

$$-\overset{\oplus}{\underset{|}{C}}-Z-H \longrightarrow -\underset{|}{C}=Z+\overset{\oplus}{H}$$

Example: Elimination of a proton from *t*-butyl cation to give *iso*-butylene.

$$H_3C-\overset{CH_3}{\underset{\oplus}{C}}-CH_2-H \longrightarrow (H_3C)_2C=CH_2+\overset{\oplus}{H}$$

(*c*) Addition to a multiple bond: A carbocation can add to a double bond generating positive charge at a new position (a new carbocation).

$$>C=C< + \overset{\oplus}{R} \longrightarrow -\overset{\oplus}{\underset{|}{C}}-\overset{|}{\underset{|}{C}}-R$$

The cation so produced may add to another double bond and such series of reactions produces poly (vinyl) compounds.

(*d*) Rearrangement: A carbocation may undergo 1, 2-shift (migration to neighbouring carbon) of hydrogen/alkyl groups/aryl groups. Some examples are:

(*i*) $$H_3C-\underset{H}{\underset{|}{CH}}-\overset{\oplus}{C}H_2 \xrightarrow[\text{shift}]{\text{1, 2-Hydride}} CH_3-\overset{\oplus}{C}H-CH_3$$

1°-Propyl carbocation (*less stable*) → 2°-Propyl carbocation (*more stable*)

(*ii*) $$H_3C-\overset{CH_3}{\underset{H}{C}}-\overset{\oplus}{C}H_2 \xrightarrow[\text{shift}]{\text{1, 2-Hydride}} H_3C-\overset{CH_3}{\underset{\oplus}{C}}-CH_3$$

iso-Butyl carbocation (2°) (*less stable*) → *t*-Butyl carbocation (3°) (*more stable*)

(*iii*) $$H_3C-\overset{CH_3}{\underset{CH_3}{C}}-\overset{\oplus}{C}H_2 \xrightarrow[\text{shift}]{\text{1, 2-Hydride}} H_3C-\overset{CH_3}{\underset{\oplus}{C}}-CH_2-CH_3$$

neo-Pentyl carbocation (1°) (*less stable*) → 2-Methyl-2-butyl carbocation (3°) (*more stable*)

The new carbocation produced can subsequently add a nucleophile or eliminate a proton from the vicinal carbocation.

2.5.2 Carbanions

Carbanions are anions of carbon which are obtained by the removal of one of the groups attached to the carbon leaving behind the bonding electrons on it. Thus organic species carrying negative charge on one of the carbon atoms is called a **carbanion.** In contrast to carbocations in which the positively charged carbon atom has six electrons, in these anions the negatively charged carbon atom has eight electrons (6 bonding and 2 non-bonding). These ions are also short lived highly reactive species. Their reactivity is believed to be due to the presence of formal negative charge on carbon atom.

Classification

Alkyl carbanions are classified into three categories, *i.e.*, primary (1°), secondary (2°) and tertiary (3°) depending upon whether the anionic carbon is primary, secondary or tertiary. Methyl carbanion with one carbon atom is a special case, e.g.,

$\overset{\ominus}{C}H_3$ — Methyl carbanion

$CH_3-\overset{\ominus}{C}H_2$ — Ethyl carbanion (1°)

$(H_3C)_2\overset{\ominus}{C}H$ — *iso*-Propyl carbanion (2°)

$H_3C-\overset{\ominus}{C}(CH_3)-CH_3$ — *tert*-Butyl carbanion (3°)

Structure

Depending upon their nature and source from which they are formed, the carbanions can have any one of the following two geometries:

(*i*) Pyramidal

(*ii*) Planar.

(*i*) Pyramidal: Simple alkyl carbanions possess one unshared pair of electrons and three pairs of bonding electrons around the central carbon atom which is sp^3-hybridized. Thus, it is expected that a carbanion would have tetrahedral structure with unshared pair of electrons occupying one apex of the tetrahedron. Actually, they have been shown to have pyramidal structure similar to ammonia or amines with which they are **isoelectronic** (Fig. 2.8)

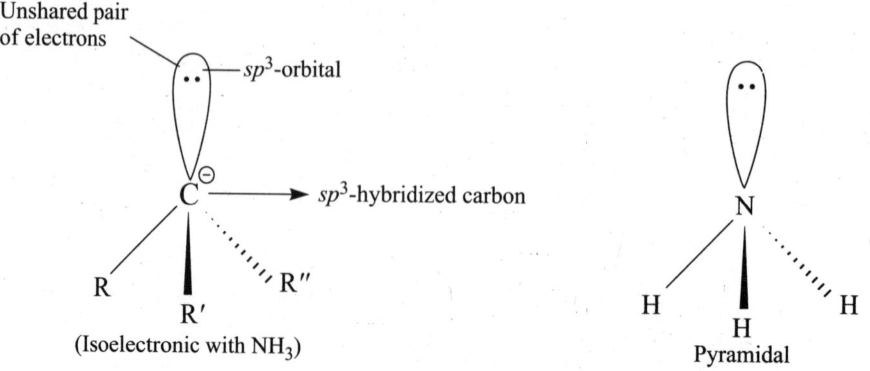

Fig. 2.8: Pyramidal structure of carbanion and ammonia

The pyramidal structure of carbanion can be explained on the basis of the fact that it possesses one pair of non-bonding electrons and three pairs of bonding electrons. Due to non-bonding—bonding repulsions, which are greater than bonding-bonding repulsions, there is a reduction of bond angles between bonding pairs to a value which is lesser than the tetrahedral value of 109.5°. Hence the simple carbanions such as methyl anion are pyramidal. (with bond angles between 97° – 100°), like ammonia with which they are isoelectronic. If this is the picture of a carbanion, then a carbanion in which three groups are different (R, R' and R" in Fig. 2.8) should be chiral and any reaction in which such a carbanion is an intermediate should proceed with retention of configuration. However, it has been shown that the reaction involving such anions result in loss of optical activity, *i.e.*, they produce racemic mixture.

Reason for loss of optical activity in reactions involving carbanions of the type ($\overset{\ominus}{C}RR'R''$).

In order to explain the above observation, it has been proposed that the unshared pair and the central carbon atom rapidly oscillate from one side of the plane to the other, *i.e.*, the two enantiomeric pyramidal structures are in a state of rapid equilibrium due to which there is a loss of optical activity.

$$R,R',R''\text{-}\overset{\ominus}{C}(:) \rightleftharpoons (:)\overset{\ominus}{C}\text{-}R,R',R''$$

Planar: Resonance stabilized carbanions, in which the orbital containing unshared pair of electrons overlaps with the π-orbital of the neighbouring multiple bond, are essentially planar. This is due to the fact that planarity is essential for resonance to occur. Thus in this type of carbanions the anionic carbon is sp^2-hybridized. For example,

$$R-\underset{H}{\overset{R'}{C}}-\overset{R''}{C}=O \xrightarrow{-\overset{\oplus}{H}} \left[R-\overset{R'}{\underset{\ominus}{C}}-\overset{R''}{C}=O \longleftrightarrow R-\overset{R'}{C}=\overset{R''}{C}-\overset{\ominus}{O} \right]$$

sp^2 sp^2

Resonance stabilized; essentially planar

Similarly allyl anion is also resonance stabilized.

$$\left[\overset{\ominus}{C}H_2-CH=CH_2 \longleftrightarrow H_2C=CH-\overset{\ominus}{C}H_2 \right]$$

The following molecular orbital depiction of the above type of overlaping makes this concept more clear:

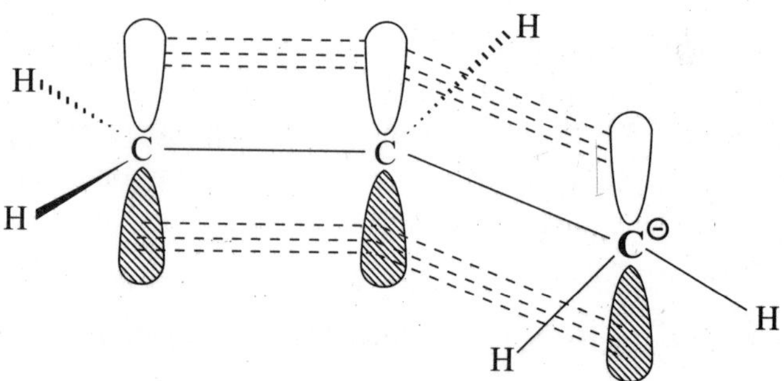

All carbon atoms are sp^2-hybridized. Hence the anion is planar.

It has been shown that if this type of anions are generated on a chiral carbon atom, there is loss of optical activity. The reason for this is that the chirality of anionic carbon is destroyed by the formation of a carbon-carbon double bond.

In conclusion it may be said that the carbon atom of an unconjugated carbanion is sp^3-hybridized and it has a pyramidal structure. On the other hand in a resonance stabilized carbanion the carbon is sp^2-hybridized and the anion is essentially planar. In both cases there is a loss of optical activity in those reactions which involve optically active reactants.

Stability

Generally speaking carbanions are stabilized by dispersal of negative charge on the neighbouring atoms or groups. Main factors which stabilize carbanions are as follows:

(*a*) Inductive effect
(*b*) Resonance effect
(*c*) Increase in *s*-character of carbanionic carbon
(*d*) Aromatization.

(*a*) Inductive effect: As expected electron-withdrawing groups (–I-effect) stabilize the carbanions by dispersal of negative charge, whereas electron-donating groups (+I-effect) destabilize them by intensifying the negative charge. For example,

(*i*) The anions $\overset{\ominus}{C}F_3$ and $\overset{\ominus}{C}(CF_3)_3$ are quite stable the latter being more stable due to –I-effect of highly electronegative F atoms (3 in first case and 9 in the second case)

$F \leftarrow \overset{\ominus}{C}{:}$ (with $\uparrow F$ above and $\downarrow F$ below)

Less stable (3 F atoms)

$F \leftarrow C(F)(F) \leftarrow \overset{\ominus}{\ddot{C}} \rightarrow C(F)(F) \rightarrow F$, with $\overset{\ominus}{\ddot{C}}$ also bonded ($\downarrow$) to $F \leftarrow C(\downarrow F) \rightarrow F$

More stable (9 F atoms)

(*ii*) The order of stability of simple alkyl anions is as follows:

$$\overset{\ominus}{C}H_3 > H_3C \rightarrow \overset{\ominus}{C}H_2 > H_3C \rightarrow \underset{\underset{CH_3}{\uparrow}}{\overset{\ominus}{C}H} > H_3C \rightarrow \underset{\underset{CH_3}{\uparrow}}{\overset{\ominus}{C}} \leftarrow CH_3$$

This is due to intensification of negative charge on anionic carbon by +I-effect of methyl groups. Here it may be mentioned than the above order of carbanion stability is just the reverse of carbocation stability.

(*b*) **Resonance effect:** If a multiple bond (say C=C, C=O, NO_2, C=N, etc.) is located α- to the anionic carbon, it stabilises the anion by resonance involving dispersal of charge. Some examples of resonance stabilized carbanions are:

(*i*) ***Allyl and benzyl anions***: They are stabilized by dispersal of negative charge due to resonance as depicted below:

$$CH_2{=}CH{-}\overset{\ominus}{C}H_2 \longleftrightarrow \overset{\ominus}{C}H_2{-}CH{=}CH_2$$

Allyl anion (resonance stabilized)

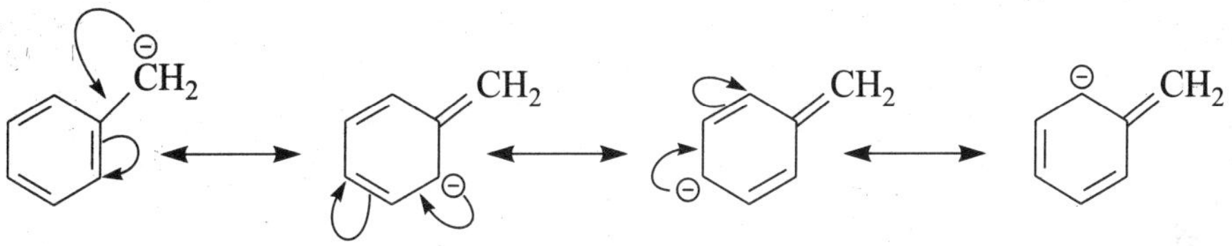

Benzyl anion (Resonance stabilized)

Effect of substituents on resonance stabilized carbanions: The presence of electron donating groups (such as alkyl, OH, NH_2 having +R-effect) at *o*-and *p*-position of the ring with respect to anionic centre results in destabilization of the anion as they intensity negative charge on anionic carbon. Whereas electron withdrawing groups (such as—NO_2, CHO, COR, COOH, COOR, etc.) on *o*-and *p*-position of the ring with respect to anionic centre stabilize the anion by –R-effect. Thus, decreasing order of stability of *p*-nitrobenzyl, benzyl and *p*-methoxy benzyl anion is

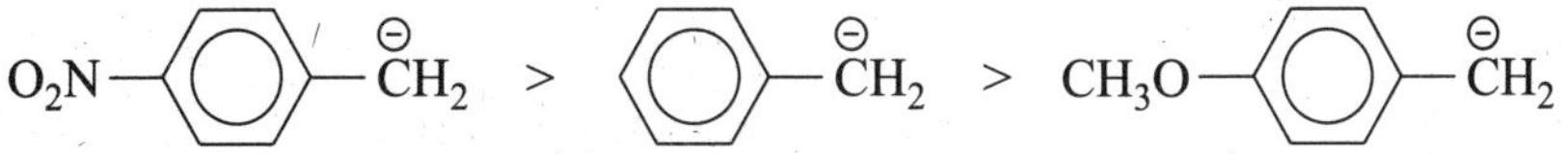

$$O_2N{-}C_6H_4{-}\overset{\ominus}{C}H_2 > C_6H_5{-}\overset{\ominus}{C}H_2 > CH_3O{-}C_6H_4{-}\overset{\ominus}{C}H_2$$

Here it may be mentioned that the order of stability of carbanions is the reverse of the stability of carbocations.

The resonance stabilization of benzyl anion suggests that the greater is the number of benzene rings attached to the anionic carbon, stabler is the anion. Thus diphenylmethyl anion is more stable than benzyl anion and triphenylmethyl anion is more stable than diphenylmethyl anion. This is due to increasing number of resonance structures.

Diphenylmethyl anion

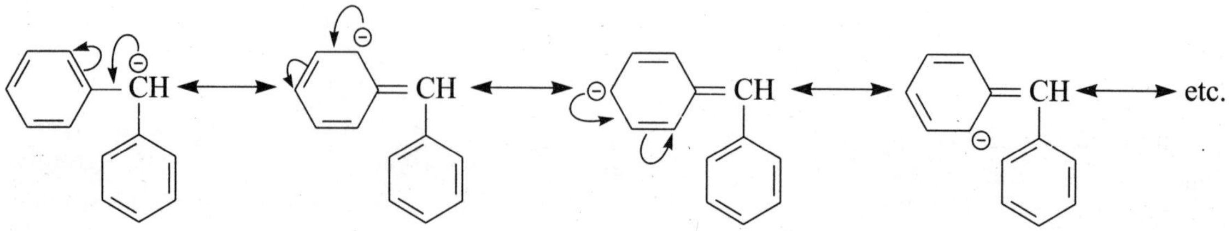

$C_6H_5-\overset{\ominus}{C}H-C_6H_5 \longleftrightarrow \ldots = CH \longleftrightarrow \ldots = CH \longleftrightarrow \ldots = CH \longleftrightarrow$ etc.

Three more contributing forms

Triphenylmethyl anion

It is more stable than diphenylmethyl anion due to larger number resonating forms:

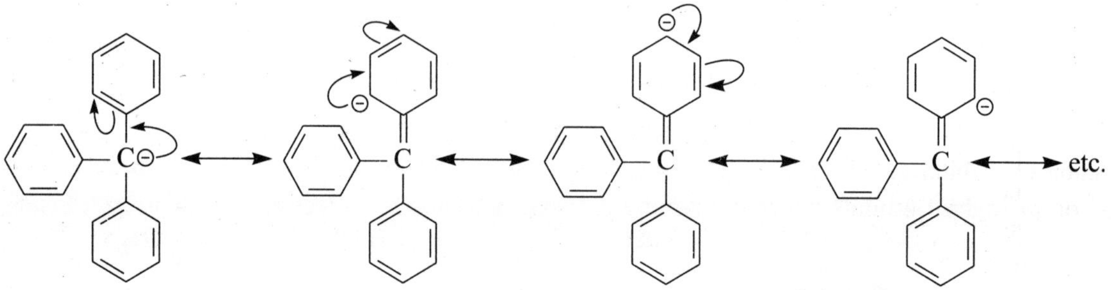

$(C_6H_5)_3C^{\ominus} \longleftrightarrow \ldots C \longleftrightarrow \ldots C \longleftrightarrow \ldots C \longleftrightarrow$ etc.

Six more contributing forms

(ii)* *Anion derived from nitromethane

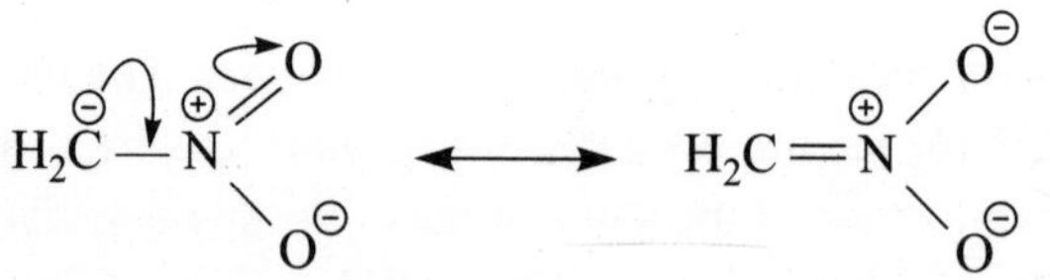

$$H_2\overset{\ominus}{C}-\overset{\oplus}{N}(=O)-O^{\ominus} \longleftrightarrow H_2C=\overset{\oplus}{N}(-O^{\ominus})-O^{\ominus}$$

(iii)* *Enolate anion derived from acetoacetic ester

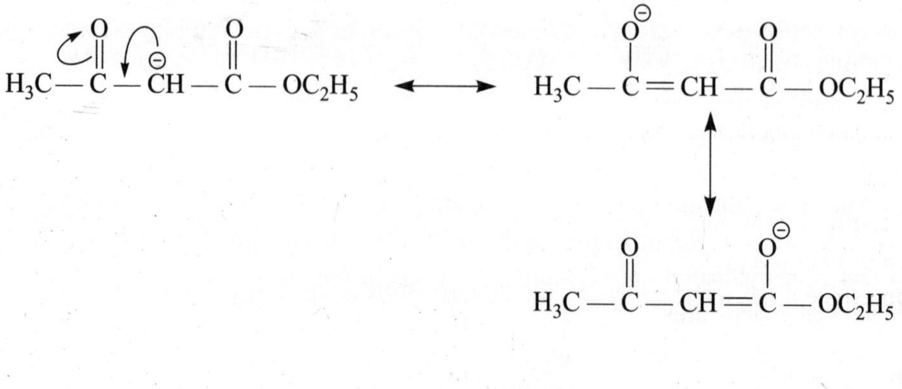

$$H_3C-\overset{O}{\overset{\|}{C}}-\overset{\ominus}{C}H-\overset{O}{\overset{\|}{C}}-OC_2H_5 \longleftrightarrow H_3C-\overset{O^{\ominus}}{\overset{|}{C}}=CH-\overset{O}{\overset{\|}{C}}-OC_2H_5$$

$$\updownarrow$$

$$H_3C-\overset{O}{\overset{\|}{C}}-CH=\overset{O^{\ominus}}{\overset{|}{C}}-OC_2H_5$$

(iv) Enolate ion of acetaldehyde

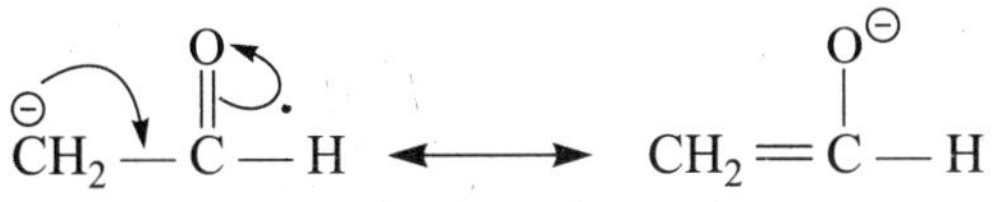

$$\overset{\ominus}{C}H_2-\overset{\overset{O}{\|}}{C}-H \longleftrightarrow CH_2=\overset{\overset{O^{\ominus}}{|}}{C}-H$$

It is due to increase in stability of anions due to resonance that diphenylmethane (pK_a 33.5), triphenylmethane (pK_a 31.5), acetaldehyde (pK_a 16.0) , nitromethane (10.0) and ethylacetate (pK_a 9.0) are acidic in nature.

(*c*) Increase in *s*-character of the carbanionic centre: Another factor which contributes to the stability of carbanions is *s*-character of the orbitals of anionic carbon. The observed carbanion stability, in decreasing order is as follows:

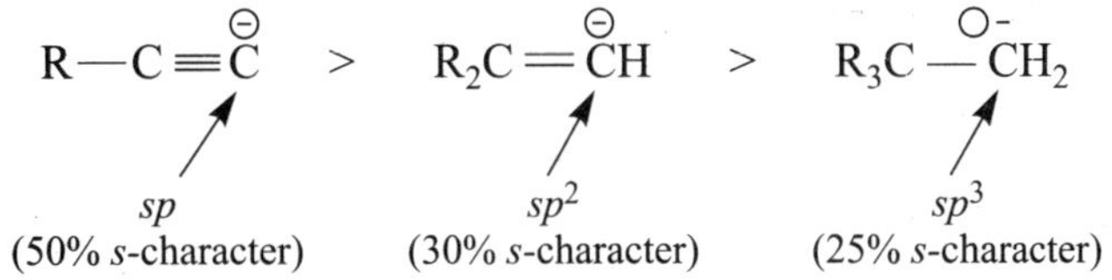

$$R-C\equiv\overset{\ominus}{C} \;>\; R_2C=\overset{\ominus}{C}H \;>\; R_3C-\overset{\ominus}{C}H_2$$

sp (50% *s*-character) — sp^2 (30% *s*-character) — sp^3 (25% *s*-character)

Highest stability of acetylide anion is attributed to the fact that in acetylide ion the electron pair occupies *sp* orbital and hence is less available for combining with acids as compared to electron pair or sp^2 or sp^3 hybrid orbitals which have more *p*-character. This is the reason why terminal alkynes (pK_a 25) are stronger acids than alkenes (pK_a 43) or alkanes (pK_a 50).

(*d*) Aromatization: Cylopentadiene readily loses a proton (pK_a 16) to form cyclopentadienyl anion which is quite stable (Resonance energy = 130 kJ mol^{-1} compared to 151 kJ mol^{-1} for benzene).

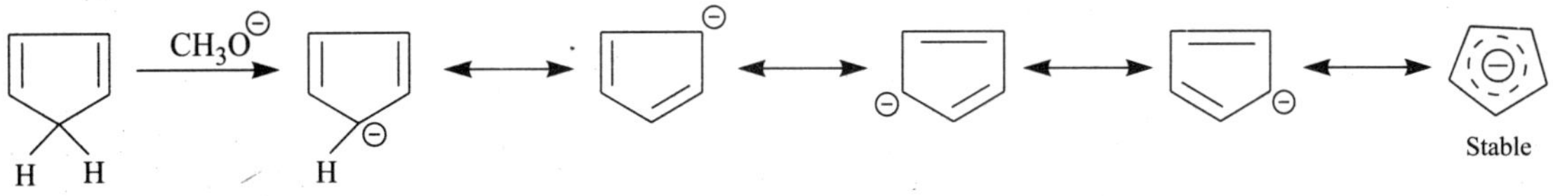

$CH_3O^{\ominus}$

H H — H — Stable

High stability of this anion is due to the fact that it constitutes an aromatic system having $4q + 2\pi$-electrons (4 from two double bonds and 2 from non-bonding pair).

Detection of Carbanions

(*a*) Spectroscopic methods for the detection of carbanions have not been found to be as successful as in the case of carbocations. Quite often an nmr spectrum of a carbanion is complicated by the covalent bonds of the particular compound which, in turn, are influenced by solvent and temperature. The chemical shift is influenced by these factors. However, the extent of negative charge localized at carbanionic carbon can be related to the degree of upfield shift experienced by the neighbouring protons.

(*b*) Chemical evidence for the existence of carbanions is provided by the base-catalyzed recemization of (+) –phenyl *sec*-butyl ketone. Here, the abstraction of a proton by the base yields the carbanion and since it has a flat structure and the reaction is reversible, the proton may get attached to either face, producing the racemic mixture. Also, the rate constants for racemization and for bromination of the above ketone are found to be identical. The rate of racemization is also found to be equal to the rate of deuteration when the reaction is carried out in presence of a base in D_2O. All these observations strongly support the formation of a *planar anion* in the rate-determining step.

H_5C_2, H, H_3C — C — C(=O) — C_6H_5 + :B

Slow

$\overset{\oplus}{B}H$ + [H_5C_2, H_3C > C ⋯ $\overset{\ominus}{C}$(O) — C_6H_5] (A) $\xrightarrow[\text{Fast}]{Br_2}$ H_5C_2, H_3C, Br — C — C(=O) — C_6H_5

Racemic

(A) —
H_2O Fast → H_5C_2, H, CH_3 — C — C(=O) — C_6H_5 + H_5C_2, H_3C, H — C — C(=O) — C_6H_5

Racemic

D_2O Fast → C_2H_5 — CD(CH_3) — C(=O) — C_6H_5

Racemic

Formation of Carbanions

As there is little difference between the electronegativity of carbon and hydrogen (2.5 and 2.1, respectively) the polarity of the C—H bond is very small, Therefore, the heterolytic fission of this bond to form an anion and a proton should be a very difficult process. In other words, a hydrogen atom bound to an sp^3 carbon atom shows negligible acidity. However, the presence of electron attracting substituents such as nitro, cyano or carbonyl groups on the same carbon renders the hydrogen relatively acidic. The increase in acidic character is not only due to the electron-withdrawing ability of these substituents, but also due to their ability to delocalize the negative charge of the anion. Thus hydrogens on the carbon atom alpha to nitro, cyano or carbonyl groups have acidic character and can be removed as protons leaving behind resonance stabilized anions having a non-bonding pair of electrons. Thus, carbanions may also be regarded as nucleophiles or Lewis bases.

Some important methods of their formation are:

(*i*) *Abstraction of hydrogens attached to the carbon atoms alpha to nitro, cyano or carbonyl groups with a strong base. For example,*

$$H\overset{\ominus}{O} + H-CH_2-\overset{\oplus}{N}(=O)-O^{\ominus} \longrightarrow \left[:\overset{\ominus}{C}H_2-\overset{\oplus}{N}(=O)-O^{\ominus} \longleftrightarrow CH_2=\overset{\oplus}{N}(-O^{\ominus})-O^{\ominus}\right] + H_2O$$

Nitromethane — Resonance stabilized anion

$$C_2H_5\overset{\ominus}{O} + H-CH_2-C\equiv N \longrightarrow \left[:\overset{\ominus}{C}H_2-C\equiv N \longleftrightarrow CH_2=C=\overset{\ominus}{N}\right] + C_2H_5OH$$

Ethoxide — Acetonitrile — Resonance stabilized anion

$$C_2H_5\overset{\ominus}{O} + H-CH_2-\overset{O}{\overset{\|}{C}}-H \longrightarrow \left[:\overset{\ominus}{C}H_2-\overset{O}{\overset{\|}{C}}-H \longleftrightarrow CH_2=\overset{\overset{\ominus}{O}}{\overset{|}{C}}-H\right] + C_2H_5OH$$

Acetaldehyde — Resonance stabilized anion

(*ii*) *Abstraction of hydrogens from terminal alkynes using a strong base.*

$$H_2\overset{\ominus}{N}: + H-C\equiv CH \longrightarrow :\overset{\ominus}{C}\equiv CH + NH_3$$

Amide ion (Stronger base) — Acetylene

(*iii*) *Addition of nucleophiles to α, β-unsaturated compounds.*

$$C_2H_5\overset{\ominus}{O} + CH_2=CH-C\equiv N \longrightarrow \left[H_5C_2OCH_2-\overset{\ominus}{C}H-C\equiv N \longleftrightarrow H_5C_2OCH_2-CH=C=\overset{\ominus}{N}\right]$$

Ethoxide ion — Vinyl cyanide — Resonance stabilized anion

Reactions of Carbanions

Carbanions undergo three basic types of reactions as follows:

(*a*) Combination with an electrophile.

(*b*) Nucleophilic substitution and addition.

(*c*) Rearrangements.

(*a*) Combination with an electrophile: The most common reaction of carbanions is combination with a positive species (electrophile) usually a proton.

$$R-\overset{\overset{\displaystyle O}{\|}}{C}-\overset{\ominus}{C}H_2 + \overset{\oplus}{E} \longrightarrow R-\overset{\overset{\displaystyle O}{\|}}{C}-CH_2-E$$

(*b*) Nucleophilic substitution and addition: (*i*) Carbanions may act as nucleophiles and push out one of the four groups on a saturated carbon atom (S_N2 reactions).

$$HC\equiv\overset{\ominus}{C} + H_3C-\underset{H}{\overset{H}{C}}-Br \xrightarrow{-\overset{\ominus}{Br}} HC\equiv C-\underset{H}{\overset{H}{C}}-CH_3$$

(*ii*) Carbanions frequently add to the carbonyl double bond (aldol condensation)

$$CH_3-\overset{\overset{\displaystyle O}{\|}}{C}-H \xrightarrow{-\overset{\oplus}{H}} \overset{\ominus}{C}H_2-\overset{\overset{\displaystyle O}{\|}}{C}-H$$

$$CH_3-\overset{\overset{\displaystyle O}{\|}}{C}-H + \overset{\ominus}{C}H_2-\overset{\overset{\displaystyle O}{\|}}{C}-H \longrightarrow CH_3-\underset{H}{\overset{O^{\ominus}}{C}}-CH_2-\overset{\overset{\displaystyle O}{\|}}{C}-H$$

$$\Big\downarrow\Big\uparrow\ H_2O$$

$$CH_3-\underset{H}{\overset{OH}{C}}-CH_2-\overset{\overset{\displaystyle O}{\|}}{C}-H$$

(Aldol)

(*c*) Rearrangements: Carbanions also undergo rearrangement reactions, though they are rare, e.g.,

$$C_6H_5-\underset{C_6H_5}{\overset{C_6H_5}{C}}-\overset{\ominus}{C}H_2 \longrightarrow C_6H_5-\underset{C_6H_5}{\overset{\ominus}{C}}-CH_2C_6H_5$$

2.5.3 Free Radicals

In addition to heterolytic cleavage of a covalent bond which produces carbocations and carbanions rupture of this bond may be *homolytic* in which each of the two fragments possesses an unpaired electron.

$$X{-}Y \longrightarrow \dot{X} + \dot{Y}$$

These fragments carrying an odd or unpaired electron have no charge and are called free radicals.

Although free radicals are electrically neutral but due to the presence of an odd electron they are weakly attracted by the magnetic field and hence are paramagnetic in nature. They are extremely reactive and yield a variety of products depending upon the experimental conditions.

Classification of Free Radicals

Free radicals are classified as 1°, 2°, or 3° similar to carbocations

$^{\bullet}CH_3$	$H_3C\dot{C}H_2$	$(H_3C)_2\dot{C}H$	$(CH_3)_3\dot{C}$
Methyl	1°	2°	3°

Structure

There are two possible structures for simple alkyl radicals.

(*i*) **Planar:** A free radical may be a planar species in which the carbon atom bearing the odd electron is sp^2-hybridized (as in carbocation) and the odd electron remains in the *p*-orbital (Fig. 2.9). The *p*-orbital containing this odd electron is always perpendicular to the plane of sp^2-carbon.

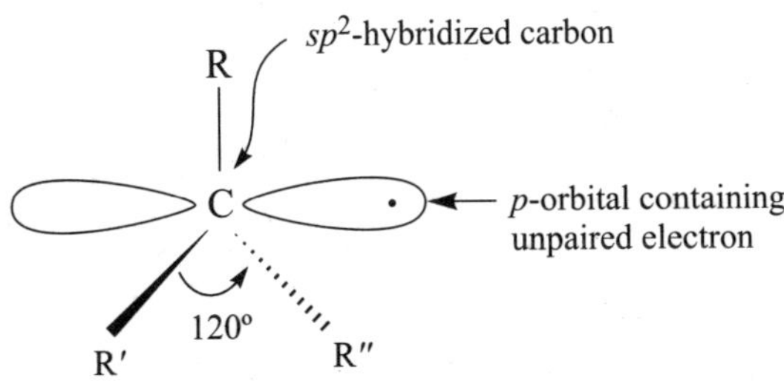

Fig. 2.9: Planar structure of free radical

(*ii*) **Pyramidal:** Free radical can also aquire a shape resembling a shallow pyramid which is neither planar nor tetrahedral and the orbital containing unpaired electron is some sort of a hybrid between a *p* and sp^3-orbital (Fig. 2.10)

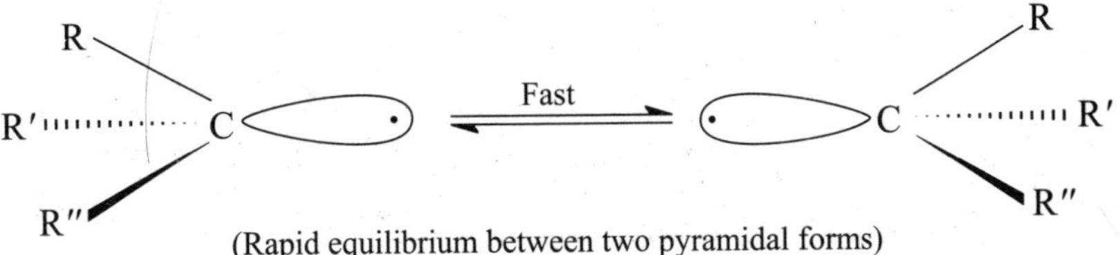

Fig. 2.10: Pyramidal structure of free radical

The question as to whether free radicals exist in planar configuration or in a rapid equilibrium between the two pyramidal forms is difficult to answer with certainty. However, available evidence indicates that the unpaired electron is in a *p*-orbital. Resonance stabilized free radicals certainly exist in planar configuration.

Stability of Free Radicals

Similar to carbocations, the stability of free radicals can be explained on the basis of hyperconjugation, resonance and steric factors.

1. Hyperconjugation: In general, it has been found that the stability of alkyl radicals is in the order: 3° > 2° > 1°. This is due to the fact that in going from methyl to *t*-butyl, there is increasing tendency for electron release due to *hyperconjugation*. As we have seen that the carbon atom carrying an unpaired electron has a half-filled orbital that gives the radical its characteristic instability; the hyperconjugative mechanisms tend to fill this orbital and thus stabilize the radical to some extent. Various hyperconjugative contributing forms of *tert*-butyl radical, for example, are given below:

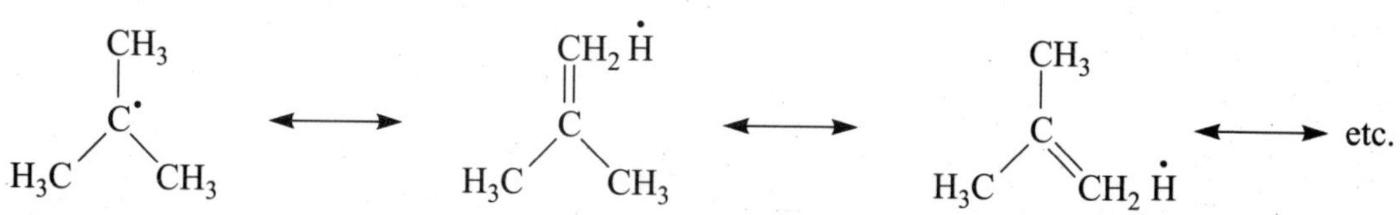

$$(CH_3)_2\dot{C}CH_3 \longleftrightarrow (H_3C)(CH_3)C{=}CH_2\ \dot{H} \longleftrightarrow (H_3C)(CH_3)C{=}CH_2\ \dot{H} \longleftrightarrow \text{etc.}$$

2. Resonance: Allyl, benzyl, diphenylmethyl and triphenylmethyl radicals are particularly stable because of resonance involving π-electrons of the double bond or aromatic ring.

$$CH_2{=}CH{-}\dot{C}H_2 \longleftrightarrow \dot{C}H_2{-}CH{=}CH_2$$

(Resonance forms of allyl radical)

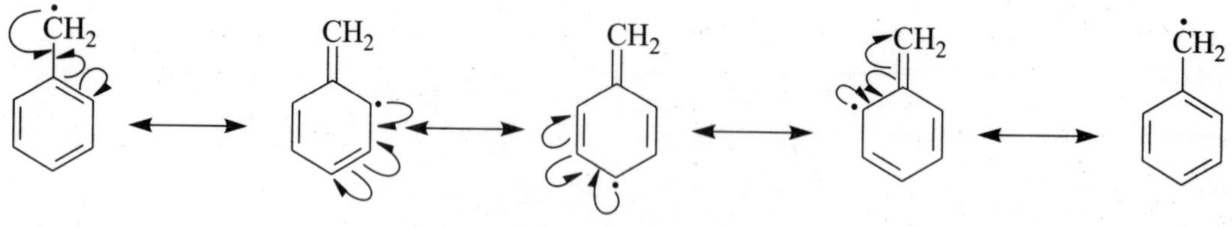

(Resonance structures of benzyl radical)

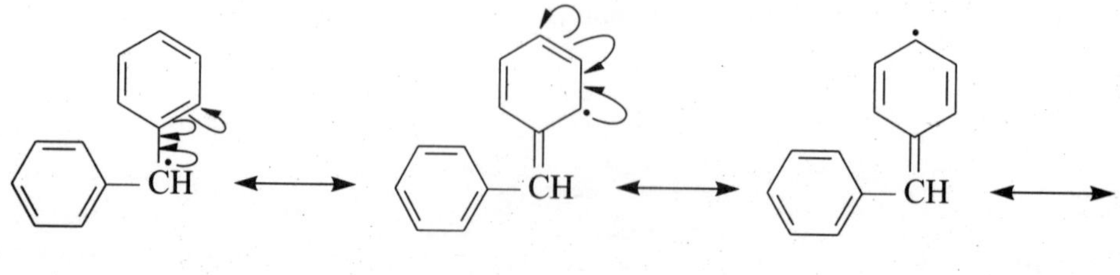

Four more such resonating structures

(Resonance structures of diphenylmethyl radical)

Triphenylmethyl free radical is still more stable due to larger number of resonating structures.

Six more such resonating structures

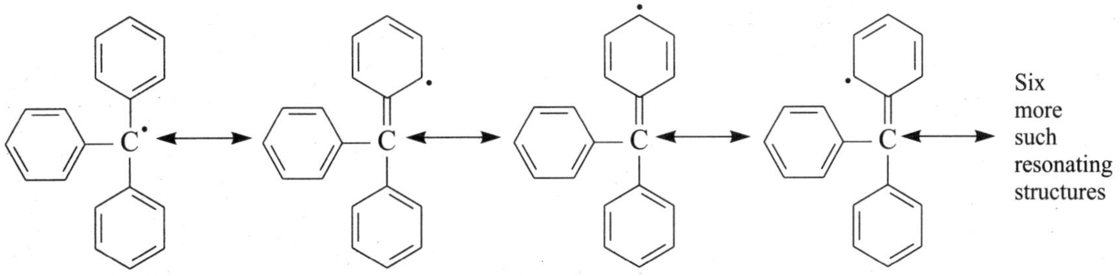

(Resonance structures of triphenylmethyl radical)

The free radicals having the odd electron on a doubly bonded carbon atom such as vinyl ($CH_2{=}\dot{C}H$) or phenyl ($\dot{C}_6H_5$) radicals, cannot be stabilized by hyperconjugation or by resonance and hence are usually highly reactive. These radicals are even less stable than methyl free radical.

3. Steric factor: Another factor that is responsible for the increased stability of tertiary radicals may be steric. Formation of a *tert*-butyl radical by removal of a hydrogen atom from *iso*-butane, $(CH_3)_3CH$, is favoured as the steric repulsion between the methyl groups is relieved to a certain extent by an increase in bond angles from 109.5° to about 120°. Clearly this steric relief should be the greatest in the formation of tertiary free radicals (cf. stability of carbocations).

Relative Stability

On the basis of all the above factors we conclude that the order of relative stability of different free radicals follows the sequence: Benzyl > Allyl > *t*-Butyl > *iso*-Propyl > Ethyl > Methyl > Phenyl > Vinyl.

Detection

Direct evidence for the presence of an odd electron is usually obtained from its magnetic properties. An earlier method made use of the magnetic susceptibility of the substances that depended on the paramagnetic contribution of the odd electron.However, this method of detecting radicals has been superceded by a technique which directly measures the paramagnetism due to an unpaired electron. This method, known as electron spin resonance (*esr*) or electron paramagnetic resonance (*epr*) spectroscopy is applicable for the study of radicals even at concentrations as low as 10^{-8}M. The basic principle of this form of spectroscopy is that an unpaired electron, like a proton, has spin and this spin has an associated magnetic moment.

When a free radical is placed in a magnetic field and exposed to electromagnetic radiation, the magnetic moment generated by the spinning electron can be lined up *with* or *against* the external magnetic field. These two orientations define two energy states which differ in energy one being, more stable than the other. However, with the absorption of light energy in the microwave region, the spin state of the electron is changed from alignment with the field to against the field (Fig. 2.11).

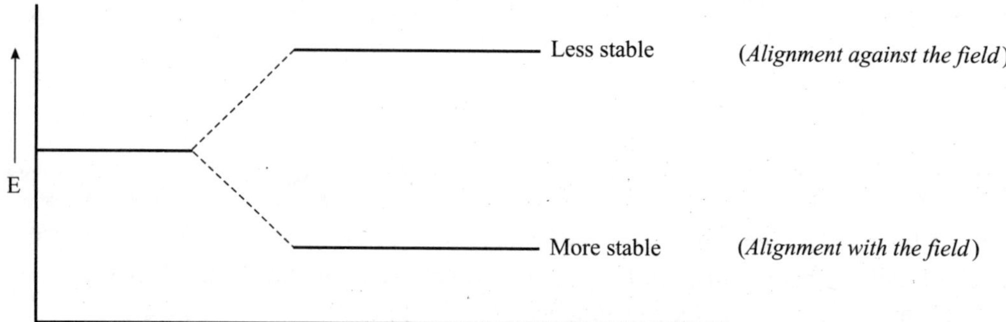

Fig. 2.11: Principle of *esr* spectroscopy

Transition between stable and less stable levels produces a line in the *esr* spectrum. Very often the *esr* spectrum of a radical has several lines instead of a single line (Fig. 2.12). This splitting of a line is due to the coupling of the odd electron most frequently with the protons situated on the same atom that carries the odd electron. If the number of coupling protons is n then, the number of lines in the *esr* spectrum is always $n + 1$. Methyl radical ($\dot{C}H_3$) for instance, shows $3 + 1 = 4$ resonance lines.

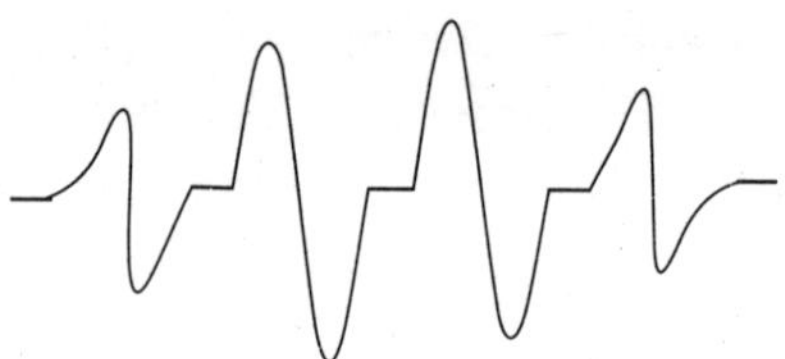

Fig. 2.12: The *esr* spectrum of a methyl radical

Formation of Free Radicals

Free radicals are often produced when a molecule is supplied with sufficient energy—thermal, photochemical or electrochemical—to cause *homolysis* of a covalent bond. Free radicals are also produced in redox reactions. Some typical reactions producing free radicals are given below:

(*a*) Thermolytic cleavage (*pyrolysis*): Any organic molecule, when heated at high temperature in the gas phase, produces free radicals. Some examples are as follows:

(*i*) *Thermolysis of tetramethyl lead.*

$$\underset{\text{(Tetramethyl lead)}}{(CH_3)_4Pb} \xrightarrow{\Delta} 4\,\dot{C}H_3 + P$$

(*ii*) *Thermolysis of azo compounds.*

$$\underset{\text{(Azomethane)}}{CH_3—N{=}N—CH_3} \xrightarrow{\Delta} 2\,\dot{C}H_3 + N_2$$

(*iii*) *Thermolysis of aryl peroxides.*

$$C_6H_5-\overset{\overset{\displaystyle O}{\|}}{C}-O-O-\overset{\overset{\displaystyle O}{\|}}{C}-C_6H_5 \xrightarrow{\Delta} 2C_6H_5-\overset{\overset{\displaystyle O}{\|}}{C}-\dot{O} \longrightarrow 2\,\dot{C}_6H_5 + 2CO_2$$

Benzoyl peroxide — Benzoyloxy radical — Phenyl radical

(*b*) Photolytic cleavage: The energy content of light ranging from 600–300 nm is 200–400 kJ mol^{-1}. This is sufficient to cleave a covalent bond in homolytic manner to produce free radicals. Some typical examples are as follows:

(*i*) *Photolysis of halogens.*

$$Cl-Cl \xrightarrow{h\nu} 2\,\dot{Cl}$$

(*ii*) *Photolysis of ketones.*

$$CH_3COCH_3 \xrightarrow{h\nu} \dot{C}H_3 + CH_3\dot{C}O$$

Acetyl radical

$$CH_3\dot{C}O \longrightarrow \dot{C}H_3 + CO$$

(*iii*) *Photolysis of azo compounds.*

$$H_3C-N{=}N-CH_3 \xrightarrow{h\nu} 2\,\dot{C}H_3 + N_2$$

(*iv*) *Photolysis of aryl peroxides.*

$$C_6H_5-\overset{\overset{\displaystyle O}{\|}}{C}-O-O-\overset{\overset{\displaystyle O}{\|}}{C}-C_6H_5 \xrightarrow{h\nu} C_6H_5-\overset{\overset{\displaystyle O}{\|}}{C}-\dot{O}$$

$$C_6H_5-\overset{\overset{\displaystyle O}{\|}}{C}-\dot{O} \xrightarrow{h\nu} \dot{C}_6H_5 + CO_2$$

(*c*) Electrolytic cleavage (Kolbe's electrolysis):

$$R-\overset{\overset{\displaystyle O}{\|}}{C}-\ddot{\underset{\cdot\cdot}{O}}{:}^{\ominus} \xrightarrow[-e^{\ominus}]{\text{at anode}} R-\overset{\overset{\displaystyle O}{\|}}{C}-\dot{O} \longrightarrow \dot{R} + CO_2$$

(*d*) Redox reactions: These reactions, involving single electron transfer, are known to generate free radicals. These reactions involve metal ions such as Fe^{2+}/Fe^{3+} and Cu^{+}/Cu^{2+}. Some examples are:

(*i*) *Decomposition of aryl peroxides.*

$$Ar-\overset{\overset{O}{\|}}{C}-O-O-\overset{\overset{O}{\|}}{C}-Ar+Cu^{+}\longrightarrow \underset{\text{Acyloxy free radical}}{Ar-\overset{\overset{O}{\|}}{C}-\dot{O}}+Cu^{2+}+\underset{\text{Acyloxy anion}}{Ar-\overset{\overset{O}{\|}}{C}-O^{\ominus}}$$

(*ii*) *Sandmeyer reaction.*

The Cu^{+} ion is also involved in generation of aryl radicals from arenediazonium salts.

$$Ar-\overset{\oplus}{N}\equiv N+Cu^{+}\longrightarrow Ar^{\bullet}+N_2+Cu^{2+}$$

(*iii*) *Auto-oxidation.*

Auto-oxidation of benzaldehyde, is catalyzed by Fe^{3+}

$$C_6H_5-\overset{\overset{O}{\|}}{C}-H+Fe^{3+}\longrightarrow C_6H_5-\overset{\overset{O}{\|}}{C^{\bullet}}+\overset{\oplus}{H}+Fe^{2+}$$

Reactions

Reactions of free radicals frequently yield stable products by a process called *termination reaction.* Some typical reactions of free radicals are as follows:

(*a*) Addition reactions

(*b*) Substitution reactions

(*c*) Disproportionation reactions.

(*d*) Rearrangements.

(*a*) Addition reactions: (*i*) Free radicals, being electrophilic in nature add to a carbon-carbon double bond forming another radical.

$$\dot{R}+\;>C=C<\;\longrightarrow R-\overset{|}{\underset{|}{C}}-\dot{C}<$$

The radical formed above may add to another double bond and this may continue further. This is one of the main route of **vinyl polymerisation**.

(*ii*) The most celebrated example of free radical addition is the *abnormal* addition of hydrogen bromide to olefins which takes place in the following steps:

$$H-Br\xrightarrow{h\nu}\dot{H}+\dot{Br}$$

$$\underset{\text{Propene}}{CH_3-CH=CH_2}+\dot{Br}\longrightarrow CH_3-\dot{C}H-CH_2Br\xrightarrow{H-Br}CH_3CH_2CH_2Br+\dot{Br}$$

(*b*) Substitution reactions: (*i*) *In aliphatic compounds,* *e.g.*, halogenation of alkanes takes place by free radical chain mechanism taking place in three stages as under with example of chlorination of methane.

Chain Initiation

$$Cl-Cl \xrightarrow{h\nu} 2\dot{Cl}$$

Chain Propagation

$$CH_4 + \dot{Cl} \longrightarrow \dot{C}H_3 + HCl$$

$$\dot{C}H_3 + Cl-Cl \longrightarrow CH_3Cl + \dot{Cl}$$

Chain Termination

$$\dot{Cl} + \dot{Cl} \longrightarrow Cl-Cl$$

$$\dot{C}H_3 + \dot{Cl} \longrightarrow CH_3-Cl$$

$$\dot{C}H_3 + \dot{C}H_3 \longrightarrow CH_3-CH_3$$

(*ii*) ***In aromatic compounds***

$$C_6H_6 + \dot{R} \xrightarrow[\text{Slow}]{} [C_6H_6R]^{\bullet} \xrightarrow[\text{Fast}]{\dot{R}} C_6H_5R + R-H$$

(*c*) **Disproportionation reactions:** Two free radicals may react with each other in such a way that one is transformed to alkane and other to alkene.

$$CH_3-\dot{C}H_2 + CH_3-\dot{C}H_2 \longrightarrow CH_3-CH_3 + CH_2{=}CH_2$$

(*d*) **Rearrangements:** Less stable radicals attain stability by a 1, 2 shift of an alkyl/aryl radical to give more stable radical.

$$R-C(R)_2-\dot{C}H_2 \longrightarrow R-\dot{C}(R)-CH_2-R$$

Although a primary radical may sometimes produce a tertiary radical by the migration of a substituent from the neighbouring carbon atom, in practice radical rearrangements have been found to be distinctly less common than carbocation rearrangements. This diminished tendency for rearrangements is due to the fact that the difference in stability between a primary and a tertiary radical is not as much as that between a primary and a tertiary carbocation.

2.5.4 Carbenes

Carbenes can be defined as *neutral, bivalent carbon intermediates in which a carbon is covalently bonded to two atoms and has two non-bonding orbitals containing two electrons between them.* These are also called methylenes e.g., **:CH_2 (methylene) and :CCl_2** (dichlorocarbene or dichloromethylene).

Types of Carbenes

Theoretical considerations suggest that there are two possible kinds of carbenes.

(*i*) Singlet carbene

(*ii*) Triplet carbene.

In the **singlet carbene**, *a carbon atom is presumed to approximate sp^2-hybridization.* Two of the three sp^2 hybrid orbitals are utilized in forming two covalent bonds whereas the third hybrid orbital contains the unshared pair of electrons. The unhybridized *p*-orbital remains vacant. Thus, *singlet* carbene resembles a carbocation very closely and is represented as ⇅ CH_2 (Fig. 2.13). The two electrons have opposite spins in singlet carbene.

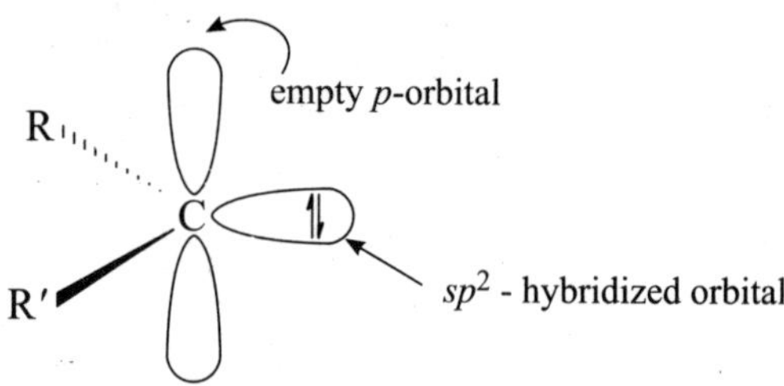

Fig. 2.13: Structure of singlet carbene

On the other hand, *carbon atom of a **triplet carbene** is sp-hybridized* and it is a *linear* or *non-linear* species. These two hybrid orbitals are involved in the bond formation with two groups and the two electrons are placed, one each, in the equivalent, mutually perpendicular p_y and p_z orbitals. Since these electrons have parallel spins, a carbene with this structure is said to be in a *triplet state* and represented as ⇈ CH_2 (Fig. 2.14).

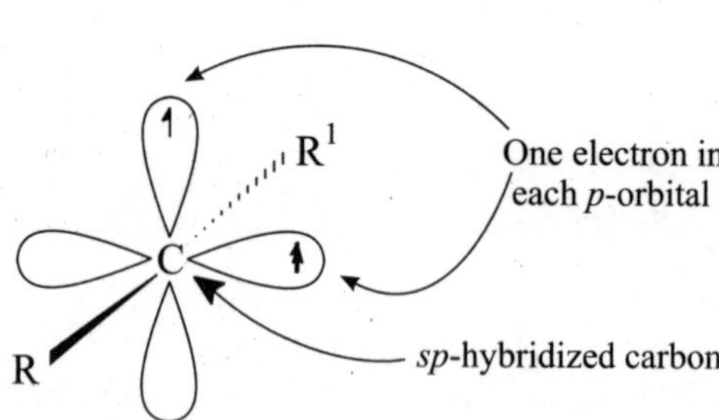

Fig. 2.14: Structure of triplet carbene

Stability of Carbenes

At the first sight, it appears that a singlet carbene has lower energy as the unshared electron pair is in a sp^2 hybrid orbital; but the consideration of the electron pair repulsion energy which must be overcome to place two electrons in a single orbital put it at the higher energy level than a triplet structure. It is, therefore, resonable to believe that the *triplet state of a carbene is more stable than singlet state and should be expected to be the ground state.* This is supported by the observation that triplet methylene is about 35.5 kJ mol^{-1} more stable than the singlet methylene.

Carbenes in which the carbene carbon atom is attached to two atoms, each bearing an unshared pair of electrons, are somewhat more stable due to resonance.

$$(R_2\ddot{N})_2C: \longleftrightarrow R_2\overset{\oplus}{N}{=}\overset{\ominus}{C}:(-\ddot{N}R_2) \longleftrightarrow R_2\ddot{N}{-}\overset{\ominus}{C}:{=}\overset{\oplus}{N}R_2$$

Detection of Carbenes

(*i*) Electron spin resonance (*esr*) studies have been useful for the detection of triplet carbenes.

(*ii*) A common method of trapping carbenes involves the insertion into the —O—H bond of an alcohol.

$$>C: + R-O-H \longrightarrow R-O-\overset{|}{\underset{|}{C}}-H$$

Formation of Carbenes

Some of the important reactions generating carbenes are as follows:

1. α-Elimination: α-Elimination is a reaction involving first a loss of group without its lone pair (usually a proton) which is followed by loss of an anion usually a halide ion. Some examples of α-elimination are:

(a) From chloroform

$$H-CCl_3 \xrightarrow[-H^{\oplus}]{\text{Base}} \overset{\ominus}{:}CCl_3 \xrightarrow{-\overset{\ominus}{Cl}} :CCl_2$$

Dichlorocarbene

(b) From sodium salt of trichloroacetic acid

$$Cl_3C-CO\overset{\ominus}{O} \xrightarrow{\Delta} :CCl_2 + CO_2 + \overset{\ominus}{Cl}$$

2. Photochemical or thermal cleavage of three membered rings

(*a*)

$$C_6H_5(H)C\langle CH_2-CH_2 \rangle \xrightarrow{\Delta \text{ or } h\nu} C_6H_5CH{=}CH_2 + :CH_2$$

Phenylcyclopropane

(*b*)

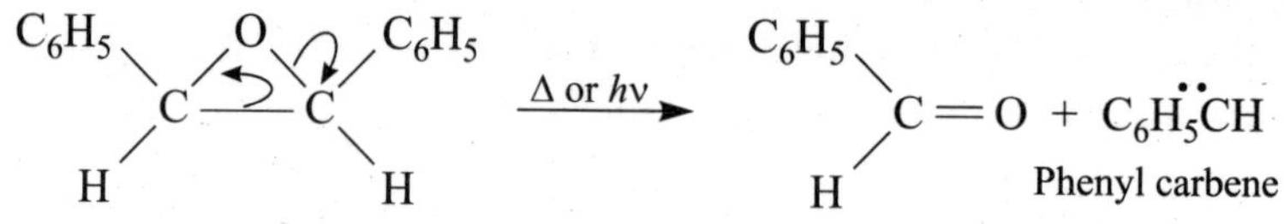

$$\underset{\text{Stilbene oxide}}{C_6H_5(H)C\overset{O}{\frown}C(H)C_6H_5} \xrightarrow{\Delta \text{ or } h\nu} C_6H_5(H)C{=}O + \underset{\text{Phenyl carbene}}{C_6H_5\ddot{C}H}$$

3. Decomposition of certain compounds having multiple bonds

Compounds such as ketenes, diazoalkanes and diazirines on decomposition give rise to carbenes. For example,

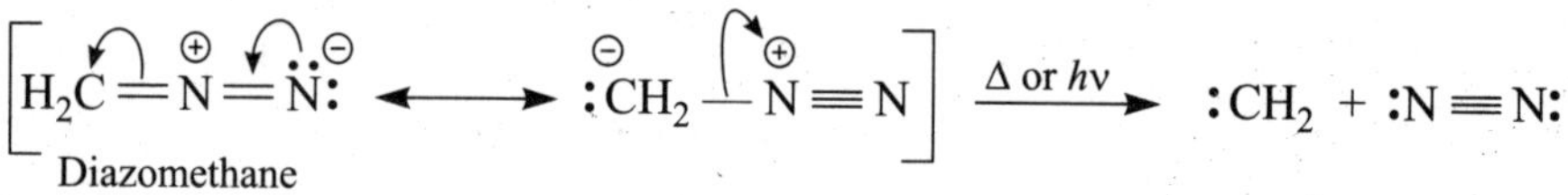

(*a*)

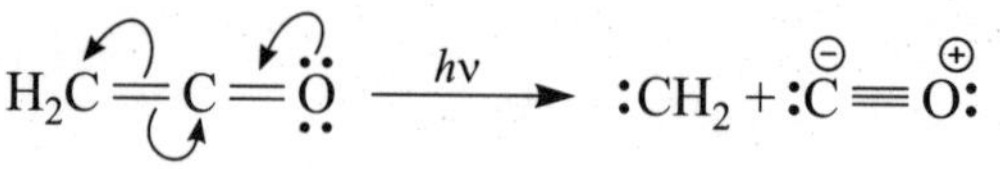

$$\underset{\text{Ketene}}{H_2C{=}C{=}\ddot{\underset{\cdot\cdot}{O}}} \xrightarrow{h\nu} :CH_2 + :\overset{\ominus}{C}{\equiv}\overset{\oplus}{O}:$$

(*b*)

$$\left[\underset{\text{Diazomethane}}{H_2C{=}\overset{\oplus}{N}{=}\overset{\ominus}{\ddot{N}}:} \longleftrightarrow :\overset{\ominus}{C}H_2{-}\overset{\oplus}{N}{\equiv}N\right] \xrightarrow{\Delta \text{ or } h\nu} :CH_2 + :N{\equiv}N:$$

(*c*)

$$R_2C\begin{matrix}N\\ \|\\ N\end{matrix} \longrightarrow R_2C: + :N{\equiv}N:$$

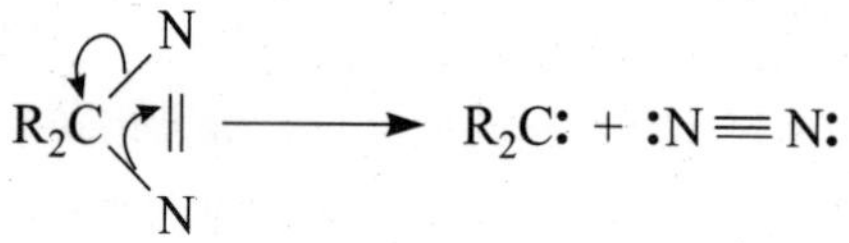

Diazirine
(*Isomeric with diazomethane*)

Reactions

Carbenes are highly reactive electrophilic species. They undergo the following reactions:

(*a*) Cycloadditions

(*b*) Insertion reactions

(*c*) Rearrangements.

(*a*) Cycloadditions: Carbenes readily add to olefinic double bond to form cyclopropane derivatives. Their reactivity is reflected in the fact that they even add to aromatic systems. In latter cases the products (norcaradiene derivatives) undergo ring expansion.

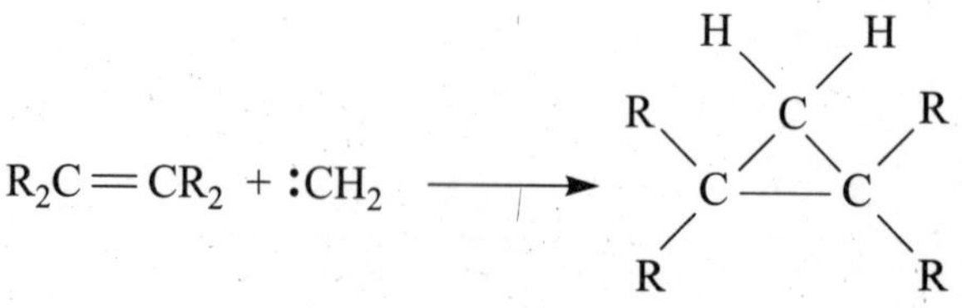

$$R_2C{=}CR_2 + :CH_2 \longrightarrow R_2C\overset{CH_2}{\triangle}CR_2$$

$$\text{Benzene} + :CH_2 \longrightarrow \text{Norcaradiene (Unstable)} \xrightarrow[\text{Expansion}]{\text{Ring}} \text{Cycloheptatriene (Stable)}$$

(*b*) **Insertion reactions:** Carbenes (in singlet state) are so reactive that they even insert themselves into σ C — H bonds.

$$R-CH_2-H + :CH_2 \text{ (Singlet carbene)} \longrightarrow R-CH_2-CH_2-H$$

These reactions are synthetically useless but are important because they reflect their extreme reactivity.

However, triplet carbenes abstract hydrogen or other atoms to produce radicals which couple to give products.

$$:CH_2 + CH_3CH_3 \text{ (Triplet carbene)} \longrightarrow \dot{C}H_3 + \dot{C}H_2CH_3 \longrightarrow CH_3CH_2CH_3$$

(*c*) **Rearrangements:** Alkyl carbenes can undergo rearrangements involving *1, 2-shift of hydrogen or alkyl group.*

$$CH_3-CH_2-CH_2-\ddot{C}H \longrightarrow CH_3-CH_2-CH=CH_2$$

$$R-\overset{O}{\overset{\|}{C}}-\ddot{C}H \text{ (Acylcarbene)} \longrightarrow R-CH=C=O \text{ (Ketene)}$$

Unlike other rearrangements most of the carbene rearrangements give rise to stable products.

2.5.5 Arynes

Arynes may be defined as *aromatic compounds containing a formal carbon-carbon triplet bond.* The best known aryne is **benzyne** which may be regarded as the aromatic counterparts of acetylene or in other words, it is benzene minus two *ortho* hydrogens and can also be called as *dehydrobenzene.* In benzyne, the newly formed π-bond is very weak and hence benzyne is a highly reactive molecule.

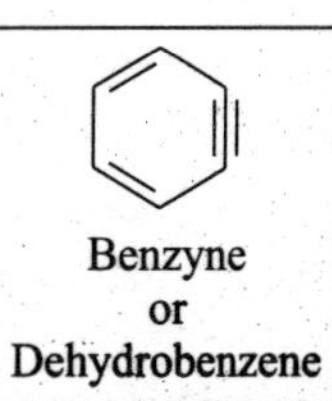
Benzyne or Dehydrobenzene

Structure

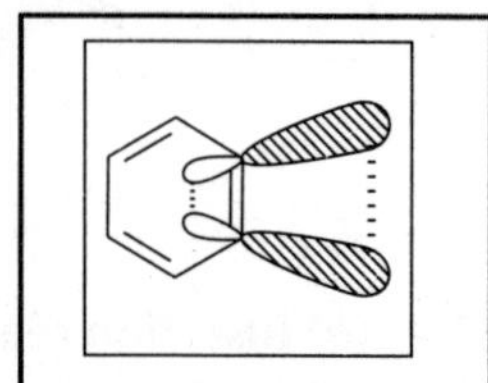

The benzyne bond is not like the triple bond of acetylene where the two carbons form a σ-bond using *sp*-orbitals and the remaining *p*-orbitals are used to form two mutually perpendicular π-bonds. Such a structure is not possible in benzyne because of the hexagonal geometry associated with the benzene ring. Most probably the new bond of benzyne is formed by the 'overlap' of sp^2-orbitals belonging to two neighbouring carbon atoms. These sp^2-orbitals are orthogonal to the π-molecular orbital of the benzene ring *i.e.*, the **triple bond** in benzyne consists of one sp^2-sp^2 σ-bond, one π-bond formed by sideways overlap of two unhybridized *p*-orbitals and another 'π-bond' by sp^2-sp^2 overlap.

Stability: The new bond of benzyne, formed by the overlap of sp^2-orbitals belonging to two neighbouring carbon atoms is very weak, and therefore, arynes are extremely reactive chemical species.

Detection: Arynes are usually detected either through spectroscopy or by their participation in such reactions as dimerization and by trapping through cycloaddition to substances such as furan.

1. Spectroscopic method: An elegant method to detect benzyne using spectroscopy involves mass spectrometry. It has been shown that when benzenediazonium-2-carboxylate is photolyzed in an apparatus which is coupled with a mass spectrometer, the ions at m/z 28 (N_2), 44 (CO_2) and 76 (benzyne) are recorded. However, with the passage of time there is decrease in the intensity of ion at m/z 76 and new ion at m/z 152 (biphenylene) starts making its appearance, obviously arising because of dimerization reaction of benzyne.

2. Trapping method

Dimerization → Biphenylene (*Self trapping*)

+ O → Diels-Alder reaction

Formation of Arynes

1. By the action of strong bases on halobenzenes: Benzyne has been shown to be intermediate in several important organic reactions. For example, in the presence of strong base, aryl halides eliminate HX from 1, 2-positions to produce benzyne which then rapidly reacts with the available nucleophile to regenerate the aromatic system. For example, treatment of bromobenzene with sodamide gives aniline.

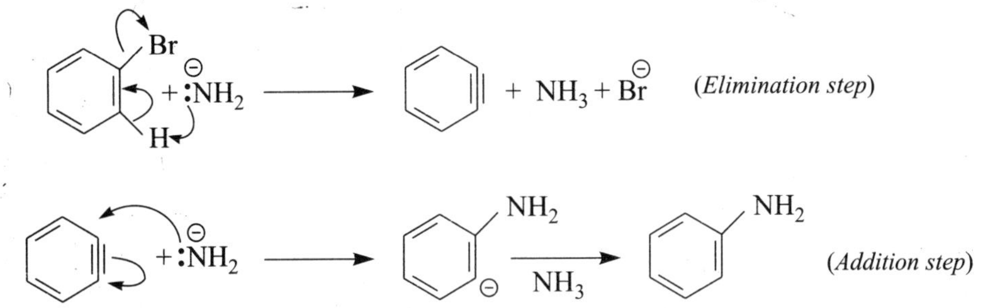

(Elimination step)

(Addition step)

The above mechanism is called *elimination-addition mechanism.*

Similarly, the intermediacy of benzyne in the Dow's method for industrial production of phenol from chlorobenzene and strong alkali has clearly been demonstrated by using (1-^{14}C) chlorobenzene.

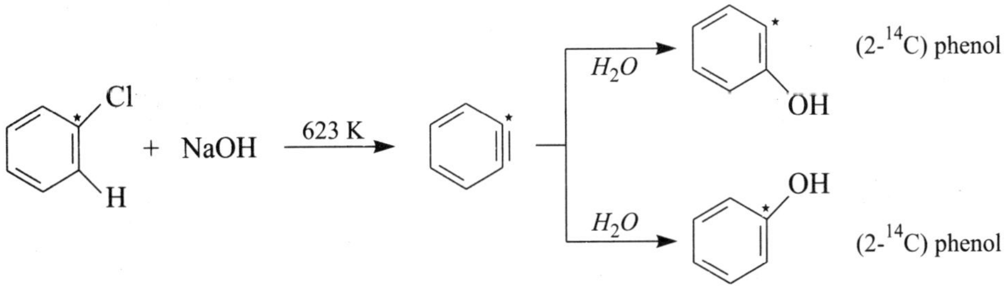

(2-^{14}C) phenol

(2-^{14}C) phenol

2. Thermal method: A convenient procedure for generating benzyne consists of heating benzenediazonium 2-carboxylate.

Anthranilic acid

Benzenediazonium-2-carboxylate

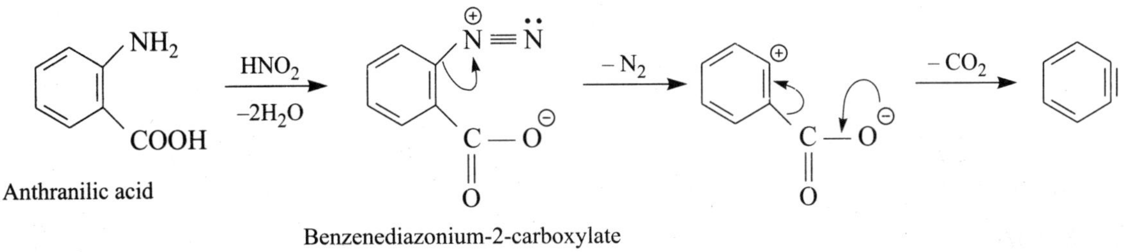

Reactions: In addition to dimerization reaction, benzyne reacts with olefins and dienes to yield different types of products. Some important reactions of benzyne are:

(*a*) Cycloaddition reactions

(*b*) Nucleophilic addition reactions.

***(a) Cycloaddition reactions*:** Benzyne reacts with both olefins and conjugated dienes to produce a variety of products, *e.g.*, (*i*) + $CH_2=CHCN$ ⟶ CN

(*ii*) + ⟶

Triptycene

(*iii*) + ⟶

(*b*) *Nucleophilic addition reactions*: Benzyne undergoes nucleophilic addition with a variety of nucleophiles.

(*i*) NHR ⟵ RNH_2, $-H^{\oplus}$ ⟶ H_2O, $-H^{\oplus}$ OH

(*ii*) + ⟶

2.5.6 Nitrenes

Nitrenes are univalent neutral nitrogen species having a sextet of electrons on nitrogen. Out of these six electrons four are present as two pairs of non-bonding electrons and two electrons are used for bonding with a monovalent atom or group. They can be said to be nitrogen analogs of carbenes. Like carbenes they are also short-lived highly reactive chemical species. Thus like carbenes, they also behave as electrophiles (Lewis acids) and are intermediates in many organic reactions. They are represented by formula.

$$R-\ddot{\underset{\cdot\cdot}{N}} \qquad (\text{cf} \quad R-\ddot{C}-R)$$

Structure of Nitrenes

Like carbenes, these intermediates also exist in two states, viz; singlet and triplet. These states are depicted below:

$$\underset{\text{Singlet}}{R-\ddot{N}:(\uparrow\downarrow)} \qquad \underset{\text{Triplet}}{R-\ddot{N}:(\uparrow\uparrow)} \quad \text{or} \quad R-\dot{\ddot{N}}\cdot$$

In **singlet nitrene**, the nitrogen is sp^2-hybridized. Out of three sp^2-hybridized orbitals one, having single electron, is utilized in *s*-bond formation with a monovalent atom or group, and the remaining two are occupied by two pairs of electrons. An unhybridized *p*-orbital is empty and this is responsible for their electrophilic character. On the other hand in triplet nitrene the nitrogen is *sp*-hybridized. Out of two *sp*-hybrid orbitals one having single electron is used in *s*-bond formation and the other is occupied by a pair of electrons. The two unhybridized *p*-orbitals are occupied by one electron each. The orbital picture of both the species is as follows:

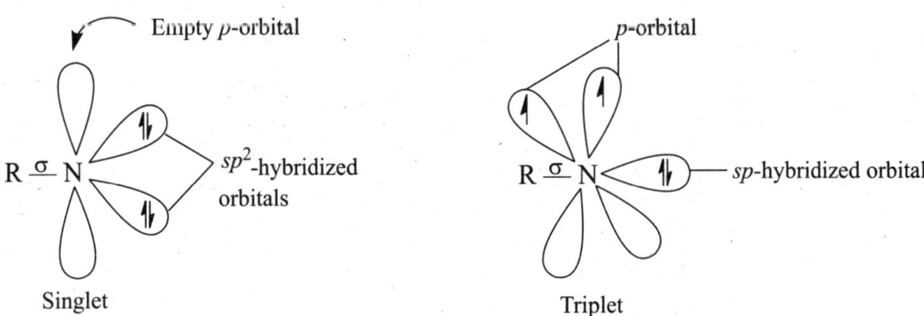

Detection

Nitrenes have been detected as reactive intermediates by applying the techniques of spectroscopy and trapping. These techniques are illustrated below:

1. Spectroscopic Techniques: The triplet nitrenes can be detected by electron spin resonance (*esr*) spectroscopy.

2. Trapping: The involvement of nitrenes as intermediates has been detected by trapping experiments. Two examples are given below:

$$Ar-\ddot{N}: + :C=O \longrightarrow Ar-\ddot{N}=C=O$$

$$EtOOC\ddot{N}: + \;{>}C=C{<}\; \longrightarrow \underset{\text{Aziridine}}{{>}C\!-\!C{<}\text{ (ring with }N-COOEt)}$$

Generation

Following methods of generation of nitrenes are analogous to those used for generation of carbenes:

1. α-Elimination: Treatment of hydroxylamine-O-sulphonates results with alkali in the formation of nitrenes by the elimination of sulphonic acid group from nitrogen.

$$R-\ddot{N}(H)(OSO_2Ar) + \overset{\ominus}{O}H \longrightarrow R-\ddot{N}: + H_2O + ArSO_3^{\ominus}$$

2. Decomposition of azides: Azides eliminate nitrogen either on heating or irradiating with light to produce nitrenes.

(*i*) $$\left[R-\ddot{N}=\overset{\oplus}{N}=\overset{\ominus}{\ddot{N}}: \longleftrightarrow R-\overset{\ominus}{\ddot{N}}-\overset{\oplus}{N}\equiv N:\right] \xrightarrow{\Delta \text{ or } h\nu} R-\ddot{N}: + N_2$$

(*ii*) $$N\equiv\overset{\oplus}{N}-\overset{\ominus}{\ddot{N}}-\overset{O}{\overset{\|}{C}}-OEt \xrightarrow{\Delta \text{ or } h\nu} N_2 + :\ddot{N}COOEt$$

Ethyl azidoformate — Ethoxycarbonyl nitrene

3. Decomposition of Isocyanates

$$R-\ddot{N}=C=\ddot{O}: \xrightarrow{h\nu \text{ or } A} R-\ddot{N}: + :\overset{\ominus}{C}\equiv\overset{\oplus}{O}:$$

Reactions

The chemical reactions of nitrenes closely resemble those of carbenes. Here it may be pointed out that many reactions where a nitrene intermediate can be proposed may not involve free nitrene. So, it is not easy to conclude that a particular reaction involves a nitrene intermediate or not. However, some of the reactions in which nitrenes may be intermediates are:

(*a*) Addition reactions

(*b*) Insertion reactions

(*c*) Rearrangements

(*d*) Abstraction of hydrogen

(*e*) Dimerization

***(a)* Addition reactions:** Both singlet and triplet nitrenes add on to $C=C$ bonds forming aziridines.

$$R_2C{=}CR_2 + R'{-}\ddot{N}: \longrightarrow \text{Aziridine (three-membered ring of } C{-}C \text{ and } N{-}R'\text{, each C bearing two R)}$$

Aziridine

(b) Insertion reactions: Some nitenes, *e.g.*, acyl nitrenes and sulphonyl nitrenes can insert into C—H or other sigma bonds, *e.g.*, formation of amides when an acyl nitrene is generated in hydrocarbon solvent.

$$R'{-}\overset{O}{\overset{\|}{C}}{-}\ddot{N}: + R_3CH \longrightarrow R'{-}\overset{O}{\overset{\|}{C}}{-}\underset{H}{\underset{|}{N}}{-}CR_3$$

(c) Rearrangements: Alkyl nitrenes readily undergo rearrangement to give imines instead of undergoing the above two reactions.

$$R{-}\underset{H}{\underset{|}{CH}}{-}\ddot{N}: \longrightarrow R{-}CH{=}NH$$

Imine

A typical example of a rearrangement: A reaction where a nitrene seems to be involved as an intermediate is Hoffmann rearrangement for which the following mechanism may be proposed.

$$R{-}\underset{O}{\underset{\|}{C}}{-}\ddot{N}H_2 \xrightarrow[NaOH]{Br_2} R{-}\underset{O}{\underset{\|}{C}}{-}\ddot{N}(Br){-}H \xrightarrow{^{\ominus}OH} R{-}\underset{O}{\underset{\|}{C}}{-}\overset{\ominus}{\ddot{N}}{-}Br \xrightarrow{-Br^{\ominus}} \left[R{-}\overset{O}{\overset{\|}{C}}{-}\ddot{N}:\right] \longrightarrow R{-}N{=}C{=}O$$

Amide

N-Bromoamide

Acylnitrene

Isocyanate

However, it may be pointed out here that a nitrene has not yet bean detected in this reaction. Hence, the following concerted mechanism may be proposed.

$$R-\overset{O}{\overset{\|}{C}}-\ddot{\underset{\cdot\cdot}{N}}^{\ominus}-Br \longrightarrow O=C=N-R$$

(d) Abstraction of hydrogen: The triplet nitrenes (like triplet carbenes) are capable of abstracting a hydrogen atom from a s C—H bond or other such group to form a pair of radicals which may couple.

$$R-\dot{\underset{\cdot\cdot}{N}}\cdot + R-H \longrightarrow R-\underset{\cdot\cdot}{\overset{H}{\overset{|}{N}}}\cdot + R^{\cdot} \longrightarrow R-\underset{\cdot\cdot}{\overset{H}{\overset{|}{N}}}-R$$

(e) Dimerisation: One of the major reaction of the parent nitrene $H-\ddot{N}:$ is its dimerisation to give diimide.

$$2H-\ddot{N}: \longrightarrow \underset{\text{Diimide}}{R-N=N-H}$$

Similarly, aryl nitrenes produce azoarene in the reactions where these nitrenes are involved.

$$2Ar-\ddot{N}: \longrightarrow Ar-N=N-Ar$$

2.5.7 Enamines

α, β-Unsaturated amines are called enamines (*ene* + *amine*). They are regarded as resonance hybrids of the following contributing forms:

$$\underset{\text{Enamine}}{\overset{\beta}{C}=\overset{\alpha}{C}-\ddot{N}\lt} \longleftrightarrow \overset{\ominus}{C}-C=\overset{\oplus}{N}\lt$$

Formation of Enamines

They are usually obtained from secondary amines and aldehydes or ketones containing α-hydrogen atom in presence of a trace amount of an acid such as *p*-toluene sulphonic acid (PTSA or *p*-TSOH).

$$R-CH_2-\overset{O}{\overset{\|}{C}}-R' + R''_2\ddot{N}H \underset{C_6H_6}{\overset{p\text{-TSOH}}{\rightleftharpoons}} \left[R-CH_2-\underset{NR''_2}{\overset{OH}{C}}-R' \right]$$

α-hydrogen

Carbinolamine (*Unstable*)

$$\xrightarrow{H^{\oplus}} R-\underset{H}{\overset{H}{C}}-\underset{NR''_2}{\overset{\overset{\oplus}{O}H_2}{C}}-R' \xrightarrow{-H_2O} R-CH=C\begin{matrix} R' \\ NR''_2 \end{matrix}$$

Enamine

The above equilibrium is shifted in the forward direction by removing H_2O as an azeotrope with benzene. The most commonly employed 2° amines in the formation of enamines are pyrrolidine, piperidine or morpholine.

Reactions of Enamines

Two important reactions of enamines include α-alkylation (and acylation) of ketones and Michael type reactions.

***(i) α-Alkylation (and α-acylation) of ketones*:** Enamines serve as useful intermediates in the α-alkylation and α-acylation of ketones when treated with an active alkyl halide such as allylic, benzylic, propargyl, α-haloether, α-haloesters, etc. They get alkylated at the α-carbon. Subsequent hydrolysis of the imine salt thus formed gives the α-alkylated ketone. For example,

Cyclohexanone + Pyrrolidine $\xrightarrow[-H_2O]{C_6H_6,\ p\text{-TsOH}}$ Enamine $\xrightarrow{BrCH_2COOEt}$ Imine salt ($Br^{\ominus}$, CH_2COOEt) $\xrightarrow{\overset{\oplus}{H}/H_2O}$ Pyrrolidinium bromide ($\overset{\oplus}{N}H_2$ $Br^{\ominus}$) + 2-(CH_2COOEt)cyclohexanone

This reaction is called *Stork enamine synthesis*. It has two advantages as a synthetic reaction. These are:

1. This procedure yields only monoalkylated ketones whereas usual base—catalyzed alkylation gives a mixture of mono and dialkylation products.
2. In the case of unsymmetrical ketones alkylation occurs only at less substituted α-carbon.

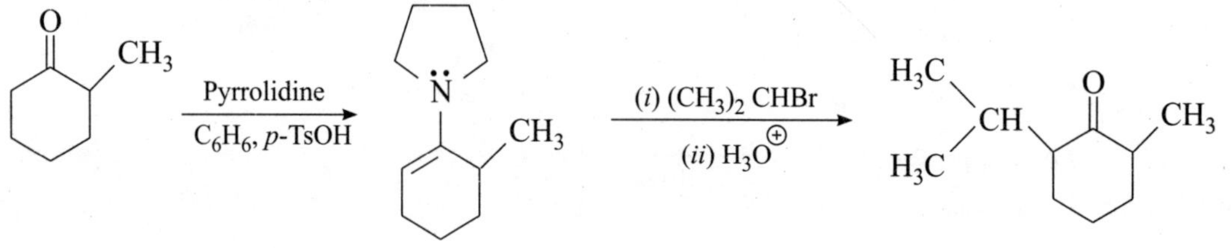

Pyrrolidine / C_6H_6, p-TsOH; (i) $(CH_3)_2CHBr$ / (ii) $H_3O^{\oplus}$

It is due to the fact that the enamine derived from such a ketone can have two structures A and B. The structure A predominates as it is stabilized by resonance more than B. This resonance involves overlapping of *p*-orbital of nitrogen atom, having non-bonding electron pair and π-orbital of carbon-carbon double bond and for this resonance to occur the two carbon atoms of the double bond and nitrogen atom should be coplanar. Now such a coplanarity is possible in structure A whereas it is not possible in B due to steric crowding between the methyl group and methylene of pyrrolidine moiety.

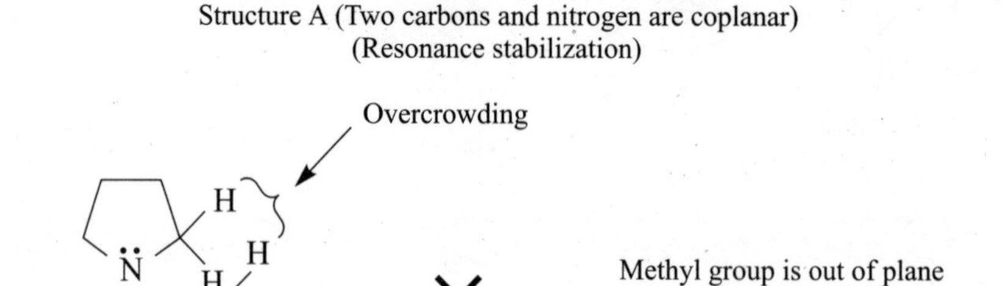

No crowding

Structure A (Two carbons and nitrogen are coplanar)
(Resonance stabilization)

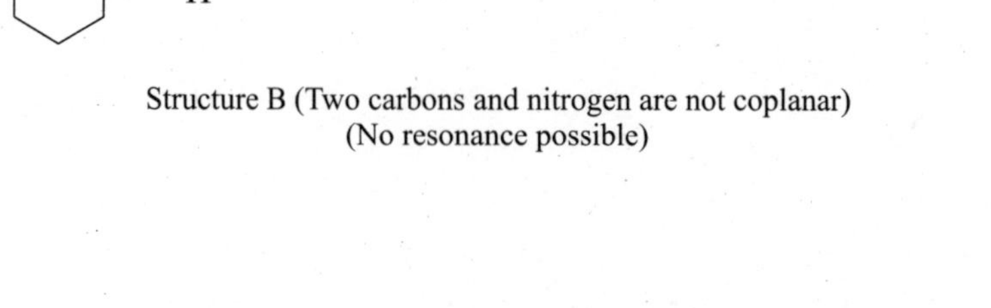

Overcrowding

Methyl group is out of plane due to overcrowding and hence no resonance can occur.

Structure B (Two carbons and nitrogen are not coplanar)
(No resonance possible)

α-Acylated ketones can also be prepared by treating enamines with acid chlorides.

Enamine + R—C(=O)—Cl (Acid chloride) ⟶ $\overset{\ominus}{Cl}$ (iminium salt) $\xrightarrow{H_3O^{\oplus}}$ 2-Acylcyclohexanone

***(ii) Micheal type reaction*:** Nucleophilic addition of enamines on α, β-unsaturated ketones first gives Micheal type adducts which upon hydrolysis yield 1, 5-diketones. For example,

Enamine + $CH_2 = CH — C(=O) — CH_3$ (MVK) ⟶ [$CH_2 — CH = C(O^{\ominus}) — CH_3$ ⟷ $CH_2 — \overset{\ominus}{C}H — C(=O) — CH_3$]

$\xrightarrow{H^{\oplus}/H_2O}$ $CH_2 — CH_2 — C(=O) — CH_3$ + $\overset{\oplus}{N}H_2$ $\overset{\ominus}{O}H$

2 - (3-Oxobutyl) cyclohe.:an - 1- one
(A 1, 5-diketone)

2.6 ASSIGNING FORMAL CHARGES TO INTERMEDIATES AND OTHER SPECIES

An organic species (a molecule or an intermediate) can be electrically neutral or positively charged or negatively charged. In a neutral species the sum of positive charges is equal to that of negative charges. For example, in nitroalkane, which is neutral, there is a unit positive and a unit negative charge.

$$R-\overset{\oplus}{N}(=O)-O^{\ominus}$$

Positive or negative charge assigned to an atom in a molecule is called formal charge which is defined *as a positive or negative charge which must be assigned to a particular atom to describe the Lewis structure completely.*

For assigning formal charge to any species we make use of **average electron theory.**

According to average electron theory, for assignment of formal charge to any atom in a molecule first we calculate *the number of average electrons* on it. This is done by making use of the following expression:

$$\text{No. of average electrons} = \frac{\text{No. of bonding electrons}}{2} + \text{No. of non-bonding electrons}$$

(**Note:** The number of bonding electrons is divided by 2 because they are equally divided between bonded atoms). The number of average electrons calculated as above is compared to the number of electrons on the neutral atom or number of electrons in the valence shell, *e.g.*, the number of electrons on the neutral atom is 4 for carbon, 5 for nitrogen, 6 for oxygen and 7 for halogens. Now if the number of average electrons on the atom concerned is less by one unit than that in the valence shell (neutral atom), the atom is assigned a formal charge of +1 and if it is two unit it is assigned a formal charge of +2. Similarly, if it is more by one unit the atom is assigned formal charge of –1 and if it is more by 2 units it is assigned formal charge of –2. Thus, Formal charge = No. of valence electrons–No. of average electrons.

Examples

1. Calculation of formal charge on methyl cation

(*i*) Lewis structure CH_3; (*ii*) No. of bonding electrons on carbon = 6; (*iii*) No. of non-bonding electrons on carbon = 0; (*iv*) No. of valence electrons on C = 4.

Now

(*i*) Average no. of electrons on carbon = (No. of bonding electrons)/2 + No. of non-bonding electrons = 6/2 + 0 = 3 + 0 = 3.

(*ii*) Formal charge on C = No. of valence electrons on carbon –Average number of electrons = 4 – 3 = +1.

So methyl cation is written as $^{\oplus}CH_3$.

2. Calculation of formal charge on methyl anion

(*i*) Lewis structure $-CH_3$; (*ii*) No. of bonding electrons = 6; (*iii*) No. of non-bonding electrons = 2; (*iv*) No. of valence electrons on carbon = 4

Now

(*i*) Average no. of electrons = 6/2 + 2 = 3 + 2 = 5

(*ii*) Formal charge on carbon = 4 – 5 = –1

So methyl anion in written as $^{\ominus}CH_2$

3. Calculation of formal charge on methyl radical

(*i*) Lewis structure $-CH_3$; (*ii*) No. of bonding electrons = 6; (*iii*) No. of non-bonding electrons = 1; (*iv*) No. of valence electrons on carbon = 4.

Now

(*i*) Average no. of electrons = 6/2 + 1 = 3 + 1 = 4

(*ii*) Formal charge on C = 4 – 4 = 0

So methyl radical is written as $\dot{C}H_3$

4. Calculation of formal charge on a molecule: The case of diazomethane

The molecule of diazomethane can be written as the hybrid of the following three contributing forms:

$$\underset{\textbf{A}}{H_2\ddot{C}-N_a\equiv N_b:} \longleftrightarrow \underset{\textbf{B}}{H_2C=N_a=\ddot{N}_b:} \longleftrightarrow \underset{\textbf{C}}{H_2\ddot{C}-\ddot{N}_a=N_b:}$$

The formal charges on carbon, nitrogen (*a*) and nitrogen (*b*) for each structure can be calculated.

Structure A

Formal charge on carbon = No. of valence electrons – Average no. of electrons
= 4 – (6/2 + 2) = 4 – 5 = –1

Formal charge on N_a = 5 – (8/2 + 0) = 5 – 4 = +1

Formal charge on N_b = 5 – (6/2 + 2) = 5 – 5 = 0

So, structure **A** is written as: $:\overset{\ominus}{C}H_2-\overset{\oplus}{N}\equiv N$

Structure B

Formal charge on carbon = 4 – (8/2 + 0) = 4 – 4 = 0

Formal charge on N_a = 5 – (8/2 + 0) = 5 – 4 = +1

Formal charge on N_b = 5 – (4/2 + 4) = 5 – 6 = –1

So, structure **B** is written as: $CH_2=\overset{\oplus}{N}=\overset{\ominus}{\ddot{N}}:$

Structure C

Formal charge on carbon = 4 – (6/2 + 2) = 4 – 5 = –1

Formal charge on N_a = 5 – (6/2 + 2) = 5 – 5 = 0

Formal charge on $N_b = 5 - (4/2 + 2) = 5 - 4 = +1$

So, structure **C** is written as: $H_2\overset{\ominus}{\ddot{C}}—\ddot{N}=\overset{\oplus}{N}:$

Finally the following picture emerges

$$\underset{A}{H_2\overset{\ominus}{\ddot{C}}—\overset{\oplus}{N}\equiv N:} \longleftrightarrow \underset{B}{H_2C=\overset{\oplus}{N}=\overset{\ominus}{\ddot{N}}:} \longleftrightarrow \underset{C}{H_2\overset{\ominus}{\ddot{C}}—\ddot{N}=\overset{\oplus}{N}:}$$

The sum of the formal charges is zero in all the three Lewis structure written above. Therefore, diazomethane as a whole is an electrically neutral molecule.

2.7 METHODS OF DETERMINATION OF REACTION MECHANISM

The mechanism of a reaction tells us the precise position of atoms, as a function of time in the reactant molecules as they are converted into products. Since many chemical reactions proceed very rapidly ($\sim 10^{-14}$ s), it is not possible to follow the reaction path directly and hence our knowledge of reaction mechanisms is mostly based on indirect evidence. The first step in the study of mechanism is characterization of products. After characterization of the product(s), the mechanism of a reaction is determined by considering all the possibilities that are consistent with the experimental observations. In most of the cases, new experiments are to be performed to exclude a good number of the proposed mechanisms. If such elimination leads to only one mechanism then it is said to be the *established mechanism* of that reaction.

The proposed mechanism of a reaction should have the following characteristics:

(*a*) be energetically reasonable,

(*b*) account for the experimental observations, and

(*c*) be in agreement with what is known about analogous reactions.

There are a number of methods for determining the mechanism of a reaction. These include:

1. Product analysis
2. Determination of the presence of intermediates
3. Cross-over experiments
4. Isotopic labelling
5. Isotope effect
6. Stereochemical evidence
7. Kinetic evidence

It should be noted that the above order does not necessarily reflect the order of experimentation. Further it may not be necessary to apply all the methods to determine the mechanism of a particular reaction.

1. Product Analysis

Identification of products, byproducts if any, and determining their relative quantities constitutes the first step in the investigation of a mechanism. To illustrate this, let us consider the following reactions:

(*i*) Hydrolysis of isomeric allylic chlorides, for instance, yields a mixture of 85 per cent of a tertiary alcohol and 15 per cent of a primary alcohol. Since product composition remains the same in either case, a common intermediate is suggested.

$(CH_3)_2C{=}CH{-}CH_2Cl$ / $(CH_3)_2C(Cl){-}CH{=}CH_2$ → [$(CH_3)_2C{=}CH{-}\overset{\oplus}{C}H_2$ ↔ $(CH_3)_2\overset{\oplus}{C}{-}CH{=}CH_2$] (Intermediate) → $(CH_3)_2C{=}CH{-}CH_2OH$ (15%) + $(CH_3)_2C(OH){-}CH{=}CH_2$ (85%)

(*ii*) Product identification has helped in establishing the correct mechanism for the **Beckmann rearrangement.** Only one geometric isomer of oxime A **(I)** can furnish the substituted amide **(III)** on treatment with phosphorus pentachloride. Earlier, chemists believed that the oxime which gives this amide has the migrating group (R′) nearer the leaving group (—OH) *i.e.*, isomer II should from

$RR'C{=}N{-}OH$ **A**

I (RR′C=N–OH, OH anti to R′) $\xrightarrow{PCl_5}$ III (RC(=O)–NRR′) $\xleftarrow[\times]{PCl_5}$ II (RR′C=N–OH, OH syn to R′)

However, it was subsequently discovered that one of the oximes of 2-bromo-5-nitroacetopenone readily undergoes ring closure to benzisoxazole (V) thus establishing its structure as (IV). The other isomer (VI) does not undergo ring closure, but readily gives the Beckmann rearrangement product, the amide (VII). The identification of (VII) thus leads to the conclusion that *rearrangement involves a trans migration of the substituted phenyl group.*

IV (oxime, O_2N, Br, CH_3, N–OH) $\xrightarrow{HO^{\ominus}}$ V (benzisoxazole with O_2N and CH_3)

VI (oxime, O_2N, Br, CH_3, N–OH) $\xrightarrow{PCl_5}$ VII ($O{=}C{-}CH_3$ on NH; aryl with Br and O_2N)

2. Determination of the Presence of Intermediates

We have already noted that many organic reactions involve formation of intermediates, which can be sometimes isolated, detected by spectroscopy, or trapped by chemical reaction with an added compound with which they are known to react and the adducts so formed are then characterised. These techniques are illustrated below:

Isolation: Occasionally, it is possible to isolate *stable* intermediates by stopping a reaction before completion , or by using relatively milder conditions. These intermediates, for example, an N-bromoamide (RCONHBr), its anion $(RCONBr)^-$ and an isocyanate (RNCO) have been isolated in the **Hoffmann rearrangement** of an amide to an amine.

$$\underset{\text{1° Amide}}{R-\overset{\overset{O}{\|}}{C}-NH_2} \xrightarrow[-KBr,\ -H_2O]{Br_2/KOH} \underset{\text{N-Bromoamide}}{R-\overset{\overset{O}{\|}}{C}-\ddot{N}H-Br} \xrightarrow[-H_2O]{KOH} \underset{\text{Anion}}{\left[R-\overset{\overset{O}{\|}}{C}-\overset{\ominus}{N}-Br\right]\overset{\oplus}{K}}$$

$$\underset{\text{Anion}}{\left[R-\overset{\overset{O}{\|}}{C}-\overset{\ominus}{N}-Br\right]\overset{\oplus}{K}} \xrightarrow{-KBr} \underset{\text{Isocyanate}}{R-N{=}C{=}O} \xrightarrow{2KOH} \underset{\text{1° Amine}}{R-NH_2} + K_2CO_3$$

Thus any proposed mechanism for this rearrangement must account for the formation of all these intermediates.

Detection: *Unstable* intermediates are often detected by spectroscopic methods. The intermediacy of carbocations in S_N1 and E1 reactions has been established by Nmr spectroscopy. Similarly, the intermediates of electrophilic substitution reactions such as VIII have been characterized by their NMR and electronic spectra.

$$\underset{\text{Mesitylene}}{C_6H_3(CH_3)_3} + HF + BF_3 \longrightarrow \underset{\text{VIII}}{[(CH_3)_3C_6H_4]^{\oplus}\ BF_4^{\ominus}}$$

Free radical intermediates are detected by the use of electron spin resonance (*esr*) spectroscopy.

Trapping: An intermediate is often trapped by running the reaction in the presence of a compound with which it is known to react. For example,

(*i*) *Carbenes* are trapped as cyclopropane derivatives.

$$\underset{\text{Diazomethane}}{\overset{\ominus}{C}H_2-\overset{\oplus}{N}{\equiv}N} \xrightarrow[-N_2]{h\nu} \underset{\text{Carbene}}{H_2C:} \xrightarrow{\text{cyclohexene}} \underset{\substack{\text{Bicyclo [4.1.0] heptane} \\ \text{(A cyclopropane derivative)}}}{H_2C\!<\!\text{(fused to cyclohexane)}}$$

(*ii*) *A benzyne* reacts with anthracene to give triptycene.

Benzenediazonium -2-caboxylate $\xrightarrow[-CO_2,\ -N_2]{h\nu}$ Benzyne $\xrightarrow{\text{anthracene}}$ Triptycene

(*iii*) *Carbocations*: Nucleophiles are commonly employed to trap carbocations. Direct evidence for the intermediacy of carbocations during the addition of a halogen to alkenes, for example, is provided by the observation that when the reaction is carried out in the presence of other nucleophiles, a mixture of products is obtained.

$$CH_2{=}CH_2 + \overset{\delta^+}{Br}{-}\overset{\delta^-}{Br} \longrightarrow \overset{\oplus}{C}H_2{-}CH_2{-}Br + \overset{\ominus}{Br}$$

$$\underset{\text{Carbocation}}{\overset{\oplus}{C}H_2{-}CH_2{-}Br} \xrightarrow{\overset{\ominus}{Br}} \underset{|\ Br}{CH_2}{-}CH_2{-}Br$$

$$\overset{\oplus}{C}H_2{-}CH_2{-}Br \xrightarrow{\overset{\ominus}{I}} \underset{|\ I}{CH_2}{-}CH_2{-}Br$$

$$\overset{\oplus}{C}H_2{-}CH_2{-}Br \xrightarrow{\overset{\ominus}{O}NO_2} \underset{|\ ONO_2}{CH_2}{-}CH_2{-}Br$$

(*iv*) *Free radicals*: These can be trapped by carrying out the reaction in presence of nitric oxide with which they react rapidly *e.g.*, free redical formed by thermal decomposition of azidoisobutyronitrile is trapped by nitric oxide to form a hydroxylamine derivative.

3. Crossover Experiments

Whether a rearrangement is a one-step or a two-step process can be decided by carrying out the reaction taking a mixture of two similarly constituted but non-identical reactants and then analyzing the products. Since the migrating group must become free in a two-step process, we should get products containing fragments of both the reactants. *Fries rearrangement,* which involves migration of an acyl group of a phenyl ester to the benzene ring, provides such an example. The two esters, IX and X, rearrange under identical conditions and at approximately the same rate to give the products XI and

XII, respectively. When the same reaction is carried out with a mixture of two esters (IX and X) under similar conditions, two possibilities arise:

(1) If the reaction is one-step the ester IX will give the ketone XI and the ester X will give the ketone XII. *i.e.*, only two ketones will be formed.

(2) If the reaction occurs in two-steps, then the migrating groups *i.e.*, C_6H_5CO and CH_3CO must become free and hence cross-over products XIII and XIV will be formed in addition to the expected products XI and XII.

In actual, rearrangement of a mixture of IX and X, however, not only gives XI and XII but also the *cross products* XIII and XIV. Isolation of cross products unequivocally demonstrates that the *Fries rearrangement is a two-step process.*

$OCOC_6H_5$ $\xrightarrow{AlCl_3}$ OH COC_6H_5

CH_3 IX CH_3 XI

Cl $OCOCH_3$ $\xrightarrow{AlCl_3}$ Cl OH $COCH_3$

CH_3 X CH_3 XII

IX + X $\xrightarrow{AlCl_3}$ XI + XII + OH $COCH_3$ CH_3 XIII + Cl OH COC_6H_5 CH_3 XIV

4. Isotopic Labelling

Mechanisms of many reactions have been investigated by using molecules in which an isotope is substituted for an atom, and then tracing the location of the same in the products. Some examples of use of isotopic labelling in elucidation of the mechanism of a reaction are given below:

(i) Hydrolysis of esters: The question whether hydrolysis of an ester involves acyl-oxygen or alkyl-oxygen bond fission has been solved by hydrolyzing amyl acetate in $H_2{}^{18}O$.

$$CH_3-\overset{\displaystyle O}{\overset{\|}{C}}-O-C_5H_{11} + H_2\overset{18}{O} \xrightarrow{HO^{\ominus}} CH_3-\overset{\displaystyle O}{\overset{\|}{C}}-\overset{18\ominus}{O} + C_5H_{11}OH$$

The fact that hydrolysis yields an unlabelled alcohol and labelled acid has been taken as evidence for *acyl-oxygen bond fission.*

(ii) Cannizzaro reaction: The mechanism of *Cannizzaro reaction*, which involves transfer of a hydride ion from one molecule to another, has been investigated by the use of deuterated benzaldehyde (C_6H_5CDO) instead of benzaldehyde.

$$2\,C_6H_5CDO + HO^{\ominus} \xrightarrow{H_2O} C_6H_5-\overset{\displaystyle O}{\overset{\|}{C}}-\overset{\ominus}{O} + C_6H_5CD_2OH$$

Since, the deuterium is exclusively located on the carbon atom of benzyl alcohol bearing the hydroxyl group, the possibility of exchange of hydrogen with hydrogen atoms in the solvent is ruled out. Thus, there is a transfer of hydrogen (or deuterium) between the carbonyl carbons of two aldehyde molecules and in the course of the reaction hydrogen (or deuterium) never gets attached to an oxygen atom. A plausible mechanism for Cannizzaro reaction can thus be *visualized* as follows:

$$C_6H_5-\overset{\displaystyle O}{\overset{\|}{C}}-H \xrightarrow{^{\ominus}OH} C_6H_5-\overset{\displaystyle O^{\ominus}}{\overset{|}{\underset{\underset{\displaystyle OH}{|}}{C}}}-H \quad + \quad C_6H_5-\overset{\displaystyle O}{\overset{\|}{C}}-H$$

$$\downarrow$$

$$C_6H_5-C\overset{\displaystyle //O}{\underset{\displaystyle \backslash O^{\ominus}}{}} + C_6H_5-CH_2OH \longleftarrow \left[C_6H_5-\overset{\displaystyle O}{\overset{\|}{\underset{\underset{\displaystyle OH}{|}}{C}}} + C_6H_5-\overset{\displaystyle O^{\ominus}}{\overset{|}{\underset{\underset{\displaystyle H}{|}}{C}}}-H \right]$$

(iii) Nucleophilic aromatic substitution: The Benzyne Mechanism: As already discussed that treatment of chlorobenzene with sodium amide ($NaNH_2$) in liquid ammonia at 240 K produces aniline. This reaction involves **benzyne** as an intermediate. Support for this mechanism comes from the observation that when [1-^{14}C]-chlorobenzene *i.e.*, chlorobenzene having ^{14}C-carbon at 1-position, is treated with sodium amide and liquid ammonia at 240 K, it produces a mixture of [1-^{14}C]-aniline and [2-^{14}C] aniline.

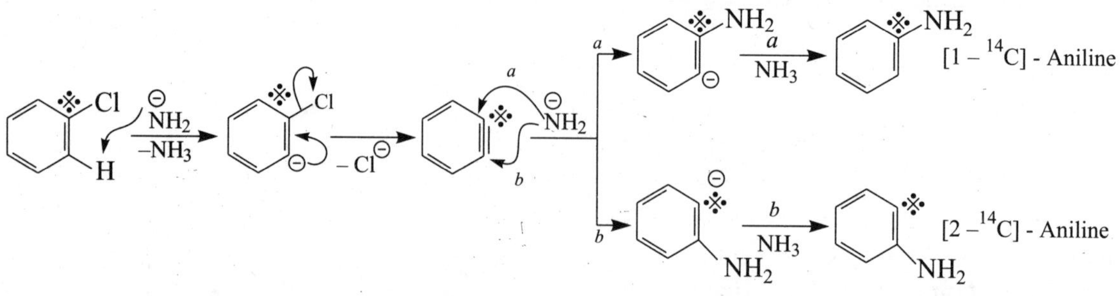

5. Isotope Effects

In the above discussion of isotopic labelling experiments, it has been presumed that there is no difference in *chemical behaviour* of different isotopes and hence the chemical properties of the compounds are not quantitatively changed when a particular atom is replaced by its isotope. However, the isotopes are not identical in their behaviour in all aspects because the rates of reactions involving two isotopes may be different.

The above difference in the rate of a reaction due to different isotopes is called the *isotope effect* and is usually small. However, if the cleavage of the bond between a particular atom (or its isotope) and rest of the substrate is involved in the rate determining step, the magnitude of isotope effect may be relatively large and so this type of isotope effect is called *primary isotope effect** or primary kinetic isotope effect.

Since, the mass of a deuterium atom is double and that of a tritium atom is triple than that of a hydrogen atom, therefore, the magnitude of the primary isotope effect is greatest when a hydrogen atom is replaced by a deuterium (D) or tritium (T). It is usually expressed in terms of the rate ratio *i.e.*, K_H/K_D or K_H/K_T, respectively. The value of this rate ratio is always greater than one which means that the compound containing hydrogen reacts faster than that containing deuterium or tritium. Typical values for K_H/K_D and K_H/K_T are 7 and 16, respectively.

The reason for the primary isotope effect, *i.e.*, when C—H and C—D bonds cleave in rate determining step, is that bond dissociation energy of the C—D bond is more than that of C—H bond. It is even higher for the C—T bond. Therefore, the energy of activation for cleaving the C—D bond would be much higher than that for C—H bond and hence the rate of the reaction involving C—D bond

---

***Secondary isotope effect.** Deuterium isotope effects have also been seen in cases where it is known that the C—H bond does not cleave at all during the reaction. Such effects are called *secondary isotope effects and their magnitude is quite small as compared to the primary isotope effects.*

These effects are further classified as α, β or γ if the deuterium atom is present respectively at α, β or γ positions with respect to the carbon at which the reaction is taking place. An example of β–secondary isotope effect is the solvolysis (*i.e.*, hydrolysis by the solvent) of isopropyl bromide, *i.e.*,

$$(CH_3)_2\,CHBr + H_2O \xrightarrow{K_H} (CH_3)_2\,CHOH + HBr$$

$$(CH_3)_2\,CHBr + H_2O \xrightarrow{K_D} (CD_3)_2\,CHOH + HBr$$

For this reaction the K_H/K_D rate ratio is 1.34. Most probably such secondary isotope effect is due to hyperconjugation.

cleavage will be slow. It may, however, be noted here that if the C—D and C—H bond cleavages are not involved in the rate-determining step of the reaction, then the rate of the reaction will be hardly affected when hydrogen is replaced by its isotope, D or T. Most of the studies of isotope effect have been carried out using D and not T because it is more easily available and also because T is radioactive.

Primary Isotope Effect and Mechanism

Primary isotope effect has been found to be a powerful tool in the elucidation of mechanisms of a number of reactions. Some examples are discussed below:

***(i)* Dehydrohalogenation of isopropyl bromide:** When in isopropyl bromide the hydrogen atoms of the two methyl groups are replaced by deuterium (D), the rate of dehydrohalogenation of this compound with sodium ethoxide is considerably lowered.

Reaction	Relative rate
$(CH_3)_2CHBr + C_2H_5O^{\ominus} \longrightarrow CH_3CH{=}CH_2 + C_2H_5OH + Br^{\ominus}$	1.0
$(CD_3)_2CHBr + C_2H_5O^{\ominus} \longrightarrow CD_3CH{=}CD_2 + C_2H_5OD + Br^{\ominus}$	0.15

This large isotopic effect definitely suggests that the rate-determining step of the reaction involves the breaking of the C—H bond. The following mechanism for the reaction is ruled out because in this case the C—H bond cleavage takes place at faster rate. The following mechanism is consistant with the observation:

$$C_2H_5O^{\ominus} + H{-}CH_2{-}CH(Br){-}CH_3 \longrightarrow \left[\begin{matrix} C_2H_5O^{\delta-} \\ \vdots \\ H \quad H \\ H{-}C{=\!=}C{-}CH_3 \\ H \quad Br^{\delta-} \end{matrix} \right]_{\text{Transition state}} \longrightarrow H_2C{=}CHCH_3 + C_2H_5OH + Br^{\ominus}$$

It may, therefore, be concluded that the C—H bond fission rather than C—Br bond fission is involved in rate-determining step of the reaction.

***(ii)* Nitration of benzene:** It is a typical example of aromatic electrophilic substitution. In this reaction the electrophile is positively charged nitronium ion ($\overset{\oplus}{N}O_2$) which in the first step adds to the carbon-carbon double bond forming cyclohexadienyl cation as intermediate which loses a proton to give nitrobenzene. This reaction does not exhibit isotope effect as is evident from the fact that the rates of nitration of benzene (C_6H_6) hexadeuterated benzene (C_6D_6) and hexatritiated benzene (C_6T_6) are the same. It means that the rate determining step does not involve the cleavage of C—H or C—D or C—T bonds. Thus, the following mechanism, which is in close agreement with the above observation can be proposed.

$$C_6H_6 + \overset{\oplus}{N}O_2 \xrightarrow{\text{Slow}} [C_6H_6NO_2]^{\oplus} \xrightarrow[\text{Fast}]{-H^{\oplus}} C_6H_5NO_2$$

Cyclohexadienyl cation

6. Stereochemical Evidence

Sometimes the proposed mechanism of a reaction can be supported by following the stereochemistry of the reactants and the products. Thus, the mechanisms of nucleophilic substitution reactions (S_N1) and (S_N2) have been investigated by carrying out reactions with optically active compounds and then checking the optical purity of the products. Following examples are illustrative:

***(i)* Mechanism of S_N1 reactions.** Hydrolysis of optically active 1-chloro-1-phenyl-ethane

$$\underset{\substack{\text{(S) - 1- Chloro -1- phenylethane} \\ \text{(optically active)}}}{(H_5C_6)(H)(H_3C)C—Cl} \xrightarrow[\text{Slow}]{\text{KOH (aq.)}} \underset{\substack{\text{Carbocation} \\ \text{(planar)}}}{(C_6H_5)(H)(CH_3)\overset{\oplus}{C}} + Cl^{\ominus}$$

$$\underset{\substack{\text{(R) - 1- Phenylethanol} \\ \text{(50\%)}}}{HO—C(C_6H_5)(H)(CH_3)} \xleftarrow[\text{(Back side attack)}]{\text{(Fast)} \overset{\ominus}{O}H} (C_6H_5)(H)(CH_3)\overset{\oplus}{C} \xrightarrow[\text{(Front side attack)}]{\overset{\ominus}{O}H \text{ (Fast)}} \underset{\substack{\text{(S) - 1- Phenylethanol} \\ \text{(50\%)}}}{(C_6H_5)(H)(CH_3)C—OH}$$

Racemic mixture (optically inactive) [Equimolar mixture of two enantiomers R and S]

Racemization of the product suggests that the reaction occurs through a planar *carbocation intermediate* since the attack of nucleophile, *i.e.*, OH^- ion on this carbocation can occur from both the faces (*front and back side*) almost equally to give a 50 : 50 mixture of the two enantiomers.

Thus it may be concluded that S_N1 reactions involve carbocation intermediates.

***(ii)* Mechanism of S_N2 reactions:** Hydrolysis of optically active 2-bromooctane gives optically active 2-octanol. This will be possible only when the nucleophile (OH^-) attacks the side of the carbon atom opposite to that of the leaving group giving first a transition state which then expels the Br^- ion to give the product.

$$HO^{\ominus} + H\text{-}C(C_6H_{13})(CH_3)\text{-}Br \longrightarrow \left[HO^{\delta-}\cdots C(C_6H_{13})(H)(CH_3) \cdots Br^{\delta-} \right] \longrightarrow HO\text{-}C(C_6H_{13})(CH_3)\text{-}H$$

(S)-2-Bromooctane (*optically active*) — Transition state — (R)-2-Octanol (*optically active*)

The reaction proceeds with *stereochemical inversion of configuration* which is usually called as **Walden inversion**.

Thus, it may be concluded that S_N2 reactions are *one-step reactions in which optically active reactants give optically active products with inversion of configuration.*

(*iii*) Elimination reactions. It has been observed that *cis*-1, 2-dichloroethane gives chloroacetylene twenty times faster than the corresponding *trans*-isomer. This suggests that E2 reactions occur more rapidly when the departing groups are *trans* to each other than when they are *cis*.

$$\text{HClC=CHCl (cis)} \xrightarrow[\text{Fast}]{\overset{\ominus}{O}H} Cl-C\equiv C-H \xleftarrow[\text{Slow}]{\overset{\ominus}{O}H} \text{HClC=CHCl (trans)}$$

cis-1, 2-Dichloroethene (*H and Cl are trans*) — *trans*-1, 2-Dichloroethane (*H and Cl are cis*)

7. Kinetic Evidence

An unambiguous (doubtless) mechanism cannot be proposed for a reaction on the basis of kinetic studies alone. However, a knowledge of the order of reaction is a very useful tool for the study of the mechanism. Thus on the basis of data pertaining to rate law, order and molecularity of a reaction, it becomes possible to rule out several of the proposed mechanistic schemes.

Perhaps only the most important examples of reactions *whose mechanisms have been established on the basis of the kinetic studies are the hydrolysis of alkyl halides, e.g.,*

(*i*) Hydrolysis of methyl bromide: It is a typical example of aliphatic nucleophilic substitution. Methyl bromide when treated with aqueous sodium hydroxide gives rise to methyl alcohol—the hydroxide ion replacing bromide ion.

$$H\overset{\ominus}{O} + CH_3-Br \longrightarrow HO-CH_3 + \overset{\ominus}{Br}$$

The reaction shows second order kinetics, meaning that both the species are involved in the rate-determining step.

$$\text{Rate} = \frac{-d\,[CH_3Br]}{dt} = [CH_3Br]\,[\overset{\ominus}{O}H]$$

Following mechanism is in close agreement with this observation:

$$HO^{\ominus} + H_3C{-}Br \longrightarrow \left[\overset{\delta^-}{HO} \cdots C(H)_3 \cdots \overset{\delta^-}{Br} \right] \longrightarrow HO{-}CH_3 + Br^{\ominus}$$

Transition state

This reaction is called as S_N2 reaction.

(*ii*) Hydrolysis of *tert*-butyl bromide

$$HO^{\ominus} + (CH_3)_3C{-}Br \longrightarrow (CH_3)_3C{-}OH + Br^{\ominus}$$

In contrast to methyl bromide this reaction shows first order kinetics involving only ($(CH_3)_3CBr$) in the rate determining step.

$$\text{Rate} = \frac{-d[(CH_3)_3CBr]}{dt} = [(CH_3)_3CBr]$$

Following mechanistic scheme is in close agreement with the above observation, the reaction taking place in two steps:

First step

$$(H_3C)_3C{-}Br \xrightarrow[\text{Ionization}]{\text{Slow}} \left[(CH_3)_3C^{\oplus} \right] + Br^{\ominus}$$

Second step

$$\left[(CH_3)_3C^{\oplus} \right] + \overset{\ominus}{O}H \xrightarrow{\text{Fast}} (CH_3)_3COH$$

This reaction is called S_N1 reaction.

A BRIEF REVIEW

- **Curved arrows** (↷) are used for *two-electron* movements and *fish hook arrows* (⇀) for one electron movements while *double headed arrows* (⟷) separate different *canonical forms* and middle head arows(→—) represent inductive effect *i.e.*, drift of electrons.
- During **homolytic fission** a covalent bond breaks in such a way that one electron of the shared pair is taken away by each atom linked by covalent bond due to which free radicals are produced. It occurs in non-polar solvents and is favoured by high temperature and UV-radiation.
- During **heterolytic fission** a covalent bond breaks in such a way that both the electrons of the shared pair are taken away by only one of the bonded atoms due to which a cation and an anion are produced. It occurs in polar solvents.
- **Electrophiles** are *electron-loving* neutral or positively charged species.
- **Nucleophiles** are *nucleus-loving* neutral or negatively charged species.
- **Substitution reactions** involve substitution of one atom or a group in an organic molecule by another atom or a group. These can be nucleophilic, electrophilic or free radical substitution reactions.
- **Addition reactions** involve combination of two molecules to give a single product. These can be nucleophilic, electrophilic and free radical addition reactions.
- **Elimination reactions** involve the loss of two atoms or groups from the same or adjacent atoms of a molecule leading to the formation of a multiple bond. These can be α-β- and γ-elimination reactions according as the two atoms or groups are lost from the α-β- and γ-positions.
- **Condensation reactions** involve combination of two or more molecules of the same or different reactants to give a product with or without the loss of some simple molecules such as H_2O, NH_3, HCl, ROH, etc. *e.g.*, aldol condensation, Claisen condensation, condensation polymerization, Claisen-Schmidt condensation, etc.
- **Isomerization reactions** involve interconversion of one isomer into another without any change in the molecular formula and the carbon skeletons of the molecules. For example,

$$\textit{trans}\text{–But-2-ene} \underset{}{\overset{h\nu}{\rightleftharpoons}} \textit{cis}\text{–But-2-ene}$$

- **Pericyclic reactions** are concerted reactions which occur through cyclic transition states. For example, Diels-Alder reaction.
- **Rearrangement reactions** involve the migration of an atom or a group from one atom to another within the same molecule *e.g.*, 1, 2-hydride shift, Beckmann rearrangement, etc.
- **Free energy** of a reaction is given by ΔG or ΔF = ΔH – TΔS. Only those reactions are feasible for which **ΔG has a negative value.**

- **Collision theory:** The rate of reaction = $PZe^{-E_a/RT}$ where P is probability factor, Z is the collision frequency, E_a is the activation energy and $e^{-Ea/RT}$ is the fraction of collision having energy greater than E_a.
- **Transition state theory:** Reactants are converted into products through a transition state (activated complex). *The energy difference between the reactants and the transition state is called the free energy of activation.* Lower is the energy of activation, faster is the reaction and *vice-versa*.
- The formation of an **intermediate** means that the reaction occurs through two transition states. The reactive intermediate cannot be isolated when the free energy of activation of second transition state is lower than that of the first. However, the intermediate can be isolated when the free energy of activation of second transition state is higher than that of the first.
- During a **step-wise reaction** the number of transition states is always equal to the number of steps of the reaction. The rate-determining step will have the transition state having highest free energy of activation.
- **Hammonds postulate** states that for any reaction the geometry of transition state resembles the side to which it is closer in free energy of activation. Therefore, the T.S. may be reactant-like, product-like or intermediate-like according as its energy lies closer to the reactant, the product or the intermediate, respectively.
- A **catalyst** increases the rate of a reaction by providing an alternate route from reactants to products with a lower energy transition state. The catalyst is never consumed in the process.
- The **reactive intermediates** are *short-lived* and *highly reactive* species through which a large number of chemical reactions occur. For example, carbocations, carbanions, free-radicals, carbenes, nitrenes, arynes and enamines are formed as intermediates during different chemical reactions.
- **Carbocations** are chemical species carrying a positive charge on the carbon atom which is sp^2-hybridized to make the carbocation a planar species. These are produced by heterolytic fission of covalent bonds and their stability follows the order: 3° > 2° > 1° > $^+CH_3$.
- **Carbanions** are chemical species carrying a negative charge on the carbon atom which is sp^3-hybridized. The carbanions may be *planar* or *pyramidal*. The order of stability follows the order: $^-CH_3$ > 1° > 2° > 3°. These are also produced by heterolysis of a covalent bond. Simple alkyl carbanions are pyramidal having sp^3-carbon but the carbanions having charged carbon next to a double bond or a benzene ring are planar having sp^2-carbon and these are stabilized by resonance.
- **Conjugated** cyclic carbocations and carbanions having 6π-electrons are **aromatic** and hence are stabilized by resonance but those having 4π-electrons are antiaromtic and hence are destabilized by resonance.
- **Free radicals** are neutral species having an odd electron. The carbon atom is sp^2-hybridized to give planar structure. The order of stability is: 3° > 2° > 1° > $^\bullet CH_3$.

- **Carbenes** are neutral divalent carbon species. There are two types of carbenes—**singlet** and **triplet**. Singlet carbene ($\uparrow\downarrow CH_2$) having sp^2-carbon is a non-linear species. However, the triplet carbene ($\uparrow\uparrow CH_2$) having sp-hybridized carbon atom is linear. *A triplet carbene is always more stable than the corresponding singlet carbene* due to lesser interelectronic repulsions.
- **Nitrenes** are nitrogen analogs of carbenes. Similar to carbenes, nitrenes also exist in *singlet* and *triplet* forms. Nitrogen atom is sp^2-hybridized in singlet form and sp-hybridized in triplet form of nitrene. *A triplet nitrene is more stable than a singlet nitrene.*
- **Arynes** are aromatic compounds containing a formal C≡C bond. **Benzyne** is best known aryne, which occurs as an intermediate in nucleophilic substitution reactions of unsubstituted aryl halides *e.g.*, **Dow's method** for production of phenol from chlorobenzene involves benzyne formation.
- **Enamines** (ene + amine) are produced when an aldehyde or a ketone having α–hydrogen reacts with a 2°-amine in presence of a trace of PTS (*i.e.*, **Stork enamine reaction**) . It is better than base-catalysed alkylation or acylation of ketones and aldehydes because it can be stopped at monoalkylation or monoacylation stage.
- **Formal change on an atom = No. of valence electrons–1/2 [No. of bonding electrons] – No. of non-bonding electrons.**
- **Hydrolysis of esters** involves cleavage of acyl-oxygen bond and not that of alkyl-oxygen bond. This mechanism has been established by carrying out the hydrolysis of an ester by OH^- in H_2O^{18}.
- **Primary isotope effect studies** have shown that the rates of nitration of C_6H_6, C_6D_6 and C_6T_6 are the same which suggests that the rate-determining step of the reaction involves attack of nitronium ion on the benzene ring to form a carbocation intermediate and not the loss of $H^{\oplus}$ ion from this intermediate.
- **S_N1 reaction** occurs through carbocation intermediates. This has been confirmed by the racemization of optically active alkyl halides during hydrolysis.

SOLVED PROBLEMS

Q.1. The carbocation, F_3C^+ is more stable than carbocation, $F_3C—C^+$. Explain.

Ans. In case of carbocation, $F_3C—C^+$, the strongly electron-withdrawing F_3C group withdraws electrons of the C—C bond towards itself thereby intensifying the positive charge and thus destabilizing the carbocation, $F_3C—C^+$ relative to F_3C^+. But in case of carbocation F_3C^+, lone pair of electrons on each of the three F-atoms overlap with the empty *p*-orbital of the carbon atom carrying the positive charge thereby dispersing the positive charge and thus stabilizing the carbocation, F_3C^+ relative to $F_3C—C^+$.

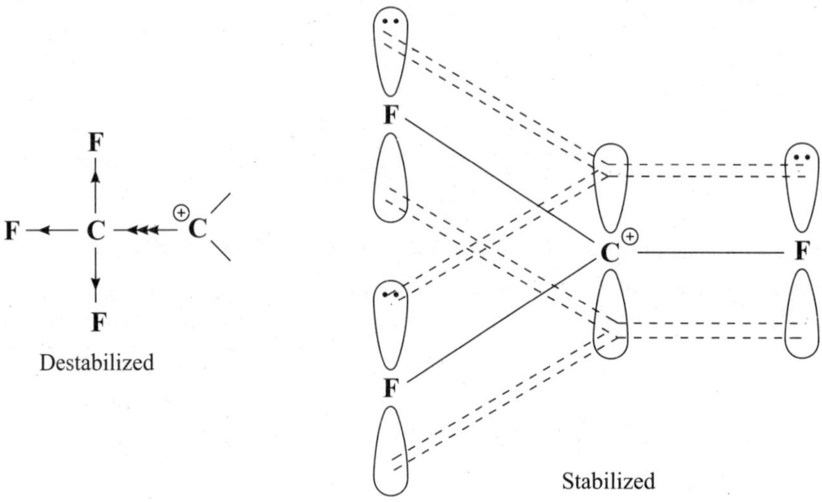

Q.2. Arrange the following in order of increasing stability and justify.

$$CH_3CH_2^-, CH \equiv C^-, CH_2 = CH^-$$

Ans. The carbon atom is *sp*-hybridized in acetylide ion ($CH \equiv C^-$) and has 50% *s*-character; the carbon atom is sp^2-hybridized in $CH_2 = \dot{C}H^-$ ion, and has 33.3% *s*-character while the carbon atom is sp^3-hybridized in $CH_3CH_2^-$ ion, and has 25% *s*-character.

We know that *s*-electrons, on an average, are closer to the nucleus than *p*-electrons, therefore, a carbon atom with greater *s*-character can accommodate or stabilize the negative charge better than a carbon atom with smaller *s*-character. In other words, the stability of the carbanion increases as the *s*-character of the carbon atom carrying the negative charge increases. Now since the *s*-character of the carbon decreases as we move from *sp* to sp^2 to sp^3-carbon, therefore, the relative stabilities of the three carbanions follows the sequence: $CH \equiv C^- > CH_2 = CH^- > CH_3CH_2^-$.

Q.3. Explain why a less stable carbocation can undergo rearrangement to a more stable carbocation, but a free radical does not?

Ans. We know that in both carbocations and free radicals, the central carbon atom is sp^2-hybridized. The unhybridized *p*-orbital is empty in carbocations but in free radicals it has one electron. During the rearrangement of carbocations, the two electrons of the migrating σ-bond can easily move into the empty *p*-orbital of catiomic carbon since in accordance with Pauli Exclusion Principle, an orbital at the maximum can accommodate two electrons. As a result, less stable carbocations get converted to more stable carbocations.

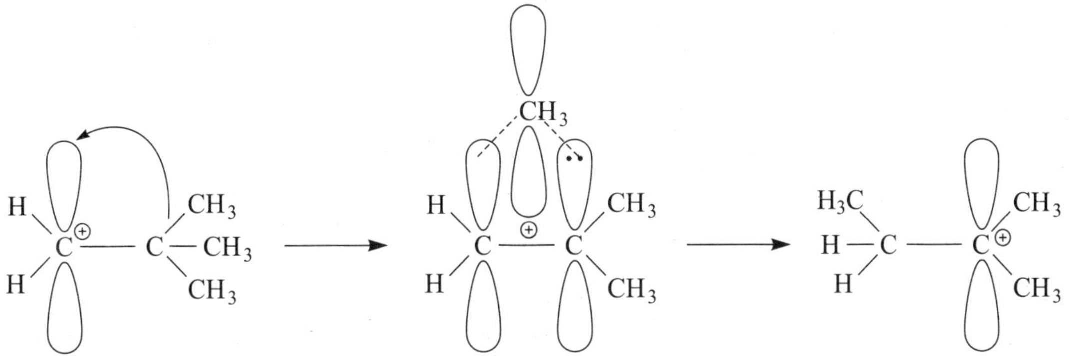

Less stable carbocation(1°)

Transition state for rearrangement

More stable carbocation (3°)

In contrast, in case of free radicals, the two electrons of the migrating σ-bond cannot move into *p*-orbital which already has one electron since by doing so the *p*-orbital will have three electrons which is against Pauli Exclusion Principle. If at all the rearrangement has to occur, the third electron has to go to the antibonding molecular orbital thereby making the rearrangement energetically unfavourable.

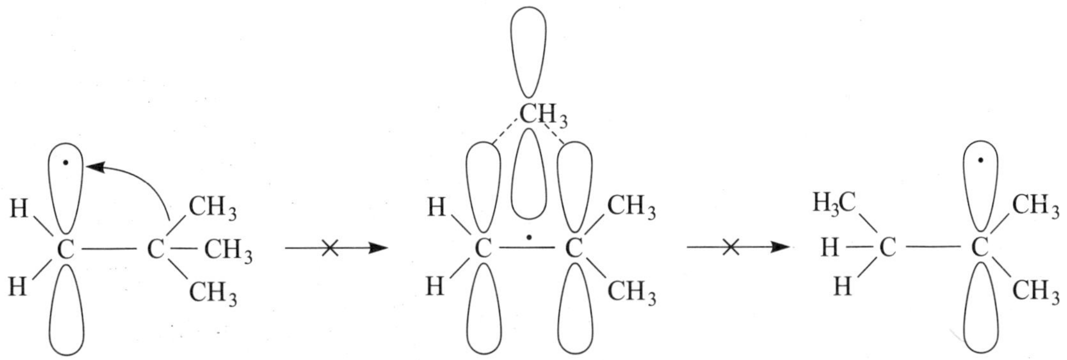

Less stable free radical (1°)

Transition state for rearrangement

More stable free radical (3°)

Therefore, we conclude that a carbocation rearrangement occurs because it is energetically favourable while free radical rearrangement does not occur since it is energetically unfavourable.

Q.4. Which of the following carbocations is more stable and why?

(*i*) $(CH_3)_3C^+$ and (*ii*) $(CD_3)_3C^+$

Ans. Both carbocations (*i*) and (*ii*) are stabilized by hyperconjugation as shown below:

$$\underset{\text{I}}{H_3C-\overset{\oplus}{C}(CH_3)_2} \longleftrightarrow \underset{\text{II}}{\overset{\oplus}{H}\;H_2C{=}C(CH_3)_2} \longleftrightarrow \underset{\text{III}}{\overset{\oplus}{H}\;H_2C{=}C(CH_3)_2} \longleftrightarrow \underset{\text{IV}}{\overset{\oplus}{H}\;H_2C{=}C(CH_3)_2} \longleftrightarrow \text{etc.}$$

$$\underset{\text{V}}{D_3C-\overset{\oplus}{C}(CD_3)_2} \longleftrightarrow \underset{\text{VI}}{\overset{\oplus}{D}\;D_2C{=}C(CD_3)_2} \longleftrightarrow \underset{\text{VII}}{\overset{\oplus}{D}\;D_2C{=}C(CD_3)_2} \longleftrightarrow \underset{\text{VIII}}{\overset{\oplus}{D}\;D_2C{=}C(CD_3)_2} \longleftrightarrow \text{etc.}$$

Contribution of structures (V–VIII) towards stability of carbocation, $(CD_3)_3C^+$ is less than those of structures (I–IV) for carbocation $(CH_3)_3C^+$ due to stronger C—D bond. Therefore, carbocation (*i*) is more stable than carbocation (*ii*) This effect is also sometimes called as ***β-secondary isotope effect.***

Q.5. Ammonium cyanate on heating gives urea. What type of reaction is this? Write its mechanism?

Ans. It is a rearrangement reaction, the mechanism of which is shown below:

$$H_3\overset{\oplus}{N}-H \quad \overset{\ominus}{O}-C\equiv N \longrightarrow H_3\ddot{N} + H-O-C\equiv N: \longrightarrow$$

$$HO-\underset{\underset{^{+}NH_3}{|}}{C}=\ddot{N}:^{\ominus} \xrightarrow[\text{transfer}]{\text{Proton}} H-O-\underset{\underset{NH_2}{|}}{C}=NH \xrightarrow{\text{Tautomerism}} O=\underset{\underset{NH_2}{|}}{C}-NH_2$$

Urea

UNSOLVED PROBLEMS

1. Define and illustrate the terms (*i*) electrophile, (*ii*) nucleophile, (*iii*) intermediate, and (*iv*) transition state.
2. In how many ways can fission of a covalent bond take place? Discuss briefly the conditions under which each type of fission occurs.
3. Give a brief account of the common type of organic reactions with suitable examples.
4. Propanol reacts with dry hydrogen chloride to produce *n*-propyl chloride and water,
   (*i*) Classify this reaction as substitution, addition, elimination or molecular rearrangement.
   (*ii*) Hydrogen chloride is a stronger acid than $CH_3CH_2CH_2{}^{\oplus}OH_2$. Will an acid-base reaction occur between propanol and HCI?
   (*iii*) What in your opinion is the intermediate in this reaction? If you think there is no intermediate, write.
5. (*i*) Differentiate between intermediate and transition state.
   (*ii*) State Hammond's postulate. Explain with a suitable example the application of Hammond's postulate in determining the shape and geometry of transition state.
6. Classify the following reactions as nucleophilic or electrophilic. In each case identify the reagent as electrophile or nucleophile:
   (*i*) $C_6H_6 + HNO_3 \longrightarrow C_6H_5NO_2 + H_2O$
   (*ii*) $(CH_3)_2C=CH_2 + HCI \longrightarrow (CH_3)_3\,CCI$

(*iii*) $(CH_3)_3CBr + H_2O \longrightarrow (CH_3)_3COH$

(*iv*) $CH_3CHO + HCN \longrightarrow CH_3CH(OH)CN$

7. Draw energy profile diagrams for the following:

(*i*) A one-step reaction.

(*ii*) A two-step reaction in which the first step is rate-determining.

(*iii*) A three-step reaction in which the second step is rate-determining.

(*iv*) A catalyzed and an uncatalyzed reaction.

8. With chosen examples, illustrate the terms kinetic and thermodynamic control of reactions.

9. What is the structure of a carbocation? Discuss the factors influencing the stability of carbocations. Listed below are pairs of carbocations which are interconvertible by a hydride shift. Tell for each pair which cation will have tendency to undergo rearrangement.

(*i*) $\underset{A}{(CH_3)_2CH\overset{\oplus}{C}H_2}$, $\underset{B}{(CH_3)_3\overset{\oplus}{C}}$

(*ii*) $\underset{A}{CH_3CH_2\overset{\oplus}{C}H_2}$, $\underset{B}{CH_3\overset{\oplus}{C}HCH_3}$

(*iii*) $\underset{A}{CH_3CH_2CH_2\overset{\oplus}{C}HCH_3}$, $\underset{B}{CH_3CH_2CH_2CH_2\overset{\oplus}{C}H_2}$

(*iv*) $\underset{A}{C_6H_5CH{=}CH-CH_2-\overset{\oplus}{C}H_2}$, $\underset{B}{C_6H_5CH{=}CH-\overset{\oplus}{C}H-CH_3}$

10. (*i*) Explain with suitable examples the role of the NMR spectroscopy in the detection of a carbocation.

(*ii*) Arrange the following carbocations in order of increasing stability:

$$C_6H_5\overset{\oplus}{C}H_2,\ (C_6H_5)_2\overset{\oplus}{C}H,\ (CH_3)_2\overset{\oplus}{C}H,\ (CH_3)_3\overset{\oplus}{C},\ CH_3\overset{\oplus}{C}H_2,\ \overset{\oplus}{C}H_3$$

11. When propyl fluoride and isopropyl fluoride are dissolved in SbF_5/SO_2 only isopropyl cation is formed. What is the relative stability of propyl and isopropyl cations ?

12. (*i*) What are carbanions? How are they formed? Comment upon the statement, "Carbanions can be either planar or pyramidal."

(*ii*) How can inductive and resonance effects be used to explain the stability of carbanions?

$$(ROOC)_2\overset{\ominus}{C}H,\ (CH_3)_3\overset{\ominus}{C},\ CH_3CO-\overset{\ominus}{C}H-CH_3$$

(*iii*) Carbanions in which the unshared pair of electrons is conjugated with a carbon-carbon, carbon-oxygen and carbon-nitrogen double bond are much more stable than the simple carbanions. Justify the statement by giving one example in each case.

13. (*i*) Discuss the planar and pyramidal structure of carbanions. Illustrate with suitable examples how these intermediates are detected?

(*ii*) Arrange the following carbanions in order of decreasing stability (most stable first):

$$\text{cyclopentadienyl anion}, (C_6H_5)_3\overset{\ominus}{C}, (CH_3)_3\overset{\ominus}{C}, (CH_3)_2\overset{\ominus}{C}H, CH_3\overset{\ominus}{C}H_2, \overset{\ominus}{C}H_3$$

14. Comment upon the structure of free radicals. How are they detected? Pick up the most stable and least stable radicals from the following:

$$\dot{C}H_3, (CH_3)_3\dot{C}, C_6H_5\dot{C}H_2, CH_3\dot{C}H_2, (C_6H_5)_3\dot{C}$$

15. What are free radicals? Discuss two methods of their generation. Comment upon the statement, "primary, secondary and tertiary" carbocations and free radicals follow the same order of stability but the former are prone to rearrangements and the latter are not.

16. Discuss briefly the various factors which affect the stability of (*i*) carbocations, (*ii*) carbanions, and (*iii*) free radicals.

17. Discuss the structure of carbenes, nitrenes and arynes. Illustrate the technique of 'trapping' for the detection of these intermediates.

18. What are enamines? Discuss briefly their synthetic utility.

19. Explain the average electron theory for assigning formal charges to the different atoms in a molecule. Explain your answer with proper justification.

20. Define and illustrate the term "isotope effect". In what way the knowledge of isotope effect is useful in ascertaining the mechanism if a reaction?

21. What is isotopic labelling? Discuss its importance in determining the mechanism of an organic reaction.

22. Illustrate the use of the following techniques in determining mechanism of reactions:

(*i*) Crossover experiments (*ii*) Stereochemical studies (*iii*) Product analysis.

23. (*i*) How are hyperconjugative effects used to explain stability of carbocations? Rearrange the carbocations in each of the following sets in order of stability (least stable first):

(*a*) $CH_3CH_2\overset{\oplus}{C}HCH_3$ (A); $CH_3CH_2CH_2\overset{\oplus}{C}H_2$ (B); $CH_3-\overset{\oplus}{\underset{CH_3}{\underset{|}{C}}}-CH_3$ (C)

(*b*) $(CH_3)_2CHCH_2\overset{\oplus}{C}H_2$ (A); $(CH_3)_2CH\overset{\oplus}{C}HCH_3$ (B); $(CH_3)_2\overset{\oplus}{C}CH_2CH_3$ (C)

(*ii*) Benzyl cation is said to be stabilized by resonance involving dispersal of charge on the benzene ring. Can you write a contributing from having positive charge on the ring at *meta* position?

24. Draw the resonance structures of benzyl free radical. Using curved arrows show how one structure is converted into the other.

25. Draw the resonance structures of the following carbocation. Using curved arrows show how one structure is converted into another.

$$H-\ddot{O}-C_6H_4-\overset{\oplus}{C}H_2$$

26. (*i*) What are electrophiles and nucleophiles? Classify the following as electrophiles and nucleophiles.

SO_3, $AlCl_3$, $C_2H_5NH_2$, NO_2^+, $CH_3CH_2^+$, $:CX_2$, $C_6H_5CH_2^-$, $(CH_3)_3\dot{C}$

(*ii*) What do you understand by electrophilic and nucleophilic reagents? Which of the two will attack a carbanion and why?

27. What are elimination reactions? What are their types? Give one example of each type.

28. What are rearrangement reactions? Discuss briefly the role of 1, 2-hydride and 1, 2-alkyl shifts in the rearrangement of carbocations.

29. Sketch the mechanism of the following reaction:

$$(CH_3)_2C(OH)-C(OH)(CH_3)_2 \xrightarrow[\text{dil. } H_2SO_4]{} (CH_3)_3\,C-COCH_3$$

30. How do rearrangement reactions differ from isomerization reactions? Illustrate your answer with suitable examples.

31. What is the state of hybridization of a carbocation, carbanion, free radical, carbene and benzyne?

32. (*i*) For which of the following mechanistic schemes is the following rate law applicable:

$$\frac{-d\,[CH_3CH_2Cl]}{dt} = k\,[CH_3CH_2Cl]\,[\overset{\ominus}{O}H]$$

(*a*) $CH_3CH_2Cl \xrightarrow{\text{Slow}} CH_3\overset{\oplus}{C}H_2 + \overset{\ominus}{Cl}$

$CH_3\overset{\oplus}{C}H_2 + \overset{\ominus}{O}H \xrightarrow{\text{Fast}} CH_3CH_2OH$

(*b*) $CH_3CH_2Cl + \overset{\ominus}{O}H \longrightarrow CH_3CH_2OH + \overset{\ominus}{Cl}$

(*ii*) For the reaction $CH_3CH_2CH_2CH_2Br + \overset{\ominus}{O}H \longrightarrow CH_3CH_2CH_2CH_2OH + \overset{\ominus}{Br}$ the kinetic rate law states that

$$\frac{-d\,[CH_3CH_2CH_2CH_2Br]}{dt} = k\,[CH_3CH_2CH_2CH_2Br]\,[\overset{\ominus}{O}H]$$

Predict the effect of following changes on the rate and rate constant of the reaction:

(*a*) Concentration of $CH_3CH_2CH_2CH_2Br$ is doubled.

(*b*) Concentration of OH^- is halved.

(*c*) The temperature is raised.

33. Comment upon the unusual stability of the following:

(*i*) Cyclopentadienyl anion and (*ii*) Triphenylmethyl anion.

34. Distinguish between carbanion and enolate ion.

35. How do you explain the acidic nature of (*i*) cyclopentadiene and (*ii*) nitromethane?

36. Explain why carbocations are always planar while carbanions and free radicals may be either planar or pyramidal?

37. How do hyperconjugation and resonance effects explain the stability of free radicals? Arrange the following free radicals in the increasing order of their stability (least stable first):

$$C_6H_5—CH—\dot{C}H{=}CH_2, \quad C_6H_5—\dot{C}H—CH_3, C_6H_5CH_2\dot{C}H_2, \quad C_6H_5\dot{C}H_2$$

38. Differentiate between singlet and triplet carbenes.

39. Explain why triplet carbenes are more stable than singlet carbenes?

40. How is dichlorocarbene generated? Name two reactions in which dichlorocarbene is a reactive intermediate.

41. Discuss the structure of benzyne. Why is the triple bond in benzyne much more reactive than that of acetylene?

42. The enamine derived from 2-methylcyclohexanone and pyrrolidine reacts with benzyl bromide to give 2-benzyl-6-methylcyclohexanone and not the isomeric compound 2-methyl-2-benzylcyclohexanone. Justify.

43. Write three resonating structures for azide ion and assign formal charge to all the nitrogen atoms in all the three resonating structures.

44. Beckmann's rearrangement involves *anti*-migration of the group. Explain.

45. In the nitration of benzene, the attack of NO_2^+ on the ring is a slow-step. Explain.

46. Primary isotope effect is quite useful in elucidating the mechanism of an organic reaction. Justify.

47. Differentiate between threshold energy and energy of activation.

48. Which of the following will generate the most stable carbocation and why?

$CH_3CH_2OH, CH_3CHOHCH_3, C_6H_5CHOHCH_3, (CH_3)_3COH$

49. Give one example of a reaction which is believed to occur through a nitrene intermediate.

50. Explain why base-catalysed dehydrohalogenation of *cis*-1, 2-dichloroethene occurs faster than *trans*-1, 2-dichloroethene?

SECTION II

CHAPTER 3

Stereochemistry of Organic Compounds

3.1 CONCEPT OF ISOMERISM

The term *isomerism* (Greek: *isos* = equal; *meros* = part) was coined by Berzelius to describe the relationship between two clearly different compounds having the same elemental composition. Such pairs of compounds differ in their physical and chemical properties and are called **isomers**. For example, ethyl alcohol (CH_3CH_2OH) and dimethyl ether (CH_3OCH_3) are isomers.

3.2 TYPES OF ISOMERISM

Isomerism is of the following two types:

1. **Structural or Constitutional Isomerism**
2. **Stereoisomerism.**

These two types are further sub-divided into various categories as shown in Fig. 3.1.

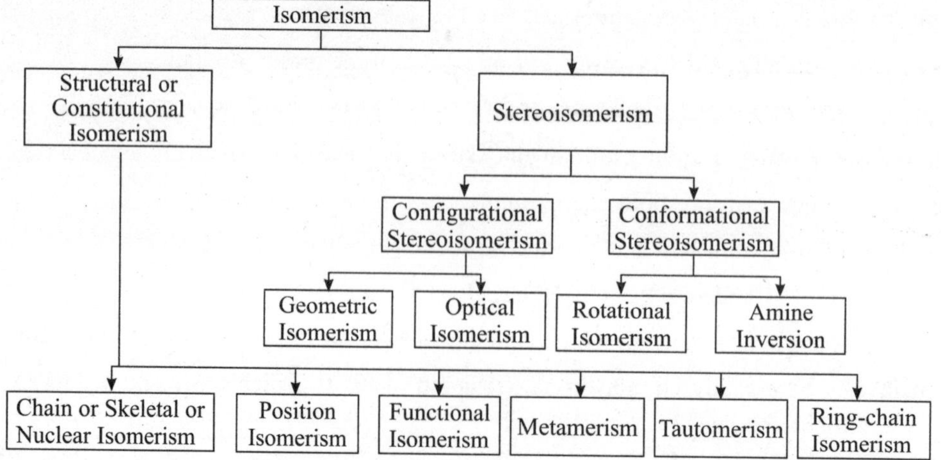

Fig. 3.1: Different types of isomerism of organic compounds

3.2.1 Structural or Constitutional Isomerism

Structural or constitutional isomers are those isomers which differ from each other in the way their atoms are connected, *i.e.*, in their structures. This phenomenon is called as **structural or constitutional isomerism.** Structural isomerism is further subdivided in following six types signifying the main difference in the structural features of the isomers.

I. Chain or Skeletal or Nuclear Isomerism

Chain or skeletal or nuclear isomers are the compounds having same molecular formula but different arrangement of carbon chain within the molecule and the phenomenon is known as **chain or skeletal or nuclear isomerism**. Following examples are illustrative:

(*i*) Molecular formula C_4H_{10} represents two chain isomers: *n*-Butane (having straight chain) and *iso*butane (having branched chain).

$H_3C—CH_2—CH_2—CH_3$

n-Butane
(*straight chain*)

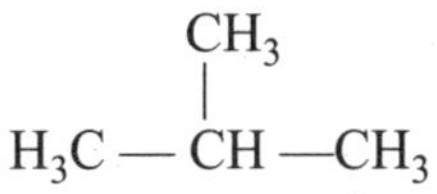

$$\begin{array}{c} \quad CH_3 \\ \quad | \\ H_3C—CH—CH_3 \end{array}$$

2-Methylpropane (*Iso*-butane)
(*branched chain*)

(*ii*) Further, molecular formula C_5H_{12} represents three isomers, one straight chain and two branched chain isomers, *e.g.*, *n*-pentane (straight chain), *iso*-pentane and *neo*-pentane (both branched chain).

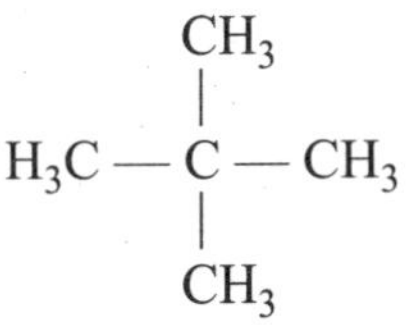

$$\begin{array}{c} CH_3 \\ | \\ H_3C—C—CH_3 \\ | \\ CH_3 \end{array}$$

2, 2-Dimethylpropane
(*neo*-pentane)

$CH_3CH_2CH_2CH_2CH_3$

n-Pentane

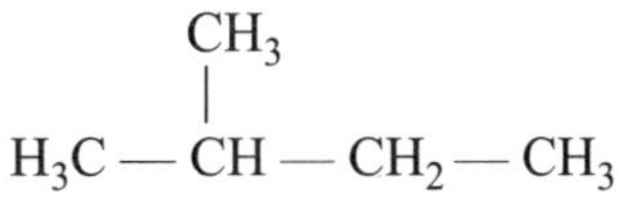

$$\begin{array}{c} CH_3 \qquad\qquad \\ | \qquad\qquad \\ H_3C—CH—CH_2—CH_3 \end{array}$$

2-Methylbutane
(*iso*-pentane)

It may be worthwhile to mention here that this type of isomerism is not confined to hydrocarbons alone. For example, formula $C_4H_{10}O$ represents three types of alcohols which are chain or skeletal isomers of each other.

$CH_3CH_2CH_2CH_2OH$

n-Butyl alcohol
(*straight chain*)

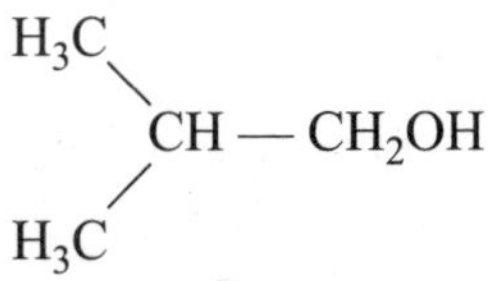

$$(H_3C)_2CH—CH_2OH$$

Iso-butyl alcohol
(*branched chain*)

$$\begin{array}{c} CH_3 \\ | \\ H_3C—C—OH \\ | \\ CH_3 \end{array}$$

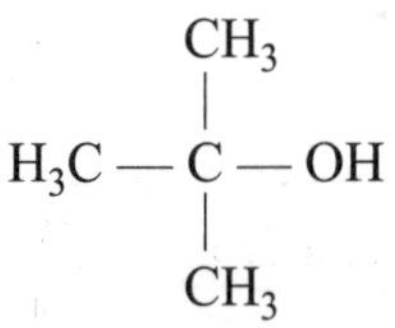

Tert-butyl alcohol
(*branched chain*)

II. Position Isomerism

Position isomers are those compounds which have the same molecular formula and same carbon skeleton but differ in the position of attached atoms or groups or in position of multiple the (double or triple) bonds. The phenomenon is known as position isomerism. For example,

(a) Different positions of functional groups

(*i*) Molecular formula C_3H_8O represents two position isomers.

$CH_3CH_2CH_2OH$ — Propan-1-ol (The OH group at C_1)

$CH_3—CH(OH)—CH_3$ — Propan-2-ol (The OH group at C_2)

(*ii*) Molecular formula $C_6H_4Cl_2$ represents three position isomers.

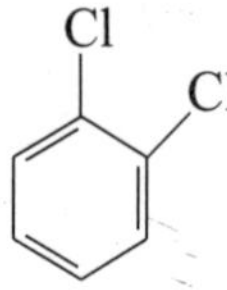

o-Dichlorobenzene

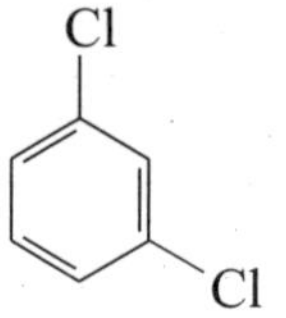

m-Dichlorobenzene

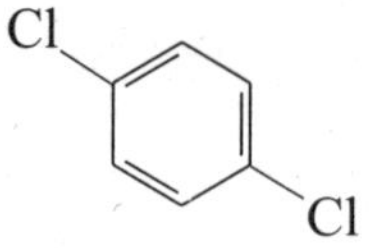

p-Dichlorobenzene

(b) Different positions of multiple bonds

(*i*) Molecular formula C_4H_8 represents two position isomers.

$\overset{1}{H_3C}—\overset{2}{C}H=\overset{3}{C}H—\overset{4}{CH_3}$

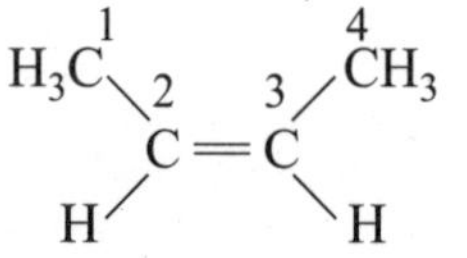

But-2-ene (*Double bond at C_2*)

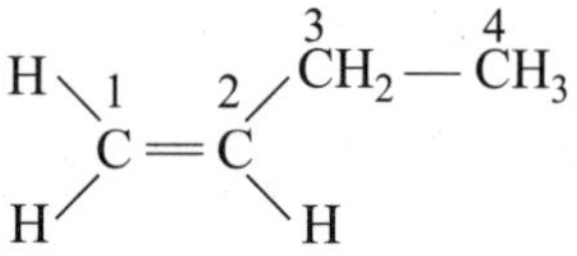

$H_2\overset{1}{C}=\overset{2}{C}H—\overset{3}{CH_2}—\overset{4}{CH_3}$

But-1-ene (*Double bond at C_1*)

(*ii*) Molecular formula C_4H_8 represents two position isomers.

$\overset{4}{C}H_3\overset{3}{C}H_2\overset{2}{C}\equiv\overset{1}{C}H$

But-1-yne (*Triple bond at C_1*)

$\overset{4}{CH_3}—\overset{3}{C}\equiv\overset{2}{C}—\overset{1}{CH_3}$

But-2-yne (*Triple bond at C_2*)

III. Functional Isomerism

Functional isomers are the compounds having same molecular formula but different functional groups. This phenomenon is known as **functional isomerism**. For example, the molecular formula C_2H_6O represents ethanol (an alcohol) and dimethyl ether (an ether) which have different functional groups.

CH_3CH_2OH — Ethanol

$CH_3—O—CH_3$ — Dimethyl ether

Some other examples of functional isomerism are as under:

***(a) Carboxylic acids and esters*:** Molecular formula $C_2H_4O_2$ represents ethanoic acid (a carboxylic acid) and methyl methanoate (an ester).

$$CH_3—\overset{\overset{\large O}{\|}}{C}—OH \qquad\qquad H—\overset{\overset{\large O}{\|}}{C}—OCH_3$$

Ethanoic acid — Methyl methanoate

***(b) Ketones and aldehydes*:** Molecular formula C_3H_6O represents acetone (a ketone) and propanal (an aldehyde).

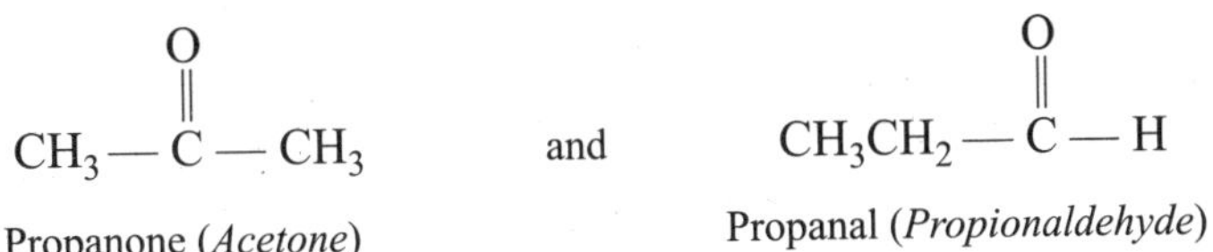

Propanone (*Acetone*) — Propanal (*Propionaldehyde*)

***(c) Primary, secondary and tertiary amines*:** Molecular formula C_3H_9N represents

$$CH_3CH_2CH_2NH_2, \qquad CH_3CH_2—N\begin{matrix}\diagup CH_3\\ \diagdown H\end{matrix} \qquad \text{and} \qquad H_3C—\overset{\overset{\large CH_3}{|}}{N}—CH_3$$

Propan-1-amine (1° *Amine*) — N-Methylethanamine (2° *Amine*) — N, N-Dimethylmethanamine (3° *Amine*)

***(d) Cyanides and isocyanides*:** Molecular formula C_2H_3N represents

$$CH_3—C \equiv N \qquad \text{and} \qquad CH_3—N \equiv C$$

Ethanenitrile — Methyl isocyanide

***(e) Dienes and alkynes*:** Molecular formula C_4H_6 represents

$$CH_2=CH—CH=CH_2 \qquad \text{or} \qquad CH_3—CH=C=CH_2$$

But-1, 3-diene (*an alkene*) — But-1, 2-diene (*an alkene*)

$$CH_3CH_2—C \equiv CH \qquad \text{or} \qquad CH_3—C \equiv C—CH_3$$

But-1-yne (*an alkyne*) — But-2-yne (*an alkyne*)

***(f) Nitroalkanes and alkyl nitrites*:** Molecular formula $C_2H_5NO_2$ represents

$$CH_3—CH_2—NO_2 \qquad \text{and} \qquad CH_3—CH_2—O—NO$$

Nitroethane — Ethyl nitrite

(*g*) ***Aromatic alcohols, phenols and ethers*:** Molecular formula C_7H_8O represents

$C_6H_5CH_2OH$ — Benzyl alcohol (*an alcohol*)

$C_6H_5OCH_3$ — Anisole (*an ether*)

o-$CH_3C_6H_4OH$ — *o*-Cresol (*a phenol*) or m-$CH_3C_6H_4OH$ — *m*-Cresol (*a phenol*) or p-$CH_3C_6H_4OH$ — *p*-Cresol (*a phenol*)

Here it may be worthwhile to mention that *o*-cresol, *m*-cresol and *p*-cresol are also position isomers.

IV. Metamerism

The isomers having different number of carbon atoms (or alkyl groups) on either side of a bifunctional group (*i.e.*, —O— , —S—, —NH—, —CO— etc.) are called **metamers**. This phenomenon is called **metamerism**. Metamerism is exhibited by members of the same family, *i.e.*, same functional groups.

Some examples of compounds showing metamerism are as follows:

(***a***) Molecular formula $C_4H_{10}O$ represents following three ethers:

$CH_3CH_2—O—CH_2CH_3$ is a metamer of $CH_3—O—CH_2CH_2CH_3$ or $CH_3—O—CH(CH_3)—CH_3$

Ethoxyethane (*Diethyl ether*); 1-Methoxypropane (*Methyl n-propyl ether*); 2-Methoxypropane (*Isopropyl methyl ether*)

(***b***) Molecular formula $C_5H_{10}O$ represents the following three ketones:

$CH_3CH_2—C(=O)—CH_2CH_3$ is a metamer of $CH_3—C(=O)—CH_2CH_2CH_3$ or $CH_3—C(=O)—CH(CH_3)—CH_3$

Pentan-3-one (*Diethyl ketone*); Pentan-2-one (*Methyl n-propyl ketone*); 3-Methylbutan-2-one (*Isopropyl methyl ketone*)

It may be mentioned here that some compounds exhibit both position isomerism as well as metamerism. For example, pentan-2-one and pentane-3-one are position isomers as well as metamers.

V. Tautomerism

Structural or constitutional isomers which exist in easy and rapid equilibrium by migration of an atom or group are called tautomers and their equilibrium is called **tautomerism**.

The most common type of tautomerism involving structures which differ in the point of attachment of hydrogen, is **keto-enol tautomerism**, *i.e.*, a rapid equilibrium between one tautomer having a *keto* group (>C = O) and the other having an *enol* ($-\overset{|}{C}=\overset{|}{C}-OH$) group. For example,

$$CH_3-\overset{\overset{O}{\|}}{C}-H \rightleftharpoons CH_2=\overset{\overset{OH}{|}}{C}-H$$

Acetaldehyde (*keto form*) — Vinyl alcohol (*enol form*) (*Negligible amount*)

$$CH_3-\overset{\overset{O}{\|}}{C}-CH_3 \rightleftharpoons CH_2=\overset{\overset{OH}{|}}{C}-CH_3$$

Acetone (*keto form*) — Prop-1-ene-2-ol (*enol form*) (*Negligible amount*)

As shown above aldehydes and ketones such as acetaldehyde and acetone, respectively, exist mainly in keto form with negligible amount of enol form.

However, in those compounds where the enol form can be stabilized by intramolecular hydrogen bonding (also called as *chelation*) the amount of enol form increases. Some examples are as follows:

(*i*) Acetoacetic ester exists as an equilibrium mixture containing 93% keto and 7% enol form.

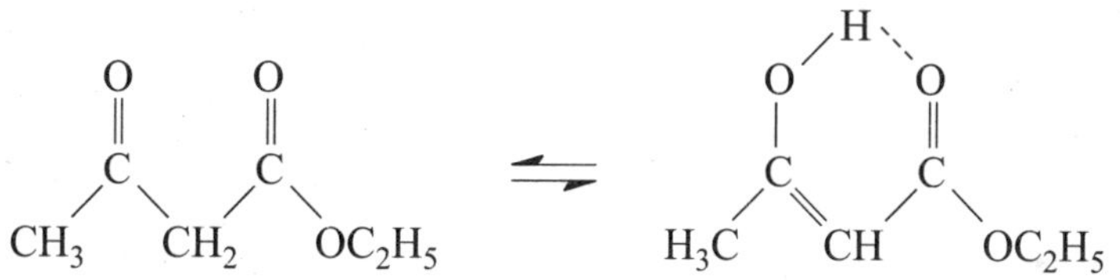

$$CH_3-\overset{\overset{O}{\|}}{C}-CH_2-\overset{\overset{O}{\|}}{C}-OC_2H_5 \rightleftharpoons H_3C-\overset{\overset{O-H}{|}}{C}=CH-\overset{\overset{O\cdots}{\|}}{C}-OC_2H_5$$

Keto form (93%) — Enol form (7%)

(*ii*) Acetylacetone exists as an equilibrium mixture containing 24% keto form and 76% enol form. This may be due to the fact that the keto carbonyl group in this compound is a strong electron withdrawing group than the ester carbonyl.

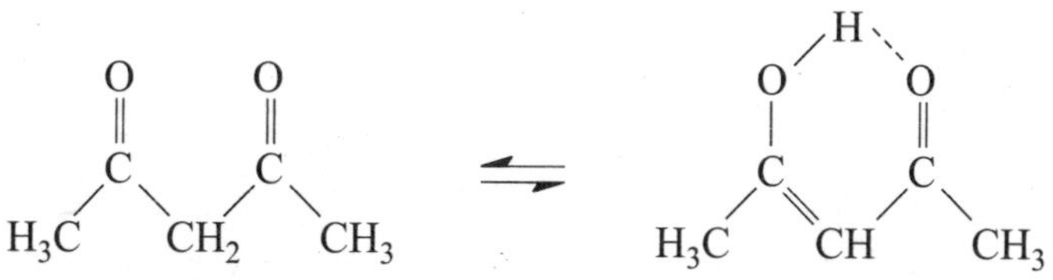

$$H_3C-\overset{\overset{O}{\|}}{C}-CH_2-\overset{\overset{O}{\|}}{C}-CH_3 \rightleftharpoons H_3C-\overset{\overset{O-H}{|}}{C}=CH-\overset{\overset{O\cdots}{\|}}{C}-CH_3$$

Keto form (24%) — Enol form (76%)

Necessary and sufficient condition for a compound to exhibit keto-enol tautomerism

Only those carbonyl compounds which contain atleast one α-hydrogen exhibit keto-enol tautomerism. For example, acetophenone, butanone, propionaldehyde which have α-hydrogen exhibit keto-enol tautomerism whereas those having no α-hydrogen such as benzaldehyde and benzophenone do not.

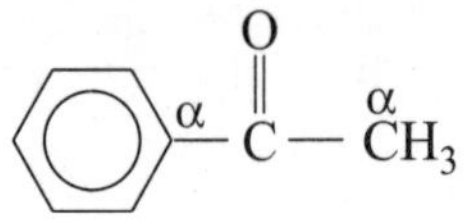

$C_6H_5 \overset{\alpha}{-} \overset{O}{\overset{\|}{C}} - \overset{\alpha}{C}H_3$

Acetophenone

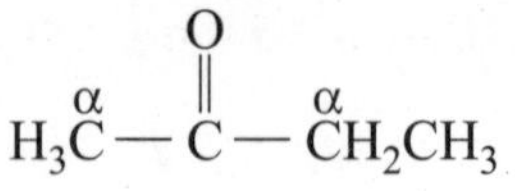

$H_3\overset{\alpha}{C} - \overset{O}{\overset{\|}{C}} - \overset{\alpha}{C}H_2CH_3$

Butan-2-one

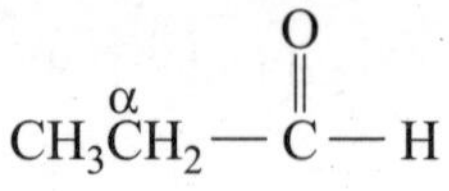

$CH_3\overset{\alpha}{C}H_2 - \overset{O}{\overset{\|}{C}} - H$

Propionaldehyde

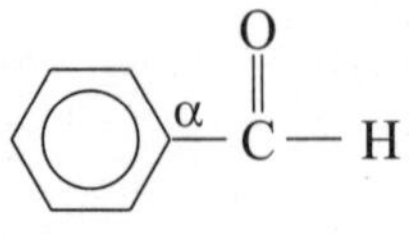

$C_6H_5 \overset{\alpha}{-} \overset{O}{\overset{\|}{C}} - H$

Benzaldehyde

$C_6H_5 \overset{\alpha}{-} \overset{O}{\overset{\|}{C}} \overset{\alpha}{-} C_6H_5$

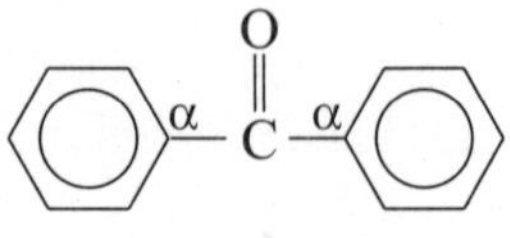

Benzophenone

Some compounds other than carbonyl derivatives also exhibit tautomerism. For example, nitromethane exists in the following equilibrium.

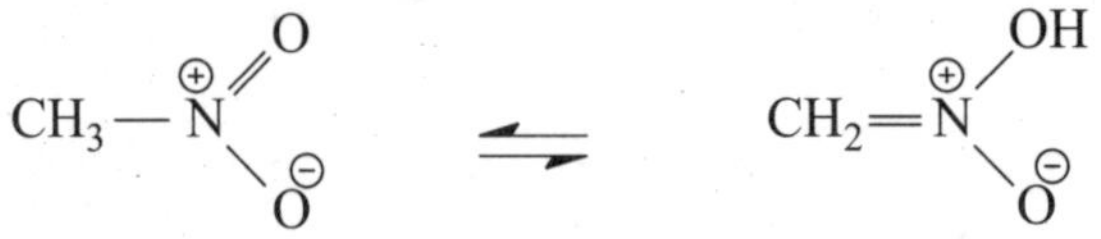

$$CH_3-\overset{\oplus}{N}(=O)-O^{\ominus} \rightleftharpoons CH_2=\overset{\oplus}{N}(-OH)-O^{\ominus}$$

Nitro form Acinitro form

This type of tautomerism is called *nitro-acinitro tautomerism.*

VI. Ring-chain Isomerism

Open chain and cyclic compounds having the same molecular formula are called **ring-chain isomers** and the phenomenon is known as **ring-chain isomerism**. For example,

(*a*) Molecular formula C_3H_6 represents propene as well as cyclopropane both of which are called ring-chain isomers.

$CH_3CH=CH_2$ and [cyclopropane]

Propene Cyclopropane

(*b*) Molecular formula C_3H_4 represents two compounds, propyne and cyclopropene both of which are ring-chain isomers.

$CH_3-C\equiv CH$ and [cyclopropene]

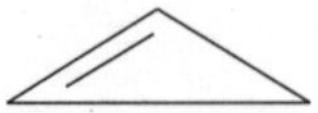

Propyne Cyclopropene

Double Bond Equivalent (DBE) or Index of Hydrogen Deficiency (IHD)

Double bond equivalent (DBE) is also called as **index of hydrogen deficiency (IHD)** and is of great utility in solving structural problems particularly structural assignments to isomers. It tells us about the number of double bonds or rings present in the molecule.

DBE (or IHD) is calculated from the expression: $\text{DBE} = \dfrac{\Sigma n(v-2)}{2} + 1$

where n = number of different kinds of atoms present

v = valency of each atom

For example, DBE (or IHD) for molecular formula C_3H_6O

$$= \frac{3(4-2)+6(1-2)+1(2-2)}{2}+1=0+1=1$$

Thus molecules having molecular formula C_3H_6O will have either one double bond or one ring.

Now if DBE (IHD) of a molecule is 2 it means that the molecule has two double bonds or one triple bond or two rings or one double bond and one ring.

3.2.2 Stereoisomerism

Isomers which have the same molecular formula and same structural formula but differ in the manner their atoms or groups are arranged in the space are called **stereoisomers** and the phenomenon is called **stereoisomerism.**

Stereoisomerism is further subdivided into the following two categories:

I Configurational Isomerism.

II. Conformational Isomerism.

I. Configurational Isomerism

The stereoisomers which cannot be interconverted unless a covalent bond is broken are called configurational isomers. This type of isomerism is called **configurational isomerism**. These isomers can be separated under normal conditions.

The configurational isomerism is again of two types:

(*a*) *Optical Isomerism or Enantiomerism.*

(*b*) *Geometrical Isomerism.*

(*a*) *Optical isomerism or enantiomerism*: The stereoisomers which are related to each other as an object and its non-superimposable mirror image are called *optical isomers* or *enantiomers* (Greek: **enantion** means **opposite**). This type of isomerism is called ***optical isomerism or enantiomerism.***

The optical isomers can also rotate the plane of polarised light to an equal degree but in opposite direction. *The property of rotating plane of polarised light is known as* ***optical activity****. The optical isomers have similar physical and chemical properties.* For example, molecular formula $C_2H_6O_3$ represents two enantiomeric lactic acids as shown below:

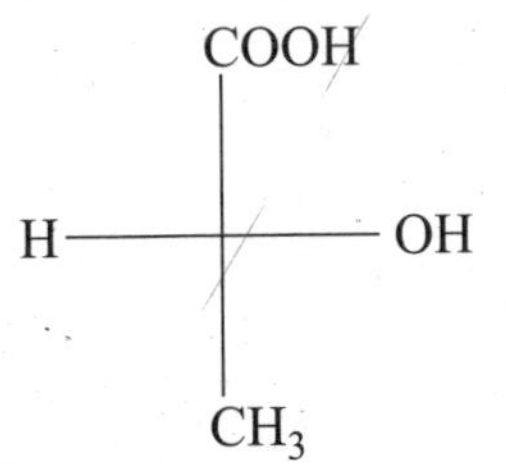

(–)Lactic acid
(Rotates the plane of polarized light towards left hand side i.e., anticlockwise)

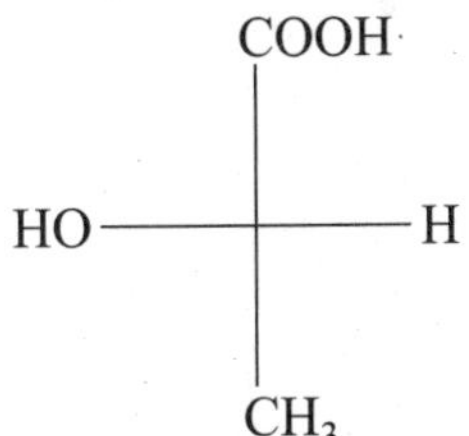

(+)Lactic acid
(Rotates the plane of polarized light towards right hand side i.e., clockwise)

(b) Geometrical isomerism*:** Geometric isomers are the stereoisomers which differ in their spatial geometry due to restricted rotation across a double bond. These isomers are also called as ***cis-trans isomers. This type of isomerism is called as ***geometrical isomerism*** or ***cis-trans isomerism***. For example, molecular formula $C_2H_2Cl_2$ corresponds to two geometric isomers as follows:

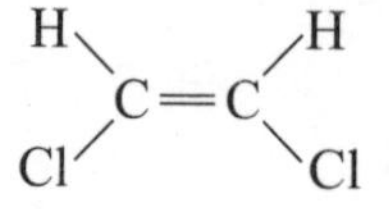

cis-1, 2-Dichloroethene 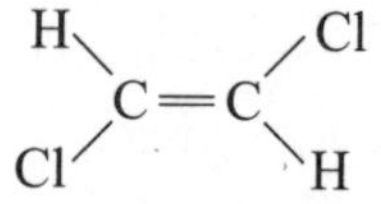 *trans*-1, 2-Dichloroethene

II. Conformational Isomerism

The stereoisomers which can be interconverted rapidly at room temperature without breaking a covalent bond are called **conformational isomers** or **conformers**. This type of isomerism is called **conformational isomerism**. Because such isomers can be readily interconverted, they cannot be separated under normal conditions.

There are the following two types of conformational isomers:

***(a) Conformational isomers resulting from rotation about single bond*:** Because the single bond in a molecule rotates continuously, the compounds containing single bonds have many interconvertible conformational isomers. For example, '*boat*' and '*chair*' forms of cyclohexane.

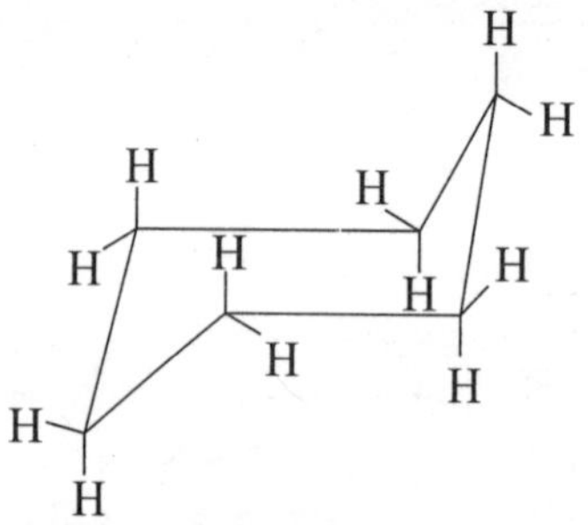

Cyclohexane (*Chair form*) Cyclohexane (*Boat form*)

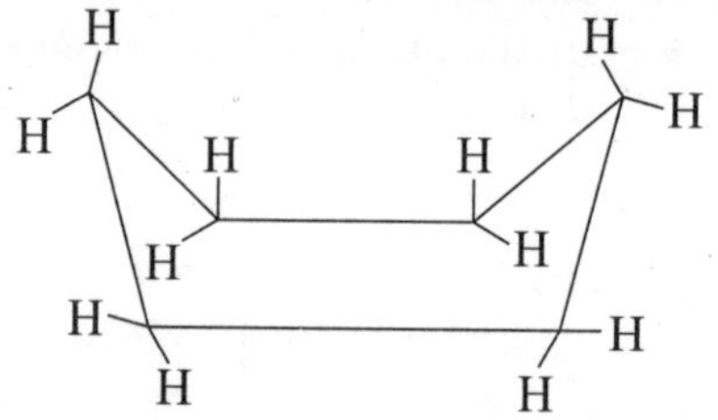

***(b) Conformational isomers arising from amine inversion*:** This type of conformational isomers exist because nitrogen atom of amines has a pair of non-bonding electrons which allow the molecule to turn "*inside out*" rapidly at room temperature. This is called **amine inversion** or **Walden inversion**.

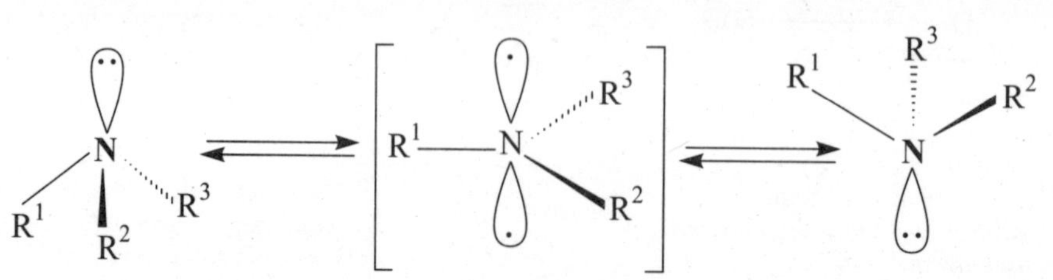

Transition state

3.3 OPTICAL ACTIVITY

Enantiomers are known to possess same physical and chemical properties but they differ from each other in the way they interact with **plane polarised light**.

What is Plane Polarised Light?

Normal light is composed of electromagnetic waves of different wavelengths which vibrate in all possible planes perpendicular to the direction of propagation. This is also true for the monochromatic light. (Monochromatic light is produced by passing the normal light either through a prism or a grating or using a source which emits light of single wavelength).

The above type of vibrations can be made to occur only in a single plane (*i.e.*, plolarised by cutting off all the vibrations occurring in other planes), by passing the beam of light through a **Nicol prism** (made of calcite, $CaCO_3$) or a **polarising lens**. Both these devices are known as polariser and such a light is called **plane polarised light**. This polarisation of light can be detected by viewing it through another **Nicol prism** or **analysing lens** which are known as analysers. When the polariser and analyser are placed in parallel positions, the polarised light emerges from analyser with undiminished intensity. However, when polariser and analyser are placed at right angles to each other, no light emerges from the analyser.

Now when a solution of a chiral compound, in suitable solvent is placed between the polariser and analyser which are kept in parallel positions, the intensity of the emergent light is diminished. This lowering of intensity of light emerging from the analyser is attributed to the fact that the plane of polarised light has rotated in passing through the chiral substance. It has been shown that such a rotation can be either in clockwise or anti-clockwise direction. If one enantiomer rotates the plane of polarised light in a clockwise direction, the other will rotate it by exactly the same amount but in anti-clockwise direction.

Substances which can rotate the plane of polarised light are said to the **optically active**. When such a substance rotates the plane of polarised light in a clockwise direction or to the right, it is called **dextrorotatory** (Latin: **dextre** means **right**) and is indicated by **(+)** sign. Whereas, the one which rotates it in an anti-clockwise (or counter clockwise) direction or to the left is called **laevorotatory** (Latin: **laeves** mean **left**) and is indicated by **(–)** sign. Those substance which do not rotate the plane of polarised light are called **optically inactive**.

The magnitude of rotation of plane polarised light by an optically active substance is measured by an instrument called **polarimeter** (Fig. 3.2).

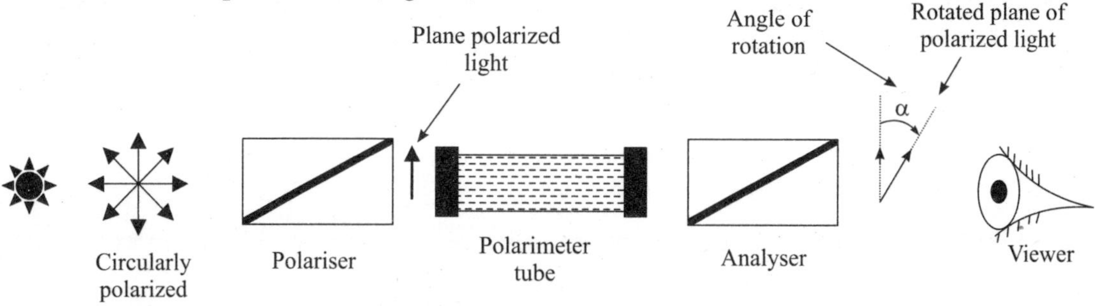

Fig. 3.2: A schematic representation of polarimeter

The essential parts of a typical polarimeter and its working are discussed below:

Parts of the polarimeter: A polarimeter consists of the following parts:

1. Light source, 2. Polariser, 3. Polarimeter tube, and 4. Analyser.

1. *Light source*: It is usually a sodium lamp which emits monochromatic light having $\lambda = 589$ nm.

2. *Polariser*: It is either a nicol prism or a polarising lens and is placed next to the light source.

3. *Polarimeter tube*: It is a glass tube and is filled with the solution of the sample to be measured or the neat liquid as the case may be and is placed in between polariser and analyser.

4. *Analyser*: It is also either a nicol prism or a polarising lens and is placed next to the polarimeter tube.

All the above parts are set in a line as shown in Fig. 3.2.

Working

Initially the empty polarimeter tube is placed between the polariser and the analyser and the light source is switched on. The polariser and analyser are then set in a position to obtain maximum intensity of emergent light. The polarimeter tube is now filled with the sample to be measured and placed in position. The intensity of emergent light is found to be diminished when viewed through the analyser. The analyser is then rotated either to the right or to the left until the intensity of emergent light again becomes maximum.

The angle by which the analyser has to be rotated to obtain maximum intensity of emergent light is called **angle of rotation**. It is denoted by α and is measured in degrees. The value of α depends upon the following factors: (*i*) Nature of the substance; (*ii*) Concentration of the solution in g/mL; (*iii*) Length of the polarimeter tube; (*iv*) Wavelength of the incident monochromatic light and (*v*) Temperature of the sample. It means that the value of α will vary with variation in any of these factors.

Specific Rotation

For quantitative studies, the optical activity is generally reported in terms of specific rotation which is denoted by the symbol $[\alpha]$. **Specific rotation** *is defined as the number of degrees of rotation caused by a solution of 1.0 g of compound per mL of solution taken in a polarimeter tube 1.0 dm (10 cm) long at a specific temperature and wavelength.*

The specific rotation is calculated from observed angle of rotation $[\alpha]$ by using the following expression; $[\alpha]_{\lambda}^{t^{\circ}} = \dfrac{\alpha}{l \times c}$

Where $[\alpha]$ = specific rotation; t° = temperature of the sample; λ = wavelength of incident light (where sodium D-line is used λ is replaced by D); α = observed angle of rotation; l = length of the polarimeter tube in decimeters; C = concentration of sample in g/mL of the solution. For example, when a compound is reported to be having $[\alpha]_{D}$293 K = + 40° (C = 0.2 g/mL; water), it means that:

(*i*) The substance is dextrorotatory (+) sign.

(*ii*) It has a specific rotation of 40°.

(*iii*) The measurements are made at 293 K, using sodium D line ($\lambda = 589$ nm) by taking the solution in a polarimeter tube which is one decimeter long.

(*iv*) The concentration of the solution used is 0.2g/mL *i.e.*, 20% (w/v) in water.

For **neat liquids** the specific rotation is calculated by using the following equation:

$$[\alpha]_{\lambda}^{t^{o}} = \frac{\alpha}{l \times d}$$

where all other symbols have the same meaning and d = density of the liquid (g/mL).

Molecular Rotation

Some times the optical activity of a given compound is also reported in terms of molecular rotation which is preferred over specific rotation. It is calculated from specific rotation by using the following formula:

$$[M]_{\lambda}^{t^{o}} = \frac{[\alpha]_{\lambda}^{t^{o}} \times M}{100}$$

where M = molecular weight of the optically active substance.

Utility of specific rotation: Just like other physical constants such as melting point, boiling point, density, refractive index, etc., the specific rotation is also a characteristic property for establishing the identity of a given optically active compound. It is an **intensive property**.

3.3.1 Chirality-Optical Activity: Discovery

As early as 1848, French chemist Louis Pasteur discovered that crystalline optically inactive sodium ammonium tartarate was a mixture of two different kinds of crystals which were mirror images of each other. On separation of these two kinds of crystals mechanically, he found that although the original mixture was optically inactive, each set of crystals when dissolved in water was optically active. The specific rotations of the two solutions were exactly equal, but of opposite sign. In other words, one solution rotated plane polarised light to the right and the other to the left to the same extent. In all other properties, the two substances were identical. Since the difference in rotation was found in the two samples in solution in which shapes of crystals disappear, Pasteur laid the foundation of stereochemisty when he proposed that like the two sets of crystals, the molecules making up the crystals were themselves mirror-images of each other and the difference in rotation was due to **'molecular dissymmetry'**.

Such substances which can rotate the plane of polarised light are called ***optically active substances*** *and the property of a substance to rotate the plane of polarised light is called* ***optical activity***.

A substance which rotates the plane of polarised light towards right (in the clockwise direction) is called dextrorotatory (Latin: **dextre = right**) and is usually represented by the prefix **'d'** or **'(+)'**. On the other hand, a substance which rotates the plane of polarised light towards left (in the anticlockwise direction) is called laevorotatory (Latin: **laevo = left**) and is usually represented by the prefix **'l'** or **'(–)'**.

3.3.2 Chirality

A hand cannot be superimposed on its mirror image. If you hold your left hand in front of a mirror, the image looks like a right hand. When you try to superimpose your hands (both palms down) you simply can not do it (Fig. 3.3). Have you ever thought why you cannot wear a left handed glove on a right hand?

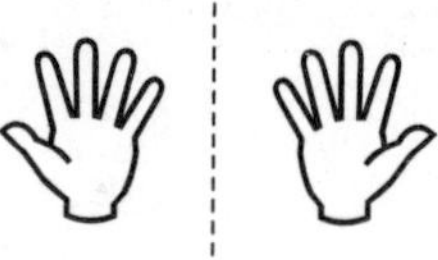

Fig. 3.3: Non-superimposable mirror image-chiral object (both hands palms down)

Such an object which cannot be superimposed on its mirror-image is said to be chiral (ky - ral) (Greek : **Cheir** 'Handedness') and the property of non-superimposability is called **chirality**. Thus, our hands are chiral.

Similarly, alphabets R, F, J are chiral and A, M, O are achiral (Fig. 3.4).

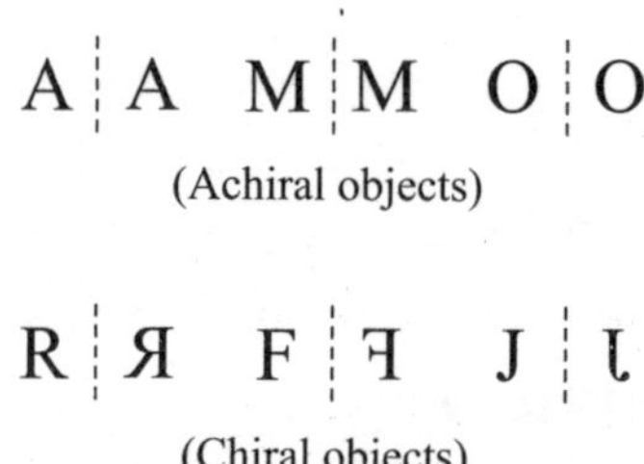

Fig. 3.4: Chiral and achiral objects

On the other hand, a molecule (or an object) which is superimposable on its mirror image is called **achiral** (non-dissymmetric or symmetric).

We can also say that objects having a plane of symmetry are achiral, whereas objects not having a plane of symmetry are usually chiral. Some examples are:

Chiral objects: Human hand, gloves, shoes, etc.

Achiral objects: A sphere, a cube, a button, socks without thumb, etc.

3.3.3 Molecular Chirality and Asymmetric Carbon

It has been shown that there is a close relationship between *chirality* and *optical activity*, and quite often the descriptions *optically active* and *chiral* are used interchangeably. It should, however, be remembered that optical activity refers to one property of chiral molecules, namely the ability to rotate the plane of polarised light.

Chirality in molecules is usually due to the presence of an sp^3 carbon atom with four different groups attached to it. Such a carbon atom is called a **chiral carbon** or a **chirality centre**. The presence of a chirality centre usually leads to **molecular chirality**. Such a molecule has no plane of symmetry and exists as a pair of enantiomers. Such a carbon atom is sometimes also referred to as asymmetric carbon atom.

It is now known through electron diffraction, X-ray diffraction and spectroscopic studies supported by quantum mechanical considerations that the four valencies of sp^3-hybridized carbon are directed towards the four corners of a regular tetrahedron with carbon atom at the centre. Let us construct a ball and stick model of a compound C_{wxyz}, a derivative of methane, where w, x, y and z are all different atoms or groups. We also construct a model of its mirror image. Thus,

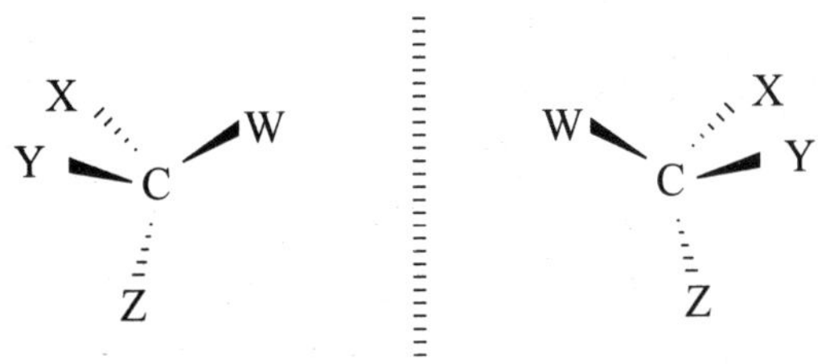

Mirror

We may twist and turn the above two representations in any way we like so long as we do not break any bond, yet we find that the two are not superimposable. Therefore, they must represent two isomers, *i.e.*, two enantiomers.

Although we have distinguished the two enantiomers of C_{wxyz} with the help of the above picture of a three-dimensional model to show the tetrahedral arrangement of the groups attached with the chiral carbon, it is not covenient to draw such picture in all the cases. It, therefore, became necessary to have a simpler convention for distinguishing between enantiomers. The following projection formulas (Fischer projection formulas) represent the two enantiomers in two dimensions with the assumption that the two horizontal bonds (C—Y and C—W) project towards us out of the plane of the paper, and the two vertical bonds (C—X and C—Z) project away from us behind the paper. The superimposability of two such flat two-dimensional structures is tested by rotating end to end without raising them (in our mind) out of the plane of the paper. The asymmetric carbon atom is sometimes not inserted in such projection formulas in which case it is assumed that it is located at the junction of the crossed lines.

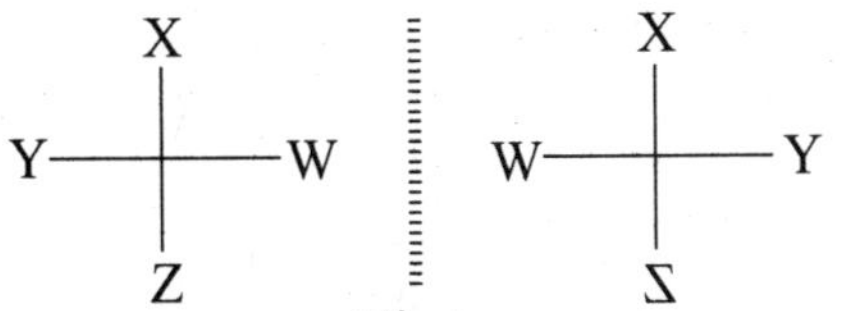

Mirror

The dextrorotatory (+) and laevorotatory (–) glyceraldehydes and lactic acids are simple examples to illustrate the non-superimposability of the enantiomers, as shown on next page.

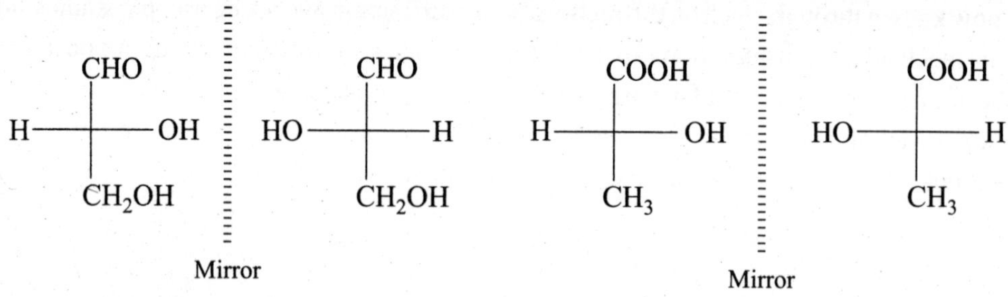

Fischer projections (mirror line between each pair):

- Glyceraldehyde, left: CHO at top, H on the left, OH on the right, CH_2OH at the bottom.
- Glyceraldehyde, right: CHO at top, HO on the left, H on the right, CH_2OH at the bottom.
- Lactic acid, left: COOH at top, H on the left, OH on the right, CH_3 at the bottom.
- Lactic acid, right: COOH at top, HO on the left, H on the right, CH_3 at the bottom.

Mirror

(+) and (–) Glyceraldehydes

Mirror

(–) and (+) Lactic acids

It may be noted that an enantiomer is changed into the other (*inversion of configuration*) when two atoms or groups about the chiral carbon are interchanged. However, if we effect two such interchanges, we get the same configuration as the first. In other words, rotation of a Fischer projection formula by 180° in the plane of the paper does not alter the configuration. It is to be noted that a 90° rotation of the projection formula about the chiral centre *or* one exchange of groups inverts the configuration of the original structure. These points are illustrated by taking glyceraldehyde as an example.

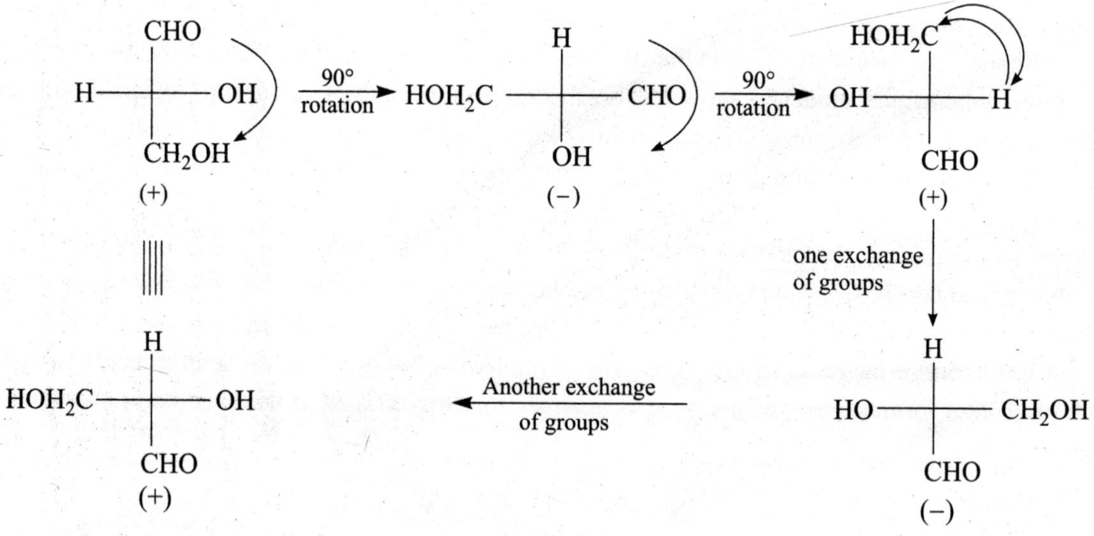

Scheme (Fischer projections):

- (+): CHO at top, H on the left, OH on the right, CH_2OH at the bottom.
- 90° rotation gives (–): H at top, HOH_2C on the left, CHO on the right, OH at the bottom.
- Another 90° rotation gives (+): HOH_2C at top, OH on the left, H on the right, CHO at the bottom.
- One exchange of groups (the HOH_2C and H positions) gives (–): H at top, HO on the left, CH_2OH on the right, CHO at the bottom.
- Another exchange of groups gives (+): H at top, HOH_2C on the left, OH on the right, CHO at the bottom. This structure is identical (≡) to the starting (+) structure.

Therefore, *chirality or molecular dissymmetry is the necessary and sufficient condition for a molecule to be optically active.*

3.4 PROJECTION FORMULAS OF CHIRAL MOLECULES

How to draw and interconvert them into each other?

Configuration of a chiral molecule is a three dimensional structure and it is not very easy to depict it on a paper having only two-dimensions. To overcome this problem the following four two-dimensional structures known as projections have been evolved.

1. Fischer projection;
2. Newman projection;

3. Sawhorse projection; and
4. Flying wedge formula.

3.4.1 Fischer Projection

Fischer projections are always generated by orienting the molecule so that the vertical bonds connected to the chirality centre are directed away from the viewer and the horizontal bonds point towards the viewer. On paper the projection of bonds is in the shape of cross with chiral centre at the centre of the cross.

Thus Fischer projection of one stereoisomer of glyceric acid HOOC — CHOH — CH_2OH is given below:

COOH
H — C — OH
CH_2OH
(I)

or

COOH
H — OH
CH_2OH
(I)

Away from the viewer

Towards the viewer

Some characteristic features of Fischer projection are as follows:

(*i*) Fischer projection of a chiral molecule can be changed into another identical Fischer projection by interchanging the position of any two pairs of substituents *i.e.*, by two interchanges). Thus, structure I after interchange of two pairs of substituents generate identical structure III.

COOH / H — OH / CH_2OH (I) —First interchange→ H / HOOC — OH / CH_2OH (II) —Second interchange→ H / HOOC — CH_2OH / OH (III)

Only one such interchange produces an enantiomer. As we will see later, structures I and III are identical and I and II are enantiomers.

Rotation of a Fischer projection by an angle of 180° about the axis which is perpendicular to the plane of the paper gives identical structure. However, similar rotation by an angle of 90° produces non-identical structure.

CH_2OH / HO — H / COOH (IV) —rotation by 180°→ COOH / H — OH / CH_2OH (I) —rotation by 90°→ H / HOH_2C — COOH / OH (V)

Thus, I and IV are identical whereas I and V are non-identical.

(**Note:** During rotation of a Fischer projection formula, it should not be lifted out of plane).

3.4.2 Newman Projection

In Newman projection, we look at the molecule down the length of a particular carbon-carbon bond. The carbon atom away from the viewer is called 'rear' carbon and is represented by a circle. The carbon atom facing the viewer is called 'front' carbon and is represented as the centre of the above circle which is shown by dot. The remaining bonds on each carbon are shown by small straight lines at angles of 120° as follows:

(*i*) Bonds joineds to 'front' carbon intersect at the central dot.

(*ii*) Bonds joined to 'rear' carbon are shown as emanating from the circumference of the circle.

The concept of Newman projection can be understood by the following drawings:

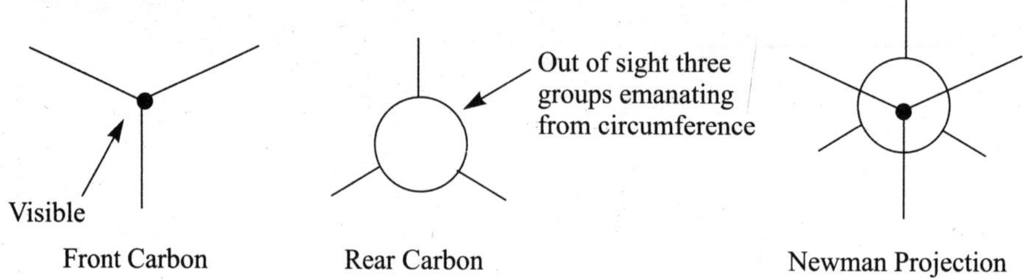

For example, Newman projections of *n*-butane.

Staggered

Eclipsed

Staggered and eclipsed conformations arise due to free rotation about the carbon-carbon single bond (front and rear carbon atoms).

3.4.3 Sawhorse Projection

The Sawhorse projection is very useful for understanding the mechanism and stereoelectronic aspects of reactions.

In this projection the bond between two carbon atoms is shown by a longer diagonal line because we are looking at this bond from an oblique angle. The bonds linking other substituents to these carbons are shown projecting above or below this line. Thus Sawhorse projection of a molecule can be drawn as follows taking *n*-butane as an example:

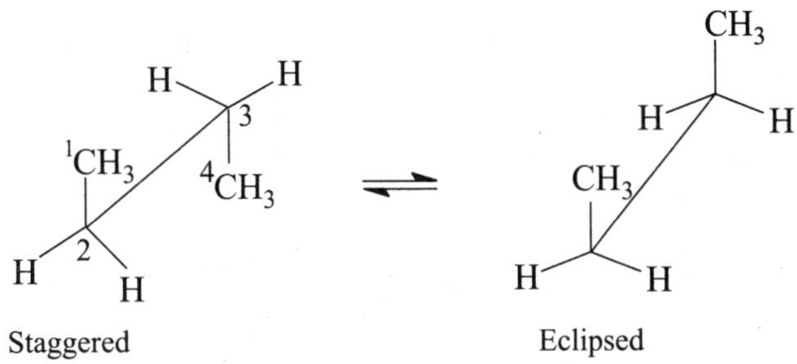

Due to free rotation along the central, bond two extreme conformations are possible — the staggered and the eclipsed (for details section 3.14 of this chapter).

3.4.4 Flying Wedge Formula

It is a three-dimensional representation and is particularly useful for chiral molecules. The procedure for drawing this formula is as follows:

Out of four bonds connecting substituents to a chiral carbon, two bonds are drawn in the plane of the paper, which are shown by normal lines, third bond lies above this plane which is shown by a solid cone (▬) and the fourth bond lies below the plane of the paper and is shown by broken lines (·········). The flying wedge formulas of two enantiomeric lactic acids are shown below:

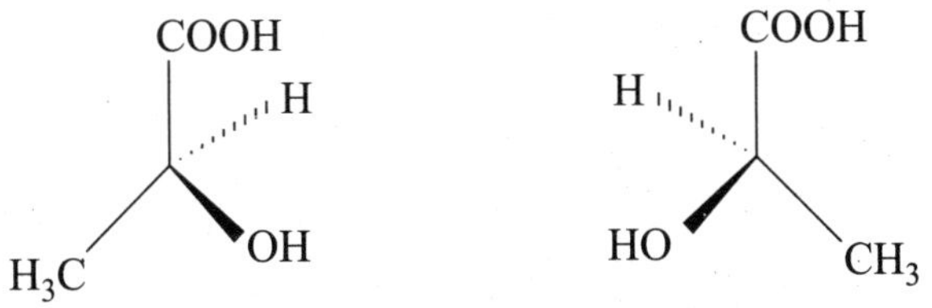

Both these structures are mirror image of each other.

(**Note:** The main functional group is generally held on the upper side in the vertical plane.)

Interconversions of Projection Formulas

It is quite important, for an understanding of stereochemical implications of the reactions, to have a clear idea of the inter-relationship between various projections of a given organic molecule. For this we should be well conversant with the methods of translating one projection into another. Given below are the methods used for converting one projection into another without changing the configuration.

1. Conversion of Fischer Projection into Sawhorse Projection and *Vice-versa*

Fischer projection of a compound can be converted into Sawhorse projection first in the eclipsed form (in Fischer projection the groups on neighbouring carbons are considered to be eclipsing each other) by holding the model in horizontal plane in such a way that the groups on the vertical line point above and the last numbered chiral carbon faces the viewer. Then one of the two carbons is rotated by an angle of 180° to get staggered form (*more stable* or *relaxed form*).

For example, conversion of Fischer projection of an optically active tartaric acid into staggered Sawhorse projection.

Rear carbon

Front carbon

Fischer Projection

Eclipsed

rotation by 180° along $C_2 - C_3$ axis

Staggered

Sawhorse Projection

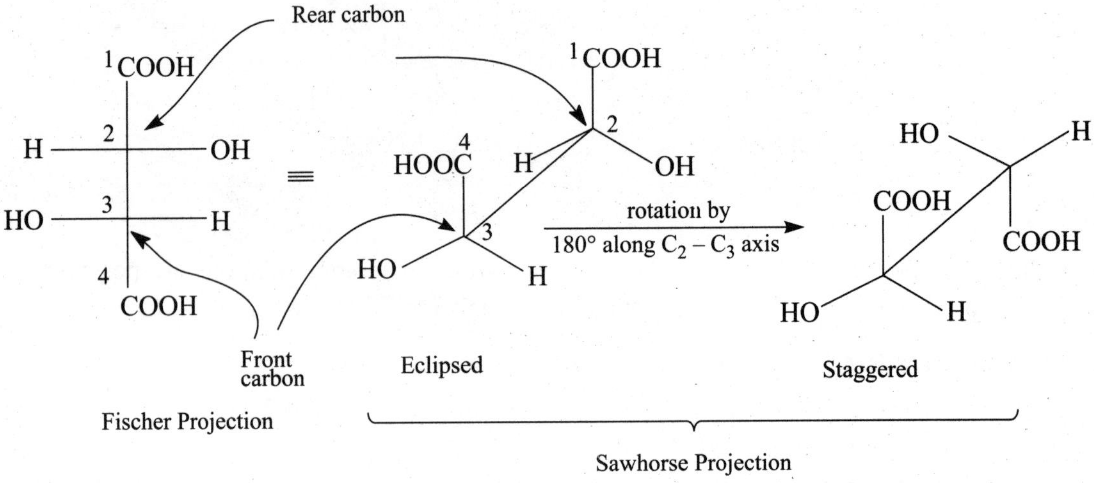

Conversion of Sawhorse Projection into Fischer Projection

First the staggered Sawhorse projection is converted in eclipsed projection. It is then held in the vertical plane in such a manner that the two groups pointing upwords are away from the viewer, *i.e.*, both these groups are shown on the vertical line. Thus, for 2, 3-dibromobutane.

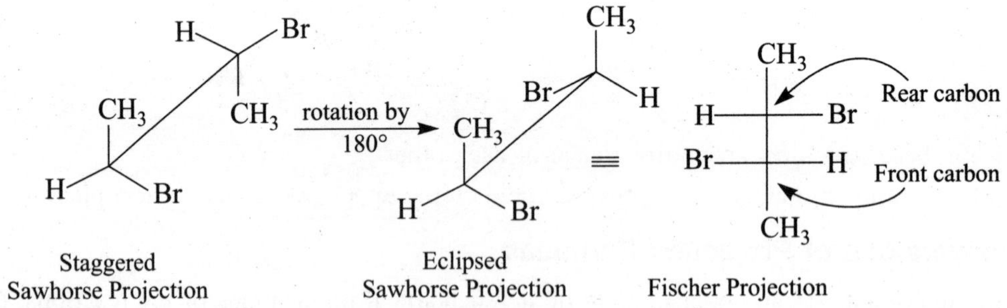

rotation by 180°

Rear carbon

Front carbon

Staggered Sawhorse Projection

Eclipsed Sawhorse Projection

Fischer Projection

2. Conversion of Sawhorse to Newman to Fischer Projection

Sawhorse to Newman projection is quite easy. View the molecule from front carbon (the central C—C bond being invisible) to get the staggered Newman projection. Rotate the front carbon by 180° to get eclipsed Newman projection. Now hold the molecule in vertical plane, *i.e.*, central bond is visible in vertical plane, in such a manner that front carbon is the lowest carbon.

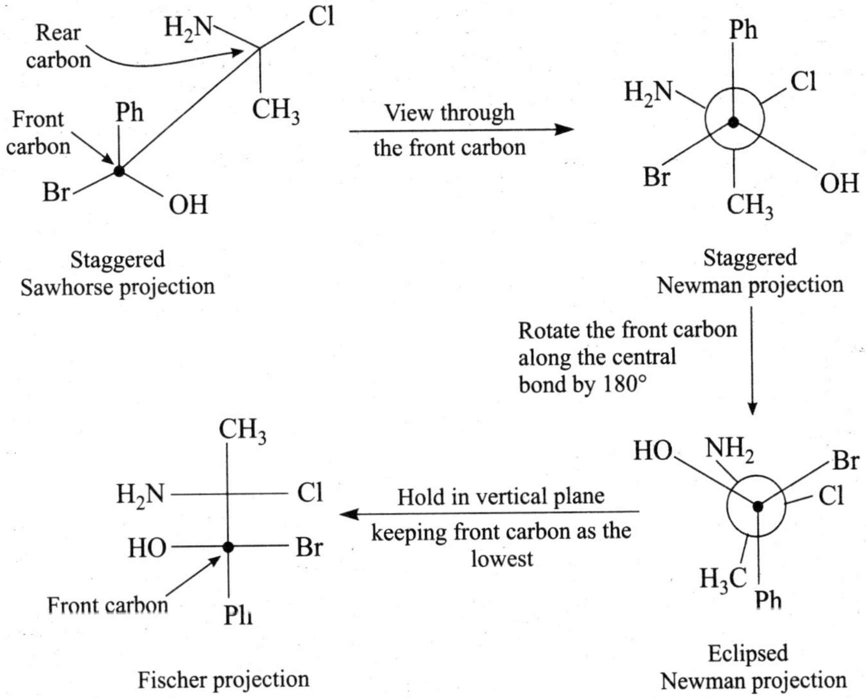

Rear carbon H_2N Cl
Front carbon Ph CH_3 View through the front carbon
Br OH

Staggered Sawhorse projection

Ph
H_2N Cl
Br OH
CH_3

Staggered Newman projection

Rotate the front carbon along the central bond by 180°

HO NH_2 Br Cl
H_3C Ph

Eclipsed Newman projection

Hold in vertical plane keeping front carbon as the lowest

CH_3
H_2N — Cl
HO — Br
Front carbon Ph

Fischer projection

Conversion of Fischer to Newman to Sawhorse Projection

View the molecule through the lowest chiral carbon which becomes front carbon and draw eclipsed Newman projection. Convert it into staggered conformation. Now view the molecule through the bond connecting front carbon with rear carbon. Thus,

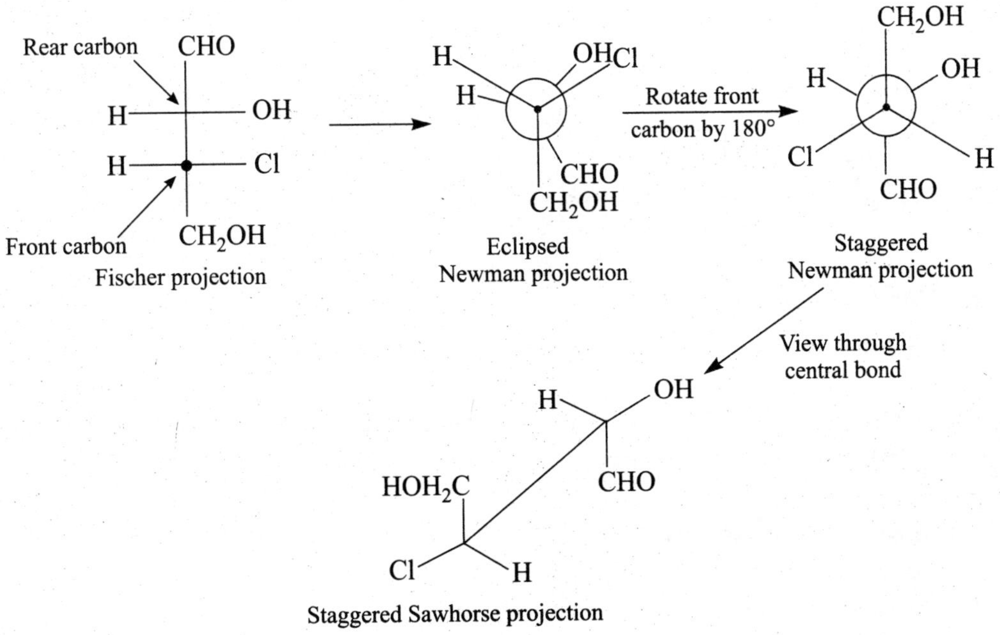

Rear carbon CHO
H — OH
H — Cl
Front carbon CH_2OH

Fischer projection

H OH Cl
H
CHO
CH_2OH

Eclipsed Newman projection

Rotate front carbon by 180°

CH_2OH
H OH
Cl H
CHO

Staggered Newman projection

View through central bond

H OH
HOH_2C CHO
Cl H

Staggered Sawhorse projection

3. Conversion of Fischer Projection into Flying Wedge Formula

The vertical bonds in the Fischer projection are drawn in the plane of the paper using simple lines (—) consequently horizontal bonds will project above and below the plane.

Example: Conversion of Fischer projection of one of the enantiomer of α-bromopropanoic acid into two flying wedge formulas (without changing the configuration) is illustrated below:

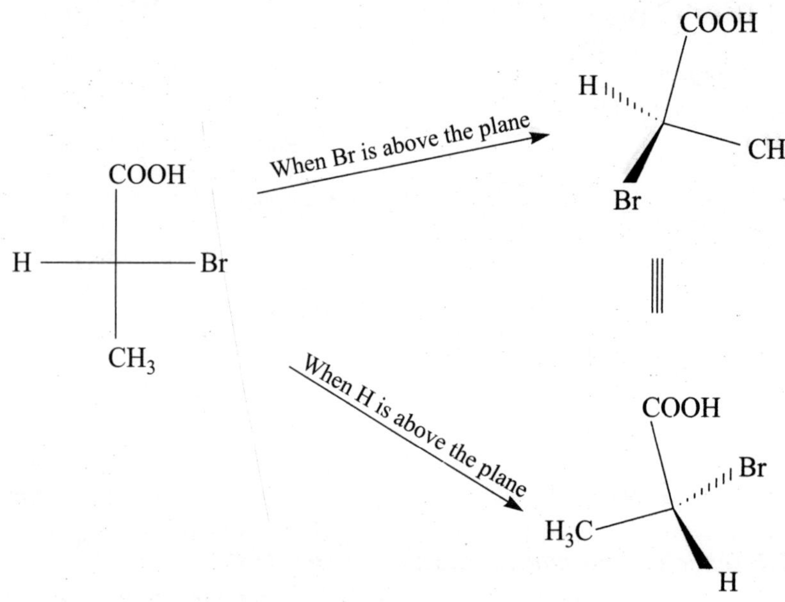

Conversion of Flying Wedge Formula into Fischer Projection

The molecule is rotated (in the vertical plane) in such a way that the bonds shown in the plane of the paper go away from the viewer and are vertical.

COOH
H
CH3
Br
Rotate the model in vertical plane so that C — COOH and C — CH_3 go away from the viewer
COOH
H
Br
CH_3

3.5 ELEMENTS OF SYMMETRY

We have seen that enantiomerism depends on the fact that a molecule is not superimposable on its mirror image. If it is superimposable, the molecule is optically inactive. If it is not superimposable, the molecule is optically active. The most convenient method of inspecting superimposability consists in determining whether the molecule has any of the four elements of symmetry. These elements of symmetry are:

1. Plane of symmetry (σ).
2. Centre of symmetry (*i*).
3. Simple or proper axis of symmetry (C_n).
4. Alternating or improper axis of symmetry (S_n).

This method can be applied to the Fischers projection formula and tetrahedral formula alike. If there is atleast one symmetry element present, the molecule will not show enantiomerism. We will now discuss these elements as follows:

3.5.1 Plane of Symmetry (σ)

A plane of symmetry is defined as an imaginary plane which divides a molecule in such a way that one-half is mirror image of the other half.

A molecule with atleast a plane of symmetry can be superimposed on its mirror image and is **achiral**. A molecule that does not have a plane of symmetry is usually **chiral**; it cannot be superimposed upon its mirror image.

A plane of symmetry may pass through atoms, between atoms or both.

For example, the lower half of the molecule of *meso*-tartaric acid is mirror image of the upper half and both are superimposable. Therefore, *meso-tartaric acid is achiral and hence optically inactive.*

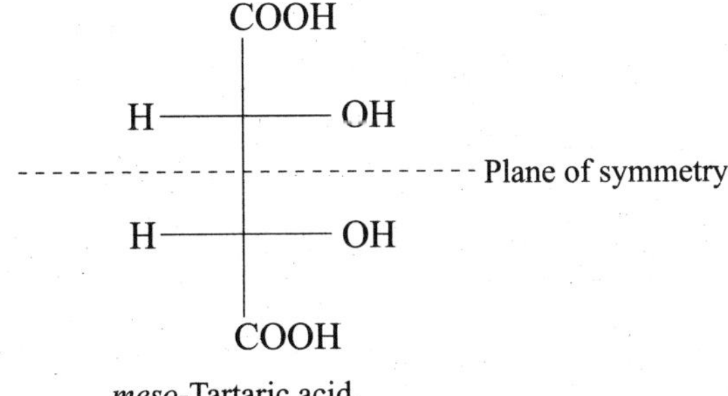

meso-Tartaric acid

3.5.2 Centre of Symmetry (*i*)

A centre of symmetry, which is usually present in an even-membered ring, is also called *centre of inversion and is defined as a point within the molecule such that if an atom is joined to it by a straight line which if extrapolated to an equal distance beyond it in opposite direction meets an equivalent atom.* In other words *it is a point at which all the straight lines joining identical points in the molecule cross each other.*

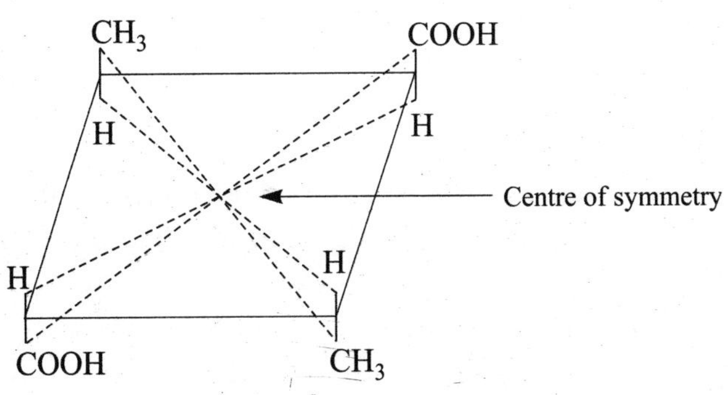

Chapter 3

For instance, the molecule of 2, 4-dimethylcyclobutane-1, 3-dicarboxylic acid has a centre of symmetry (centre of ring) and is, therefore, optically inactive.

Another interesting example is that of dimethyl-diketopiperazine which exists as two geometric isomers, *cis* and *trans*, depending on whether the methyl groups are on the same side or opposite side. The *cis*-isomer has no centre of symmetry and can exist in optically active forms. The *trans*-isomer, on the other hand, is optically inactive because it has a centre of symmetry.

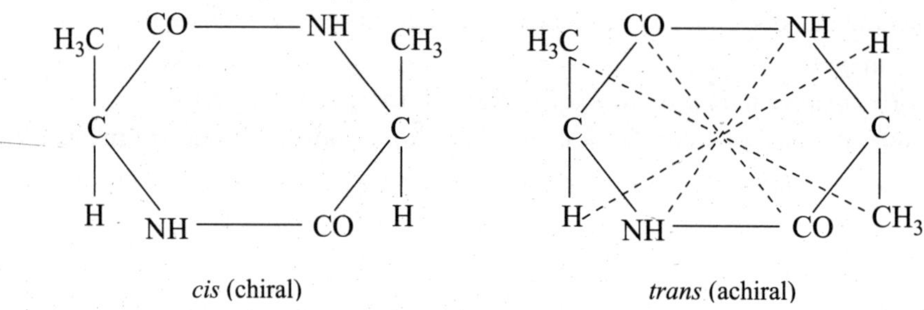

cis (chiral) *trans* (achiral)

3.5.3 Proper Axis of Symmetry (C_n)

An axis of symmetry is defined as an imaginary line passing through the molecule in such a way that when the molecule is rotated about it by an angle of 360°/n, an arrangement indistinguishable from the original is obtained. Such an axis is called *n*-fold axis of symmetry. For example, *cis*-1, 3-dimethylcyclobutane has a **two** fold axis of symmetry (C_2) *i.e.*, rotation by 180° gives indistinguishable appearance.

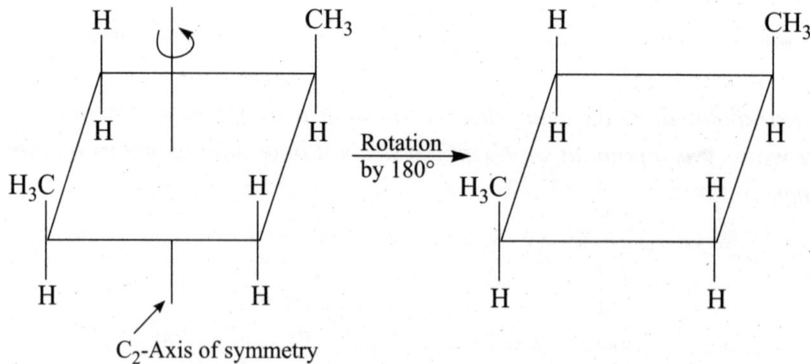

3.5.4 Alternating or Improper or Rotation Reflection Axis of Symmetry (S_n)

Alternating axis of symmetry, is rather rare. It is also called *improper or rotation-reflection axis of symmetry*. To determine whether an *n*-fold improper axis of symmetry is present, the molecule is rotated through an angle of 360°/*n* about this axis followed by reflection in a plane perpendicular to the axis. If then we arrive at the original molecule, it is said to have an *n*-fold improper axis of symmetry and will be optically inactive. For instance, 1, 2, 3, 4-tetramethylcyclobutane (***a***) has a four fold axis of symmetry because its rotation by 90° about the axis AB (passing through the centre of the ring perpendicular to its plane) gives **(*b*)**. The form **(*a*)** is the mirror image of **(*b*)** by its reflection in the plane parrallel to the ring *i.e.*, perpendicular to the axis AB. We should note that it so happens the molecule in this case has two vertical planes of symmetry through the diagonals of the ring.

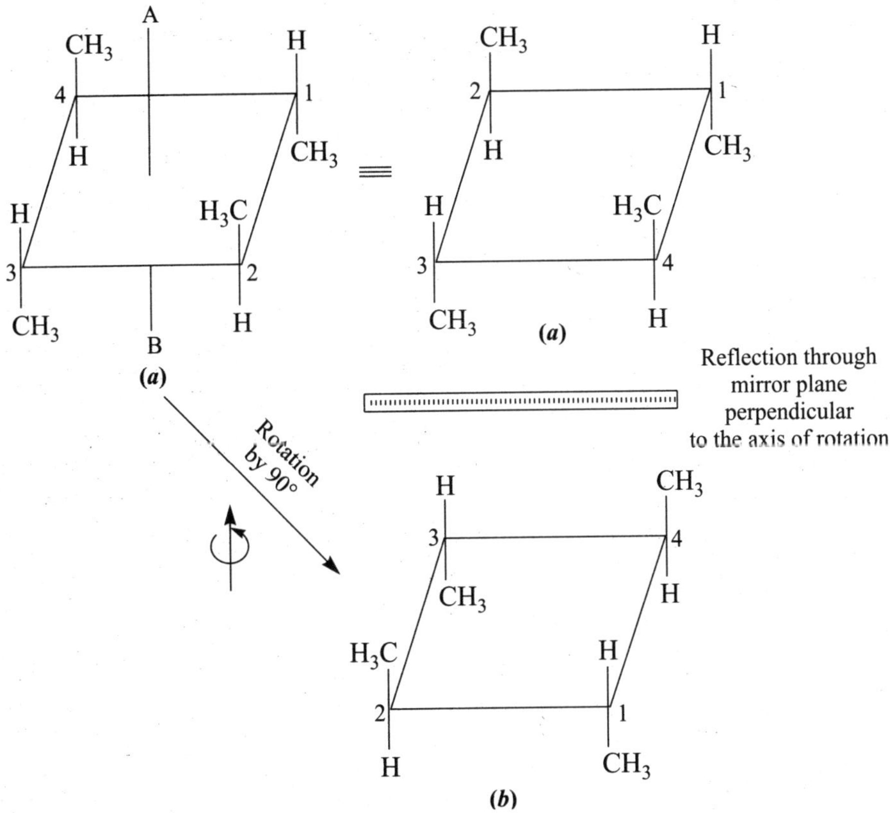

From the *foregoing*, it can be concluded that the necessary and sufficient condition for a molecule to show optical activity is that such a molecule is not superimposable on its mirror image. In other words, *a molecule can exist in enantiomeric forms only when it has molecular dissymmetry.* The concept of molecular dissymmetry is important since the phenomenon of optical rotation is not related to chiral carbon atoms only. Even when a chiral carbon atom is present, a compound need not necessarily have molecular dissymmetry. For instance, some compounds, such as *meso*-compounds which have two or more chiral carbons, do not exhibit optical activity. On the other hand, there are optically active compounds having no chiral carbon. For instance, allenes having no chiral carbons show optical activity due to dissymmetry of the molecule.

Asymmetry *vs* Dissymmetry

We know that, only those substances which are related to each other as non-superimposable mirror images are optically active. Such substances may be dissymmetric or asymmetric. So far these two terms have been used interchangeably (*i.e.*, without distinction) to describe optical acitivity. But this is not so and there exists a subtle difference between them. In general the term *asymmetry is used for those optically active compounds which have none of the four elements of symmetry.* In contrast the term *dissymmetry is used for all stereoisomeric compounds which are capable of existing as pairs of non-superimposable mirror images despite the presence of some elements of symmetry.* In other words the term dissymmetry is applicable to all stereoisomers, which are related to each other as non-superimposable mirror images of each other. The following example is illustrative.

The molecule of 2, 3-dibromobutane possesses a C_2 axis of symmetry passing through the centre of the molecule at right angle to the plane of the paper.

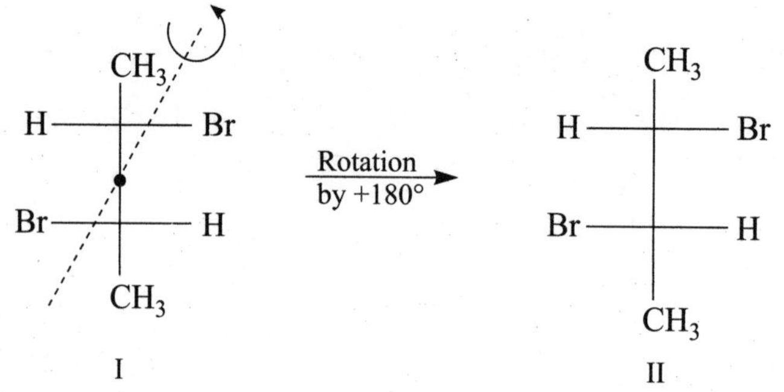

CH_3
H — Br
Br — H
CH_3
I

Rotation by +180° →

CH_3
H — Br
Br — H
CH_3
II

Since structures I and II are indistinguishable, the molecule has C_2 axis of symmetry. But it is non-superimposable on its mirror image so it is dissymmetric and not asymmetric and exhibits optical activity.

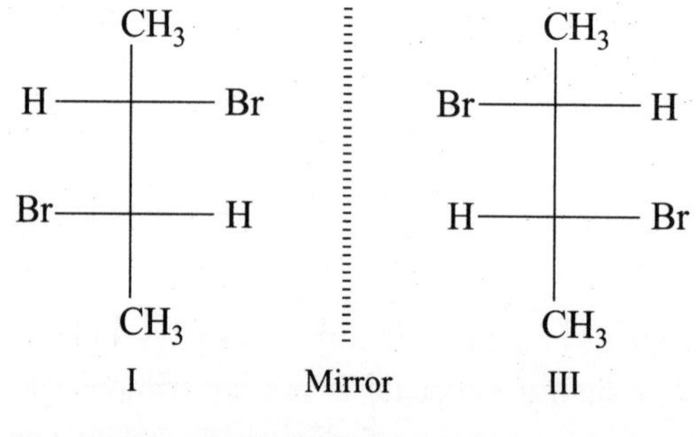

CH_3
H — Br
Br — H
CH_3
I

Mirror

CH_3
Br — H
H — Br
CH_3
III

I and III are non-superimposable mirror images.

From the above discussion it becomes quite clear that all asymmetric molecules are dissymmetric whereas all dissymmetric molecules are not asymmetric. However, both these types of molecules show optical activity.

In the end it may be interesting to mention that both asymmetric and dissymmetric molecules are chiral. So in order to avoid any confusion, in using these terms, — asymmetry or dissymmetry — the term chirality is used.

3.6 STEREOGENIC CENTRE OR CHIRALITY CENTRE

As we have seen molecules having tetrahedral carbon atoms attached to four different atoms or groups are called asymmetric carbon atoms or chiral carbon atoms. However, in 1996, the IUPAC recommended that a tetrahedral carbon atom bearing four different atoms or groups may be called a **chirality centre**. Several earlier terms including asymmetric centre, asymmetric carbon, chiral centre, stereogenic centre and stereocentre are still widely used.

Here it may be mentioned that if a molecule has one chirality centre it is very easy to know that it is chiral. For example, 2-butanol is chiral having C-2 as the chirality centre. But 1-butanol is achiral in which no carbon bears four different atoms.

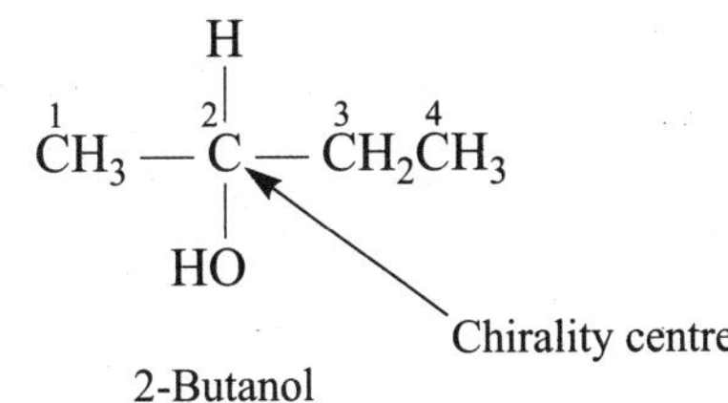

2-Butanol

Further, any mutual exchange of any two atoms or groups on a stereogenic carbon atom gives another configuration which is non-equivalent (or gives a new stereoisomer).

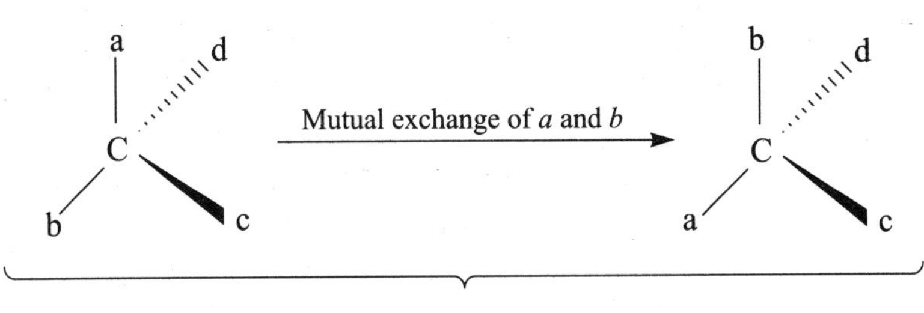

Two non-equivalent configurations
(Two enantiomers)

3.7 CHARACTERISTICS OF ENANTIOMERS

The term enantiomers and enantiomerism have already been introduced in Section 3.3. Here we shall discuss the necessary and sufficient conditions for enantiomerism and properties of enantiomers.

3.7.1 Necessary and Sufficient Condition for Enantiomerism

The necessary and sufficient condition for a compound to exhibit enantiomerism is that it should be chiral or dissymmetric *i.e.*, the molecule and its mirror image should be non-superimposable, even if it may not have an assymmetric carbon or stereocentre.

In general it has been observed that compounds having one or more chirality centre show enantiomerism and, therefore, optically active. However, this statement does not hold good for all such molecules. For example,

(i) Compounds having chirality centre(s) but not enantiomeric

Meso-2,3-dibromobutane contains two chirality centres (marked with *) but it does not exhibit enantiomerism and hence is optically inactive.

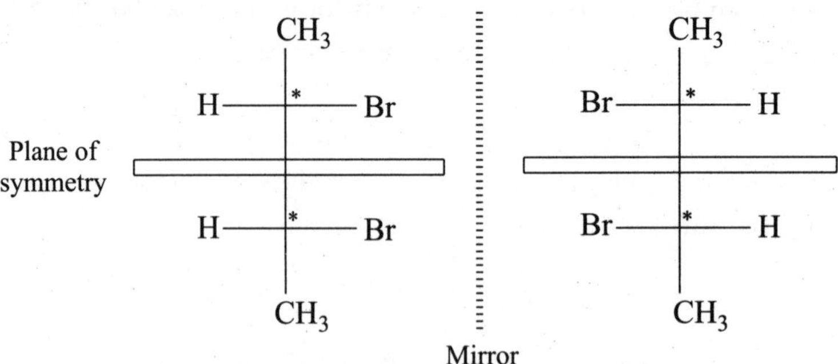

Meso-2, 3-dibromobutane (optically inactive)

Reason: The *meso* molecule possesses a plane of symmetry which divides the molecule into two equal halves which are superimposable mirror images of each other. Thus it is an achiral, non-dissymmetric molecule. Actually what happens is that the optical activity due to one half is cancelled by equal and opposite sign of rotation of the other half. This is called *internal compensation.* So *meso*-2, 3-dibromobutane is an example of an optically inactive compound containing two chirality centers.

(ii) Compounds having no chirality centre but are enantiomeric

Such compounds are also known which do not have a chirality centre but still they are optically active, *i.e.*, their molecules show chirality or dissymmetry and hence enantiomerism. Examples of such compounds are *o*-substituted biphenyls and allenes having even number of double bonds.

Biphenyls: The chirality and hence the optical activity of biphenyls such as 2, 2'-difluoro-6, 6'-dinitrobiphenyl is due to the fact that the substituents in the *ortho* positions prevent the free rotation of benzene rings about the single bond which joins them. Hence, these rings do not lie in the same plane and the molecule is devoid of symmetry.

NO_2 F F O_2N

F O_2N NO_2 F

Mirror

2, 2'-Difluoro - 6, 6'-dinitrobiphenyl (*No chirality centre*)
(Non-superimposable mirror images hence enantiomeric and optically active)

Allenes: Chirality of allenes is due to *sp*-hybridization of central carbon which forms two π-bonds perpendicular to each other and thus the two groups attached to terminal carbon atoms are also *orthogonal.* Due to this arrangement the molecule of allene is devoid of symmetry and hence is chiral.

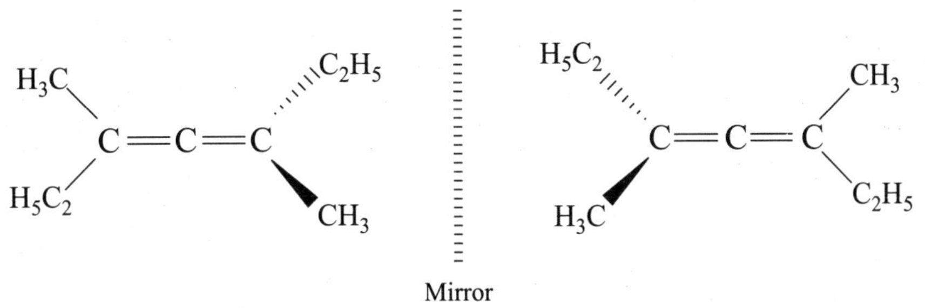

H_3C, H_5C_2 C=C=C C_2H_5, CH_3 | Mirror | H_5C_2, H_3C C=C=C CH_3, C_2H_5

1, 3-Diethyl-1, 3-dimethylallene (*No chirality centre*)
(Non-superimposable mirror images, hence enantiomeric and optically active)

Finally, the above two types of examples prove that the *necessary and sufficient condition for compounds to exhibit enantiomerism is that they should possesses chirality or dissymmetry rather than asymmetry.*

3.7.2 Properties of Enantiomers

As mentioned earlier, the enantiomers have same physical and chemical properties. They differ in their sign of rotation of plane polarised light and their rates of reactions with chiral compounds. Some important properties of enantiomers are discussed below:

(i) They have identical physical properties but differ in the direction of the rotation of plane polarized light. Although the two enantiomers rotate the plane-polarised light in two opposite directions, the magnitude of rotation is the same. The direction of rotation is usually denoted by the prefix (+) for dextrorotatory and (–) for laevorotatory enantiomers. The following table shows the physical properties of (+) and (–) forms of some familiar compounds:

Tartaric acids

Enantiomer	*Specific Rotation*	*M.Pt.*	*Density*
(+)	+11.98°	443 K	1.76
(–)	–11.98°	443 K	1.76

2-Methyl-1-butanols

Enantiomer	*Specific Rotation*	*B.Pt.*	*Refractive Index*
(+)	+5.75°	402 K	1.41
(–)	–5.75°	402 K	1.41

It is clear that two enantiomers have the same melting points, boiling points, refractive indices, etc. The magnitude of rotation of polarized light is also the same, but in opposite direction. The molecules being so similar, that they are expected to have same degree of rotation. But being mirror images of each other, their rotations should also be the mirror images of each other. Indeed, clockwise and anti-clockwise rotations are the mirror images of each other.

(ii) The enantiomers have identical chemical properties towards optically inactive reagents because the structural environment in the two enantiomers is same and thus the optically inactive reagents such as H_2SO_4, HBr and CH_3COOH encounter the same environment while approaching either enantiomer, *e.g.*, enantiomers such as those of 2-methyl-1-butanol are both converted to 2-methylbutene when treated with strong H_2SO_4, to 2-methylbutyl bromide on treatment with HBr and give esters on esterification with CH_3COOH.

$$\underset{\substack{(+\text{ and }-)\\ \text{2-Methylbutyl bromide}}}{CH_3CH_2-\overset{\overset{\displaystyle CH_3}{|}}{\underset{\underset{\displaystyle H}{|}}{C}}-CH_2Br} \xrightarrow{HBr} \underset{\substack{(+\text{ and }-)\\ \text{2-Methyl-1-butanol}}}{CH_3CH_2-\overset{\overset{\displaystyle CH_3}{|}}{\underset{\underset{\displaystyle H}{|}}{C}}-CH_2OH} \xrightarrow{H_2SO_4} \underset{\substack{(+\text{ and }-)\\ \text{2-Methylbutene)}}}{(CH_3)(H_3CH_2C)C=CH_2}$$

$$\downarrow CH_3COOH$$

$$\underset{\substack{(+\text{ and }-)\\ \text{Esters}}}{CH_3CH_2-\overset{\overset{\displaystyle CH_3}{|}}{\underset{\underset{\displaystyle H}{|}}{C}}-CH_2-O-\overset{\overset{\displaystyle O}{\|}}{C}-CH_3}$$

The rates of the above transformations are also the same.

(iii) The enantiomers have different chemical properties towards optically active reagents. If we use an optically active reagent, the reaction rates will be different. If we esterify the two enantiomers of 2-methyl-1-butanol with (–)-lactic acid, the influence exerted by the reagent will not be identical due to the different spatial disposition of the OH group in the two enantiomers in relation to the groups attached to the chirality centre of (–)-lactic acid. It, therefore, follows that rate of esterification of (+)-2-methyl-1-butanol will be different form that of (–)-2-methyl-1-butanol.

(iv) The enantiomers have different biological properties. In biological systems this difference is a rule to the extent that more often reaction with one of the enantiomers does not take place at all. This is quite understandable as the biological catalysts, enzymes, and most of the compounds with which react, are *optically active*. The following examples are illustrative:

1. (+)-Glucose plays an important role in animal metabolism and fermentation, but (–)-glucose is not metabolized by animals, and furthermore cannot be fermented by yeasts.
2. *Penicillium glaucum*, consumes only the (+)-enantiomer when fed with a mixture of equal quantities of (+)-and (–)-tartaric acids.
3. Hormonal activity of (–)-adrenaline is many times more than that of its enantiomer.

We can also understand the differential activity of optically active reagents towards enantiomers by transition-state approach. The transition states in the reaction of two enantiomers (+)-A and (–)-A with an optically inactive reagent, B, will be mirror images of each other and hence will be of the same energy. On the other hand, the transition states in the reaction of the enantiomers with an optically active reagent (*e.g.*, (–)-B) will not be mirror images and, therefore, will not have the same energy. Thus the energy of activation will be different and so will be the rates of reaction.

3.8 COMPOUNDS WITH SEVERAL CHIRALITY CENTRES

In general the optical rotation of asymmetric compounds depends upon the presence of a chirality centre. Therefore, the number of optically active forms increases with increase in the number of chiral carbon atoms. If there are *n* chiral carbons, the compound will exist in 2^n optically active forms, provided the chiral atoms are not identically substituted. Thus 2-bromo-3-hydroxybutanedioic acid, HOOC—CH(OH)—CH(Br)—COOH, in which the two chiral carbon atoms are dissimilar, exists in $2^2 = 4$ optically active forms. The two chiral carbon atoms of tartaric acid, HOOC—CHOH—CHOH—COOH, on the other hand, are identically substituted (similar) and hence the total number of optically active isomers cannot be predicted by using 2^n formula. We will now discuss few examples of such molecules.

3.8.1 Compounds with Two Dissimilar Stereogenic Centres (Chirality Centres): Diastereomers

Compounds such as 2-bromo-3-hydroxybutanedioic acid have two dissimilar chiral carbon atoms and hence exist in four stereoisomeric forms *i.e., as two pairs of enantiomers*

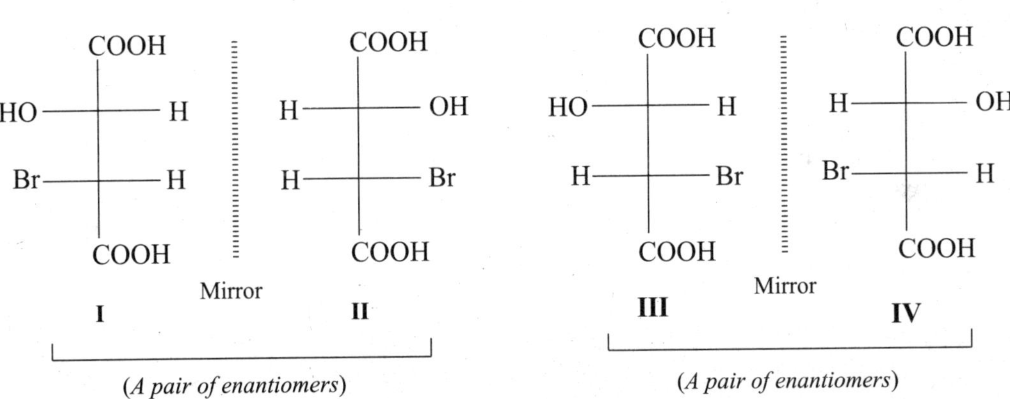

The two enantiomers of any pair (I, II and III, IV) have equal and opposite rotations and if mixed in equal amounts, I and II will give rise to one racemic modification, while III and IV will give another. Now the question arises as to what is the relationship between I and III or I and IV or between II and III or II and IV. They are optically active, but are not the mirror images. Such stereoisomers are referred to as **diastereomers**. For example, stereoisomers I or II is a diastereomer of III or IV.

Diastereomers are stereoisomers which have the same configuration at one chirality centre but different configuration at the other. In other words *diastereomers are stereoisomers which are not mirror images of each other.*

3.8.2 Properties of Diastereomers

As diastereomers are not related to each other as mirror images, they have different properties. Some of the properties are:

1. Physical Properties

(*i*) *Melting points, boiling points, etc.* The melting points, boiling points, water solubility, relative densities of diastereomers etc. are different. The physical properties of tartaric acids as given in Table 3.1 are illustrative.

Table 3.1: Physical properties of tartaric acid

	(+)	(–)	(±)	*Meso*
$[\alpha]_D^{20°}$	+12°	–12°	0°	0°
Melting points (K)	443	443	478	413
Water solubility (g/100 mL)	147	147	25	120
Relative density	1.760	1.760	1.687	1.666

(*ii*) Diastereomers having different solubilities, boiling points and Rf values can be easily separated using techniques such as fractional crystallization, fractional distillation and chromatography.

2. Behaviour Towards Plane-Polarised Light

If the diastereomers happen to be optically active they have different values of $[a]_D^{+0}$.

3. Chemical Properties

Diastereomers have similar but non-identical chemical properties. In particular they react with chiral or achiral reagents at different rates. This may be due to the fact that neither the reagents nor the transition states are mirror images of each other. Thus diastereomers differ in their energies of activation for a particular reaction and hence react at different rates.

3.8.3 Threo and Erythro Diastereomers

Another example of compounds with two dissimilar chiral carbon atoms are the four carbon sugars, *erythrose* and *threose*.

(–)-Erythrose	(+)-Erythrose	(–)-Threose	(+)-Threose
CHO	CHO	CHO	CHO
H–C–OH	HO–C–H	HO–C–H	H–C–OH
H–C–OH	HO–C–H	H–C–HO	HO–C–H
CH_2OH	CH_2OH	CH_2OH	CH_2OH

By analogy, diastereomers of compounds possessing two dissimilarly substituted chiral carbon atoms are commonly distinguished from each other by using the perfix *threo* or *erythro*. These prefixes are used to designate those compounds which contain two stereocentres having two identical and a third non-identical groups attached. *The erythro diastereomer is the one in which two like groups on the chiral carbon atoms are on the same side in the Fischer projection formula as in erythrose.* In contrast, the *threo* diastereomer resembles threose as the two like groups are on opposite sides. This nomenclature is illustrated by taking the example of 3-bromo-2-butanol.

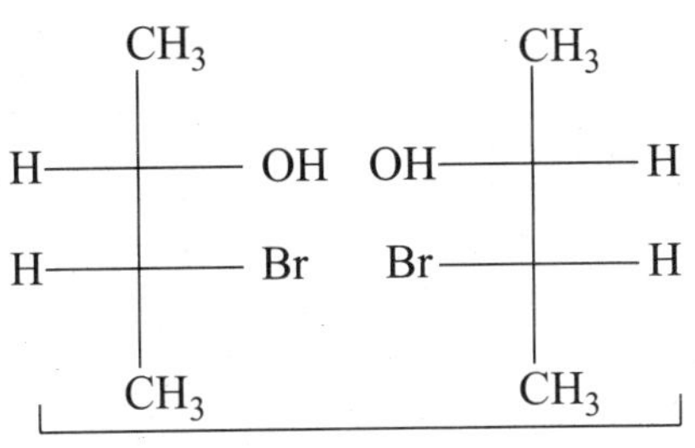

Erythro-3-bromo-2-butanol

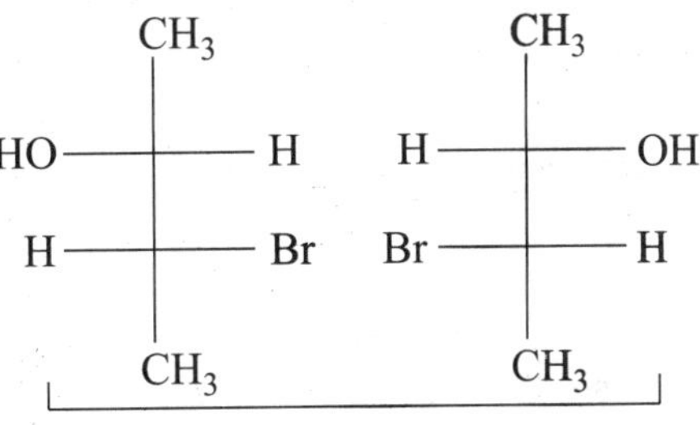

Threo-3-bromo-2-butanol

Although the Fischer projection formulas are conveniently constructed and show at a glance whether the given compound is *erythro* or *threo*, they however, give the impression that the molecule exists in the *eclipsed form.* Actually it exists in the *staggered form* in which the bulky substituents are as far apart as possible

We can thus say that an erythro isomer corresponds to that diastereomer, which when viewed along the bond connecting the chiral carbons has a rotational isomer in which all similar groups are eclipsed. The threo diastereomers, on the other hand, does not have an isomer in which all similar groups are eclipsed.

3.8.4 Compounds with Two Similar Chiral Carbon Atoms: *Meso* Compounds

When a compound contains chiral carbon atoms that are similar, the number of optical isomers cannot be predicted by the application of 2^n formula. The best known example of this class of compounds is tartaric acid, HOOC—*CH(OH)—*CH(OH)—COOH, with two chiral carbon atoms denoted by asterisks. We can write four Fischer projection formulas for tartaric acid.

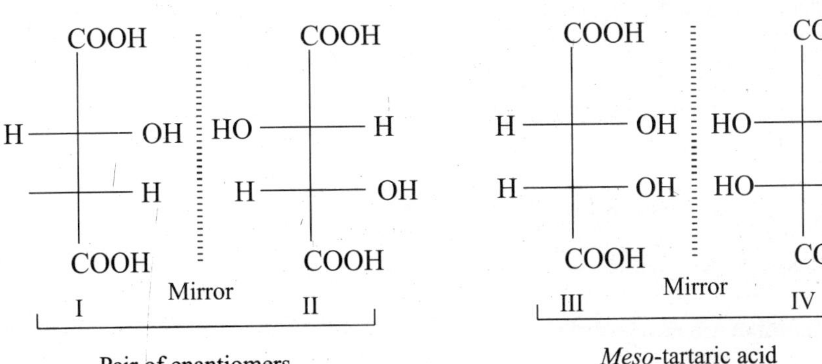

Pair of enantiomers

Meso-tartaric acid

The isomers I and II represent a pair of enantiomers and each is diastereomeric to III or IV. The isomers III and IV appear to be another pair of enantiomers by a casual inspection, but they are really identical as rotation of one (III or IV) gives the other. A closer examination of III (or IV) reveals that it has a plane of symmetry that bisects the molecule between the two chiral carbon atoms. Such compounds are called *meso* compounds. Thus, the combined activity of each chiral centre in III (or IV) is zero by *internal compensation*. This is in contrast with *external compensation* where the combined optical activity of equimolar proportions of an optically active molecule and its enantiomer is zero. This is commonly referred to as *recemization*.

The isomers having two similar chirality centres such as III are optically inactive due to the presence of a plane of symmetry and are termed meso-compounds. Hence, meso-compounds are optically inactive compounds whose molecule is superimposable on its mirror image.

3.8.5 Relationship between Enantiomers, *Meso*-Compound and Racemic Modification

A look at the three stereoisomers of tartaric acid. (I, II and III) *i.e.*, (+)-tartaric acid or (–)-tartaric acid on one hand and *meso*-tartaric acid on the other, reveals that they are not mirror images of each other. Hence (+)-tartaric acid and *meso*-tartaric acid are diastereomers. Similarly, (–)-tartaric acid and *meso*-tartaric acid are also diastereomers. Being diastereomers, the properties of *meso*-tartaric acid are different from those of (+)-tartaric acid and (–)-tartaric acid.

Racemic modification (±) of tartaric acid can be obtained by mixing equimolar amounts of (+) and (–)-tartaric acids (*i.e.*, I and II).

(±)-Tartaric acid is a diastereomer of *meso*-tartaric acid and hence will have different physical properties.

Recemic modification being a mixture of two enantiomeric compounds can be resolved. In contrast, a *meso*-compound is only one stereoisomer so it cannot be resolved.

Prediction of the number of stereoisomers i.e., optical isomers and meso-forms

The number of optical isomers and *meso*-forms in compounds containing chirality centres depend upon the following:

(*i*) Number of chirality centres, and

(*ii*) Whether the chirality centres are similar or dissimilar.

Following formulas are useful for making predictions about the number of optical isomers and *meso*-forms.

1. For molecules having dissimilar chirality centres

Suppose the number of chirality centres $= n$

Now, (*a*) Number of optical isomers $= 2^n$

(*b*) Number of *meso*-forms $= 0$

Total number of stereoisomers = No. of optical isomers + No. of *meso*-forms

$= 2^n + 0 = 2^n$

Example

(*i*) Butan –1,2,3-triol (CH_3*CHOH—*CHOH—CH_2OH) has two dissimilar chiral carbon atoms.Here $n = 2$

Now, 1. Number of optical isomers $= 2^2 = 4$

2. Number of *meso*-forms $= 0$

Total number of stereoisomers= $4 + 0 = 4$

In fact also there are four stereoisomers as shown below:

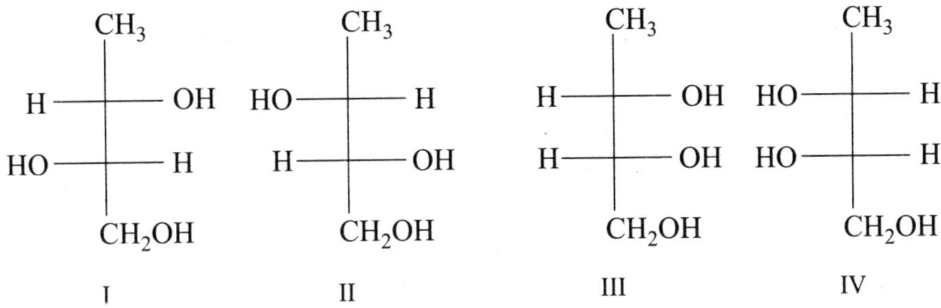

The compounds I and II as well as III and IV are enantiomers where as compounds I and III (or IV) as well as II and III (or IV) are diastereomers.

2. For molecules having similar chirality centres

These molecules are of two types:

(*a*) Molecules having even number of chiral carbons.

(*b*) Molecules having odd number of chiral carbons.

For molecules having even number of chiral carbons the number of stereoisomers is calculated as follows:

(*a*) Number of optical isomers $= 2^{(n-1)}$

(*b*) Number of *meso*-forms $= 2^{(n/2-1)}$

Total number of stereoisomers $= 2^{n-1} + 2^{n/2-1}$

Example

Tartaric acid (HOOC— *CHOH —*CHOH — COOH) has two similar chiral carbon atoms, *i.e.*, $n = 2$

Now, (*a*) Number of optical isomers $= 2^{n-1} = 2^{2-1} = 2^1 = 2$

(*b*) Number of *meso*-forms $= 2^{n/2-1} = 2^{1-1} = 2^0 = 1$

Total number of stereoisomers $= 2 + 1 = 3$

In fact also there are three stereoisomers as shown below:

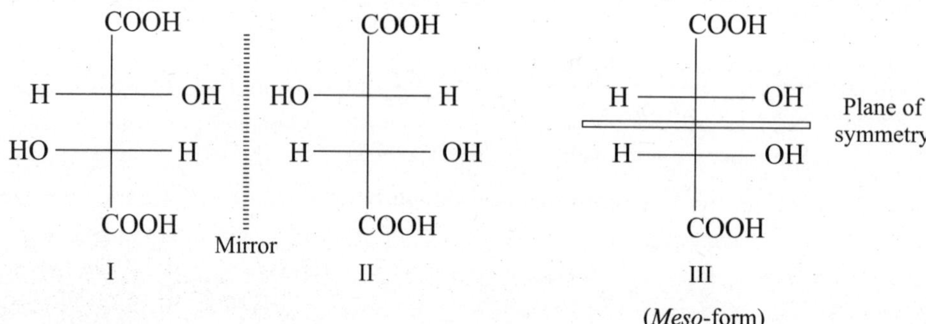

The compounds I and II are related as enantiomers where as compounds I and III as well as II and III are related as diastereomers.

For molecules having odd number of chiral carbon atoms the number of stereoisomers is calculated as follows:

(*a*) Number of optical isomers $= 2^{n-1} - 2^{(n-1)/2}$

(*b*) Number of *meso* forms $= 2^{(n-1)/2}$

Total number of stereoisomers $= [2^{n-1} - 2^{(n-1)/2}] + [2^{(n-1)/2}] = 2^{n-1}$

Example

Trihydroxyglutaric acid, (HOOC— *CHOH —*CHOH—*CHOH—COOH), has three chiral carbon atoms. *i.e.*, $n = 3$.

Now, (*a*) Number of optical isomers $= 2^{3-1} - 2^{(3-1)/2}$

$= 2^2 - 2^1 = 4 - 2 = 2$

(*b*) Number of *meso*-forms $= 2^{(3-1)/2} = 2^1 = 2$

Total number of stereoisomers $= 2 + 2 = 4$ (or $2^{3-1} = 2^2 = 4$)

In fact also there are four stereoisomers as shown below:

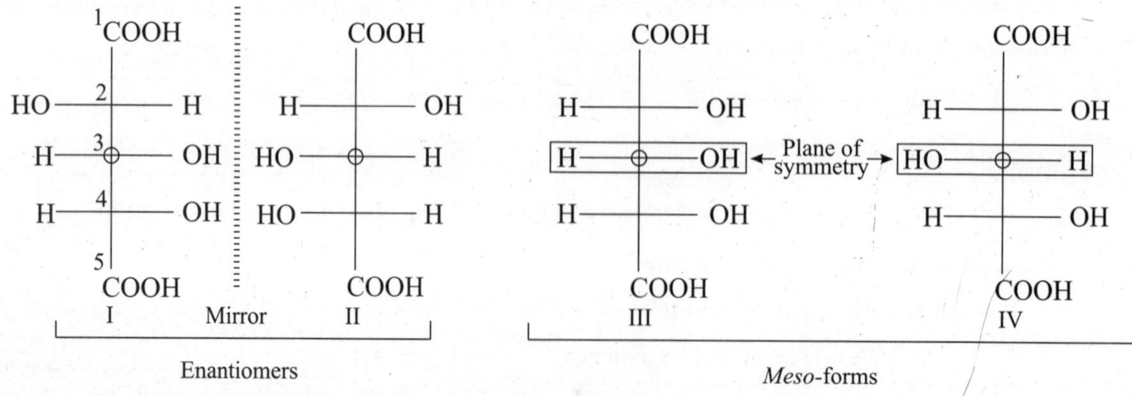

Here I and II are optically active and are enantiomers whereas III and IV are *meso* compounds having a plane of symmetry passing through central carbon C_3. In I and II the carbon C_3 is called a **pseudo chiral centre** or **pseudoasymmetric centre**, where *C_3 is a stereogenic centre but not a chirality centre.*

3.9 PROCHIRALITY

A carbon atom in a molecule is said to be **prochiral** when the replacement of one of its hydrogen atoms by another atom or group generates a chirality centre at the carbon. When replacement of one hydrogen at a time gives an enantiomer, such a hydrogen atom is called **enantiotopic hydrogen**. For example, C_2 carbon atom in ethanol is prochiral and the two hydrogen atom H_R and H_S are enantiotopic because replacement of one of them by deuterium would produce a chirality centre *i.e.*, giving rise to one enantiomer.

Pro-S; Pro-R; CH_3; H_S—C—H_R; OH; Prochiral centre; Ethanol

Replacement of H_R or H_S by D →

CH_3; H—+—D; OH; (R-isomer) or CH_3; D—+—H; OH; (S-isomer)

The carbon atom to which the two hydrogen atoms are attached is called **prochirality centre** and the moelcule of ethanol is called **prochiral molecule**.

Further, that enantiotopic hydrogen, the replacement of which gives R-configuration is called **pro-R** and the other which give S-configuration is called **pro-S** (For details of R-S configuration see Section 3.12).

3.10 RETENTION AND INVERSION OF CONFIGURATION

The stereochemical consequences of the reactions of chiral molecules may be as follows:

(*a*) The configuration of the product may be the same as that of the reactant. In such a case the reaction is said to proceed with **retention of configuration**.

(*b*) The configuration of the product may be opposite to that of the reactant. In such a case the reaction is said to proceed with **inversion of configuration**.

Retention or inversion of configuration in the reaction of a chiral molecule depends upon:

(*i*) *The side of the molecule from which the reagent attacks the reactant.*

(*ii*) *Manner of bond cleavage in the reaction i.e., whether the bond between the substituent and chirality centre is broken or not.*

We shall now discuss the stereochemical consequences of some of the important reactions of chiral molecules particularly in the light of the manner of bond cleavage.

1. If the bond between the chirality centre and the substituent remains intact throughout the course of the reaction, the reagent will attack at some other site and the reaction will proceed with retention of configuration. For example, conversion of optically active 2-butanol to 2-butyl tosylate.

$$\underset{\text{Optically active 2-Butanol}}{H_3C(H)(H_3CH_2C)C-O-H} + Cl-Ts \xrightarrow{-HCl} \underset{\text{2-Butyl tosylate}}{H_3C(H)(H_3CH_2C)C-O-Ts}$$

Not broken so retention of configuration

2. If the bond between the chirality centre and the substituent is cleaved, the consequence may be either *inversion* or *retention* of configuration. For example,

(*a*) *Inversion of Configuration*

All S_N2 reactions proceed with inversion of configuration at the chirality centre. For instance, conversion of 2-butyl chloride into 2-butanol by treating with *aq.*NaOH.

$$HO^{\ominus} \quad H\text{''''}C(CH_3)(H_5C_2)-Cl \xrightarrow[\text{(Inversion)}]{S_2N} HO-C(CH_3)(C_2H_5)\text{''''}H + Cl^{\ominus}$$

(Inverted configuration)

The reaction involves cleavage of C—Cl bond with simultaneous backside attack of $H\overset{\ominus}{O}$ on the chirality centre. Since the process of this inversion was discovered by Walden, it is also called as **Walden inversion**.

(*b*) *Retention of Configuration*

Retention of configuration may be the result of the following two mechanisms:

(*i*) The S_N2 reaction takes place twice on the same chirality centre. This will result in inversion followed by another inversion which means no inversion *i.e.*, retention of configuration. For example, treatment of 2-bromopropionate with dilute *aq.*NaOH produces lactate ion with retention of configuration. Various steps involved are as follows:

$$\underset{\text{2-Bromopropionate}}{{}^{\ominus}O(O=)C-C(H)(Br)-CH_3} \xrightarrow[\text{(1st inversion)}]{\text{Intramolecular } S_N2} O=C\overset{O}{\triangle}C\text{''''}H(CH_3)$$

$\overset{\ominus}{O}H$ (2nd Inversion)

Intermolecular S_N2

$$\underset{\text{Lactate (retained configuration)}}{{}^{\ominus}O(O=)C-C(H)(OH)-CH_3}$$

(*ii*) Bond breaking and bond making may occur at the front side of the molecule. For example, reaction of a chiral alcohol with thionyl chloride to give a chiral alkyl halide with retention of configuration. The reaction involves a chlorosulphite as intermediate, which decomposes to supply Cl^- to produce alkyl chloride as shown below:

$$H_3C(H)(H_5C_2)C-\ddot{O}-H + Cl-S(=O)-Cl \longrightarrow H_3C(H)(H_5C_2)C-O-S(=O)Cl \longrightarrow H_3C(H)(H_5C_2)C-Cl$$

2-Butanol — Chlorosulphite intermediate — 2-Chlorobutane (retention of configuration)

3.11 RACEMIC MODIFICATION OR RACEMIC MIXTURE

An equimolar mixture of two enantiomers does not possess optical activity and is called ***racemic mixture*** *or* ***racemic modification*** *or* ***conglomerate****. The loss of optical activity is due to the fact that the rotation due to the molecules of one enantiomer is cancelled by an equal and opposite rotation due to the same number of molecules of the other enantiomer.*

Prefixes such as (dl) or (±) or (RS) are used before the name of the compound to specify that it is racemic. For example, if equal amount of (+) and (–) tartaric acids are mixed together, it gives rise to racemic mixture designated as (±) tartaric acid, or dl-tartaric acid or RS-tartaric acid.

[50% (+)-Tartaric acid] + [50% (–)-Tartaric acid] = (±)-Tartaric acid (*A racemic mixture*).

$$[\alpha]_D^{293} = +12^\circ \qquad [\alpha]_D^{293} = -12^\circ \qquad [\alpha]_D^{293} = 0^\circ$$

The optical rotation as well as other physical properties of the racemic mixture such as melting point, boiling point, solubility in a given solvent etc., are also different from those of enantiomers.

3.11.1 Racemisation and Resolution of Enantiomers

Definitions

***Racemisation*:** Conversion of an enantiomer into a racemic mixture (usually as a result of chemical reaction) is known as **racemisation**. In this process the optical activity is diminished to zero.

***Resolution of enantiomers*:** The process of separation of a racemic mixture into its constituent enantiomers is called **resolution of enantiomers**. In the process the optical activity is said to be enhanced and becomes equal to that of pure enantiomers (a phenomenon known as **optical activation**) in the case of complete resolution.

Here it may be mentioned that racemisation and resolution are complementary to each other.

Racemisation: It has been found that attempts to synthesis chiral molecules using achiral reactants and achiral reagents result in the formation of a racemic mixture. In addition, many reactions of chiral molecules also produce racemic mixture. Thus, racemisation is an undesirable process and if the product molecule is chiral, it becomes necessary to avoid the use of reaction conditions and reagents which bring about racemisation.

Mechanism of Racemisation

Racemisation can occur under a wide variety of conditions. During racemisation an enantiomer has to pass through a symmetric species such as a symmetric transition state or a symmetric intermediate so that when the molecule is formed again the two enantiomers are formed in equal amounts and with equal ease.

Methods of Racemisation

Some of the more important methods of racemisation alongwith their mechanisms are discussed below using suitable examples:

1. *Racemisation involving a carbanion as an intermediate*

When an optically active aldehyde or ketone having a hydrogen atom on the α-carbon, which is chiral, is treated with an acid or a base it undergoes racemisation. For example, (–)-lactic acid is racemised in aqueous sodium hydroxide. The mechanism of racemisation is sketched below:

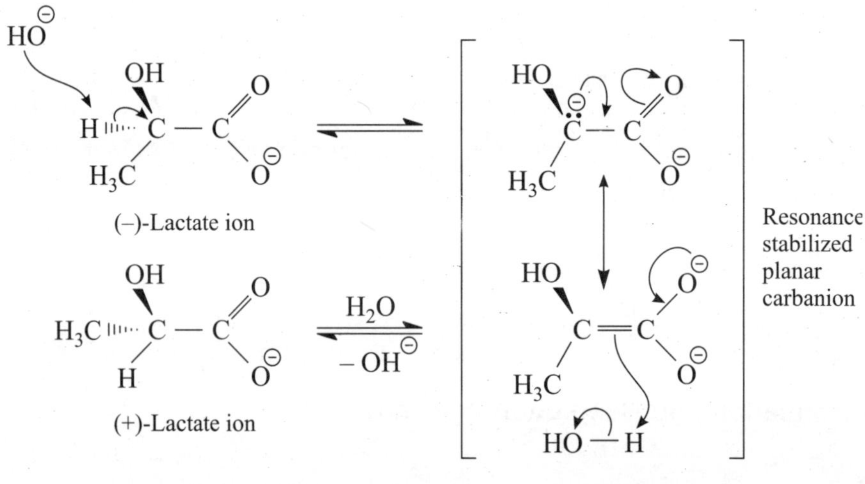

(–)-Lactate ion

(+)-Lactate ion

H_2O

$-OH^{\ominus}$

Resonance stabilized planar carbanion

In this reaction, the initial abstraction of proton by base produces a resonance stabilized carbanion which, being planar, is attacked by water on either face with equal ease to give equimolar mixture of (–) and (+)-lactate ion.

2. *Racemisation involving a carbocation as an intermediate* (S_N1 *mechanism*)

When a group leaves from a chirality centre alongwith electron pair, it leaves behind a carbocation. We know that, carbocations are planar and hence achiral. Recombination of an anion can take place from either side of the carbocation with equal ease thereby leading to racemisation. For example, when (+) or (–) form of α-chloroethylbenzene is treated with $SbCl_5$ in liquid SO_2 it racemises spontaneously. The mechanism of this transformation is given below:

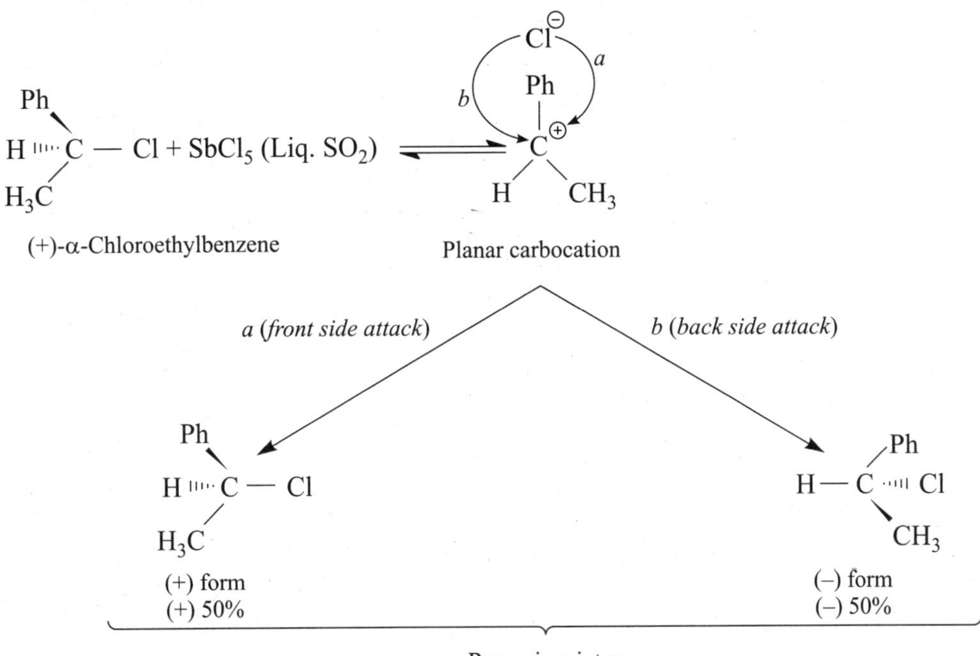

3. *Racemisation involving Walden inversion (S_N2 mechanisms)*

Enantiomers having the halogen at chirality centre can undergo racemisation by S_N2 mechanism. For instance, a solution of (+) or (–)-2-iodobutane on treatment with NaI in acetone produces (±)-2-iodobutane.

$$I^{\ominus} \quad H\text{—}C(CH_3)(C_2H_5)\text{—}I \;\underset{inversion}{\overset{Walden}{\rightleftharpoons}}\; I\text{—}C(CH_3)(C_2H_5)\text{—}H \quad {}^{\ominus}I$$

(One enantiomer) (Other enantiomer)

Any one enantiomer of 2-iodobutane can undergo Walden inversion when treated with sodium iodide to give a 1 : 1 mixture of the two enantiomers *i.e.*, a racemic mixture.

4. *Racemisation involving rotation about carbon-carbon single bond*

We have already discussed that optical activity of biphenyls arises due to restricted rotation. It is, therefore, reasonable to believe that if the rings of such biphenyl derivatives become planar their optical activity should be lost. In agreement with this it has been found that a number of optically active compounds can be racemised under suitable conditions, *e.g.*, heating which overcomes the energy barrier between two enantiomers. As a result of this a racemic modification results, due to rotation about the carbon-carbon single bond between the two rings.

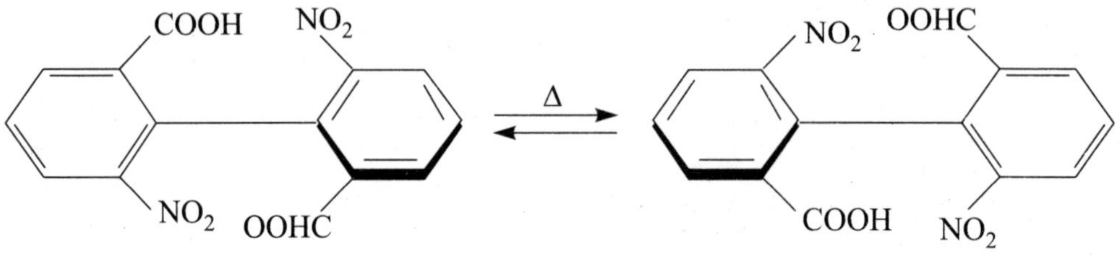

COOH NO$_2$ NO$_2$ OOHC

Δ

NO$_2$ OOHC COOH NO$_2$

6, 6′-Dinitrodiphenic acid
(Racemic mixture)

Resolution

It is well known that enantiomers do not differ in their physical properties such as boiling point, solubility, adsorptivity etc. Therefore, usual methods of separation such as fractional distillation, fractional crystallization or adsorption techniques cannot be used for the separation of enantiomers. Therefore, some special procedures are needed for resolution of racemic mixtures. Some of the more important methods are discussed below:

1. Mechanical Separation

Pasteur (1948) used this method to prove that the compound called **"racemic acid"** is actually an equimolecular mixture of (+) and (–) tartaric acids. He found that when racemic sodium ammonium tartarate was crystallized below 300 K, two types of crystals, were obtained. One type had all the (+) molecules and other type had all (–) molecules. These crystals had distinguishable hemihedral faces and were non-superimposable. They were separated with the help of tweezers and a magnifying glass.

Limitations

(*i*) This method is painstaking and time consuming.

(*ii*) It is of limited use being applicable to only to those compounds which can crystallize to give two well defined types of crystals.

2. Preferential Crystallization

Preferential crystallization is closely related to mechanical separation of crystals. In this technique a supersaturated solution of the racemic mixture is inoculated with a crystal of one of the enantiomers or an isomorphous crystal of another chiral compound. For example, when the saturated solution of (±) sodium ammonium tartarate is seeded with the crystal of one of the pure enantiomer or a crystal of (–) asparagine, (–) sodium ammonium tartarate crystalises out first.

This method is also called as **entrainment** and the seed crystal is called **entrainer**.

3. Biochemical Method

Microorganisms or enzymes are known to be highly stereoselective in their reactions. Pasteur first found that when a solution of ammonium salt of (±) tartaric acid is fermented in presence of yeast or a mold, *e.g.*, *Pencillium glaucum*, the (+) tartaric acid is completely consumed leaving behind (–) tartaric acid. This method has also found wide applicability in the resolution of (±) amino acids. Thus when acetyl derivative of a (±) amino acid is treated with ***hog-kidney acylase*** until half of acetyl groups are hydrolysed away, only acetyl derivative of L-amino acid is hydrolysed leaving behind acetyl derivative of D-amino acid.

This method has the following **limitations**:

(*i*) These reactions are to be carried out in dilute solutions, so isolation of products becomes difficult.

(*ii*) There is loss of one enantiomer which is consumed by the microorganism. Hence only half of the compound is isolated. In other words this procedure is *partially destructive*.

4. Resolution through the Formation of Diastereomers: The Chemical Method

Basic Principle

This method was also discovered by Pasteur. It is based upon the principle that the treatment of a racemic mixture (±)-A with an optically pure reagent (+) or (–)-B results in the formation of a mixture of two products which are diastereomers. The reagent (+) or (–)-B is called the resolving agent.

$$(\pm)\text{-A} + (+)\text{-B} \rightarrow (+)\text{ A }(+)\text{ B} + (-)\text{ A }(+)\text{ B}$$

We know that, the physical properties such as solubility, boiling point and adsorption coefficient of the diastereomers are different because they are not mirror images of each other.

Thus, the mixture of diastereomers obtained above can be separated by using the methods of fractional distillation, fractional crystallization, etc. The pure diastereomers are then decomposed each into the corresponding enantiomer and the original optically active reagent, which are then separated.

Method

In this method the racemic mixture is treated with an optically active pure resolving agent. The product so obtained is separated into diastereomeric derivatives by fractional crystallization if it is a solid and by fractional distillation if it is a liquid. The pure diastereomers are then decomposed to get pure enantiomers and optically active resolving agent which are finally separated.

To illustrate, let us take the case of (±) tartaric acid. When treated with (–)-cinchonidine, (±)-tartaric acid gives a mixture of two diastereomeric salts: (+)-tartarate of (–) chinchonidine and (–)-tartarate of (–)-cinchonidine, which are diastereomers and have widely different solubility. The latter separates out first when the mixture is subjected to fractional crystallization. After the separation and purification of the two diastereomeric salts, each is treated with dilute sulphuric acid which decomposes the salt into the (+)-or (–)-tartaric acid, as the case may be; the base cinchonidine remains in solution as its soluble sulphate.

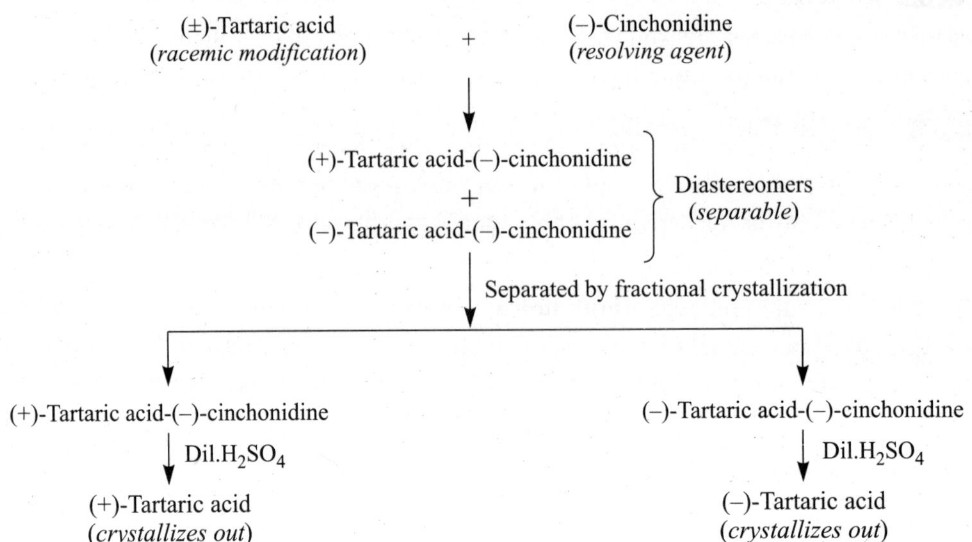

The same procedure may be adopted for the resolution of a (±) base with an optically active acid.

(+)-base	+ (+)-acid ⟶	(+)-(+)-salt	$\xrightarrow{HO^{\ominus}}$	(+)-base
(–)-base		(–)-(+)-salt	$\xrightarrow{HO^{\ominus}}$	(–)-base
Racemic modification		Diastereomers (*separable*)		

The choice of the reagent will depend upon the reactive group in the racemic compound. For example, a racemic alcohol can be resolved as follows:

(+)-alcohol	+ (+)-acid ⟶	(+) - (+) -ester	$\xrightarrow{HO^{\ominus}}$	(+)-alcohol
(–)-alcohol		(–) -(+) -ester	$\xrightarrow{HO^{\ominus}}$	(–)-alcohol
Racemic modification		Diastereomers (*separable*)		

The chemical method of resolution is widely used and has the advantage that both the enantiomers are obtained. This method will be successful if the following conditions are fulfilled:

(*i*) The resolving agent should be optically pure.

(*ii*) The substrate (racemic mixture) and the resolving agent should have suitable functional groups for reaction with each other.

(*iii*) The resolving agent should be cheap and should be capable of regeneration and recycling.

(*iv*) The resolving agent should be such which produces easily crystallizable diastereomeric products.

(*v*) The resolving agent should be easily separable from pure enantiomers.

5. Chromatographic Method

When a racemic mixture is placed on a chromatographic column having a chiral adsorbent, then in principle the rates of movement of the two enantiomers through the column should be different (due to difference in the extent of adsorption). They should thus be separable by elution with suitable solvent.

This method has an advantage over chemical separation as the enantiomers need not be converted into diastereomers.

The techniques used include paper, column, thin layer, gas and liquid chromatography. Some examples are given in Table 3.2:

Table 3.2: Some examples of chromatographic separation of racemic mixtures

S. No.	*Chiral packing material for filling the column*	*Compound whose racemic mixtures are separated*
1.	Microcrystalline triacetyl cellulose (MCTC)	Alcohols, olefins, ketones, acids, esters, biphenyls
2.	Starch	Mandelic acid
3.	Lactose	Chloromycetin, Troger base
4.	Brucine	2-Octyl hydrogen phthalate

3.11.2 Optical Purity

We can easily determine the amount of both enantiomers present in a mixture. It has been shown that for an enantiomerically pure sample (*i.e.*, only one enantiomer) the value of specific rotation [α] is the highest. Any contamination by the other enantiomer lowers the value of specific rotation proportionately.

For example, the observed specific rotation of enantiomerically pure (S) -(+) -2-bromobutane has been found to be +23.1°. Now if a give sample of the compound has [α] = +23.1° it can be concluded that the sample is 100% (S) -(+) -enantiomer. If the specific rotation of the sample is zero we will say that it is a racemic mixture. But if the value of [α] lies between 0 and +23.1° it means that the sample is a mixture of unequal amounts of the two enantiomers. The positive sign of the observed specific rotation means that the mixture has some excess of (S) -enantiomer over (R) -enantiomer. This excess is known as **enantiomeric excess (ee).**

The amount of each enantiomer present in the mixture can be calculated in two steps from the observed specific rotation.

Step I

The **optical purity** of the sample is determined using the following formula:

$$\text{Optical purity (OP)} = \frac{\text{Observed specific rotation, } [\alpha]_{obs}}{\text{Specific rotation of the pure enantiomer } [\alpha]_{max}}$$

Step II

Now suppose a sample of 2-bromobutane has observed specific rotation of +9.2°. We know that, for the pure sample $[\alpha]_{max}$ is +23.1°.

$$\therefore \qquad \text{Optical purity} = \frac{+9.2°}{+23.1°} = 0.4 \text{ or } 40\%$$

The positive sign of observed specific rotation indicates that the mixture has some excess of S-enantiomer. The amount is calculated from optical purity as follows:

As calculated above, the optical purity of the sample is 40%. It means that 40% of the mixture is excess of S-isomer and 100 – 40 = 60% is racemic mixture.

∴ Total amount of enantiomer -S in the mixture will be 40 + 60/2 = 40 + 30 = 70% and the amount of enantiomer -R in the mixture is, therefore, 30%.

3.12 ABSOLUTE AND RELATIVE CONFIGURATIONS

For describing the configurations of different stereoisomers the following two terms are used:

I. Absolute configuration.

II. Relative configuration.

I. Absolute Configuration

It denotes the actual arrangement of atoms or groups of atoms in the space of a particular stereoisomer of a compound. Absolute configuration can be ascertained by X-ray studies of the crystals of pure compound.

II. Relative Configuration

It denotes the arrangement of atoms or groups of atoms in the space of a particular stereoisomer relative to the atoms or groups of atoms of another compound chosen as arbitrary standard for comparison. For example, the configuration (A) was arbitrarily assigned to designate the configuration of (+)-glyceraldehyde. Taking this as standard, the relative configuration of (–) lactic acid was assigned as shown below:

CHO
H—C—OH
CH_2OH
A
(+)-Glyceraldehyde
(Arbitrarily assigned that OH group is on right hand and H on left hand side of the viewer)

—[O], Bromine water→

COOH
H—C—OH
CH_2OH
(–)-Glyceric acid

—P/Br_2→

COOH
H—C—OH
CH_2Br
1-Bromo-2-hydroxypropanoic acid

—Redn.→

COOH
H—C—OH
CH_3
(–)-Lactic acid

In the above reactions the bond between chiral carbon and the substituents has not broken in any step. Hence, the configuration of (–)-lactic acid is the same as that of (+)-glyceraldehyde. Thus, this is the relative configuration of (–)-lactic acid.

Now the question arises to define the configuration of the given enantiomer of a chiral substance. Two commonly used conventions are as follows: 1. D-L System and 2. R-S System.

3.12.1 D-L System

This is one of the oldest and the most commonly used system for assigning configuration to a given enantiomer. It is based upon the comparison of the projection formula of one enantiomer to which the name is to be assigned, with that of a standard substance arbitrarily chosen for comparison.

The following two conventions are used for this purpose:

(*i*) Hydroxy acid or amino acid convention and

(*ii*) Sugar convention.

(*i*) Hydroxy acid or amino acid convention

According to this convention the prefix D- and L- refer to the configuration of α-hydroxy or α-amino acids (*i.e.*, the lowest numbered chirality centre) in the Fischer projection formula. If the α-OH or α-NH_2 group is on the right hand side (of the viewer), the prefix D is used whereas if these groups are on the left hand side the prefix L- is used.

COOH
H — ⓧ → On the right hand side of the viewer

D-Hydroxy acid (X = OH)
D-Amino acid (X = NH_2)

COOH
On the left hand side of the viewer ← ⓧ — H

L-Hydroxy acid (X = OH)
L-Amino acid (X = NH_2)

Examples

COOH
H — OH
CH_3

D-(–)-Lactic acid

COOH
HO — H
CH_3

L-(+)-Lactic acid

COOH
H — OH
HO — H
COOH

D-(+)-Tartaric acid

COOH
HO — H
H — OH
COOH

L-(–)-Tartaric acid

$$\begin{array}{c} COOH \\ | \\ H — C — NH_2 \\ | \\ CH_3 \end{array} \qquad \begin{array}{c} COOH \\ | \\ H_2N — C — H \\ | \\ CH_3 \end{array}$$

D-(–)-Alanine L(+)-Alanine

(**Note:** Here it should always be kept in mind that the sign of rotation (+ or –) has nothing to do with configuration D- or L-).

(*ii*) Sugar convention

This convention was evolved by **Emil Fischer** who arbitrarily assigned D- and L- configurations to (+) and (–)-glyceraldehydes, respectively. He assigned D-configuration (OH on the right) to (+)-glyceraldehyde and L-configuration (OH on the left) to (–)-glyceraldehyde.

$$\begin{array}{c} CHO \\ | \\ H — C — OH \\ | \\ CH_2OH \end{array} \qquad \begin{array}{c} CHO \\ | \\ HO — C — H \\ | \\ CH_2OH \end{array}$$

D-(+)-Glyceraldehyde L-(–)-Glyceraldehyde

The relative configurations of a large number of compounds were determined by correlating them with D (+) or L (–)-glyceraldehyde, *e.g.*, relative configuration of (–)-lactic acid which was designated as D-(–) -lactic acid as it had the same configuration as D (+) glyceraldehyde.

For compounds containing several chiral carbon atoms, the configuration at one centre is related to glyceraldehyde and the configuration at other carbon atoms are determined relative to the first. In the case of glucose, this carbon atom is C_5 which is next to the CH_2OH group *i.e.*, *the highest numbered chiral carbon*. Since naturally occurring glucose was assumed to have the OH group of this carbon projecting at right hand side of the viewer, it belongs to the D series of compounds and hence designated D-glucose. In case of the compounds having the OH group on the highest numbered chiral carbon on left side of the viewer, notation L- is used.

In 1951, however, **Bijvoet** determined the absolute configuration of (+) sodium rubidium tartarate using X-ray crystallography. It was found to be the same as assigned arbitrarily. Before Bijvoet's research, configurations of several thousand compounds were assigned relative to (+)-tartaric acid. So after this discovery all these relative configurations became absolute configurations.

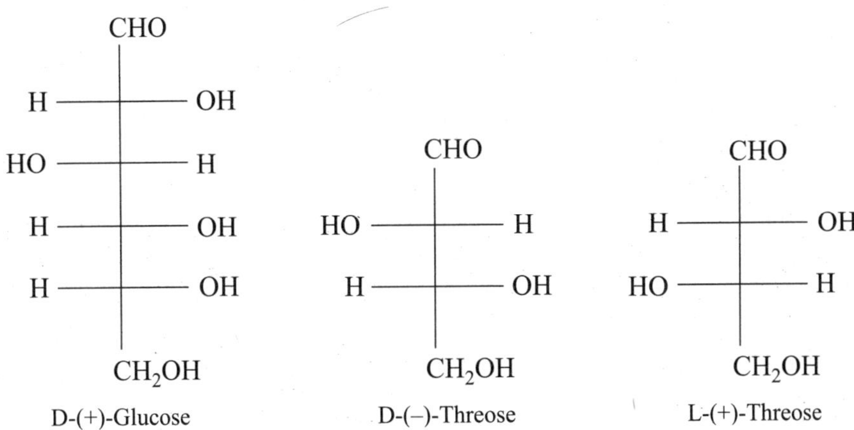

D-(+)-Glucose D-(–)-Threose L-(+)-Threose

Limitations of sugar convention

1. Sugar molecules have more than one chirality centres and the configuration of only the highest numbered chirality centre is assigned and that of the other centres are not shown (hidden in their names).

2. Another complication is encountered when compound of one series is converted into that of the other. For example, saccharic acid which is obtained by oxidation of D-glucose is assigned D-configuration, as the configuration of highest numbered chiral carbon is D.

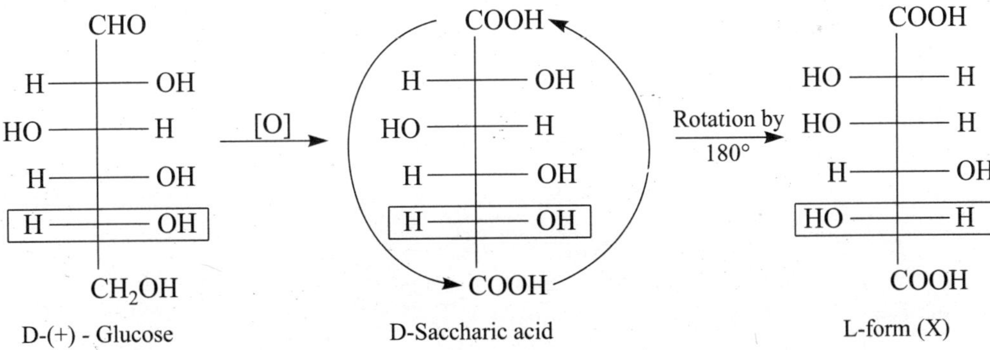

D-(+) - Glucose D-Saccharic acid L-form (X)

Now if we rotate the molecule by 180° in the plane of the paper we get formula (X) which should be designated as L-. This is very serious drawback in the system *i.e.*, the same molecule can have both D- and L-configurations.

3. Cases of (+) - Tartaric Acid and (–) - Threonine:
Both these compounds be designated as D-as well as L-depending upon whether the reference compound is glyceraldehyde (highest numbered chiral carbon) or hydroxy or amino acid (lowest numbered chiral carbon).

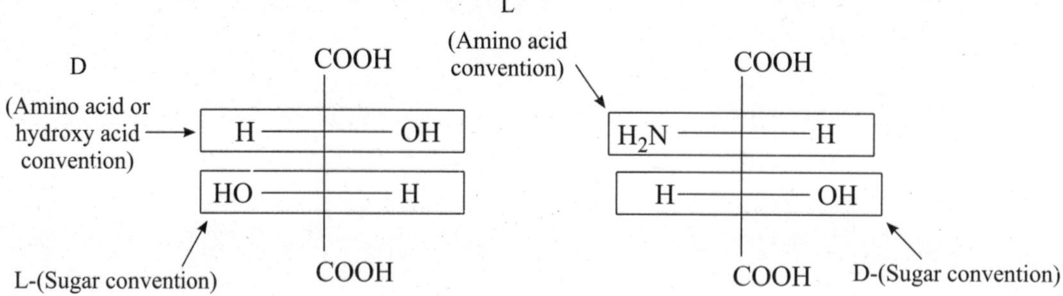

In conclusion it may be added that this system is of limited use as it is confined only to:
1. Sugars, 2. Hydroxy acids and 3. Amino acids.

3.12.2 R-S System

We have seen above that the greatest drawback of the D and L system for specification of configuration around a chiral carbon was not found to be suitable for molecules which are not related with a particular glyceraldehyde. To overcome this problem **R.S. Cahn** (England), **Sir Christopher Ingold** (England), and V. **Prelog** (Zürich) evolved a new and unambiguous system for assigning absolute configuration to chiral molecules. This system is named as CIP (**C**ahn, **I**ngold, **P**relog) system after their names. It is called as R-S system as the prefixes R-and S-are used to designate the configuration at a particular chirality centre. A racemic mixture is named as (RS). This system is based on certain rules called as **sequence rules** and also as **CIP rules**.

The designation of a chirality centre is done in the following steps:

***Step I*:** Identify the atoms or groups directly attached to the chirality centre and assign them a sequence of priority by using greek numerals 1, 2, 3 and 4 where number 1 is assigned to atom or group of highest priority and 4 is assigned to the group of lowest priority.

The priorities of the groups are assigned as per CIP rules (***vide infra***), *e.g.*,

H

H_3C — C — OH

H_3CH_2C

2-Butanol

The priorities of the substituents as determined by CIP rules are —OH is **1**, CH_3CH_2— is **2**, CH_3— is **3** and H is **4** *i.e.*, —OH has the highest priority and H has the lowest priority.

***Step II*:** View the molecule in such a way that the lowest ranked group (priority 4) points away from you.

In the above formulation the molecule has already been appropriately oriented. H-atom having lowest priority (4) is away from us.

***Step III*:** Draw the remaining three groups as they appear to you in the following manner when the molecule is oriented as specified in step II.

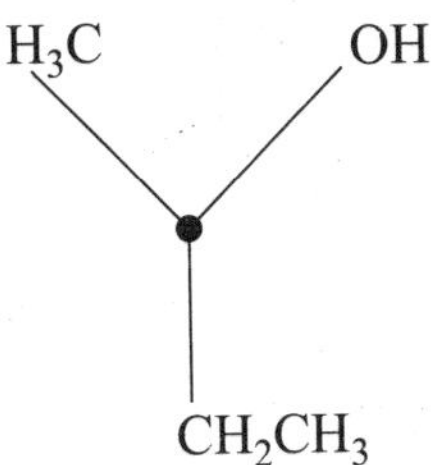

Step IV: Move your eye from the group of priority number 1 to group of priority number 3 *via* the group of priority number 2 *i.e.*, in the following order:

$$1 \rightarrow 2 \rightarrow 3$$

If during this movement your eye travels in the clockwise direction, the molecule under examination is designated as R (Latin: **Rectus** meaning **right**) and if it moves in the anticlockwise direction it is designated as S (Latin: **Sinister** meaning **left**). The letters R and S are written in parenthesis.

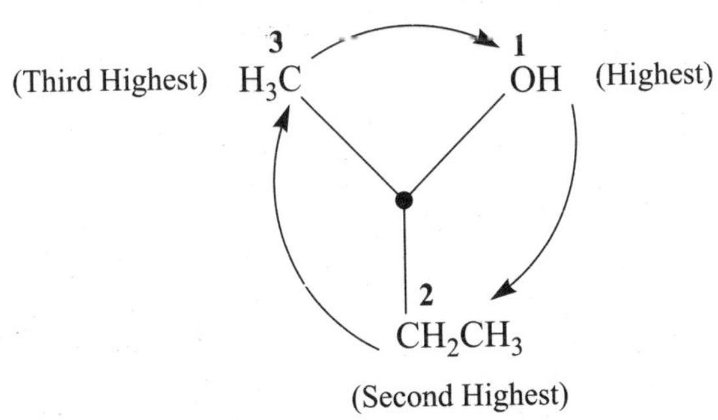

In the above arrangement our eye moves in clockwise direction, so the absolute configuration of (–)-2-butanol is **R**. Here it may be mentioned that the complete name of the enantiomer specifies both configuration and direction of rotation. Thus, the full name of the above molecule is (R)-(–)-2-butanol.

A comprehensive list of some common groups in increasing order of priority is given in the box below:

> ***Priority sequence order of various groups***
> **Lowest:** Non-bonding electrons on N (At. No. = 0) —H, —D, —CH_3, —CH_2CH_3, —$CH_2(CH_2)_nCH_3$, —CH_2—CH=CH_2, —CH_2—C≡CH, —CH_2—C_6H_5, —$CH(CH_3)_2$, —CH=CH_2, —$C(CH_3)_3$ —C≡CH, —C_6H_5, —CH_2OH, —CHO, —COR, —$CONH_2$, —COOH, —COOR, —NH_2, —$NHCH_3$, —$N(CH_3)_2$, —NO, —NO_2, —OH, —OCH_3, —OC_6H_5, —OCOR, —F, —SH, —SR, —SOR, —Cl, —Br, —I **Highest**.

Sequence Rules

For fixing priorities of various atoms/groups attached to a chirality centre, the following sequence rules have been evolved:

***Sequence rule* 1:** *If four atoms/groups attached to the chirality centre are all different, the atom with highest atomic number is given the highest priority.* For example, in bromochlorofluoroiodomethane, the priorities are:

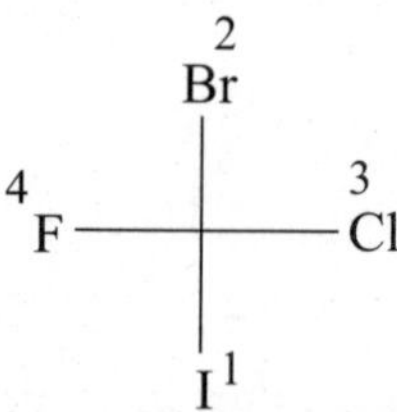

I = 1; Br = 2; Cl = 3; F = 4 *i.e.*, the sequence of priorities is I > Br > Cl > F.

However, if *two isotopes of the same element are attached to the chirality centre, the atom with higher mass number is given higher priority.* For example in the α-deuterioethyl bromide.

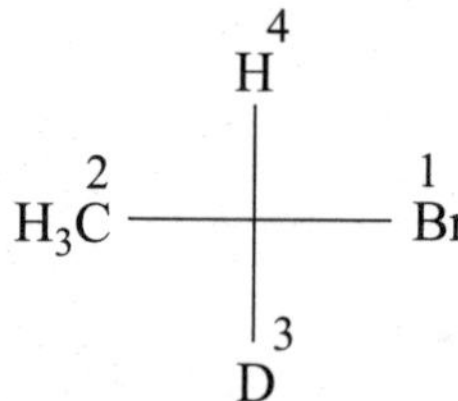

Where two isotopes of hydrogen *i.e.*, H and D are attached to the chiral carbon, D having higher mass number is given higher priority than H.

Hence, the sequence of priority of the four groups will be: Br > CH_3 > D > H

***Sequence rule* 2:** *If on the basis of the* **sequence rule 1** *the priorities of two groups cannot be decided, it can be determined by a similar comparison of the next atoms, in both groups. If by doing so the priority cannot be decided, one goes to 'next' atom and continues moving outwards commencing with the chiral atom till one reaches the first point of difference.*

Some examples are as follows:

(*i*) $—CH_2CH_3$ and $—CH_3$

In C_2H_5 the atoms are H, H, C and in CH_3 they are H, H, H. So, CH_3CH_2 gets higher priority than CH_3.

(*ii*) $—CH_2OH$ and $—CH_2NH_2$

In $—CH_2OH$ the atoms are H, H, O and in $—CH_2NH_2$ they are H, H, N. So, the priority order is $—CH_2OH > —CH_2NH_2$

(*iii*) $—CH_2CHF$ Br and $—CH_2CHFCl$

No decision can be taken as the first groups are CH_2 in both. Now going to next C the groups are H, F, Br in the first case and in the second case they are H, F, Cl. So, the priority order is $—CH_2CHFBr > —CH_2CHFCl$.

(*iv*) $—CH_2\,CH_3$ $—CH(CH_3)_2$

A   B

In **A** the groups are H, H, C and in **B** they are H, C, C. So, —$CH(CH_3)_2$ > —CH_2CH_3

(*v*) —$CH(CH_3)_2$ —CH_2Cl

A B

In **A** atoms are H, C, C and in **B** they are H, H, Cl. So, —CH_2Cl > —$CH(CH_3)_2$

(**Note :** The decision about priority should be made at the very first point of difference, and should not be effected from the consideration of substituents further along the chain.)

***Sequence rule* 3:** *In case the group attached to the chiral carbon contains a double bond or a triple bond, both atoms joined by multiple bonds are considered to be duplicated* (*in case of a double bond*) *and triplicated* (*in case of a triple bond*). For example,

1. Group containing a double bond say >C=X is considered to be equal to —C(X)—X(C) [C bearing a duplicate X; X bearing a duplicate C]

2. Group containing a triple bond say —C≡X is considered to be equal to —C(X)(X)—X(C)(C) [C bearing two duplicate X; X bearing two duplicate C]

3. The phenyl substituent (benzene ring) is considered equal to HC C CH with C above and C below (each ring carbon carrying duplicate C atoms)

This rule is illustrated by the following examples:

(*i*) **Glyceraldehyde**

H—C=O
|
H—C—OH
|
CH_2OH

The problem is to decide the priorities of CH_2OH and CHO.

Now — CHO is regarded as equal to —C(H)(O)—O(C) and CH_2OH is —C(H)(H)—O—H

Here carbon atom of —CHO having H, O, O takes priority over that of CH_2OH having H, H, O.

(*ii*) $C_6H_5-CH(NH_2)-CH(CH_3)-CH_3$

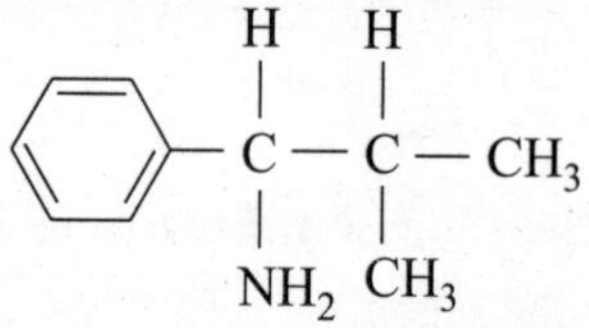

In this case priorities of phenyl ring and isopropyl group are to be decided.

Now – phenyl group 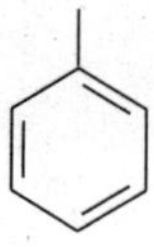 is regarded as equal to H—Ⓒ Ⓒ Ⓒ — C

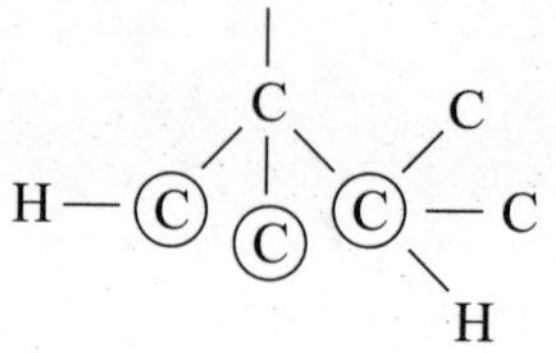

and isopropyl group is $-C(C)(H)-C$ so phenyl ring having C, C, C takes priority over isopropyl group having C, C, H. So, the sequence of priority of the four groups is $—NH_2$ > Ph > Isopropyl > H.

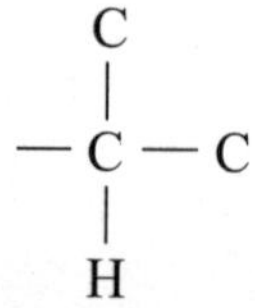

Examples

A. Tetrahedral wedge formulas

(*i*) 2 COOH, H, 3 CH_3, HO 1

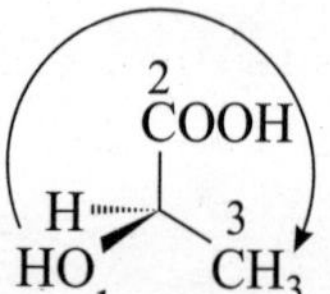

1. The priorities are: OH > COOH > CH_3 > H.
2. The group of lowest priority (H) is away from us.
3. Movement of eye from 1 → 2 → 3 takes place in clockwise direction. So, the configuration is R.

(*ii*) 2 COOH, 3 H_3C, H, OH 1

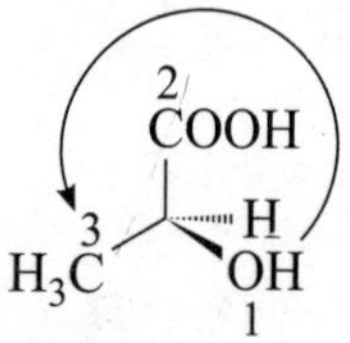

Similar operations as in (*i*) above reveal that it has S-configuration.

B. Fischer projection formulas

Step I: Assign priorities to four atoms attached to the chiral carbon. The assignment of configuration will depend upon the fact whether the group of lowest priority falls on the vertical line or on horizontal line. Further steps are as follows:

(*a*) **Group of lowest priority on vertical line:** It does not matter whether upward or downward because in Fischer projection both the groups on the vertical line are always away from the viewer.

The eye is moved in the 1 → 2 → 3 direction. If it is clockwise the configuration then is R and if it is counter clockwise the configuration then is S. For example,

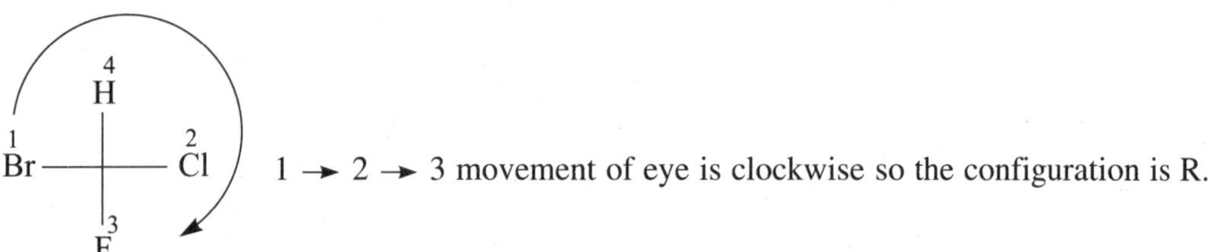

3 F, 2 Cl, 1 Br, 4 H — 1 → 2 → 3 movement of eye is clockwise so the configuration is R.

(*b*) **The group of lowest priority on the horizontal line:** In such a case make two interchanges of the atoms or groups in such a way that the atom or group of lowest priority falls on vertical line either upward or downward.

This procedure is illustrated by the following examples:

D (+) Glycozaldehyde

2 CHO, 4 H, 1 OH, 3 CH_2OH ≡ (2 top, 4 left, 1 right, 3 bottom) —First interchange of 4 and 3→ (2 top, 3 left, 1 right, 4 bottom) —Second interchange of 2 and 3→ (3 top, 2 left, 1 right, 4 bottom)

Group of lowest priority on the horizontal line

Clockwise
So it has R configuration

D (–) Lactic acid

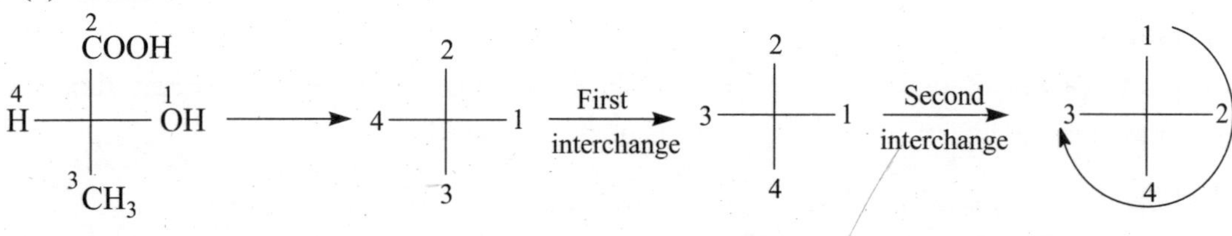

2-Butanol

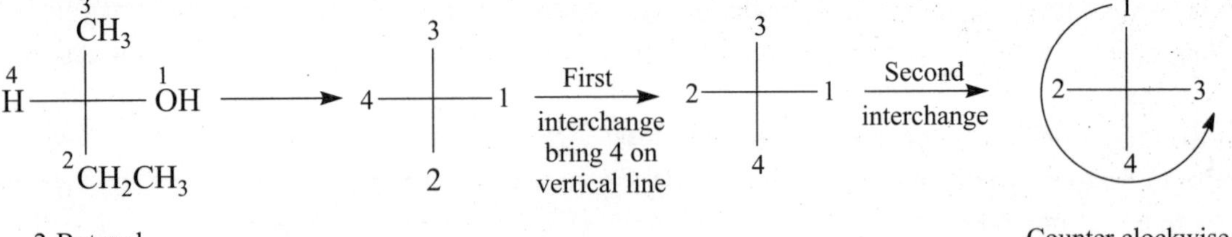

2-Butanol

Counter clockwise
So it has S configuration

Alternate procedure for assigning configuration

Out of a number of alternative procedures for assigning R- and S-configurations on the basis of Fischer projection, the simplest and most widely accepted one was suggested by Epliel. who coined the term **"very good"** for the expression **"vertical-good!"** for his procedure which consists of two operations:

Operation I: Fix up priorities of the groups and move your eye from $1 \rightarrow 2 \rightarrow 3$ ignoring 4. Now, if

(*i*) Group of lowest priority (4) is on the vertical line (whether on top or bottom), and the sequence $1 \rightarrow 2 \rightarrow 3$ is in clockwise direction the configuration is R and if it is in counter clockwise direction the configuration is S.

(*ii*) Group of lowest priority (4) is on the horizontal line assign the configuration which is opposite to what you see *i.e.*; if the movement of the eye from $1 \rightarrow 2 \rightarrow 3$ is in clockwise direction, assign S-configuration and if it moves in counter clockwise direction assign R-configuration. For example,

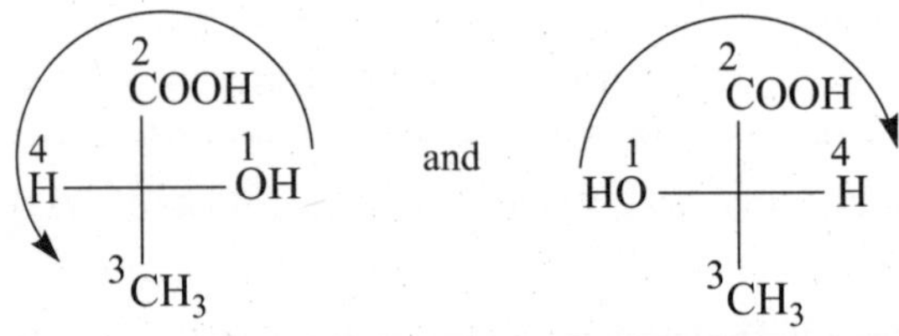

Movement of eye from $1 \rightarrow 2 \rightarrow 3$ is in counter clockwise direction so it has R-configuration.
Movement of eye from $1 \rightarrow 2 \rightarrow 3$ is clockwise direction so it has S-configuration.

R-S Nomenclature of Compounds having more than one Chiral Carbon

When a compound possesses two different chiral carbons, each of these chirality centres may be either (R) or (S). Four different stereoisomers are possible having the following configurations; (1R, 2R); (1R, 2S); (1S, 2R) and (1S, 2S). Each chiral carbon is considered separately for assigning (R) or (S) configurations. Let us consider the configuration around chiral carbon atoms number 2 and 3 in the 2, 3, 4-trihydroxybutanal, commonly referred to as D-erythrose.

For carbon 2

Movement of eye from $1 \rightarrow 2 \rightarrow 3$ counter clockwise so as per "very good" device carbon 2 has R-configuration as the lowest priority (4) group is on horizontal line.

For carbon 3

Movement of eye from $1 \rightarrow 2 \rightarrow 3$ counter clockwise so as per "very good" device carbon 3 also has R-configuration.

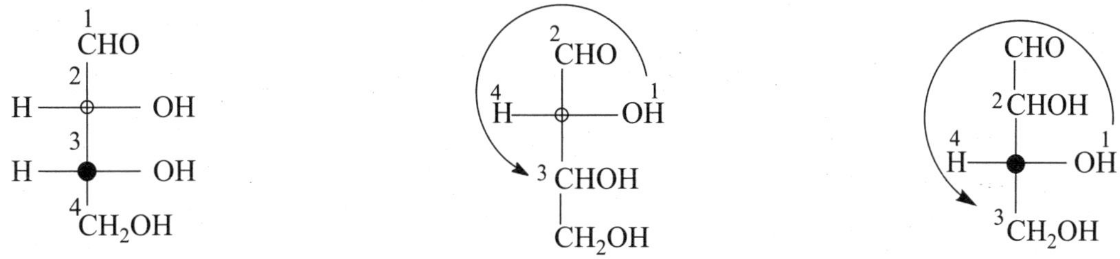

The configuration of the above compound will, therefore, be (2R, 3R) -2, 3-4-trihydroxybutanal or 2R,3R- erythrose.

We can similarly assign configurations to the other compounds.

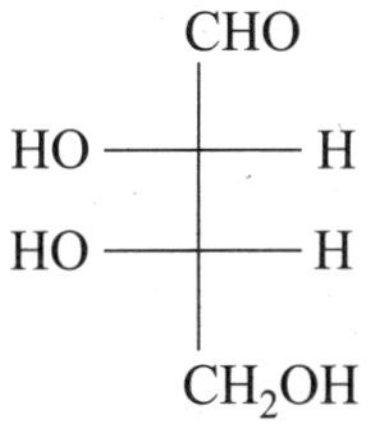

L-Erythrose (2S, 3S) | D-Threose (2S, 3R) | L-Threose (2R, 3S)

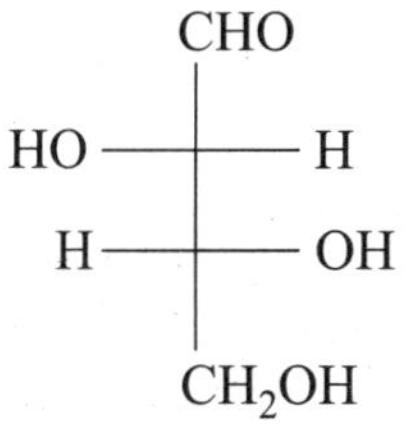

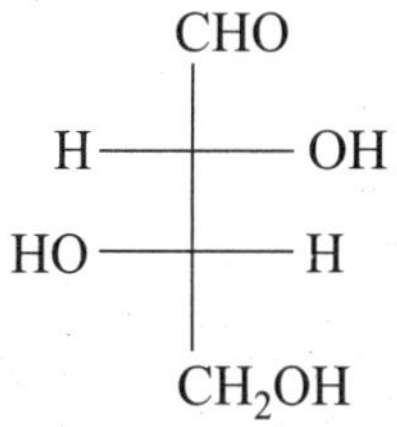

3.13 GEOMETRICAL ISOMERISM

Definitions and General Aspects

Isomers having the same structural formula but different relative spatial arrangement of atoms or groups around a double bond are called **geometric isomers**, ***cis-trans*** **isomer** or **E-Z isomers** and this phenomenon is known as **geometric isomerism**, *cis-trans* or **E-Z isomerism**. This type of isomerism arises if we assume that there is no free rotation about the double bond.

Due to different arrangement of atoms or groups in the space, geometric isomerism is designated as stereoisomerism. The geometric isomers belong to the category of configurational isomers because they cannot be interconverted without breaking two covalent bonds. Further, geometric isomers are examples of diastereomers because they are not mirror images of each other.

Due to rigidity of double bond towards free rotation, these isomers have two spatial arrangements. In one arrangement, called *cis*-(Latin: ***cis*** **= on this side**), two similar or identical atoms or groups are on the same side of the double bond. In the other arrangement called *trans*-(Latin : ***trans*** **= across**), two similar or identical atoms or groups are on the opposite side of the double bond. For example, the two geometric isomers of butenedioic acid are:

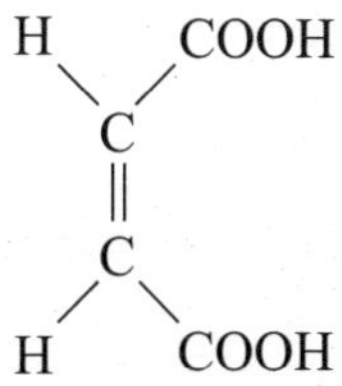

cis-Butenedioic acid (*Maleic acid*)

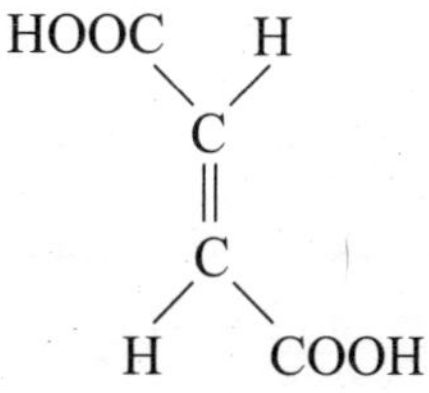

trans-Butenedioic acid (*Fumaric acid*)

In general geometric isomerism is exhibited by the compounds of the type **A** having non-identical groups attached to doubly bonded carbon atoms. It is not exhibited by compounds of the type **B** because in this type of molecules at least one of the doubly bonded carbon atom bears two identical groups or atoms.

$$\underset{\text{A}}{\begin{matrix} a & & a \\ & C=C & \\ b & & b \end{matrix}} \qquad \underset{\text{B}}{\begin{matrix} a & & a \\ & C=C & \\ a & & b \end{matrix}}$$

Geometric isomerism is not confined only to the compounds having carbon-carbon double bonds. In fact the following compounds exhibit this type of isomerism:

(*i*) Compounds having a double bond, *i. e.*, olefins (C = C), imines (C = N) and azo compounds (N = N).

(*ii*) Cyclic compounds.

(*iii*) Compounds exhibiting geometric isomerism due to restricted rotation about carbon-carbon single bond.

3.13.1 Cause of Geometric Isomerism—*Hindered Rotation*

In contrast to cylindrically symmetrical single bond (σ-orbital) joining two carbon atoms, overlap of two *p*-atomic orbitals to give a π-molecular orbital completely prevents rotation of the groups about the double bond under normal conditions. Carbon atoms involved in double bond formation and all the atoms attached to these doubly bonded carbon atoms must lie in the same plane because π-bond can be formed only by parallel overlap of the two *p*-orbitals. There will be decrease in the overlap of *p*-orbitals if an attempt is made to destroy this coplanarity. In other words, neither of the doubly bonded carbon atom can be rotated about the double bond without destroying the π-orbital (Fig 3.5).

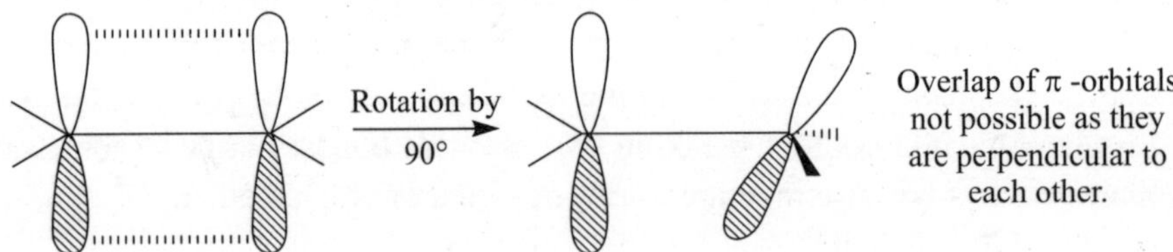

Fig. 3.5: Rotation about a C = C bond

This process which is really a transfer of electrons from the *p*-molecular orbital to the *p*-atomic orbital is associated with high energy about 271.7 kJ mol^{-1}. Thus at ordinary temperatures, rotation about a double bond is prevented and hence compounds such as $CH_3CH = CHCH_3$ exist in isolable and stable geometrical isomeric forms.

$$\underset{\textit{cis}\text{-But-2-ene}}{\begin{matrix} H & & H \\ & C=C & \\ H_3C & & CH_3 \end{matrix}} \qquad \underset{\textit{trans}\text{-But-2-ene}}{\begin{matrix} H_3C & & H \\ & C=C & \\ H & & CH_3 \end{matrix}}$$

The two compounds are normally quite stable since interconversion requires high energy and can only be achieved at high temperatures or by irradiation with ultraviolet light. This may be contrasted to conformational isomers in which the groups rotate about σ bonds because of relatively low energy barriers between the conformers which are, therefore, not isolable at ordinary temperatures.

3.13.2 Necessary and Sufficient Condition for Geometric Isomerism

As stated above the existence of geometric isomerism depends upon the substitution pattern at the double bond. *Geometrical isomerism will not be possible if one of the unsaturated carbon atoms is bonded to two identical groups.* For instances no two stereoisomers are possible for compounds such as $CH_3HC = CH_2$, $(CH_3)_2C = CH_2$ and $Cl_2C = CHCl$.

A few examples of compounds existing in two stereoisomeric forms are listed below:

```
H           H              H            COOH
 \         /                \          /
  C  =  C                    C  =  C
 /         \                /          \
H5C6        COOH          H5C6          H
```

cis-Cinnamic acid *trans*-Cinnamic acid

```
        H           H                 H            CH3
         \         /                   \          /
          C  =  C                       C  =  C
         /         \                   /          \
H3C — H2C           CH3       H3C — H2C            H
```

cis-Pent-2-ene *trans*-Pent-2-ene

3.13.3 Determination of the Configuration of the Geometric Isomers

Methods used for determination of configuration of geometric isomers can be broadly classified in the following two categories:

I. Physical methods

II. Chemical methods

I. Physical Methods

***(a) Melting points and boiling points*:** In general the *trans* isomer has a **higher m. pt.** and a **lower b. pt.** than that of the corresponding *cis*-isomer.

Reason: *Trans*-isomers being symmetrical get closely packed in the crystal lattice due to which they have strong inter-molecular forces of attraction and hence higher melting points.

As we shall see below, *cis*-isomers have higher dipole moments than *trans*-isomers due to which *cis*-isomers have higher b.pt. than those of *trans*-isomers. This may be due to dipole-dipole attractive forces holding the molecules together.

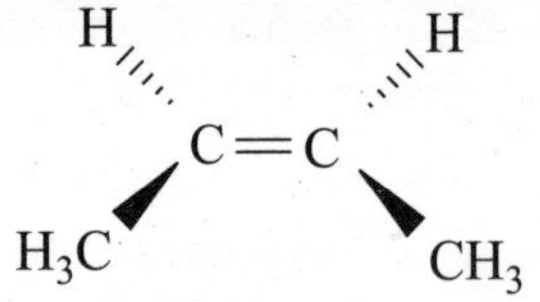

H, H / H_3C, CH_3 (C=C)

b.pt. 277 K

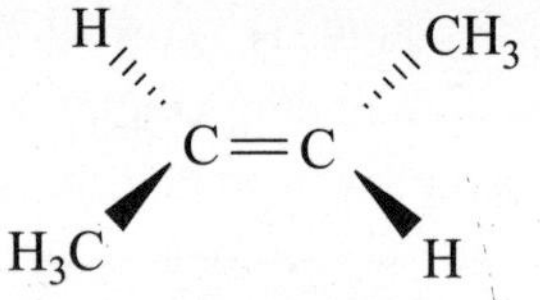

H, CH_3 / H_3C, H (C=C)

b.pt. 274 K

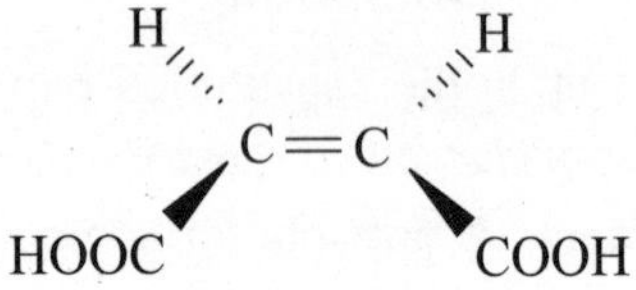

H, H / $HOOC$, $COOH$ (C=C)

m.pt. 403 K

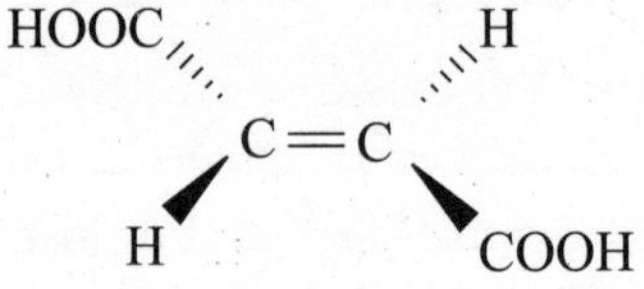

$HOOC$, H / H, $COOH$ (C=C)

m.pt. 575 K

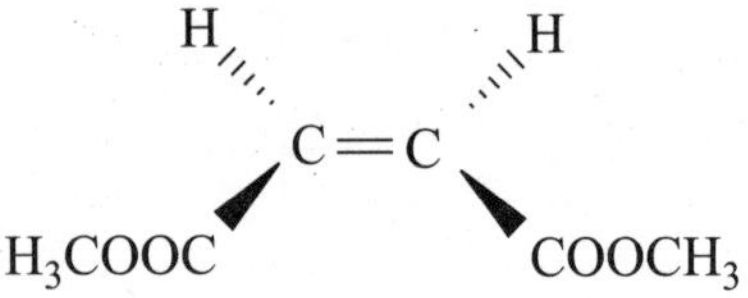

H, H / H_3COOC, $COOCH_3$ (C=C)

m.pt. 264.3 K

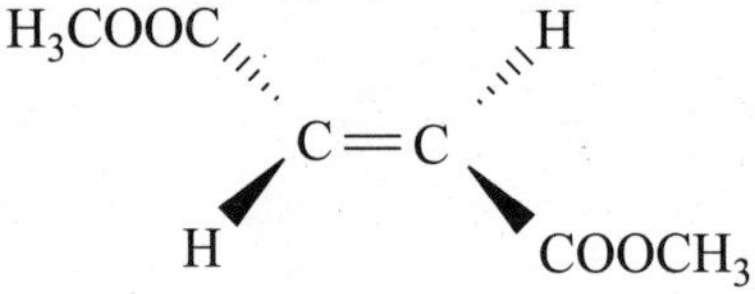

H_3COOC, H / H, $COOCH_3$ (C=C)

m.pt. 374 K

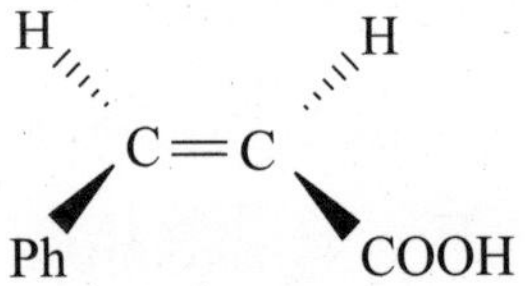

H, H / Ph, $COOH$ (C=C)

m.pt. 341 K

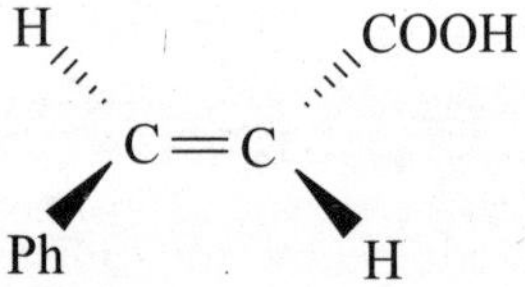

H, $COOH$ / Ph, H (C=C)

m.pt. 406 K

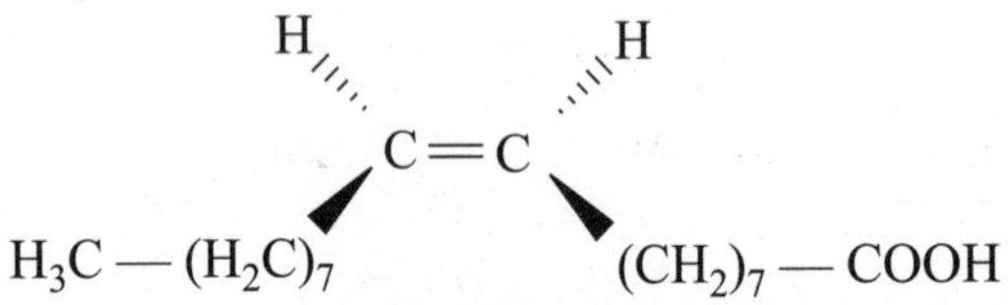

H, H / $H_3C-(H_2C)_7$, $(CH_2)_7-COOH$ (C=C)

m.pt. 289 K

H, $(CH_2)_7-COOH$ / $H_3C-(CH_2)_7$, H (C=C)

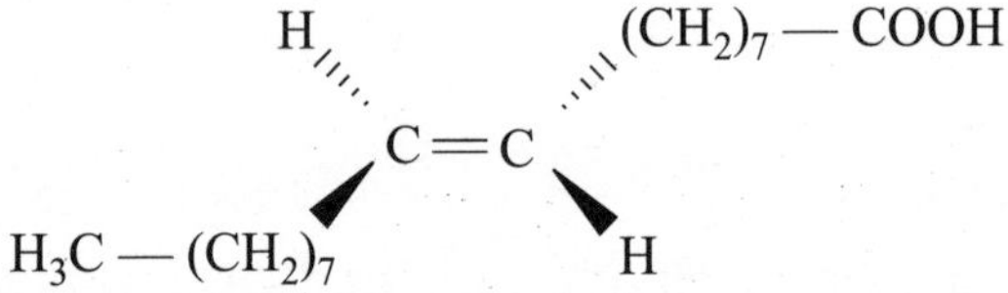

m.pt. 324 K

***(b) Solubility*:** The *cis*-isomers have higher solubilities. The solubilities of maleic and fumaric acids for example, are noteworthy.

Maleic acid: 79.0 g/100 mL at 293 K

Fumaric acid: 0.7g /100 mL at 293 K

***(c) Dipole moment*:** In general, *cis*-isomers have the greater dipole moment. The *cis*-isomer of 2-butene, for instance, has a significant dipole moment because the two methyl groups are attached to the same side of the double bond and the combined inductive effects are additive. *Trans*-2-butene, on the other hand, has one methyl group and one hydrogen on each side of the double bond and hence the bond moments cancel each other. Similarly, the *cis*-isomer of 1, 2-dichloroethene has a significant dipole moment while that for the *trans*-isomer is zero.

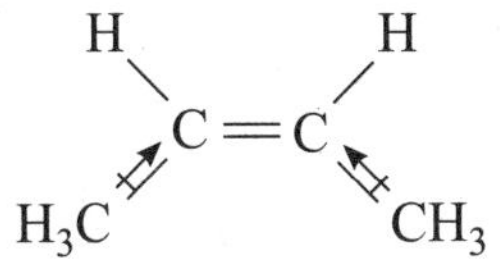

$\mu = 0.4$ D

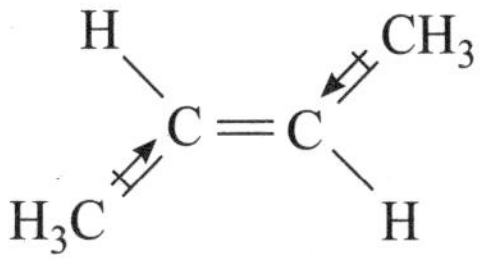

$\mu = 0$

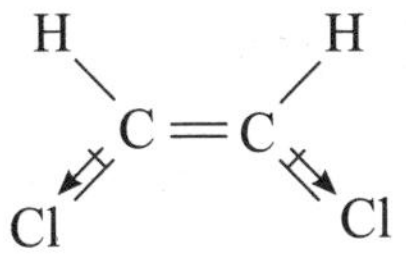

$\mu = 1.85$ D

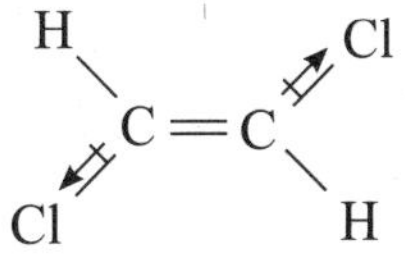

$\mu = 0$

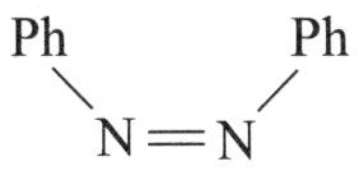

$\mu = 3.0$ D

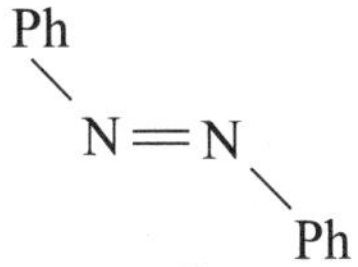

$\mu = 0$

(d) Spectroscopic data:

IR: Infrared spectra are sometimes used to distinguish between *cis*- and *trans*-isomers of the type RCH $=$ CHR′. Isomer having a *trans* configuration is readily identified by the appearance of a characteristic band near 970–960 cm^{-1}. No such band is observed in the spectrum of the *cis*-isomer.

NMR: Nuclear magnetic resonance spectroscopy provides another powerful tool for distinction between *cis*- and *trans*-isomers of the above type. The protons in the two isomers have different coupling constants *e.g.*, *trans*-vinyl protons have a larger value of the coupling constant than the *cis*-isomer. This difference is well exemplified by the *cis*- and trans-cinnamic acids.

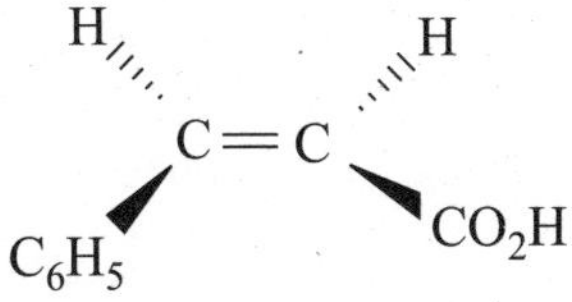

cis-Cinnamic acid (J_H, H = 12 Hz)

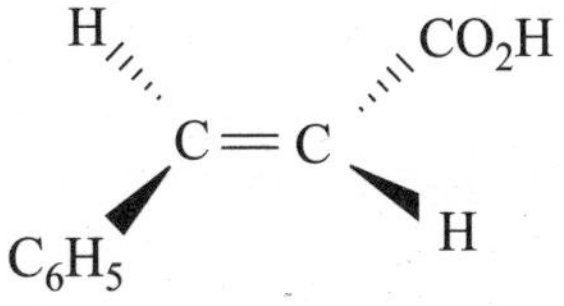

trans-Cinnamic acid (J_H, H = 16 Hz)

II. Chemical Methods

***(a) Methods of formation from cyclic compounds*:** Suppose we oxidize benzene or quinone to an unsaturated dicarboxylic acid of m.pt. 403 K. From the structure of benzene or quinone, it becomes clear that the two carboxyl groups must be on the same side (*cis*).

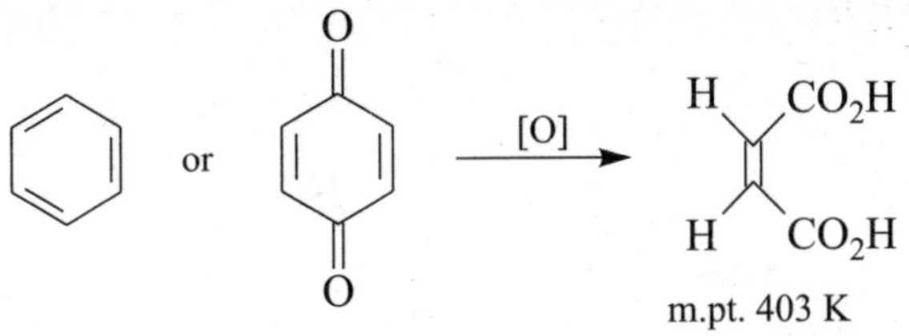

Benzene or p-benzoquinone $\xrightarrow{[O]}$ H(CO$_2$H)C=C(CO$_2$H)H (m.pt. 403 K)

Therefore, maleic acid *i.e.*, the isomer having m.pt. 403 K, must be *cis* and the other isomer fumaric acid (m.pt. 575 K) must be *trans*.

(*b*) *Methods of formation of cyclic compounds*

(*i*) *Maleic and fumaric acids:* Very often two geometric isomers are distinguished through reactions that lead to the formation of rings. It is expected that a *cis*-isomer will undergo ring closure much more readily than the *trans*-isomer because in *cis*-isomer the two reacting groups are in close proximity of each other and it is almost impossible to prepare a ring compound containing a *trans* double bond.

Maleic and fumaric acids represent two most interesting compounds in this respect. Both the acids have the molecular formula $C_4H_4O_4$. The structures of these acids are established by the fact that they are obtained by the dehydration of maleic acid and their catalytic reduction yields succinic acid.

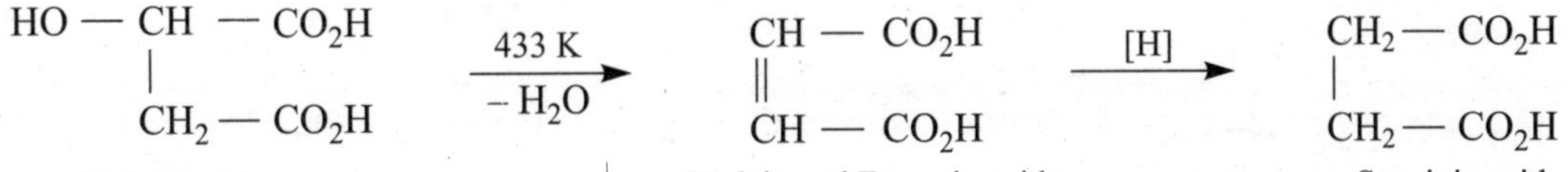

$$\begin{array}{c} HO-CH-CO_2H \\ | \\ CH_2-CO_2H \\ \text{Maleic acid} \end{array} \xrightarrow[-H_2O]{433\ K} \begin{array}{c} CH-CO_2H \\ \| \\ CH-CO_2H \\ \text{Maleic and Fumaric acids} \end{array} \xrightarrow{[H]} \begin{array}{c} CH_2-CO_2H \\ | \\ CH_2-CO_2H \\ \text{Succinic acid} \end{array}$$

In spite of this similarity, both the acids differ in their physical properties as becomes evident from the following:

	M.pt.	Water solubility	Density	Heat of combustion (kJ mol^{-1})
Maleic Acid	403 K	79.0	1.59	327
Fumaric Acid	575 K	0.7	1.64	320

Now the question arises whether the configuration of maleic acid is *cis* or *trans*?

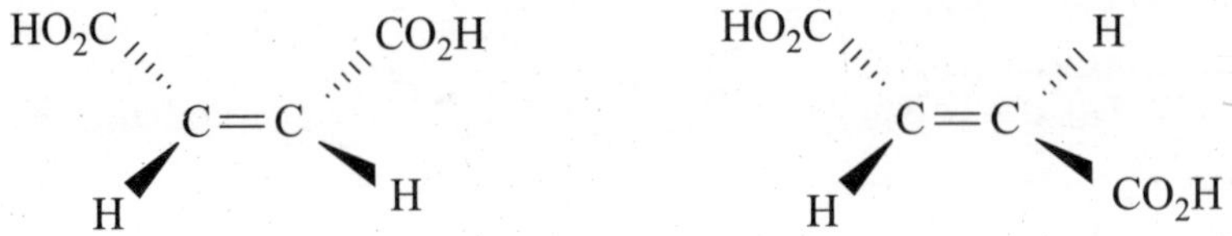

$$\begin{array}{c} HO_2C\quad\quad CO_2H \\ C{=}C \\ H\quad\quad\quad H \\ \textit{Cis} \end{array} \qquad \begin{array}{c} HO_2C\quad\quad H \\ C{=}C \\ H\quad\quad\quad CO_2H \\ \textit{Trans} \end{array}$$

It was found that maleic acid readily loses water when heated to about 423 K to yield a cyclic anhydride, whereas fumaric acid does not give the anhydride at this temperature. Rather it must be heated to 573 K to get the same anhydride. Hydrolysis of the anhydride yields only maleic acid.

H CO$_2$H C=C H CO$_2$H (Maleic acid) $\xrightarrow[-H_2O]{423\ K}$ H CO C=C H CO, O (Maleic anhydride) $\xleftarrow[-H_2O]{573\ K}$ H CO$_2$H C=C HO$_2$C H (Fumaric acid)

Maleic acid

Maleic anhydride

Fumaric acid

Maleic anhydride $\xrightarrow{\text{Hydrolysis}}$ H CO$_2$H C=C H CO$_2$H

Maleic acid

It is, therefore, reasonable to conclude that maleic acid is the *cis*-isomer and fumaric acid is the *trans*-isomer. The latter forms the anhydride *via* the formation of maleic acid at high temperature which involves rupture of π-bond and rotation of the acid groups followed by reformation of the π-bond and loss of water.

H CO$_2$H C=C HO$_2$C H $\xrightarrow{\Delta}$ H CO$_2$H Ċ—Ċ HO$_2$C H $\xrightarrow{\text{Rotation}}$ H CO$_2$H Ċ—Ċ H CO$_2$H

Fumaric acid

Diradical

Diradical

↓

H CO$_2$H C=C H CO$_2$H $\xrightarrow{-H_2O}$ H CO C=C H CO, O

Maleic anhydride

Maleic acid

Chapter 3

(*ii*) *Ortho-aminocinnamic acids*: Let us take another instance. The Basalt of an isomer of ortho-aminocinnamic acid on treatment with CO_2 at room temperature gives carbostyril. This shows that the carboxyl group and the substituted phenyl group must be *cis* in this isomer. On the other hand, the Basalt of the other isomer of *ortho*-aminocinnamic acid does not give carbostyril under the same condition and, therefore, it must have the *trans* configuration.

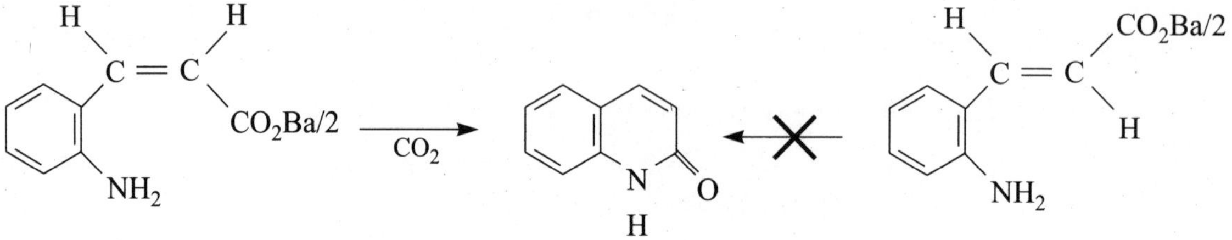

Cis $\xrightarrow{CO_2}$ Carbostyril $\nleftarrow$ *Trans*

(**c**) ***Method of chemical correlation***: Suppose we have already known the configuration of a geometric isomer, say A. Let A be converted by chemical reactions under mild conditions to a geometric isomer A′, of another compound. Since under mild conditions interconversion of the geometric isomers will not take place, we can conclude that the configuration of A′ will be the same as that of A. For instance, under (***ii***) we have determined which of the isomers of *ortho*-aminocinnamic acids has the *cis* configuration and which has *trans*. Now if we treat these isomers (separately) with nitrous acid (diazotization) followed by reduction with hypophosporous acid, we get the isomer of cinnamic acids having m.pt. 341 K and 406 K, respectively. Thus, the isomer having m.pt. 341 K must be *cis* and the other (m.pt. 406 K) *trans*.

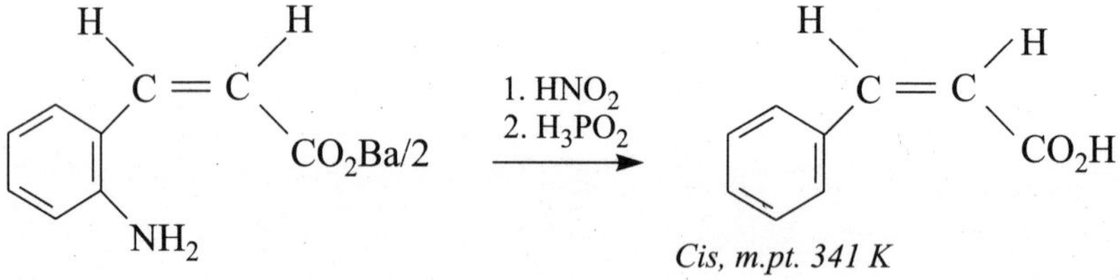

Cis $\xrightarrow{1.\ HNO_2\quad 2.\ H_3PO_2}$ *Cis, m.pt. 341 K* Allocinnamic acid

Trans $\xrightarrow{1.\ HNO_2\quad 2.\ H_3PO_2}$ *Trans, m.pt. 406 K* Cinnamic acid

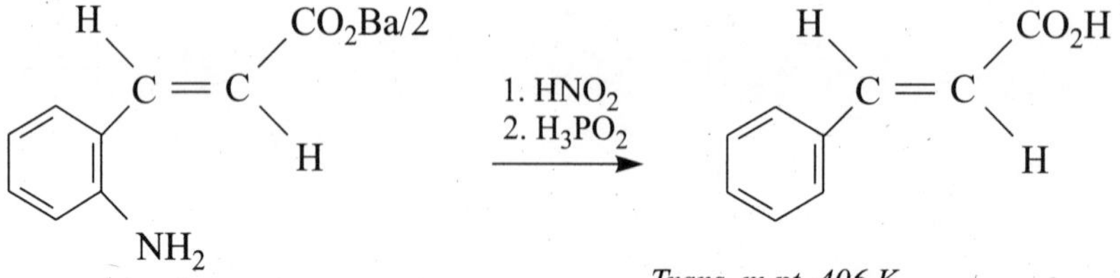

(d) Method of stereoselective addition reactions

(i) Hydroxylation of double bond: We shall see (Chapter 6), hydroxylation of alkenes by *aq.* $KMnO_4$ or OsO_4 is stereospecifically *cis*. Thus *cis*-isomer will give *meso*-compound and *trans*-will form racemic mixture, *e.g.*, hydroxylation of maleic acid produces *meso*-tartaric acid and that of fumaric acid gives racemic mixture. Hence, maleic acid is *cis*-isomer and fumaric acid is *trans*-isomer.

Maleic acid (H, COOH / C=C / H, COOH) $\xrightarrow{aq.\ KMnO_4 \text{ or } OsO_4}$ *meso*-Tartaric acid (COOH; H–C–OH; H–C–OH; COOH)

Fumaric acid (H, COOH / C=C / HOOC, H) $\xrightarrow{aq.\ KMnO_4 \text{ or } OsO_4}$ (+)-Tartaric acid (50%) (COOH; HO–C–H; H–C–OH; COOH) + (–)-Tartaric acid (50%) (COOH; H–C–OH; HO–C–H; COOH)

Racemic mixture

(ii) Addition of bromine to double bond: In contrast to hydroxylation, addition of bromine to alkenes is stereospecifically *trans*. Therefore, addition of bromine to *trans*-isomer will give rise to *meso*-compound and the same addition to *cis*-isomer will produce racemic mixture. Thus the stereoisomer of 2-butene which gives *meso*-product is *trans* and the one which gives racemic mixture is *cis*.

Cis-But-2-ene (H_3C, H / C=C / H_3C, H) $\xrightarrow{Br_2/CCl_4}$ (CH_3; H–C–Br; Br–C–H; CH_3) + (CH_3; Br–C–H; H–C–Br; CH_3)

Racemic mixture

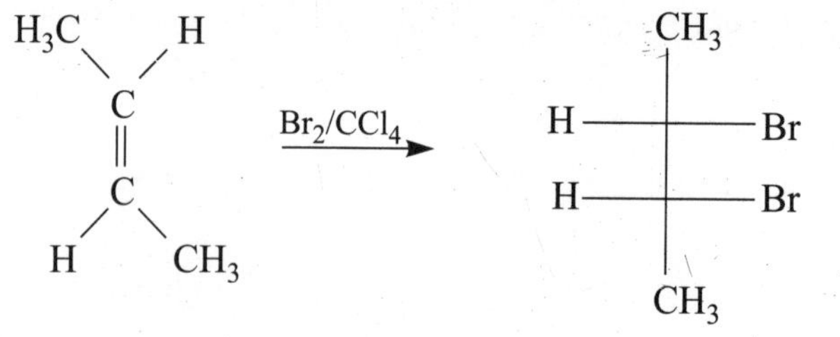

$$\underset{\textit{trans}\text{-But-2-ene}}{H_3C(H)C{=}C(H)CH_3} \xrightarrow{Br_2/CCl_4} \underset{\textit{meso}\text{-2,3-Dibromobutane}}{CH_3\text{–}CH(H)(Br)\text{–}CH(H)(Br)\text{–}CH_3}$$

trans-But-2-ene *meso*-2,3-Dibromobutane

3.13.4 E and Z System of Nomenclature

The *cis* and *trans* designations have been used for many years to indicate the spatial relations of atoms or groups of atoms around a double bond. For compounds of the type $ab\text{C}=\text{C}ab$ or $ab\text{C}=\text{C}ac$, the prefixes *cis*- and *trans*-are quite unambiguous. However, when there are three or four different atoms or groups attached to the carbon atoms of a double bond *cis-trans* nomenclature leads to ambiguity and sometimes creates a lot of confusion. Let us consider a molecule in which the two carbon atoms of a double bond are attached with four different halogens.

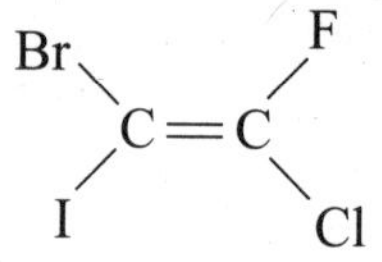

$$(Br)(I)C=C(F)(Cl)$$

When we say that Br and CI are *trans* to each other we can also say with equal degree of confidence that I and CI are *cis* to each other. It is thus difficult to name such a substance either the *cis* or the *trans* isomer. To eliminate this confusion, a more general and easy system for designating configuration about a double bond has been adopted. This method, which is called the *E and Z system*, is based on *a priority system* originally developed by Cahn, Ingold and Prelog for use with optically active substances (Section 3.12). In olefins we assign a first (1) and a second (2) priority to the atoms or groups attached to each carbon of the double bond. Priority is then compared at one carbon relative to the other. If the higher-priority atoms or groups are on opposite side of the double bond, the configuration is designated as **E** (German: ***entgegen*** **= across**). If the higher-priority atoms or groups are on the same side of the double bond, the configuration is designated as **Z** (German: ***zusammen*** **= together**). Thus,

$$\underset{E}{(1)(2)C=C(2)(1)} \qquad \underset{Z}{(1)(2)C=C(1)(2)}$$

E Z

Note : Rules for fixing the priorities of atoms or groups attached to doubly bonded carbon atoms are the same as discussed earlier. Some examples are as follows:

(2) Br, (1) I on one carbon; F (2), Cl (1) on the other: $C=C$

Z-1-Bromo-2-chloro-2-fluoro-1-iodoethene

(2) Br, (1) I on one carbon; Cl (1), F (2) on the other: $C=C$

E-1-Bromo-2-chloro-2-fluoro-1-iodoethene

(1) H_3C, (2) H on one carbon; CH_3 (1), H (2) on the other: $C=C$

Z-2-Butene

(1) H_3C, (2) H on one carbon; H (2), CH_3 (1) on the other: $C=C$

E-2-Butene

(1) H_3C, (2) H on one carbon; $CH_2\,CH_2\,CH_3$ (1), CH_3 (2) on the other: $C=C$

Z-3-Methylhex-2-ene

(2) ClH_2C, (1) BrH_2C on one carbon; CD_3 (1), CH_3 (2) on the other: $C=C$

E-isomer

(1) HOOC, (2) H on one carbon; COOH (1), H (2) on the other: $C=C$

Z-But-2-ene-1, 4-dioic acid
(*Maleic acid*)

(2) H, (1) HOOC on one carbon; COOH (1), H (2) on the other: $C=C$

E-But-2-ene-1, 4-dioic acid
(*Fumaric acid*)

If there are two or more double bonds prefixes E and Z are used for designating configuration around each double bond. For example,

H_3C(1), H on C2; H on C3; $C_2=C_3$; C3–C4; H on C4; CH_3(6), H on C5; $C_4=C_5$

2E, 4Z-2, 4-Hexadiene

H_3C(1), H on C2; H on C3; $C_2=C_3$; C3–C4; H on C4; H, CH_3(6) on C5; $C_4=C_5$

2Z, 4Z-2, 4-Hexadiene

Number of Geometrical isomer of compounds containing two or more Double Bonds with Non-equivalent terminii

Dienes in which the two terminii are different (*i.e.*, XHC=CH—CH=CHY), exit as four geometrical isomers.

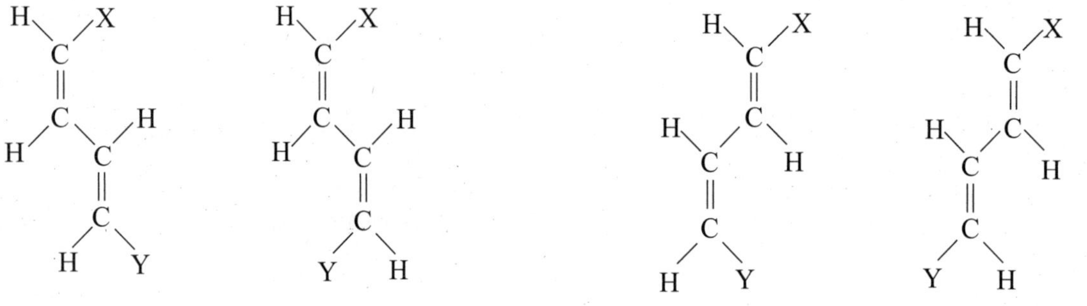

Z, E, or *cis-trans* Z, Z, or *cis-cis* E, Z or *trans-cis* E, E or *trans-trans*

Thus it can be concluded that for a given molecule the number of geometrical isomers is 2^n, where *n* is the number of double bonds. It may be noted here that this formula applies only to those molecules in which the two terminii are different. However, in identical terminii (*i.e.*, XHC = CH — CH = CHX) the number of geometrical isomers is given by the following expressions:

(*i*) **If *n* is even**, the number of geometric isomers = $2^{(n-1)} + 2^{(n/2-1)}$

(*ii*) **If *n* is odd,** the number of geometric isomers = $2^{(n-1)} + 2^{\frac{n-1}{2}}$

3.13.5 Geometric Isomerism of Oximes

Oximes are a class of compounds obtained by treating aldehydes and ketones with hydroxylamine in acidic medium (pH 4.5). The oximes obtained from aldehydes are called aldoximes and those obtained from ketones are called ketoximes. They have a formal carbon-nitrogen double bond.

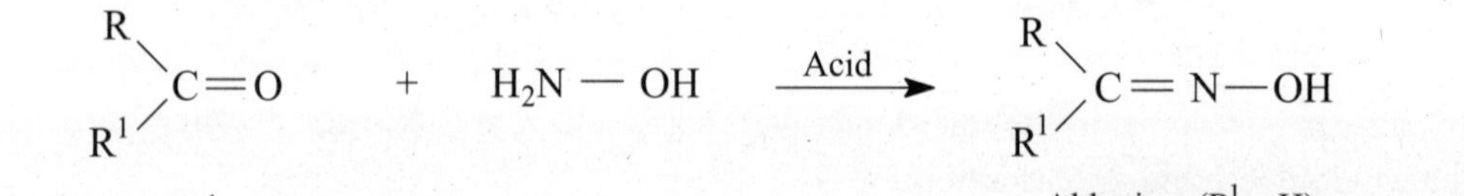

$$\text{R(R}^1\text{)C=O} + \text{H}_2\text{N}-\text{OH} \xrightarrow{\text{Acid}} \text{R(R}^1\text{)C=N}-\text{OH}$$

Aldehyde (R^1 = H)
Ketone (R^1 = alkyl or aryl)

Aldoxime (R^1 = H)
Ketoxime (R^1 = alkyl or aryl)

The carbon and nitrogen atoms of oximes are sp^2-hybridized, as in alkenes. Forther, like alkenes this double bond consists of a σ-bond and a π-bond. Out of three sp^2-orbitals of nitrogen, two are involved in forming σ-bonds with C and OH. The third orbital is occupied by a lone pair of electrons.

Thus all groups in oximes lie in the same plane and hence they should also exhibit geometric isomerism if groups R and R^1 are different. Accordingly, Beckmann (1889) observed that benzaldoxime existed in two isomeric forms and Hantzsh and Werner (1890) suggested that these oximes exist as the following two geometric isomers (I and II).

H_5C_6, H, C=N, OH — I
H_5C_6, H, C=N, HO — II

Considering the lone pair of electrons also as one of the substituents these isomers can be written as follows:

H_5C_6, H, C=N, OH — I
H_5C_6, H, C=N, HO — II

Nomenclature of Oximes

For specifying the configuration of oximes the prefixes *syn*- and *anti*- are used instead of *cis*-and *trans*. The prefixes *syn*-and *anti* are used in different context for aldoximes and ketoximes.

Aldoximes

In the case of aldoximes *syn*-form is the one in which both the hydrogen atom and the hydroxyl group are on the same side of the double bond and *anti*-is used if they are on opposite side of the double bond. Thus **I** is *syn*-benzaldoxime and **II** is *anti*-benzaldoxime.

H, OH, C=N, H_5C_6 — I *syn*-Benzaldoxime
H, C=N, H_5C_6, OH — II *anti*-Benzaldoxime

Ketoximes

In the case of ketoximes, the prefix *syn* indicates that the first-named group and hydroxyl groups are on the same side of the double bond whereas prefix *anti*-indicates that they are on the opposite side of the double bond. For Example,

H_3C, OH, C=N, H_5C_2 — *anti*-Ethylmethylketoxime
H_3C, C=N, H_5C_2, OH — *syn*-Ethylmethylketoxime

As in the case of *cis-trans* isomerism, this nomenclature is ambiguous and often creates confusion. To avoid this, the system of E-Z nomenclature has been adopted. For fixing priority the lone pair of electrons on nitrogen is taken as the group of lowest priority. Some examples are given below:

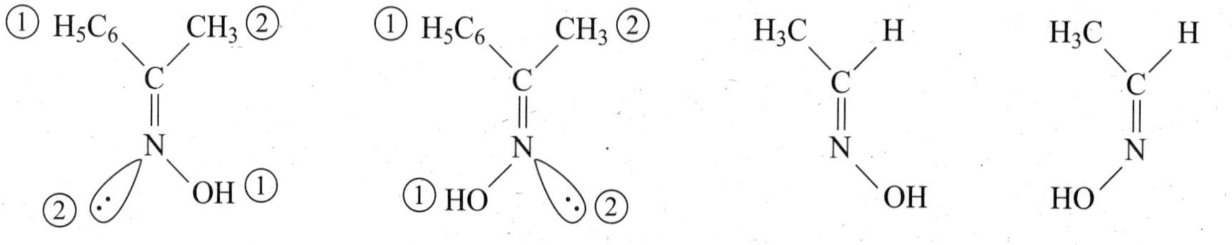

E-Methylphenylketoxime Z-Methylphenylketoxime E-Acetaldoxime Z-Acetaldoxime

Determination of Configuration of Oximes

(*a*) *Aldoximes*

It has been shown that aromatic aldoximes exist in both *syn-* and *anti-*forms. In contrast aliphatic aldoximes exist only in more stable form which is *anti*-form. The method used for determining the configuration of aromatic aldoxides is based upon the difference in behaviour of their acetyl derivatives when treated with alkali. Accordingly, the given oxime is first acetylated by treatment with acetic anhydride and the acetyl derivative so obtained is heated with *aq.* Na_2CO_3. The acetyl derivative of one isomer regenerated the original oxime whereas that of the other isomer eliminated acetic acid by E2 mechanism to form aryl cyanide. As we shall see later that in E2 reaction, the groups eliminated should be *trans-* to each other. Therefore, this isomer is *anti* and by implication the other isomer is *syn.*

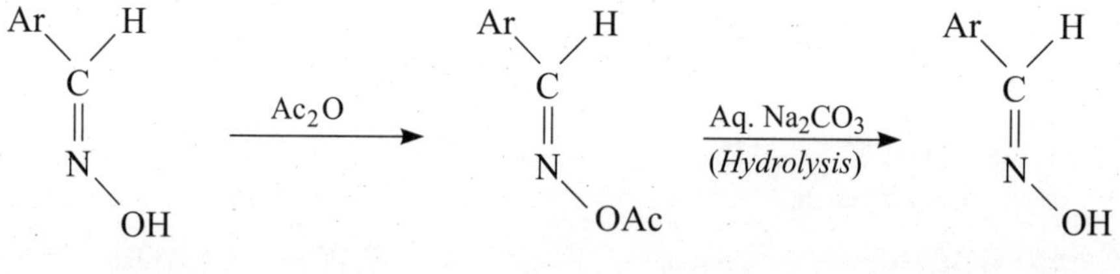

E or *syn*-Oxime $\xrightarrow{Ac_2O}$ $\xrightarrow[\textit{(Hydrolysis)}]{Aq.\ Na_2CO_3}$ *Original oxime*

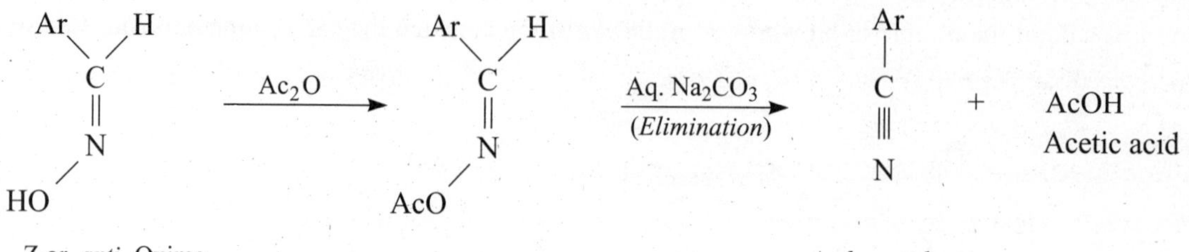

Z or *anti*-Oxime $\xrightarrow{Ac_2O}$ $\xrightarrow[\textit{(Elimination)}]{Aq.\ Na_2CO_3}$ *Aryl cyanide* + AcOH Acetic acid

(*b*) *Ketoximes*

The configurations of the geometric isomers of the unsymmetrical ketoximes are determined by *Beckmann rearrangement* which consists in treating ketoxime with acidic reagents such as PCl_5, H_3PO_4, P_2O_5, etc., when the oxime isomerizes to a substituted amide by migration of the group (R_1 or R_2) which is *anti* to the hydroxyl group.

$$\underset{R^2}{\overset{R^1}{>}}C=N\diagdown_{OH} \longrightarrow R^2-CO-NHR^1 \longrightarrow R^2COOH + R^1NH_2$$

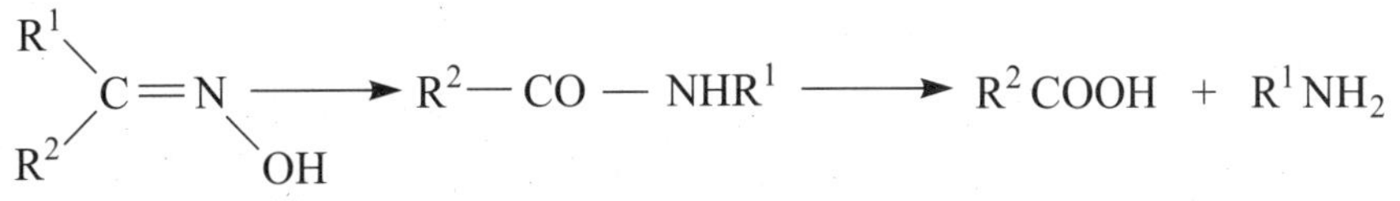

The structures of the amides are determined by their hydrolysis to the corresponding acids and amines. The determination of the structure of amine formed in the above sequence of reactions plays a key role in deciding which group has migrated during Beckmann rearrangement. That the group *anti* to the hydroxyl group actually migrates was shown by the following reactions:

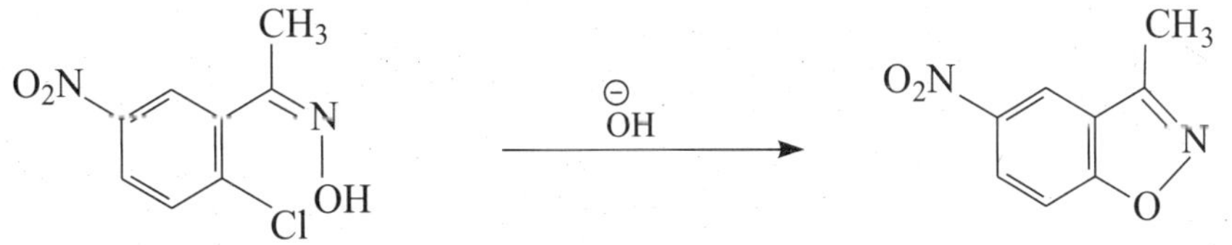

A (2-chloro-5-nitrophenyl, CH_3, N–OH syn) $\xrightarrow{\overset{\ominus}{O}H}$ B (benzisoxazole with O_2N, CH_3)

C (2-chloro-5-nitrophenyl, CH_3, N–OH anti) $\xrightarrow{NaOH}$ D (oxime salt, $\overset{\ominus}{O}\overset{\oplus}{Na}$, Cl)

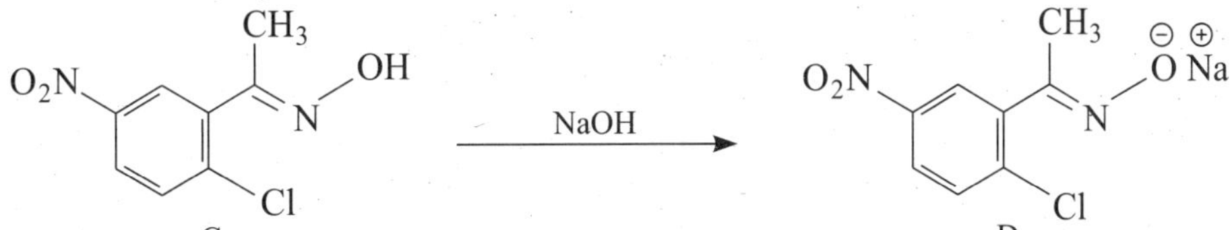

The above experiment shows that **A** has the structure in which substituted phenyl group is *syn* to the hydroxyl group. The other geometric isomer **C** then must have these groups *anti* to each other. When C was treated with PCl_5 it gave the amide **D** which produced 2-chloro -5-nitroaniline upon hydrolysis. Thus, it is proved beyond doubt that the migrating group is *anti*- to OH.

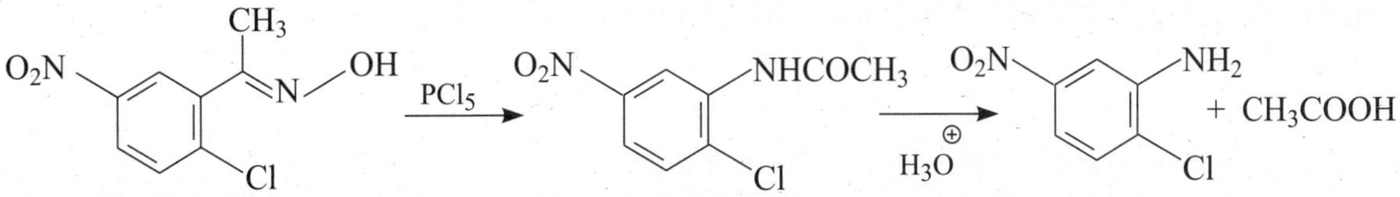

C (2-chloro-5-nitrophenyl, CH_3, N–OH) $\xrightarrow{PCl_5}$ D (O_2N, $NHCOCH_3$, Cl) $\xrightarrow{H_3\overset{\oplus}{O}}$ (O_2N, NH_2, Cl) + CH_3COOH

3.13.6 Geometric Isomerism in Alicyclic Compounds

We have seen that geometric isomerism arises only when there is hindered rotation about a bond. Such hindrance is not confined to double bonds only. For instance, cyclic compounds such as the disubstituted derivatives of cyclopropane, cyclobutane, cyclopentane and cyclohexane can also show *cis-trans* isomerism, because the basic condition for such isomerism—that there should be sufficient hindrance to rotation about a linkage between atoms—is also satisfied in these systems. Atoms joined in a ring are not free to rotate around the sigma bond.

To understand geometric isomerism in cyclic compounds, we presume that the carbon atoms of a cyclic structure such as cyclohexane form a plane (this assumption is really not correct, as we will see later in this Chapter). The plane of the ring is considered being almost horizontal with the edge of the ring projecting towards us being shadeds more heavily. In a cyclic compound, each carbon atom is joined to its neighbouring ring carbon atoms and also to two other atoms or groups which are shown by vertical lines. When an atom (or a group) is placed at the top of a vertical line. It is said to be *above the plane of the ring*, and when an atom (or a group) is attached to the bottom of a vertical line, it is said to be *below the plane of the ring*. Usually, hydrogen atoms attached to the ring are not shown. For example,

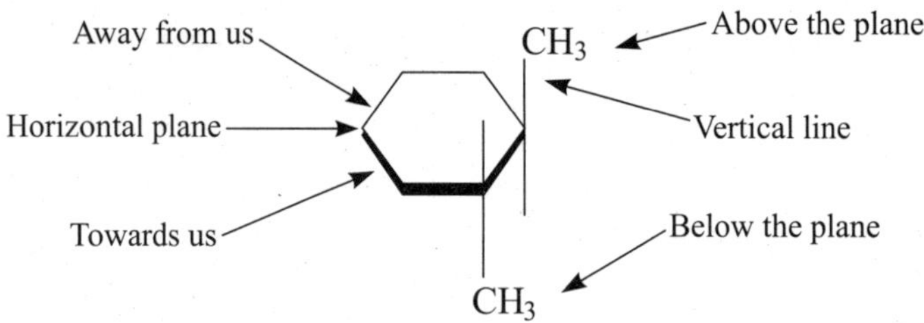

1, 2-Dimethylcyclohexane

Sometimes, a broken wedge is used to indicate a group below the plane of the ring, and a solid line to represent a group above the plane.

Above the plane

CH_3

Below the plane

CH_3

The main point to be noted in both the above structures is that the two methyl groups are on *opposite sides* of the plane of the ring. Just as we have seen in the case of *cis* and *trans*-alkenes, when the two groups are on opposite sides of the ring they are *trans* and when they are on the same side they are *cis*. Thus, these compounds are also geometric isomers of each other.

trans-1, 2-Dimethylcyclohexane *cis*-1, 2-Dimethylcyclohexane

It has been shown that 1, 2-disubstituted alicyclic compounds exhibit position isomerism in addition to geometric isomerism. The geometric isomers of various disubstituted alicyclic compounds along with their structures are illustrated below:

(*a*) *Disubstituted cyclopropanes*: Only two isomers exist since position isomerism is not possible.

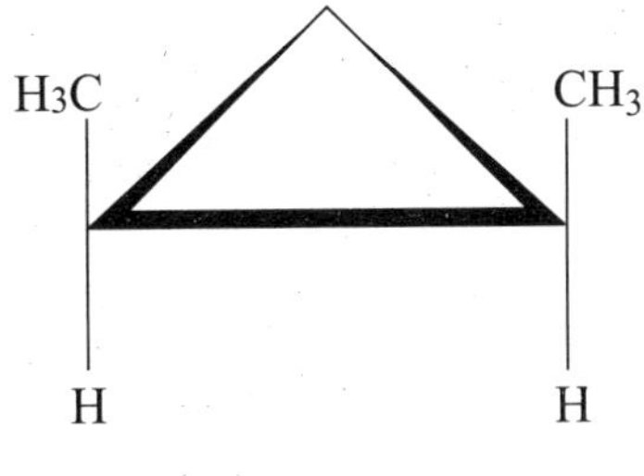

cis-1, 2-Dimethylcyclopropane

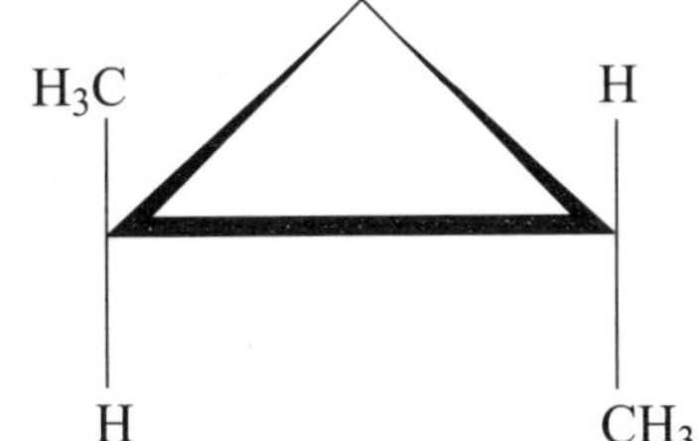

trans-1, 2-Dimethylcyclopropane

(*b*) *Disubstituted cyclobutanes* and *cyclopentanes* exist as two position isomers *i.e.*, 1, 2 and 1, 3 and each one of them exists as *cis*- and *trans*-isomers. So, in all four isomers are possible as shown below:

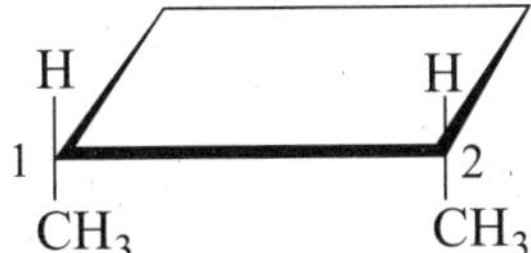

cis-1, 2-Dimethylcyclobutane

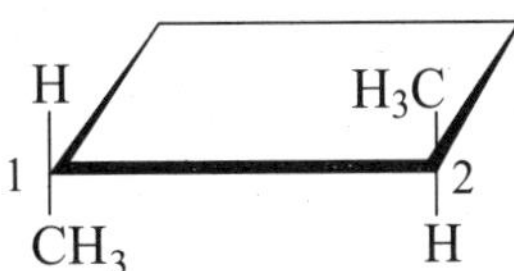

trans-1, 2-Dimethylcyclobutane

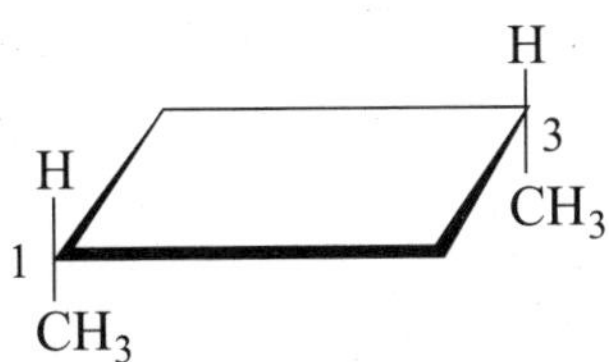

cis-1, 3-Dimethylcyclobutane

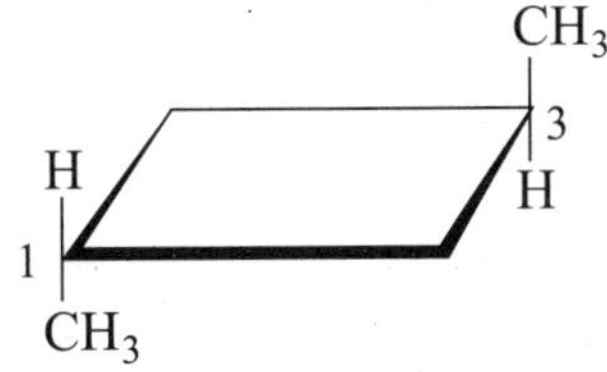

trans-1, 3-Dimethylcyclobutane

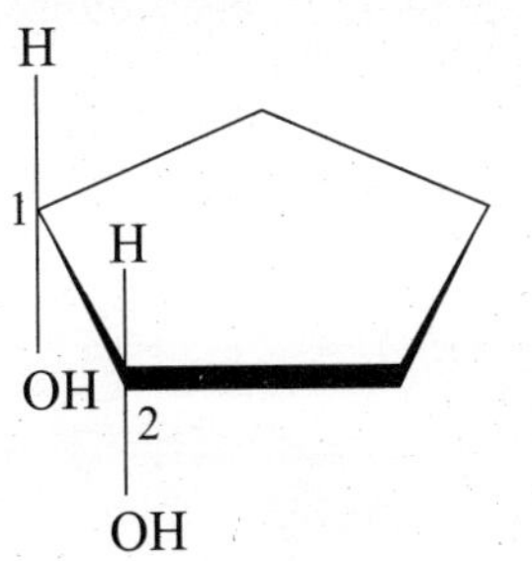

cis-Cyclopentane-1, 2-diol

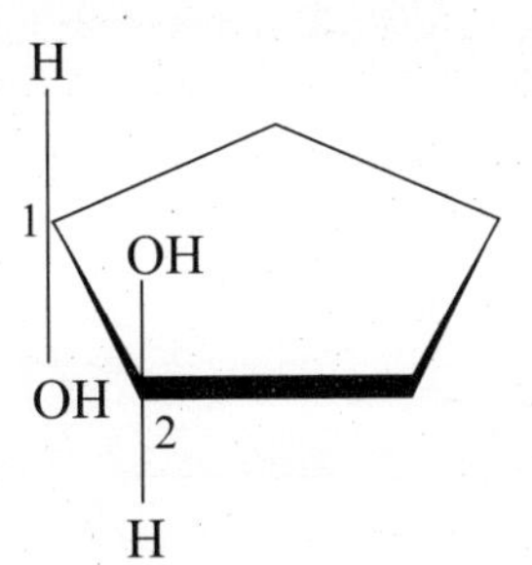

trans-Cyclopentane-1, 2-diol

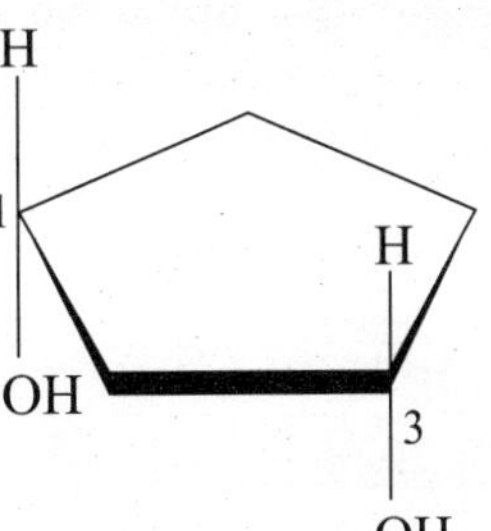

cis-Cyclopentane-1, 3-diol

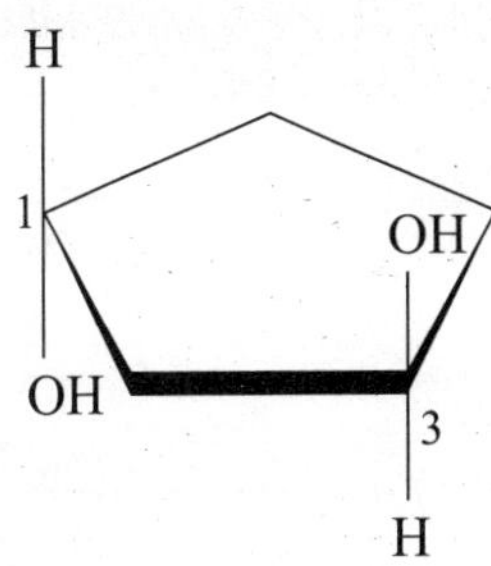

trans-Cyclopentane-1, 3-diol

(*c*) *Disubstituted cyclohexane* exists as three position isomers *i.e.*, 1, 2-, 1, 3- and 1, 4- and each of them exists as *cis*- and *trans*-isomers. So in all six isomers are possible as shown below:

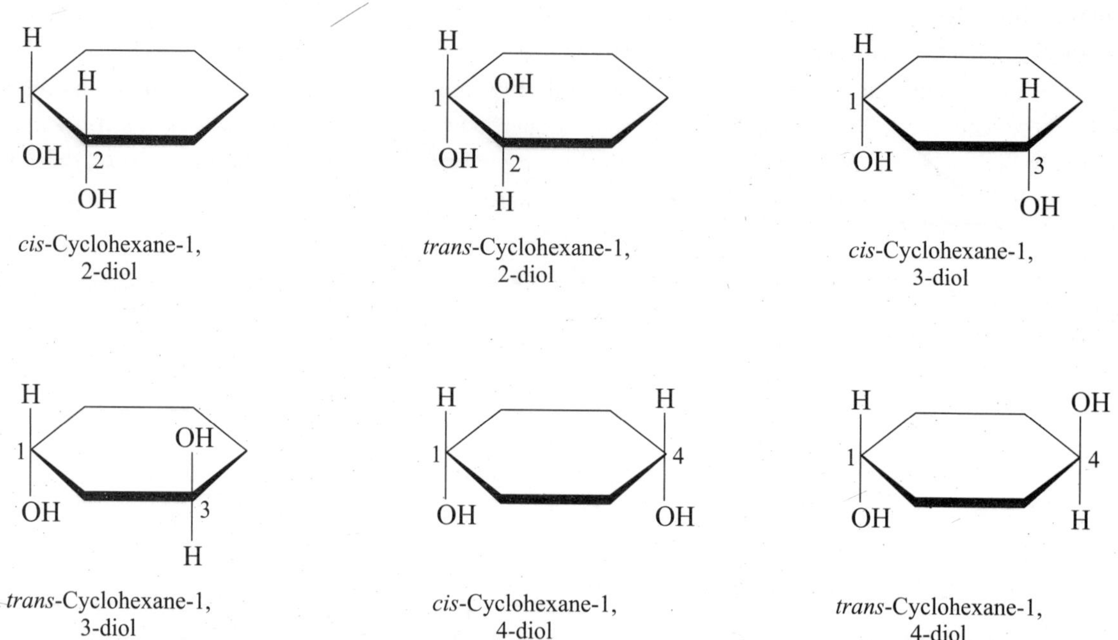

3.14 CONFORMATIONAL ISOMERISM

We know that a carbon-carbon σ-bond (a single bond) is formed by an **end-on** overlapping of sp^3-orbitals of the two carbon atoms. This bond is cylindrically symmetrical about the axis (Fig. 3.6) and has the highest electron density along the bond axis (this is an imaginary line connecting the centres of two atoms joined by a σ-bond). Due to cylindrical symmetry, free rotation about a carbon-carbon single bond is possible without effecting the amount of orbital overlap. Almost an infinite number of spatial arrangements of atoms about the cabon-cabon single bond exist. All those arrangements which result from free rotation about a single bond are called **conformations** or **conformers** or **rotational isomers** or **simply rotamers.**

Unlike configurational isomers, the isolation of conformers in non-rigid systems is not feasible at ordinary temperature. This is due to very low potential energy barrier (~10–12 kJ mol^{-1}) between two conformers.

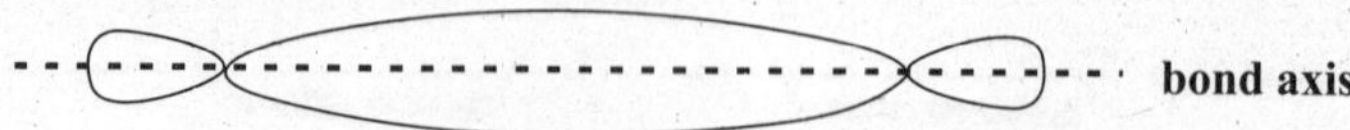

Fig. 3.6: A cylindrically symmetric MO of a single bond obtained by sp^3-sp^3 overlap of two carbon atoms

3.14.1 Conformations of Ethane

In acyclic hydrocarbons, such as ethane, it was originally assumed that there is complete freedom of rotation about the carbon-carbon single bond. But **Pitzer (1936)** postulated that there exists a potential energy barrier which causes restriction in free rotation.

Let us consider the rotation of one of the methyl groups about the central C—C bond in ethane keeping the other methyl group stationary. As a consequence of such rotation two extreme conformations can result. They are an **eclipsed conformation (I)** and a **staggered conformation (II)**. In addition to these, an infinite number of conformations between these extremes are also possible.

I
Eclipsed conformation

II
Staggered conformation

In eclipsed conformation (I) all the hydrogen atoms on one carbon are directly opposed to and eclipsed by vicinal hydrogen atoms on the other carbon atom. In this arrangement the electrons of one carbon-hydrogen bond will repel the electrons of the other carbon-hydrogen bond because they are too close to each other (internuclear H—H distance 229 pm), the dihedral angle (angle between two planes, ϕ) between two such C—H bonds being 0°. So, this conformation is the least stable one because only in this conformation, the C—H bonds are nearest to each other. This extra energy of eclipsed conformation is called **torsional strain**. *The term torsional strain is used for the repulsion felt by bonding electrons on one substituent when it passes close to the bonding electrons of another substituent.*

In contrast to eclipsed conformation, the staggered conformation is the most stable one because the distance between C—H bonds on neighbouring carbons is maximum (internuclear H—H distance is 310 pm), dihedral angle between two C—H bonds being 60°. Experimentally also it has been found to be true. The staggered conformation of ethane is more stable by about 12 kJ mol^{-1} than eclipsed conformation. In fact this energy difference (12 kJ mol^{-1}) between eclipsed and staggered conformations is the energy barrier to free rotation about the C—C single bond in ethane.

The variation of energy of conformations of ethane with rotation about C—C single bond is shown in Fig. 3.7. A look at the figure reveals that at the initial stage *i.e.*, the eclipsed conformation with dihedral angle 0°, the molecule has maximum energy. A rotation of front carbon by 60° generates a staggered conformation having minimum energy. Another rotation by 60° (*i.e.*, the rotation by 120° from initial conformation) generates another eclipsed form and so on. Thus, every rotation of 60° alternatively generates eclipsed and staggered conformations.

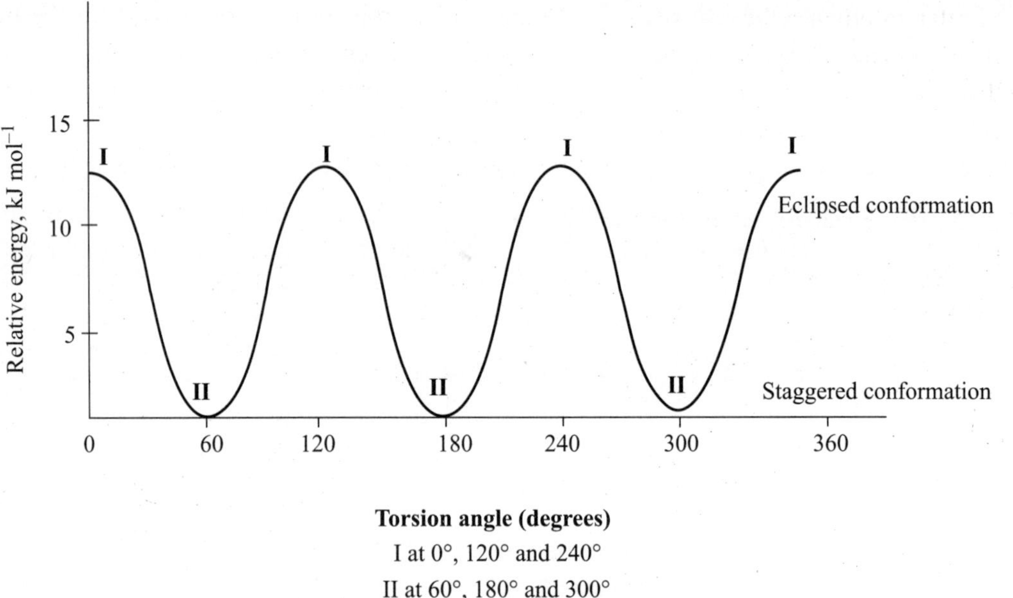

Fig. 3.7: Rotational or torsional energy in ethane

It may be noted here that all the staggered conformations are of equal energy. Similarly all eclipsed conformations are of equal energy.

The investigation of the various conformations of a compound and their relative stabilities is called **conformational analysis.**

3.14.2 Conformations of Propane

For its conformational analysis, propane can be considered as a derivative of ethane in which one hydrogen atom of ethane is replaced by a methyl group as shown below:

```
         H    H
        2|   1|
  CH3—C—C—H
         |    |
         H    H
```

In case of propane also an infinite number of conformations are possible due to rotation of one methyl group around the C—C single bond, through a complete rotation of 360°. Out of these only two extreme conformations, *i.e.*, eclipsed and staggered are important which are shown as below:

CH_3 60° H H H H H

Staggered conformation

CH_3 H H H H H

Eclipsed conformation

Similar to ethane, the staggered conformation of propane is more stable than the eclipsed conformation. The energy difference (14.2 kJ mol^{-1}) between these two conformations is slightly more than that in ethane (12 kJ mol^{-1}). Even then this energy barrier to free rotation about C—C single bond is so small that it cannot prevent rotation completely. Like ethane, at room temperature the two conformations are not readily isolable, and are inter-converted by rotation through an angle of 60°.

3.14.3 Conformations of *n*-Butane

n-Butane can be assumed to be derived from ethane by replacing one hydrogen atom of each carbon by a methyl group.

$$\overset{1}{H_3C} - \overset{2}{CH_2} - \overset{3}{CH_2} - \overset{4}{CH_3}$$

The molecule of *n*-butane reveals that each of the two central carbon atoms C_2 and C_3 has one methyl group and two hydrogen atoms attached to it. Thus *n*-butane has three C—C single bonds, and the molecule can rotate about each of them. However, we will consider here only those conformations which result from rotation about C_2—C_3 bond. Various conformations derived due to rotation about this bond are shown below in their Newman projections:

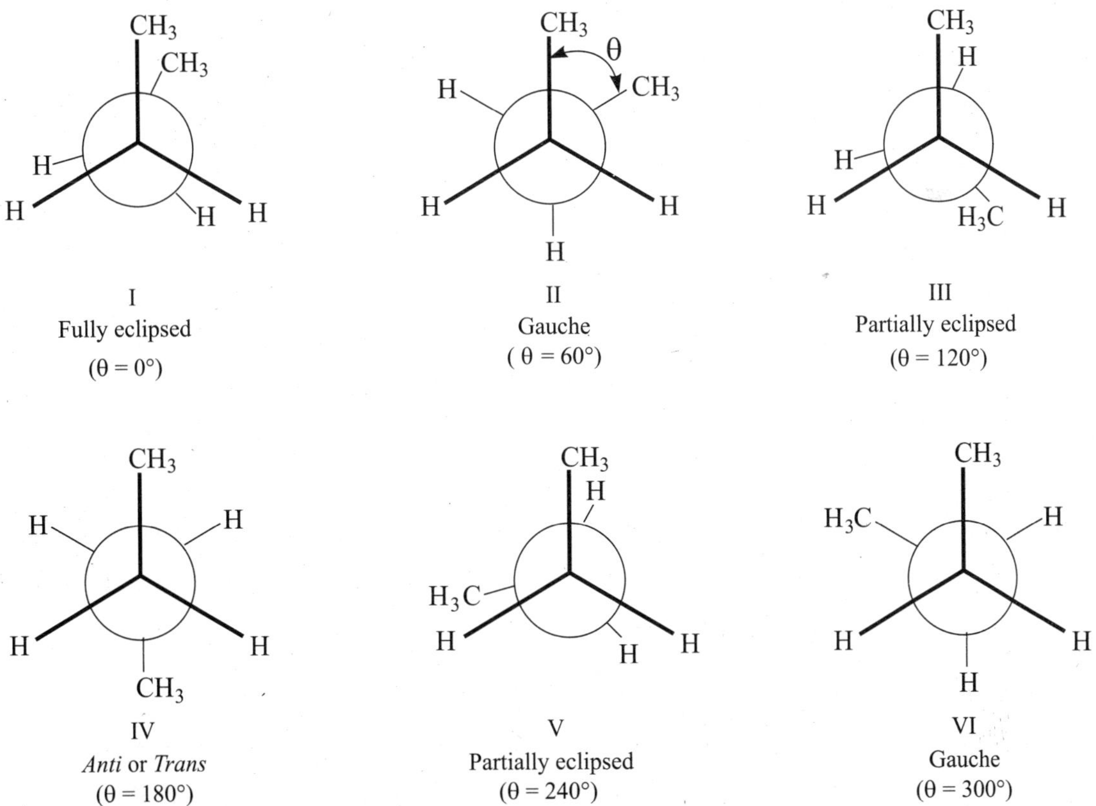

I
Fully eclipsed
(θ = 0°)

II
Gauche
(θ = 60°)

III
Partially eclipsed
(θ = 120°)

IV
Anti or *Trans*
(θ = 180°)

V
Partially eclipsed
(θ = 240°)

VI
Gauche
(θ = 300°)

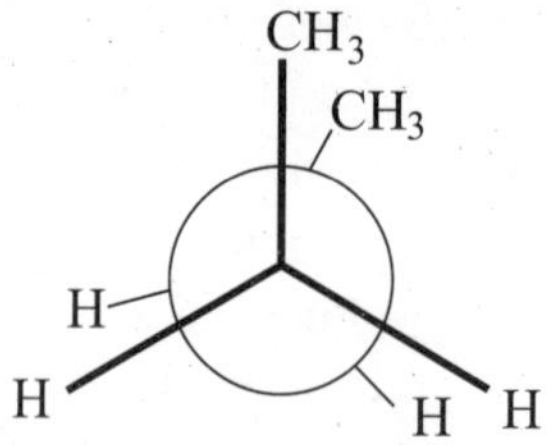

CH_3 CH_3 H H H H

I

Fully eclipsed

($\theta = 0°$ or $360°$)

Thus unlike ethane in which the staggered and eclipsed conformations are equivalent, two different staggered and two different eclipsed conformations occur in *n*-butane. These are discussed below:

(*i*) Two different kinds of staggered conformations are as follows:

(*a*) *Anti* (or *trans*) conformation (*Greek*: ***anti*** **= opposite**): In this conformation the two methyl groups are farthest apart (IV, Q =180°) and is the most stable.

(*b*) *Gauche* (or skew) conformation (*French*: ***Gauche*** **= left**): Both these conformations (II and VI) are staggered, so, they are free from torsional strain. However, the hydrogens of methyl groups in *gauche* conformation lie within 210 pm of each other. This distance is less than the sum of their van der Waals radii of 240 pm. Due to this congestion in space a repulsive force acts between the methyl groups. This type of destabilization is called **van der Waals strain** or **steric hindrance**. In the case of butane, due to van der Waals strains the *gauche* conformation is less stable than *anti*-conformation by 4 kJ mol^{-1}. This is also called as ***n*-butane gauche (or skew) intraction.** Further, if these two forms could be isolated they would show different physical properties. However, their isolation is not possible because the energy barrier of 4 kJ mol^{-1} is too small to permit isolation of the two forms at normal temperature. This may, however, be possible to separate them at very low temperature (43 K).

Finally, it may be mentioned that conformations II and VI are non-superimposable mirror images of each other. Such pairs of conformations are called as **conformational enantiomers**. However, due to low energy barrier these enantiomers cannot be resolved.

(*ii*) Two different kinds of eclipsed conformations are as follows:

(*a*) Partially eclipsed conformation: The first type of eclipsed conformation, also called partially eclipsed conformation, is reached during rotation from *anti*-to *gauche* conformation. In this eclipsed conformation there are two CH_3 : H and one H : H eclipsed interactions (structures III and V). This eclipsed conformation in less stable than anti conformation by approximately 16 kJ mol^{-1}.

(*b*) Fully eclipsed conformation: The second type of eclipsed conformation, also called fully eclipsed conformation, is reached during rotation from one *gauche* conformation to the other. In this eclipsed conformation there is one CH_3 : CH_3 and two H : H eclipsed interactions (structure I). This conformation is least stable and has higher energy than anti-conformation by approximately 20 kJ mol^{-1}. Thus the molecule in this conformation possesses highest potential energy.

The potential energy relationship of various conformations of *n*-butane is illustrated in Fig. 3.8. The staggered conformations, representing energy minima, are more stable than eclipsed representing energy maxima.

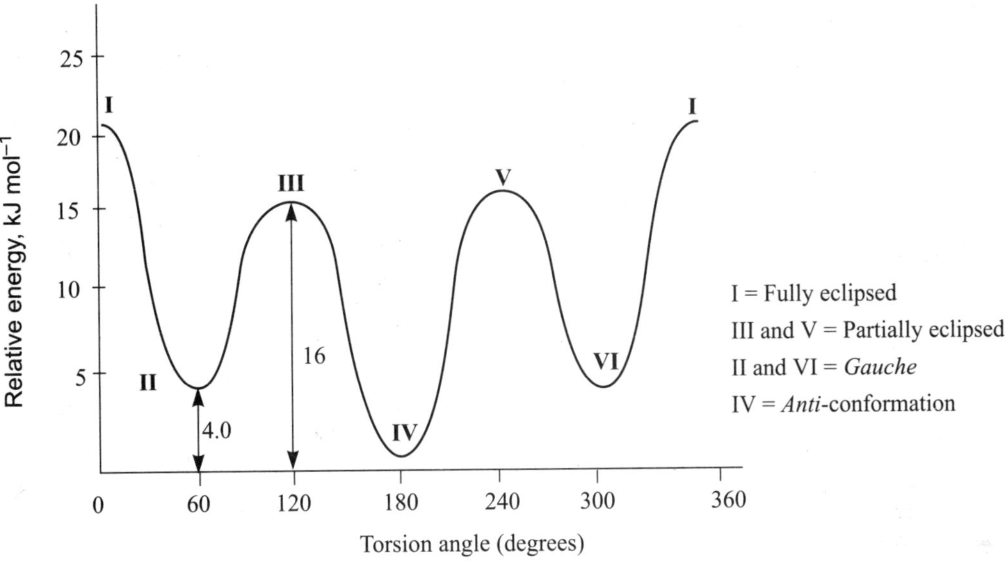

Fig. 3.8: Rotational or torsional energy in *n*-butane

At room temperature, almost all molecules exist in staggered conformation and amongst staggered conformations 78% exist in *anti* and 22% in *gauche* conformations. The point of maximum potential energy (fully eclipsed) lies about 20 kJ mol^{-1} above the *anti*-conformation. The total strain of this form is divided between torsional strains and *van der Waals* strain between two methyl groups.

3.14.4 Conformations of 1, 2-Dibromoethane

Like *n*-butane, 1, 2-dibromoethane can exist in four different conformations as shown below:

Fully eclipsed — Partially eclipsed — Gauche — Anti

On the basis of *torsional strain* and *van der Waals steric hindrance*, staggered (*anti*) conformation of 1, 2-dibromoethane is the most stable followed by *gauche*.

A look at these conformations shows that the dipole moment of *anti*-conformation is zero because the two C—Br dipoles are equal but opposite. However, the *gauche* conformation has some finite dipole moment since the two C—Br dipoles are at an angle of 60° to each other.

Experimentally it has been observed that the dipole moment of 1, 2-dibromoethane is 1.0 D, therefore, the molecule cannot exist entirely in the *anti* form. Further, the molecule also cannot exist entirely in the *gauche* form as the dipole moment varies with temperature. Therefore, to justify these observations, it is assumed that there is an equilibrium between the staggered and the gauche conformations and with an increase in temperature, the equilibrium shifts in favour of the gauche conformation and hence the dipole moment increases with temperature.

$$\textit{Anti} \rightleftharpoons \textit{Gauche}$$

3.14.5 Conformations of 1, 2-Glycols: *Ethylene Glycol*

Similar to *n*-butane and 1, 2-dibromoethane, ethylene glycol can exist in four different conformations as shown below:

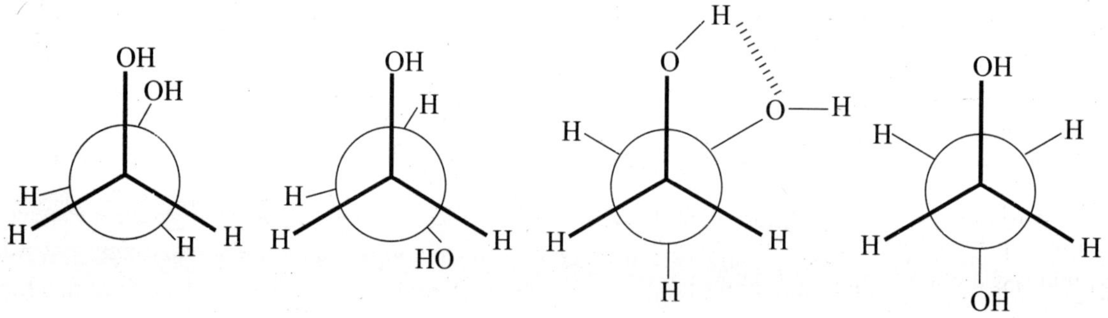

Fully eclipsed   Partially eclipsed   Gauche   *Anti*

We know that, *anti*-conformation is the most stable conformation followed by *gauche*. But in case of ethylene glycol due to intramolecular H-bonding the *gauche* form becomes more stable than *anti*-conformation because there will be no such H-bonding possible in *anti*-conformation. The formation of such H-bond stabilizes the molecule by approximately 20–30 kJ mol^{-1}. Therefore, *gauche form of ethylene glycol is the most stable conformation.*

Similarly, due to intramolecular H-bonding ethylene chlorohydrin, ($CH_2Cl — CH_2OH$), exists in *gauche* conformation which is more stable than its *anti*-conformation.

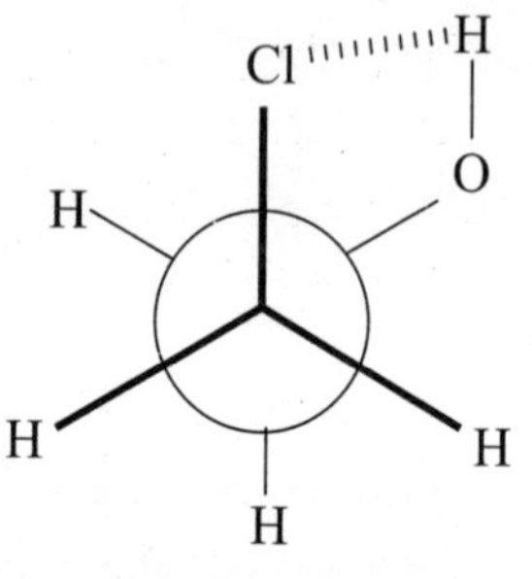

Gauche conformation (*more stable*)
Ethylene chlorohydrin

3.14.6 Alicyclic System: Cyclohexane

Cyclohexane can have two conformations free from **Baeyer or angle strain**, called the *chair* form (I) and the *boat* form (II), respectively.

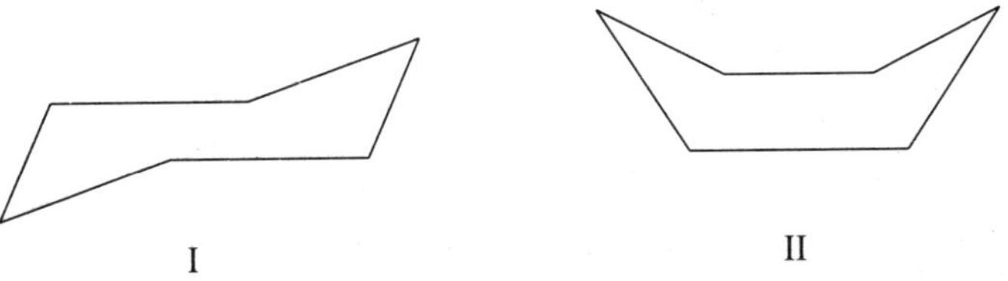

Determination of the heats of combustion of alicyclic hydrocarbons shows absence of substantial Baeyer strain in rings of six or more carbon atoms. Such strain when present should have a direct effect on the thermodynamic stability of a molecule and will be reflected in the heat of combustion. For instance, heat of combustion per CH_2 group in acyclic alkanes is 660 kJ mol^{-1}, whereas in various cycloparaffins it is as follows: Cyclopropane, 697.5 kJ mol^{-1}; cyclobutane, 686.3 kJ mol^{-1}; cyclopentane, 665 kJ mol^{-1}; cycloheptane, 663 kJ mol^{-1}; cyclooctane, 664 kJ mol^{-1}; cyclononane, 665 kJ mol^{-1}.

The first experimental evidence confirming the concept of Sachse and Mohr was made available by Kohlrausch (1936). He found that the Raman spectrum of cyclohexane was best explained by the chair structure. A similar conclusion was reached by a study of infrared spectra. It was in 1947 that Hassel, through an extensive study of cyclohexane by the electron diffraction method, confirmed that practically all the molecules were in symmetrical chair forms and that all C—C bond lengths were normal and all bond angles were approximately tetrahedral.

The above experimental findings were further supported by the consideration of *n*-butane skew and eclipse interactions as discussed earlier. Cyclohexane may be regarded as composed of six *n*-butane segments. Thus the chair form has six *n*-butane skew interactions and the boat form has four *n*-butane skew and two eclipse interactions (2, 3 and 5, 6). The energy difference between the boat and chair forms will therefore be $4 \times 3.77 + 2 \times 18.85 — 6 \times 3.77 = 30.16$ kJ mol^{-1}. In other words, cyclohexane chair form is stabler than the boat form by about 30.16 kJ mol^{-1}. Thus cyclohexane will spend most of its time in the chair conformation.

Axial and Equatorial Bonds

The twelve carbon-hydrogen bonds in the chair form of cyclohexane may be divided into two classes. The six C—H bonds parallel to the axis in the structures shown below are called axial (*a*) bonds, and the six C—H bonds which extend radially outward at an angle of 109° 28′ to the axis are called equatorial (*e*) (Fig. 3.9).

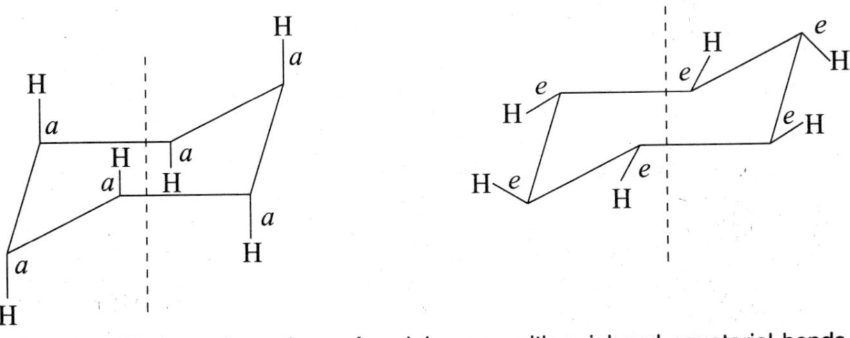

Fig. 3.9: Chair conformations of cyclohexane with axial and equatorial bonds

In the chair form the carbon atoms are in two parallel planes each containing three carbon atoms and the distance between the planes is about 0.5Å. But the chair form of cyclohexane can easily flip at ordinary temperature and the molecule will assume a second chair form in which all the C—H bonds which were axial in the former will be equatorial in the latter, and *vice versa*.

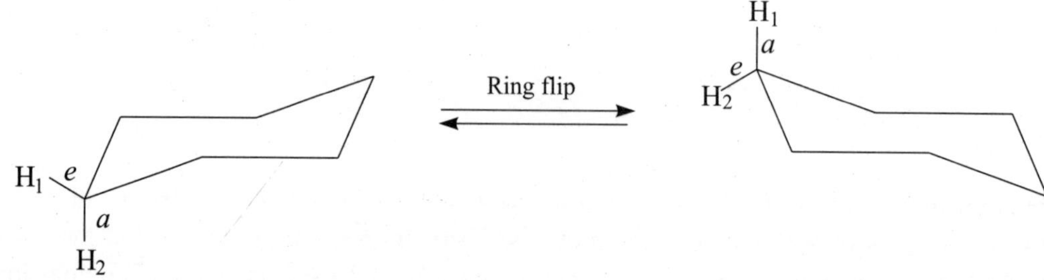

The regular boat form of cyclohexane has the hydrogen atoms disposed as shown below:

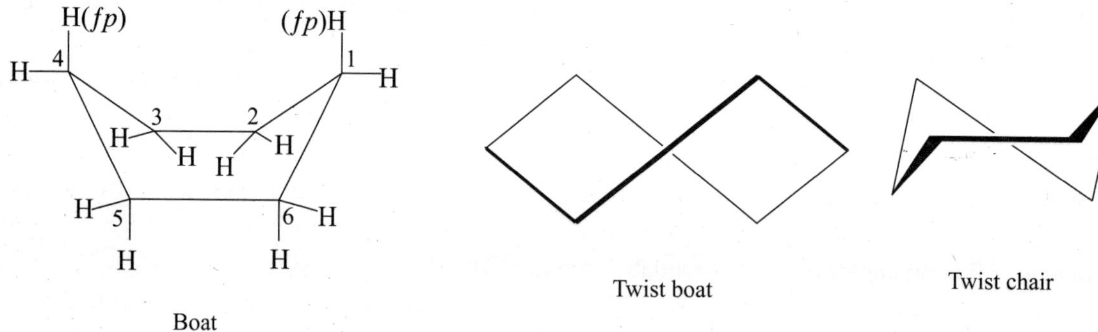

Boat

Twist boat

Twist chair

There is considerable non-bonded interaction between the flagpole (fp) hydrogens and also between the other eight eclipsed hydrogens. As a consequence, this form of cyclohexane can flex into what is known as twist boat conformation (flexible form) which is stabler by about 7.54 kJ mol^{-1} than the regular boat because none of the *n*-butane units in the twist form is either exactly eclipsed or exactly staggered (skew). This can be shown with the help of a model. If we hold the two bonds $C_2—H_a$ and $C_5—H_b$ and gently press them nearer, we find that C_2 and C_5 go down, C_3 and C_6 go up and the two flagpole hydrogens move diagonally apart. When the distances between the flagpole hydrogens and that between H_a and H_b become equal, we have brought to the minimum the flagpole-flagpole interaction as well as the eclipse interaction (torsional strain) at the $C_2—C_3$ and $C_5—C_6$ bonds, which makes the twist boat more stable than the regular boat form.

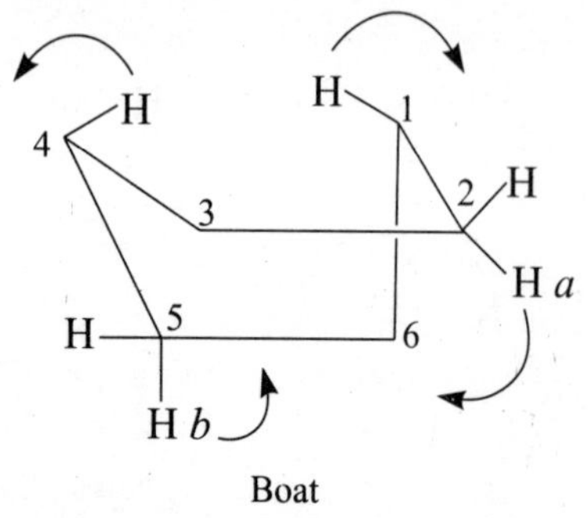

Boat

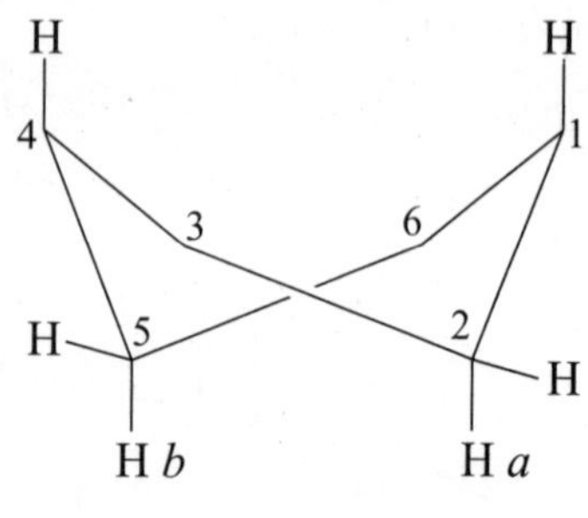

Twist boat

The chair form, on the other hand, is energized into the twist chair conformation which is of higher energy than the chair form by about 41.9 kJ mol^{-1} (Fig. 3.10).

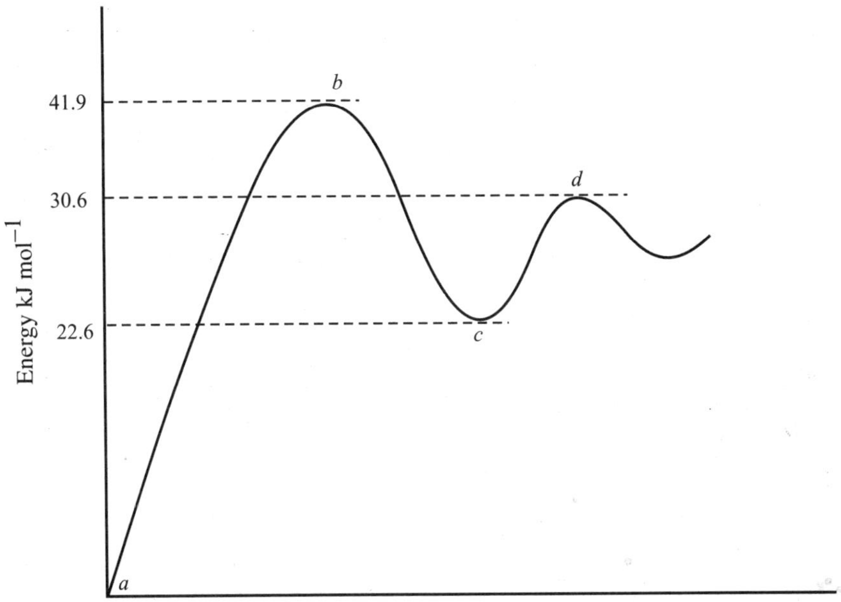

Fig. 3.10: Potential energy of cyclohexane, *(a)* chair; *(b)* twist chair; *(c)* twist boat; *(d)* boat

Cyclohexane Derivatives

As long as the axial and equatorial bonds in the chair form are all bound to hydrogen or the same atom or group, nothing new will be obtained by the flip from one chair to another chair form. But with the substitution of one of the hydrogens by another group or atom a kind of isomerism is possible. In methylcyclohexane, for instance, there exists an equilibrium in which two conformations are present, one with the methyl group axial and the other in which it is equatorial. The axial conformer will have two more *n*-butane skew interactions (7.54 kJ mol^{-1}) whereas in the equatorial conformer no additional interaction or torsional strain is introduced since the two new *n*-butane segments in it are both fully staggered (*anti*).

CH_3 *a* H

Axial form

e CH_3 H

Equatorial form

The two new skew (gauche) interactions in the axial conformer are best demonstrated by drawing the Newman projection formula for the *n*-butane segment, CH_3, C_1, C_2, C_3 and CH_3, C_1, C_6, C_5.

Axial CH_3

Newman projection for axial conformer along C_1—C_2 bond

Gauche (skew) *n*-butane

It is clear from the above projection formula that there is *n*-butane skew interaction between the axial methyl and the atoms at C_3. The same situation is obtained if we draw the Newman projection along C_1—C_6, the *n*-butane skew interaction being in this case between CH_3 and the atoms at C_5. On the other hand, Newman projection for the equatorial conformer, as shown below, clearly shows the absence of any additional skew interaction involved in introducing the equatorial methyl group.

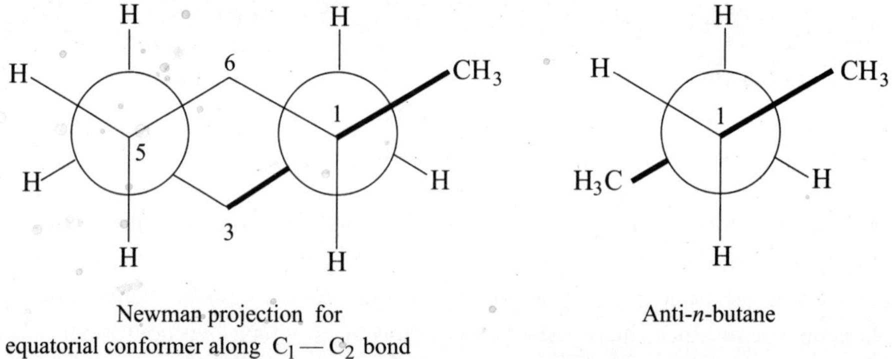

Newman projection for equatorial conformer along C_1—C_2 bond

Anti-*n*-butane

We reach the same conclusion if we consider that in the axial conformer the two axial hydrogens on C_3 and C_5 are closer to the axial than to the equatorial methyl group. Thus, there will be more repulsive interaction due to crowding of the two axial hydrogens at C_3 and C_5 with the large axial group than among the three axial hydrogens in the equatorial isomer.

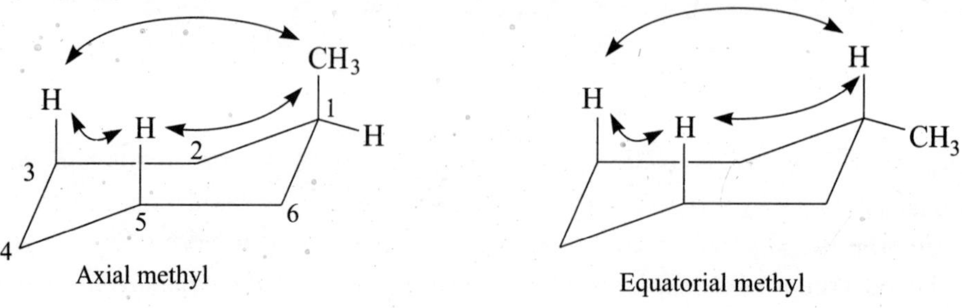

Axial methyl

Equatorial methyl

It is therefore, clear that the equilibrium will have a predominance of the equatorial conformer (93%). *As a rule, between two conformers that conformer will be more stable and therefore more abundant in the equilibrium which has the maximum number of equatorially bound larger groups.* This predominance of the equatorial conformer can be raised to 100 per cent if the substituent group is so large, that it cannot be accommodated in the axial position due to appreciable non-bonded interaction between the large group and the axial hydrogen at the 3- and 5-positions. In other words, flipping of the ring is prevented and the ring is 'frozen'. The interactions between the axial atoms or groups at 1- and 3- or 5-positions are called 1, 3-*diaxial interactions* and in the case of 1, 3-dimethylcyclohexane, the 1, 3-diaxial interaction has been assigned the value of 22.6 kJ mol^{-1}. Thus *cis*-1, 3-dimethylcyclohexane exists at ordinary temperatures almost wholly in the diequatorial conformation.

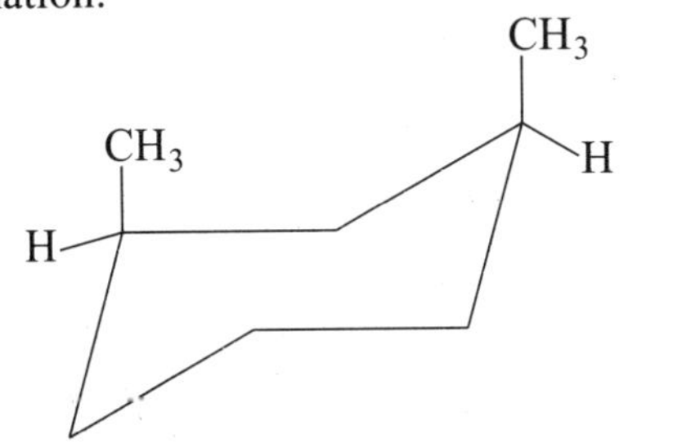

cis-1, 3-Dimethylcyclohexane (diaxial conformer; much less stable)

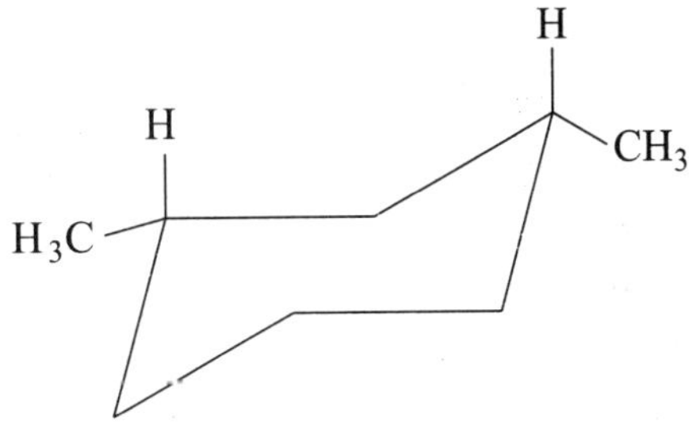

cis-1, 3-Dimethylcyclohexane (diequatorial conformer; much more stable)

We can, therefore, visualize that 1, 3, 5-trimethylcyclohexane would have all the three methyl groups 100 per cent in the equatorial disposition.

tert-Butylcyclohexane exists 100 per cent in the equatorial conformation (A), the ring being frozen due to the prevention of the flip to a conformation (B) in which the non-bonded 1,3-diaxial interactions between the axially bound *tert*-butyl group and the two axial hydrogens at the 3-and 5-positions will be forbiddingly large.

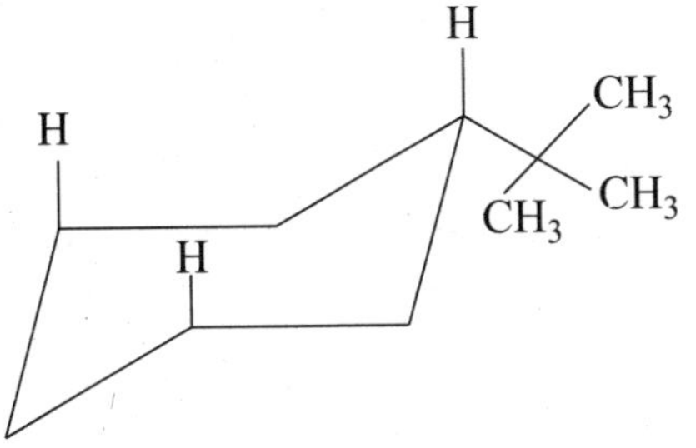

A

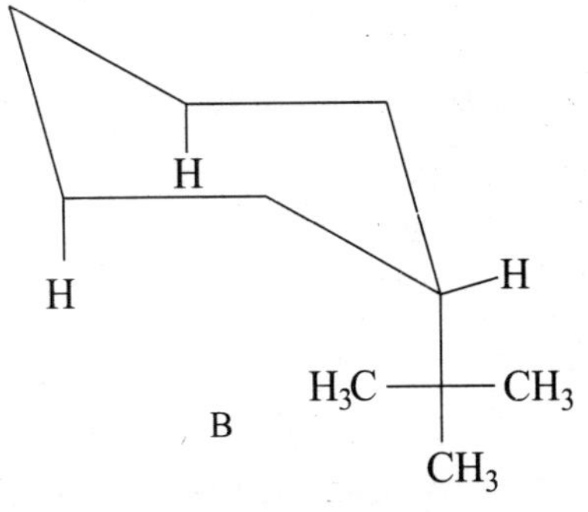

B

It is clear from the above considerations that the axial bonds experience non-bonded interactions with other axial bonds at 3- and 5-positions whereas the equatorial bonds are free from such steric interactions. In other words, the axially bound groups will experience more steric crowding than the equatorially bound groups. *This explains why in most of the cases the equatorially bound groups in cyclohexane derivatives are more reactive than the axially bound ones.* For instance, equatorially bound hydroxyl groups are more easily esterified than the axial ones. Similarly, the equatorial acetoxy group undergoes hydrolysis faster than the axial group. The same is true for esterification of the esters.

3.15 DIFFERENCE BETWEEN CONFORMATION AND CONFIGURATION

Although both the terms represent spatial arrangement of atoms or groups in a molecule, they are used in different contexts, so their difference should be clearly understood.

As discussed earlier the term conformations is used for various spatial isomers which can be easily inter-converted by simple rotation about a single bond. In contrast the term configurations is used for various spatial isomers which can be interconverted only by breaking and making of covalent bonds. This difference can be attributed to the fact that the energy difference between two conformers is very small due to which they can be interconverted by molecular collisions even at room temperature.

Since conformational isomers are readily interconvertible, they cannot be separated. However, due to high energy needed for interconversion of configurational isomers, they cannot be readily interconverted and hence can be separated easily.

Examples

Conformational Isomers

(*a*) Boat and chair forms of cyclohexane.

(*b*) Staggered and eclipsed forms of *n*-butane.

Configurational Isomers

(*a*) *Cis* and *trans*-2-butenes.

(*b*) R(–)-lactic acid and S (+)-lactic acid.

A BRIEF REVIEW

I. Structural Isomerism

- **Structural isomers** are the compounds having the same molecular formula but different arrangement of atoms within the molecule.
- **Chain or nuclear isomers** have the same molecular formula but different carbon skeleton *e.g.*, pentane, isopentane and neopentane.
- **Position isomers** have the same molecular formula but different positions of functional groups, multiple bonds or the substituents on the ring. *e.g.*, (*i*) but-1-ene and but-2-ene, (*ii*) butan-1-ol and butan-2-ol (*iii*) *o, m* and *p*-dichlorobenzenes.
- **Functional isomers** have the same molecular formula but different functional groups, *e.g.*, acetone, propionaldehyde, allyl alcohol and methyl vinyl ether are all functional isomers having molecular formula. C_3H_6O.

- **Metamers** have the same molecular formula but different number of carbon atoms on either side of the divalent functional groups such as —S—, $-\overset{O}{\overset{\|}{C}}-$, —O—, —NH—, etc.

  e.g., $CH_3CH_2OCH_2CH_3$ and $CH_3CH_2CH_2OCH_3$;

$$CH_3CH_2CH_2-\overset{O}{\overset{\|}{C}}-CH_3 \quad \text{and} \quad CH_3CH_2-\overset{O}{\overset{\|}{C}}-CH_2CH_3$$

- *Tautomers* are *keto-enol* isomers which involve 1, 3-intramolecular shift of a proton from one polyvalent atom to the other. Only those aldehydes or ketones which have one or more α-hydrogen can undergo *keto-enol* tautomerism. *e.g.*

$$CH_3-\overset{O}{\overset{\|}{C}}-CH_2-\overset{O}{\overset{\|}{C}}-CH_3 \quad \text{and} \quad CH_3-\overset{O}{\overset{\|}{C}}-CH=\overset{OH}{\overset{|}{C}}-CH_3$$

(*Keto Form*) (*Enol Form*)

  The relative amount of enol form increases when enol form in stabilized by intramolecular *H*-bonding (*chelation*).

- **Ring-chain isomers** have the same molecular formula but differ in having the open chain and cyclic structures *e.g.*, but-1-ene and cyclobutane.

$$CH_3CH_2CH=CH_2 \quad \text{and} \quad \square$$

II. Stereoisomerism

- **Stereoisomers** have the same molecular formula but different relative spatial arrangements of atoms or groups in the molecule.
- **Conformational isomers** can be interconverted by simple rotation about the sigma bond, *e.g.*, eclipsed and staggered conformations of *n*-butane. Such isomers are not separated easily because they have low energy barrier.
- **Configurational isomers** can be interconverted only by breaking and making of covalent bonds and not by simple rotation about the sigma bonds, *e.g.*, *cis*-but-2-ene and *trans*-but-2-ene. Configurational isomers are of two types:

  (*i*) Optical Isomers (*ii*) Geometrical Isomers

- **Optical isomers** are optically active substances having same molecular formula and structural formula but rotate the plane of polarised light in opposite directions in equal magnitude.

- **Dextrorotatory** substances rotate the plane of polarised light towards right hand side and are represented by the prefix '*d*' or (+).
- **Leavoratory** substances rotate the plane of polarised light towards the left hand side and are represented by the prefix '*l*' or (–).
- **Asymmetric** molecules have no element of symmetry and such molecules are generally optically active.
- **Dissymmetric** molecules are optically active molecules which have no improper axis of rotation. Such molecules may have proper axis of symmetry. *e.g.*, tartaric acid having C_2 axis of symmetry is an optically active substance. *All asymmetric molecules are dissymmetric but all dissymmetric molecules are not asymmetric.*
- **Chiral** molecules are not superimposable on their mirror images. If a carbon atom is attached to four different atoms or groups it is called **chiral** or **asymmetric carbon**. Chirality of the *molecule is the necessary and sufficient condition for a molecule to be optically active and not only the presence of a chirality centre. e.g.*, some allenes and biphenyls are optically active even though they do not have any chiral atom.
- **Enantiomers** are the optically active compounds which are non-superimposable mirror images of each other. They have identical physical and chemical properties but rotate the plane of polarized light in opposite direction to the same extent. These are represented by the prefixes *d* or *l* (+ or –).
- **Racemic mixture** or **racemic modification** is represented by (±) or '*dl*' which means an equimolar mixture of two enantiomers. Racemic mixtures are optically inactive due to *external compensation.* Such mixtures can be resolved by different methods to get pure enantiomers.
- **Diastereomers** are stereoisomers which are not mirror images of each other. They have different physical properties such as m.pt., b.p. and solubility. These can be separated easily by fractional distillation or fractional crystallization. They may or may not be optically active.
- **Meso compounds** are optically inactive stereoisomers of a compound. These are optically inactive due to the presence of a plane of symmetry which divides the molecule in two parts which are mirror images of each other. One part of the molecule cancels the optical rotation due to the other and this phenomenon is called **internal compensation.**
- **Geometric isomers** have same molecular formula but different spatial arrangement of atoms around the double bond. This type of isomerism is shown by the compounds of the type $abC = Cab$ or $abC = Cad$ or $abC = Cde$ but not by the compounds in which any carbon of the double bond has same atoms/groups *e.g.*, $aaC = Cab$ or $aaC = Cbd$ etc. This isomerism arises due to restricted rotation about C = C double bond. Compounds with C = N or N = N also exhibit geometric isomerism.
- ***Cis*-isomers** have two similar atoms or groups on the same side of the double bond while in ***trans*-isomer** the two similar groups or atoms are on the opposite sides of the double bond. *Cis*-isomer has higher dipole moment, b.pt. and solubility than corresponding *trans*-isomer.

- **Geometric isomerism** is also exhibited by disubstituted alicyclic compounds such as 1, 2-disubstituted cyclopropane, 1, 2 and 1,3-disubstituted cyclobutanes and cyclopentanes.
- **Conformational or rotational isomers or rotamers** are the infinite number of arrangements of atoms in space which arise due to rotation about a single bond. Two such extreme conformations are **eclipsed** and **staggered**. These isomers cannot be separated due to lower energy barrier between them.
- Eclipsed conformations are less stable than staggered ones due to the presence of **torsion strain** and **van der Waals strains** in eclipsed form. Staggered conformation of ethane is more stable than eclipsed conformation by 12 kJ mol^{-1}.
- The relative stability of conformations of *n*-butane is: Staggered > gauche > partially eclipsed > fully eclipsed.
- Cyclohexane exists in **chair** and **boat conformations**. Chair form is more stable than the boat form. Both these conformations are free from angle strain but in boat form there are *n*-butane eclipsed interactions which make it less stable than chair form which has no such eclipsed interactions but have *n*-butane skew-interactions. Both chair and boat form are interconvertible and cannot be separated.
- In case of substituted cyclohexanes, the substituent group prefers to remain in the equatorial position than axial.

Chapter 3

SOLVED PROBLEMS

Q. 1. Explain why the dipole moment of meso form is much lower (μ =1.27 D) than that of the optically active form (μ = 2.75 D) of stilbene dichloride.

Ans. The three important conformations of *meso*-form of stilbene dichloride (μ = 1.27 D) are

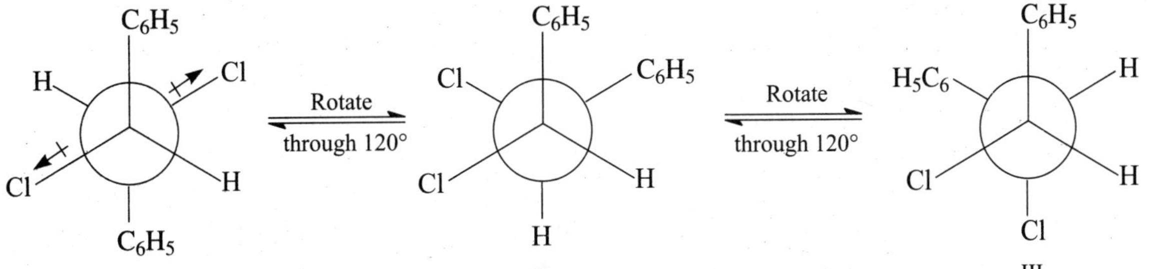

C_6H_5 H Cl Cl H C_6H_5 — I

Rotate through 120°

C_6H_5 Cl C_6H_5 Cl H H — II

Rotate through 120°

C_6H_5 H_5C_6 H Cl H Cl — III

The conformations of optically active (+ or –) form of stilbene dichloride (μ = 2.75 D) are

IV: front — C_6H_5 (top), Cl (left), H (right); back — C_6H_5 (upper right), Cl (lower right), H (bottom)
$\underset{\text{through } 120°}{\overset{\text{Rotate}}{\rightleftharpoons}}$
V: C_6H_5 (top), H, H, Cl, Cl, C_6H_5 (bottom)
$\underset{\text{through } 120°}{\overset{\text{Rotate}}{\rightleftharpoons}}$
VI: C_6H_5 (top), H_5C_6, H, H, Cl, Cl (bottom)

IV V VI

The most favoured conformation of *meso*-stilbene dichloride is (I). This is due to the reason that in this conformation not only the two bulky phenyl groups are *anti* to each other, the two C—Cl dipoles also oppose each other. Since this particular conformer has zero dipole moment, therefore, the overall dipole moment of the *meso*-form is quite low.

The preferred conformation of (+) or (–) stilbene dichloride is (V) in which the two bulky phenyl groups are *anti* to each other. Since in this conformation, the two C—Cl dipole are inclined to each other at angle of 60°, therefore, the overall dipole of (+) or (–) stilbene dichloride is relatively large.

Q. 2. How many asymmetric carbon atoms are created during the complete reduction of benzil (PhCOCOPh) with $LiAlH_4$? Also write the number of possible stereoisomers of the product.

Ans. Benzil on reduction with $LiAlH_4$ gives 1, 2-diphenylethane-1, 2-diol which contains two asymmetric carbon atoms as shown by asterisks.

$$\underset{\text{Benzil}}{Ph—\overset{O}{\overset{\|}{C}}—\overset{O}{\overset{\|}{C}}—Ph} \xrightarrow{LiAlH_4} \underset{\text{1, 2-Diphenylethane-1, 2-diol}}{Ph—\overset{OH}{\overset{|}{\underset{H}{\underset{|}{\overset{*}{C}_1}}}}—\overset{OH}{\overset{|}{\underset{H}{\underset{|}{\overset{*}{C}_2}}}}—Ph}$$

Since, 1, 2-diphenylethane-1, 2-diol contains two similar asymmetric carbons (cf. tartaric acid) it exists in the following three stereoisomeric forms (I, II and III):

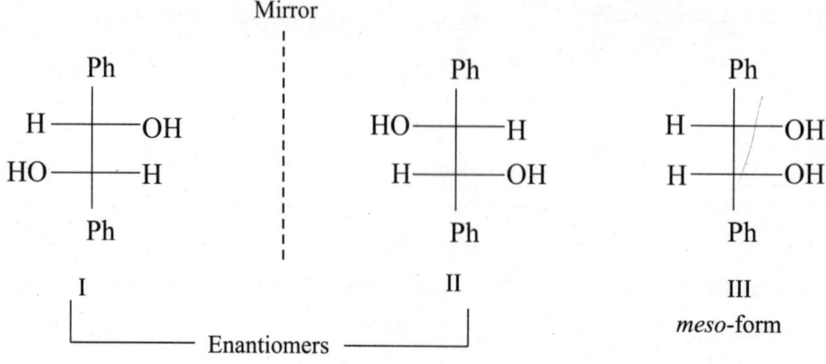

UNSOLVED PROBLEMS

1. (*i*) What is meant by isomerism? Discuss its various types giving atleast one example in each case.

(*ii*) Label each of the following pairs of structures as structural isomers of each other, as geometric isomers of each other or as the same compound:

(*a*) $$\begin{array}{ccc} H & & H \\ & C=C & \\ H_9C_4 & & H \end{array} \quad \text{and} \quad \begin{array}{ccc} H & & H \\ & C=C & \\ H & & C_4H_9 \end{array}$$

(*b*) $$\begin{array}{ccc} H & & H \\ & C=C & \\ H_7C_3 & & CH_3 \end{array} \quad \text{and} \quad \begin{array}{ccc} H & & C_3H_7 \\ & C=C & \\ H_3C & & H \end{array}$$

(*c*) $$\begin{array}{ccc} H & & Cl \\ & C=C & \\ Cl & & H \end{array} \quad \text{and} \quad \begin{array}{ccc} Cl & & H \\ & C=C & \\ H & & Cl \end{array}$$

(*d*) $H_3C—C{\equiv}C—C_2H_5$ and $H_5C_2—C{\equiv}C—CH_3$

(*e*) $$\begin{array}{ccc} CH_3H_2C & & CH_2CH_3 \\ & C=C & \\ H & & H \end{array} \quad \text{and} \quad \begin{array}{ccc} CH_3 & & H \\ & C=C & \\ H & & CH_2CH_2CH_3 \end{array}$$

2. (*i*) Discuss briefly keto-enol tautomerism in aldehydes and ketones. Also discuss the conditions under which enol form predominates.

(*ii*) Write the keto-enol forms of acetoacetic ester and pentane-2, 4-dione. Which of the two has higher enol content? Explain.

(*iii*) Draw the structures of all the isomeric alcohols and ethers having the molecular formula, $C_5H_{12}O$.

3. Define optical activity. Discuss various methods for the resolution of a racemic mixture.

4. (*a*) What are the requirements for a compound to be optically active? Comment upon the observation that the presence of a chiral carbon is not always essential for a compound to exhibit optical activity.

(*b*) Predict which of the following compounds are chiral:

(*i*) 2,6-dinitrophenyl ring bonded to a second benzene ring carrying COOH groups (one hashed, one wedged)

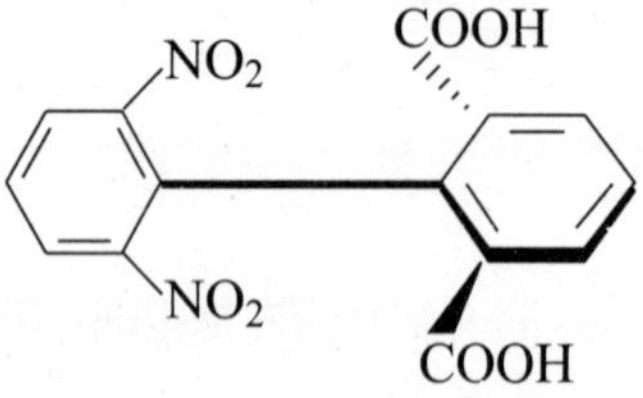

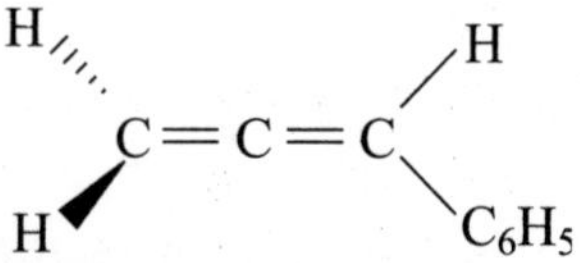

(*ii*) H, H (hashed, wedged) C=C=C H, C_6H_5

(*iii*) H, HOOC (hashed, wedged) C=C=C H, C_6H_5

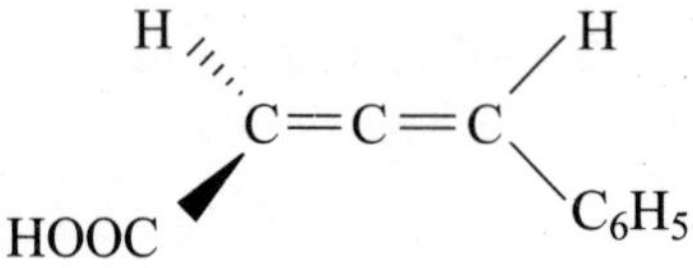

5. (*i*) A carboxylic acid having the molecular formula $C_3H_5O_2Br$ is optically active. Write its structure.

(*ii*) Certain ester of tartaric acid is optically active. But it yields an optically inactive acid on hydrolysis. Give its structure.

(*iii*) What is the structure of an optically active alcohol having the molecular formula $C_4H_{10}O$.

(*iv*) Write structures for all the possible olefins of molecular formula C_5H_{10} that exhibit geometric isomerism.

(*v*) An acid having molecular formula $C_5H_{10}O_2$ is optically active. What is its structure?

6. (*i*) What is the lowest molecular weight alkane which is chiral? Is there another alkane of the same molecular formula that is also chiral? If it is so, then give its structure also.

(*ii*) Listed below are several chiral and achiral molecules. Which is which?

(*a*) $CH_3—C(Cl)(Br)—CH_2—CH=CH_2$

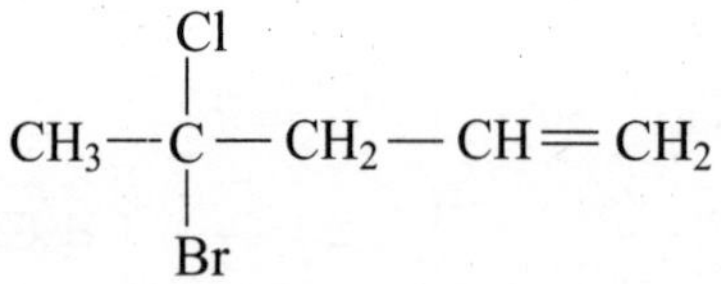

(*b*) $CH_3—CH(CH_2CH_3)—C(=O)—CH_3$

(*c*) $CH_3—CHBr—CH_3$

(*d*) $CH_3CH_2—CH(CH_3)—Cl$

(*e*) $C_6H_5—CH(Cl)—CH_3$

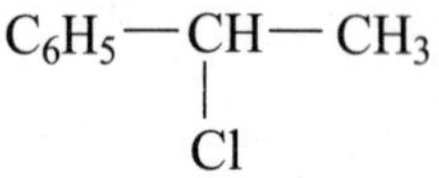

(*f*) $H_2N—CH_2—COOH$

(*g*) $H_2NCH(CH_3)COOH$

(*h*) $CH_3—CH(Br)—CH(Br)—CH_2—CH_3$

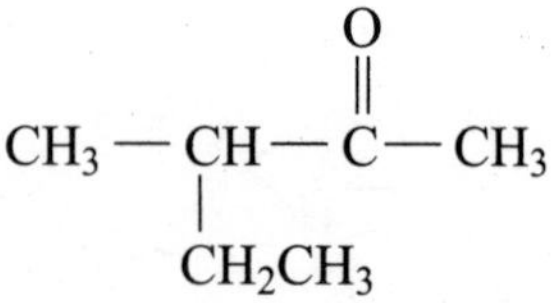

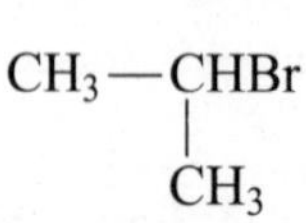

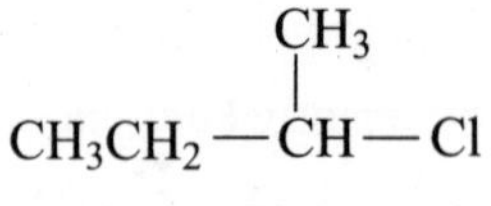

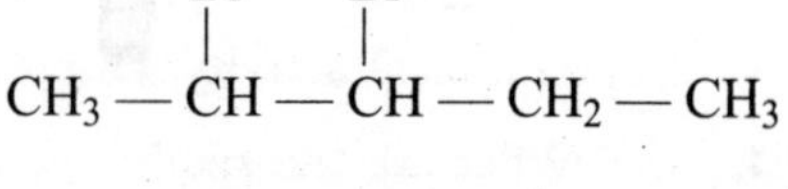

(*i*) $(CH_3)_2CH—CH(Br)—CH_3$

(*j*) $CH_3—CH(CH_3)—OH$

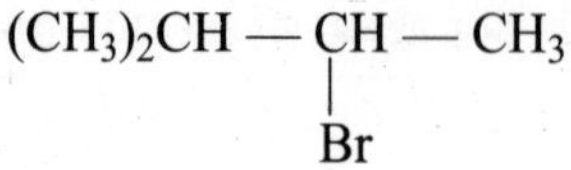

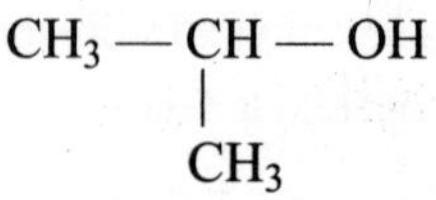

7. (*i*) What is meant by the terms:

(*i*) Chirality, (*ii*) Diastereomers, (*iii*) Configurational isomers, (*iv*) Conformational isomers, (*v*) R and S convention, (*vi*) Plane polarised light, and (*vii*) Specific rotation.

(*ii*) Differentiate between asymmetry, dissymmetry and chirality. Explain why chirality is the essential condition for optical activity.

8. What are sequence rules? How are these used for assigning R and S-configurations? Discuss with suitable examples.

9. Explain D and L system of assigning absolute configuration to a pair of stereoisomers. What are its limitations and how are these overcome in R and S system?

10. Assign R or S designation to each of the following molecules:

(*i*) H; Cl — CH_3; Br

(*ii*) CH_2CH_3; Br — CH_2OH; CH_3

(*iii*) OH; H — Cl; CH_3

(*iv*) C_4H_9; H_3C — C_2H_5; C_3H_7

(*v*) CH_3; H — Cl; NH_2

(*vi*) CH_3; H — OH; H — OH; CH_3

(*vii*) CH_3; H — Cl; Cl — H; CH_3

(*viii*) F; H — Cl; Br

11. Draw Fischer projections for the two enantiomers of each of the non-cyclic isomers of:

(*i*) $C_5H_{11}Cl$ and (*ii*) C_6H_{12} which possess a chiral carbon. Designate the configuration of each enantiomer by R and S convention.

12. (*i*) What is racemization? Discuss briefly various mechanisms by which racemization can occur.

(*ii*) Discuss briefly the characteristics of enantiomers towards chiral and achiral reagents.

13. Draw wedge and dash and Fisher projections of all the possible stereoisomers of 2-aminopropanoic acid and 3-iodopentane-2-ol. Comment upon the stereochemical relationships between their respective stereoisomers.

14. Discuss optical isomerism in 3, 4-dichlorohexane.

15. Draw the Newman projection formulas for the compounds having the following stereochemistry:

(*i*) (2R,3S)–Tartaric acid (*ii*) (2S,3R)-3-bromobutan-2-ol.

16. What is Walden inversion? Give one example.

17. Discuss briefly the cause of geometrical isomerism.

18. Define and explain the term geometrical isomerism. State the necessary and sufficient condition for a molecule to exhibit geometrical isomerism. How can you explain the fact that derivatives of cyclohexane exhibit geometrical isomerism even though they do not contain a double bond?

19. Discuss briefly the properties of geometrical isomers taking maleic acid and fumaric acid as examples.

20. Discuss briefly the geometrical isomerism in oximes and alicyclic compounds.

21. What do the symbols E and Z stand for? Discuss briefly the E and Z system of naming a pair of geometrical isomers. What are advantages of E and Z system over conventional *cis-trans* system?

22. Give a brief account of the various methods which have been used to ascertain the configuration of a pair of geometrical isomers.

23. How is the configuration of geometrical isomers of oximes established?

24. Indicate whether each of the following compound is E or Z:

(*i*) $\mathrm{(Cl)(H)C{=}C(Cl)(I)}$

(*ii*) $\mathrm{(H)(CH_3)C{=}C(CH_2Cl)(Cl)}$

(*iii*) $\mathrm{(F)(H)C{=}C(CH_2CH_2CH_3)(Cl)}$

(*iv*) $\mathrm{(D)(H)C{=}C(H)(D)}$

(*v*) $\mathrm{(CH_3)(H)C{=}C(CH_2CH_2CH_3)(CH_2CH_3)}$

(*vi*) $\mathrm{(I)(H)C{=}C(CH_3)(CH_2CH_3)}$

25. Which of the following can exhibit geometrical isomerism?

(*i*) 2, 3-Dimethyl-2-butene (*ii*) 2-Hexene

(*iii*) 2, 3-Dimethyl-1-butene (*iv*) 4-Methyl-2-pentene.

26. Discuss how geometrical isomers affect melting points and boiling points.

27. Briefly discuss various factors which affect the stability of a conformation. Draw the potential energy diagram for the various conformations of *n*-butane and explain their relative stability.

28. (*i*) Write structures of two geometric isomers of 1, 2-cyclopentanediol, one of which exists as a pair of enantiomers, while the other is a *meso* compound. Give a suitable reason for their existence.

(*ii*) Draw the potential energy diagram for various conformations of cyclohexane and explain their relative stability. Why is the chair conformation of cyclohexane more stable than the boat conformation?

29. What are axial and equatorial bonds in cyclohexane? Discuss briefly the conformations of mono-substituted cyclohexanes and their relative stability.

30. Why is twist boat conformation of cyclohexane obtained? Explain why it is more stable than boat conformation.

31. Label each of the following compounds as *cis, trans* or neither:

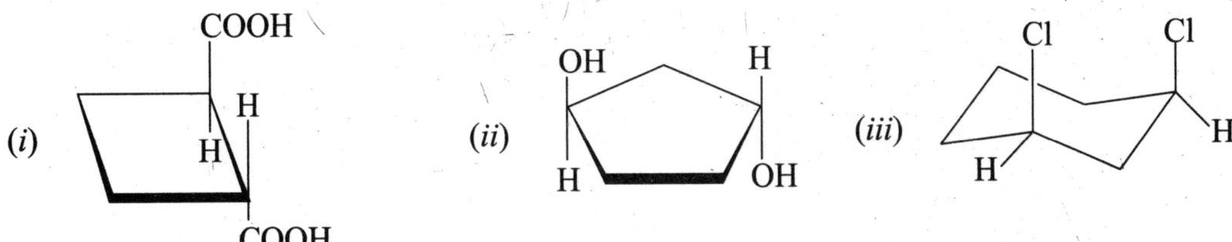

(iv) H, CH_3, H, CH_3 (v) CH_3, CH_3, H, H

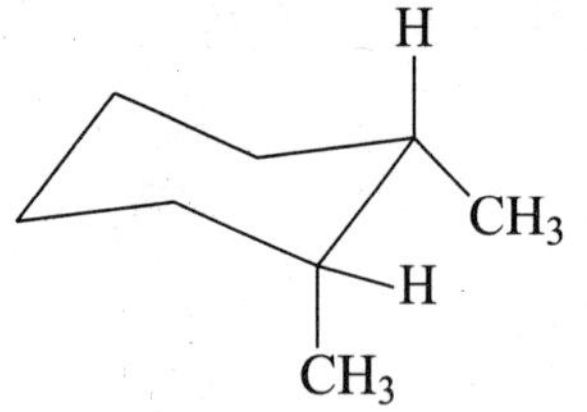

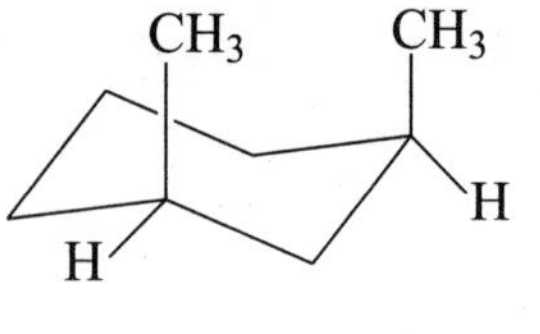

(vi) H_3C— —CH_3 (vii) CH_3, CH_3 (viii) H_3C, CH_3, H, H

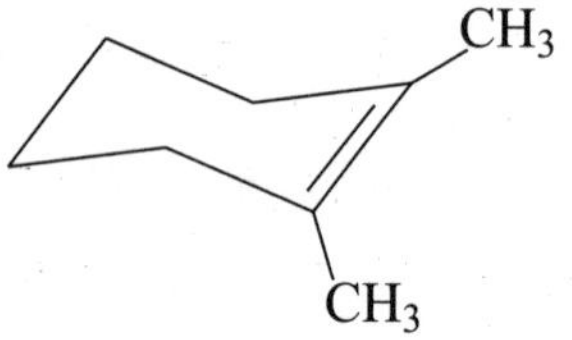

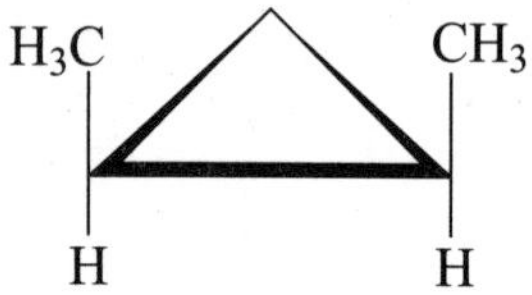

32. (*i*) Why is it not possible under ordinary conditions to resolve amines even if three different groups are attached to nitrogen atoms?

(*ii*) Discuss the stereochemistry of oximes.

33. Draw Newman projections of axial and equatorial methylcyclohexanes and comment upon their relative stability.

34. Explain the following terms with suitable examples:

(*i*) Angle strain (*ii*) Torsional strain (*iii*) Steric strain.

35. Draw various conformations of *n*-butane and compare their relative stability.

36. Draw Sawhorse, Newman, Fischer and Flying Wedge representations of 3-bromo-2-butanol.

37. Write Fischer projection formulas for all the different optical isomers of 3-bromo-2, 5-hexanediol (A).

$$CH_3CH(OH)CH_2CH(Br)\underset{\displaystyle CH_3}{\underset{|}{C}}HOH$$

(A)

38. Distinguish between the following terms:

(*i*) Configuration and conformation.

(*ii*) Absolute and relative configuration.

(*iii*) Observed rotation and specific rotation.

(*iv*) Enantiomers and diastereomers.

(*v*) Meso and racemic forms.

39. Examine each of the following structure pairs, and find out whether or not they are identical. If they are different, indicate the relationship they have to each other (enantiomers, diastereomers or conformational isomers).

(*i*) CH_3, H, Br, C_2H_5 and Br, H, CH_3, C_2H_5

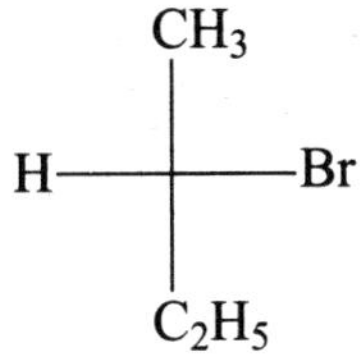

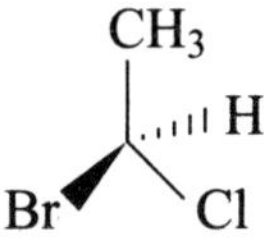

(*ii*) CH_3, Cl, H, Br and CH_3, H, Br, Cl

(*iii*) Br, H_3C, OH, Cl, F, H and Cl, H, CH_3, Br, F, OH

(*iv*) Cl, H, Br, H and H, Cl, Br, H

(*v*) CH_3, OH and OH, CH_3

(*vi*) NH_2, H, CH_2CH_3, H_3C and NH_2, H_3C, CH_2CH_3, H

(*vii*) CH_3, H, H, CH_3 and H, CH_3, CH_3, H

(*viii*) CH_3, H, $N^{\oplus}$, CH_2CH_3, Ph and CH_2CH_3, Ph, $N^{\oplus}$, CH_3, H

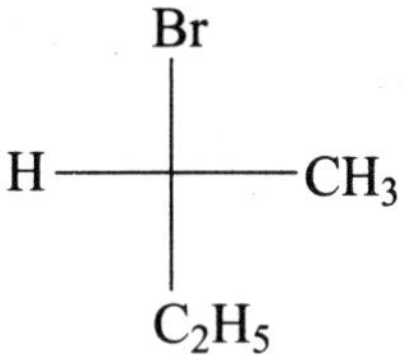

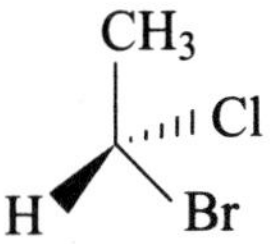

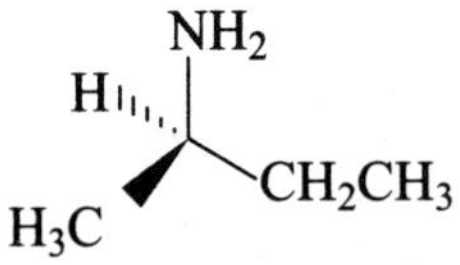

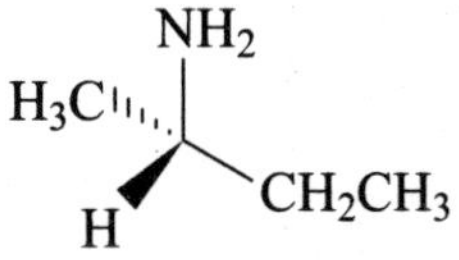

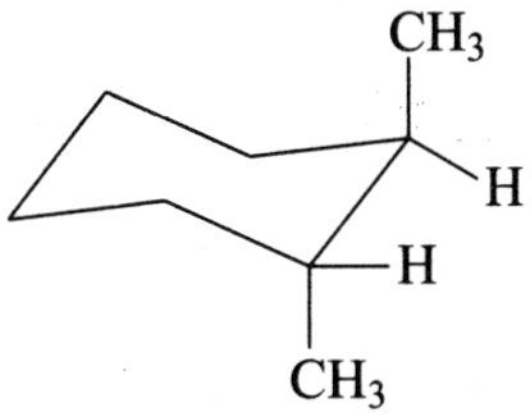

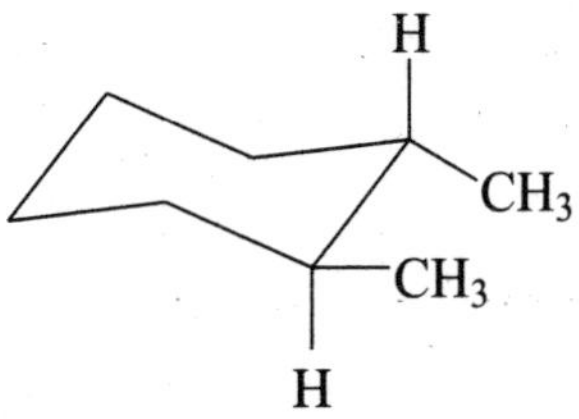

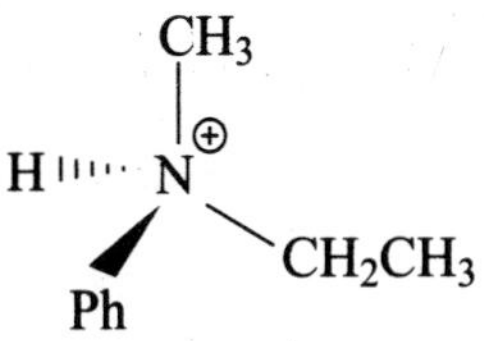

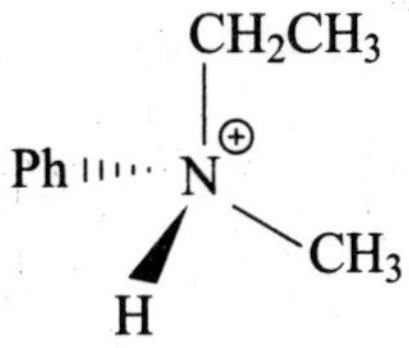

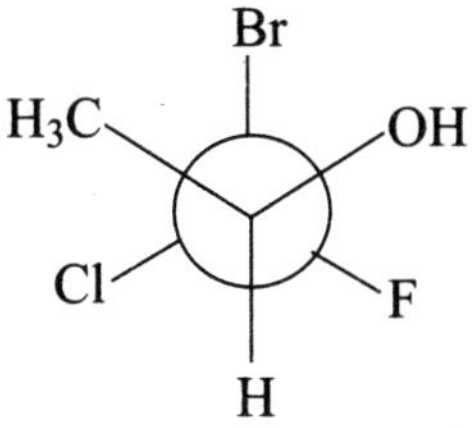

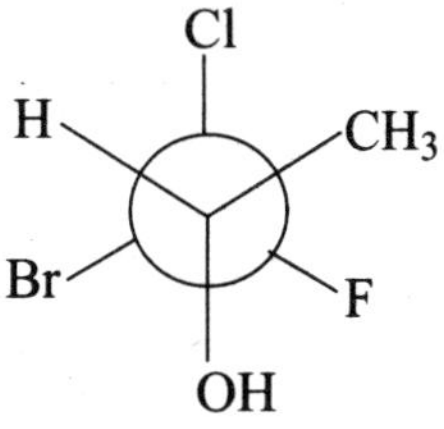

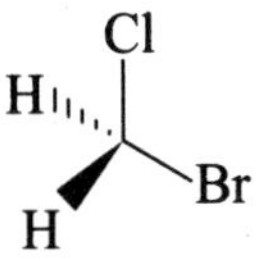

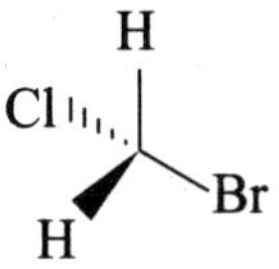

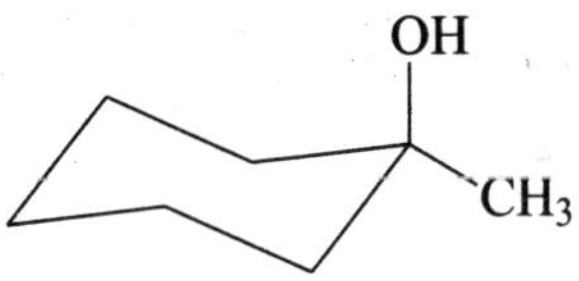

SECTION III

Alkanes and Cycloalkanes – I (Alkanes)

4.1 HYDROCARBONS

The term hydrocarbons is used for the organic compounds containing only carbon and hydrogen.

4.1.1 Classification

Broadly speaking the hydrocarbons are classified into various categories as shown in Fig. 4.1.

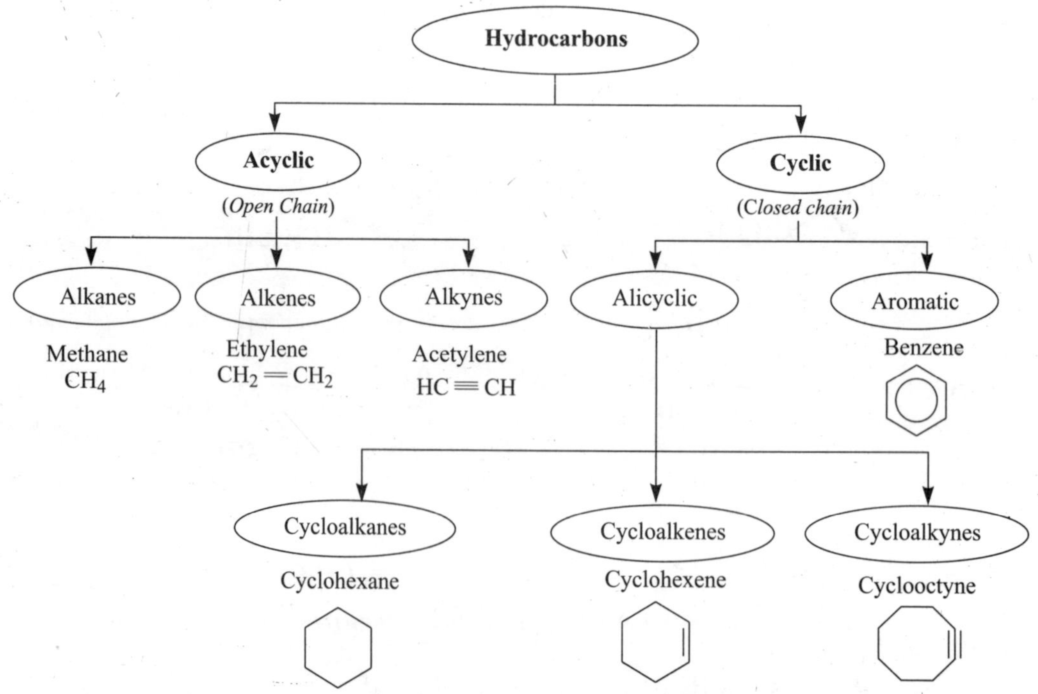

Fig. 4.1: Classification of hydrocarbons

We will now discuss these classes in more details.

1. Acyclic Hydrocarbons (*Open Chain Hydrocarbons*)

They are generally referred to as aliphatic hydrocarbons. They are further divided into three categories as follows:

(*i*) The *alkanes* are saturated hydrocarbons having general formula C_nH_{2n+2}, where n must be an integer. An alkane is, therefore, characterized by the presence of all carbon-carbon and carbon-hydrogen *single bonds*. They possess low chemical activity. Therefore, they are sometimes referred to as paraffins (*L. parum* = little, *affinis* = affinity or reactivity). For example, methane (CH_4), ethane (C_2H_6), propane (C_3H_8), etc.

(*ii*) *The alkenes* are hydrocarbons which have general formula C_nH_{2n} and are characterized by the presence of *one double bond* between two carbon atoms. For example, ethylene ($H_2C = CH_2$), propylene ($CH_3CH = CH_2$), etc. They are also called olefins (*L. oleum = oil, fiant = forming*) because the simplest compound, ethylene, produces an oily liquid on treatment with chlorine.

(*iii*) *The alkynes* are hydrocarbons which are designated by the general formula C_nH_{2n-2} and are characterized by the presence of *one triple bond* between two carbon atoms. For example, acetylene ($HC \equiv CH$), propyne ($CH_3C \equiv CH$) etc. Sometimes hydrocarbons of this class are referred to as acetylenes being the derivatives of the simplest hydrocarbon acetylene.

The alkenes and alkynes are also called *unsaturated hydrocarbons* since they contain lesser percentage of hydrogen compared to that of corresponding alkanes. The latter type of compounds are called *saturated hydrocarbons*.

2. Cyclic Hydrocarbons (*Closed Chain Hydrocarbons*)

These are the hydrocarbons containing carbocyclic rings *i.e.*, rings made up of carbon atoms only. They are further sub-divided into two categories:

(*a*) The alicyclic hydrocarbons.

(*b*) The aromatic hydrocarbons.

(*a*) The alicyclic hydrocarbons: With a few exceptions, they resemble aliphatic hydrocarbons in their chemical behaviour, so they are called alicyclic (*aliphatic cyclic*). Like acyclic hydrocarbons they are of the three types as follows:

(*i*) *Cycloalkanes*: They are cyclic hydrocarbons having *all single bonds* and hence resemble alkanes in their chemical properties. Their general formula is C_nH_{2n}. For example,

Cyclopentane, Cyclohexane etc.

(*ii*) *Cycloalkenes*: They are cyclic hydrocarbons having a carbon-carbon *double bond* and hence resemble alkenes in their chemical behaviour. Their general formula is C_nH_{2n-2}. For Example, Cyclopentene, Cyclohexene etc.

***(iii) Cycloalkynes*:** They are cyclic hydrocarbons having a carbon-carbon triple bond and hence resemble alkynes in their chemical behaviour. Their general formula is C_nH_{2n-4}. For example,

Cyclooctyne

(*b*) The aromatic hydrocarbons: These are polyunsaturated carbocyclic compounds containing a six-membered ring with *three conjugated double bonds*. Despite high degree of unsaturation they are quite stable and prefer to undergo substitution rather than addition reactions. For example,

Benzene, Toluene CH_3 etc.

In this chapter, we will discuss the chemistry of alkanes in details.

4.2 ALKANES

4.2.1 Nomenclature

The different systems used for naming these compounds are as follows:

1. The common or trivial names.
2. As methane derivatives.
3. IUPAC system.

1. The common or trivial names: The lower members of the series are known by their common names which are generally based on their source or some other facts, *e.g.*,

CH_4	$CH_3—CH_3$	$CH_3—CH_2—CH_3$
Methane	Ethane	Propane

Alkanes having more than three carbon atoms can be represented by more than one structure. Broadly speaking the following two types of structures are possible.

(*a*) Structures having straight chain or unbranched chain of carbon atoms (straight chain alkanes): These are the compounds having unbranched chain of carbon atoms in which a given carbon is attached to not more than two carbon atoms. Such compounds are also called straight chain hydrocarbons. Their common names are the same as IUPAC names except that the prefix *n* (meaning **normal**) is used in the common names. For example,

$CH_3CH_2CH_2CH_3$	$CH_3CH_2CH_2CH_2CH_3$	$CH_3CH_2CH_2CH_2CH_2CH_3$
n- Butane	*n*-Pentane	*n*-Hexane

(*b*) *Branched chain alkanes*: These are the compounds in which at least one carbon atom is linked to three or four carbon atoms. In writing common names of these hydrocarbons prefixes *iso*- and *neo*- are used when a carbon atom is attached to three and four carbons, respectively. For example,

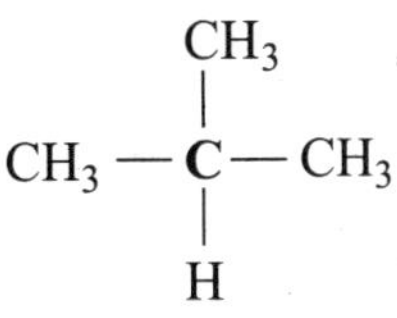

$$CH_3-\overset{\overset{CH_3}{|}}{\underset{\underset{H}{|}}{\mathbf{C}}}-CH_3$$
Isobutane

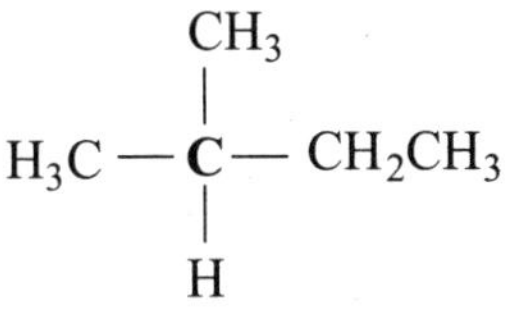

$$H_3C-\overset{\overset{CH_3}{|}}{\underset{\underset{H}{|}}{\mathbf{C}}}-CH_2CH_3$$
Isopentane

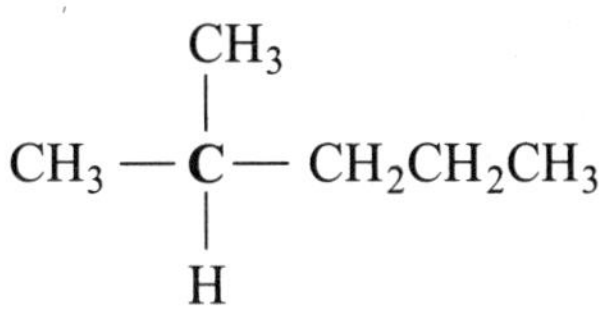

$$CH_3-\overset{\overset{CH_3}{|}}{\underset{\underset{H}{|}}{\mathbf{C}}}-CH_2CH_2CH_3$$
Isohexane

(Carbon atom in bold is attached to three carbons)

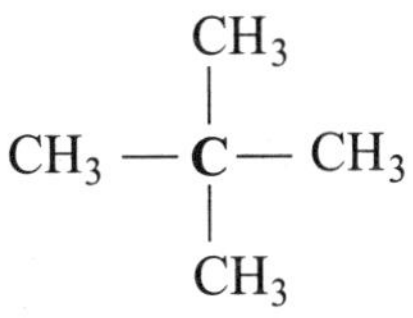

$$CH_3-\overset{\overset{CH_3}{|}}{\underset{\underset{CH_3}{|}}{\mathbf{C}}}-CH_3$$
Neopentane

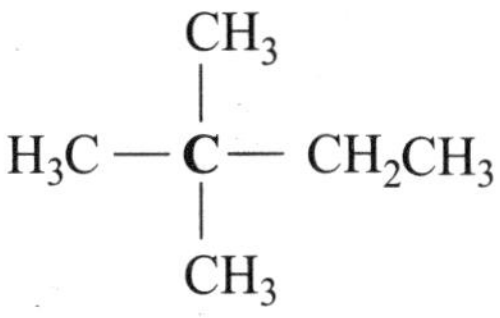

$$H_3C-\overset{\overset{CH_3}{|}}{\underset{\underset{CH_3}{|}}{\mathbf{C}}}-CH_2CH_3$$
Neohexane

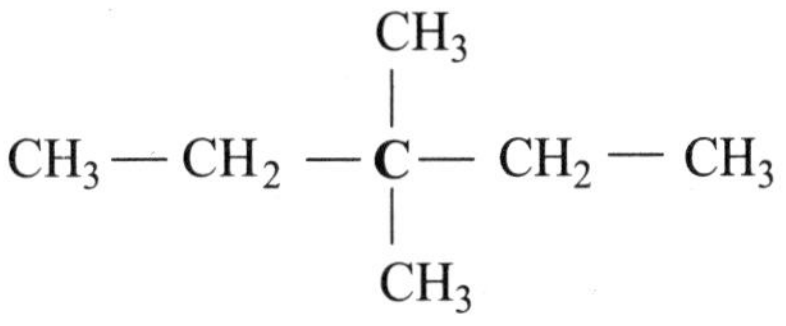

$$CH_3-CH_2-\overset{\overset{CH_3}{|}}{\underset{\underset{CH_3}{|}}{\mathbf{C}}}-CH_2-CH_3$$
Neoheptane

(Carbon atom in bold is attached to four carbons)

The first four members are given common names whereas higher members are named systematically by putting the Latin *prefix* denoting the number of carbon atoms (pent for 5 carbons, hex for 6 carbons, hept for 7 carbons) to the *suffix-ane*.

2. As methane derivatives: In the second system of nomenclature, the branched chain hydrocarbons are named as derivatives of methane. *The carbon atom having the maximum branching is considered to be the methane nucleus.* The names of the alkyl groups attached to this nucleus are arranged in alphabetical order as prefixes before the word methane, *e.g.*,

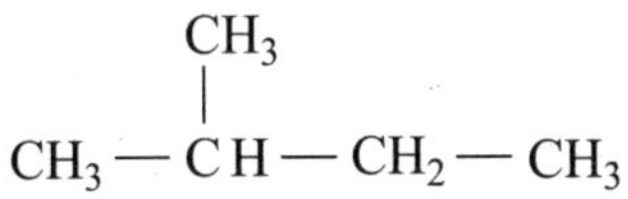

$$CH_3-\overset{\overset{CH_3}{|}}{CH}-CH_2-CH_3$$
Ethyl dimethyl methane

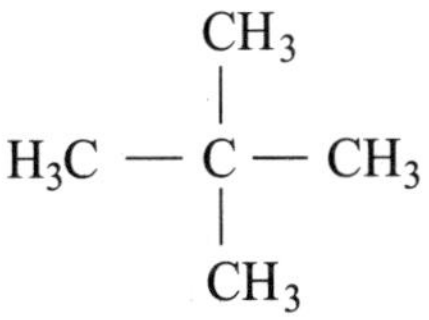

$$H_3C-\overset{\overset{CH_3}{|}}{\underset{\underset{CH_3}{|}}{C}}-CH_3$$
Tetramethyl methane

3. IUPAC system: The above two systems have not been found suitable for naming the higher branched chain alkanes. The more scientific IUPAC **(International Union of Pure and Applied Chemistry)** system has been found to be quite satisfactory for the nomenclature of such compounds. The IUPAC name of an alkane is obtained using the following rules (steps):

(i) Rule of longest chain: First of all the longest continuous carbon chain is chosen. Such a carbon chain is called the 'parent chain'. Next the number of carbon atoms are counted in this 'parent chain' and the name of the alkane corresponding to the number of carbon atoms in this chain becomes the last name of the alkane.

Other carbon atoms not included in the 'parent chain' are regarded as prefixes commonly referred to as alkyl groups/side chains/substituents, *e.g.*,

Chapter 4

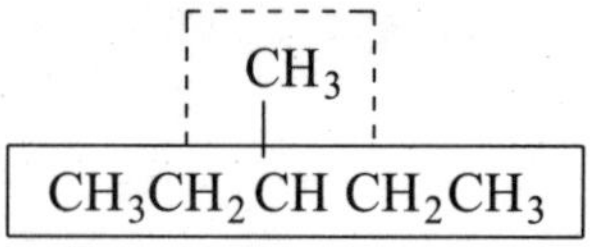

$$\begin{array}{c} CH_3 \\ | \\ CH_3CH_2CH\,CH_2CH_3 \end{array}$$

Longest continuous chain has five carbon atoms so its parent name is **pentane**.

Here it may be noted that longest continuous chain is to be chosen regardless of how the molecule is written, *i.e.*, the longest continuous chain may not always be in horizontal plane, it may have 'bends' or 'turns'. For example, the molecule of one of the isomers of hydrocarbon C_9H_{20} can be written in different ways.

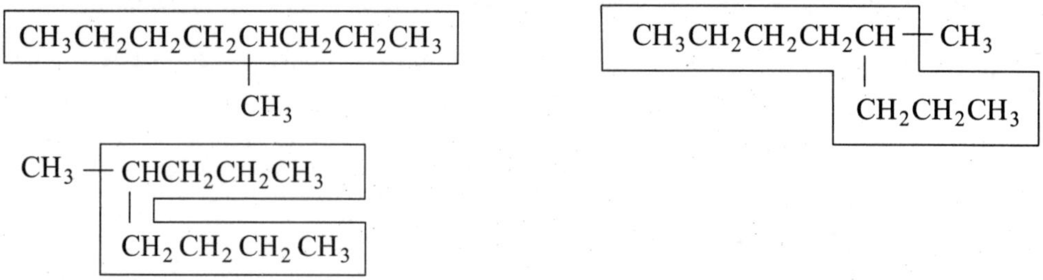

$$\begin{array}{c} CH_3CH_2CH_2CH_2CHCH_2CH_2CH_3 \\ | \\ CH_3 \end{array}$$

$$\begin{array}{l} CH_3CH_2CH_2CH_2CH - CH_3 \\ \qquad\qquad\qquad\qquad | \\ \qquad\qquad\qquad\quad CH_2CH_2CH_3 \end{array}$$

$$\begin{array}{l} CH_3 - CHCH_2CH_2CH_3 \\ \qquad\quad | \\ \qquad\quad CH_2\,CH_2\,CH_2\,CH_3 \end{array}$$

In all the formulas the longest continuous chain has eight carbons, so its parent name will be **octane**.

***(ii) Larger number of side chains (substituents) rule*:** In case two different chains of equal length, having different number of side chains, are possible then the one with larger number of side chains is selected. For example,

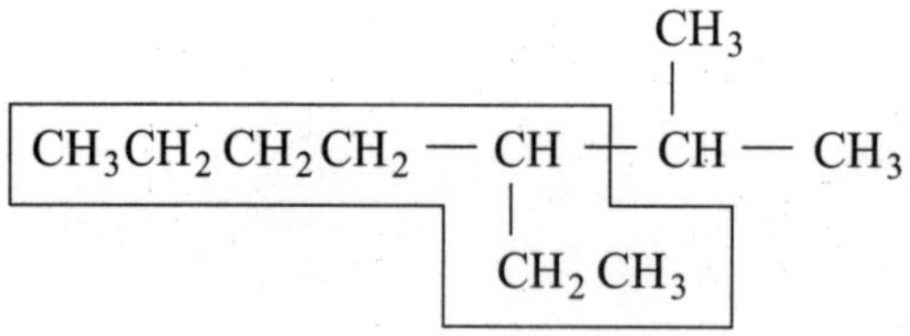

$$\begin{array}{l} \qquad\qquad\qquad\qquad\qquad\quad CH_3 \\ \qquad\qquad\qquad\qquad\qquad\quad | \\ CH_3CH_2\,CH_2CH_2 - CH - CH - CH_3 \\ \qquad\qquad\qquad\qquad | \\ \qquad\qquad\qquad\quad CH_2\,CH_3 \end{array}$$

Longest chain of seven carbon atoms with one alkyl group (side chain) — ***Incorrect***

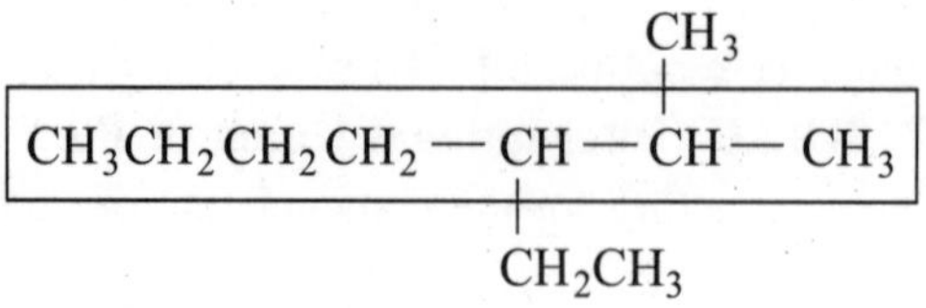

$$\begin{array}{l} \qquad\qquad\qquad\qquad\qquad\quad CH_3 \\ \qquad\qquad\qquad\qquad\qquad\quad | \\ CH_3CH_2CH_2\,CH_2 - CH - CH - CH_3 \\ \qquad\qquad\qquad\qquad | \\ \qquad\qquad\qquad\quad CH_2CH_3 \end{array}$$

Longest chain of seven carbon atoms with two alkyl groups (side chains) — ***Correct***

1. Lowest Locant or Lowest Number Rule

(*a*) *For alkane with one substituent*: The carbon atoms of the longest continuous chain, selected as above, are numbered as 1, 2, 3,..... etc., starting from the end which assigns the lowest number to the carbon atom carrying the substituent. For example, numbering of carbon atoms in structure (I) can be done in two ways:

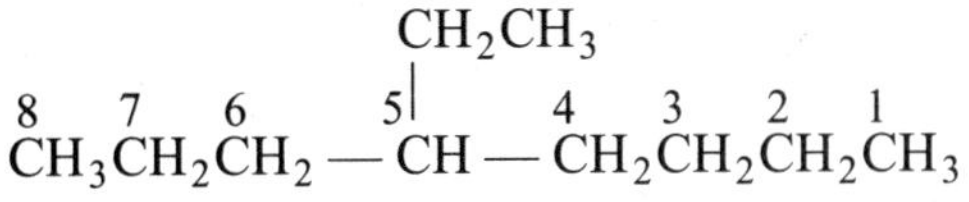

$$\overset{8}{C}H_3\overset{7}{C}H_2\overset{6}{C}H_2-\underset{\overset{|}{CH_2CH_3}\ 5}{CH}-\overset{4}{C}H_2\overset{3}{C}H_2\overset{2}{C}H_2\overset{1}{C}H_3$$

I A (Incorrect, higher number to carbon carrying the substituent)

$$\overset{1}{C}H_3\overset{2}{C}H_2\overset{3}{C}H_2-\overset{CH_2CH_3}{\overset{|\ 4}{CH}}-\overset{5}{C}H_2\overset{6}{C}H_2\overset{7}{C}H_2\overset{8}{C}H_3$$

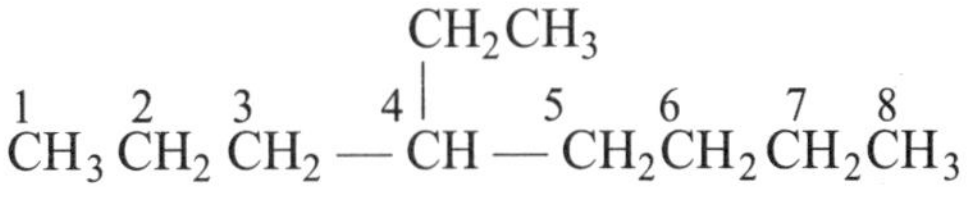

I B (Correct, lower number to carbon carrying the substituent)

The number which is used to indicate the position of the side chain (or substituent) on the 'parent chain' in called **positional number** or also the **locant**. Hence in **I** the correct positional number or locant for ethyl side chain is 4 and not 5.

(b) For alkanes with more than one substituents: *Rule of lowest set of locants*: Earlier this rule was also called as **lowest sum rule**. The term lowest sum is now replaced by lowest set of locants. This rule states that when the 'parent chain' has two or more substituents it is numbered starting from the end which gives the lowest set of locants (or lowest sum of the numbers used to indicate the position of the substituents).

For determining the lowest set of locants the parent chain is numbered commencing with all ends and a locant is assigned to each substituent in every set of numbering of the parent chain. The different sets of locants so obtained are compared term-by-term till we arrive at first point of difference. This is the reason why this rule is sometimes also called as *first point of difference rule.*

Examples

(*a*) The parent chain of alkane **II** can be numbered in two ways as follows:

$$H_3C-\underset{5}{\overset{\overset{6\quad 7}{CH_2CH_3}}{\overset{|}{CH}}}-\underset{\underset{\underset{3\quad 2\quad 1}{CH_2CH_2CH_3}}{4|}}{CH}-CH_2CH_3$$

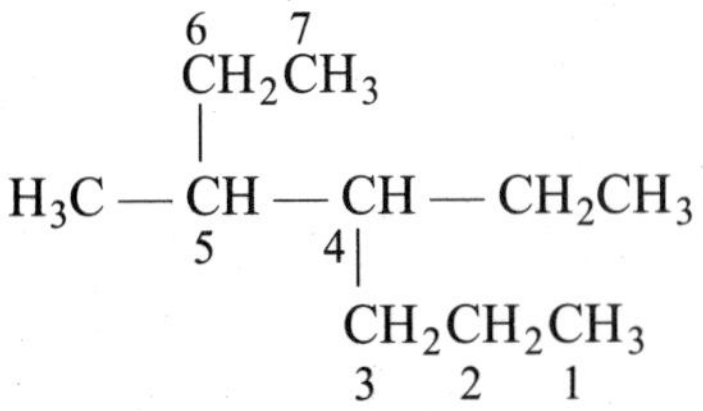

II A set of locants = 4, 5
(*Incorrect*)

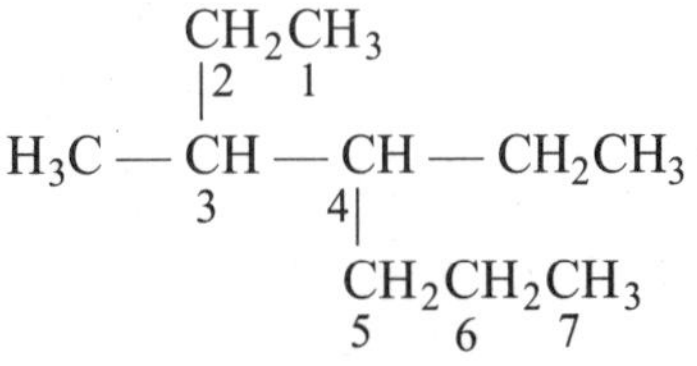

$$H_3C-\underset{3}{\overset{CH_2CH_3}{\overset{|2\quad 1}{CH}}}-\underset{\underset{\underset{5\quad 6\quad 7}{CH_2CH_2CH_3}}{4|}}{CH}-CH_2CH_3$$

II B set of locants = 3, 4
(*Correct*)

(*b*) The parent chain of alkane **III** can be numbered in two ways as follows:

$$\underset{1}{H_3C}-\underset{2}{\overset{CH_3}{\overset{|}{CH}}}-\underset{3}{CH_2}-\underset{\underset{CH_3}{4|}}{\overset{CH_3}{\overset{|}{C}}}-\underset{5}{CH_3}$$

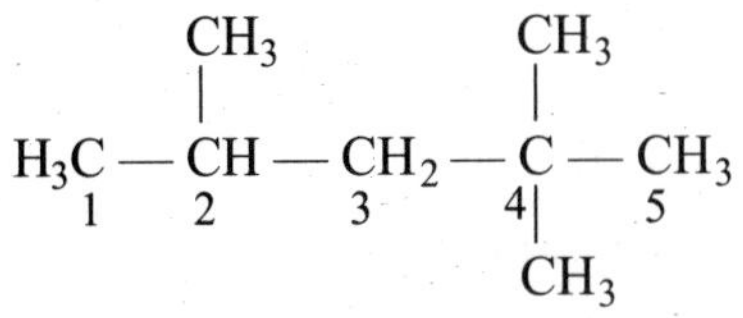

III A set of locants 2, 4, 4
second term 4 (*Incorrect*)

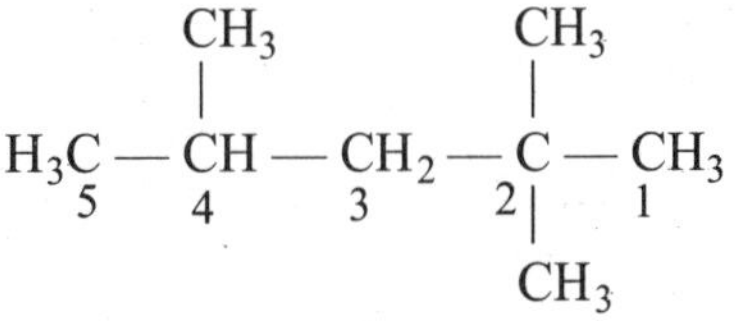

$$\underset{5}{H_3C}-\underset{4}{\overset{CH_3}{\overset{|}{CH}}}-\underset{3}{CH_2}-\underset{\underset{CH_3}{2|}}{\overset{CH_3}{\overset{|}{C}}}-\underset{1}{CH_3}$$

IV B set of locants 2, 2, 4
second term 2 (*Correct*)

Chapter 4

2. Naming the Alkane

The name of the substituent, prefixed with a number to indicate the carbon atom to which it is attached, is written before the name of the parent alkane. *The name of the alkane is then written as one word using hyphen (–) between the number and name of the substituent. For example,*

$$\begin{array}{ccccc} & & CH_3 & & \\ & & | & & \\ \underset{1}{CH_3} & — & \underset{2}{CH} & — & \underset{3}{CH_3} \end{array}$$

2-Methylpropane

$$\begin{array}{ccccccccc} & & & & CH_3 & & & & \\ & & & & | & & & & \\ \underset{1}{H_3C} & — & \underset{2}{CH_2} & — & \underset{3}{CH} & — & \underset{4}{CH_2} & — & \underset{5\ \ 6}{CH_2CH_3} \end{array}$$

3-Methylhexane

3. Naming the Alkane with more than One Substituent: Alphabetical Order of the Simple Substituents

If more than one substituents is attached to the parent chain, the substituents prefixed by their locant number are written in an alphabetical order irrespective of its locant before the name of alkane which is again written as one word using hypens. For example,

$$\begin{array}{l} \qquad\qquad\qquad\quad CH_2CH_3 \\ \quad 6\ 5\quad 4\quad 3\ \ | \\ H_3CCH_2CH_2CHCHCH_3 \\ \qquad\qquad\quad |\ \ 2\quad 1 \\ \qquad\qquad\ CH_3 \end{array}$$

2-Ethyl-3-methylhexane

$$\begin{array}{ccccccc} & & \overset{2}{CH_2} & — & \overset{1}{CH_3} & & \\ & & | & & & & \\ CH_3 & — & \underset{3}{CH} & — & \underset{4}{CH} & — & CH_2CH_3 \\ & & & & | & & \\ & & & & \underset{5}{CH_2} & — \underset{6}{CH_2} — & \underset{7}{CH_3} \end{array}$$

4-Ethyl-3-methylheptane

In arranging the substituents in alphabetical order prefixes such as *iso*- and *neo*- are taken into account while prefixes such as *sec*-and *tert*- are ignored. For example,

$$\begin{array}{ccccc} CH_3CHCH_2CH_3 & & & & \\ | & & & & \\ \underset{1\ 2\quad 3\quad 4}{H_3CCH_2CH_2CH} & — & \underset{5}{CH_2} & — & \underset{6\ \ 7\ \ 8\ \ 9}{CHCH_2CH_2CH_3} \\ & & & & | \\ & & & & H_3C — CH — CH_3 \end{array}$$

4-*sec*-Butyl-6-*iso*-propylnonane

4. Substituents at Interchangeable Positions

When during numbering of 'parent chain' two substituents are found at interchangeable (equivalent) positions, the numbering is done from the end which gives lower number to the substituent coming first in alphabetical order. For example,

$$\begin{array}{ccccc} \quad CH_3 & & & & \\ \overset{1\ 2\quad\ \ 3\ 4\quad 5}{|} & & \overset{6}{} & & \overset{7}{} \\ H_3CCH_2CHCH_2CH & — & CH_2 & — & CH_3 \\ \underset{7\ 6\quad 5\ \ 4\quad 3}{|} & & \underset{2}{} & & \underset{1}{} \\ \qquad CH_2CH_3 & & & & \end{array}$$

Methyl and ethyl groups are at interchangeable positions 3 and 5

Correct name: 3-Ethyl-5-methylheptane
Incorrect name: 5-Ethyl-3-methylheptane

5. Same Substituent Repeated

When a particular substituent is repeated on the parent chain at different positions, the positions assigned to it are separated by comma as (,) and suitable prefix such as **di** (for substituent repeated twice), **tri** (for substituent repeated thrice), **tetra** (for substituent repeated **four times**), etc., are written after the sequence of numbers. In case a substituent is repeated twice at the same carbon atom, the number of this carbon is also repeated twice.

Examples

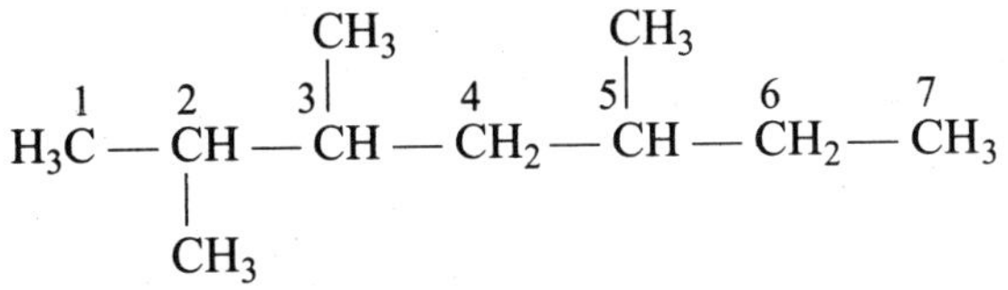

$$\begin{array}{ccccccccccccc} & & & & CH_3 & & & & CH_3 & & & & \\ 1 & & 2 & & 3| & & 4 & & 5| & & 6 & & 7 \\ H_3C & - & CH & - & CH & - & CH_2 & - & CH & - & CH_2 & - & CH_3 \\ & & | & & & & & & & & & & \\ & & CH_3 & & & & & & & & & & \end{array}$$

2, 3, 5-Trimethylheptane

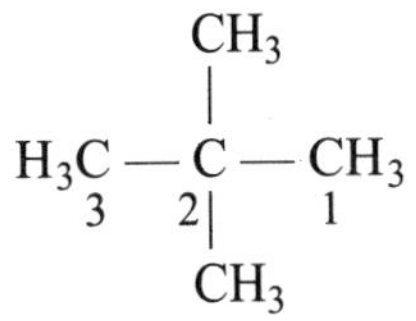

$$\begin{array}{ccccc} & & CH_3 & & \\ & & | & & \\ H_3C & - & C & - & CH_3 \\ 3 & & 2| & & 1 \\ & & CH_3 & & \end{array}$$

2, 2-Dimethylpropane

$$\begin{array}{ccccccccc} & & & & CH_3 & & CH_3 & & \\ 1 & & 2 & & 3| & & 4| & & 5\quad 6\quad 7\quad 8 \\ H_3C & - & CH_2 & - & C & - & C & - & CH_2CH_2CH_2CH_3 \\ & & & & | & & | & & \\ & & & & CH_3 & & CH_2CH_3 & & \end{array}$$

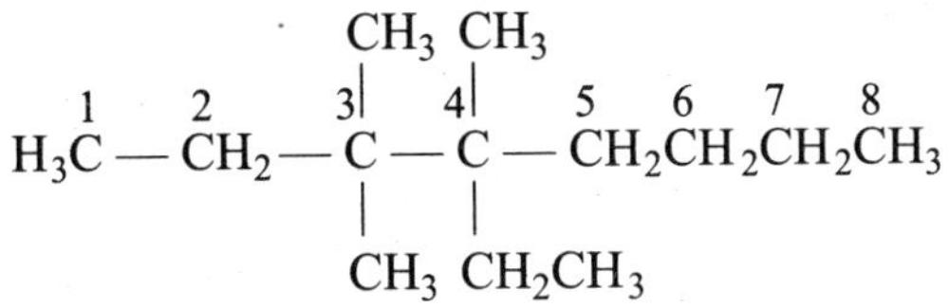

4-Ethyl-3, 3, 4-trimethyloctane

6. The Case of Branched (or Substituted) Substituents

The general guidelines, as under, are followed depending upon the nature and type of such substituents.

Guideline 1: ***For a single branched substituent***

Such a substituent is named as a substituted alkyl group by separately numbering the carbon atoms of this assigning number 1 to the carbon attached to parent chain: the name of such substituent along with the names and positions of its substituents is written in parenthesis.

Examples

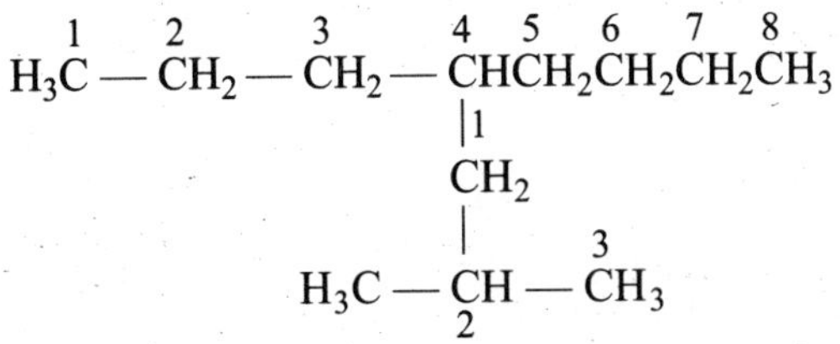

$$\begin{array}{ccccccccc} 1 & & 2 & & 3 & & & 4\quad 5\quad 6\quad 7\quad 8 & \\ H_3C & - & CH_2 & - & CH_2 & - & & CHCH_2CH_2CH_2CH_3 & \\ & & & & & & & |1 & \\ & & & & & & & CH_2 & \\ & & & & & & & | & 3 \\ & & & & H_3C & - & & CH & - CH_3 \\ & & & & & & & 2 & \end{array}$$

4-(2-Methylpropyl)octane

$$
\begin{array}{cccccccccc}
1 & 2 & 3 & 4 & 5 & 6 & 7 & 8 & 9 & 10 \\
H_3C - & CH - & CH - & CH_2 - & CH_2 - & CH - & CH_2 - & CH_2 - & CH_2 - & CH_3 \\
 & | & | & & & |_1 & & & & \\
 & CH_3 & CH_3 & & & CH_2 & & & & \\
 & & & & & | & & & & \\
 & & & & H_3C - & {}_2C - CH_3 & & & & \\
 & & & & & | & & & & \\
 & & & & & CH_3 & & & & \\
 & & & & & 3 & & & &
\end{array}
$$

2, 3-Dimethyl-6-(2, 2-dimethyl propyl)decane

Guideline 2: ***For more than one branched substituent having the same chain length***

In such a case the substituent with larger number of alkyl groups attached is taken as a part of the **'parent chain'** and the other is considered as the branched substituent.

Example

$$
\begin{array}{cccccccccc}
 & & & & & & & CH_3 & & \\
 & & & & & & & | & & \\
H_3C - & CH_2 - & CH_2 - & CH_2 - & CH_2 - & CH - & CH_2 - & C - & CH_3 & \\
10 & 9 & 8 & 7 & 6 & 5| & 4 & 3| & & \\
 & & & & & {}_1CH_2 & & CH_2 - & CH_3 & \\
 & & & & & | & & 2 & 1 & \\
 & & & & CH_3CH_2 - & CH - & CH_2CH_3 & & & \\
 & & & & & 2 & 3 \;\; 4 & & &
\end{array}
$$

5-(2-Ethylbutyl)-3, 3-dimethyldecane

Guideline 3: ***Alphabetical order of the substituent***

The first letter of the branched substituent is chosen while arranging all the substituent in alphabetical order.

$$
\begin{array}{ccccccccc}
 & CH_2CH_3 & & & CH_3 & CH_3 & & & \\
1\,2\;\;3\;\;4 & 5| & 6 & 7 & 1| & 2| & 3 & 4 & \\
H_3CCH_2CH_2CH_2 - & CH - & CH_2 - & CH - & CH - & CH - & CH_2 - & CH_3 & \\
 & & & | & & & & & \\
 & & & CH_2 - & CH_2CH_2CH_2CH_3 & & & & \\
 & & & 8 & 9 \;\; 10 \;\; 11 \;\; 12 & & & &
\end{array}
$$

7-(1, 2-Dimethylbutyl)-5-ethyldodecane

Guideline 4: ***Lowest locant at the first point of difference***

When the names of two or more branched substituents have the same words (*e.g.*, methylbutyl), the substituent having lowest locant of difference is given priority.

Example

$$\begin{array}{l}
\qquad\qquad\qquad\qquad\qquad\qquad\qquad\qquad\qquad\qquad\quad CH_3 \\
\qquad\qquad\qquad\qquad\qquad\qquad\qquad\qquad\quad \overset{1}{CH_2} - \overset{2}{CH} - \overset{3}{CH_2} - \overset{4}{CH_3} \\
\qquad\qquad\qquad\qquad\qquad\qquad\qquad\qquad\quad | \, 9 \\
\overset{1\;\;2\;\;3\;\;4\;\;5\;\;6}{CH_3CH_2CH_2CH_2CH_2CH_2} - \overset{7}{CH} - \overset{8}{CH_2} - CH - \overset{10\;\;11\;\;12\;\;13\;\;14}{CH_2CH_2CH_2CH_2CH_3} \\
\qquad\qquad\qquad\qquad\qquad\quad | \, 1 \\
\qquad\qquad\qquad\qquad H_3C - CH - \overset{2\;\;3\;\;4}{CH_2CH_2CH_3}
\end{array}$$

7-(1-Methylbutyl)-9-(2-methylbutyl)tetradecane

Note: 1-Methylbutyl is given priority over 2-methyl butyl

4.3 THE ALKYL GROUP AND ITS NOMENCLATURE

Removal of an atom of hydrogen from a molecule of alkane produces an **alkyl group**. It is generally written as R. An alkyl group constitutes a structural unit of the aliphatic compounds. The alkyl groups are represented by the general formula C_nH_{2n+1} and their names are derived from those of the corresponding alkanes, *e.g.*, methyl (CH_3—), ethyl (C_2H_5—) and propyl (C_3H_7—) from methane (CH_4), ethane (C_2H_6) and propane (C_3H_8), respectively. The names of these alkyl groups are retained while naming the compounds belonging to different classes. Thus CH_3—Br is methyl bromide and CH_3—OH is methyl alcohol.

In case of alkanes from propane onwards it is possible to abstract the hydrogen atom from different carbon atoms thus giving rise to isomeric alkyl groups. Alkyl groups formed by removing terminal hydrogen of the straight chain alkane is designated as *n* (or normal) alkyl group. Alkyl groups with *iso* units are designated as *iso*alkyl groups. Thus two alkyl groups derived from propane are *n*-propyl and *iso*-propyl groups.

$$CH_3 - CH_2 - CH_2 - \qquad\qquad \begin{array}{l} H_3C \diagdown \\ \qquad\quad CH - \\ H_3C \diagup \end{array}$$

***n*-Propyl** — *iso*-Propyl

Similarly, four isomeric butyl groups are derived from isomeric butanes.

$$CH_3 - CH_2 - CH_2 - CH_2 - \qquad \begin{array}{c} CH_3 \\ | \\ CH_3 - CH - CH_2 - \end{array} \qquad \begin{array}{c} | \\ CH_3 - CH - CH_2 - CH_3 \end{array} \qquad \begin{array}{c} CH_3 \\ | \\ H_3C - C - \\ | \\ CH_3 \end{array}$$

n-Butyl — *iso*-Butyl — *sec*-Butyl — *tert*-Butyl

4.4 CLASSIFICATION OF CARBON ATOMS IN ALKANES

The carbon atoms of alkanes are designated as primary, secondary, tertiary or quaternary depending upon their position in the molecule. This classification is not only useful in the study of alkanes but has also found utility in the study of other classes of compounds. A carbon (methyl carbon) which is linked to only one other carbon atom is called a *primary carbon*, *secondary carbon* if linked to two other carbon atoms, a *tertiary* carbon in case it is attached to three other carbon atoms and a *quaternary carbon* when it is linked to four other carbon atoms. All the four types of carbon atoms can be recognized in the following structure:

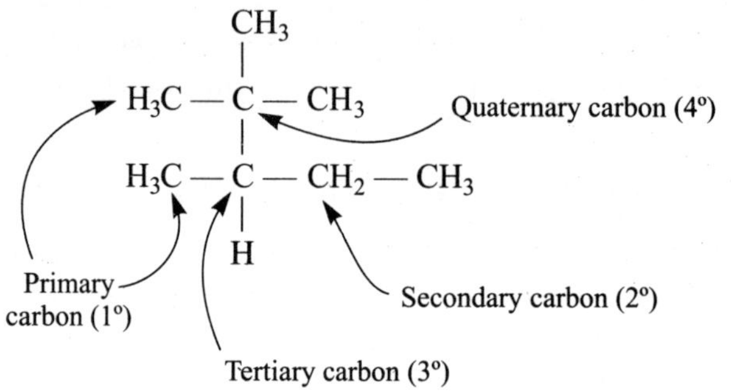

$$\begin{array}{c} CH_3 \\ | \\ H_3C - C - CH_3 \\ | \\ H_3C - C - CH_2 - CH_3 \\ | \\ H \end{array}$$

Quaternary carbon (4°)

Primary carbon (1°)

Secondary carbon (2°)

Tertiary carbon (3°)

Consequently, hydrogen atoms attached to 1°, 2° and 3° carbon atoms are called 1°, 2° and 3° hydrogen atoms, respectively.

4.5 ISOMERISM IN ALKANES

In the homologous series of alkanes the first three members can have only one structure. However, two different structures are possible for the molecular formula C_4H_{10} (butanes) and three structures are possible for the molecular formula C_5H_{12} (pentanes). These compounds differ in their melting point, boiling point, density, solubility and all other physical properties.

$$CH_3 - CH_2 - CH_2 - CH_3$$

n-Butane

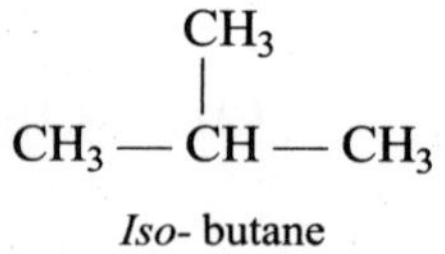

$$\begin{array}{c} CH_3 \\ | \\ CH_3 - CH - CH_3 \end{array}$$

Iso- butane

$$CH_3 - CH_2 - CH_2 - CH_2 - CH_3$$

n-Pentane

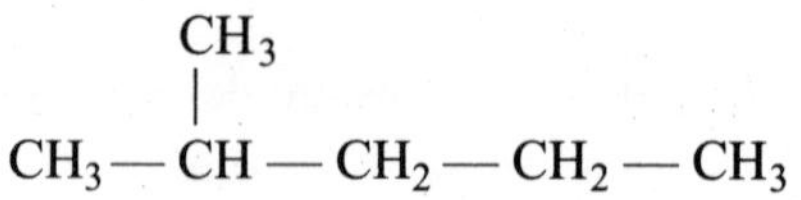

$$\begin{array}{l} \quad\;\; CH_3 \\ \quad\;\;\; | \\ CH_3 - CH - CH_2 - CH_2 - CH_3 \end{array}$$

Iso-pentane

$$\begin{array}{c} CH_3 \\ | \\ CH_3 - C - CH_3 \\ | \\ CH_3 \end{array}$$

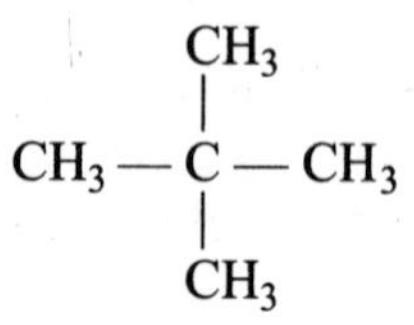

Neo-pentane

Such compounds which have same molecular formula but different structures are called isomers and the phenomenon is known as isomerism. Isomers having different positions of atoms or groups are called **structural isomers.** Further *since these isomers differ from each other in the structure of carbon chain they are called chain isomers and the phenomenon is called* **chain isomerism.**

Two isomeric butanes can be considered to have been derived by replacing hydrogen (1° or 2°) of the propane by a methyl group as illustrated below:

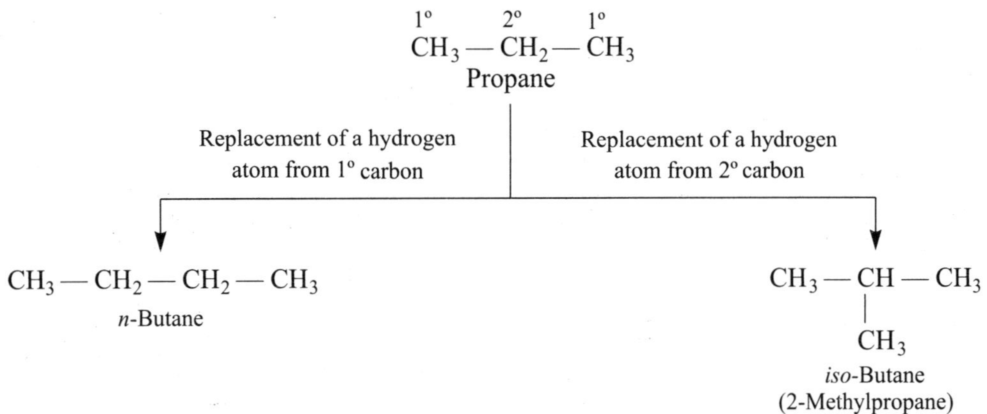

The two isomeric butanes have three types of carbon atoms (1°, 2° and 3°) and may give rise to three isomeric pentanes when a hydrogen of butane is replaced by a methyl group.

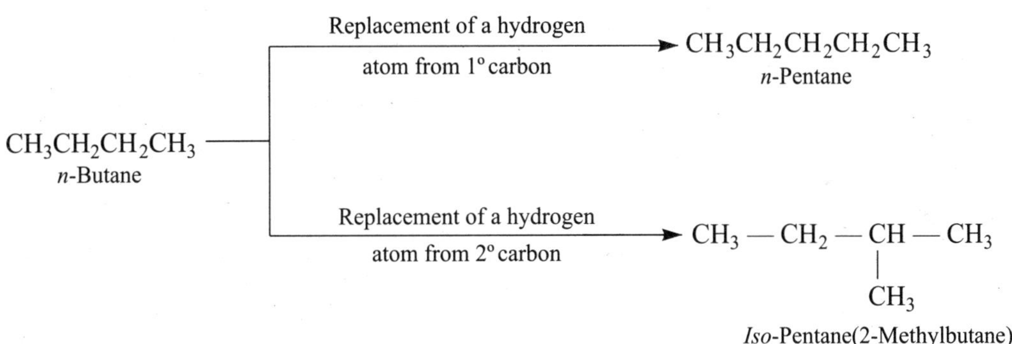

1° 3° 1°
$CH_3—CH(1°CH_3)—CH_3$
Iso-Butane

Replacement of a hydrogen atom from 1° carbon → $CH_3CH_2CH(CH_3)CH_3$
Iso-Pentane (2-Methylbutane)

Replacement of a hydrogen atom from 3° carbon → $CH_3—C(CH_3)_2—CH_3$
Neo-pentane

It is obvious that in a given homologous series, the number of isomers goes on increasing rapidly with the increase in the number of carbon atoms. Proceeding in a similar manner, it can be shown that hexane (C_6H_{14}) can be represented by five isomeric structures, heptane (C_7H_{16}) by nine structures, nonane (C_9H_{20}) by 35 structures and decane ($C_{10}H_{22}$) by 75 structures.

4.6 SOURCES OF ALKANES

The chief commercial sources of alkanes are petroleum and natural gas. Petroleum is a complex mixture of liquid alkanes and higher solid hydrocarbons. Ozokerites formed in large deposits in petroleum areas is a mixture of higher solid alkanes. Natural gas chiefly contains methane. The major constituents of petroleum are C_5—C_{50} hydrocarbons. Methane is the principal constituent of the 'marsh gas' produced by anaerobic fermentation of cellulose.

Some plant waxes also contain saturated hydrocarbons. The wood turpentine of some pine trees has been shown to contain *n*-heptane. It is interesting to note that all the paraffins present in the natural waxes have odd number of carbon atoms (C_{27}—C_{37}). This is probably due to the fact that natural fatty acids which contain even number of carbon atoms give rise to paraffins by decarboxylation (loss of CO_2). The chief constituents of beewax are $C_{27}H_{56}$, and $C_{31}H_{64}$. Lower members like methane are also produced by natural decay of animals and plants.

4.7 METHODS OF FORMATION

The characteristic feature of a homologous series is that the members can be synthesized by the general methods of preparation. The alkanes are prepared by the following methods:

1. Hydrogenation of alkenes: The addition of hydrogen to a double bond is called *hydrogenation*. It is best carried out by molecular hydrogen in the presence of finely divided metal catalysts such as platinum, palladium and nickel. Special techniques are required to prepare these highly 'activated' metal catalysts. The reaction temperature varies from room temperature to 473–573 K. Nickel catalyst is used in well-known **Sabatier-Senderens reduction**. Raney nickel* can be used at ordinary temperature. This is highly versatile synthetic method and the yields are nearly quantitative.

$$\underset{\text{Alkene}}{R-CH=CH_2+H_2} \xrightarrow[\text{Catalyst}]{\text{Metal}} \underset{\text{Alkane}}{R-CH_2-CH_3}$$

Evidence

The evidence available so far suggests that hydrogen combines with the catalyst at its surface exothermally ($\Delta H = -126$ kJ mol^{-1}) thus making the H—H bond quite weak. This becomes evident from the following equilibrium that is established between hydrogen and deuterium molecules on the nickel surface.

$$H-H+D-D \underset{}{\overset{Ni}{\rightleftharpoons}} 2H-D$$

Similarly, alkenes also combine with the catalyst at its surface (adsorption) in an exothermic ($\Delta H = -252$ kJ mol^{-1}) and reversible manner. It is possible that π-electron cloud in the alkene may be

*Raney nickel is prepared by treating Ni—Al allory with sodium hydroxide. Aluminum dissolves in sodium hydroxide leaving behind nickel in finely divided from. Nickel in this form is highly active.

responsible for this sort of combination, as this results in the weakening of double bond. Alkanes (having no π-electrons) do not combine with metals. With the simultaneous weakening of both, H—H and C = C bond, it may be conceived that hydrogenation involves the simultaneous addition of hydrogen atoms to the two carbon atoms of the Ni-alkene 'complex'. The alkane produced in this way immediately leaves the reaction site thus vacating the catalyst surface for the fresh alkene and hydrogen molecules to be adsorbed (Fig. 4.2).

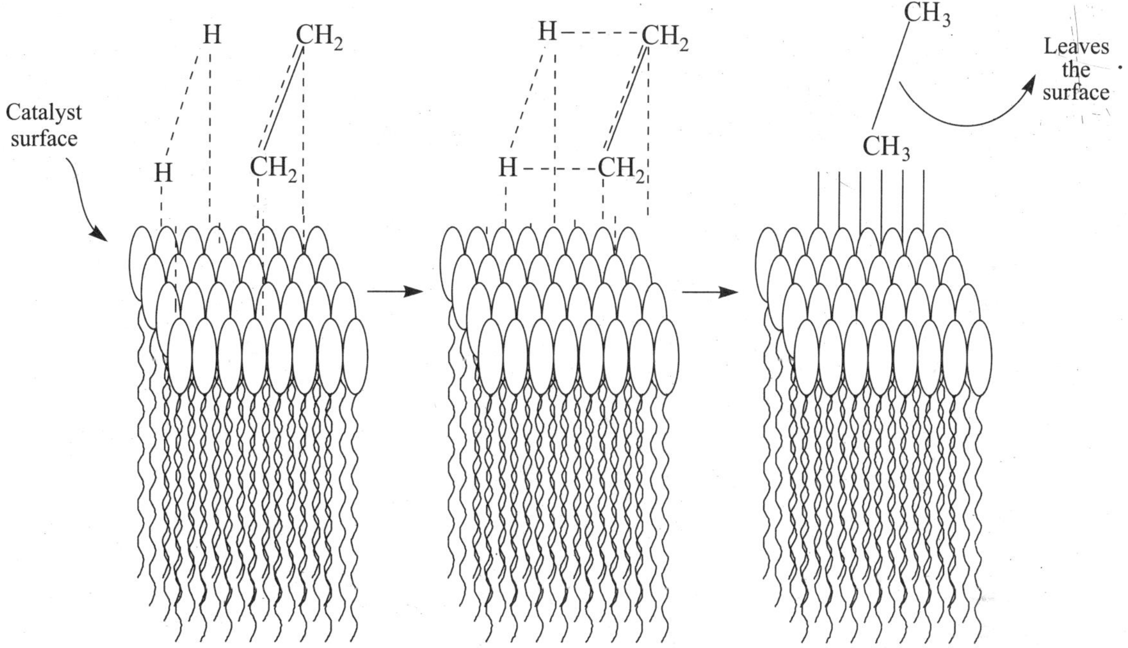

Fig. 4.2: Catalytic hydrogenation of alkenes

2. Kolbe's electrolytic method: Electrolysis of the concentrated aqueous solutions of sodium or potassium salts of saturated carboxylic acids afford alkanes.

$$R-\overset{\overset{\displaystyle O}{\|}}{C}-\overset{\ominus\oplus}{ONa} \xrightarrow{\text{Electrolysis}} \underbrace{R-R+CO_2}_{\text{at anode}} + \underbrace{NaOH+H_2}_{\text{at cathode}}$$

Mechanism

This is an example of *radical coupling* (or combination) reaction, involving formation of a covalent bond by the combination of two radicals. In this reaction the metal ion goes to cathode where it gains an electron forming metal atom which upon reaction with water forms the hydroxide and hydrogen is liberated. The carboxylate anion, on the other hand, migrates to the anode and transfers an electron to it thus generating a carboxylate radical which immediately loses CO_2 producing an alkyl radical. Coupling of two alkyl radicals yields alkanes.

At the Cathode

$$\overset{\oplus}{Na} + e \longrightarrow \dot{Na}$$

$$\dot{Na} + H_2O \longrightarrow NaOH + 1/2\ H_2$$

At the Anode

$$R-\overset{\overset{\displaystyle O}{\|}}{C}-\ddot{\underset{\cdot\cdot}{O}}{:}^{\ominus} \xrightarrow{-e} R-\overset{\overset{\displaystyle O}{\|}}{C}-\dot{\underset{\cdot\cdot}{O}}{:} \longrightarrow \dot{R} + CO_2$$

Carboxylate ion — Carboxylate radical

$$\dot{R} + \dot{R} \xrightarrow{\text{coupling}} R-R$$

Limitations

(*i*) Methane cannot be prepared by this method.

(*ii*) Only alkanes with even number of carbon atoms can be obtained.

(*iii*) When a mixture of two salts of different carboxylic acids is electrolyzed, a mixture of three alkanes is obtained.

(*iv*) Many side products are formed as a result of other reactions of free radicals.

3. Würtz reaction: Heating of an ethereal solution of alkyl halides (other than fluorides) with sodium metal affords alkanes.

$$2\ RX + 2\ Na \xrightarrow{\text{Dry ether}} R-R + 2\ NaX$$

Some other metals such as silver and copper have also been used in place of sodium. Attempts to synthesize alkanes from a mixture of two alkyl halides yield a mixture of products, therefore the reaction has limited synthetic application. Hence, only a few alkanes such as ethane, *n*-butane and *n*-hexane, etc., can be synthesized using this reaction.

$$CH_3-Br \xrightarrow{\text{Na, ether}} CH_3-CH_3$$

$$C_2H_5-Br \xrightarrow{\text{Na, ether}} CH_3-CH_2-CH_2-CH_3$$

$$CH_3-Br + C_2H_5-Br \xrightarrow{\text{Na, ether}} CH_3-CH_3 + CH_3-CH_2-CH_2-CH_3 + CH_3-CH_2-CH_3$$

Mechanism

Two mechanisms, depending upon the conditions of the reaction, have been proposed.

(*a*) *Ionic mechanism*: It operates when the reaction is carried out in solution. It has been suggested that an atom of sodium reacts with an alkyl halide to form an alkylsodium compound having carbanion character. This is followed by an S_N2 type displacement on another molecule of alkyl halide.

$$RX + 2\,\dot{Na} \longrightarrow RNa + NaX$$

$$\overset{\ominus}{R} + R-\dot{X} \longrightarrow R-R + \overset{\ominus}{X} \qquad S_N2 \text{ reaction}$$

Evidence

This mechanism also explains the formation of ***disproportionation*** products by an elimination process (E2), *e.g.*, when two molecules of ethyl halide are reacted, besides *n*-butane, ethene and ethane are also formed in small amounts.

$$CH_3-\overset{\ominus}{C}H_2 + H-CH_2-CH_2-X \longrightarrow CH_3-CH_3 + CH_2=CH_2 + \overset{\ominus}{X}$$

(b) Free radical mechanism: It has been shown that when the reaction is carried out in vapour phase (a condition suitable for the generation of free radicals) it can take place *via* a free radical mechanism. Sodium metal may react with alkyl halide forming the alkyl radicals which may couple or disproportionate to form the observed products.

$$R-X + \dot{Na} \longrightarrow \dot{R} + NaX$$

$$\dot{R} + \dot{R} \longrightarrow R-R$$

For example, if R = $\dot{C}_2H_5$

Combination (or coupling) reaction

$$\dot{C}_2H_5 + \dot{C}_2H_5 \longrightarrow C_2H_5-C_2H_5$$

Disproportionation reaction

$$\dot{C}_2H_5 + \dot{C}_2H_5 \longrightarrow CH_3-CH_3 + CH_2=CH_2$$

Limitations

(*i*) Methane cannot be prepared by this method.

(*ii*) *Tert*-alkyl halide cannot be used for this reaction.

(*iii*) Disproportionation of the intermediate leads to the formation of side products.

4. Reduction of alkyl halides: Alkanes can also be prepared by the reduction of alkyl halides in a variety of ways.

(a) Treatment of alkyl halides with reducing agents such as zinc and acid, zinc and alkali, zinc-copper couple and ethanol gives alkanes.

$$R-X \xrightarrow{\text{Reducing agent}} R-H$$

Mechanism

Until recently, it was believed that the metal reacts with an acid or alkali to produce "nascent" hydrogen, which brings about the reduction. However, the true picture is that the metal transfers two electron to the alkyl halide ultimately generating alkyl *carbanion* which, being a very strong base, abstracts a proton from the acid or any potential proton donor (may be even a solvent).

$$\text{Zn}: \longrightarrow \overset{\oplus}{\text{Zn}} + 2\overset{\ominus}{e}$$

$$\text{R}-\ddot{\underset{\cdot\cdot}{\text{X}}}: + \overset{\ominus}{e} \longrightarrow \dot{\text{R}} + :\overset{\ominus}{\ddot{\underset{\cdot\cdot}{\text{X}}}}:$$

$$\dot{\text{R}} + \overset{\ominus}{e} \longrightarrow \overset{\ominus}{\text{R}}:$$

$$\overset{\ominus}{\text{R}}: + \overset{\oplus}{\text{H}} \longrightarrow \text{R}-\text{H}$$

(From acid or solvent)

(*b*) ***Reduction of alkyl halides with hydride donors*** such as lithium aluminum hydride ($LiAlH_4$) or sodium borohydride ($NaBH_4$) also yields alkanes. It has been shown that $LiAlH_4$ reduces primary and secondary alkyl halides while $NaBH_4$ reduces only secondary and tertiary alkyl halides.

$$4\,\text{RX} + \underset{(\text{or NaBH}_4)}{\text{LiAlH}_4} \longrightarrow 4\,\text{RH} + \underset{(\text{or NaBX}_4)}{\text{LiAlX}_4}$$

The reduction takes place in four independent steps, each step involving a nucleophilic displacement by hydride ions, $\overset{\ominus}{\text{H}}$

(*c*) ***Grignard reaction*****:** On treatment with magnesium metal in dry ether, alkyl halides produce alkyl magnesium halides known as *Grignard reagents.*

$$\text{R}-\text{X} + \text{Mg} \xrightarrow{\text{Ether}} \text{RMgX}$$

The high reactivity of these reagents makes them versatile tools for synthesizing a large variety of organic compounds.

On treatment with water Grignard reagents yield alkanes.

$$\text{RMgX} + \text{H}_2\text{O} \longrightarrow \text{R}-\text{H} + \text{Mg(OH)X}$$

The whole sequence may also be considered a method for reducing alkyl halides.

(*d*) ***Alkyl iodides undergo*** *reduction on treatment with hydroiodic acid* at high temperature in presence of red phosphorus.

$$\text{R}-\text{I} + \text{HI} \xrightarrow{423\text{ K}} \text{R}-\text{H} + \text{I}_2$$

Some other classes of compounds such as alcohols, aldehydes, ketones and carboxylic acids can also be reduced to alkanes with hydroiodic acid. The reduction of aldehydes and ketones can also be achieved by **Clemmensen and Wölff-Kishner reduction**.

5. Corey-house reaction: A useful extension of the **Würtz reaction** involves the treatment of a suitable lithium dialkylcuperate with an alkyl halide to form alkanes in excellent yields. This is known as **Corey-house reaction**.

$$\underset{\text{Lithium dialkyl cuperate}}{R_2CuLi} + R'—X \xrightarrow[\text{or THF}]{\text{Dry ether}} R—R' + R—Cu + LiX$$

(R and R′ may be same or different)

The required lithium dialkylcuperate is prepared from alkyl halide according to the following reaction sequence:

$$R—X + 2\,Li \xrightarrow{\text{Diethyl ether}} R—Li + LiX$$

$$2\,R—Li + CuI \longrightarrow R_2CuLi + LiI$$

Good yields are obtained when the alkyl halide is *methyl halide, a primary alkyl halide or secondary cycloalkyl halide* and the alkyl groups in lithium dialkylcuperate are methyl, *primary*, *secondary* or *tertiary*. Following examples depicting the whole sequence of reactions are illustrative:

(*a*) $$CH_3—I \xrightarrow[-LiI]{\text{Li/ether}} CH_3Li \xrightarrow[-LiI]{CuI} \underset{\text{Lithium dimethylcuperate}}{(CH_3)_2CuLi} \xrightarrow{CH_3CH_2CH_2CH_2I}$$

$$\underset{n\text{-Pentane}}{CH_3CH_2CH_2CH_2CH_3} + \underset{\text{Methyl copper}}{CH_3—Cu} + LiI$$

(*b*) $$CH_3CH_2—\underset{\underset{Cl}{|}}{CH}—CH_3 \xrightarrow[-LiI]{\text{Li/ether}} CH_3CH_2—\underset{\underset{CH_3}{|}}{CH}—Li \xrightarrow[-LiI]{CuI} \underset{\text{Lithium di-}sec\text{ butylcuperate}}{(CH_3CH_2—\underset{\underset{CH_3}{|}}{CH})_2CuLi}$$

$$\xrightarrow[-LiBr]{CH_3CH_2CH_2CH_2Br} \underset{\text{3-Methylheptane}}{CH_3CH_2—\underset{\underset{CH_3}{|}}{CH}—CH_2CH_2CH_2CH_3} + CH_3CH_2—\underset{\underset{CH_3}{|}}{CH}—Cu + LiBr$$

Importance: Corey-house reaction is better than Würtz reaction since it can be used for the synthesis of straight chain or branch chain alkanes and also those alkanes which contain either even or odd number of carbon atoms. This reaction is particularly useful for the synthesis of higher alkanes.

6. Decarboxylation of carboxylic acids: *Decarboxylation* is the process of removal of a molecule of CO_2 from carboxylic acids.

Sodium salt of a monocarboxylic acid on fusion with sodalime (NaOH + CaO in the ratio 3 : 1) at about 630 K looses a molecule of CO_2 producing an alkane with one carbon atom less than the carboxylic acid.

$$R-COONa + NaOH \xrightarrow{CaO,\ 630\ K} R-H + Na_2CO_3$$

e.g.,

$$CH_3COONa + NaOH \xrightarrow{CaO,\ 630\ K} CH_4 + Na_2CO_3$$

Carboxylic acids can also be used for this purpose, for example,

$$R-COOH + 2\ NaOH \xrightarrow{CaO,\ 630\ K} R-H + Na_2CO_3 + H_2O$$

e.g.,

$$CH_3COOH + 2\ NaOH \xrightarrow{CaO,\ 630\ K} CH_4 + Na_2CO_3 + H_2O$$

However, sodalime is preferred because of the following reasons:

(*a*) The reaction can be carried out at a relatively higher temperature so that complete decarboxylation is ensured,

(*b*) Sodium hydroxide is hygroscopic in nature but CaO present in sodalime keeps it dry.

Mechanism

Following mechanism may be proposed for this reaction:

$$H\ddot{O}^{\ominus} + R-C(=O)-O^{\ominus} \rightleftharpoons H-O-C(O^{\ominus})(O^{\ominus})-R \rightleftharpoons R:^{\ominus} + H-O-C(=O)-O^{\ominus}$$

$$\downarrow$$

$$\underset{\text{Alkane}}{R-H} + \underset{\text{Carbonate anion}}{^{\ominus}O-C(=O)-O^{\ominus}}$$

Limitations

Only methane can be prepared in good yields. In all other cases such as decomposition of sodium propionate, butyrate, caproate, a mixture of products is obtained. For example,

$$\underset{\text{Sodium propionate}}{CH_3CH_2COONa} \xrightarrow[630\ K]{NaOH + CaO} \underset{44\%}{CH_3-CH_3} + \underset{20\%}{CH_4} + \underset{33\%}{H_2} + \underset{3\%}{\text{Unsaturated hydrocarbons}}$$

7. Hydroboration of alkenes: Diborane (B_2H_6), a versatile synthetic reagent, which behaves as its monomer, BH_3 and adds to an olefinic double bond forming mono-, di-and trialkylboranes. These alkyl boranes on treatment with acetic acid or propionic acid yield alkanes.

$$R-CH=CH_2 \xrightarrow{B_2H_6} \underset{\text{Alkyl borane}}{(R-CH_2-CH_2)_3B} \xrightarrow{CH_3COOH} \underset{\text{Alkane}}{R-CH_2-CH_3}$$

Thus, the whole sequence can be regarded as a method of reducing the olefinic bond.

8. Reduction of various functional groups using concentrated hydroiodic acid and red phosphorous: Reduction of alochols, aldehydes, ketones and carboxylic acids with red phosphorus and HI produces alkanes. Some examples are as follows:

(*a*)

$$\underset{\text{Alcohol}}{R-OH} + 2\,HI \xrightarrow[423\,K]{P} \underset{\text{Alkane}}{R-H} + I_2 + H_2O$$

$$CH_3-OH + 2\,HI \xrightarrow[423\,K]{P} CH_4 + I_2 + H_2O$$

(*b*)

$$\underset{\text{Aldehyde}}{RCHO} + 4\,HI \xrightarrow[423\,K]{P} \underset{\text{Alkane}}{RCH_3} + 2I_2 + H_2O$$

$$CH_3CHO + 4\,HI \xrightarrow[423\,K]{P} CH_3-CH_3 + 2I_2 + H_2O$$

(*c*)

$$\underset{\text{Ketone}}{R-\overset{\overset{O}{\|}}{C}-R'} + 4\,HI \xrightarrow[423\,K]{P} \underset{\text{Alkane}}{RCH_2R'} + 2I_2 + H_2O$$

$$\underset{\text{Acetone}}{CH_3-\overset{\overset{O}{\|}}{C}-CH_3} + 4\,HI \xrightarrow[423\,K]{P} \underset{\text{Alkane}}{CH_3CH_2CH_3} + 2I_2 + H_2O$$

(*d*)

$$\underset{\text{Carboxylic acid}}{R-\overset{\overset{O}{\|}}{C}-OH} + 6\,HI \xrightarrow[423\text{–}473\,K]{P} \underset{\text{Alkane}}{RCH_3} + 3I_2 + 2H_2O$$

$$\underset{\text{Acetic acid}}{CH_3-\overset{\overset{O}{\|}}{C}-OH} + 6\,HI \xrightarrow[423\text{–}473\,K]{P} \underset{\text{Ethane}}{CH_3-CH_3} + 3I_2 + 2H_2O$$

In this reaction, the effective reducing agent is HI. The role of red phosphorus is to react with I_2 to regenerate HI, which is used again.

$$2\,P + 3\,I_2 + 6\,H_2O \xrightarrow[423\ K]{P} 6\,HI + 2\,H_3PO_3$$

This method is specially useful for the preparation of higher alkanes from carboxylic acids obtained by the hydrolysis of fats.

9. Reduction of ketones with zinc amalgam and conc. HCl: ***Clemensen reduction.*** Ketones upon treatment with zinc amalgam (Zn/Hg) and conc. HCl produce hydrocarbons by reduction of C=O into CH_2. This is known as **Clemensen reduction**.

$$RCOR \xrightarrow[\text{conc. HCl}]{\text{Zn/Hg}} RCH_2R$$

$$\underset{\text{Acetone}}{CH_3COCH_3} \xrightarrow[\text{conc. HCl}]{\text{Zn/Hg}} \underset{\text{Propane}}{CH_3CH_2CH_3}$$

10. Wölff-Kishner reduction: Ketones can also be reduced to produce hydrocarbons upon treatment with hydrazine and KOH. This is known as ***Wölff-Kishner reduction.***

$$\underset{\text{Ketone}}{RCOR} \xrightarrow[\text{KOH}]{NH_2-NH_2} \underset{\text{Alkane}}{RCH_2R}$$

$$\underset{\text{Acetone}}{CH_3COCH_3} \xrightarrow[\text{KOH}]{NH_2-NH_2} \underset{\text{Propane}}{CH_3CH_2CH_3}$$

4.8 GENERAL PHYSICAL PROPERTIES

4.8.1 Physical State

The first four members (C_1—C_4), *i.e.*, methane to butane are colourless gases. The compounds C_5—C_{17}, *i.e.*, pentane to heptadecane ($C_{17}H_{36}$) are liquids and those with C_{18} and above are waxy solids. The gaseous and liquid hydrocarbons possess the typical 'Petroleum odour'. The solid hydrocarbons are generally odourless.

4.8.2 Solubility

Being non-polar in nature, the alkanes are almost insoluble in water and soluble in non-polar solvents such as ether, benzene and carbon tetrachloride. However, their solubility decreases with increase in their molecular weights.

4.8.3 Boiling Points

In the case of *n*-alkanes the boiling points rise more or less steadily with the increase of each CH_2 unit (with a difference of approximately 20–30° per CH_2 unit). The difference between the boiling points of two homologues becomes smaller and smaller with increase in the molecular size and consequently the separation of mixture of higher hydrocarbons by ordinary distillation is quite difficult. This could be understood by considering the fact that the difference in boiling point or melting point is due to the difference in the intermolecular attractive forces. As the alkanes are non-polar in nature, the only attractive forces are the weak van der Waals attractive forces (Chapter 1). For *n*-alkanes this difference is 4 kJ/CH_2 unit. In other words, these attractive forces increase with the increase in the molecular size and hence there is more or less regular gradation in boiling points in the case of *n*-alkanes. In the case of isomeric alkanes the boiling points decrease with the increase in the branching of the chain. Thus *n*-pentane has a boiling point of 36.1°; isopentane, 27.9°; neopentane, 9.5°. The higher boiling points of *n*-alkanes among isomeric alkanes can be understood on the basis of the fact that the magnitude of the van der Waals attractive forces depends upon the extent of the molecules being close together. Straight chain hydrocarbons with larger surface area can get much closer and so have higher boiling points compared with the branched ones having lesser surface area which is responsible for lesser attractive forces and hence the lower boiling points.

4.8.4 Melting Points

Unlike boiling points, the melting points of alkanes do not show regular gradation. *An alkane with even number of carbon atoms melts at a higher temperature than that with an odd number of carbon atoms.* A plausible explanation for this originates from the fact that because of the zigzag shape of the chains the terminal atoms of the alkane with odd number of carbon atoms will be on the same side while those of the alkane with even number of carbon atoms will be on the opposite side. There is thus a closer packing of the molecules with even number of carbon atoms thereby increasing the intermolecular attractive forces which are responsible for higher melting points.

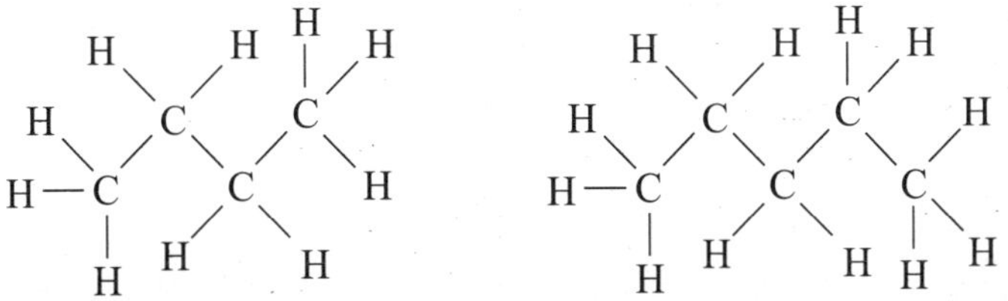

Even number of carbon atoms

Odd number of carbon atoms

This property of alkanes is known as **alternation effect**. The melting points and boiling points of first twenty members are listed in **Table 4.1**.

4.8.5 Density

The density of alkanes increases with the increase in the molecular mass till a limiting value of 0.8 g cm^{-3} is reached. It means that all alkanes are lighter than water.

4.8.6 Separation of Isomeric Alkanes from a Mixture

n-Alkanes can be separated from branched chain alkanes when present in a mixture. The procedure is as follows:

The mixture of alkanes is treated with urea preferably in the presence of a little methanol. This mixture is then allowed to crystallize. *n*-Alkane-urea inclusion compounds (Chapter 1) crystallize out leaving behind branched chain alkanes. The solid after filtration is decomposed with water when urea dissolves leaving behind *n*-alkane.

Table 4.1: Melting and boiling points of some alkanes

Name	Melting point (°C)	Boiling point (°C)
Methane	–183	–162
Ethane	–172	–88
Propane	–187	–42
n-Butane	–138	0
iso-Butane	–159	–12
n-Pentane	–130	36
iso-Pentane	–160	28
neo-Pentane	–170	10
n-Hexane	–95	69
n-Heptane	–90	98
n-Octane	–57	126
n-Nonane	–54	151
n-Decane	–30	174
n-Undecane	–26	196
n-Dodecane	–10	216
n-Tridecane	–6	234
n-Tetradecane	5.5	252
n-Pentadecane	10	271
n-Hexadecane	18	280
n-Heptadecane	22	292
n-Octadecane	28	308
n-Nonadecane	32	326
n-Bicosane	36	343

4.9 CHEMICAL REACTIONS

Under ordinary conditions, the alkanes do not react with the common reagents. However, they undergo substitution reactions involving free-radical chain mechanism in the presence of UV-light, at higher temperature or in the presence of certain radical initiators such as peroxides, etc. Some of the most common reactions are described under following categories:

1. *Substitution reactions*
2. *Isomerization reactions*
3. *Alkylation reactions*
4. *Pyrolysis or cracking*

5. *Aromatization reactions* 6. *Oxidation reactions*
7. *Insertion of carbene.*

4.9.1 Substitution Reactions

Some important substitution reactions are halogenation, nitration and sulphonation which are discussed below:

(i) **Halogenation:** One of the important chemical reactions of alkanes is halogenation which is defined as *the replacement of a hydrogen atom of the molecule by a halogen atom.*

(a) ***Chlorination:*** A mixture of an alkane and chlorine remains unaffected in the dark but a vigorous reaction occurs when it is exposed to light or heated at a higher temperature.

$$R—H + Cl_2 \xrightarrow{h\nu \text{ or } \Delta} R—Cl + HCl$$

(*Mechanism* of chlorination is discussed in Sec. 4.9).

(b) ***Bromination:*** *Bromination of alkane has close similarity to chlorination except for the fact that the rate of the reaction is slow.* This is mainly due to difference in bond dissociation energies of the molecules involved in the reaction (Section 4.9).

$$R—H + Br_2 \xrightarrow{h\nu \text{ or } \Delta} R—Br + HBr$$

(*Mechanism* of halogenation is discussed in Sec. 4.9).

(ii) **Nitration:** The displacement of an atom of hydrogen by a nitro group ($-NO_2$) is called *nitration.* Alkanes can be nitrated with nitric acid in the gas phase generally at a temperature range of 673–773 K.

$$\underset{\text{Alkane}}{R—H + HNO_3} \xrightarrow{673–773\text{ K}} \underset{\text{Nitroalkane}}{R—NO_2} + H_2O$$

Mechanism

The mechanism of this reaction is quite uncertain except for the fact that it involves *free radicals* as intermediates. Kinetic studies have revealed that this reaction is first order in nitric acid. The following mechanism, which gives a rather simple picture, is consistent with the kinetic data:

Chain initiation

$$HO—NO_2 \xrightarrow{\text{Slow}} H\dot{O} + \dot{N}O_2$$

Chain propagation

$$R—H + \dot{O}H \xrightarrow{\text{Fast}} \dot{R} + H_2O$$

$$\dot{R} + HO—NO_2 \xrightarrow{\text{Fast}} R—NO_2 + \dot{O}H$$

Chapter 4

Chain termination

$$\dot{R} + HO - NO_2 \longrightarrow R - OH + \dot{N}O_2$$

Inspite of the fact that a free radical substitution at a tertiary carbon is preferred over that of a secondary or primary carbon, a mixture of all the possible isomers is formed in the nitration of higher alkanes. Apart from this, nitroalkanes containing fewer number of carbon atoms are also formed by a carbon-carbon bond fission. For example, nitration of propane gives 1-nitropropane, 2-nitropropane, nitromethane and nitroethane.

$$CH_3CH_2CH_3 \xrightarrow[673–773K]{HNO_3} \begin{cases} \underset{\text{1-Nitropropane}}{CH_3 - CH_2 - CH_2 - NO_2} + \underset{\text{2-Nitropropane}}{CH_3 - \overset{\overset{NO_2}{|}}{CH} - CH_3} \\ \underset{\text{Nitroethane}}{CH_3 - CH_2 - NO_2} + \underset{\text{Nitromethane}}{CH_3 - NO_2} \end{cases}$$

(iii) **Sulphonation:** Substitution of a hydrogen atom of an alkane by sulphonic acid group ($-SO_3H$) is called *sulphonation* Branched chain and higher normal alkanes undergo sulphonation on heating with fuming sulphuric acid to give alkylsulphonic acids. The ease of substitution is $3° > 2° > 1°$.

$$\underset{iso\text{-Butane}}{H_3C - \overset{\overset{CH_3}{|}}{\underset{\underset{CH_3}{|}}{C}} - H} + \underset{\text{Fuming sulphuric acid}}{H_2SO_4(SO_3)} \xrightarrow{675\ K} \underset{tert\text{-Butylsulphonic acid}}{H_3C - \overset{\overset{CH_3}{|}}{\underset{\underset{CH_3}{|}}{C}} - SO_3H} + H_2O$$

Mechanism

Like nitration, sulphonation also involves *free-radical mechanism.*

$$HO - SO_3H \xrightarrow{675\ K} H\dot{O} + \dot{S}O_3H$$

$$R - H + \dot{O}H \longrightarrow \dot{R} + H_2O$$

$$\dot{R} + \dot{S}O_3H \longrightarrow \underset{\text{Alkanesulphonic acid}}{R - SO_3H}$$

4.9.2 Isomerization Reactions

Heating of straight chain alkanes with anhydrous aluminium chloride (a Lewis acid) affords branched chain alkanes. This process is termed *isomerization. n*-Butane for instance, is isomerized to *iso*-butane on heating with $AlCl_3$.

$$\underset{n\text{-Butane}}{CH_3-CH_2-CH_2-CH_3} \xrightarrow[HCl]{AlCl_3,\,\Delta} \underset{\text{iso-Butane}}{CH_3-\overset{\overset{CH_3}{|}}{\underset{\underset{CH_3}{|}}{C}}-H}$$

The process of isomerization has been of immense utility in petroleum industry for raising the octane number of a particular petroleum fraction.

Mechanism

The reaction is believed to be initiated by olefin which is formed by dehydrogenation of alkane during the transformation. Protonation of this alkane gives a carbocation. The carbocation in turn undergoes a series of 1, 2-hydride and /or alkyl shifts to produce more stable carbocation. This then abstracts hydride from the alkane. The whole sequence of these steps is shown below:

$$\underset{n\text{-Butane}}{CH_3CH_2CH_2CH_3} \xrightarrow{\text{Dehydrogenation}} \underset{\text{But-2-ene}}{CH_3CH=CHCH_3} + H_2$$

$$AlCl_3 + HCl \rightleftharpoons \overset{\oplus}{H}\overset{\ominus}{AlCl_4}$$

$$\underset{\text{But-2-ene}}{CH_3CH=CHCH_3} + \overset{\oplus}{H}\overset{\ominus}{AlCl_4} \rightleftharpoons \underset{2^\circ\text{Butyl cation}}{CH_3\overset{\oplus}{C}H-CH_2CH_3} + \overset{\ominus}{AlCl_4}$$

$$CH_3-\overset{\oplus}{C}H-CH_2-CH_3 \xrightleftharpoons{\text{1, 2-Methyl shift}} CH_3-\overset{\overset{H}{|}}{\underset{\underset{CH_3}{|}}{C}}-\overset{\oplus}{C}H_2 \xrightleftharpoons{\text{1, 2-Hydride shift}}$$

$$\underset{3^\circ\text{Butyl cation}}{CH_3-\overset{\oplus}{\underset{\underset{CH_3}{|}}{C}}-CH_3} + CH_3-\overset{\overset{H}{|}}{C}H-CH_2-CH_3 \xrightleftharpoons{\text{Hydride abstraction}} \underset{iso\text{-Butane}}{H_3C-\overset{\overset{H}{|}}{\underset{\underset{CH_3}{|}}{C}}-CH_3} + \underset{2^\circ\text{Butyl cation}}{H_3C-\overset{\oplus}{C}H-CH_2-CH_3}$$

4.9.3 Alkylation

Alkanes containing a tertiary hydrogen atom such as *iso*-butane react with alkenes when heated in the presence of acidic catalysts such as conc. H_2SO_4 or HF to produce larger branched chain alkanes. For example, *iso*-butane reacts with *iso*-butylene in presence of conc. H_2SO_4 to produce 2, 2, 4-trimethylpentane (*iso*-octane)

$$\underset{iso\text{-Butane}}{(CH_3)_3C{-}H} + \underset{iso\text{-Butylene}}{H_2C{=}C(CH_3)_2} \xrightarrow[\Delta]{\text{conc. } H_2SO_4} \underset{\substack{\text{2, 2, 4-Trimethylpentane} \\ (iso\text{-Octane})}}{CH_3{-}C(CH_3)_2{-}CH_2{-}CH(CH_3){-}CH_3}$$

Mechanism

Like isomerization, alklyation occurs in three steps involving carbocation intermediates:

Step I: Protonation of alkene

$$CH_3{-}C(CH_3){=}CH_2 + \overset{\oplus}{H} \longrightarrow \underset{\text{3°Butyl cation}}{CH_3{-}\overset{\oplus}{C}(CH_3){-}CH_3}$$

Step II: Addition of tertiary carbocation to alkene molecule.

$$(H_3C)_2C{=}CH_2 + \overset{\oplus}{C}(CH_3)_2{-}CH_3 \longrightarrow \underset{\text{3°cation}}{H_3C{-}\overset{\oplus}{C}(CH_3){-}CH_2{-}C(CH_3)_2{-}CH_3}$$

Step III: *Hydride* transfer from is-obutane to the 3° cation.

$$H_3C{-}C(CH_3)_2{-}H + H_3C{-}\overset{\oplus}{C}(CH_3){-}CH_2{-}C(CH_3)_2{-}CH_3 \xrightarrow[\text{transfer}]{\text{Hydride}} \underset{\substack{\text{2, 2, 4-Trimethylpentane} \\ (iso\text{-Octane})}}{H_3C{-}CH(CH_3){-}CH_2{-}C(CH_3)_2{-}CH_3} + \underset{\text{3° Butyl cation}}{H_3C{-}\overset{\oplus}{C}(CH_3){-}CH_3}$$

It may be noted that steps (*ii*) and (*iii*) are repeated again and again.

4.9.4 Pyrolysis or Cracking or Fragmentation

When high boiling hydrocarbons are heated strongly they give rise to a mixture of low boiling hydrocarbons. This process of thermal degradation is called *pyrolysis* or *cracking* or *fragmentation*. It involves cleavage of C—C and C—H bonds and occurs by a free radical mechanism and is mainly of two types: (*i*) Thermal cracking and (*ii*) Catalytic cracking

(*i*) *Thermal cracking*: As the name suggests *higher hydrocarbons are converted into lower ones by heat alone. Thermal cracking involves random cleavage of C—C bonds.* Therefore, cracking of a hydrocarbon such as dodecane can give a number of hydrocarbons such as decane, nonane, octane, heptane, hexane, pentane, butane, ethene, propene, butene, pentene, hexene etc., as shown below:

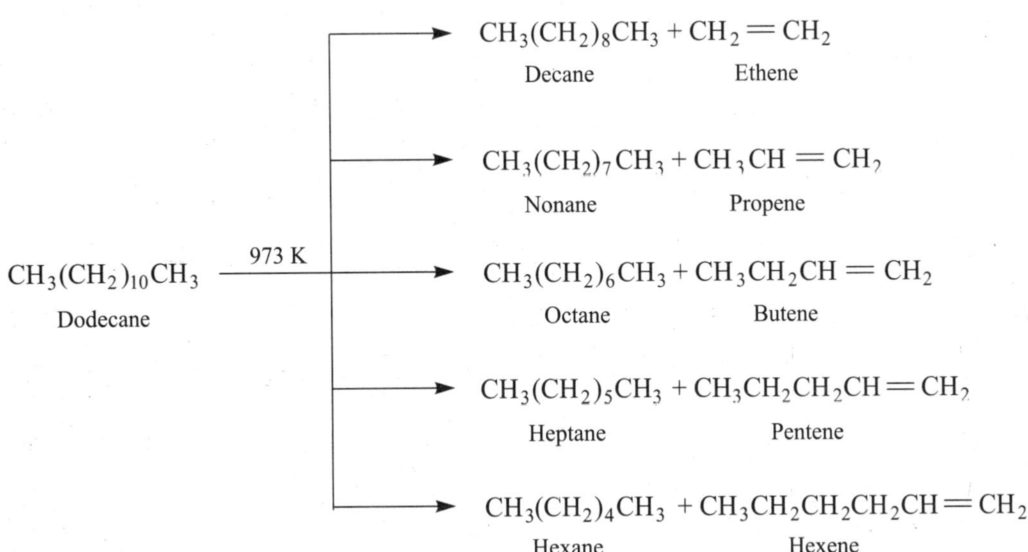

Thus, it may be concluded that thermal cracking is difficult to control and hence a complex mixture of products is always obtained. It is because of this limitation that now-a-days, catalytic cracking in used instead of thermal cracking.

(*ii*) *Catalytic cracking*: *In fact, catalytic cracking is, also, thermal cracking which takes place in presence of a catalyst.* The most commonly used catalyst is a mixture of silica and alumina (4 : 1). Catalytic cracking is generally carried out at a lower temperature (600–650 K) than thermal cracking.

$$\underset{}{CH_3(CH_2)_{10}CH_3} \xrightarrow[\text{Catalyst}]{600\text{–}650\text{ K}} \underset{\substack{\textbf{Heptane}\\ \text{b.pt. 371 K}}}{C_7H_{16}} + \underset{\substack{\textbf{Pentene}\\ \text{b.pt. 309 K}}}{C_5H_{10}}$$

Cracking process is very widely used in petroleum industry. About 85% of the total world supply of gasoline is obtained by this process. Further, this process is also useful for the manufacture of alkenes such as ethene, propene and butene which are widely used in industry.

4.9.5 Aromatization or Reforming Reactions

When *n*-alkanes containing C_6 to C_8 are heated to 873 K under a pressure of 10–20 atm. and in presence of a catalyst, they produce aromatic hydrocarbons. This process is called **aromatization**. The catalysts used are oxides of chromium, molybdenum and vanadium supported over alumina. Other commonly used catalysts are Pd, Pt or Ni.

$$\underset{n\text{-Hexane}}{CH_3(CH_2)_4CH_3} \xrightarrow[\text{Cyclization } (-H_2)]{873\text{ K, Pt}} \underset{\text{Cyclohexane}}{C_6H_{12}} \xrightarrow[\text{Aromatization } (-3\,H_2)]{873\text{ K, Pt}} \underset{\text{Benzene}}{C_6H_6}$$

$$\underset{n\text{-Heptane}}{CH_3(CH_2)_5CH_3} \xrightarrow[\text{Cyclization } (-H_2)]{873\text{ K, Pt}} \underset{\text{Methylcyclohexane}}{C_6H_{11}CH_3} \xrightarrow[\text{Aromatization } (-3\,H_2)]{873\text{ K, Pt}} \underset{\text{Toluene}}{C_6H_5CH_3}$$

This process is also used in petroleum industry to improve the quality of gasoline*, therefore, it is also called **reforming**. Further, it has been found that platinum is the best catalyst. This is the reason why reforming is also called **platforming**.

4.9.6 Oxidation Reactions

Oxidation of alkanes gives different products under different conditions as follows:

***(a) Complete oxidation or complete combustion*:** Alkanes are readily burnt in air or excess of oxygen producing CO_2 and H_2O, and a lot of heat, *e.g.*,

$$CH_4 + 2\,O_2 \longrightarrow CO_2 + 2\,H_2O \quad \Delta H^\circ = -\,890\text{ kJ mol}^{-1}$$

$$2\,CH_3—CH_3 + 7\,O_2 \longrightarrow 4\,CO_2 + 6\,H_2O \quad \Delta H^\circ = -1555\text{ kJ mol}^{-1}$$

---

*Cycloalkanes and aromatic hydrocarbons have higher octane number than *n*-alkanes and hence a higher percentage of these hydrocarbons improves the quality of gasoline.

This energy is so much that alkanes which are the constituents of LPG, gasoline, kerosine oil and diesel are widely used as fuel. The generation of heat in this process has been exploited by using these alkanes as fuel in the internal combustion engines.

It has been shown that in such an engine, in addition to carbon dioxide and water, some other poisonous gases such as unburnt hydrocarbons, CO and oxides of nitrogen are also formed. These gases cause a lot of atmospheric pollution (which is a topic of current interest throughout the world) and in order to check this, attempts are being made to develop **'Converters'** where hydrocarbons and CO can be completely oxidized, and oxides of nitrogen can be broken down to nitrogen and oxygen.

(b) Incomplete combustion

(*i*) If the combustion of the alkanes is carried out in a limited supply of air or oxygen, CO is produced along with unburnt carbon in the form of carbon black or soot.

$$CH_4 + 3\,O_2\text{ (limited)} \longrightarrow 2\,CO + 4\,H_2O$$

$$CH_4 + O_2\text{ (limited)} \longrightarrow C + 2\,H_2O$$

Carbon black is used in the preparation of black inks, paints, polishes, etc.

(*ii*) *Reactions with steam:* Methane reacts with steam at 1273 K in presence of Ni as a catalyst forming CO and H_2.

$$CH_4 + H_2O \xrightarrow{\text{1273 K, Ni}} CO + 3\,H_2$$

***(c) Catalytic or controlled oxidation*:** Alkanes give different products under different conditions of temperature, pressure and catalyst. Some examples are:

(*i*) When a mixture of methane and oxygen [9 : 1(v/v)] is passed through a copper tube at 100 atm. and 573 K, it forms methanol.

$$2\,CH_4 + O_2 \xrightarrow[\text{Cu tube}]{\text{100 atm, 573 K}} \underset{\text{Methanol}}{2\,CH_3OH}$$

(*ii*) A mixture of methane and oxygen upon heating under pressure in presence of molybdenum oxide, produces methanol,

$$CH_4 + O_2 \xrightarrow[\text{High pressure}]{Mo_2O_3,\,\Delta} \underset{\text{Methanol}}{HCHO + H_2O}$$

(*iii*) Although alkanes are inert towards usual oxidising agents such as $KMnO_4$, $K_2Cr_2O_7$, etc. those with a tertiary hydrogen are oxidized easily to the corresponding alcohols.

$$H_3C-\underset{CH_3}{\overset{CH_3}{\underset{|}{\overset{|}{C}}}}-H + [O] \xrightarrow{\text{Alk. } KMnO_4} CH_3-\underset{CH_3}{\overset{CH_3}{\underset{|}{\overset{|}{C}}}}-OH$$

tert-Butyl alcohol

4.9.7 Insertion of Carbenes

Carbene generated by the irradiation of ketene is inserted into a σ C—H bond of an alkane. A mixture of propane and ketene, for instance, on irradiation produces *n*-butane and *iso*-butane.

$$\underset{\text{Ketene}}{CH_4 = C = O} \xrightarrow{h\nu} \underset{\text{Carbene}}{:CH_2 + CO}$$

$$CH_3 - CH_2 - CH_3 + :CH_2 \longrightarrow \underset{n\text{-Butane}}{CH_3 - CH_2 - CH_2 - CH_3} + \underset{iso\text{-Butane}}{CH_3 - \underset{\underset{H}{|}}{\overset{\overset{CH_3}{|}}{C}} - CH_3}$$

4.10 MECHANISM OF HALOGENATION OF ALKANES

This reaction takes place by a free radical chain mechanism which proceeds in the following three steps: (*i*) *Chain initiation* (*ii*) *Chain propagation* (*iii*) *Chain termination.*
Free radical chlorination of methane constitutes a typical example.

(*i*) *Chain Initiation*

This is the first step of the reaction. When a mixture of methane and chlorine is either heated in dark at 525–675 K or irradiated with UV-light at 487.5 nm, chlorine molecule undergoes *homolysis* to produce chlorine atoms. The reaction is endothermic requiring 242.5 kJ mol^{-1} of energy.

$$\underset{242.5\ \text{kJ mol}^{-1}}{Cl - Cl} \xrightarrow[\text{or } h\nu\ (487.5\ \text{nm})]{525\text{–}675\ \text{K}} 2\,\dot{Cl} \quad (\Delta H = +242.5\ \text{kJ mol}^{-1})$$

(*ii*) *Chain Propagation*

Chain propagation is a two step process.

(*a*) ***Step* 1:** The chlorine atom produced in the above chain initiation step collides with methane molecule and abstracts an atom of hydrogen from it forming methyl radical.

$$\dot{Cl} + \underset{(+426.36\ \text{kJ mol}^{-1})}{H - CH_3} \longrightarrow \underset{\text{Transition state}}{\left[\overset{\delta^{\bullet}}{Cl} \cdots H \cdots \overset{\delta^{\bullet}}{CH_3}\right]} \longrightarrow \underset{(-430.54\ \text{kJ mol}^{-1})}{\dot{C}H_3 + H - Cl}$$

$$(\Delta H_1 = -4.18\ \text{kJ mol}^{-1})$$

(*b*) ***Step 2:*** The methyl radical collides with another chlorine molecule forming methyl chloride and another chlorine atom.

$$\dot{C}H_3 + Cl-Cl \longrightarrow \left[\overset{\delta^\bullet}{C}H_3 \cdots Cl \cdots \overset{\delta^\bullet}{Cl}\right] \longrightarrow CH_3-Cl + \dot{Cl}$$

(+242.5 kJ mol^{-1}) Transition state (– 338.58 kJ mol^{-1})

(ΔH_2 = – 96.08 kJ mol^{-1})

Both these steps involve their respective transition states as shown in the energy profile diagram (Fig. 4.3). The overall energy released in the two steps is given by

$$\Delta H = \Delta H_1 + \Delta H_2 = -4.18 - 96.08 = -100.26 \text{ kJ mol}^{-1}.$$

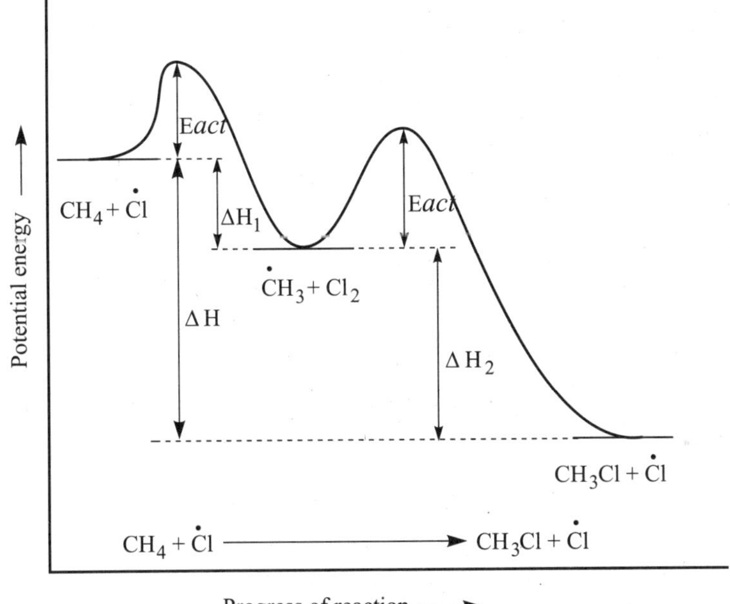

Fig. 4.3: Energy profile diagram for chlorination of methane

A look at Fig. 4.3 reveals that E*act* for first step is high so this is the rate determining step. It means that the formation of $\dot{C}H_3$ is slow but once formed it reacts quickly with Cl_2 molecule to form CH_3Cl.

Both the above steps (1 and 2) are repeated over and over again. In other words, a single photon of light absorbed by Cl_2 can transform a large number of methane molecules into methyl chloride. Such reactions in which excitation of one molecule results in the transformation of a large number of molecules are called *chain reactions*. Both the steps are exothermic and hence the overall energy balance (ΔH) of the reaction is negative.

After a sufficient quantity of CH_3Cl is produced, the chlorine atoms produced in initial step have a greater probability to collide with CH_3Cl rather than CH_4. Such collision leads to the formation of $\dot{C}H_2Cl$ radicals which abstract $\dot{Cl}$ from fresh Cl_2 molecule to give CH_2Cl_2 and another $\dot{Cl}$. Similarly,

CH_2Cl_2 produces $CHCl_3$ which in turn produces CCl_4. The whole sequence of events is shown below:

$$CH_3Cl + \dot{C}l \longrightarrow \dot{C}H_2Cl + HCl$$
$$\dot{C}H_2Cl + Cl_2 \longrightarrow CH_2Cl_2 + \dot{C}l$$
$$CH_2Cl_2 + \dot{C}l \longrightarrow \dot{C}HCl_2 + HCl$$
$$\dot{C}HCl_2 + Cl_2 \longrightarrow CHCl_3 + \dot{C}l$$
$$CHCl_3 + \dot{C}l \longrightarrow \dot{C}Cl_3 + HCl$$
$$\dot{C}Cl_3 + Cl_2 \longrightarrow CCl_4 + \dot{C}l$$

(iii) Chain Termination

The above chain may be terminated by the coupling of any two radicals.

$$\dot{C}l + \dot{C}l \longrightarrow Cl—Cl$$
$$\dot{C}H_3 + \dot{C}l \longrightarrow CH_3—Cl$$
$$\dot{C}H_3 + \dot{C}H_3 \longrightarrow CH_3—CH_3$$

Looking at the bond dissociation energies it becomes clear that although *energy must be supplied for the chain initiation, the two propagation steps are exothermic and are repeated several times. Thus a small amount of energy brings about conversion of several alkane molecules into halides. In other words, we may say that the quantum yield of the reaction in very high.*

The ease of substitution at various carbon atoms is of the order: 3° > 2° > 1°, which is the same as the stability of various alkyl radicals. For instance, the chlorination of propane yields a mixture of 1-chloropropane and 2-chloropropane. Since there are six *primary* hydrogens as against two *secondary* hydrogens, it can be expected that the product ratio would be 3 : 1, respectively. But actually, the ratio has been found to be 1 : 1. It can, therefore, be inferred that the ease of formation of a secondary radical is three times more than that of a primary radical.

$$CH_3—CH_2—CH_3 \xrightarrow{\dot{C}l} \begin{cases} CH_3—CH_2—\dot{C}H_2 \text{ (1° Radical, } \textit{Less stable}) \xrightarrow{Cl_2} CH_3—CH_2—CH_2—Cl \\ CH_3—\dot{C}H—CH_3 \text{ (2° Radical, } \textit{More stable}) \xrightarrow{Cl_2} CH_3—CH(Cl)—CH_3 \end{cases} \text{ 1 : 1 ratio}$$

Bromination of alkanes has close similarities to chlorination except for the fact that the rate of the reaction is slow. This is mainly due to difference in bond dissociation energies of the molecules involved in the reaction.

$$\underset{(192.7\ \text{kJ mol}^{-1})}{Br-Br} \xrightarrow{h\nu \text{ or } \Delta} 2\,Br^{\bullet} \qquad (\Delta H = +192.7\ \text{kJ mol}^{-1})$$

$$Br^{\bullet} + \underset{(427\ \text{kJ mol}^{-1})}{H-CH_3} \longrightarrow \underset{(363.6\ \text{kJ mol}^{-1})}{Br-H} + \dot{C}H_3 \qquad (\Delta H = +63.4\ \text{kJ mol}^{-1})$$

$$\dot{C}H_3 + \underset{(192.7\ \text{kJ mol}^{-1})}{Br-Br} \longrightarrow \underset{(280.7\ \text{kJ mol}^{-1})}{CH_3-Br} + Br^{\bullet} \qquad (\Delta H = -88\ \text{kJ mol}^{-1})$$

Although in this reaction the last step is exothermic, the hydrogen abstraction by bromine radical is endothermic because of the fact that H—C bond is stronger than H—Br bond. Thus it is relatively difficult for the bromine atom to acquire sufficient energy for abstracting a hydrogen atom in competition with various chain terminating steps such as coupling with another bromine atom or with an alkyl radical. Hence bromination is slower than chlorination. An important consequence of low reactivity of bromine atom is that it becomes more selective for tertiary and secondary positions compared to chlorine atom (which is less selective). Thus bromination of isobutane affords *tert*-butyl bromide as the major product.

$$(CH_3)_3C-H \xrightarrow[\text{or } \Delta]{Br_2,\ h\nu} \underset{99\%}{(CH_3)_3C-Br} + \underset{1\%}{H_3C-CH(CH_3)-CH_2Br}$$

Fluorination of alkanes is so violent that it results in the cleavage of C—C as well as C—H bonds. This can be readily understood by a consideration of the bond dissociation energies of various species involved in the chain initiation and propagation steps.

$$\underset{(158\ \text{kJ mol}^{-1})}{F-F} \xrightarrow{h\nu \text{ or } \Delta} 2\,F^{\bullet} \qquad (\Delta H = +158\ \text{kJ mol}^{-1})$$

$$F^{\bullet} + \underset{(427\ \text{kJ mol}^{-1})}{H-CH_3} \longrightarrow \dot{C}H_3 + \underset{(565.8\ \text{kJ mol}^{-1})}{H-F} \qquad (\Delta H = -138.8\ \text{kJ mol}^{-1})$$

$$\dot{C}H_3 + \underset{(158\ \text{kJmol}^{-1})}{F-F} \longrightarrow \underset{(450.6\ \text{kJ mol}^{-1})}{CH_3-F} + F^{\bullet} \qquad (\Delta H = -292.6\ \text{kJ mol}^{-1})$$

The above equations make it abundantly clear that the reaction is highly exothermic and the release of such a huge amount of energy explains why fragmentation (of C—C and C—H bonds) is an important side reaction.

Although the chain initiation in the case of **iodination** is the easiest of all the halogenation processes (ΔH = +150 kJ mol^{-1}), the propagation steps are endothermic and so the overall reaction is also endothermic. Furthermore, the reaction is reversible resulting in the regeneration of the alkane:

$$R—H + I_2 \rightleftharpoons R—I + HI$$

The reaction is, therefore, not synthetically useful and direct iodination is usually achieved by using a strong oxidizing agent to destroy the HI formed in the process.

Evidence in Support of Free Radical Mechanism of Halogenation

Following observations fully support the free radical chain mechanism in halogenation of alkanes:

(*i*) When methane and chlorine are mixed together, they do not react at all at room temperature. However, when this mixture is heated above 525 K (indark) or irradiated with ultraviolet light at room temperature they react very readily. We know that, high temperature or ultraviolet light favour the formation of free radicals. Hence, the above reaction also involves free radicals as intermediates. The chain initiation step involving homolytic fission of Cl_2 molecule into $\dot{C}l$ atoms, being endothermic, requires a lot of energy in the form of heat or light to initiate this process.

(*ii*) **Quantum yield*:** The quantum yield of chlorination of methane is 10^6 to10^7at room temperature. Absorption of one photon of light by a chlorine molecule, results in the formation of several thousand molecules of CH_3Cl. This observation supports the occurrence of chain propagation steps which are exothermic and repeat themselves over and over again, thereby setting up a chain of events.

(*iii*) There is practically no reaction between ehtane and chlorine at 410 K. However, in the presence of a trace amount of tetraethyllead (0.002 mole percent) the reaction takes place vigorously. The role of tetraethyllead is as follows: Tetraethyllead is well-known as free radical initiator. At reaction temperature it decomposes readily to produce ethyl radicals which react with chlorine molecule to produce $\dot{C}l$ atoms. The chorine atoms, so produced, enter the propagation step to produce methyl radical and the chain propagates.

$$(C_2H_5)_4Pb \xrightarrow{410\ K} 4\ CH_3\dot{C}H_2 + Pb$$

Hence, the involvement of free radicals in the reaction is evident.

(*iv*) The rate of the reaction is retarded considerably in presence of a small amount of oxygen. The reason for this is that oxygen exists as a diradical $\dot{O}—\dot{O}$ in the ground state. As such it reacts with $\dot{C}H_3$ radical (produced in Step 2) to give methylperoxy radical which is relatively less reactive. Hence, the involvement of free radicals in the reaction.

$$\dot{C}H_3 + O_2 \longrightarrow CH_3—O—\dot{O}$$

***Note*:** Substances such as $\dot{O}—\dot{O}$, which retard the rate of a free radical reaction are called **radical inhibitors**[+].

*Quantum yield of a photochemical reaction is defined by the following equation:

$$\text{Quantum yield } \phi = \frac{\text{No. of molecules reacting or formed in unit time}}{\text{No. of quanta of light absorbed in unit time}}$$

[+] Oxygen is known to exist as a diradical, $\dot{O}—\dot{O}$, in the grounds state. It may act as free radical inhibitor probably because of this fact.

(*v*) Well-known free radical precursors such as acyl peroxides start the halogenation of alkanes. In presence of such an initiation, the reaction temperature is also lowered considerably. The role of acyl peroxides is illustrated in the folllowing reactions:

$$\underset{\text{Benzoylperoxide}}{C_6H_5-\overset{\overset{O}{\|}}{C}-O-O-\overset{\overset{O}{\|}}{C}-C_6H_5} \longrightarrow \underset{\text{Benzoyloxy radical}}{2\,C_6H_5-\overset{\overset{O}{\|}}{C}-\dot{O}}$$

$$C_6H_5-\overset{\overset{O}{\|}}{C}-\dot{O} \longrightarrow \dot{C}_6H_5 + CO_2$$

$$\dot{C}_6H_5 \;+\; Cl-Cl \longrightarrow C_6H_5Cl \;+\; \dot{C}l$$

Relative Reactivities of Halogens Toward Halogenation of Alkanes

The order of relative reactivities of halogens is: $F_2 > Cl_2 > Br_2 > I_2$.

The above order is in agreement with ΔH values of these reactions which are given below:

$$\underset{(158.5\text{ kJ mole}^{-1})}{F-F} + \underset{(434.7\text{ kJ mole}^{-1})}{H-CH_3} \longrightarrow \underset{(451.5\text{ kJ mole}^{-1})}{F-CH_3} + \underset{(568.5\text{ kJ mole}^{-1})}{H-F}$$

$$\Delta H = -426.8\text{ kJ mole}^{-1}$$

$$\underset{(242.5\text{ kJ mole}^{-1})}{Cl-Cl} + \underset{(434.7\text{ kJ mole}^{-1})}{H-CH_3} \longrightarrow \underset{(351.5\text{ kJ mole}^{-1})}{Cl-CH_3} + \underset{(430.5\text{ kJ mole}^{-1})}{H-Cl}$$

$$\Delta H = -104.4\text{ kJ mole}^{-1}$$

$$\underset{(192.3\text{ kJ mole}^{-1})}{Br-Br} + \underset{(434.7\text{ kJ mole}^{-1})}{H-CH_3} \longrightarrow \underset{(292.6\text{ kJ mole}^{-1})}{Br-CH_3} + \underset{(367.8\text{ KJ mole}^{-1})}{H-Br}$$

$$\Delta H = -33.4\text{ kJ mole}^{-1}$$

$$\underset{(150.5\text{ kJ mole}^{-1})}{I-I} + \underset{(434.7\text{ kJ mole}^{-1})}{H-CH_3} \longrightarrow \underset{(234.1\text{ kJ mole}^{-1})}{I-CH_3} + \underset{(296.8\text{ kJ mole}^{-1})}{H-I}$$

$$\Delta H = +54.3\text{ kJ mole}^{-1}$$

A look at the above values of ΔH reveals that fluorination, chlorination and bromination are exothermic whereas iodination is endothermic. Further since the values of ΔH follow the order: Fluorination > chlorination > bromination > iodination, it is reasonable to believe that the rates of these reactions should follow the same order.

Let us now examine the energy changes (ΔH) for each of the three steps of halogenation.

Steps				ΔH (kJ mole^{-1}) F	Cl	Br	I
1. Initation	X_2	⟶	$2\dot{X}$	+158	+242.5	+192.3	+150.3
2. Propagation	$\dot{X} + CH_4$	⟶	$HX + \dot{C}H_3$	–138.8	–4.18	+63.4	+137.9
3. Propagation	$\dot{C}H_3 + X_2$	⟶	$CH_3X + \dot{X}$	–292.6	–96.08	–88	–83.6

Because step 1 involves only dissociation of molecules into atoms, it can be easily assumed that for this step ΔH = E*act*. Thus chlorine with highest value of E*act* (+242.5 kJ mole^{-1}) should dissociate most slowly and iodine with lowest value of E*act* (+150.3 kJ mole^{-1}) should dissociate most rapidly. So, the halogen reactivity should follow the order $I_2 > F_2 > Br_2 > Cl_2$. But this order is not in agreement with the observed order of reactivity. Hence, we conclude that the ease of dissociation of halogens into atoms cannot determine the observed order of reactivities.

Coming to step 3 which involves attack of methyl redicals on hologen. It is exothermic for all the four halogens, with ΔH for chlorine, bromine and iodine being quit close to each other. The E*act* for these halogens is expected to be quite low, *i.e.* only a fraction of kJ. Thus step 3 can also not account for the observed order of halogen reachinity.

Thus, we are left with step 2 which involves the abstraction of hydrogen atom from methane by a halogen atom. This shows a wide variation of ΔH ranging from **highly exothermic** reaction with F° (ΔH = –138.8 kJ mole^{-1}) to highly **endothermic reaction** with I° (ΔH = +137.9 kJ mole^{-1}).

The experimental values of E*act*, for all the four halogens, are listed below:

Experiment Values of E*act*

reaction	E*act* (kJ mole^{-1}) F	Cl	Br	I
$\dot{X} + CH_4 \longrightarrow HX + \dot{C}H_3$	4.18	16.72	75.24	137.9

Thus, the increasing E*act* for halogens follows the order: Fluorination < chlorination < bromination < iodination which is in agreement with the observed order of reactivities, *i.e.*, Fluorination > chlorination > bromination > iodination.

Thus, it may be concluded that the reactivity of halogens is controlled by step 2 *i.e.*, abstraction of hydrogen by halogen.

4.11 ORIENTATION OF HALOGENATION OF ALKANES

It is well-known that monochlorination of methane and ethane gives only one product, whereas that of propane, *n*-butane or *iso*-butane gives two products in each case. This is due to the fact that all the hydrogens in methane and ethane are only of one type *i.e.*, they are primary. On the other hand, the hydrogens of propane, n-butane and *iso*-butane are of two types, in the first two they are 1° and 2° and in the third they are 1° and 3°. Further, as expected three different types of hydrogens (1°, 2° and 3°) will be abstracted at different rates. Therefore, the different isomeric monohalogen derivatives will be formed in different amounts. *Thus in case of compounds which can give rise to two or more isomeric products, the process of ascertaining at which site in molecule the reaction is most likely to occur (and what are the relative amounts of products formed) is called* ***orientation****.*

To understand the concept of orientation clearly, let us take the monochlorination of propane as an example. It is known that the reaction gives rise to a mixture of n-propyl chloride and *iso*-propyl chloride.

$$\underset{\text{Propane}}{CH_3-CH_2-CH_3} \xrightarrow[h\nu]{Cl_2} \xrightarrow{\text{Abstraction of } 1^\circ H} \underset{n\text{-Propyl radical}}{CH_3-CH_2-\dot{C}H_2} \xrightarrow{Cl_2} \underset{n\text{-Propyl chloride}}{CH_3-CH_2-CH_2-Cl}$$

$$\underset{\text{Propane}}{CH_3-CH_2-CH_3} \xrightarrow[h\nu]{Cl_2} \xrightarrow{\text{Abstraction of } 2^\circ H} \underset{Iso\text{-propyl radical}}{CH_3-\dot{C}H-CH_3} \xrightarrow{Cl_2} \underset{Iso\text{-propyl chloride}}{CH_3-CHCl-CH_3}$$

The relative amounts of n-propyl chloride and *iso*-propyl chloride produced depend upon the relative rates of formation of n-propyl radicals and *iso*-propyl radicals, respectively. Now, if it is presumed that *iso*-propyl chloride will be formed faster and will constitute a larger fraction of the product mixture, on similar lines we can say that if n-propyl radical is formed faster then n-propyl chloride will constitute a larger fraction of the product mixture. Thus it can be said that orientation can be determined by relative rates of the two competing reactions.

Now the question arises as to what factor(s) govern the observed orientation.

We know that, the rate of a given reaction depends upon the following factors:

(*a*) Frequeny of collisions between reacting species.

(*b*) Probability factor.

(*c*) Energy factor.

We will now examine which of these factors govern the formation of different amounts of n-propyl chloride and *iso*-propyl chloride.

(*a*) Frequency of Collisions Between Reacting Species

Since both the reactions involve collisions between the same particles, *viz*., a propane molecule and a $\dot{C}l$ atom, the collision frequency will not have any effect on rate of the two competing reactions.

(*b*) Probability Factor

If the $\dot{C}l$ atom has to abstract a primary hydrogen then at the time of collision the propane molecule should be so oriented that $\dot{C}l$ atom strikes only a primary hydrogen. Similarly, if it has to abstract a secondary hydrogen than at the time of collision the propane molecule should be so oriented that $\dot{C}l$ atom strikes only secondary hydrogens. Since in propane molecule there are 6 primary and 2 secondary hydrogens, it expected that the probability factor will favour abstraction of primary hydrogen over secondary hydrogens by a ratio 6 : 2 or 3 : 1, *i.e.*, the ratio of *n*-propyl chloride and *iso*-propyl chloride formed should be 3 : 1. But in actual practice both the isomers are formed in almost equal amounts (45% *n*-propyl chloride and 55% *iso*-propyl chloride). Hence probability alone does not govern the observed ratios of products formed.

(*c*) Energy Factor

We have seen that the observed ratio of isomeric propyl chlorides formed cannot be accounted for on the basis of frequencies of collisions and probability factor. Thus, it is quite reasonable to believe that E*act* for the abstraction is responsible for this ratio. As expected E*act* for abstraction of 2° hydrogen is lower than the E*act* for abstraction of 1° hydrogen. This is in close agreement with the fact that a 2° radical is more stable than a 1° radical.

Another illustrative example is the chlorination of *iso*-butane. In this case, abstraction of one out of nine 1° hydrogen gives *iso*-butyl chloride and the abstraction of the single 3° hydrogen produces *tert*-butyl chloride.

If we work on the lines of probability factor alone, we would predict that chlorination of *iso*-butane will produce *iso*-butyl chloride and *tert*-butyl chloride in the ratio 9 : 1. But actual ratio of these products is 64 : 32 or 2 : 1 (or 9 : 4.5). It means that the tertiary hydrogen is abstracted about 4.5 times faster than expected. In other words, the successful collisions with tertiary hydrogen are 4.5 more than with primary hydrogen. This may be explained on the basis of the fact that E*act* for abstraction of 3° hydrogen is far less than that for 1° or 2° hydrogens. As a result of studies on free radical chlorination of a large number of hydrocarbons, it has been found that the ease of abstraction of hydrogens by $\dot{C}l$ atom follows the order 3° > 2° > 1°.

Further, on the basis of the amount of each chlorination product obtained from various hydrocarbons, chemists have been able to calculate that at room temperature the relative rates of abstraction (or relative reactivities) of various types of hydrogens are 5.0 for 3° H, 3.8 for 2° H and 1.0 for 1° H.

These values can be used to predict the ratio of isomeric chlorination products formed from a given alkane. For example, monochlorination of *n*-butane.

$$\underset{n\text{-Butane}}{CH_3CH_2CH_2CH_3} \xrightarrow[h\nu]{Cl_2} \underset{n\text{-Butyl chloride}}{CH_3CH_2CH_2CH_2Cl} + \underset{sec\text{-Butyl chloride}}{CH_3CH_2\overset{\overset{\displaystyle Cl}{|}}{C}HCH_3}$$

The expected ratio of two products can be calculated with the help of the following expression:

$$\frac{\text{Amount of } n\text{-butyl chloride}}{\text{Amount of } iso\text{-butyl chloride}} = \frac{\text{No of } 1^\circ\text{H}}{\text{No of } 2^\circ\text{H}} \times \frac{\text{Reactivity of } 1^\circ\text{H}}{\text{Reactivity of } 2^\circ\text{H}}$$

Here No. of 1° H = 6   Reactivity of 1° H = 1
No. of 2° H = 4   Reactivity of 2° H = 3.8

Putting these values in the above equation, we get

$$\frac{\text{Amount of } n\text{-butyl chloride}}{\text{Amount of } iso\text{-butyl chloride}} = \frac{6 \times 1}{4 \times 3.8} = \frac{6}{15.2} = \frac{28\%}{72\%}.$$

This ratio has been verified experimentally.

For bromination the relative rates of abstraction (or reactivity) of hydrogens has been found to be 1600 for 3°, 82 for 2° and 1 for 1°. Let us use these values to predict the relative amounts of isomeric products from the following reaction:

$$\underset{\text{Propane}}{CH_3CH_2CH_3} \xrightarrow[h\nu]{Br_2} \underset{n\text{-Propyl bromide}}{CH_3CH_2CH_2Br} + \underset{iso\text{-Propyl bromide}}{CH_3-\overset{\overset{\displaystyle Br}{|}}{C}H-CH_3}$$

Here No. of 1° H = 6   Reactivity of 1° H = 1
No. of 2° H = 2   Reactivity of 2° H = 82

$$\frac{\text{Amount of } n\text{-propyl chloride}}{\text{Amount of } iso\text{-propyl chloride}} = \frac{6 \times 1}{2 \times 82} = \frac{3}{82} = \frac{3.5\%}{96.5\%}$$

Actual amounts = 3% *n*-propyl bromide and 97% *iso*-propyl bromide.

4.12 REACTIVITY *vs.* SELECTIVITY

These terms are applied *to specify the rates and orientations of halogenation of alkanes.* The term reactivity refers to relative rates of chlorination and bromination of alkanes without any reference to the type of hydrogen abstracted by the halogen. On the other hand, selectivity refers to relative rates of abstraction of different types of hydrogens (1°, 2°, 3°) in alkanes by halogen. It has been observed that at 548 K out of 15 million collisions 375000 are effective in the case of chlorine and only one in the case of bromine, so chlorine is more reactive than bromine. Further, the relative rates of abstraction of different types of hydrogen are 1600 for 3° H, 82 for 2° H and 1 for 1° H in the case of bromine against 5.0 for 3° H, 3.8 for 2° H and 1 for 1° H in case of chlorine, so bromine is more selective. Thus we can say that in a set of similar reactions the less reactive reagent is more selective and more reactive reagent is less selective. This is called as **reactivity-selectivity principle**.

To find a suitable explanation for the reactivity- selectivity relationship, we will now consider the nature of transition states involved during abstraction of hydrogen by $\dot{Cl}$ and $\dot{Br}$ atoms.

In the case of the attack by highly reactive $\dot{Cl}$ atom, the reaction is exothermic só the E*act* is low, under these conditions the transition state leading to the formation of alkyl radical is reached early *i.e.*, when the alkyl group has developed only slight radical character (*i.e.*, it is reactant like). Thus, the stability of incipient (in the process of forming) radical has little role to play. Hence in the case of chlorination the selectivity is low.

$$R—H + \dot{Cl} \longrightarrow [\overset{\dot{\delta}}{R} \text{--------} H \text{----------} \overset{\dot{\delta}}{Cl}] \longrightarrow \dot{R} + H—Cl$$

Transition state reached early
so little radical character

On the other hand, the attack of less reactive $\dot{Br}$ atom is endothermic so its E*act* is quite high. Under these conditions the transition state leading to the formation of alkyl radical is reached late, *i.e.*, when alkyl group has developed sufficient radical character (it is intermediate-like). Thus, the relative stability of incipient alkyl radicals, determines the relative energies of three transition states involved in hydrogen abstraction. So, it is this order of stabilites of transition states which determines the rate of abstraction of three types of hydrogens (1°, 2°, 3°). Now since the relative stabilities of alkyl radicals follow the order 3° > 2° > 1° therefore, the relative stabilities of the corresponding transition states also follow the same order (**Hammond's postulate**). The difference in energies of the three transition states is quite high. Hence, the reaction prefers to take up the path involving the lowest energy transition state. Hence in the case of bromination of alkanes the selectivity is high.

A BRIEF REVIEW

- **Alkanes** are the acyclic saturated hydrocarbons having general formula C_nH_{2n+2}.
- **Catalytic hydrogenation** of alkenes and alkynes is a cis-addition process which occurs readily in presence of Pt, Pd or Raney Ni at room temperature to give the corresponding alkanes.
- **Sabatier-Sanderens reduction** is the reduction of alkenes which occurs in presence of Ni at 523–573 K.
- **Reduction of alkyl halides to alkanes** can be carried out in the following ways:
  (*a*) Using dissolved zinc metal such as Zn/CH_3COOH, Zn/HCI or Zn/NaOH.
  (*b*) $LiAlH_4$ can reduce 1° and 2° alkyl halides to alkanes.
  (*c*) $NaBH_4$ can reduce 2° and 3° alkyl halides to alkenes.
  (*d*) H_2/Pd-C is used for catalyic hydrogenolysis of alkyl halides.
  (e) HI/Red P reduces alkyl iodides to alkanes.
  (*f*) Mg in dry ether forms **Grignard reagents** with alkyl halides which react with active hydrogen containing substances such as H_2O, ROH, H_2S, NH_3 RC ≡ CH etc., to give alkanes corresponding to alkyl group of the **Grignard reagent**.
- **Würtz reaction:** Symmetrical alkanes *i.e.*, alkane having even number of carbon atoms are obtained when alkyl halides react with metallic sodium in presence of dry ether.
- **Corey-house reaction:** Lithium dialkylcuperate and a suitable alkyl halide react to give alkanes of all types whether branched or unbranched containing odd or even number of carbon atoms.

- **Kolbe's electrolysis** of sodium or potassium salts of fatty acids gives alkanes.
- **Decarboxylation:** Sodium salts of fatty acids react with sodalime to give alkanes containing one carbon atom less than the corresponding original acid. Here one carbon is lost as CO_2.
- The b. pt.of straight chain alkanes increases regularly with an increase in the number of carbon atoms.
- B. pt. decreases with branching amongst isomeric alkanes. For instance, b. pt. decreases in the order: *n*-Pentane > *iso*-pentane > *neo*-pentane.
- **Alternation effect:** *n*-Alkanes having even number of carbon atoms possess much higher m. pt. than those having odd number of carbon atoms.
- Alkanes are relatively less reactive molecules due to the presence of strong C—C and C—H bonds.
- **Halogenation** of alkanes takes place when a mixture of an alkane and chlorine is heated to 520–627 K in dark or in presence of UV radiations.

  The order of reactivity of halogens is: $F_2 > Cl_2 > Br_2 > I_2$

  The order of reactivity of alkanes is: $3° > 2° > 1°$
- **Nitration** of alkanes takes place on heating a mixture of an alkane with fuming nitric acid at 623–673 K at high pressure.
- **Sulphonation** of alkanes takes place when branched chain alkanes or alkanes having six or more carbon atoms are heated with oleum ($H_2SO_4 + SO_3$).
- **Isomerization:** An *n*-alkane undergoes isomerization to give branched chain alkane on heating with anhyd. $AlCl_3$ and HCI. For instance,

$$\underset{n\text{-Butane}}{CH_3CH_2CH_2CH_3} \xrightarrow{\text{Anhyd. } AlCl_3,\ HCl} \underset{iso\text{-Butane}}{CH_3-\underset{\underset{CH_3}{|}}{CH}-CH_3}$$

- **Alkylation:** In presence of conc. H_2SO_4 or HF, 3° alkanes react with alkenes to give larger branched chain alkanes. For example, isobutane reacts with *iso*-butylene to give *iso*-octane.
- **Cracking:** High boiling hydrocarbons on heating strongly in absence of air get converted into low boiling hydrocarbons. This process is called cracking and involves C—C bond cleavage and is mainly of two types: (*a*) Thermal cracking, and (*b*) Catalytic cracking.
- **Thermal cracking** involves random cleavage of C—C bonds to give a complex mixture of low boiling alkanes and alkenes.
- **Catalytic cracking** gives less complex mixture and hence is widely used in petroleum industry for manufacture of about 85% of the total world production of gasoline at present. Catalytic cracking is carried out at lower temperature (600–650 K) in presence of a catalyst (Silica : Alumina :: 4 : 1).

- **Aromatization:** Aromatic hydrocarbons are obtained, from *n*-alkanes having six to eight carbon atoms, by heating them in presence of a catalyst (metal oxides or Pd or Pt or Ni). This process is also called **reforming** which involves cyclization, isomerization and dehydrogenation to improve the quality of gasoline by increasing its octane number.
- **Oxidation:** Alkanes rapidly burn in air to produce CO_2 and H_2O. Alkanes give different products on oxidation under different conditions. This process is called **combustion** during which a large quantity of heat is produced. Due to this generation of heat energy alkanes in the form of LPG, gasoline, kerosene oil and diesel are used as fuels.
- **Mechanism of halogenation** involves a **chain reaction** which occurs by a **free radical chain mechanism**. This reaction is an exothermic process.
- **Reactivity-selectivity principle:** *Bromine is less reactive but more selective while chlorine is more reactive but less selective.* Therefore, free-radical bromination rather than chlorination is a useful synthetic procedure for the preparation of alkyl halides.
- **Orientation of halogenation:** Isomeric monohalogen derivatives are formed by alkanes containing more than one type of H-atoms (*i.e.*, 3°, 2°, 1°). Their relative amount depends upon:

  (*a*) The number of each type of H-atoms, and

  (*b*) The relative reactivity of different types of H-atoms.

  Orientation is the process of determining the relative amounts of various isomeric products formed in a reaction.

SOLVED PROBLEMS

Q.1. How will you prepare 2-deuteriopropane?

Ans.

(*i*) $$\underset{\textit{iso}\text{-Propyl iodide}}{(CH_3)_2CH—I} \xrightarrow{\text{Mg/ether}} (CH_3)_2CH—MgI \xrightarrow{D_2O} \underset{\text{2-Deuteriopropane}}{(CH_3)_2CH—D} + Mg(OD)I$$

(*ii*) $$(CH_3)_2CH—I \xrightarrow[\text{–LiI}]{\text{Li/ether}} (CH_3)_2CH—Li \xrightarrow{D_2O} (CH_3)_2CH—D + LiOD$$

Q.2. Outline two conceivable synthesis of 2-methylpentane from three carbon compounds. Which of the two would you actually use? Why?

Ans. 2-Methylpentane contains two three carbon components, *i.e.*, *iso*-propyl group and the *n*-propyl group. Therefore, either of these two groups can act as a source of lithium dialkylcuperate or the alkyl halide. Thus, the two conceivable syntheses are

(*i*) $CH_3-\underset{|}{\overset{CH_3}{CH}}-Br \xrightarrow[-\,LiBr]{Li/ether} CH_3-\underset{|}{\overset{CH_3}{CH}}-Li \xrightarrow{CuI} \left[CH_3-\overset{CH_3}{\overset{|}{CH}}\right]_2 CuLi$

Iso-Propyl bromide; Lithium diisopropylcuperate

$\xrightarrow[S_N2]{CH_2CH_2CH_2Br} CH_3-\overset{CH_3}{\overset{|}{CH}}-CH_2CH_2CH_3$

2-Methylpentane

(*ii*) $CH_2CH_2CH_2Br \xrightarrow[-\,LiBr]{Li/ether} CH_3CH_2CH_2-Li \xrightarrow{CuI} (CH_3CH_2CH_2)_2CuLi$

n-Propyl bromide; Lithium di -*n*-propylcuperate

$\xrightarrow[S_N2]{(CH_3)_2CH-Br} CH_3CH_2CH_2-\overset{CH_3}{\overset{|}{CH}}-CH_3$

2-Methylpentane

Since S_N2 reactions occur readily on 1°than on 2° alkyl halides, therefore second method which uses a primay alkyl halide, *i.e.*, *n*-propyl bromide is preferred.

Q.3. Electrolysis of an aqueous solution of sodium propanoate gives butane, ethane, ethylene and ethyl propanoate. Propose a suitable mechanism to account for the formation of these products.

Ans. Propanoate ion discharges at the anode to form first a propanoate free radical which subsequently decomposes to from ethyl free radical and carbon dioxide.

$$C_2H_5-\overset{O}{\overset{\|}{C}}-\ddot{\underset{..}{O}}{:}^{\ominus} \xrightarrow{-e^-} C_2H_5-\overset{O}{\overset{\|}{C}}-\dot{\underset{..}{O}}{:} \longrightarrow \dot{C}_2H_5 + \overset{O}{\overset{\|}{C}}{=}O$$

Propanoate ion; Propanoate free radical; Ethyl radical

The formation of products from ethyl radical is explained below:

(*i*) $\dot{C}_2H_5 + \dot{C}_2H_5 \xrightarrow{\text{Coupling}} C_2H_5-C_2H_5$

n -Butane

(*ii*) $CH_3-\dot{C}H_2 + H-CH_2-\dot{C}H_2 \xrightarrow{\text{Disproportionation}} CH_3-CH_3 + CH_2{=}CH_2$

Ethane; Ethylene

Chapter 4

$$(iii)\quad C_2H_5-\overset{\overset{\displaystyle O}{\|}}{C}-\dot{O} + \dot{C}_2H_5 \longrightarrow C_2H_5-\overset{\overset{\displaystyle O}{\|}}{C}-OC_2H_5$$

Ethyl propanoate

Q.4. Out of isomeric pentanes, neopentane has the lowest boiling point but the highest melting point. Explain.

Ans. The b.pt. of an alkane depends upon its surface area. Since out of all the isomeric pentanes, neopentane has the lowest surface area (due to its spheroidal shape), therefore, it has the lowest b.pt.

The m.pt. on the other hand, depends upon the packing of the molecules in the crystal lattice. Since out of all the isomeric pentanes, neopentane is most symmetrical, therefore, it is most closely packed in the crystal lattice and hence has the highest m.pt.

Q.5. Although free radical chlorination of alkanes is not a good method for the laboratory preparation of alkyl chlorides yet neopentyl chloride is generally prepared by free radical chlorination of neopentane. Explain.

Ans. Free radical halogenation is not a suitable method for the laboratory preparation of alkyl chlorides because of the following two reasons:

(*i*) It gives a mixture of isomeric monochlorinated products whose boiling points are so close that they cannot be easily separated in the laboratory.

(*ii*) Polychlorination can also occur to some extent thereby making the mixture more complex and hence more difficult to separate.

Neopentyl chloride is generally prepared by free radical chlorination because neopentane has only one type of H-atoms and hence no isomeric products are formed. Further since free radicals do not rearrange, there is no danger of any rearrangement products being formed during the reaction.

$$CH_3-\underset{\underset{\displaystyle CH_3}{|}}{\overset{\overset{\displaystyle CH_3}{|}}{C}}-CH_3 \xrightarrow[-\,HCl]{\dot{C}l} CH_3-\underset{\underset{\displaystyle CH_3}{|}}{\overset{\overset{\displaystyle CH_3}{|}}{C}}-\dot{C}H_2 \xrightarrow[-\,\dot{C}l]{Cl_2} CH_3-\underset{\underset{\displaystyle CH_3}{|}}{\overset{\overset{\displaystyle CH_3}{|}}{C}}-Cl$$

Neopentane — Neopentyl free radical (does not rearrange) — Neopentyl chloride

Q.6. Bromination and not chlorination of alkanes is a usual synthetic method for preparation of alkyl halides. Explain.

Ans. Bromination of alkanes is more selective than chlorination. For example, chlorination of butane gives a mixture of 1-chlorobutane and 2-chlorobutane in the ratio 28 : 72 while bromination under similar conditions gives a mixture of 1-bromobutane and 2-bromobutane approx. in the ratio 2 : 98. In other words, the product of bromination of butane *i.e.*, 2-bromobutane is almost pure while the product of chlorination contains only 78% 2-chlorobutane. Therefore, bromination is preferred to chlorination for preparation of alkyl halides. The main reason of this selectivity is that the rates of abstraction of 2° H by Cl and Br atoms are in the ratio 3 : 82. Thus, the energy of activation for abstraction of a 2° H by $\dot{Br}$ atom is much lower than that by $\dot{Cl}$ atom.

Q.7. Though iodine-iodine bond is very weak, the iodination of alkanes does not readily take place.

OR

Explain why iodination of methane is much slower as compared to chlorination inspite of the fact that iodine-iodine bond is weaker than chlorine-chlorine bond.

Ans. Since, the bond dissociation energy of I—I bond is the lowest (151 kJ mol^{-1}) amongst halogens, iodine radicals are more readily available for iodination than bromine radicals for bromination and chlorine radicals for chlorination. But the rate-determining step of the reaction is not the availability of these free radicals but their attack, on the methane molecule to form the methyl radical.

$$\dot{I} + H—CH_3 \longrightarrow HI + \dot{C}H_3;\ \Delta H = 137.8 \text{ kJ mol}^{-1}$$

This step is most endothermic (+137.8 kJ mol^{-1}) in case of iodination as compared to bromination (+66.9 kJ mol^{-1}) and chlorination (+4.18 kJ mol^{-1}). Therefore, iodination does not occur readily even though the iodine radicals are easily available.

Q.8. Halogenation of alkanes in presence of tetramethyllead proceeds at a lower temperature than when it is done in its absence. Explain.

Ans. Halogenation of alkanes is usually carried out between 525–675 K since at this temperature range, halogens undergo homolytic fission to form halogen free radicals which are needed for halogenation.

But tetramethyl lead (TML) decomposes at a much lower temperature (*i.e.*, 415 K) to produce first methyl free radicals, which, in turn, react with halogens to from halogen free radicals.

$$(CH_3)_4\ Pb \xrightarrow{415\text{ K}} Pb + 4\dot{C}H_3\ ; \qquad \dot{C}H_3 + X—X \longrightarrow CH_3—X + \dot{X}$$

Since in presence of TML, halogen radicals are produced at a lower temperature, therefore, halogenation also occurs at a lower temperature.

Q.9. An equimolar mixture of neopentane and ethane yields neopentyl chloride and ethyl chloride in the ratio 2.3 : 1. How does the reactivity of a primary hydrogen in neopentane compare with that of primary hydrogen in ethane?

Ans. The reactions involved are:

$$\underset{\text{Ethane}}{CH_3-CH_3} + Cl_2 \longrightarrow \underset{\text{Ethyl chloride}}{CH_3-CH_2Cl} + HCl$$

$$\underset{\text{Neopentane}}{CH_3-\overset{\overset{CH_3}{|}}{\underset{\underset{CH_3}{|}}{C}}-CH_3} + Cl_2 \longrightarrow \underset{\text{Neopentyl chloride}}{CH_3-\overset{\overset{CH_3}{|}}{\underset{\underset{CH_3}{|}}{C}}-CH_2Cl} + HCl$$

No. of 1° H-atoms in ethane = 6

No. of 1°H-atoms in neopentane = 12

Ratio of $(CH_3)_3CCH_2Cl : CH_3CH_2Cl$ = 2.3 : 1(given)

The reactivity of their primary hydrogen atoms can be compared as follows:

$$\frac{\text{Neopentyl chloride}}{\text{Ethyl chloride}} = \frac{\text{No. of 1° H in neopentane}}{\text{No. of 1° H in ethane}} \times \frac{\text{Reactivity of 1° H in neopentane}}{\text{Reactivity of 1° H in ethane}}$$

$$\therefore \quad \frac{2.3}{1.0} = \frac{12}{6} \times \frac{\text{Reactivity of 1° H in neopentane}}{\text{Reactivity of 1° H in ethane}}$$

or $$\frac{\text{Reactivity of 1° H in neopentane}}{\text{Reactivity of 1° H in ethane}} = \frac{2.3}{1.0} \times \frac{6}{12} = 1.15 : 1.$$

Q.10. Tell the number and write the structures of all the haloalkanes that could be produced by the free radical chlorination of ethane.

Ans. Nine products are possible. These are:

Monochloro; **one,** CH_3CH_2Cl; Dichloro, **two,** CH_3CHCl_2, $CH_2Cl—CH_2Cl$

Trichloro, **two,** $CH_3—CCl_3$, $CH_2Cl—CHCl_2$

Tetrachloro, **two,** $CHCl_2—CHCl_2$, $CH_2Cl—CCl_3$

Pentachloro, **one,** $CCl_3—CHCl_2$; Hexachloro, **one,** $CCl_3—CCl_3$.

Q.11. The boiling points and melting points of three isomeric alkanes: 2, 2, 3, 3-tetramethylbutane, *n*-octane and 2, 2-dimethylhexane are given below. Match each compound with its boiling and melting point.

(*i*) Boiling points: 379.5 K, 298.7 K, 363 K. (*ii*) Melting points: 216.2 K, 154.7 K, 172.3 K.

Ans.	2, 2, 3, 3-Tetramethylbutane	2, 2-Dimethylhexane	*n*-Octane
B.pt.	363 K	379.5 K	398.7 K
M.pt.	172.3 K	154.7 K	216.2 K

Q.12. Of the three isomeric pentanes, which one do you think will produce four isomers of monochloro derivatives. Draw their structures.

Ans. 2-Methylbutane or isopentane gives four monochloro derivatives as shown below:

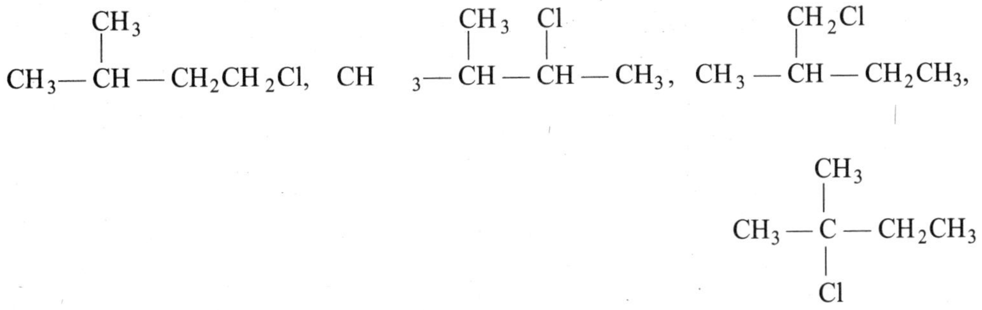

$$\begin{array}{l} \quad\;\; CH_3 \\ \quad\;\;\; | \\ CH_3-CH-CH_2CH_2Cl, \end{array} \quad \begin{array}{l} \quad\;\;\; CH_3 \;\; Cl \\ \quad\;\;\;\; | \quad\;\; | \\ CH_3-CH-CH-CH_3, \end{array} \quad \begin{array}{l} \quad\;\;\;\; CH_2Cl \\ \quad\;\;\;\;\; | \\ CH_3-CH-CH_2CH_3, \end{array}$$

$$\begin{array}{c} CH_3 \\ | \\ CH_3-C-CH_2CH_3 \\ | \\ Cl \end{array}$$

UNSOLVED PROBLEMS

1. (*i*) Write the structural formula for 2, 2, 3-trimethylhexane and indicate 1°, 2°, 3° and 4° carbon atoms in it.

(*ii*) Write IUPAC names for the following compounds:

(*a*)
$$\begin{array}{lccl} & & CH_3 & \\ & & | & \\ CH_3CH_2CH_2CH_2CH_2- & CH - & C - & CH_2CH_3 \\ & | & | & \\ & CH_3 & CH_2CH_2CH_3 & \end{array}$$

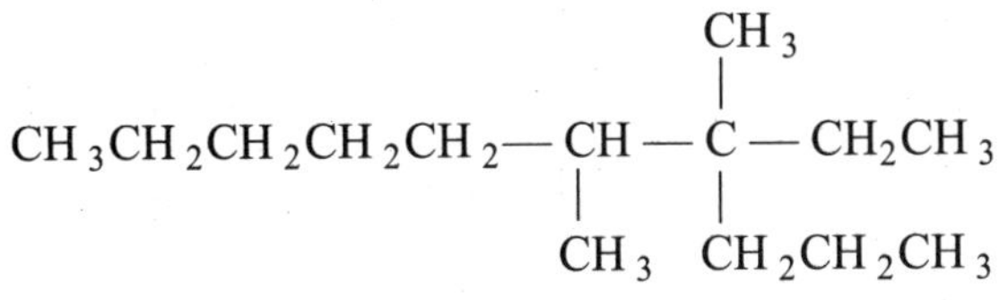

(*b*)
$$\begin{array}{ll} CH_3CH_2CH_2CH_2 & CH-CH_3 \\ & | \\ & CH_2 \\ & | \\ & CH_3 \end{array}$$

(*c*)
$$\begin{array}{lccccl} & & & CH_3 & CH_3 & \\ & & & | & | & \\ CH_3CH_2 & CH - & CH - & C - & CH - & CH_3 \\ & | & | & | & & \\ & CH_3 & CH_3 & CH_2CH_3 & & \end{array}$$

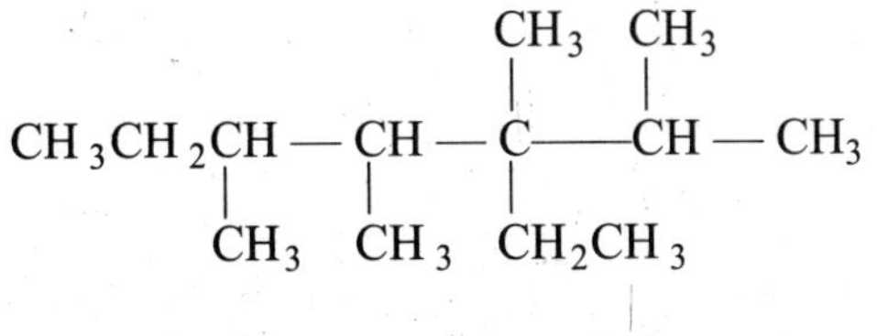

(*d*)
$$\begin{array}{lllccl} CH_3 & CHCH_2 & CHCH_2 & CH - & CH - & CH_3 \\ & | & | & | & | & \\ & CH_3 & CH_3 & CH_3 & CH_3 & \end{array}$$

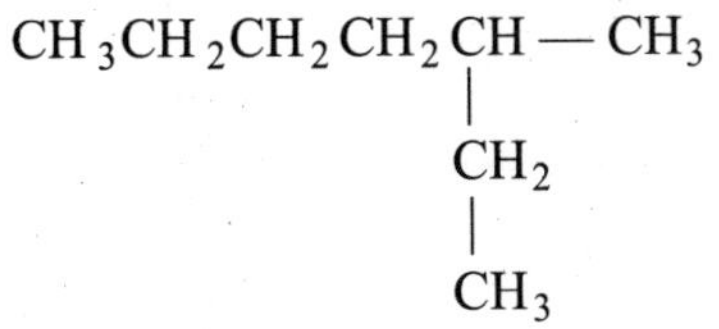

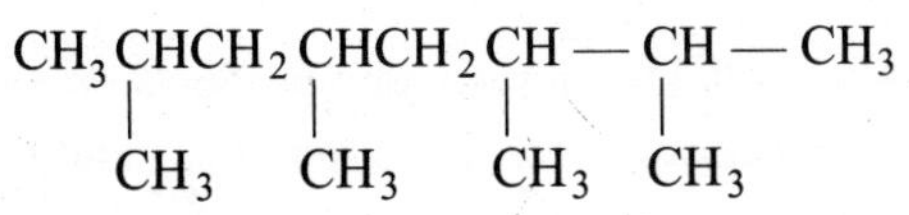

2. Write the structure and IUPAC name of an alkane or cycloalkane having the formula:

(*i*) C_6H_{12} (only secondary hydrogens) (*ii*) C_5H_{12} (only primary hydrogens)

(*iii*) C_5H_{10} (only secondary hydrogens) (*iv*) C_5H_{12}(only one tertiary hydrogen)

(*v*) C_6H_{14} (only primary and tertiary hydrogens).

3. The following names are incorrect. Write the correct names:

(*i*) 6-Methyloctane (*ii*) 3, 5, 6-Trimethyl-4-propylheptane

(*iii*) 2, 4, 5-Trimethylhexane (*iv*) 3-Propylpentane

(*v*) *t*-Butylmethane (*vi*) 3-Methyl-3-ethyl-4-methyldecane

(*vii*) 4, 5, 5-Trimethylhexane.

4. (*i*) Write structures of all the isomers of the following:

(*a*) Monochloro derivatives of pentane (normal isomer).

(*b*) Monochloro derivatives of 2, 2-dimethylbutane.

(*c*) Mono and dichloro derivatives of 2, 2-dimethylpropane.

(*ii*) How many isomers, regarded as derivatives of pentane, are possible for the formula C_8H_{18}? Give their structures and arrange them in order of increasing melting points and boiling points. Give reasen for your answer.

5. (*i*) Offer suitable explanation for the following:

(*a*) The boiling points of isomeric alkanes decrease with the increase in the branching.

(*b*) Alkanes with even number of carbon atoms melt at higher temperatures than those having odd number of carbon atoms.

(*c*) The b.pt. of *n*-alkanes increase as the molecular mass increases.

(*ii*) Select from each of the following sets the hydrocarbons having lowest and highest boiling point:

(*a*) 3, 3-Dimethylpentane, 2-methylhexane, *n*-heptane.

(*b*) *n*-Butane, *n*-hexane, *n*-pentane.

6. Discuss various methods of formation of alkanes with particular mention to the mechanisms of the reactions involved.

7. What is the role of light (or heat) in the halogenation of alkanes? Discuss the mechanism of the reaction with particular reference to the ease of substitution of different types of hydrogens.

8. What do you mean by orientation in halogenation of alkanes? On what factors does it depend?

9. Discuss briefly the relative reactivities of halogens in the halogenation of alkanes.

10. (*i*) Tell the number and write the structures of all the chloroalkanes that could be produced by the free radical chlorination of propane.

(*ii*) The free radical chlorination of methane yields four organic products. We wish to prepare only monochloromethane. Which of the following reaction conditions would be most suitable for this purpose:

(*a*) Take equimolar mixture of methane and chlorine.

(*b*) Take an excess of methane.

(*c*) Take an excess of chlorine.

11. Comment upon the following:

(*a*) Würtz reaction (*b*) Kolbe's electrolysis

(*c*) Corey-house reaction (*d*) Decarboxylation of monocarboxylic acids

(*e*) Reduction of alkyl halides by Zn / HCl.

12. (*a*) An alkane M.F. C_5H_{12} gives only one monochloro derivative. Write its structure.

(*b*) With the help of equations, depict various steps involved in the free radical chlorination of ethane.

(*c*) Why are alkanes less reactive in majority of organic reactions?

13. Halogenation of alkanes in presence of tetraethyllead proceeds at a much lower temperature than without it. Explain why it is so? What is orientation of halogenation?

14. Comment on the statement, 'Bromination but not chlorination of an alkane is a useful synthetic method for the preparation of alkyl halides. But both bromination and chlorination can be used to prepare neopentyl halide in the laboratory'.

15. Justify the following statements:

(*i*) Chlorination of alkanes is an exothermic reaction and yet it requires either high temperature or exposure to light for its success.

(*ii*) Iodination of methane does not take place at all under conditions under which chlorination and bromination take place in spite of the fact that iodine-iodine bond in iodine molecule is the weakest.

(*iii*) The iodination of alkanes is carried out in the presence of HIO_3 but chlorination and bromination is done as such.

16. How will you justify that the formation of H—X bond determines the rates of various halogenation processes in methane?

17. Discuss the relative rates of halogenation of alkanes in terms of bond dissociation energy.

18. In the chlorination and bromination of propane, the percentage of 2-haloderivatives are 56 and 97, respectively. Explain.

19. Bromine is less reactive but more selective whereas chlorine is more reactive but less selective. Explain.

CHAPTER **5**

Cycloalkanes

A large number of carbocyclic or homocyclic compounds, which are chemically similar to aliphatic compounds are known. They are called **alicyclic** (meaning aliphatic cyclic) compounds. The saturated hydrocarbons of this family are known as **cycloalkanes** or **cycloparaffins** and are designated by the general formula $C_n H_{2n}$. Hydrocarbons having medium size rings (C_5 or C_6) are quite stable and closely resemble alkanes. But the compounds having rings with carbon atoms fewer than five are specially interesting because their chemical properties are somewhat intermediate between alkanes and alkenes.

5.1 NOMENCLATURE

There are the following two systems of naming these compounds.

5.1.1 Common System

As the saturated alicyclic hydrocarbons have methylene units (CH_2) joined together in the form of a ring, these compounds are called polymethylenes. According to this system of nomenclature these compounds are named by putting a prefix such as tri, tetra, penta, etc., before the word methylene. Thus,

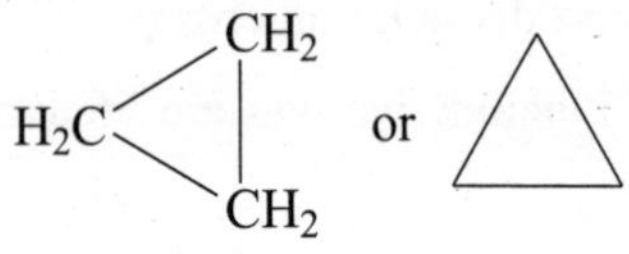

H_2C, CH_2, CH_2 (three-membered ring) or (triangle); $H_2C—CH_2$, $H_2C—CH_2$ (four-membered ring) or (square); CH_2, H_2C, CH_2, H_2C, CH_2, CH_2 (six-membered ring) or (hexagon)

Trimethylene 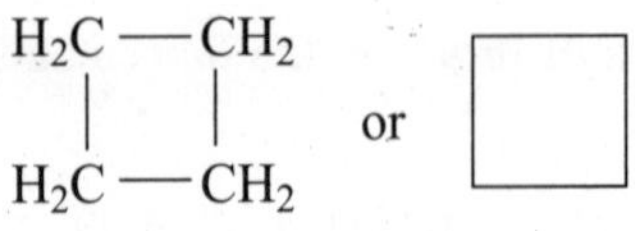 | Tetramethylene 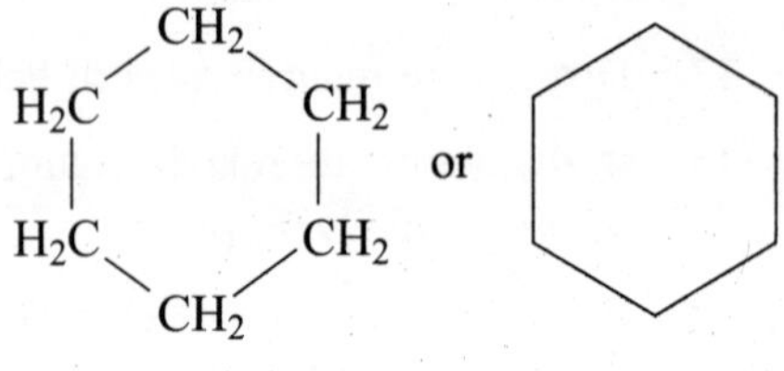 | Hexamethylene

It may be noted that for the sake of convenience the structures of these compounds are drawn as regular polygons whose each corner depicts a CH_2 group.

5.1.2 IUPAC System

(*a*) Parent Hydrocarbons

These compounds are regarded as alkanes whose carbon atoms are involved in forming a ring. Their names are derived by prefixing the term **cyclo**- to the name of the **alkanes** having the same number of carbon atoms as are present in the ring. Thus, they are called as **cycloalkanes** for example,

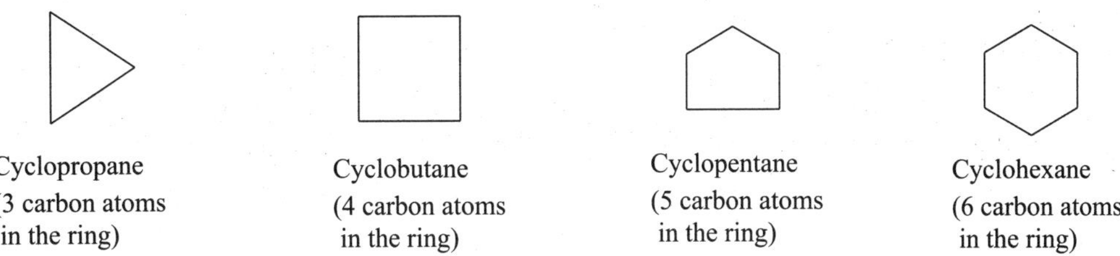

Cyclopropane (3 carbon atoms in the ring) | Cyclobutane (4 carbon atoms in the ring) | Cyclopentane (5 carbon atoms in the ring) | Cyclohexane (6 carbon atoms in the ring)

(*b*) Substituted Cycloalkanes

Rule for assigning IUPAC names to different substituted cycloalkanes vary with the number and types of the substituents present. Following examples are illustrative:

(*i*) **Monosubstituted cycloalkanes in which the substituent has fewer number of carbon atoms than those in the ring:** The cycloalkane is chosen as the **parent hydrocarbon.** The name of the substituent (also called as **side chain**) is prefixed to the name of the parent hydrocarbon which is written as one word. For instance,

CH_3 — Methylcyclohexane

CH_2CH_3 — Ethylcyclobutane

CH_2CH_3 — Ethylcyclohexane

Further, in cases where the side chain can have isomeric structures (*e.g.*, propyl onwards), carbon atoms of the side chain are numbered starting from that end which gives lowest number to the carbon atom which is attached to the ring. The name of the side chain is then written in the parenthesis along with the number of carbon atom through which it is attached to the ring. This name is then prefixed to the name of the parent cycloalkane. Here also the final name is written as one word as illustrated below:

$\overset{1}{C}H_2\overset{2}{C}H_2\overset{3}{C}H_2\overset{4}{C}H_3$

(1-Butyl) cyclopentane

$H_3\overset{1}{C}—\overset{2}{C}H—\overset{3}{C}H_2\overset{4}{C}H_3$

(2-Butyl) cyclopentane

***(ii)* Monosubstituted cycloalkanes in which the substituent has more number of carbon atoms than those in the ring:** The substituent side chain is chosen as the parent hydrocarbon (alkane) and the (cycloalkyl) ring is named as the substituent. Thus, the names of such hydrocarbons are written as **cycloalkylalkanes**. The position of the cycloalkyl group is shown in the alkanes. For example,

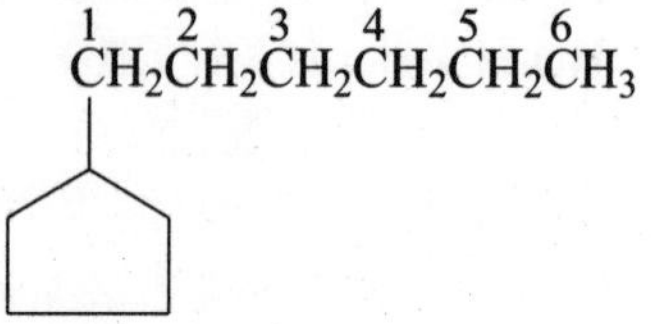

$\overset{1}{C}H_2\overset{2}{C}H_2\overset{3}{C}H_2\overset{4}{C}H_2\overset{5}{C}H_2\overset{6}{C}H_3$ (C-1 attached to cyclopentyl ring)

1-Cyclopentylhexane
(Ring has 5 carbons and side chain has 6 carbons)

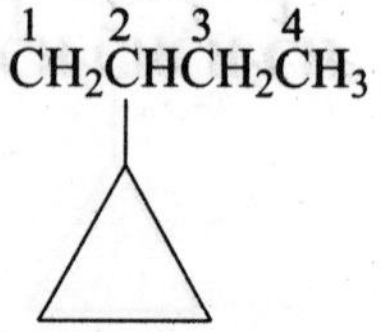

$\overset{1}{C}H_2\overset{2}{C}H\overset{3}{C}H_2\overset{4}{C}H_3$ (C-2 attached to cyclopropyl ring)

2-Cyclopropylbutane
(Ring has 3 carbons and side chain has 4 carbons)

***(iii)* Cycloalkanes having two or more substituents (side chains):** The names of the side chains, along with their positions are prefixed to the name of the parent compound in alphabetical order. The ring carbons are numbered 1, 2, 3, 4, ...etc., starting from the carbon atom to which substituent coming first in alphabetical order is attached and so on, keeping in mind the lowest locant rule. For example,

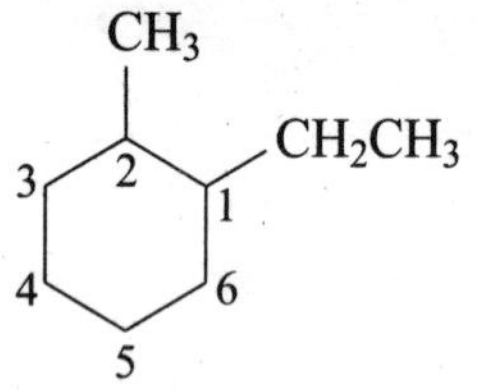

CH_3, CH_2CH_3 (ring positions 1–6)

1-Ethyl-2-methylcyclohexane

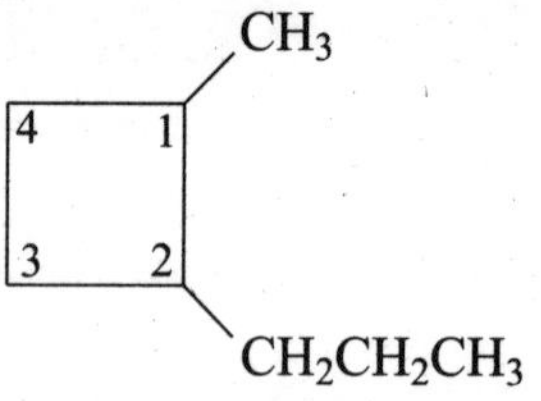

CH_3, $CH_2CH_2CH_3$ (ring positions 1–4)

1-Methyl-2-(1-propyl) cyclobutane

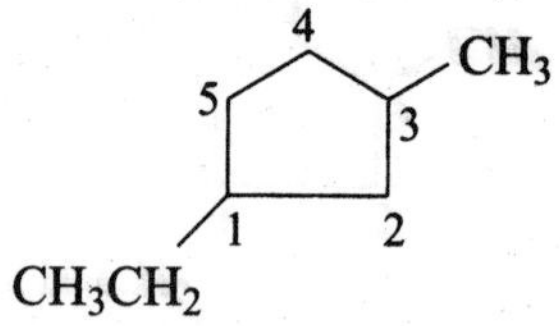

CH_3, CH_3CH_2 (ring positions 1–5)

1-Ethyl-3-methylcyclopentane

(Ethyl coming first in alphabetical order is shown at carbon-1 and methyl coming next at carbon -2)

Note: If the same side chain is repeated then the prefixes **di-, tri-, tetra-**, etc., are used before the name of the side chain along with the number of ring carbons to which they are attached. For example,

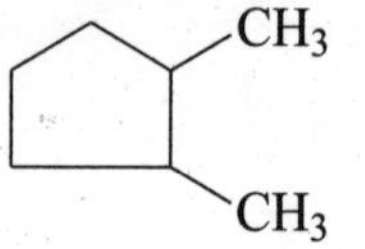

CH_3, CH_3

1, 2-Dimethylcyclopentane

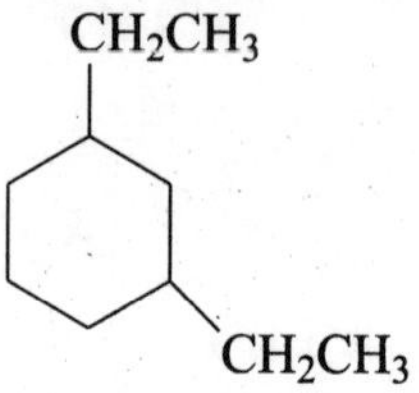

CH_2CH_3, CH_2CH_3

1, 3-Diethylcyclohexane

(*iv*) Cycloalkanes having unsaturated side chain and/or a functional group on the chain: The side chain is named as **"parent chain"** and cycloalkyl ring as the substituent. For example,

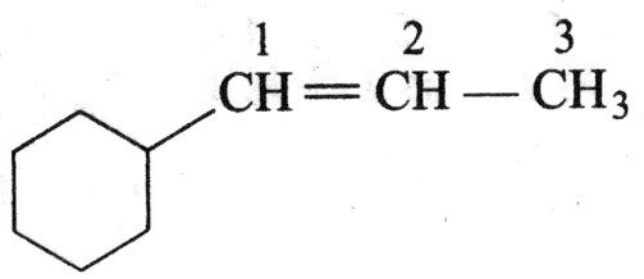

$C_6H_{11}-\overset{1}{C}H=\overset{2}{C}H-\overset{3}{C}H_3$

1-Cyclohexylpropene

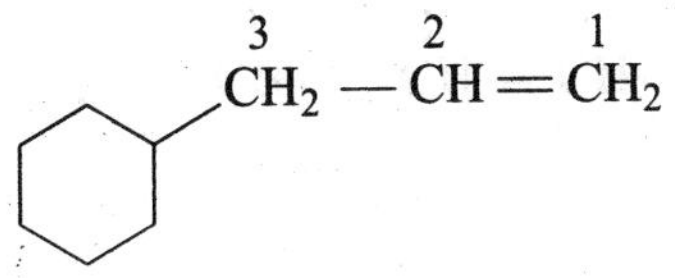

$C_6H_{11}-\overset{3}{C}H_2-\overset{2}{C}H=\overset{1}{C}H_2$

3-Cyclohexylpropene

$C_6H_{11}-\underset{4}{C}H=\underset{3}{C}H-\underset{2}{C}(=O)-\underset{1}{C}H_3$

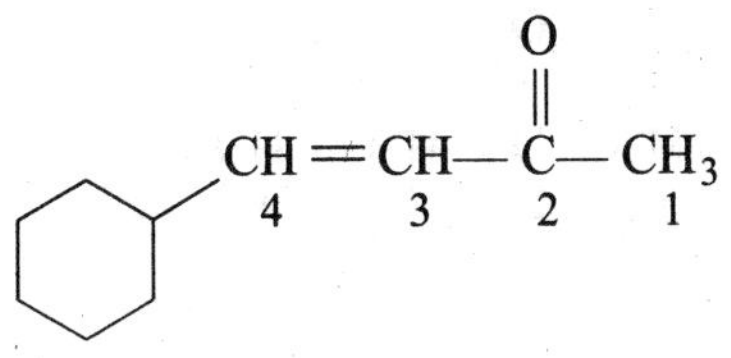

4-Cyclohexylbut-3-en-2-one

(*v*) Cycloalkanes having an alkyl side chain and a functional group or two simple functional groups: To assign position to such substituents, the ring carbons are numbered 1-, 2-, 3-, etc., starting from the carbon atom to which the functional group is attached (lowest number to carbon bearing the functional group which gets priority). For example,

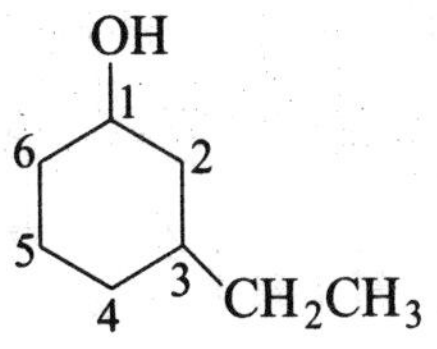

3-Ethylcyclohexan-1-ol

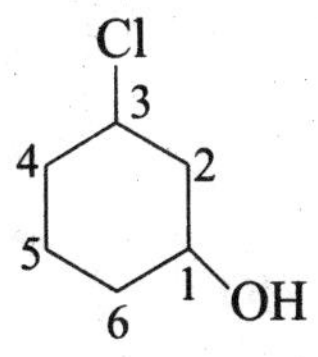

3-Chlorocyclohexan-1-ol

(*vi*) Cycloalkanes having same functional group attached to ring as well as to the side chain: The ring and the side chain are named as substituents depending upon the relative number of carbon atoms in them [Rule 2(*b*) (*i*) and 2(*b*) (*ii*)]. The name of the substituent is written in the parenthesis. For example,

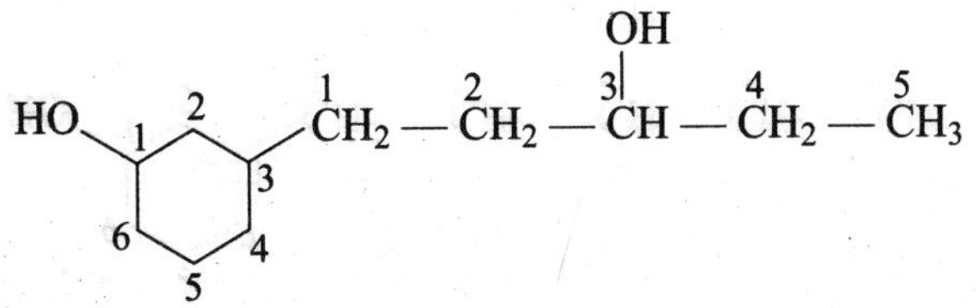

$\overset{1}{C}H_2-\overset{2}{C}H_2-\overset{3}{C}H(OH)-\overset{4}{C}H_2-\overset{5}{C}H_3$ (attached at C-3 of ring bearing HO at C-1)

3-(3-Hydroxypent-1-yl) cyclohexan-1-ol

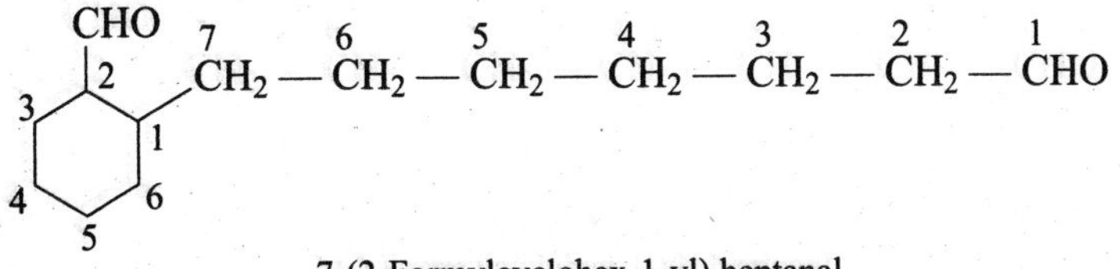

$\overset{7}{C}H_2-\overset{6}{C}H_2-\overset{5}{C}H_2-\overset{4}{C}H_2-\overset{3}{C}H_2-\overset{2}{C}H_2-\overset{1}{C}HO$ (attached at C-1 of ring bearing CHO at C-2)

7-(2-Formylcyclohex-1-yl) heptanal

Chapter 5

(*vii*) **Cycloalkanes having cycloalkane ring directly attached to an aromatic ring:** The cycloalkane ring is regarded as a substituent and the compound is named as a substituted benzene derivative. The carbon atoms of both the rings are numbered in such a way that number 1 is assigned to carbon atoms of both the rings which are directly attached to each other. For example,

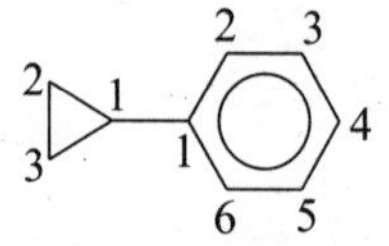

Cyclopropylbenzene

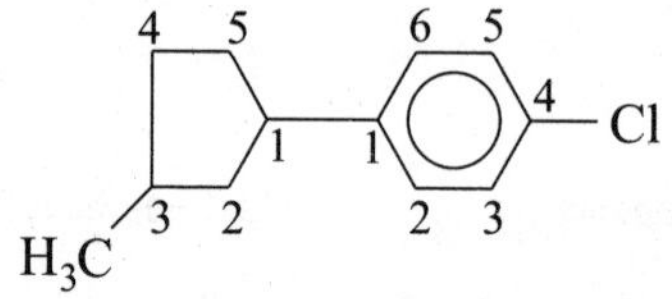

1-(3-Methylcyclopentyl)-4-chlorobenzene

(*viii*) **Cycloalkanes having more than one ring attached to the same carbon or a carbon chain:** The open carbon chain is regarded as **'parent chain'** and ring is regarded as the substituent. For example,

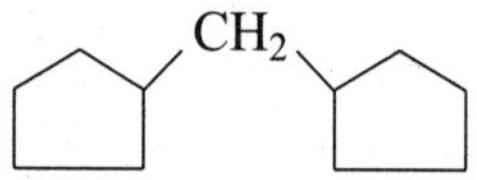

Dicyclopentylmethane

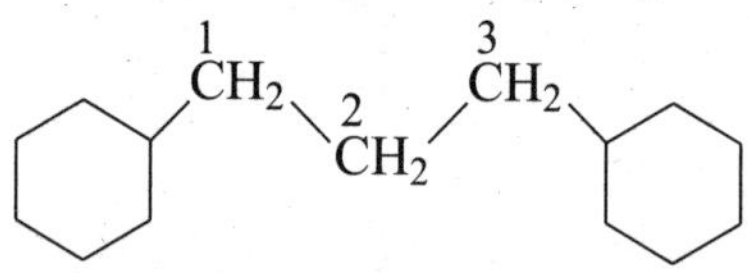

1, 3-Dicyclohexylpropane

(*ix*) **Cycloalkanes having functional groups other than those mentioned in rule 2 (b, v):** The ring carbons bearing these functional groups are numbered according to the sequence of priorities specified under IUPAC rules for naming poly functional aliphatic compounds (Chapter 4). Here the number of carbon atom to which the functional group of highest priority is attached is not shown in the name (therefore, it is always No.1). Names of these functional groups as prefixes and suffixes are listed in Table 5.1.

Table 5.1: Names of functional groups as prefixes and suffixes

S. No.	Functional group	Name as prefix	Name of suffix
1.	—CN	Cyano	Carbonitrile
2.	—COOR	Alkoxycarbonyl or carbalkoxy	Alkyl carboxylate
3.	—$CONH_2$	Carbamoyl	Carboxamide
4.	—CHO	Formyl	Carboxaldehyde
5.	—COOH	Carboxy	Carboxylic acid
6.	— COX (X = Cl, Br, I)	Halocarbonyl	Carbonyl halide

Some examples are:

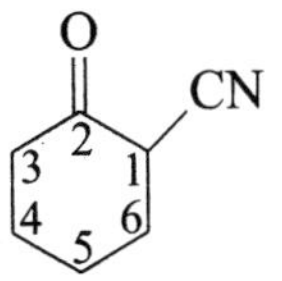

2-Oxocyclohexane - carbonitrile

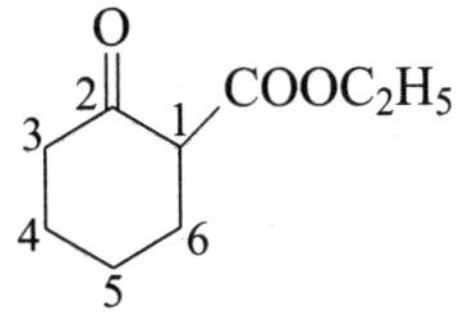

Ethyl (2-oxocyclohexan) carboxylate

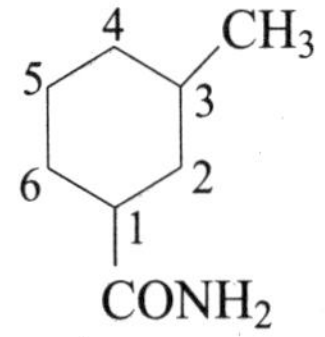

3-Methylcyclohexanecarboxamide

***(x)* Polycyclic compounds:** In addition to monocyclic compounds, a large number of polycyclic compounds are also known. They contain two or more rings having two or more carbon atoms in common. Systematic naming of compounds having more than one rings is done by putting the prefixes **bicyclo, tricyclo** etc., before the name of alkane denoting the total number of carbon atoms in the ring. The ring carbons are now numbered commencing with any one of the carbon atoms at the junction and is continued to the next junction first along the **longest path** (maximum number of carbon atoms), then along the shorter and shorter paths, successively, and finally along the shortest path. The number of carbon atoms involved in each of these 'paths' is written in a square bracket in descending order which is inserted in between the words **bicyclo** and **alkane**. The following examples are illustrative:

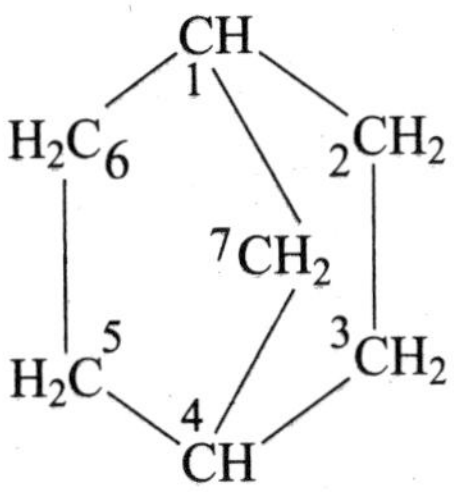

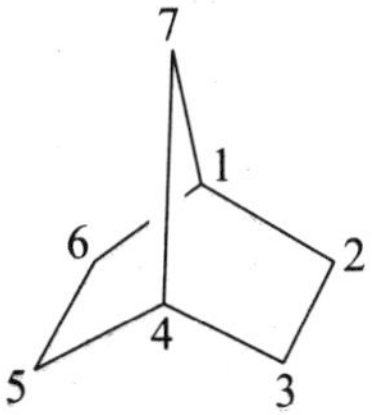

Bicyclo[2.2.1] heptane

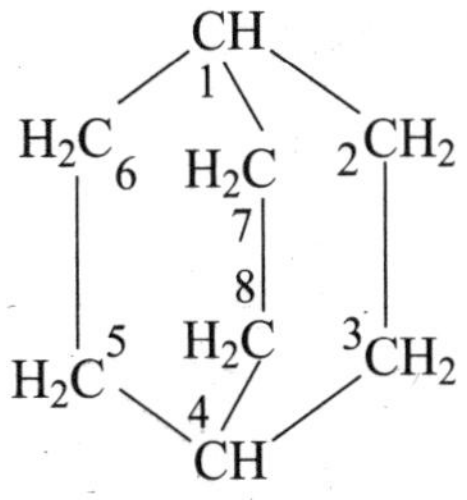

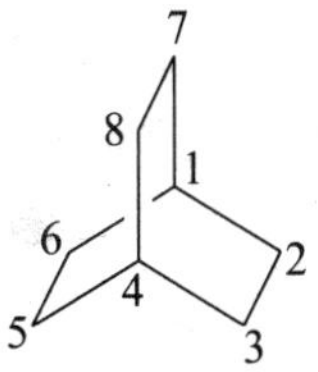

Bicyclo [2.2.2] octane

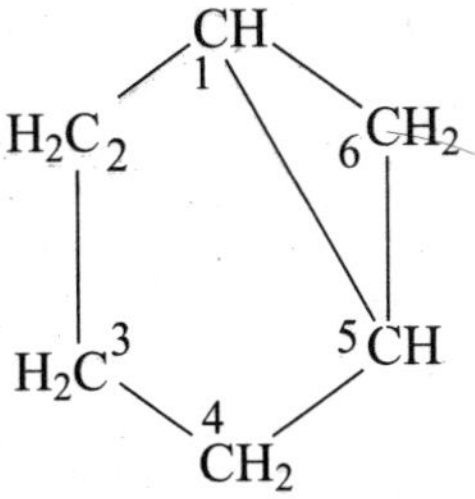

Bicyclo [3.1.0] hexane

The positions of the substituents are indicated by the number of carbon atoms to which they are attached for instance

2-Bromo-3-methylbicyclo [2.2.2] octane

Quite recently certain polycyclic hydrocarbons having fascinating structures have been synthesized. They are generally given common names, sometimes indicating their shapes, *e.g.*,

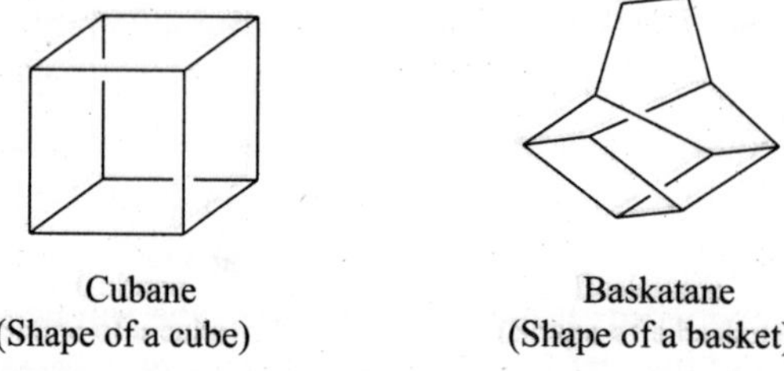

Cubane (Shape of a cube) Baskatane (Shape of a basket)

5.2 METHODS OF FORMATION

Cycloalkanes are synthesized from the aliphatic compounds by employing suitable reactions that bring about cyclization. As we shall see below, derivatives of cycloalkanes are produced in most of the reactions. For obtaining hydrocarbons, suitable reactions which replace these groups by hydrogen are employed. They are also prepared from aromatic compounds. Following are some of the more common methods used for the synthesis of these compounds.

***(i)* From aromatic compounds:** On catalytic hydrogenation aromatic compounds yield the corresponding alicyclic compounds, *e.g.*, cyclohexane from benzene.

$$\text{Benzene} + 3\,H_2 \xrightarrow{\text{Catalyst}} \text{Cyclohexane}$$

Phenol is almost quantitatively reduced to cyclohexanol which can be converted into cyclohexane by dehydration followed by the hydrogenation of the intermediate cyclohexene.

$$\text{Phenol} + 3\,H_2 \xrightarrow{\text{Catalyst}} \text{Cyclohexanol} \xrightarrow[-H_2O]{\overset{\ominus}{H}, \Delta} \text{Cyclohexene} \xrightarrow{H_2/\text{Catalyst}} \text{Cyclohexane}$$

Cyclohexanol Cyclohexene Cyclohexane

(*ii*) **From α, ω-dihaloalkanes**: *(a)* ***Freunds method***: Cycloalkanes are produced by treating appropriate dihaloalkanes with sodium or zinc. It can be regarded as **intramolecular Würtz reaction**. For example,

$$H_2C\begin{cases}CH_2Br\\ CH_2Br\end{cases} + Zn \text{ (or Na)} \longrightarrow H_2C\begin{cases}CH_2\\ |\\ CH_2\end{cases} + ZnBr_2$$

1, 3-Dibromopropane

$$H_2C\begin{cases}CH_2-CH_2Br\\ CH_2-CH_2\end{cases}\!\!\!\diagup CH_2Br + Zn \text{ (or Na)} \longrightarrow H_2C\begin{cases}CH_2-CH_2\\ CH_2-CH_2\end{cases}\!\!\!\diagup CH_2 + ZnBr_2$$

1, 6-Dibromohexane

If in a dihalide, X—$(CH_2)_n$—X, $n > 6$, the products of intermolecular Würtz reaction are obtained.

(*b*) A large number of alicyclic compounds have been synthesized by the action of disodiomalonic ester or disodioacetoacetic ester on dihaloalkanes as shown in the following reaction sequence:

$$H_2C\begin{cases}CH(COOEt)_2\\ CH(COOEt)_2\end{cases} \xrightarrow{Na} H_2C\begin{cases}\overset{\ominus}{C}(COOEt)_2\\ \overset{\ominus}{C}(COOEt)_2\end{cases} \xrightarrow{CH_2I_2} \begin{matrix}EtOOC & & COOEt\\ & C & \\ H_2C & & CH_2\\ & C & \\ EtOOC & & COOEt\end{matrix}$$

$$\xrightarrow[2.\,CO_2]{1.\,HCl} \begin{matrix} & COOH & \\ & | & \\ & CH & \\ H_2C & & CH_2\\ & CH & \\ & | & \\ & COOH & \end{matrix}$$

(*iii*) **From dicarboxylic acids** — ***Wislicenus method:*** Calcium, barium or thorium salts of dicarboxylic acids yield cyclic ketones on heating. These cyclic ketones are converted into cycloalkanes by **Clemmensen reduction** (Zn-Hg/HCl). For instance, cyclopentanone is obtained from barium adipate through the following reaction sequence:

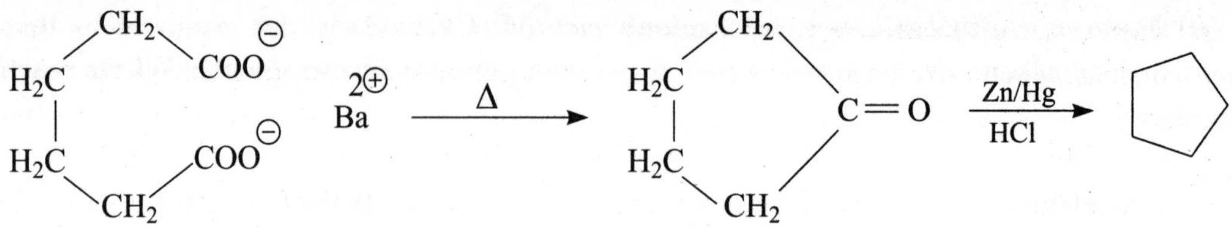

Macrocyclic rings having as many as 34 carbon atoms have been prepared by the pyrolysis of the thorium salts of the corresponding dicarboxylic acids.

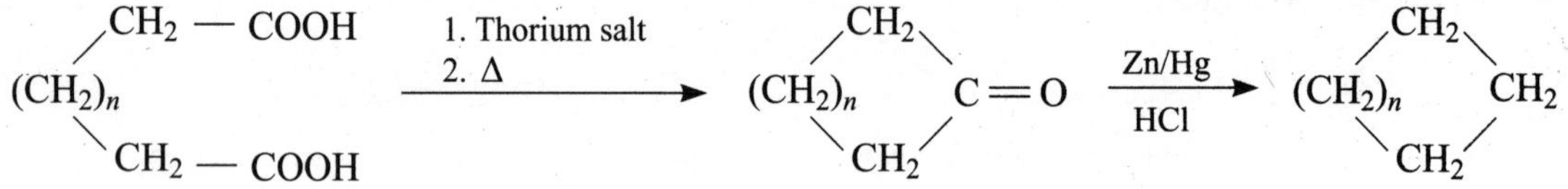

(n + 4) Carbon atoms

(n + 3) Carbon atoms

In accordance with the **Blanc's rule***, this method is suitable for the syntesis of 5-and 6-membered ring ketones.

Mechanism

The reaction takes place according to the following mechanistic scheme:

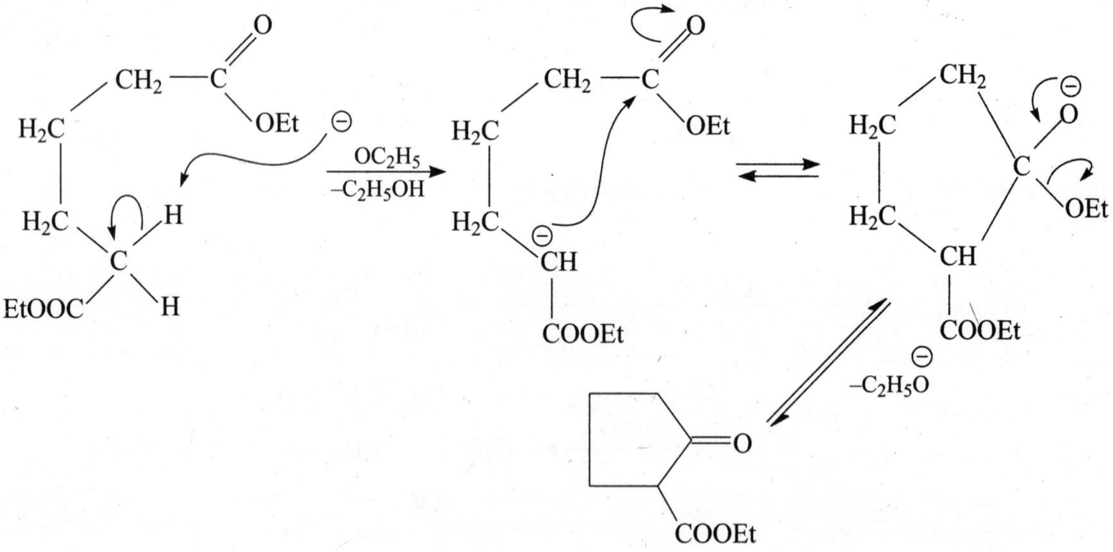

* Blanc's rule is an empirical rule based on the experimental observation of the ease of formation of cyclic anhydrides and ketones. It states that pyrolysis of 1,4- and 1,5-dicarboxylic acids yields cyclic anhydrides while 1,6 and 1,7-dicarboxylic acids afford cyclic ketones; the 1,8-and higher dicarboxylic acids remaining uneffected. This rule is useful in determining the size of the ring and also the chain length of the dicarboxylic acids. In the compounds of unknown ring size, a double bond is introduced by some known methods(s) followed by oxidative ring fission with the help of $KMnO_4$. This gives a dicarboxylic acid, the pyrolysis of which can tell us about the size of the ring.

(*iv*) **Dieckmann cyclization or condensation:** Esters of dicarboxylic acids undergo *intramolecular Claisen condensation* in presence of sodium dust or sodium ethoxide to give β-ketoesters. These β-ketoesters on subsequent hydrolysis form β-keto acids which undergo decarboxylation on heating to produce cycloalkanones.

These compounds can be converted into cycloalkanes by the method described above. Thus cyclopentane can be synthesized from diethyl adipate through the following sequence of reactions:

Diethyl adipate ($EtOOC-CH_2CH_2CH_2CH_2-COOEt$) $\xrightarrow[\text{Na or NaOEt}]{\text{Condensation}}$ Cyclopentanone-2-carboxylic ester $\xrightarrow[H_3O^{\oplus}]{\text{Hydrolysis}}$ Cyclopentanone-2-carboxylic acid $\xrightarrow[\Delta,\ -CO_2]{\text{Decarboxylation}}$ Cyclopentanone $\xrightarrow[\text{Reduction}]{\text{Zn/Hg, HCl}}$ Cyclopentane

Further, **intermolecular Claisen condensation** can also yield cyclic products by selection of appropriate reactants. For instance, condensation of oxalic ester with glutaric ester affords a cyclopentadione derivative which can be converted into cycloalkane by usual procedure.

$$\begin{matrix} COOEt \\ | \\ COOEt \end{matrix} + \begin{matrix} CH_2-COOEt \\ CH_2 \\ CH_2-COOEt \end{matrix} \xrightarrow{\text{Na or NaOEt}} \text{cyclopentane-1,2-dione-3,5-dicarboxylic ester (C(=O)–C(=O)–CH(COOEt)–CH}_2\text{–CH(COOEt))}$$

(*v*) **Thrope reaction —*Ziegler's method*:** This method is based on reaction of a dicyano (α, ω-dinitrile) compound with lithium diethylamide ($LiNEt_2$) in a large volume of benzene or toluene which produces cyclic imino compound. Hydrolysis of imine gives the corresponding ketone in good yield. The sequence of reaction is shown below:

Chapter 5

$(CH_2)_n(CH_2CN)_2 \xrightarrow[\text{Condensation}]{LiNEt_2}$ cyclic imine $(CH_2)_n$, $C{=}NH$, $CH{-}CN$ $\xrightarrow{\overset{\oplus}{H}/H_2O \text{ Hydrolysis}}$ [cyclic ketone $(CH_2)_n$, $C{=}O$, $CH{-}COOH$] $\xrightarrow[\text{Decarboxylation}]{-CO_2}$ cyclic ketone $(CH_2)_n$, $C{=}O$ $\xrightarrow[\text{HCl, Clemmensen reduction}]{Zn/Hg}$ cycloalkane $(CH_2)_n$, CH_2

The mechanism of the reaction is similar to that of Dieckmann cyclization, $LiNEt_2$ acting as a strong base. Ziegler Large Dilution Technique* is required to ensure intramolecular reaction instead of intermolecular reaction. At present this is the best method for preparing large macrocyclic compounds in good yield.

***(vi)* Acyloin condensation:** This consists in treating a dicarboxylic ester with metallic sodium in a large volume of benzene or toluene followed by protic solvents. Ziegler large dilution is required to ensure intramolecular condensation instead of intermolecular reaction. The mechanism of the reaction is as follows:

$(CH_2)_n(CH_2{-}COOEt)_2 \xrightarrow{2\dot{N}a}$ $(CH_2)_n(CH_2{-}\dot{C}(OEt){-}\overset{\ominus}{O}\overset{\oplus}{Na})_2 \xrightarrow{\text{Radical coupling}}$ $(CH_2)_n$ ring with $CH_2{-}C(OEt)(\overset{\ominus}{O}\overset{\oplus}{Na}){-}C(OEt)(\overset{\ominus}{O}\overset{\oplus}{Na}){-}CH_2$ $\xrightarrow{-2\,NaOEt}$ $(CH_2)_n(CH_2{-}C{=}O)_2$ $\xrightarrow{2\dot{N}a}$ $(CH_2)_n(CH_2{-}C(\overset{\ominus}{O}\overset{\oplus}{Na}){=})_2$ $\xrightarrow{\overset{\oplus}{H}}$ $(CH_2)_n(CH_2{-}C(OH){=})_2$ $\xrightarrow{\text{Keto enol Tautomerism}}$ $(CH_2)_n$ ring with $CH_2{-}C{=}O$, $CH_2{-}CH{-}OH$ $\xrightarrow[\text{Zn-Hg/HCl}]{\text{Reduction}}$ $(CH_2)_n$ ring with $CH_2{-}CH_2{-}CH_2{-}CH_2$

Cycloalkane with same number of carbon atoms

It may be noted that unlike other methods, this method gives the rings having the same number of carbon atoms as in the dicarboxylic ester.

*Ziegler large dilution means a very dilute solution of the dicyano compound in benzene is added to the solution of basic catalyst in large excess of ether.

***(vii)* Cycloaddition reactions:** A reaction, which involves addition of a system of $m\pi$-electrons to another system of $n\pi$-electrons to produce a ring compound, is called **cycloaddition**. Depending upon the number of π-electrons involved in the process, these reactions are designated as $[m + n]$ cycloaddition reactions. The number of π-electrons of the species involved in the reaction are written in a descending order as shown below: [2 + 2], [4 + 2], [6 + 4] cycloaddition reactions. An important characteristic of these reactions is that they all are concerted reactions *i.e.*, one step reactions involving *cyclic transition state**. Another notworthy feature of these reactions is that they may take place either under thermal or photochemical conditions and do not need any catalyst. Following cycloaddition reactions are usually employed for the synthesis of cycloalkanes.

***(a)* [2 + 2] *Cycloaddition reactions*:** A very important route for the synthesis of cycloalkanes is the addition of one alkene molecule to the other. In general these reactions take place under **photochemical** conditions and are also called **photodimerisations**. They proceed in a concerted manner involving four-membered cyclic transition state and produce cyclobutane derivatives. For example,

$$R_2C{=}CR_2 + R_2C{=}CR_2 \xrightarrow{h\nu} [\text{Transition state}] \longrightarrow \text{cyclobutane (R}_8\text{)}$$

Transition state

Addition of *cis*- and *trans*-alkenes to ethylene produces a mixture of two stereoisomeric cyclobutanes. For example, irradiation of *cis*- or *trans*-2-butene with ethylene produces a mixture of *cis*- and *trans*-cyclobutane derivatives in the same ratio.

$$H_2C{=}CH_2 + \text{cis-2-Butene (or trans-2-Butene)} \xrightarrow{h\nu} \text{cis-1,2-Dimethyl cyclobutane} + \text{trans-1,2-Dimethyl cyclobutane}$$

cis-2-Butene

or

trans-2-Butene

cis -1, 2-Dimethyl cyclobutane

trans -1, 2-Dimethyl cyclobutane

---

* The reactions proceeding through a cyclic transition state are called **pericyclic reactions**. Cycloaddition is, therefore, a type of pericyclic reactions.

This is due to two different modes of concerted cyclisation, viz., *suprafacial* and *antarafacial.* A detailed discussion of these modes is beyond the scope of this book. Here it may be worthwhile to mention that substituted olefins having electron-withdrawing groups such as —CN, —COR, —F, —COOR etc., undergo [2 + 2] cycloaddition under thermal conditions.* These reactions proceed *via* free radical mechanism involving a diradical intermediate. For example,

(*i*) $CF_2{=}CR_2 + CF_2{=}CR_2 \xrightarrow{\Delta} F_2C{-}CF_2$ ($R_2C^\bullet$ $^\bullet CR_2$) Diradical $\longrightarrow$ $F_2C{-}CF_2$ / $R_2C{-}CR_2$ (cyclobutane)

(*ii*) $F_2C{=}CR_2$ (R=Cl) $+ CH_2{=}CH{-}C_6H_5 \xrightarrow{\Delta} F_2C{-}CH_2$ ($R_2C^\bullet$ $^\bullet CH{-}C_6H_5$) $\longrightarrow$ $F_2C{-}CH_2$ / $R_2C{-}CH{-}C_6H_5$ (cyclobutane)

(*b*) **[4 + 2] Cycloaddition reactions—Diels—Alder reaction:** This reaction involves the addition of an olefin molecule [2π-electrons], called a dienophile, to a conjugated **diene** [4π-electrons]. The product formed is a cyclohexene derivative. For example, cyclohexene is formed by heating a mixture of butadiene and ethylene at 473 K.

$CH_2{=}CH{-}CH{=}CH_2$ (Diene) $+ CH_2{=}CH_2$ (Dienophile) $\xrightarrow{473\ K}$ Cyclohexene (*adduct*)

This reaction takes place readily if:

(*i*) Dienophile has an electron-withdrawing group.

(*ii*) Diene has an electron-releasing group.

Some examples are as follows:

(*i*) *Dienophile having an electron-withdrawing group*

1, 3-Butadiene + $CH_2{=}CH{-}CHO$ (Acrolein) $\xrightarrow{473\ K}$ Cyclohex-3-en-1-carbaldehyde

1, 3-Butadiene + Maleic anhydride $\xrightarrow{473\ K}$ 1, 2, 5, 6-Tetrahydrophthalic anhydride

*According to Woodward-Hoffmann rules for conservation of orbital symmetry, concerted [2 + 2] cycloadditions are symmetry allowed only under photochemical conditions. However, if any [2 + 2] cycloaddition reaction takes place under thermal condition it will be non-concerted and would involve diradical intermediates.

Tetracyanoethylene

It may be noted that tetracyanoethylene has been found to be the most reactive dienophile because of four strongly electron-withdrawing CN groups on dienophile.

(*ii*) *Diene having electron-donating group*

Isoprene + Dimethyl acetylene-dicarboxylate → Dimethyl 3, 6-dihydro-4-methylphthalate

Stereochemistry of Diels-Alder reaction: Diels-Alder reaction is stereospecific in nature, *i.e.*, with *cis*-dienophile we get *cis*-product and with *trans*-dienophile we get *trans*-product. This gives support to the fact that the reaction is concerted involving a six-membered cyclic transition state.

(*viii*) Addition of carbenes to olefins: The formation of cyclopropanes by the 1, 2-addition of carbenes to alkenes was first reported by **Doering** (1954). This process is now widely used for the synthesis of cyclopropanes. Some examples are given below:

(a) Addition of methylene to olefins: Photolysis or thermolysis of diazomethane in presence of an alkene produces cyclopropane derivatives according to the reaction sequence given below:

$$H_2\ddot{C}-\overset{\oplus}{N}\equiv N: \xrightarrow{h\nu \text{ or } \Delta} :CH_2 + :N\equiv N:$$

Diazomethane → Methylene

$$H_3C-CH=CH_2 + :CH_2 \longrightarrow H_3C-CH-CH_2 \text{ (ring closed by } CH_2\text{)}$$

Propene + Methylene → Methylcyclopropane

Methylene, being a highly reactive species, produces a mixture of products which are formed as a result of side reactions such as insertion, etc.

(b) Addition of dichlorocarbene to olefins: Dichlorocarbene, being more stable than methylene produces the corresponding cyclopropane derivatives. For example, when chloroform is treated with potassium *tert*-butoxide in presence of an alkene it produces a cyclopropane derivative in higher yield.

For example,

$$(H_3C)_3CO^{\ominus}K^{\oplus} + H—CCl_3 \underset{}{\overset{\text{Fast}}{\rightleftharpoons}} (H_3C)_3COH + {:}\overset{\ominus}{C}Cl_3$$

$${:}\overset{\ominus}{C}Cl_3 \xrightarrow{\text{Slow}} {:}CCl_2 + Cl^{\ominus}$$

Dichlorocarbene

$$\text{Cyclohexene} + {:}CCl_2 \longrightarrow$$

7, 7-Dichlorobicyclo [4.1.0] heptane

***Note*:** The formation of dichlorocarbene from chloroform is an example of α-elimination taking place by 1, 1-ElcB mechanism.

***(ix)* Simons-Smith procedure:** As discussed above free methylene is not very useful for addition to alkenes because it produces many side products. The Simons-Smith procedure gives good yields of cyclopropanes without a free carbene intermediate and without side reactions. In this procedure the alkene is treated with methylene iodide (CH_2I_2) and Zn-Cu couple*. The actual species which reacts with the olefin is an organozinc intermediate (I—CH_2—ZnI) called as a **carbenoid** or **Simons-Smith reagent**. The entire sequence along with the mechanism of reaction with a carbenoid is as follows:

$$CH_2I_2 + Zn \longrightarrow ICH_2—Zn—I$$

Carbenoid or Simon-Smith reagent

$$>C{=}C< + CH_2(I)(ZnI) \longrightarrow [\text{Transition state}] \longrightarrow \text{cyclopropane} + ZnI_2$$

Transition state

***(x)* Demjanov rearrangement:** Demjanov rearrangement, which in an example of carbocation rearrangement, has been used for converting cycloalkyl rings into larger (*Ring expansion*) or smaller (*ring contraction*) ones. For this purpose appropriate cycloalkyl amines are treated with nitrous acid. Whereas cycloalkylmethyl amines undergo ring expansion, the cycloalkyl amines undergo ring

---

* It is prepared by heating CuCl with Zn in ether in the atmosphere of nitrogen.

contraction. The reaction sequences for both the processes are discussed below:

1. *Ring expansion*: Cyclobutylmethyl amine upon treating with nitrous acid produces a mixture containing cyclobutylmethyl alcohol, methylenecyclobutane, cyclopentanol and cyclopentene.

CH_2NH_2 $\xrightarrow{HNO_2}$ CH_2 + CH_2OH

Cyclobutylmethyl amine — Methylenecyclobutane (A) — Cyclobutylmethyl alcohol (B)

OH +

Cyclopentanol (C) — Cyclopentene (D)

The last two products are formed as a result of ring expansion. The mechanism of this reaction is shown in the following scheme:

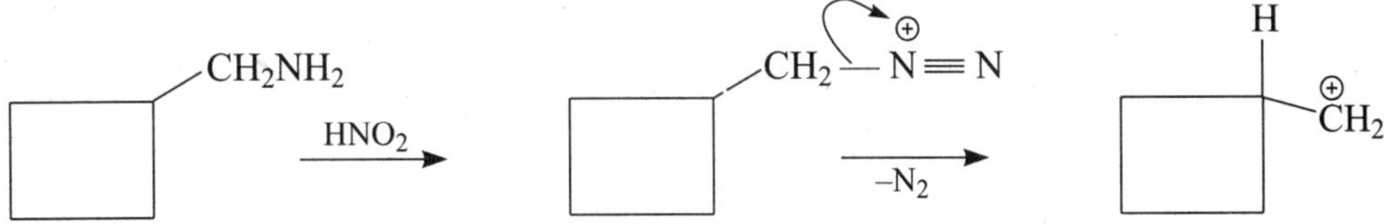

Cyclobutylmethyl cation

The cyclobutylmethyl cation undergoes the following reactions:

(*a*) *Loss of proton to produce methylene cyclobutane.*

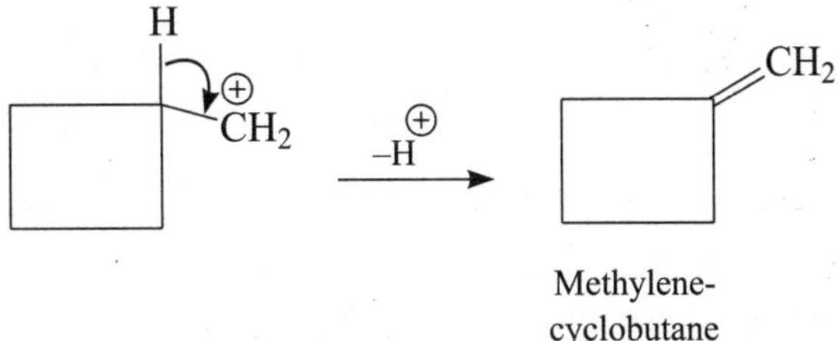

Methylene-
cyclobutane

(*b*) *Nucleophilic addition of water followed by the loss of proton to produce cyclobutylmethyl alcohol.*

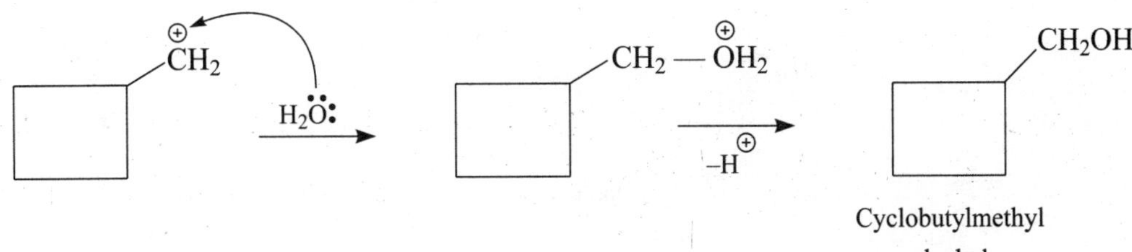

Cyclobutylmethyl
alcohol

(*c*) *Rearrangement to produce cyclopentyl carbocation.*

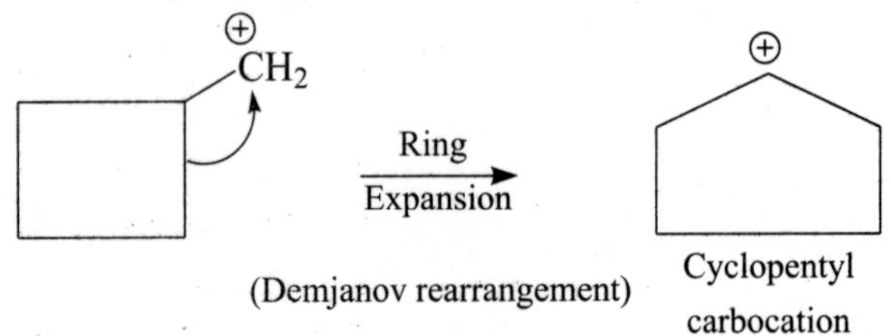

Cyclopentyl carbocation

This carbocation may then undergo following reactions:

(*i*) *Loss of proton to give cyclopentene.*

$$\xrightarrow{-H^{\oplus}}$$

(*ii*) *Nuclecophilic addition of water followed by loss of proton to produce cyclopentanol.*

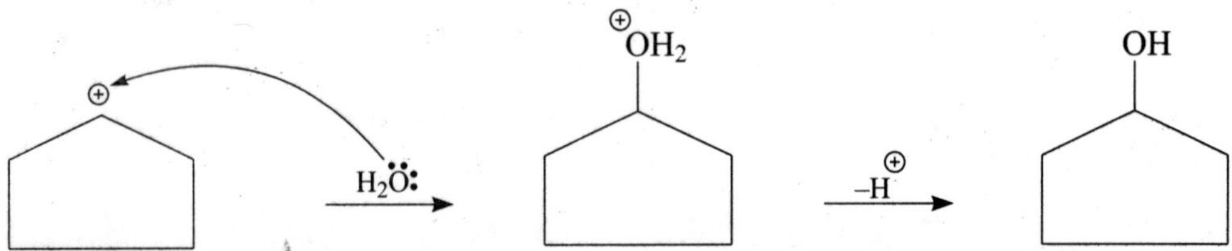

2. *Ring contraction*: Cyclobutylamine upon treatment with nitrous acid produces a mixture containing cyclobutyl alcohol and cyclopropylmethyl alcohol.

$$\text{Cyclobutylamine (}NH_2\text{)} \xrightarrow{HNO_2} \text{Cyclobutyl alcohol (OH)} + \text{Cyclopropyl-}CH_2OH$$

Cyclobutyl alcohol

Cyclopropyl-methyl alcohol

Here cyclopropylmethyl alcohol is produced as a result of ring contraction. The mechanism of the reaction is shown in the following scheme:

NH_2 $\xrightarrow{HNO_2}$ $\overset{\oplus}{N}\equiv N$ $\longrightarrow$ $\oplus$ + $N\equiv N$

Cyclobutyl cation

The cyclobutyl cation formed in the reaction may undergo the following reactions:

(*i*) *Nucleophilic addition of water followed by the loss of proton to produce cyclobutyl alcohol.*

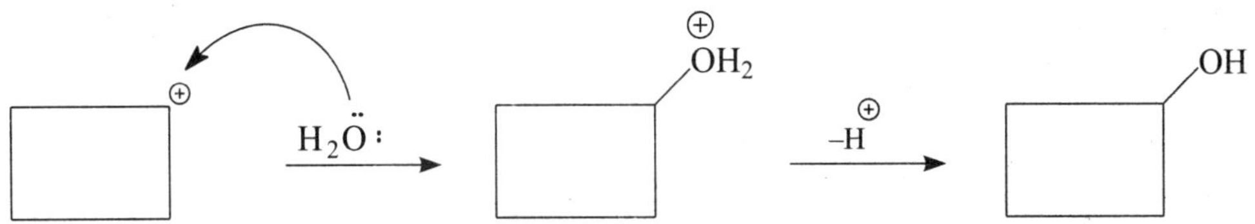

(*ii*) *Demjanov rearrangement to produce ring contracted cyclopropylmethyl cation* which undergoes nucleophilic addition of water followed by loss of a proton to give cyclopropylmethyl alcohol.

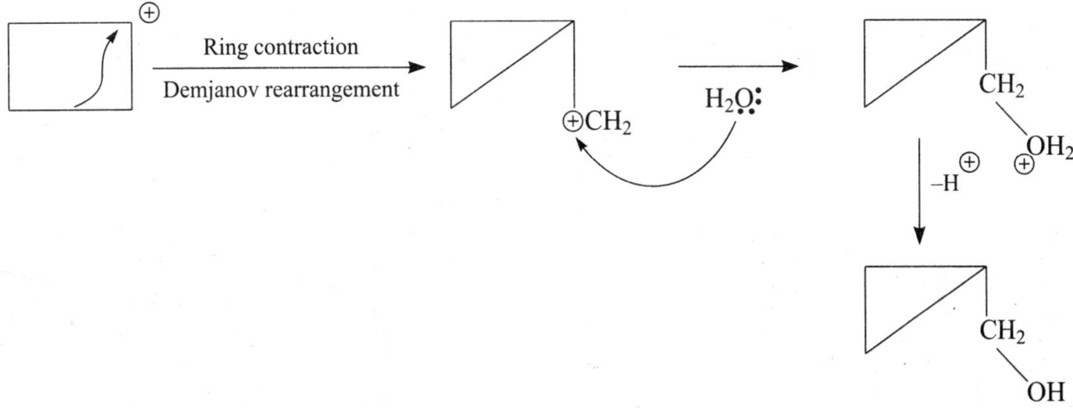

Note**:** The cyclobutyl and cyclopropyl methyl cations do not eliminate protons because the expected products cyclobutene and methylenecyclopropane are quite unstable due to strain.

(*xi*) From active methylene compounds: Active methylene compounds such as malonic ester and acetoacetic ester upon treatment with suitable dihaloalkane in presence of a base produce cycloalkyl carboxylic esters which are generally hydrolyzed. The acids so produced are decarboxylated to give cycloalkanes. However, as shown in the following sequence of reactions, cyclobutane carboxylic acid cannot be directly decarboxylated because on heating the ring undergoes fission. Hence a multistep reaction sequence is employed. Various steps involved in the synthesis of cyclobutane from malonic ester and cyclopentane from acetoacetic ester are shown below:

(*i*) *Cyclobutane from malonic ester.*

$$\overset{\ominus\ \oplus}{C_2H_5ONa} + H{-}CH(COOC_2H_5)_2 \longrightarrow \overset{\oplus}{Na}\ \overset{\ominus}{:CH}(COOC_2H_5)_2 + C_2H_5OH$$

Sodiumethoxide; Malonic ester; Monosodium salt of malonic ester

$$H_2C(CH_2Br)_2 + \overset{\oplus}{Na}\ \overset{\ominus}{CH}(COOC_2H_5)_2 \xrightarrow{-NaBr} BrCH_2{-}CH_2{-}CH_2{-}CH(COOC_2H_5)_2$$

$$\xrightarrow[-C_2H_5OH]{\overset{\oplus}{Na}\overset{\ominus}{O}C_2H_5} BrCH_2{-}CH_2{-}CH_2{-}\overset{\ominus}{C}(COOC_2H_5)_2 \longrightarrow \text{cyclo-}C_4H_6(COOC_2H_5)_2 \xrightarrow{\text{Hydrolysis}} \text{cyclo-}C_4H_6(COOH)_2$$

$$\xrightarrow[\Delta]{-CO_2} \underset{\text{Cyclobutane carboxylic acid}}{\text{cyclo-}C_4H_7{-}COOH} \xrightarrow{PCl_5} \underset{\text{Cyclobutane carboxy chloride}}{\text{cyclo-}C_4H_7{-}COCl} \xrightarrow{NH_3} \underset{\text{Cyclobutane carboxamide}}{\text{cyclo-}C_4H_7{-}CONH_2}$$

$$\xrightarrow[\text{Hoffmann bromamide degradation}]{Br_2/KOH} \text{cyclo-}C_4H_7{-}NH_2 \xrightarrow[\text{Hoffmann exhaustive methylation}]{CH_3I\ (\text{Excess})} \text{cyclo-}C_4H_7{-}\overset{\oplus}{N}(CH_3)_3\ \overset{\ominus}{HO}$$

$$\xrightarrow[AgOH]{\Delta} \text{Cyclobutene } (H_2C{-}CH{=}CH{-}CH_2) \xrightarrow[H_2/Ni]{373\ K} \text{Cyclobutane}$$

The above procedure is called as **Perkin's method** and can be used for the preparation of cyclopropane as well as cyclopentane using 1, 2-dibromoethane, 1, 4-dibromobutane, etc.

(*ii*) *Cyclopentane from acetoacetic ester.*

$$H_3C-CO-CH_2-COOC_2H_5 + \overset{\oplus}{Na}\overset{\ominus}{O}C_2H_5 \xrightarrow{-C_2H_5OH} H_3C-CO-\overset{\ominus}{C}H\overset{\oplus}{Na}-COOC_2H_5$$

Monosodium salt of ethyl acetoacetate

$$H_3C-CO-\overset{\ominus}{C}H\overset{\oplus}{Na}-COOC_2H_5 + BrCH_2-CH_2-CH_2-CH_2-Br \xrightarrow{-NaBr} (H_3COC)(H_5C_2OOC)CH-CH_2-CH_2-CH_2-CH_2-Br$$

$$\xrightarrow[-C_2H_5OH]{\overset{\oplus}{Na}\overset{\ominus}{O}C_2H_5} (COOH_3)(COOH_2H_5)\overset{\ominus}{C}-CH_2-CH_2-CH_2-CH_2-Br \longrightarrow \text{cyclo-}(H_2C-CH_2-CH_2-CH_2)C(COCH_3)(COOH_2H_5)$$

$$\xrightarrow[(ii)\ \overset{\oplus}{H}/H_2O]{(i)\ \text{dil. KOH}} \text{cyclo-}(H_2C-CH_2-CH_2-CH_2)C(COCH_3)(COOH) \xrightarrow[-CO_2]{\Delta} \text{cyclo-}(H_2C-CH_2-CH_2-CH_2)CH-COCH_3$$

$$\xrightarrow[\text{Clemmensen reaction}]{\text{Zn-Hg / Conc. HCl}} \text{cyclo-}(H_2C-CH_2-CH_2-CH_2)CHCH_2-CH_3$$

Ethylcyclopentane

1. conc. KOH
2. $\overset{\oplus}{H}$ / H_2O
Acidic hydrolysis

$$\text{cyclo-}(H_2C-CH_2-CH_2-CH_2)C(COCH_3)(COOH_2H_5) \longrightarrow \text{cyclo-}(H_2C-CH_2-CH_2-CH_2)C(COCH)(COOH) \xrightarrow[\Delta]{-CO_2} \text{cyclo-}(H_2C-CH_2-CH_2-CH_2)CHCOOH$$

Cyclopentane carboxylic acid

5.3 PHYSICAL PROPERTIES

With a few exceptions, the boiling and melting points of cycloalkanes show some gradation (Table 5.2). A comparison of melting and boiling points of cycloalkanes and alkanes (Table 4.1) reveals that former compounds melt and boil at higher temperatures than the latter ones. This may probably be due to the fact that zig-zag alkane molecules are constantly fluttering and this makes it difficult for the molecules, in the solid state, to be closely packed thereby lowering their melting points as compared to cycloalkanes where the more rigid ring structure prevents this fluttering to a great extent. Similarly, this fluttering makes the alkane molecules, in the liquid state, more hostile towards their neighbours. In other words, the intermolecular attractions in alkanes are less than those in corresponding cycloalkanes hence lower boiling points of the former.

Table 5.2: Boiling and melting points of cycloalkanes

Compound	Boiling point (K)	Melting point (K)	Density (g/ml) at 293 K
Cyclopropane	240	154	—
Cyclobutane	286	193	0.746
Cyclopentane	322.5	179	0.746
Cyclohexane	354	279.4	0.778
Cycloheptane	390	271	0.810
Cyclooctane	420	287	0.830

5.4 CHEMICAL REACTIONS

Although their general formula, C_nH_{2n} suggests that they should be unsaturated, cycloalkanes closely resemble alkanes in chemical properties. Their similarity to alkanes can be attributed to the fact that all the carbon atoms of these compounds are sp^3-hybridized forming σ-bonds with two neighbouring carbons and two hydrogens. Also it has been shown that cycloalkanes from cyclopentane onwards show high degree of stability and remain uneffected by acids, alkalis and common oxidizing agents, such as $KMnO_4$, under usual conditions. They, however, undergo free radical substitution reactions. But cyclopropane and cyclobutane are exceptions as they show a high degree of reactivity, cyclopropane being particularly reactive. Under suitable conditions both of them undergo free radical substitutions (like alkanes) and also undergo addition reactions (like alkenes) involving ring fission to form open chain saturated compounds. Some reactions are:

***(i)* Free radical substitution (*halogenation*):** Halogenation takes place in the presence of sunlight at room temperature or on heating up to 525–575 K in dark.

$$\underset{\text{(cyclopropane: } H_2C—CH_2,\ CH_2)}{} + Cl_2 \xrightarrow{h\nu} \underset{\text{(chlorocyclopropane: } H_2C—CH—Cl,\ CH_2)}{} + HCl$$

$$\text{cyclopentane} + Br_2 \xrightarrow{h\nu} \text{bromocyclopentane (ring–Br)} + HBr$$

***(ii)* Addition reactions:** Due to high strain involved (See Sections 5.5 and 5.6) the cyclopropane ring behaves like a double bond. Therefore, like alkenes cyclopropane undergoes addition reactions.

Some of the examples are:

$$\underset{\text{cyclopropane}}{H_2C{-}CH_2{-}CH_2} \xrightarrow{Br_2,\ FeBr_3} BrH_2C{-}CH_2{-}CH_2Br$$

$$\xrightarrow{HBr} H_3C{-}CH_2{-}CH_2Br$$

$$\xrightarrow{(i)\ \text{conc. } H_2SO_4\ (ii)\ H_2O} H_3C{-}CH_2{-}CH_2OH$$

$$\xrightarrow{H_2/Ni,\ 353\ K} H_3C{-}CH_2{-}CH_3$$

However, unlike alkenes, cyclopropane is not hydroxylated with aqueous solution of $KMnO_4$ and also does not add ozone.

Cyclobutane is not as reactive as cyclopropane as it does not undergo addition reactions under ordinary conditions. However, it can be hydrogenated to *n*-butane at higher temperature.

$$\text{(cyclobutane) } CH_2{-}CH_2{-}CH_2{-}CH_2 \text{ (ring)} + H_2 \xrightarrow{Ni,\ 393\ K} CH_3{-}CH_2{-}CH_2{-}CH_3$$

Thus cyclopropane is less reactive than alkenes and cyclobutane is still less reactive than cyclopropane, while cyclopentane and higher homologues are much less reactive than these two. Now question arises how to account for this special behaviour *(less stability)* of three and four membered ring compounds?

5.5 BAEYER'S STRAIN THEORY

Adolf von Baeyer (1885) proposed his famous *Baeyer's strain theory* in order to account for the above behaviour of three and four membered ring compounds *vis-à-vis* higher cycloalkanes. The main assumptions of Baeyer strain theory, which is based on a purely mechanical concept for bonding, are as follows:

(*i*) Since carbon atom is tetrahedral in nature with all the four valencies directed towards four corners of a regular tetrahedron, the angle between any two bonds should be 109.5°. *Any deviation from this angle would result in internal strain in the molecule.* Thus a double bond in an olefin causes considerable deviation from the tetrahedral angle (if we regard the angle between two carbon-carbon valencies of a double bond to be 0°) and consequently the molecule is highly strained and hence the olefins are highly reactive.

(*ii*) All the carbon atoms constituting a cycloalkane ring are *planar* meaning thereby that cyclopropane ring is an *equilateral triangle;* cyclobutane ring is a *square* and other cycloalkane rings are *regular polygons.*

(*iii*) When these molecules assume the above mentioned shapes there is deviation from the normal tetrahedral angle causing a lot of strain in the molecule. This strain, commonly referred to as *angle strain*, would depend upon the extent of such deviation.

(*iv*) *Greater the deviation of bond angle from 109.5° the greater is the strain and consequently more unstable* (*or reactive*) *is the ring*. Thus, the bond angles of cyclopropane and cyclobutane must be compressed to 60° and 90°, respectively.

(*v*) *The greater the stability of a ring, more easily it is formed.*

Baeyer showed that for cycloalkanes these deviations (listed in Table 5.3) from the normal tetrahedral angle can be calculated as follows:

$$d = \frac{1}{2}(109.5 - \text{bond angle in the planar ring})$$

(deviation from 109.5° for one bond)

Table 5.3: Deviations from the normal tetrahedral angle

Compound and its geometry	Bond angle in planar ring	Deviation
Cyclopropane (equilateral triangle)	60°	½(109.5–60) = 24.75°
Cyclobutane (square planar)	90°	½(109.5–90) = 9.75°
Cyclopentane (regular pentagon)	108°	½(109.5–108) = 0.75°
Cyclohexane (regular hexagon)	120°	½(109.5–120) = –5.25°
Cycloheptane (regular heptagon)	128.5°	½(109.5–128.5) = –9.5°
Cyclooctane (regular octagon)	135°	½(109.5–135) = –12.75°
Alkanes (chain)	109.5°	½(109.5–109.5) = 0°

The positive values of deviation mean that the bond angles are compressed while negative values mean expansion of the bond from the tetrahedral angle. In both the situations, the molecule would be strained. From the Table 5.3 it is clear that the deviation from the normal tetrahedral angle falls rapidly in going from C_3 to C_5 ring system and then they rise slowly. This is in agreement with the fact that cyclopropane, being more strained, is more reactive than cyclobutane. As the deviation in cyclopentane is the least it should be expected to be quite stable. This is also in agreement with the observed facts. Baeyer predicated, of course wrongly, that cyclohexane should be less stable than cyclopentane (because of increase in deviation) and that rings larger than C_7 should not exist and if at all they exist, they should be highly strained. This was partly because of the fact that such compounds were not synthesized at that time.

5.5.1 Some Observations Supporting the Baeyer Strain Theory

1. Reactivity of the Double Bond

A carbon-carbon double bond can be regarded as a two-membered ring with a carbon-carbon bond angle of 0°. The angle strain in this case = ½ [109.5° – 0°] = +54.75°. This angle strain is the highest compared to that of all the cycloalkanes. Therefore, it can be said that *alkenes should be highly reactive*. This is known to be true.

2. Reactivity of Cyclopropane and Cyclobutane

The heats of combustion of cyclopropane and cyclobutane show that they possess more energy (38.5 kJ and 27.6 kJ, respectively) per CH_2 group than the corresponding straight chain alkanes. Hence, they should be less stable than the corresponding *n*-alkanes —a fact in agreement with Baeyer's strain theory. It can, therefore, be said that their tendency to undergo addition reactions through ring fission is related to this instability.

5.5.2 Limitation of Baeyer Strain Theory

Baeyer strain theory successfully accounts for the relative stabilities of small ring compounds (cyclopropane and cyclobutane). But it fails to explain the relative stabilities of cyclopentane onwards. For example, it predicts (**incorrectly**) that cyclopentane (*lowest value for angle of deviation*) to be more stable than cyclohexane (*having higher value of the angle of deviation than cyclopentane*). Further, according to this theory molecules such as cycloheptane onwards should be more strained due to larger degree of angle of deviation. In other words, rings smaller and larger than cyclopentane should be strained and hence less stable. However, the observed stabilities of cycloalkanes do not fully agree with the prediction of this theory.

Some of the anomalies in the Bacyer strain theory become clear by using the heat of combustion data (Table 5.4) rather than angle of deviation for comparing their relative energies of this class of compounds. The most important figure in the table is heat of combustion per CH_2 unit which is obtained by dividing the heat of combustion of the molecule by number of carbon atoms in the ring.

The heat of combustion of alkanes have been found to be almost the same, *i.e.*, 657.5 kJ/CH_2 unit and has been taken as a standard for comparing the relative stability of cycloalkanes. Thus if a cycloalkane has higher value of heat of combustion per CH_2 unit it should be less stable.

With this background now we will examine some of the limitations of Baeyer's strain theory and try to explain the observed facts on the basis of heat of combustion data.

1. According to Baeyer's strain theory cyclopentane with least angle of deviation (+ 0.75°) should be more stable than cyclohexane and cyclocheptane having more angles of deviation *i.e.*, –5.25° and –9.5°, respectively. But their heats of combustion per CH_2 unit are in the range comparable to those of alkanes. Hence, these molecules should be stable. This has been found to be true. However, *cyclopentane which is predicted to be most stable according to Baeyer's strain theory has been found to be little less stable than cyclohexane, etc.*

2. According to this theory cycloheptane (angle of deviation = 9.5°) should be as unstable as cyclobutane (angle of deviation = 9.75°). But their heats of combustion per CH_2 unit show appreciable difference in their stabilities with cyclobutane having the value of 685.5 kJ per CH_2 unit as against cycloheptane having the value of 661.7 kJ per CH_2 unit.

3. In contrary to the predictions of Baeyer's strain theory in cycloalkanes with C_7 onwards there is practically negligible rise in heats of combustion per CH_2 unit which means that these compounds should be almost equally stable.

4. Baeyer's strain theory fails to explain the ease of formation of a carbon-carbon double bond.

Table 5.4: Heats of combustion of cycloalkanes

S. No.	Cycloalkane $(CH_2)_n$	Ring size (n)	Heat of combustion (kJ mole^{-1})	Heat of combustion per CH_2 unit (kJ mole^{-1})
1.	Cyclopropane	3	2089.2	696.4
2.	Cyclobutane	4	2742.0	685.5
3.	Cyclopentane	5	3317.0	663.4
4.	Cyclohexane	6	3947.4	657.9
5.	Cycloheptane	7	4613.9	661.7
6.	Cyclooctane	8	5303.2	662.9
7.	Cyclononane	9	5974.2	663.8
8.	Cyclodecane	10	6617.0	661.7
9.	Cycloundecane	11	7283.1	662.1
10.	Cyclododecane	12	7905.6	658.6
11.	Cyclotridecane	13	8574.8	659.6
12.	Cyclotetradecane	14	9210.6	657.9
13.	Cyclopentadecane	15	9876.0	658.4

For alkanes (taken as strandard for comparison) heat of combustion /CH_2 unit = 657.9.

5.6 SACHSE-MOHR THEORY OF STRAINLESS RINGS

The main drawback in Baeyer's strain theory is that Baryer presumed all cycloalkanes rings to be **planar,** *i.e.*, these compound were regarded to be having the shapes of regular polygons, *e.g.*, cyclohexane, was assumed to be a regular hexagon with C—C—C bond angle of 120°, etc. But the fact is that neither cyclohexane is a hexagon nor cyclodecane a regular decagon.

To account for the stability of six-membered and larger rings Sachse proposed that, contrary to Baeyer's postulate, carbon atoms in a cycloalkane ring are not planar. They actually assume a strain-free puckered form in which all the ring carbon atoms retain their normal tetrahedral angle (109.5°). On this basis he constructed two models of cyclohexane and called them as **boat** and **chair forms**.

Boat form Chair form

Sachse's proposal was initially discarded as it was based on the assumption that above two types cyclohexanes should exist. But at that time only one type of cyclohexane was known. A little later Mohr revived the proposal of Sachse and proposed that the two forms of cyclohexane readily undergo interconversion by rotation across single bonds involving no deviation from the tetrahedral angles. After Mohr's proposal this theory is known as **Sachse-Mohr theory** of strainless rings and can be stated as follows:

Five-membered or smaller rings can be planar, but six membered and large rings are non-planar assuming puckered structures. Various forms are readily interconvertible by rotation about single bonds and hence represent the **conformations** of cycloalkanes.

Recent works on physical measurements and thermodynamic calculations has revealed that a molecule of cyclohexane does undergo interconversion but at room temperature, it exists entirely in the chair conformation. The postulation of Sachse and Mohr was responsible for the synthesis and isolation of a number of multi-membered ring compounds shown in Fig. 5.1.

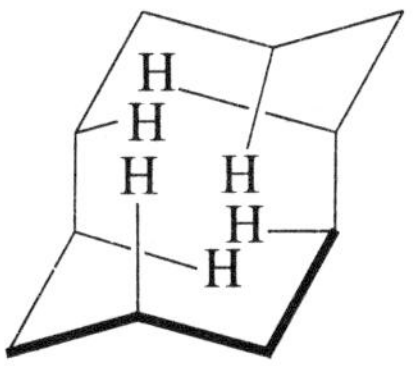

Cyclodecane

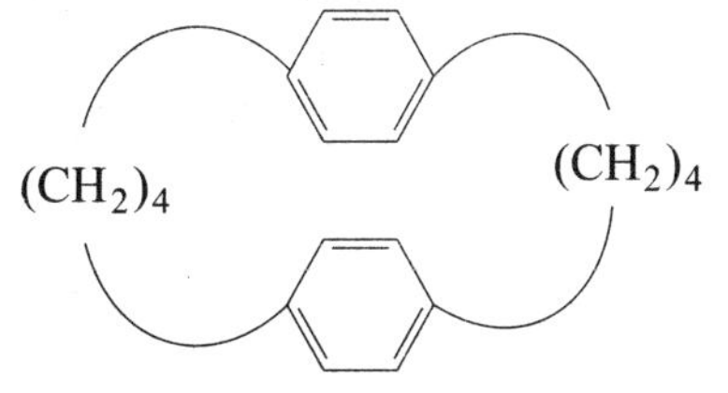

$(CH_2)_4$ $(CH_2)_4$

Paracyclophane

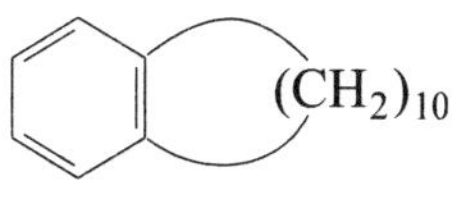

$(CH_2)_{10}$

Cyclophane

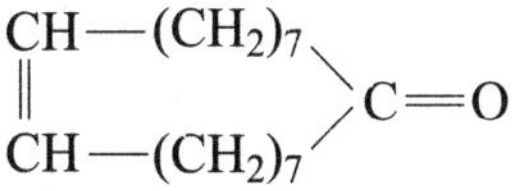

$CH—(CH_2)_7$ / $CH—(CH_2)_7$ $C=O$

Civetone
(From civet cat)

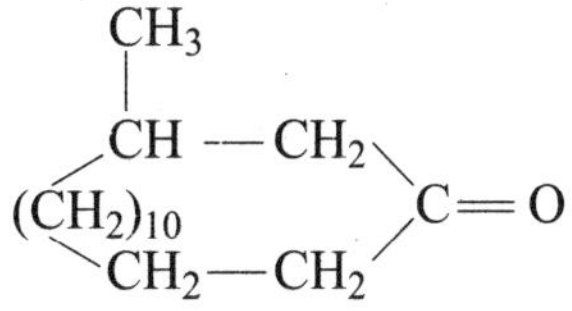

CH_3 / $CH—CH_2$ / $(CH_2)_{10}$ $C=O$ / $CH_2—CH_2$

Muscone
(Fragrant principle of musk, found in musk deer)

Fig. 5.1: Some multi-membered ring compounds

It is interesting that the difficulty encountered in the synthesis of large-ring compounds was wrongly ascribed by Baeyer to their unstability. Rather it arises out of the fact that for the ring closure to take place it is essential that the coupling functional groups at the terminii of a bifunctional compound should come sufficiently close so as to react with each other. But in the synthesis of large-membered rings there is a problem involving competition between *inter-* and *intramolecular* condensations with the former predominating. Thus if two such groups are situated at the terminii of a very long chain there is less possibility of these two groups coming close enough to react with each other and hence intermolecular condensation between functional groups of different molecules may result in polymerization (Fig 5.2). Recently, this difficulty in synthesizing multi-membered rings has been overcome with the help of high dilution techniques, as mentioned earlier, which minimize collisions between two molecules so that due to **'flopping'** the two ends of the molecule come closer enough to undergo intramolecular condensation to give closed ring product instead of intermolecular condensation product. However, no difficulty was encountered in the case of 1, 5 and 1, 6 disubstituted compounds, because the two terminii are quite close to react with each other and form five and six membered ring.

CH_2X $\xrightarrow[\text{condensation}]{\text{Intramolecular}}$ CH_2

CH_2X CH_2

CH_2X XH_2C CH_2X $\xrightarrow[\text{condensation}]{\text{Intermolecular}}$ $CH_2—H_2C$ CH_2X

CH_2X CH_2X CH_2X CH_2X $CH_2—CH_2$

Fig. 5.2: Intramolecular (upper) and intermolecular (lower) condensation

In conclusion, it can be said that Baeyer's strain theory can be acceptable, though in some different form for three and four membered rings only, but it does not apply to higher members of the series. Thus cyclopropane ring must be planar, cyclobutane molecule may or may not be planar, cyclopentane ring can assume a puckered structure which is envelope shape but nearly normal tetrahedral angle is retained even in planar form and all other cycloalkane rings assume strain-free puckered structures.

5.7 THE CASE OF CYCLOPROPANE RING: BANANA BOND

Cyclopropane offers an interesting case of a bent C—C covalent bond. It is planar with the internal bond angles θ of 60°. It means a considerable deviation from the tetrahedral angle of 109.5°. But carbon cannot have sp^3-hybridized orbitals oriented at less than 90° to each other, and since there are four bonds on each carbon, the angles have to be more than even 90°. This would mean that in cyclopropane the ring σ-bonds cannot be formed by the bonding atomic orbitals pointing directly at one another. It has been shown that the orbitals are hybridized in such a way that the ring-forming orbitals have a slightly greater *p*-character than those forming the normal sigma bonds between carbon atoms in saturated molecules. This explains why the angle ϕ between two such orbitals of one carbon atom in cyclopropane is 104°, somewhat less than the tetrahedral angle. Thus, the major axes of symmetry of the orbitals forming the C—C bonds in cyclopropane do not point towards one another but outwards from the molecule with the consequence that the C—C bonding MO's are not symmetrical about the internuclear axes but are bent giving the appearance of bananas (Fig. 5.3).

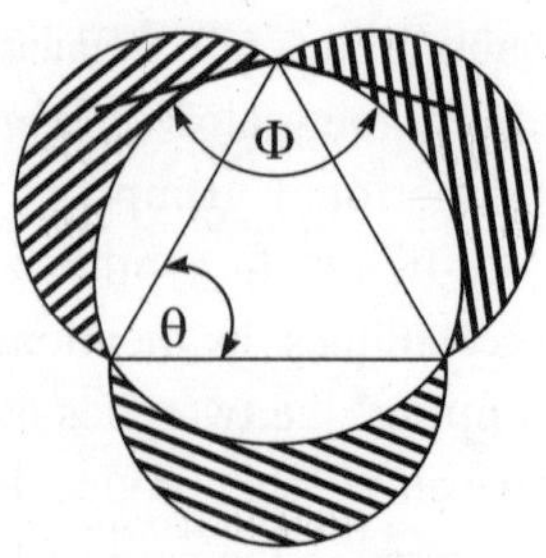

Fig. 5.3: Shape of cyclopropane: The banana bond

A significant point arising out of the higher *p*-character of atomic orbitals of the bonded C—C atoms is that there is a small but appreciable degree of pi delocalization around the ring, which occurs, unlike in an aromatic system, in the plane of the ring and not above and below the carbon framework. This in-plane pi delocalization adds some stability to the ring.

A BRIEF REVIEW

- **Alicyclic compounds** are aliphatic cyclic compounds which are also called **carbocyclic compounds** as their rings are made up of only carbon atoms.
- **Cycloalkanes** are saturated alicyclic compounds having sp^3-hybridized carbon atoms. They have general formula C_nH_{2n}. These are also called polymethylenes because their rings are obtained by joining a number of methylene (CH_2) units.
- **Diels-Alder reaction** ([4 + 2] cycloaddition) gives six membered rings while [2 + 2] cycloaddition gives four membered rings *i.e.*, cyclobutane and its derivatives.
- **Cyclopropane** and its derivatives are obtained by photolysis of diazomethane in presence of alkenes which involve addition of free carbene *i.e.*, methylene to carban-carbon double bond.
- **Simmons-Smith reaction** is the addition of **carbenoids** (carbene like species) to alkene to give cyclopropane derivatives. **Iodomethylene zinc**, which is formed by reaction of CH_2I_2 and Zn-Cu couple, acts as an important source of carbenoid.
- **Freunds method**, which involves heating of a suitable α,ω-dihaloalkane in presence of active metals such as Zn or Na, is used to prepare cyclopropane in good yields.
- **Dieckmann cyclization** of esters of α,ω-dicarboxylic acids in presence of sodium ethoxide leads to the formation of 2-carbethoxy-cycloalkanones which upon hydrolysis and subsequent decarboxylation give cycloalkanones.
- **Clemmensen reduction** is the reduction of cycloalkanones to cycloalkanes in presence of Zn/Hg and HCI.
- **Wislicenus method** involves *pyrolysis* of calcium or barium salts of dicarboxylic acids to give cycloalkanones which upon subsequent *Clemmensen reduction* (Zn- Hg/HCI) give cycloalkanes.
- **Blanc's rule** shows the effect of heat on dicarboxylic acids. It states that
  (*i*) 1, 3-dicarboxylic acids on heating alone or on distillation with acetic anhydride undergo decarboxylation to give monocarboxylic acids;
  (*ii*) 1, 4- and 1, 5-dicarboxylic acids give cyclic anhydrides;
  (*iii*) 1, 6-and 1, 7-dicarboxylic acids give cyclic ketones with one carbon atom less due to loss of CO_2 and
  (*iv*) 1, 8- and higher dicarboxylic acids remain unaffected.

- **Thorpe reaction** involves interaction of α,ω-dinitrile with lithium diethylamide in benzene or toluene to from first a cyclic β-iminonitrile which upon hydrolysis followed by Clemmesen reduction gives cycloalkanes.

- **Ziegler high dilution technique** (*i.e.*, use of dilute solution) is used to prepare macrocyclic rings in good yields. In case of **Thorpe-Ziegler reaction** a dilute solution of dinitrile in benzene is added to the basic solution of catalyst in large excess of ether.

- **Demjanov rearrangement** is the expansion or contraction of alicyclic amines with HNO_2 which involve carbocation intermediates.

  Ring expansion occurs in case of cycloalkyl methylamines whereas *ring contraction* occurs in case of cycloalkylamines.

- **Perkin's method** of preparation of cycloalkanes involves interaction of **active methylene** compounds such as malonic ester or acetoacetic ester with a suitable α, ω-dihalide in presence of a strong base.

- **Free-radical substitution** of cycloalkanes with Cl_2 or Br_2 takes place in presence of light to give corresponding halocycloalkanes.

- **Addition reactions:** Cyclopropane is most reactive and undergoes addition reactions ($Cl_2/FeCl_3$, $Br_2/FeBr_3$, HBr etc.) through the cleavage of ring to give propane derivatives.

- According to **Baeyer's strain theory** all the cycloalkanes are planar molecules due to which their bond angles are either more or less than the normal tetrahedral angles. This deviation of bond angle value causes a strain in the molecule which is called **angle strain**.

  Higher the angle strain lesser is the stability of the cycloalkane and hence more is the reactivity.

  Baeyer's strain theory fails to explain the stability of six and higher membered rings.

- **Sachse and Mohr theory** of strainless ring explains the stability of six and higher membered rings on the basis of the fact that larger rings are non-planar *i.e.*, puckered having no deviation of angle from the tetrahedral value, due to which there is no strain into the molecule.

- It is very difficult to synthesize six or higher membered rings due to intermolecular condensation instead of intramolecular condensation which is required for ring closure. However, the probability of intramolecular condensation increases under high dilution conditions which minimises the intermolecular condensations therefore, six or higher membered rings can be synthesized by increasing the dilution.

- Cyclopropane bonds are called **banana bonds**. These bonds appear as bananas because C—C bonding molecular orbitals are not symmetrical about the internuclear axis.

SOLVED PROBLEMS

Q.1. How will you prepare 1,1-dimethylcyclopropane from malonic ester?

Ans.

$$CH_2(CO_2C_2H_5)_2 \xrightarrow[(ii)\ CH_3I]{(i)\ NaOC_2H_5} CH_3—CH\ (CO_2C_2H_5)_2 \xrightarrow[(ii)\ CH_3I]{(i)\ NaOC_2H_5}$$

$$(CH_3)_2C(CO_2C_2H_5)_2 \xrightarrow{LiAlH_4} (CH_3)_2C(CH_2OH)_2 \xrightarrow{PBr_3} (CH_3)_2C(CH_2Br)_2$$

$$(CH_3)_2C(CH_2Br)_2 \xrightarrow[\text{(Freund reaction)}]{\text{Zn dust, }\Delta} (CH_3)_2C{<}\begin{matrix}CH_2\\|\\CH_2\end{matrix} + ZnBr_2$$

1, 1-Dimethylcyclopropane

Q.2. Complete the following reactions:

(*i*) $ClCH_2CH_2CH_2Cl \xrightarrow{\text{Mg, ether}}$ (*ii*) $ClCH_2CH_2CH_2CH_2Cl \xrightarrow{\text{Mg, ether}}$

Ans. (*i*) Due to favourable entropy over thermal stability factor, three-membered rings are readily formed, *i.e.*,

$$CH_2{<}\begin{matrix}CH_2Cl\\CH_2Cl\end{matrix} \xrightarrow{Mg} CH_2{<}\begin{matrix}CH_2—MgCl\\CH_2—Cl\end{matrix} \xrightarrow{\text{Freund reaction}} CH_2{<}\begin{matrix}CH_2\\|\\CH_2\end{matrix} + MgCl_2$$

(*ii*) The formation of four-membered rings is very slow since both the thermal stability factor and the entropy factors are unfavourable. The thermal stability is low because of the angle strain. The entropy factor is unfavourable since the two ends of the ***zig-zag*** chain of the substrate always adopt conformation (I) with the reactive groups far apart and even in the best conformation (II) the two ends are still far away.

$$\underset{\text{I}}{^{\ominus}Nu\text{/\\/\\}E^{\oplus}} \rightleftarrows \underset{\text{II}}{\overset{\ominus}{Nu}\text{\\\_\_/}\overset{\oplus}{E}} \longrightarrow \begin{matrix}Nu—E\\|\quad\ \ |\\\text{—}\text{—}\end{matrix}$$

It is because of these two reasons that $ClCH_2CH_2CH_2Cl$ on treatment with Mg in ether forms only the Grignard reagent and not the four-membered ring.

$$\begin{array}{l} CH_2CH_2—MgCl \\ | \\ CH_2CH_2—MgCl \end{array} \xleftarrow{Mg} \begin{array}{l} CH_2—CH_2—Cl \\ | \\ CH_2—CH_2—Cl \end{array} \xrightarrow{Mg} \begin{array}{l} CH_2—CH_2—MgCl \\ | \\ CH_2—CH_2—Cl \end{array} \not\longrightarrow \begin{array}{l} CH_2—CH_2 \\ | \quad\quad\;\; | \\ CH_2—CH_2 \end{array}$$

Q.3. Cyclopropanes have the greatest ring strain yet they are readily prepared.

Ans. Formation of cyclic compounds mainly depends upon the following two factors:

(*i*) Rates of reactions, *i.e.*, kinetics of the reaction.

(*ii*) Stability of the ring, *i.e.*, themodynamic factor.

Out of all the cycloalkanes, cyclopropane ring is least stable since it has the highest Baeyer's angle strain (24. 44°). However, the two ends of the reagent leading to the formation of a cyclopropane are always close together as shown below:

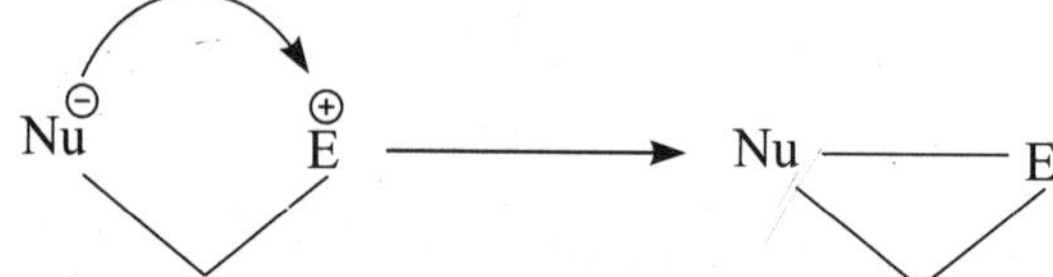

Therefore, the rate of formation of three-membered ring is quite high due to favourable entropy factor which more than compensates the higher thermal instability.

Q.4. Give the two possible products of the reaction of 1,1-dimethylcyclopropane with Br_2 at room temperature and predict the major product and account for its formation.

Ans. Since during ring opening, a C—C bond cleavage occurs, therefore, two carbocations (III and IV) can be formed. Each one of them will react with Br^- to afford products (I and II). Since carbocation (III) being 3° is more stable than the 1° carbocation (IV), therefore, the major product of the reaction is product (I).

$(CH_3)_2C(CH_2)_2$ (bonds *a* and *b*)

Path '*a*': $\xrightarrow[\text{Path }'a']{Br_2}$ $CH_3—C(Br)(CH_3)—CH_2—CH_2Br$
1, 3-Dibromo-3-methylpropane (I) (major product)

Path '*b*': $\xrightarrow[\text{Path }'b']{Br_2}$ $Br—CH_2—C(CH_3)_2—CH_2Br$
1, 3-Dibromo-2, 2-dimethylpropane (II) (minor product)

Path '*a*' $\longrightarrow$ $CH_3—\overset{\oplus}{C}(CH_3)—CH_2—CH_2Br$
(III)
3° Carbocation (*more stable*) $\xrightarrow{+Br^\ominus}$ (I)

Path '*b*' $\longrightarrow$ $\overset{\oplus}{C}H_2—C(CH_3)_2—CH_2Br$
(IV)
1° Carbocation (*less stable*) $\xrightarrow{+Br^\ominus}$ (II)

UNSOLVED PROBLEMS

1. (*i*) What are cycloalkanes? How are they named? Discuss briefly the general methods of preparation of cycloalkanes.

   (*ii*) What are cycloparaffins? Describe four methods of their preparation.

2. Write structures for the following:

   (*i*) Bicyclo [2.2.0] hexane
   (*ii*) 1-Iodo-2-methyl-3-isopropyl-bicyclo[2.2.2]octane
   (*iii*) Adamantane
   (*iv*) Cyclodecane
   (*v*) Muscone
   (*vi*) Civetone.

3. How will you make use of the following in synthesis of alicyclic hydrocarbons?

   (*i*) Dieckmann condensation. (*ii*) Perkin method.
   (*iii*) Thorpe-Ziegler reaction.

4. Write suitable steps to indicate the synthesis of cyclopentane from ethyl adipate.

5. Commencing with adipic acid how will you synthesise cyclopentane?

6. What happens when:

   (*i*) 1, 3-Dibromopropane is heated with sodium.
   (*ii*) 1, 6-Hexadinitrile is treated with lithium diethylamide.
   (*iii*) Diethyl adipate is treated with sodium ethoxide in toluene.
   (*iv*) Diazomethane is irradiated in presence of isobutylene.

7. Discuss briefly Simmon's-Smith reaction for the preparation of cyclopropane derivatives.

8. What is Blanc's rule? Discuss its applications.

9. Discuss briefly the mechanism of addition of singlet and triplet carbenes to alkenes.

10. What are cycloaddition reactions? Give two example.

11. (*a*) Complete the following reactions:

(*i*) $\begin{array}{l} CH_2CH_2COOC_2H_5 \\ | \\ CH_2CH_2COOC_2H_5 \end{array} \xrightarrow{C_2H_5ONa} ?$ (*ii*) $H_2C\begin{array}{l} \diagup CH_2Br \\ \diagdown CH_2Br \end{array} + Zn \xrightarrow{\Delta} ?$

(*iii*) $H_2C{=}CH_2 + CH_2N_2 \xrightarrow{h\nu} ?$ (*iv*) $CH_3CH{=}CH_2 + CH_2I_2 \xrightarrow{Cu\text{-}Zn} ?$

What is the reactive intermediate in each reaction?

12. Why are intermolecular cyclizations of rings with more than six carbon atoms achieved at low concentration of reactants?

Chapter 5

13. Discuss briefly Ziegler-high dilution technique for the synthesis of macrocyclic rings.

14. Why large rings with more than six carbon atoms are stable but difficult to prepare ?

15. Write a short note on banana bonds.

16. Give plausible explanation for the following:

(*i*) The melting points, boiling points and densities of cycloalkanes are higher than those of the corresponding straight chain analogues.

(*ii*) Cyclopropane and cyclobutane undergo addition reactions but higher cycloalkanes do not.

17. Give the structures and the names of the principal products, if any, expected from each of the following reactions:

(*i*) Cyclopropane +Cl_2, $FeCl_3$ (*ii*) Cyclopropane + CI_2(300 K)

(*iii*) Cycloproane + HI (*iv*) Cyclobutane + H_2/Ni(393 K)

(*v*) Cyclopentane +Br_2(575 K) (*vi*) Cyclohexane + Br_2/$FeBr_3$

(*vii*) 1,1-Dimethylcyclopropane + HBr.

18. What is Baeyer's strain theory? Show how Baeyer's theory fails for rings large than four membered?

19. Write a short note on Baeyer's strain theory.

20. Cyclopropane is the least stable member of cycloalkanes. Justify the statement.

21. Write a brief note on strainless rings.

22. Discuss briefly the difficulty encountered in the synthesis of large membered rings. How can this problem be overcome?

23. Which member of each of the following pairs of compounds is more stable?

(*i*) *cis* or *trans* 1, 2-dimethylcyclohexane.

(*ii*) *cis* or *trans* 1, 3-dimethylcyclohexane.

(*iii*) *cis* or *trans* 1, 4-dimethylcyclohexane.

24. (*i*) Write the structural formulas for the two chair conformations of *cis*-1-*t*-butyl-4-methylcyclohexane.

(*ii*) Which of these two would be more stable? Why?

25. Write the structure and give IUPAC name of the product obtained by treating 1,2,3-tri-methylcyclopropane with bromine in the absence of light.

CHAPTER **6**

Alkenes

In the preceding chapters, we have already mentioned briefly that in addition to saturated hydrocarbons (alkanes and cycloalkanes) there are other types of hydrocarbons which contain less hydrogen atoms per carbon atom. Because of this hydrogen deficiency, they are referred to as ***unsaturated hydrocarbons*** and are characterized by the presence of carbon-carbon multiple bonds. As such the term unsaturated hydrocarbons includes *alkenes, cycloalkenes, polyenes* and *alkynes*. The present chapter is devoted to the chemistry of alkenes.

6.1 ALKENES

Alkenes are open-chain hydrocarbons having carbon-carbon double bonds. They have the general formula C_nH_{2n} and are also called as *alkylenes* or *olefins*.

6.1.1 Nomenclature

There are two systems governing the nomenclature of this class of compounds.

1. Common system.   2. IUPAC system.

1. Common System

Simplest alkenes are known by their common names such as ethylene ($H_2C == CH_2$), propylene ($CH_3CH == CH_2$), α-butylene ($CH_3CH_2CH == CH_2$), β-butylene ($CH_3 - CH == CH — CH_3$) and *iso*-butylene [$(CH_3)_2 - C == CH_2$], etc. Here α, β, etc., as prefixes indicate the position of double bond. However, this system is seldom used for the nomenclature of higher alkenes. In general, name of alkene can be derived from the name of corresponding alkane as below:

Common name of alkene = **Alkane – ane + ylene** = Alkylene.

2. The IUPAC System

It is the most useful system for the nomenclature of higher alkenes. The class name of **alkenes** is derived by replacing the suffix-**ane** from alkane by-**ene** (**Alkene = alkane – ane + ene**). The following general rules are observed for writing systematic names for alkenes.

***(i) Longest chain rule*:** First of all, longest continuous (straight) chain of carbon atoms containing the double bond is selected and the compound under reference is named as the derivative of the **alkane** having the same number of carbon atoms by replacing the suffix - **ane** by - **ene**. Thus, if the longest chain having a double bond contains 6 carbon atoms the alkene will be named as hexene.

***(ii) Assigning position to the double bond*:** The carbon atoms of the longest chain are numbered 1, 2, 3,......, etc., from that end which is nearest to the double bond. The position of the double bond is indicated by the number of the first of the two doubly bonded carbon atoms. This number is written before the name of the alkene with a hyphen (-) in between, *e.g.*,

$$\overset{5}{CH_3}-\overset{4}{CH}=\overset{3}{CH}-\overset{2}{CH_2}-\overset{1}{CH_3}$$

3-Pentene (*Incorrect*)

$$\overset{1}{CH_3}-\overset{2}{CH}=\overset{3}{CH}-\overset{4}{CH_2}-\overset{5}{CH_3}$$

2-Pentene (*Correct*)

$$\overset{1}{CH_3}-\overset{2}{CH_2}-\overset{3}{CH_2}-\overset{4}{CH}=\overset{5}{CH_2}$$

4-Pentene (*Incorrect*)

$$\overset{5}{CH_3}-\overset{4}{CH_2}-\overset{3}{CH_2}-\overset{2}{CH}=\overset{1}{CH_2}$$

1-Pentene (*Correct*)

***(iii) Assigning position to the substituents*:** If substituents such as alkyl groups, halogens, nitro, etc., are present which do not constitute the root name, the chain is still numbered from the end which is nearest to the double bond. For example

$$\overset{5}{H_3C}-\overset{4}{CH}=\overset{3}{CH}-\overset{CH_3}{\overset{|}{\underset{2}{CH}}}-\overset{1}{CH_3}$$

2-Methyl-3-pentene
(*Incorrect*)

$$\overset{1}{H_3C}-\overset{2}{CH}=\overset{3}{CH}-\overset{CH_3}{\overset{|}{\underset{4}{CH}}}-\overset{5}{CH_3}$$

4-Methyl-2-pentene
(*Correct*)

$$\begin{array}{l}\overset{6}{CH_2}-\overset{7}{CH_3}\\ \ |_{5}\quad 4 \qquad 3 \qquad\ 2 \qquad\ 1\\ Cl-C=CH-CH_2-CH_2-CH_3\end{array}$$

5-Chloro-4-heptene

(*Incorrect*)

$$\begin{array}{l}\qquad CH_2-CH_3\\ \qquad |^{2}\qquad\ 1\\ Cl-\underset{3}{C}=\underset{4}{CH}-\underset{5}{CH_2}-\underset{6}{CH_2}-\underset{7}{CH_3}\end{array}$$

3-Chloro-3-heptene

(*Correct*)

***(iv) Assigning positions to the substituents if same number is assigned to the double bond from both sides*:** In such a case, end is chosen for numbering which gives lowest number to the substituent. For example

$$\overset{1}{H_3C}\ \overset{2}{CH_2}\overset{3}{CH_2}\underset{CH_3}{\underset{|}{\overset{4}{C}}}=\overset{5}{CH}\overset{6}{CH_2}\underset{CH_3}{\underset{|}{\overset{7}{CH}}}\overset{8}{CH_3}$$

4, 7-Dimethyl-4-octene
(*Incorrect*)

$$\overset{8}{H_3C}-\overset{7}{CH_2}-\overset{6}{CH_2}-\underset{CH_3}{\underset{|}{\overset{5}{C}}}=\overset{4}{CH}-\overset{3}{CH_2}-\underset{CH_3}{\underset{|}{\overset{2}{CH}}}-\overset{1}{CH_3}$$

2, 5-Dimethyl-4-octene
(*Correct*)

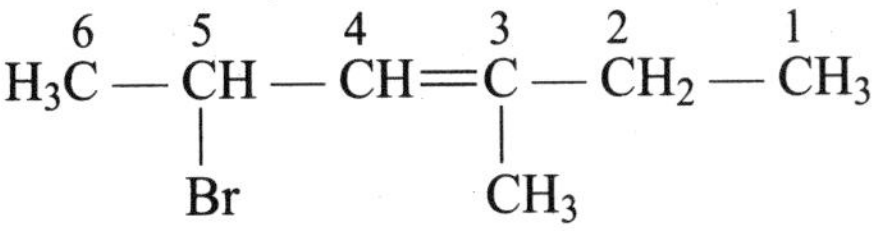

$$\overset{6}{H_3C}-\underset{\underset{Br}{|}}{\overset{5}{CH}}-\overset{4}{CH}=\underset{\underset{CH_3}{|}}{\overset{3}{C}}-\overset{2}{CH_2}-\overset{1}{CH_3}$$

5-Bromo-3-methyl-3-hexene
(*Incorrect*)

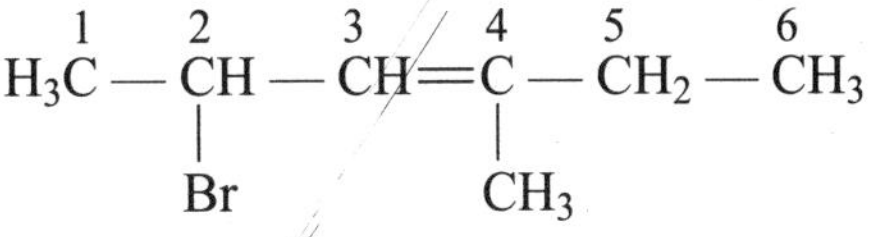

$$\overset{1}{H_3C}-\underset{\underset{Br}{|}}{\overset{2}{CH}}-\overset{3}{CH}=\underset{\underset{CH_3}{|}}{\overset{4}{C}}-\overset{5}{CH_2}-\overset{6}{CH_3}$$

2-Bromo-4-methyl-3-hexene
(*Correct*)

(*v*) *Naming the compounds with more than one substituent*: In such a cases [also see (*iv*) above] the names of the substituents are prefixed before the name of the parent alkene in alphabetical order in the same manner as in the case of alkanes.

Here, it may be noted that the prefixes *di-*, *tri-*,*tetra-*, etc., are ignored while prefixes such as *iso.*, *neo-*, *cyclo-* are not. For example,

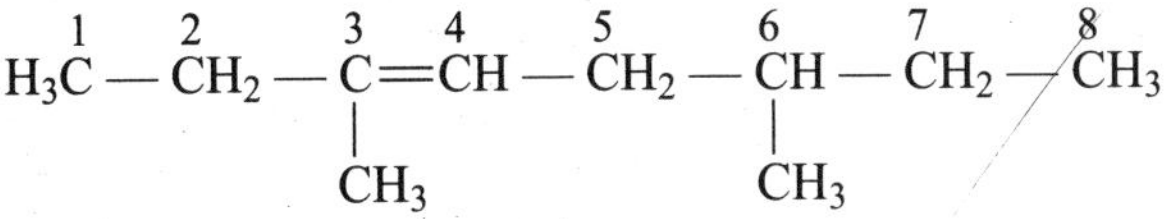

$$\overset{1}{H_3C}-\overset{2}{CH_2}-\underset{\underset{CH_3}{|}}{\overset{3}{C}}=\overset{4}{CH}-\overset{5}{CH_2}-\underset{\underset{CH_3}{|}}{\overset{6}{CH}}-\overset{7}{CH_2}-\overset{8}{CH_3}$$

3, 6-Dimethyl-3-octene

$$\overset{7}{H_3C}-\overset{6}{CH_2}-\underset{\underset{Br}{|}}{\overset{5}{CH}}-\underset{\underset{Cl}{|}}{\overset{4}{CH}}-\overset{3}{CH_2}-\overset{2}{CH}=\overset{1}{CH_2}$$

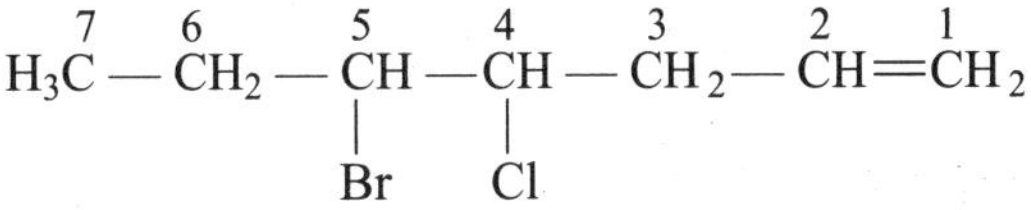

5-Bromo-4-chloro-1-heptene

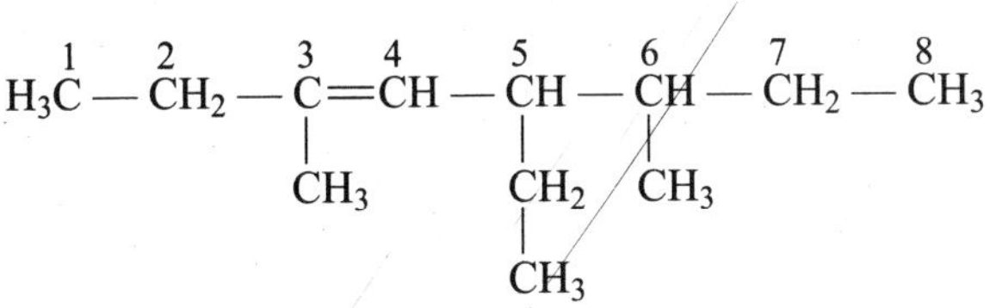

$$\overset{1}{H_3C}-\overset{2}{CH_2}-\underset{\underset{CH_3}{|}}{\overset{3}{C}}=\overset{4}{CH}-\underset{\underset{\underset{CH_3}{|}}{CH_2}}{\overset{5}{CH}}-\underset{\underset{CH_3}{|}}{\overset{6}{CH}}-\overset{7}{CH_2}-\overset{8}{CH_3}$$

5-Ethyl-3, 6-dimethyl-3-octene

(*vi*) *Naming the compound having a functional group which constitutes the root name*: If the compound has both a double bond and any one of the functional group such as —OH,—CHO, —COR,—COOH, —COOR, etc., then the numbering of the chain starts from the end which is nearest to these groups and not the double bond.

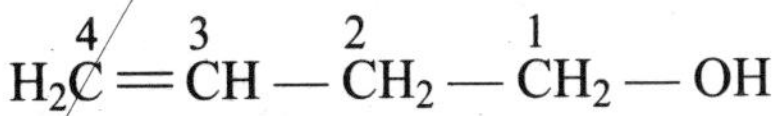

$$\overset{4}{H_2C}=\overset{3}{CH}-\overset{2}{CH_2}-\overset{1}{CH_2}-OH$$

But-3-en-1-ol
(*Correct*)

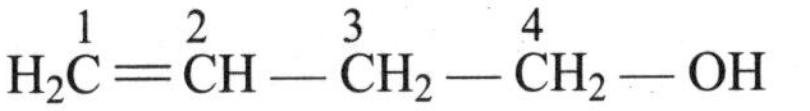

$$\overset{1}{H_2C}=\overset{2}{CH}-\overset{3}{CH_2}-\overset{4}{CH_2}-OH$$

But-1-en-4-ol
(*Incorrect*)

$$\overset{4}{H_3C}-\overset{3}{CH}=\overset{2}{CH}-\overset{1}{COOH}$$

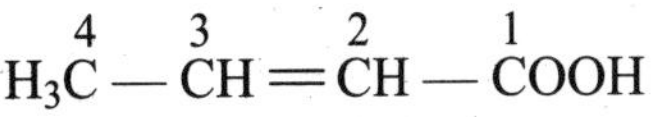

But-2-enoic acid
(*Correct*)

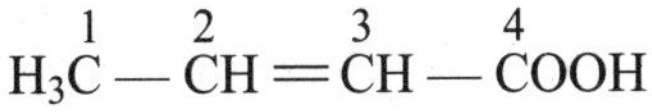

$$\overset{1}{H_3C}-\overset{2}{CH}=\overset{3}{CH}-\overset{4}{COOH}$$

But-2-enoic acid
(*Incorrect;* wrong end chosen)

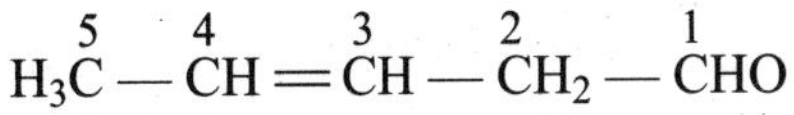

$$\overset{5}{H_3C}-\overset{4}{CH}=\overset{3}{CH}-\overset{2}{CH_2}-\overset{1}{CHO}$$

Pent-3-enal
(*Correct*)

$$\overset{1}{H_3C}-\overset{2}{CH}=\overset{3}{CH}-\overset{4}{CH_2}-\overset{5}{CHO}$$

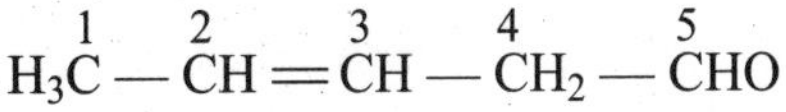

Pent-2-enal
(*Incorrect*)

Chapter 6

6.2 THE ALKENYL GROUP

The monovalent group obtained by removing one hydrogen atom of an alkene is called as **alkenyl group**. Such a group derived from ethene is called **vinyl group** or **ethenyl group**. The functional derivatives of ethene are, therefore called **vinyl compounds**.

$$\mathrm{H_2C{=}CH{-}}$$

The vinyl group

$$\mathrm{H_2C{=}CHCl}$$

Vinyl chloride

On the other hand, removal of a hydrogen atom from the methyl group of propene gives another monovalent alkenyl group called the **allyl group** and consequently its derivatives are referred to as **allyl compounds**.

$$\mathrm{H_2C{=}CH{-}CH_2{-}}$$

The allyl group

$$\mathrm{H_2C{=}CH{-}CH_2OH}$$

Allyl alcohol

Other **alkenyl groups** are

$$\mathrm{CH_3CH{=}CH{-}}$$

1-Propenyl group

$$\mathrm{CH_3CH{=}C(CH_3){-}}$$

2-Butenyl group

6.3 METHODS OF FORMATION

Lower alkenes, up to pentenes, can be obtained by the cracking of petroleum and are isolated in pure form by fractionation. The higher alkenes, formed during cracking, cannot be separated and are used as components of gasoline. Alkenes are also prepared in pure state by introducing a double bond in the molecule by the process of **elimination** of atoms or groups from the vicinal carbon atoms. In most of the cases, a hydrogen is lost as a proton, from one carbon (designated as β-carbon) and a nucleophile is lost from the adjacent carbon atom (designated as α-carbon).

$$\mathrm{{-}\overset{H}{\underset{|}{\overset{|}{C}}}_{\beta}{-}\underset{X}{\underset{|}{\overset{|}{C}}}_{\alpha}{-}} \longrightarrow \mathrm{{>}C{=}C{<}}$$

This type of elimination reaction, which is very common, is generally referred to as β-*elimination* or 1, 2-*elimination* in contrast to another type of elimination reaction where both the atoms or groups are lost from the same carbon atom. The latter type of elimination which generates carbenes, is called α-*elimination*. In the following pages, we shall discuss the formation of alkenes *via* β-*elimination* reactions.

(*i*) Dehydrohalogenation of Alkyl Halides: Mechanism of Elimination Reactions

Elimination of a hydrogen halide molecule from an alkyl halide is called *dehydrohalogenation*. The alkyl halides on refluxing with concentrated alcoholic solution of potassium hydroxide yield alkenes. For example,

$$\underset{\text{Ethyl Chloride}}{CH_3-CH_2-Cl} + \underset{\text{(Alcoholic)}}{KOH} \xrightarrow{\Delta} \underset{\text{Ethene}}{CH_2{=}CH_2} + H_2O + KCl$$

$$\underset{\text{2-Bromopropane}}{CH_3-\underset{\displaystyle Br}{\underset{|}{C}H}-CH_3} + \underset{\text{(Alcoholic)}}{KOH} \xrightarrow{\Delta} \underset{\text{Propene}}{CH_3-CH{=}CH_2} + H_2O + KBr$$

Mechanism

Depending upon the structure of alkyl halide and some other factors as detailed below, the reaction can take place by either of the two mechanistic pathways commonly referred to as E2 (**E**limination **B**imolecular) and E1 (**E**limination **U**nimolecular) mechanisms.

E2 Reaction: Bimolecular Mechanism: Many eliminations can be successful only when they are carried out in the presence of a strong base. The formation of ethene on treatment of chloro-or bromoethane with concentrated alcoholic solution of potassium hydroxide is an example of this type. Kinetic studies have revealed that the rate of the reaction is proportional to the concentration of both the reactants, the haloalkane and potassium hydroxide.

$$CH_3-CH_2-Br + \overset{\ominus}{O}Et \longrightarrow CH_2{=}CH_2 + EtOH + \overset{\ominus}{Br}$$

$$\text{Rate} = k\,[CH_3CH_2Br]\,[\overset{\ominus}{O}Et]$$

The following two alternative mechanisms can be proposed on the basis of this kinetic evidence.

1. One-step process

In the first mechanism, which is a one-step process, the base abstracts a proton from the β-carbon atom and simultaneously halide ion departs from the α-carbon along with a pair of bonding electrons as shown below:

$$C_2H_5\overset{\ominus}{O} + H-\underset{H}{\overset{H}{\underset{|}{\overset{|}{C}}}}{}_{\beta}-{}_{\alpha}\underset{Br}{\overset{H}{\underset{|}{\overset{|}{C}}}}-H \longrightarrow \left[H-\underset{H}{\overset{H\cdots\overset{\delta^-}{O}C_2H_5}{\underset{|}{\overset{|}{C}}}}\overset{---}{=}\underset{\overset{\delta^-}{Br}}{\overset{H}{\underset{|}{\overset{|}{C}}}}-H\right] \longrightarrow C_2H_5OH + \underset{H}{\overset{H}{}}{>}C{=}C{<}\underset{H}{\overset{H}{}} + \overset{\ominus}{Br}$$

Transition state

2. Two-step process

In the second mechanism, which is a two-step process, the first step involves rapid removal of a proton from the α-carbon forming a carbanion which loses the halide ion in a subsequent rate-determining step.

First step

$$C_2H_5\overset{\ominus}{O} + H-\underset{H}{\overset{H}{C}}-\underset{Br}{\overset{H}{C}}-H \underset{}{\overset{\text{Fast}}{\rightleftharpoons}} H-\underset{H}{\overset{\ominus}{C}}-\underset{Br}{\overset{H}{C}}-H + C_2H_5OH$$

Second step

$$H-\underset{H}{\overset{\ominus}{C}}-\underset{Br}{\overset{H}{C}}-H \xrightarrow{\text{Slow}} H_2C{=}CH_2 + \overset{\ominus}{Br}$$

In this mechanism the overall rate of the reaction is limited to the slower second step whose rate depends only on the concentration of the conjugate base of the alkyl halide.

Because of this, the mechanism is referred to as E1CB mechanism (**E**limination **U**nimolecular **C**onjugate **B**ase). The problem is how to decide which of the two mechanisms is operating? This problem has been solved by isotopic labelling experiments. Let us consider the formation of styrene by the action of sodium ethoxide on 2-phenylethyl bromide. This reaction follows second order kinetics. The following two alternative mechanisms consistent with the kinetic data, may be proposed.

E2CB Mechanism

$$C_2H_5\overset{\ominus}{O} + H-\underset{C_6H_5}{\overset{H}{C}}-CH_2-Br \longrightarrow \left[\begin{array}{c} \overset{\delta^-}{C_2H_5O} \\ \vdots \\ H \\ \vdots \\ H-\underset{C_6H_5}{C} \overset{---}{=} CH_2 --- \overset{\delta^-}{Br} \end{array} \right] \longrightarrow C_2H_5OH + \underset{\text{Styrene}}{C_6H_5CH{=}CH_2} + \overset{\ominus}{Br}$$

Transition state

E1CB Mechanism

$$C_2H_5\overset{\ominus}{O} + H-\underset{C_6H_5}{\overset{H}{\overset{|}{C}}}-CH_2-Br \xrightleftharpoons[]{\text{Fast}} C_2H_5OH + C_6H_5\overset{\ominus}{C}HCH_2Br$$

$$H-\underset{C_6H_5}{\overset{\ominus}{C}}-CH_2-Br \xrightarrow{\text{Slow}} C_6H_5CH{=}CH_2 + \overset{\ominus}{Br}$$

Evidence in favour of E2CB Mechanism

The first step of the second mechanism(E1CB) is reversible. So, when we carry out this reaction in C_2H_5OD instead of C_2H_5OH, the intermediate carbanion (conjugate base) should pick up deuterium from the solvent in the reverse step.

$$H-\underset{C_6H_5}{\overset{\ominus}{C}}-CH_2Br + C_2H_5OD \rightleftharpoons H-\underset{C_6H_5}{\overset{D}{C}}-CH_2Br + C_2H_5\overset{\ominus}{O}$$

In case E1CB mechanism is correct one would expect to recover labelled 2-phenylethyl bromide ($C_6H_5CHD\ CH_2Br$) if the reaction is stopped before completion. On the other hand, there should be no deuterium uptake if E2 mechanism is operating. *In actual experiment, it was observed that there is no deuterium incorporation under the above conditions.* Hence, it can be concluded that the reaction proceeds through E2 CB mechanism. However, E1CB mechanism does operate in some special cases. For instance, 1,1,1-trifluoro-2,2-dichloroethane undergoes base-catalysed exchange of β-hydrogen atom with the solvent deuterium faster than dehydrofluorination.

$$F_3C-CHCl_2 \xrightleftharpoons{C_2H_5\overset{\ominus}{O}} F_3C-\overset{\ominus}{C}Cl_2 \xrightleftharpoons{C_2H_5OD,\ \text{Fast}} F_3C-CDCl_3$$

$$F_3C-\overset{\ominus}{C}Cl_2 \xrightarrow{\text{Slow}} F_2C{=}CCl_2$$

A stronger C—F bond coupled with –I-effect of halogens explains the formation of carbanion before elimination.

Stereochemistry of E2 Reactions

It has been shown that elimination by E2CB mechanism is quite facile if both the departing groups are *trans* with respect to each other rather than *cis*. Thus, for instance, *trans*-2-bromobutene forms dimethylacetylene more rapidly than the *cis*-isomer.

Chapter 6

Trans-2-bromobutene $\xrightarrow{\text{Fast}}$ $H_3C—C\equiv C—CH_3$ (Dimethylacetylene) $\xleftarrow{\text{Slow}}$ *Cis*-2-bromobutene

Similarly, the stereo-chemistry of products of E2 reactions of compounds possessing two chiral carbon atoms depends upon the stereochemistry of the starting isomers. Thus, dehydrobromination of *meso*-stilbene dibromide yields the *cis*-alkene whereas the *trans*-alkene is obtained from the *dl*-isomer.

Meso-stilbene dibromide $\xrightarrow{-H_2O}$ *Cis*-diphenylethylene $+ Br^{\ominus}$

dl-stilbene dibromide $\xrightarrow{-H_2O}$ *Trans*-diphenylethylene $+ Br^{\ominus}$

Why *trans* elimination is more facile than *cis*? A possible explanation may be that an elimination reaction produces *p*-orbitals from what were the bonds of the leaving group and the hydrogen to be eliminated. In order to achieve maximum overlap of these *p*-orbitals in the transition state they must be coplanar, and for the minimum energy of the transition state the two groups to be eliminated must be *anti-parallel* to each other across the central bond, since there will be no crowding of the groups as well as the reagent in such an arrangement.

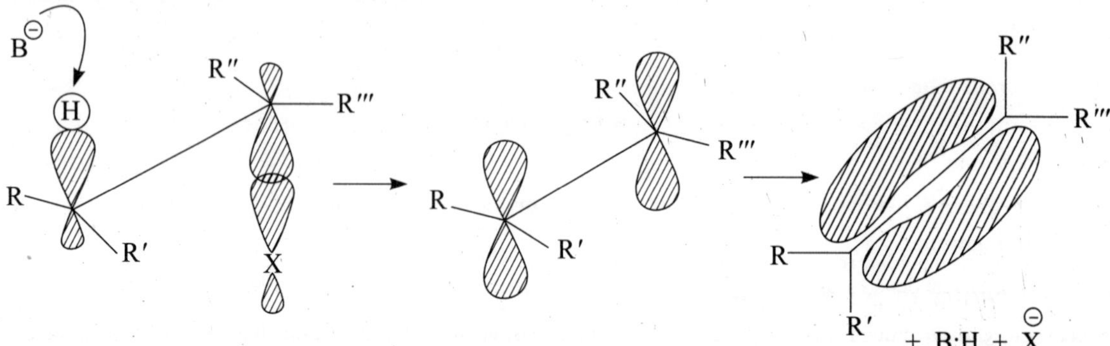

Thus, *trans*-elimination is a general rule. However, recent evidence indicates occurrence of *cis*-elimination in some rigid ring systems involving higher activation energy.

E1 Reaction: Unimolecular Mechanism: It has been shown that many secondary and tertiary halides undergo elimination reactions in neutral or acidic media. These reactions follow first order kinetics with respect to alkyl halides.

$$H_3C-\overset{\overset{CH_3}{|}}{\underset{\underset{CH_3}{|}}{C}}-X \longrightarrow H_3C-\underset{\underset{CH_3}{|}}{C}=CH_2 + HX$$

$$X = Cl, Br, I$$

$$\text{Rate} = k\,[(CH_3)_3CX]$$

Such reactions, called E1 reactions, are often accompanied by substitution (Chapter-10) as well. The main feature of E1 reactions is that under the influence of *solvation forces** the electron-attracting halogen breaks away along with the bonding electrons forming a carbocation which subsequently loses a proton to a proton acceptor which may sometimes be even the solvent.

$$H_3C-\overset{\overset{CH_3}{|}}{\underset{\underset{CH_3}{|}}{C}}-X \xrightarrow[298\ K]{C_2H_5OH} H_3C-\overset{\overset{CH_3}{|}}{\underset{\underset{CH_3}{|}}{C^{\oplus}}} + X^{\ominus}$$

$$H-CH_2-\overset{\oplus}{C}(CH_3)_2 \longrightarrow H_2C=C(CH_3)_2$$

To a proton acceptor (may be solvent)

It may be noted that the *ease of E1 reactions depends directly upon the stability of the carbocations.* Further, unlike E2 reactions these reactions have no stereochemical preferences because the intermediate carbocation has a planar structure.

Regiochemistry or Orientation of E2 and E1 reactions: Saytzeff Rule

An interesting feature of dehydrohalogenation is that if an alkyl halide is capable of forming the double bond in any one of the two possible directions it invariably yields the alkene which has the greater number of substituents across the double bond. In other words, more substituted alkene is preferably

*Solvents play an important role in many organic reactions which take place via a stepwise mechanism. The ionizing capability of a solvent depends upon its dielectric constant. Thus, the separation of ions is easier in a solvent having high dielectric constant. Secondly, the ionization capability also depends upon the ability of the solvent to solvate the ions, which may take place through hydrogen bonding between the ions and the solvent, thus stabilizing the ions. Such forces which are responsible for the formation and stability of ions are called solvation.

formed. This is in accordance with the **Saytzeff rule** which states that "neutral substrates undergoing E1or E2 elimination to produce more substituted alkenes. Dehydrobromination of 2-bromobutane, for instance, gives 2-butene as the major product.

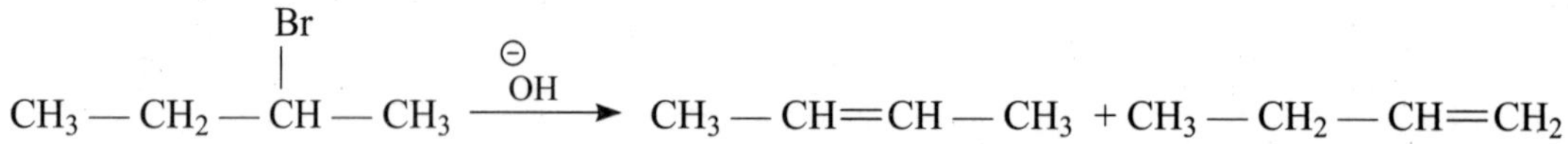

$$\underset{\text{2-Bromobutane}}{CH_3-CH_2-\overset{\overset{\displaystyle Br}{|}}{CH}-CH_3} \xrightarrow{\overset{\ominus}{OH}} \underset{\text{2-Butene (80\%)}}{CH_3-CH{=}CH-CH_3} + \underset{\text{1-Butene (20\%)}}{CH_3-CH_2-CH{=}CH_2}$$

The preferential formation of more substituted alkenes in the above reactions can be correlated with the relative stability of various alkenes. Calculations from heats of combustion and hydrogenation establish that the stability of a double bond is increased by alkyl substitution on the doubly-bonded carbon atoms. Thus the stabilities of alkenes follow the order:

$$R_2C{=}CR_2 > RCH{=}CR_2 > RCH{=}CHR > RCH{=}CH_2 > CH_2{=}CH_2$$

The greater stability of the more substituted alkenes is sometimes explained on the basis of hyperconjugation.

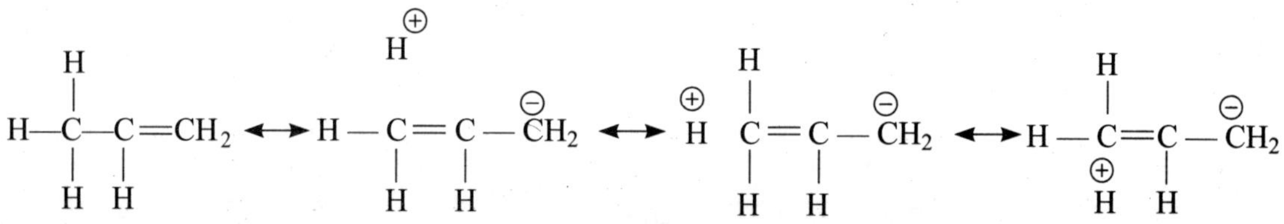

$$H-\underset{H}{\overset{H}{C}}-\underset{H}{C}{=}CH_2 \longleftrightarrow \overset{\oplus}{H}\;\; H-\underset{H}{C}{=}\underset{H}{C}-\overset{\ominus}{C}H_2 \longleftrightarrow \overset{\oplus}{H}\;\; \underset{H}{\overset{H}{C}}{=}\underset{H}{C}-\overset{\ominus}{C}H_2 \longleftrightarrow H-\underset{\overset{\oplus}{H}}{\overset{H}{C}}{=}\underset{H}{C}-\overset{\ominus}{C}H_2$$

The number of resonating structures of this hybrid increases with the increase in the number of alkyl substituents and this consequently increases the stabilization. However, it should be remembered that actually it is the relative stability of the *transition state* leading to the formation of alkene rather than that of the alkene itself which is responsible for Saytzeff type of elimination. What actually happens is that in the transition state sigma electrons of γ-carbon-hydrogen bond become hyperconjugated with the partly formed α, β-double bond, thereby reducing its energy and facilitating the reaction.

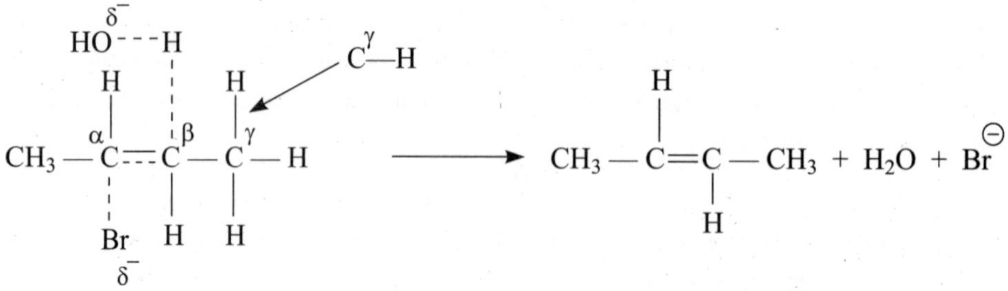

$$CH_3-\underset{\underset{\overset{\delta^-}{Br}}{\vdots}}{\overset{H}{\overset{|}{\overset{\alpha}{C}}}}{\cdots}\underset{H}{\overset{\overset{\delta^-}{HO}\cdots H}{\overset{\beta}{C}}}-\underset{H}{\overset{H}{\overset{\gamma}{C}}}-H \;\;(\overset{\gamma}{C}-H) \longrightarrow CH_3-\overset{H}{C}{=}\underset{H}{C}-CH_3 + H_2O + \overset{\ominus}{Br}$$

(*ii*) Dehydration of Alcohols

Upon heating with an acid, alcohols eliminate a molecule of water to produce alkenes. This process is known as **dehydration**.

$$\underset{\text{An alcohol}}{-\overset{\overset{H}{|}}{\underset{|}{C}}-\overset{|}{\underset{\underset{OH}{|}}{C}}-} \xrightarrow[\Delta]{H^{\oplus}} \underset{\text{An alkene}}{>C=C<} + H_2O$$

The above reaction can be carried out in the two ways:

(*i*) Heating a mixture of alcohol and conc. sulphuric acid or phosphoric acid.

(*ii*) Passing alcohol vapours on preheated catalyst such as Al_2O_3.

The ease of dehydration of various types of alcohols follows the order : **3° > 2° > 1°**. The following examples illustrate this fact :

$$\underset{\text{Ethanol (1° Alcohol)}}{CH_3CH_2OH} \xrightarrow[443\ K]{95\%\ H_2SO_4} CH_2{=}CH_2$$

$$\underset{\text{1-Butanol (1° Alcohol)}}{CH_3CH_2CH_2CH_2OH} \xrightarrow[413\ K]{75\%\ H_2SO_4} \underset{\text{Major product}}{CH_3-CH{=}CH-CH_3}$$

$$\underset{\text{2-Butanol (2° Alcohol)}}{CH_3CH_2CHOHCH_3} \xrightarrow[373\ K]{60\%\ H_2SO_4} \underset{\text{Major product}}{CH_3-CH{=}CH-CH_3}$$

$$\underset{\text{2-Methyl-2-propanol}}{H_3C-\overset{\overset{CH_3}{|}}{\underset{\underset{CH_3}{|}}{C}}-OH} \xrightarrow[360\ K]{20\%\ H_2SO_4} \underset{\text{1-Methyl-1-propene}}{H_3C-\underset{\underset{CH_3}{|}}{C}{=}CH_2}$$

$$\underset{\text{Ethanol}}{CH_3CH_2OH} \xrightarrow[630\ K]{Al_2O_3} H_2C{=}CH_2 + H_2O$$

Mechanism

This reaction is another example of an E1 elimination and involves three distinct steps as follows:

First step: In this step a proton from an acid coordinates with the —OH group of the alcohol forming the protonated alcohol (oxonium ion).

$$\underset{\text{Ethanol}}{CH_3-CH_2-\ddot{\underset{..}{O}}-H} + \underset{\text{(From } H_2SO_4)}{\overset{\oplus}{H}} \rightleftharpoons \underset{\text{The oxomium ion}}{CH_3-CH_2-\overset{\oplus}{\underset{\underset{H}{|}}{O}}-H} + \overset{\ominus}{HSO_4}$$

Second step: The protonated alcohol undergoes heterolysis to form a carbocation and water.

$$CH_3-CH_2-\overset{\oplus}{\underset{\underset{H}{|}}{O}}-H \rightleftharpoons \underset{\text{Ethyl cation}}{CH_3-\overset{\oplus}{C}H_2} + H_2O$$

Third step**:** The carbocation, being a highly reactive species, eliminates a proton from the β-position to form alkene. The removal of proton may also be facilitated by HSO_4^- which is the conjugate base of H_2SO_4.

$$\underset{\beta}{H_2\overset{H}{\overset{|}{C}}} - \underset{\alpha}{\overset{\oplus}{C}H_2} \underset{}{\overset{\text{Fast}}{\rightleftharpoons}} H_2C{=}CH_2 + \overset{\oplus}{H}$$

Here, it may be mentioned that unlike E2 elimination of alkyl halides, the dehydration of alcohols is reversible and to drive the reaction in the forward direction besides strong acids, high temperatures are needed to derive away the volatile alkene from the reaction mixture.

Evidence in support of the proposed mechanism

As discussed above, this reaction involves the formation of a carbocation as intermediate. The following facts are in agreement with the formation of a carbocation as an intermediate:

(*a*) Ease of dehydration of alcohol

Step 2, *i.e*., formation of the carbocation is the rate determining step. It is , therefore, reasonable to believe that stabler cation will be formed more readily. It means that the alcohol producing such a carbocation will dehydrate more readily. We know that the order of stability of carbocations are 3° > 2° > 1°. *The ease of dehydration of alcohols also follows the order* 3° > 2° > 1°. *Hence, the formation of carbocation as an intermediate is proved.*

(*b*) Formation of rearranged product

Dehydration of some alcohols give rise to that alkene, as a major product, in which the position of double bond is different from the expected position. In other words, in such cases the alkene is formed as a result of rearrangement. For example, acid-catalyzed dehydration of 1-butanol.

$$\underset{\substack{\text{1-Butene (20\%)}\\ (\textit{Normal})}}{CH_3 - CH_2 - CH{=}CH_2} \xleftarrow[\Delta]{\text{conc. } H_2SO_4} CH_3CH_2CH_2CH_2OH \xrightarrow[\Delta]{\text{conc. } H_2SO_4} \underset{\substack{\text{2-Butene (80\%)}\\ (\textit{Rearranged})}}{CH_3 - CH{=}CH - CH_3}$$

The formation of products such as 2-butene supports the view that these reactions involve the formation of carbocations as intermediates. It is well known that carbocations readily undergo rearrangements to produce more stable carbocations. If the structures permit, these rearrangements may take place either by a 1,2-*hydride shift* or a 1,2-*alkyl shift*. Thus, the formation of rearranged alkene supports the involvement of carbocation intermediates. Some examples are as under:

***(i) Rearrangement involving* 1,2-*hydride shift*:** In the case of *n*-butyl alcohol, the initially formed *n*-butyl cation being *primary* in nature undergoes a 1, 2-hydride shift to form *secondary* butyl cation before loosing a proton. Thus,

$$CH_3CH_2CH_2CH_2OH \xrightarrow[-H_2O]{+\overset{\oplus}{H}} CH_3CH_2-\underset{\underset{H}{|}}{C}H-\overset{\oplus}{C}H_2 \xrightarrow[\text{shift}]{\text{1, 2-hydride}} CH_3CH_2-\overset{\oplus}{C}H-CH_3$$

Butan-1-ol; 1° Cation (less stable); 2° Cation (more stable)

$$CH_3-\underset{\underset{H}{|}}{C}H-\overset{\oplus}{C}H-CH_3 \xrightarrow[-\overset{\oplus}{H}]{} CH_3-CH{=}CH-CH_3$$

But-2-ene

(ii) *Rearrangement involving* 1, 2-*alkyl shift*: Acid catalyzed dehydration of 3,3-dimethylbutan-2-ol gives 2,3-dimethylbutan-2-ene as the major product which is a rearranged product. The normal product *i.e.*, 3,3-dimethylbut-1-ene is obtained as a minor product.

$$CH_3-\underset{\underset{CH_3}{|}}{\overset{\overset{CH_3}{|}}{C}}-\underset{\underset{OH}{|}}{C}H-CH_3 \xrightarrow[\Delta]{\text{Conc. } H_2SO_4} \underset{CH_3}{\overset{CH_3}{>}}C{=}C\underset{CH_3}{\overset{CH_3}{<}} + CH_3-\underset{\underset{CH_3}{|}}{\overset{\overset{CH_3}{|}}{C}}-CH{=}CH_2$$

3, 3-Dimethylbutan-2-ol; 2, 3-Dimethylbut-2-ene (*Rearranged product*) (*major*); 3, 3-Dimethylbut-1-ene (*Normal product*) (*minor*)

The formation of rearranged product may be explained by rearrangement of the initially formed carbocation which is *secondary* to the more stable *tertiary* carbocation by 1,2-methyl shift as shown below:

$$H_3C-\underset{\underset{CH_3}{|}}{\overset{\overset{CH_3}{|}}{C}}-\underset{\underset{OH}{|}}{C}H-CH_3 \underset{}{\overset{\overset{\oplus}{H}}{\rightleftharpoons}} CH_3-\underset{\underset{CH_3}{|}}{\overset{\overset{CH_3}{|}}{C}}-\underset{\underset{\overset{\oplus}{O}H_2}{|}}{C}H-CH_3 \xrightarrow[-H_2O]{\text{Slow}} CH_3-\underset{\underset{CH_3}{|}}{\overset{\overset{CH_3}{|}}{C}}-\overset{\oplus}{C}H-CH_3$$

2°-Carbocation (*less stable*)

$$\xrightarrow[\text{shift}]{\text{1, 2 - methyl}} CH_3-\overset{\overset{CH_3}{|}}{\underset{\oplus}{C}}-\underset{\underset{H}{|}}{\overset{\overset{CH_3}{|}}{C}}-CH_3 \xrightarrow[-\overset{\oplus}{H}]{\text{Fast}} \underset{CH_3}{\overset{CH_3}{>}}C{=}C\underset{CH_3}{\overset{CH_3}{<}}$$

3°-Carbocation (*more stable*); 2, 3-Dimethylbut-2-ene

(c) Regiochemistry or Orientation in dehydration: When the formation of isomeric alkenes is possible *during the dehydration of an alcohol, the more stable, i.e., the more highly substituted alkene is always the major product.* For instance,

Chapter 6

$$\underset{\text{Butan-2-ol}}{CH_3-CH_2-\underset{OH}{\underset{|}{CH}}-CH_3} \xrightarrow[\Delta]{H_2SO_4} \underset{\substack{\text{But-2-ene}\\(major\ product)}}{CH_3CH{=}CHCH_3} + \underset{\substack{\text{But-1-ene}\\(minor\ product)}}{CH_3CH_2CH{=}Ch_2}$$

$$\underset{\text{2-Methylbutan-1-ol}}{CH_3CH_2-\underset{CH_3}{\underset{|}{CH}}-CH_2OH} \xrightarrow[\Delta]{H_2SO_4} \underset{\substack{\text{2-Methylbut-2-ene}\\(major\ product)}}{CH_3-CH{=}C(CH_3)_2} + \underset{\substack{\text{2-Methylbut-1-ene}\\(minor\ product)}}{CH_3CH_2\overset{CH_3}{\overset{|}{C}}{=}CH_2}$$

The formation of these products also supports the involvement of carbocation which rearranges to more stable carbocation followed by loss of a proton.

***(iii) Pyrolysis of quaternary ammonium hydroxides*: Hoffmann Elimination:** Many alkenes are conveniently prepared by heating tetraalkylammonium hydroxides. This is the basis of the Hoffmann's method of elimination of nitrogen from a base by *exhaustive methylation* followed by pyrolysis of the quaternary salt in presence of silver oxide.

$$CH_3CH_2-\overset{CH_3}{\overset{|\oplus}{\underset{CH_3}{\underset{|}{N}}}}-CH_2CH_2CH_3\overset{\ominus}{O}H \xrightarrow[\Delta]{Ag_2O} \underset{(96\%)}{CH_2{=}CH_2} + CH_3CH_2CH_2N(CH_3)_2 + H_2O$$

Mechanism

Mechanism of this E2 reaction involves abstraction of a proton from the β-carbon atom by hydroxide ion with the simultaneous expulsion of a tertiary base from the α-carbon atom. This reaction, thus results in the formation of a double bond between and β-carbon atoms.

$$CH_3-\overset{H}{\overset{|}{\underset{H}{\underset{|}{C}}}}-CH_2-\overset{\oplus}{N}(CH_3)_2-\overset{\alpha}{C}H_2-\overset{\beta}{\overset{H}{\overset{|}{\underset{H}{\underset{|}{C}}}}}-H \quad \overset{\ominus}{O}H \xrightarrow{\Delta} CH_3CH_2CH_2\underset{CH_3}{\underset{|}{N}}-CH_3 + CH_2{=}CH_2 + H_2O$$

Regiochemistry

These eliminations are governed by the well-known **Hoffmann rule** which states that '*the charged substrates yield the least substituted olefins*'.

$$\left[CH_3-CH_2-\overset{CH_2CH_3}{\overset{|\oplus}{N}}(H_2C-CH_2CH_3)-CH_2-CH_2-CH_3\right]\overset{\ominus}{O}H \xrightarrow{\Delta} \underset{(96\%)}{CH_2{=}CH_2} + \underset{(4\%)}{CH_3-CH{=}CH_2}$$

The following explanation seems to be reasonable for Hoffmann elimination:

In the presence of strong electron withdrawing leaving group, positive charge is induced on all the neighbouring atoms. This renders β-hydrogen atoms acidic so that they can be abstracted easily. But if an electron releasing alkyl group is attached to this carbon, the abstraction of β-hydrogen becomes a bit difficult. In other words, if the ammonium salt has ethyl and *n*-propyl attached to nitrogen, ethyl group is lost more readily to form ethene.

More acidic (H) — Less acidic (H)

$$H{-}CH_2{-}CH_2{-}\overset{\oplus}{N}(CH_3)_2{-}CH_2{-}CH_2 \leftarrow CH_3$$

$$\xrightarrow{\overset{\ominus}{O}H} CH_2{=}CH_2 + H_2O + CH_3CH_2CH_2N(CH_3)_2$$

Hoffmann *vs* Saytzeff Elimination

Here it may be worthwhile to compare regiochemistry of Hoffmann and Saytzeff eliminations which are quite different. For example,

Saytzeff elimination

$$CH_3CH_2{-}\underset{Br}{CH}{-}CH_3 \xrightarrow[\Delta]{NaOEt,\ EtOH} \underset{\substack{\text{But-2-ene} \\ (80\%)}}{CH_3{-}CH{=}CH{-}CH_3} + \underset{\substack{\text{But-1-ene} \\ (20\%)}}{CH_3CH_2CH{=}CH_2}$$

Hoffmann elimination

$$CH_3CH_2{-}\underset{\overset{\oplus}{N}(CH_3)_3\ \overset{\ominus}{O}H}{CH}{-}CH_3 \xrightarrow{\Delta} \underset{\substack{\text{But-2-ene} \\ (5\%)}}{CH_3{-}CH{=}CH{-}CH_3} + \underset{\substack{\text{But-1-ene} \\ (95\%)}}{CH_3CH_2CH{=}CH_2}$$

The reasons for this selectivity are not well known. There are two probable explanations as below:

(*a*) *The transition states are different.*

(*b*) *Steric factors.*

(*iv*) *Dehalogenation of vicinal dihalides*

The dihalogen compounds in which the two halogen atoms are attached to the adjacent carbon atoms are called vicinal(vic) dihalides. These, when heated with zinc dust, lose a molecule of halogen as zinc

Chapter 6

halide to form alkenes. The reaction is known as **dehalogenation**. Thus:

$$Br-CH_2-CH_2-Br + Zn \xrightarrow{\Delta/\text{Alcohol}} CH_2{=}CH_2 + ZnBr_2$$

1, 2-Dibromoethane (*Ethylene bromide*) → Ethene

$$CH_3-\underset{\underset{Br}{|}}{CH}-\underset{\underset{Br}{|}}{CH_2} + Zn \xrightarrow{\Delta/\text{Alcohol}} CH_3CH{=}CH_2 + ZnBr_2$$

1, 2-Dibromopropane (*Propylene bromide*) → Propene

This reaction is not a useful method for the preparation of alkenes since, the dihalides needed for the purpose are themselves prepared from alkenes by the addition of halogens. However, this method is very suitable for *purification of compounds containing double bond and for protection of carbon-carbon double bond in chemical reactions.*

Mechanism

Dehalogenation of vicinal dihalides occurs by an E2 mechanism :

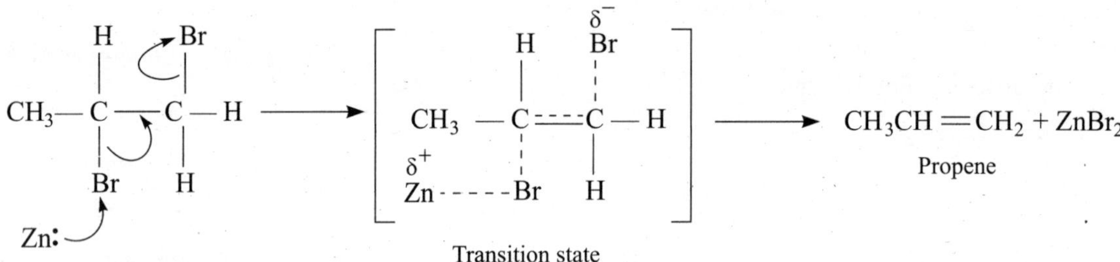

6.4 PHYSICAL PROPERTIES OF ALKENES

1. Physical State and Smell

Lower alkenes, *i.e.*, ethene, propene and butene are colourless gases; higher ones are liquids and those with C_{16} onwards are solids. Except ethene which has a pleasant smell, all other alkenes are odourless gases.

2. Melting Points

The melting points of alkenes are higher than those of the corresponding alkanes. This may be due to close packing of alkene molecules having more polarisable π-electrons than alkane molecules having less polarisable σ-electrons. Due to this intermolecular forces of attraction are stronger in alkenes than in alkanes. Hence, alkenes have higher *m.pt.* than the corresponding alkanes.

Amongst isomeric alkenes, *trans-alkenes have higher m. pt. than their corresponding cis-isomers.* For example,

$$\underset{\textit{cis}\text{-But-2-ene (m.pt. 134 K)}}{\begin{matrix} H & & H \\ & C{=}C & \\ H_3C & & CH_3 \end{matrix}} \qquad \underset{\textit{trans}\text{-But-2-ene (m.pt. 167 K)}}{\begin{matrix} H & & CH_3 \\ & C{=}C & \\ H_3C & & H \end{matrix}}$$

cis-But-2-ene (m.pt. 134 K) *trans*-But-2-ene (m.pt. 167 K)

It may be explained on the basis of the fact that the molecules of *trans*-alkene being symmetrical have more closely packed crystal lattice than the molecules of *cis*-alkenes which are less symmetrical. Thus, more energy is needed to break the crystal lattice of *trans*-alkene than that of the corresponding *cis*-alkene.

3. Boiling Points

Like those of alkanes, the boiling points of alkenes show a regular gradation with the increase in the number of carbon atoms. Except for the first few members, the b.pt. of alkenes increase by about 20°–30° for each additional carbon atom. Further, the branched chain alkenes have lower b.pt. than the corresponding straight chain alkenes. *cis-Alkenes have higher b.pt. than the corresponding trans-isomers.*

$$\begin{matrix} H & & H \\ & C{=}C & \\ H_3C & & CH_3 \end{matrix} \qquad \begin{matrix} H & & CH_3 \\ & C{=}C & \\ H_3C & & H \end{matrix}$$

cis-But-2-ene (b.pt. 277 K) *trans*-But-2-ene (b.pt. 274 K)

This may be due to the fact that *cis-alkenes are polar and have stronger dipole-dipole interactions while trans-alkenes are non-polar or weakly polar and have weak induced dipole interactions.* So, in the case of *cis*-isomer some extra energy is needed to overcome these attractive forces.

4. Dipole Moments

Symmetrical trans-alkenes, though may have polar bonds but have zero dipole moment and hence are non-polar. This is due to the fact that the dipole moments of individual groups being equal and opposite exactly cancel out each other. *Unsymmetrical trans-alkenes, however, have small but finite dipole moments.* This is because of the fact that though the two dipoles are opposite but they do not cancel out each other since they are unequal.

$$\begin{matrix} H & & CH_3 \\ & C{=}C & \\ H_3C & & H \end{matrix} \qquad \begin{matrix} H & & CH_3 \\ & C{=}C & \\ CH_3CH_2 & & H \end{matrix} \qquad \begin{matrix} H_3C & & CH_3 \\ & C{=}C & \\ H & & H \end{matrix}$$

trans-But-2-ene ($\mu = 0$) *trans*-Pent-2-ene ($\mu > 0$) *cis*-But-2-ene ($\mu = 0.25$ D)

Chapter 6

Similarly, unsymmetrical terminal alkenes such as propene and but-1-ene have finite dipole moments since the polarity due to the alkyl group is not cancelled out by corresponding polarity in the opposite direction. Therefore,

$$\underset{\substack{\text{Prop-1-ene} \\ (\mu = 0.35\ \text{D})}}{(H_3C)(H)C{=}C(H)(H)} \qquad \underset{\substack{\text{But-1-ene} \\ (\mu > 0.37\ \text{D})}}{(CH_3CH_2)(H)C{=}C(H)(H)}$$

5. Solubility

Alkenes are practically insoluble in water and other polar solvents but are highly soluble in non-polar solvents such as benzene, carbon tetrachloride, petroleum ether etc., and weakly polar solvents such as ether.

6. Density

The densities of alkenes increase with the increase in molecular mass till they reach a limiting value of 0.89 g cm^{-3}. Thus, *all alkenes are lighter than water.*

6.5 RELATIVE STABILITIES OF ALKENES : HEATS OF HYDROGENATION

We know that when alkenes are treated with molecular hydrogen in presence of finely divided metals such as Pt, Pd or Raney Ni; they form alkanes. This reaction is called **hydrogenation** and is exothermic. For example,

$$H_2C{=}CH_2 + H_2 \xrightarrow{\text{Raney Ni}} H_3C-CH_3 + \text{Heat}$$

$$H_3C-CH{=}CH_2 + H_2 \xrightarrow{\text{Pt}} CH_3-CH_2-CH_3 + \text{Heat}$$

Hydrogenation of alkenes has been found to be useful in determining the following:

(*a*) **Number of carbon-carbon double or triple bonds** (one mole of hydrogen is consumed per double bond and two moles of hydrogen are consumed per triple bond).

(*b*) **Relative stabilites of alkenes:** Here, we will discuss the use of hydrogenation for determining relative stabilities of alkenes. We have seen above that hydrogenation of alkenes produces heat. The amount of heat evolved on hydrogenation of an alkene is called **heat of hydrogenation** (ΔH°). The value of heat of hydrogenation provides very useful information about the relative stabilities of alkenes.

Heats of hydrogenation of various alkenes are given in Table 6.1. In general *lower is the heat of hydrogenation stabler is the alkene.*

Evidently, an alkene, having lower heat of hydrogenation must possess lower inherent energy and hence will be more stable. For example, both *cis*-and *trans*-2-butenes consume one mole of hydrogen per mole of alkene and produce the same product butane. Whereas, the *cis*-isomer evolves 119 kJ $mole^{-1}$ of heat, the *trans*-isomer evolves 114 kJ $mole^{-1}$ of heat. Since in this reaction the *trans*-isomer evolves 5 kJ $mole^{-1}$ less heat than the *cis*-isomer, it means that *trans*-isomer has 5 kJ $mole^{-1}$ less energy. Thus, *trans-isomer is more stable than cis-isomer* as shown in Fig. 6.1.

Table 6.1: Heats of Hydrogenation of Alkenes

S. No.	Name of alkene	Structure	Heat of hydrogenation (kJ mole^{-1})
1.	Ethene	$H_2C{=}CH_2$	136
	Monosubstituted Alkenes		
2.	Propene	$CH_3{-}CH{=}CH_2$	125
3.	1-Butene	$CH_3{-}CH_2{-}CH{=}CH_2$	126
4.	1-Pentene	$CH_3CH_2CH_2CH{=}CH_2$	126
5.	1-Hexene	$CH_3CH_2CH_2CH_2CH{=}CH_2$	126
6.	3-Methyl-1-butene	$(H_3C)_2\,CHCH{=}CH_2$	126
	Disubstituted Alkenes		
7.	*Cis*-2-butene	$(H_3C)(H)C{=}C(CH_3)(H)$ (H₃C and CH₃ on same side)	119
8.	*Trans*-2-butene	$(H_3C)(H)C{=}C(H)(CH_3)$ (H₃C and CH₃ on opposite sides)	114
9.	*Cis*-2-pentene	$(H_3C)(H)C{=}C(CH_2CH_3)(H)$	117
10.	*Trans*-2-pentene	$(H_3C)(H)C{=}C(H)(CH_2CH_3)$	113
11.	2-Methyl-1-butene	$CH_3CH_2C(CH_3){=}CH_2$	119
	Trisubstituted Alkenes		
12.	2-Methyl-2-pentene	$(H_3C)_2C{=}CH{-}CH_2{-}CH_3$	112
13.	2, 3-Dimethyl-2-butene	$(H_3C)_2C{=}C(CH_3)_2$	110
14.	2-Methyl-2-butene	$H_3C{-}C(CH_3){=}CHCH_3$	112

Chapter 6

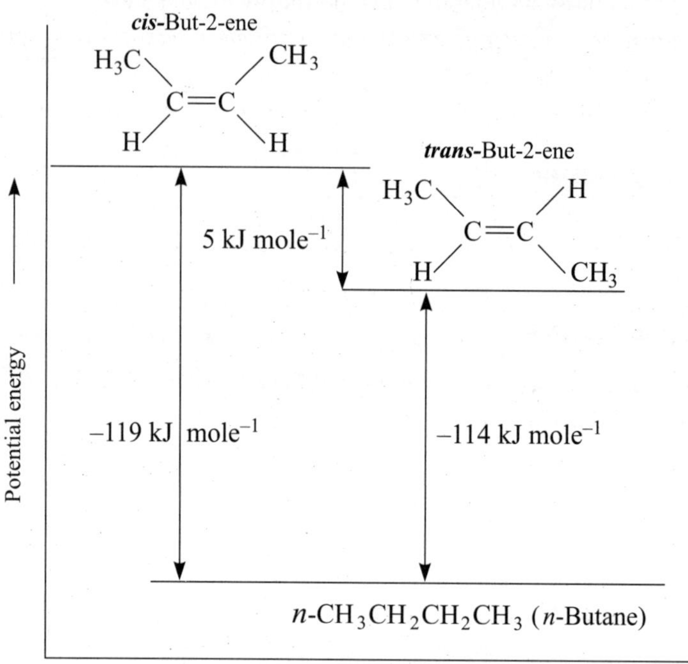

Fig. 6.1 : Stability of isomeric alkenes

Similary, *trans*-2-pentene (ΔH° = 113 kJ $mole^{-1}$) must be more stable by 4 kJ $mole^{-1}$ than the *cis*-isomer (ΔH° = 117 kJ $mole^{-1}$).

Thus, in general *trans*-alkene should be more stable than *cis*-alkene. The reason for this is that in *trans*-isomer the two larger substituents are located farther apart from each other compared to *cis*-isomer. This increase in distance results in less crowding and less Van der Waals strain.

The heat of hydrogenation data shows that the stability of alkenes also depends upon the position of double bond. For example,

$CH_3CH_2—CH=CH_2$
1-Butene
(ΔH° = – 126 kJ $mole^{-1}$)

$CH_3—CH=CH—CH_3$
2-Butene
cis (ΔH° = – 119 kJ $mole^{-1}$)
trans (ΔH° = – 114 kJ $mole^{-1}$)

$H_3CCH_2CH_2=CH_2$
1-Pentene
(ΔH° = – 126 kJ $mole^{-1}$)

$CH_3—CH_2—CH=CH—CH_3$
2-Pentene
cis (ΔH° = – 117 kJ $mole^{-1}$)
trans (ΔH° = – 113 kJ $mole^{-1}$)

$H_3C—CH(CH_3)—CH=CH_2$
3-Methyl-1-butene
(ΔH° = –126 kJ $mole^{-1}$)
(*Monosubstituted*)

$CH_3—CH_2—C(CH_3)=CH_2$
2-Methyl-1-butene
(ΔH° = –119 kJ $mole^{-1}$)
(*Disubstituted*)

$H_3C—CH=C(CH_3)—CH_3$
2-Methyl-2- butene
(ΔH° = –112 kJ $mole^{-1}$)
(*Trisubstituted*)

Based upon the heats of hydrogenation data the following generalizations can be made:

(*i*) *An unsubstituted alkene such as ethene has the highest heat of hydrogenation.*

(*ii*) *Greater the number of alkyl groups attached to the doubly bonded carbon atoms, more stable is the alkene. In general, alkenes have the following decreasing order of stability:*

$$R_2C=CR_2 > R_2C=CHR > RCH=CHR > R_2C=CH_2 > RCH=CH_2 > CH_2=CH_2$$

(*iii*) *Out of geometrical isomers, the trans-isomer is slightly more stable than the corresponding cis-isomer.*

6.6 CHEMICAL REACTIONS

The typical reactions of olefins are addition to the carbon-carbon double bond. These reactions may be initiated by an electrophile, a nucleophile, or a free radical. All the three reactions take place by a two-step mechanism, second step consisting of the combination of the resultant intermediate with a negative species, a positive species or a neutral entity respectively.

Electrophilic addition

$$>C=C< \; + \; E^{\oplus} \longrightarrow >\overset{\oplus}{C}-C(E)< \xrightarrow{Y^{\ominus}} >C(Y)-C(E)<$$

Nucleophilic addition

$$Nu^{\ominus} \; + \; >C=C< \longrightarrow >C(Nu)-\overset{\ominus}{C}< \xrightarrow{X^{\oplus}} >C(Nu)-C(X)<$$

Free-radical addition

$$>C=C< \; + \; X^{\cdot} \longrightarrow >\dot{C}-C(X)< \xrightarrow{X:Y} >C(Y)-C(X)< \; + \; X^{\cdot}$$

Since, a multiply bonded carbon atom can readily supply the comparatively losely held π-electrons, electrophilic addition is the characteristic reaction of alkenes. In contrast, the π-electron cloud tends to shield the molecule from attack by nucleophilic reagents. Such reactions, however, may be initiated in the presence of substituents in the molecule which have a high electron attracting power and these are able to polarize the π-electrons in such a manner that the β-carbon becomes electron deficient and hence enters into bonding interaction with an approaching nucleophile. We shall discuss in this section the detailed mechanisms of these two types of reactions, in addition to some special cases such as epoxidation and hydroxylation where the attack at the two carbons of the double bond is simultaneous. Free radical additions to alkenes is quite general. Some of the reactions are:

Chapter 6

(*i*) Reduction with Di-imide

The double bond of an alkene (or any other multiple bond) can be reduced by using hydrazine in conjunction with hydrogen peroxide, oxygen or potassium ferricyanide. The active reducing agent in this reaction has been shown to be *cis*-di-imide which transfers two hydrogens to the double bond *via* a cyclic transition state.

$$>C=C< \;+\; H-N=N-H \longrightarrow \left[\begin{array}{c} >C\cdots C< \\ H \qquad\quad H \\ N\cdots N \end{array} \right] \longrightarrow \begin{array}{c} >C-C< \\ |\quad\; | \\ H\quad H \end{array} + N\equiv N$$

This reduction is stereospecific and the addition of hydrogen is invariably *cis*.

(*ii*) Addition of Halogens: Electrophilic Additions

Treatment of alkenes with halogens yields 1, 2-dihalogenated alkanes (*the adducts*). For instance,

$$CH_2=CH_2 + Br_2 \xrightarrow{CCl_4} \underset{\substack{|\\ Br}}{CH_2}-\overset{\substack{Br\\ |}}{CH_2}$$

1, 2-Dibromoethane

$$CH_3-CH=CH_2 + Br_2 \xrightarrow{\text{Ether}} CH_3-\overset{\substack{Br\\ |}}{CH}-\underset{\substack{|\\ Br}}{CH_2}$$

1, 2-Dibromopropane

This reaction is used as a test for unsaturation in organic compounds where the orange red colour of bromine is discharged without evolution of H — Br. This is due to the fact that the dibromo compound formed is colourless.

Mechanism

This reaction is a typical example of a two-step electrophilic addition. The following mechanism may be proposed :

***Step* 1:** Addition of an electrophile (generally a positively charged species) to the carbon-carbon double bond resulting in the formation of an open-chain carbocation intermediate.

$$>C=C< + Br-Br \longrightarrow >\overset{\oplus}{C}-\underset{\substack{|\\ Br}}{\overset{|}{C}}- + Br^{\ominus}$$

***Step* 2:** Addition of a nucleophile (Br^- in this case) to the intermediate carbocation to produce 1, 2-dihalide.

$$Br^{\ominus} + \overset{\oplus}{C}-\underset{Br}{\overset{|}{C}}- \longrightarrow -\underset{|}{\overset{Br}{C}}-\underset{Br}{\overset{|}{C}}-$$

Evidence for the formation of the carbocation intermediate: The best evidence for the intermediacy of the carbocation comes from the trapping experiments. Thus, when ethylene is treated with bromine in presence of some other added nucleophiles ($\overset{\ominus}{I}$, $\overset{\ominus}{O}NO_2$, $\overset{\ominus}{O}-\underset{O}{\underset{\|}{C}}-H$ etc.), a mixture of products is obtained as shown below:

$$\overset{\oplus}{C}H_2-CH_2-Br \xrightarrow{\overset{\ominus}{Br}} Br-CH_2-CH_2-Br$$

$$\overset{\oplus}{C}H_2-CH_2-Br \xrightarrow{\overset{\ominus}{I}} I-CH_2-CH_2-Br$$

$$\overset{\oplus}{C}H_2-CH_2-Br \xrightarrow{\overset{\ominus}{O}NO_2} O_2NO-CH_2-CH_2-Br$$

$$\overset{\oplus}{C}H_2-CH_2-Br \xrightarrow{\overset{\ominus}{O}-\overset{O}{\overset{\|}{C}}-H} H-\overset{O}{\overset{\|}{C}}-O-CH_2-CH_2-Br$$

However, the above mechanism fails to account for the following observations:

(*a*) How does a symmetrical bromine molecule act as an electrophile?

(*b*) The addition is stereospecific, giving rise to *trans*-addition product (*i.e.*, the two bromine atoms add on the opposite side of the double bond).

(*c*) Rearranged products are not formed.

To account for the above observations the following mechanistic scheme involving a **cyclic halonium ion** is proposed.

As the bromine molecule comes near π-electron cloud of alkene, the unshared pair of electrons on the bromine atom near the double bond is repelled because of electron-electron repulsion making this atom of bromine molecule *partially positive* and the other atom *partially negative*. In other words, because of this 'contact' the bromine molecule gets polarized. Now partially positive end of the bromine molecule adds to the double bond as an electrophile forming a carbocation and bromide ion which subsequently combine to form dibromide.

$$CH_2=CH_2 + \overset{\delta^+}{Br}-\overset{\delta^-}{Br} \longrightarrow \overset{\oplus}{C}H_2-CH_2-Br + \overset{\ominus}{Br} \longrightarrow Br-CH_2-CH_2-Br$$

There is enough evidence to show that the open chain carbocation envisaged above for sake of simplicity is not the actual intermediate, rather a *cyclic bromonium* ion is formed in the process. Thus, the initial step of an electrophilic addition is the formation of a π-complex by the attack of π-electron cloud of the alkene on the vacant or partially vacant orbital of the electrophile. Such a complex does not actually involve a formal bond and may be regarded as an association in which bromine molecule (electrophile) gets embeded in one lobe of the π-orbital leading to the formation of a cyclic intermediate called a bromonium ion in which positive bromine is bonded to both the carbon atoms of the double bond.

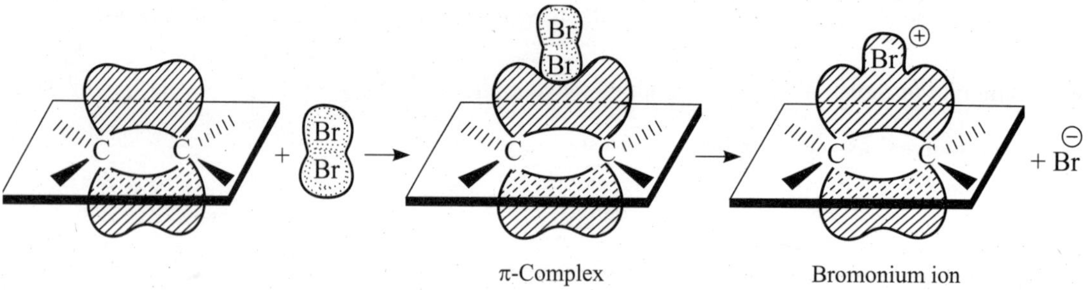

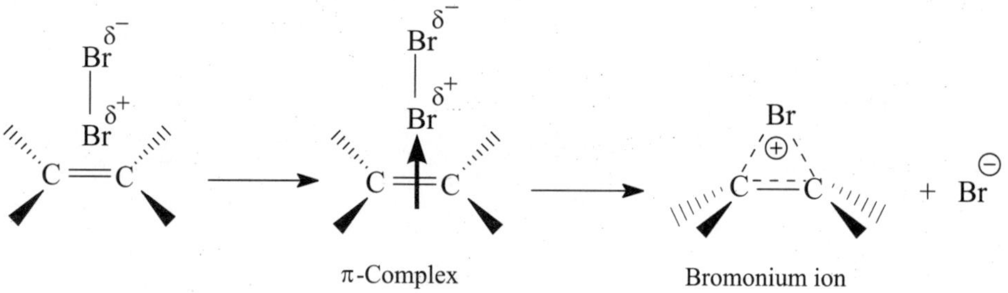

π-Complex Bromonium ion

π-Complex Bromonium ion

The above mechanism clearly explains how symmetrical molecule acts as an electrophile. As discussed below, it also explains the other facts also.

1. Formation of *trans*-products

Due to intervention of a cyclic bromonium ion, the attack of the bromide ion in the second step can only occur from the opposite side resulting exclusively in the *trans*-addition.

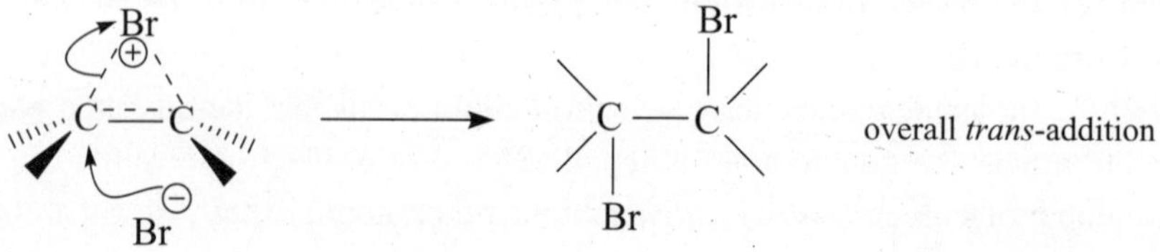

overall *trans*-addition

Direct evidence for the intermediacy of the bromonium ion has been provided by *nmr* studies.

2. Addition of bromine to *cis* and *trans*-butenes

Further evidence for the intervention of the bromonium ion comes from the stereospecific addition of bromine to isomeric 2-butenes. Thus, *cis*-2-butene exclusively forms *dl*-2,3dibromobutane and the *trans*-2-butene results in the exclusive formation of *meso*-2,3-dibromobutane.

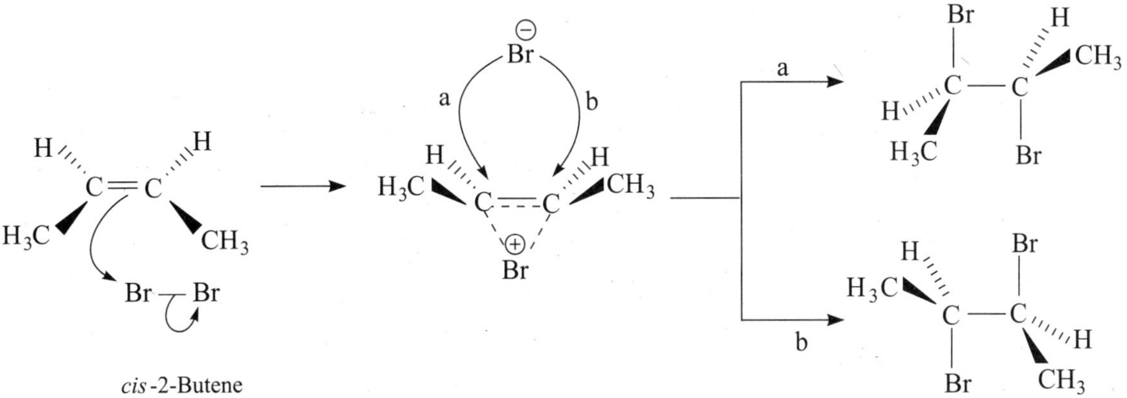

cis-2-Butene

dl-2, 3-Dibromobutane

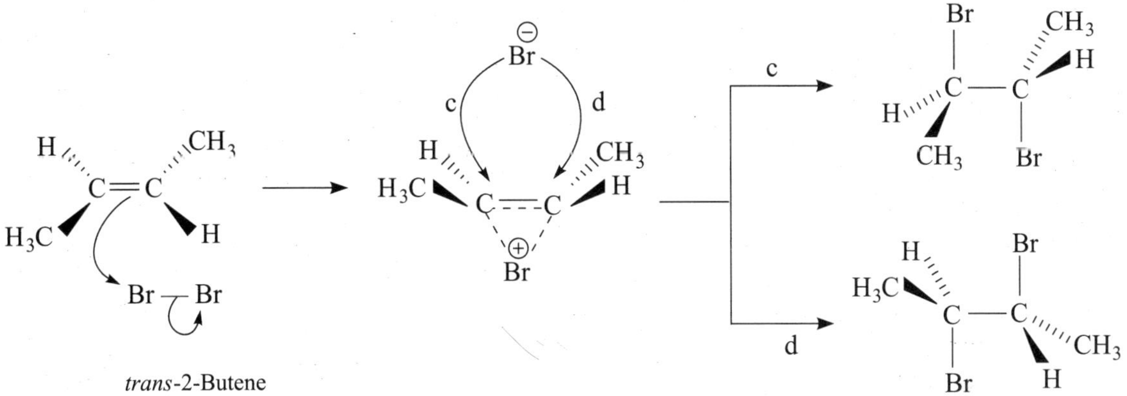

trans-2-Butene

meso-2, 3-Dibromobutane

3. Addition of bromine to cyclopentene

Addition of bromine to cyclopentene gives exclusively *trans*-1, 2-dibromocyclopentane. If this addition involves open chain carbocation as intermediate, both *cis* and *trans*-products should have been formed. Because the carbocations are planar species which can be attacked from either side to the same extent. But the formation of only *trans*-product confirms the intermediacy of *cyclic bromonium ion.*

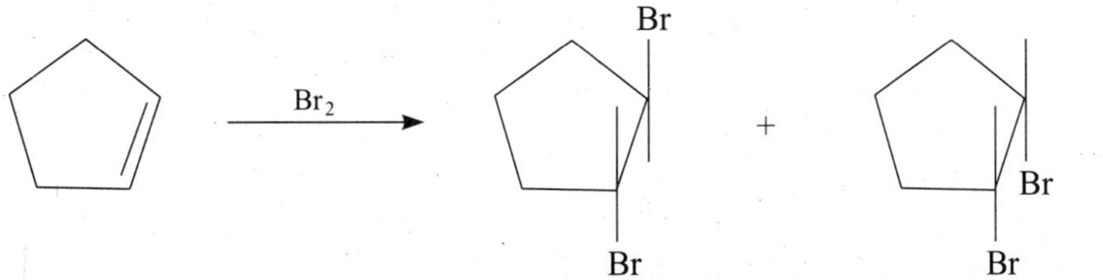

Br_2

Trans -adduct
(Exclusive product)

Cis-adduct
(Not formed)

Mechanism

δ^+ δ^-
+ Br — Br

(Bromine approaches from bottom)

$Br^{\ominus}$

Bromine ion approaches from top as the bottom side is blocked

Br

Br

4. Rearranged products are not formed

The carbocations formed as intermediates in this reaction do not rearrange. For example, 3, 3-dimethyl-1-butene upon treatment with bromine gives rise to 1, 2-dibromo-3, 3-dimethylbutane as an exclusive product and isomeric 1, 3-dibromo-2, 3-dimethylbutane is not formed at all.

$H_3C-C(CH_3)_2-CH=CH_2 \xrightarrow{\overset{\delta+}{Br}-\overset{\delta-}{Br}} H_3C-C(CH_3)_2-CH\overset{\oplus}{\underset{Br}{\triangledown}}CH_2 \xrightarrow{Br^{\ominus}} H_3C-C(CH_3)_2-CH(Br)-CH_2Br$

1, 2-Dibromo-3, 3-dimethylbutane (100%)

$H_3C-C(CH_3)_2-\overset{\oplus}{C}H-CH_2Br \xrightarrow{\text{1, 2 methyl}} H_3C-\overset{\oplus}{C}(CH_3)-CH(CH_3)-CH_2Br \xrightarrow{Br^{\ominus}} H_3C-C(CH_3)(Br)-CH(CH_3)-CH_2Br$

Rearranged product (Not formed at all)

Reactivity of Alkenes Towards Electrophilic Addition Reactions

As discussed above, the first step in the addition of an electrophile to an alkene involves the formation of a carbocation intermediate (may be even cyclic). Thus, it may be said that the rate of electrophilic addition will depend upon the stability of the carbocation involved. In other words, *stabler is the carbocation, the more reactive is the alkene*.

Let us consider the following examples:

$$H_3C-CH=CH_2 + \overset{\oplus}{E} \longrightarrow H_3C-\overset{\oplus}{C}H-CH_2E \quad ...(i)$$

Propene → 2° Carbocation

$$H_3C-CH_2-CH=CH_2 + \overset{\oplus}{E} \longrightarrow H_3C-CH_2-\overset{\oplus}{C}H-CH_2E \quad ...(ii)$$

1-Butene → 2° Carbocation

$$H_3C-CH=CH-CH_3 + \overset{\oplus}{E} \longrightarrow H_3C-\overset{\oplus}{C}H-\underset{\displaystyle E}{\underset{|}{C}H}-CH_3 \quad ...(iii)$$

2-Butene → 2° Carbocation

$$H_3C-\overset{\displaystyle CH_3}{\overset{|}{C}}=CH_2 + \overset{\oplus}{E} \longrightarrow H_3C-\underset{\oplus}{\overset{\displaystyle CH_3}{\overset{|}{C}}}-CH_2E \quad ...(iv)$$

2-Methylpropene → 3° Carbocation

$$H_2C=CH_2 + \overset{\oplus}{E} \longrightarrow H_2\overset{\oplus}{C}-CH_2E \quad ...(v)$$

Ethene → 1° Carbocation

$$Cl-CH=CH_2 + \overset{\oplus}{E} \longrightarrow Cl-\overset{\oplus}{C}H-CH_2E \quad ...(vi)$$

Chloroethene → 2° Carbocation

It has been found that 2-methylpropene is more reactive than propene, 1-butene or 2-butene which in turn are more reactive than ethene. This may be explained on the basis of the fact that 2-methylpropene gives a 3° carbocation which is more stable than 2° carbocation obtained from propene, 1-butene or 2-butene and the corresponding cations in turn are more stable than 1° carbocation obtained in the case of ethene.

The case of chloroethene is different where the secondary carbocation formed is destabilized due to – I-effect of adjacent chlorine atom. Hence, $Cl-\overset{\oplus}{C}H-CH_2E$, 2° carbocation (destabilized due to – I-effect of Cl) is less reactive than ethene.

In general, the order of reactivity of alkenes towards electrophilic addition reactions decreases in the following order:

$$R_2C=CR_2 > R_2C=CHR > R_2C=CH_2 > RHC=CHR > RHC=CH_2 > CH_2=CH_2 > CH_2=CHX$$

where R = groups having + I-effect and
X = groups having – I-effect.

(*iii*) Addition of Hydrogen Halides: Markownikoff's Rule

Hydrogen halides (HCl, HBr, HI) add on a carbon-carbon double bond forming alkyl halides. The order of reactivity of hydrogen halides is HI > HBr > HCl which can be explained on the basis of bond dissociation energy. For example,

$$\rangle C{=}C\langle \; + \; HBr \longrightarrow \rangle \underset{}{\overset{H}{C}} - \underset{Br}{C}\langle$$

The orientation of addition depends upon whether the alkene is symmetrical or unsymmetrical.

(*a*) *Addition to symmetrical alkenes*: This reaction is initiated by the addition of a proton (electrophile) forming a hydrogen *bridged intermediate* called a ***protium ion*** which is attacked by the halide ion from the back side resulting in *trans*-addition.

Mechanism

$$\rangle C{=}C\langle + H{-}Br \longrightarrow \left[\rangle \underset{H}{C} - \overset{\oplus}{C}\langle \longleftrightarrow \rangle \overset{\oplus}{C} - \underset{H}{C}\langle \longleftrightarrow \rangle \underset{H^{\oplus}}{C{=}C}\langle \right] \equiv \rangle C \overset{}{\underset{\overset{\oplus}{H}}{=\!=}} C\langle \; + \; Br^{\ominus}$$

Protium ion

$$\overset{\ominus}{Br} + \rangle C \underset{\overset{\oplus}{H}}{=\!=} C\langle \longrightarrow \overset{Br}{C} - \underset{H}{C}$$

Backside attaek

(*b*) *Addition to unsymmetrical alkenes*: *Markownikoff's rule*: In the case of unsymmetrical alkenes the position of attachment of the nucleophile is governed by the nature of substituents. Addition of H—Br to propene, for example, is expected to yield two products—1-bromopropane and 2-bromopropane.

$$H_3C - CH{=}CH_2 + HBr \longrightarrow \begin{cases} CH_3 - CH_2 - CH_2 - Br & \text{1-Bromopropane} \\ CH_3 - \underset{}{\overset{Br}{CH}} - CH_3 & \text{2-Bromopropane} \end{cases}$$

However, in actual practice 2-bromopropane is the exclusive product. To explain the exclusive formation of this type of products Markownikoff, after an intensive study of such reactions, formulated the famous *Markownikoff's rule,* which states that *addition of compounds of the type HX to unsymmetrical alkenes proceeds in such a way that the hydrogen of HX adds to the carbon carrying greater number of hydrogen atoms.*

Theoretical Explanation — *Validity of Markownikoff's rule*

Markownikoff's rule was purely empirical when it was formulated. However, now a theoretical explanation of orientation of addition to olefins can be offered on the basis of relative stability of carbocations. If there is a probability of formation of more than one carbocations, the addition of the electrophile yields the more stable one. For example, in the addition of hydrogen bromide to propene, there is a possibility of the formation of either a primary or a secondary carbocation. Since, the latter is more stable, the activation energy required for its formation will be lowest as compared to activation energy required for the formation of less stable 1° carbocation (Fig 6.2). Hence, 2-bromopropane is formed as an exclusive product. Therefore, *Markownikoff's addition occurs through more stable carbocation.*

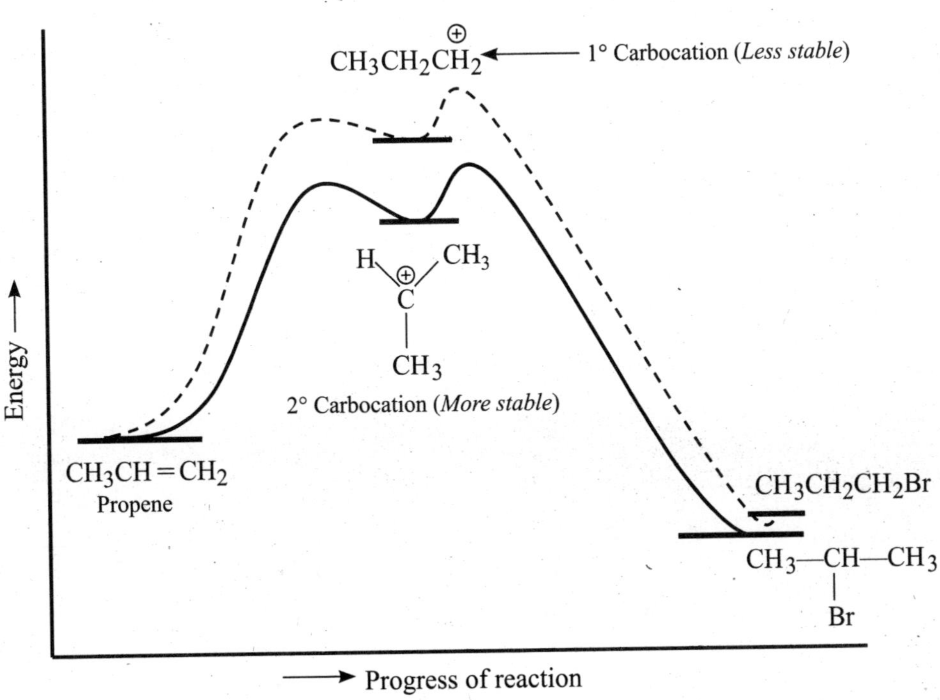

Fig. 6.2 : Potential energy diagram for the addition of HBr to propene

Chapter 6

$$CH_3-CH=CH_2 \xrightarrow{H^{\oplus}} \left[CH_3-\overset{\oplus}{C}H-CH_3\right] \longrightarrow CH_3-\underset{Br}{\underset{|}{CH}}-CH_3$$

2° Carbocation (*More stable*) — (100%)

$$CH_3-CH=CH_2 \xrightarrow{H^{\oplus}} \times \left[CH_3-CH_2-\overset{\oplus}{C}H_2\right] \longrightarrow CH_3-CH_2-CH_2Br$$

1° Carbocation (*Less stable*) — Not formed

Now, we can conclude that *validity of Markownikoff's rule depends upon the stabilities of the intermediate carbocations i.e., the energies of the transition states leading to their formation involved.*

Formation of Rearranged Product

The intermediacy of a carbocation in such reactions has also been proved by the isolation of rearranged* products on treating appropriate alkene with hydrogen halides. For examples, addition of hydrogen chloride to 3, 3-dimethyl-1-butene yields a mixture of 2-chloro-3,-3-dimethylbutane (normal product) and 2-chloro-2, 3-dimethylbutane (rearranged product).

$$H_3C-\underset{CH_3}{\overset{CH_3}{C}}-CH=CH_2 \xrightarrow{H-Cl} H_3C-\underset{CH_3}{\overset{CH_3}{C}}-\overset{\oplus}{C}H-CH_3 \xrightarrow{Cl^{\ominus}} H_3C-\underset{CH_3}{\overset{CH_3}{C}}-\underset{Cl}{CH}-CH_3$$

2° Carbocation (*Less stable*) — 2-Chloro-3,3-dimethylbutane (*Normal product*)

1, 2-Methyl shift ↓

$$(H_3C)_2\overset{\oplus}{C}-\overset{CH_3}{CH}-CH_3 \xrightarrow{Cl^{\ominus}} H_3C-\underset{Cl}{\overset{CH_3}{C}}-\underset{H}{\overset{CH_3}{C}}-CH_3$$

3° Carbocation (*More stable*) — 2-Chloro-2,3-dimethylbutane (*Rearranged product*)

The formation of rearranged product can be rationalized by considering the fact that the less stable 2° carbocation changes to more stable 3° carbocation before the addition of the nucleophile takes place.

---

*Rearrangement is a characteristic feature of reactions involving carbocations.

We can also understand the *Markownikoff's rule* by considering the electronic effect of the substituents on the ethylenic bond. In propene, for instance, the polarization of the double bond will be as depicted below due to the electron-repelling inductive effect of the methyl group, which will be transmitted through the π-bond.

$$CH_3 \rightarrow \overset{\delta^+}{C}H = \overset{\delta^-}{C}H_2$$

Thus, the positive and negative ends of the dipole H—X will add to the negative and positive ends of double bond respectively giving rise to the isopropyl derivative. Similarly, Markownikoff addition of hydrogen halide can be explained by the electron-donating resonance effect of halogen attached to a double bond dominating over its electron-attracting inductive effect. Thus, 1, 1-dihaloethane is formed from vinyl halide.

$$CH_2{=}CH{-}\ddot{X}: \xrightarrow{H-X} \left[CH_3{-}\overset{\oplus}{C}H{-}\ddot{X}: \longleftrightarrow CH_3{-}CH{=}\overset{\oplus}{X}: \right] \xrightarrow{X^{\ominus}} CH_3{-}CH(X){-}X$$

Peroxide effect: Anti-Markownikoff addition: After an extensive study on the addition of hydrogen bromide to alkenes in the presence of peroxides, **Kharasch and Mayo** found that under these conditions the addition involves *free radical mechanism*. Further, the orientation of addition in the case of asymmetrical alkenes was also shown to be contrary to Markownikoff's rule. Such additions are sometimes referred to as **anti-Markownikoff's additions**. For instance, the addition of hydrogen bromide to propene in presence of peroxides gives 1-bromopropane rather than 2-bromopropane. Since, the orientation of addition is reversed due to peculiar effect of peroxides, this is also termed as **peroxide effect** or **Kharasch effect**.

$$\underset{\text{(Not formed)}}{CH_3{-}CH(Br){-}CH_3} \xleftarrow[\text{Peroxide}]{HBr} CH_3{-}CH{=}CH_2 \xrightarrow[\text{Peroxide}]{HBr} CH_3{-}CH_2{-}CH_2{-}Br$$

Mechanism of free-radical addition of HBr to alkene: It consists of following three steps.

(a) Chain initiation: The reaction is initiated by the alkoxy radicals, produced by ***homolysis*** of peroxides which abstract an atom of hydrogen from hydrogen bromide generating bromine atoms.

$$\underset{\text{A peroxide}}{ROOR} \longrightarrow \underset{\text{Alkoxy radical}}{2\,R\dot{O}}$$

$$R\dot{O} + H{-}Br \longrightarrow ROH + \dot{Br}$$

***(b) Chain propagation*:** The bromine atom generated in this way can add to either of the two carbon atoms of the double bond in an asymmetrical alkene producing either a primary or a secondary radical. The preferential formation of the latter is in accordance with the order of stability of radicals which is tertiary > secondary > primary.

$$(i)\quad CH_3-CH=CH_2 + \dot{B}r \longrightarrow CH_3-\dot{C}H-CH_2Br$$

2°-Radical (stable)

$$(ii)\quad CH_3-\dot{C}H-CH_2Br + H-Br \longrightarrow CH_3CH_2CH_2Br + \dot{B}r$$

The secondary radical obtained above abstracts a hydrogen atom from hydrogen bromide thus producing 1-bromopropane and bromine atom which propagates the chain further.

In addition to peroxides, such free radical additions are also initiated by light, high temperature or even oxygen.

***(c) Chain termination*:** Chain termination steps involve coupling of two similar or dissimilar free radicals as shown below:

$$(i)\quad \dot{B}r + \dot{B}r \longrightarrow Br_2$$

$$(ii)\quad CH_3-\dot{C}H-CH_2Br + \dot{B}r \longrightarrow CH_3-CHBr-CHBr$$

$$(iii)\quad 2\,H_3C-\dot{C}H-CH_2Br \longrightarrow BrCH_2-\underset{CH_3}{\underset{|}{CH}}-\underset{CH_3}{\underset{|}{CH}}-CH_2Br$$

Here, it may be mentioned that peroxide effect is observed only in the case of HBr and not HF, HCl or HI.

This can be understood by considering the ΔH values (Table 6.2) of the two propagation steps involved in the addition of HF, HCl, HBr and HI.

Table 6.2: Addition of hydrogen halides to propene

Radical	ΔH (kJ mol⁻¹)	
$\dot{X}$	*First step*: Addition of halogen to double bond. $\dot{X} + H_2C=CH-CH_3 \longrightarrow H_2C(X)-\dot{C}H-CH_3$	*Second step*: Abstraction of hydrogen $H_2C(X)-\dot{C}H-CH_3 + H-X \longrightarrow H_2C(X)-CH_2-CH_3 + \dot{X}$
$F^{\bullet}$	−210	+160
$\dot{C}l$	−100	+30
$\dot{B}r$	−40	−37
$\dot{I}$	+10	−100

A look at the above table reveals that in case of HCl and HF the second step is *endothermic* and in case of HI the first step in *endothermic*. But in case of HBr, both these steps are *exothermic*. Hence, only *HBr shows peroxide effect.*

(iv) Other Free Radical Additions

Kharasch has shown that in addition to hydrogen bromide there are other compounds which may also add to carbon-carbon double bond in presence of peroxides. The most common example of such reactions is addition of carbon tetrachloride to olefins in presence of peroxides or sunlight.

$$>C=C< + Cl-CCl_3 \xrightarrow[\text{Peroxides}]{h\nu \text{ or}} >C(CCl_3)-C(Cl)<$$

The generally accepted mechanism of this reaction is analogous to what is described above. Thus,

$$\dot{R} + Cl-CCl_3 \longrightarrow R-Cl + \dot{C}Cl_3$$

$$CH_3-CH=CH_2 + \dot{C}Cl_3 \longrightarrow CH_3-\dot{C}H-CH_2-CCl_3$$

$$CH_3-\dot{C}H-CH_2-CCl_3 + Cl-CCl_3 \longrightarrow CH_3-CH(Cl)-CH_2-CCl_3 + \dot{C}Cl_3 \text{ and so on.}$$

(v) Addition of Hypohalous Acids: *Halohydrin Formation*

Hypohalous acids (HOCl, HOBr, HOI, etc.), produced by the action of halogens with water, add on carbon-carbon double bond to give halohydrins*. The order of reactivity of hypohalous acids is: HOCl > HOBr > HOI.

$$X_2 + H_2O \longrightarrow HOX + HX \qquad (X = Cl, Br)$$

$$\underset{\text{Alkene}}{>C=C<} + \underset{\text{Hypohalous acid}}{HOX} \longrightarrow \underset{\text{Halohydrin}}{-C(OH)-C(X)-}$$

For example,

$$\underset{\text{Ethene}}{CH_2=CH_2} + HOCl \longrightarrow \underset{\text{2-Chloroethanol}}{\overset{1}{C}H_2(OH)-\overset{2}{C}H_2(Cl)}$$

$$\underset{\text{Propene}}{CH_3-CH=CH_2} + HOBr \longrightarrow \underset{\text{1-Bromopropan-2-ol}}{CH_3-CH(OH)-CH_2(Br)}$$

*Compounds containing halogen atom and the hydroxyl group on adjacent carbon atoms are called halohydrins.

Mechanism

The hypohalous acids are highly polarized species with halogen as the cationoid part of the molecule ($\overset{\delta^-}{HO} - \overset{\delta^+}{X}$). Thus, these additions are also electrophilic in nature initiated by the addition of positive halogen to the carbon-carbon double bond. The addition of these reagents to unsymmetrical alkenes takes place according to Markownikoff's rule. The product is formed by the initial addition of X^+ to give a more stable carbocation regardless of the fact that whether the intermediate is an open chain carbocation or a cyclic ion. Thus, for instance, the addition of hypochlorous acid to propene can be rationalized as under:

$$CH_3-CH=CH_2 + \overset{\delta^+}{Cl}-\overset{\delta^-}{OH} \xrightarrow{\text{Slow}} CH_3-\underset{\overset{\oplus}{Cl}}{CH\text{---}CH_2} + \overset{\ominus}{O}H$$

$$[\equiv CH_3-\overset{\oplus}{C}H-CH_2Cl]$$

$$CH_3-\underset{\overset{\oplus}{Cl}}{CH\text{---}CH_2} + :\ddot{O}H_2 \xrightarrow{\text{Fast}} CH_3-\underset{\overset{|}{\overset{\oplus}{O}H_2}}{CH}-CH_2-Cl \xrightarrow[-H^{\oplus}]{} CH_3-\underset{\overset{|}{OH}}{CH}-CH_2-Cl$$

As expected these reactions also yield *trans*-addition products.

(*vi*) Addition of Sulphuric Acid—*Indirect Hydration of Alkenes*

Addition of cold conc. $H_2\,SO_4$ across a carbon-carbon double bond produces alkyl hydrogen sulphates.

$$\underset{\text{Ethene}}{CH_2{=\!=}CH_2} + \underset{\text{Sulphuric acid}}{H-O-SO_2-OH} \longrightarrow \underset{\text{Ethyl hydrogen sulphate}}{C_3H-CH_2-OSO_2OH}$$

In case of unsymmetrical alkenes, the addition takes place in accordance with Markownikoff's rule.

$$\underset{\text{Propene}}{CH_3-CH=CH_2} + H-OSO_2OH \longrightarrow \underset{\textit{iso}\text{-Propyl hydrogen sulphate}}{CH_3-\underset{\overset{|}{OSO_2OH}}{CH}-CH_3}$$

Mechanism

This is also an electrophilic addition to $C=C$ double bond and takes place by initial addition of a proton to produce a carbocation.

$$CH_3-CH{=}CH_2 + H-O-SO_2OH \xrightarrow{\text{Step I}} CH_3-\overset{\oplus}{C}H-CH_3 + \overset{\ominus}{O}-SO_2OH$$

$$\xrightarrow{\text{Step II}} CH_3-\underset{\displaystyle OSO_2OH}{\underset{|}{C}H}-CH_3$$

Applications of the Reaction

1. *Separation of alkenes from alkanes*: Alkanes are not absorbed by cold conc. H_2SO_4. This difference in behaviour of alkanes and alkenes can be used as a method for separating an alkane from an alkene. When a mixture of an alkane and an alkene is bubbled through cold conc. H_2SO_4, alkene is absorbed forming alkyl hydrogen sulphate whereas alkane escapes. The alkyl hydrogen sulphate on heating regenerates the original alkene.

$$CH_3-\underset{\displaystyle OSO_2OH}{\underset{|}{C}H}-CH_3 \xrightarrow{\Delta} \underset{\text{Alkene}}{CH_3-CH{=}CH_2} + H_2SO_4$$

2. *Formation of alcohols*: The alkyl hydrogen sulphate solution, on dilution with water and heating, undergoes hydrolysis to an alcohol having the same alkyl group as in the original alkyl hydrogen sulphate. This sequence, therefore, provides an elegant method for the conversion of an alkene into an alcohol.

$$CH_3-\underset{\displaystyle OSO_2OH}{\underset{|}{C}H}-CH_3 + H_2O \xrightarrow{\Delta} \underset{\textit{iso}\text{-Propyl alcohol}}{CH_3-\underset{\displaystyle OH}{\underset{|}{C}H}-CH_3}$$

The overall two-step conversion of an alkene into alcohol is called ***indirect hydration of alkene.***

(*vii*) Addition of Water: *Direct Hydration of Alkenes*

Many alkenes are known to add elements of water in presence of acids to yield alcohols. This addition also follows Markownikoff's rule. The mechanism of acid catalysed addition of water to *iso*-butene, for instance, can be visualized as under:

$$H_2O + H_2SO_4 \rightleftharpoons H-\overset{\oplus}{O}H_2 + \overset{\ominus}{HSO_4}$$

$$\underset{\textit{iso}\text{-Butylene}}{(H_3C)_2C{=}CH_2} + H-\overset{\oplus}{O}H_2 \underset{-H^{\oplus}}{\overset{+H^{\oplus}}{\rightleftharpoons}} (H_3C)_2\overset{\oplus}{C}-CH_3 \underset{-H_2O}{\overset{+H_2O}{\rightleftharpoons}}$$

$$(H_3C)_2\underset{\displaystyle {}^{\oplus}OH_2}{\underset{|}{C}}-CH_3 \underset{+H^{\oplus}}{\overset{-H^{\oplus}}{\rightleftharpoons}} (H_3C)_2\underset{\displaystyle OH}{\underset{|}{C}}-CH_3 + H^{\oplus}$$

It may be noted that direct hydration of alkenes is rather a simpler and commercially feasible process.

(*viii*) Hydroboration-Oxidation: *Indirect Hydration of Alkenes*

A useful synthetic procedure in organic chemistry involves addition of borane (BH_3) to a carbon-carbon double bond forming alkylboranes. This reaction discovered by H.C. Brown* is referred to as hydroboration or hydroboronation. The process cannot be stopped with addition of one molecule of olefine to BH_3 because initially formed adduct RBH_2 adds to another molecule of olefin to give R_2BH which in turn adds to third molecule of olefin to give R_3B. The whole sequence of reactions is as under:

$$CH_2{=}CH_2 + BH_3 \longrightarrow CH_3CH_2BH_2 \xrightarrow{CH_2=CH_2} (CH_3CH_2)_2\,BH \xrightarrow{CH_2=CH_2} (CH_3CH_2)_3B$$

Mechanism

Borane itself is unknown but its dimer B_2H_6 (diborane) behaves as if it were the hypothetical monomer. Because of vacant orbitals, the boron atom has a strong tendency to accept a pair of electrons. Hence, borane may be regarded as an electrophile. Thus, hydroboration is also a typical electrophilic addition. But unlike other electrophilic additions it does not involve intermediacy of carbocations and takes place in a concerted manner involving a four centered cyclic transition state. With the close approach of borane molecule towards π-electrons, the boron atom starts bonding with one of the doubly bonded carbon atoms with a positive charge developing on that carbon, which is best able to support it. Intramolecular hydride transfer then completes the reaction. Thus, the overall addition of BH_3 to the double bond takes place in accordance with *Markownikoff's rule*. For example,

$$CH_3{-}CH{=}CH_2 + H{-}BH_2 \longrightarrow \left[\begin{array}{c} \overset{\delta-}{H} \cdots\cdots \overset{\delta+}{B}H_2 \\ \vdots \qquad\quad \vdots \\ CH_3{-}\underset{\delta+}{CH} \cdots\cdots \underset{\delta-}{CH_2} \end{array}\right] \longrightarrow CH_3CH_2CH_2BH_2$$

Four centered cyclic T.S.

Because of the four centered mechanism in hydroboration the addition of borane is *cis*. As these reactions do not yield rearranged products, it can be inferred that a free carbocation is not an intermediate in the process.

The rate of addition of borane to alkenes decreases with the increase in the alkyl substituents on the double bond. For example, trisubstituted alkene gives dialkyl boranes and tetrasubstituted alkenes yield only monoalkylboranes.

* H. C. Brown shared the Nobel Prize in chemistry in 1979 for this work.

$$(CH_3)_2C{=}CH{-}CH_3 \xrightarrow{(BH_3)_2} [(CH_3)_2CH{-}\underset{}{\overset{CH_3}{\overset{|}{C}H}}]_2BH$$

$$(CH_3)_2C{=}C(CH_3)_2 \xrightarrow{(BH_3)_2} [(CH_3)_2CH{-}C(CH_3)_2]BH_2$$

Applications

The hydroboration has been found to be of immense synthetic utility as the alkylboranes can undergo a variety of useful transformations. Some important examples are:

1. *Reduction of double bond*: The alkylboranes are readily cleaved to alkanes on boiling with acetic or propionic acid. Thus, hydroboration-acid hydrolysis provides a valuable method for effecting the reduction of a double bond.

$$n\text{-}C_4H_9{-}CH{=}CH_2 \xrightarrow{(BH_3)_2} (n\text{-}C_4H_9{-}CH_2{-}CH_2)_3B \xrightarrow{C_2H_5COOH} n\text{-}C_4H_9{-}CH_2{-}CH_3$$

2. *Oxidation of trialkylboranes to alcohols*: It is a two stage conversion of alkene into alcohol. Trialkylboranes, obtained by addition of borane to alkene, on oxidation with an alkaline solution of hydrogen peroxide, give alcohols as shown below:

$$\underset{\text{Triethylborane}}{(CH_3CH_2)_3B} + 3H_2O_2 \xrightarrow{NaOH} \underset{\text{Ethanol}}{3CH_3CH_2OH} + \underset{\text{Boric acid}}{H_3BO_3}$$

$$\underset{\text{Tri-}n\text{-propylborane}}{(CH_3CH_2CH_2)_3B} + 3H_2O_2 \xrightarrow{NaOH} \underset{\text{1-Propanol}}{3CH_3CH_2CH_2OH} + H_3BO_3$$

Thus, the overall result of these two-steps—addition of borane to carbon-carbon double bond followed by oxidation of alkylborane appears to be **anti-Markownikoff's addition** of water to carbon-carbon double bond of alkene although hydroboration takes place according to *Markownikoff's rule*.

Mechanism

The oxidation of alkylboranes with alkaline hydrogen peroxide proceeds in the following steps:

Step I*: *Formation of hydroperoxide ion

$$H{-}O{-}O{-}H + \overset{\ominus}{O}H \longrightarrow \underset{\text{Hydroperoxide ion}}{H{-}O{-}\overset{\ominus}{O}} + H_2O$$

Step II*: *Nucleophilic addition of hydroperoxide ion on boron atom of trialkylborane.

$$R{-}\overset{R}{\underset{R}{B}} + \overset{\ominus}{O}{-}O{-}H \longrightarrow R{-}\overset{R}{\underset{R}{\overset{\ominus}{B}}}{-}O{-}O{-}H \quad (R = CH_3CH_2CH_2)$$

Chapter 6

Step III: *The adduct formed in step II undergoes 1, 2-alkyl shift from boron to oxygen. The dialkylalkoxyborane so produced reacts further with H_2O_2/OH^- to form dialkoxy and finally trialkoxyborane.*

$$R_2\overset{\ominus}{B}(R)-O-O-H \xrightarrow{\text{1, 2-Alkyl shift}} R_2B-OR \xrightarrow{H_2O_2/\overset{\ominus}{O}H} R(OR)\overset{\ominus}{B}(R)-O-O-H \xrightarrow{\text{1, 2-Alkyl shift}} R-B(OR)-OR \xrightarrow{\text{1, 2-Alkyl shift}} (RO)_2B-OR$$

Dialkylalkoxy-borane; Alkyldialkoxy-borane; Trialkoxyborane

Step IV: *Hydrolysis of trialkoxyborane with NaOH to form alcohol. Trialkoxyborane on subsequent hydrolysis gives the corresponding alcohol*

$$B(OR)_3 + 3\,NaOH \longrightarrow \underset{\text{Alcohol}}{3ROH} + NaBO_3$$

Stereochemistry

Another significant feature of the whole process is the stereospecificity of both steps, which result in *cis*-addition of water across a double bond.

$$\text{1-Methylcyclopentene} \xrightarrow{BH_3} \text{Alkylborane intermediate} \xrightarrow{H_2O_2\,/\,\overset{\ominus}{O}H} \textit{Trans}\text{-2-Methylcyclopentanol}$$

3. *Oxidation to carbonyl compounds*: Alkylboranes can be converted into carbonyl compounds.

(*ix*) Oxymercuration-Reduction

Mercuric acetate adds across a carbon-carbon double bond to form an adduct, which upon reduction with sodium borohydride gives rise to the corresponding alcohol.

$$CH_3CH_2CH{=}CH_2 \xrightarrow[\text{THF}-H_2O]{(CH_3COO)_2Hg} CH_3CH_2-\underset{OH}{\underset{|}{CH}}-\underset{HgOCOCH_3}{\underset{|}{CH_2}} \xrightarrow{NaBH_4} CH_3CH_2-\underset{OH}{\underset{|}{CH}}-CH_3$$

But-1-ene (Oxymercuration adduct) Butan-2-ol

The above two stage process is known as **oxymercuration-reduction** or **oxymercuration-demercuration**. Regiochemistry of overall addition of water is in accordance with *Markownikoff's rule*.

Mechanism

It is also an example of electrophilic addition reaction. The mechanism is quite similar to halohydrin formation. A cyclic *mercurinium ion* is an intermediate in this reaction. The whole sequence of reactions is shown below:

$$\underset{\text{Alkene}}{R-CH{=}CH_2} + Hg(OCOCH_3)_2 \xrightarrow[-CH_3COO^{\ominus}]{} R-CH\overset{\oplus}{\cdots}CH_2 \ (\text{bridged by } Hg-OCOCH_3) \xrightarrow{H_2\ddot{O}:}$$

(*Mercurinium ion*)

$$CH_3CO\overset{\ominus}{O} \quad R-\underset{\overset{|}{\overset{\oplus}{O}H_2}}{CH}-CH_2-Hg\,OCOCH_3 \xrightarrow[-CH_3COOH]{} R-\overset{OH}{\overset{|}{CH}}-CH_2-Hg-OCOCH_3 \xrightarrow[\text{Hydride transfer}]{[H-BH_3]^{\ominus}Na^{\oplus}} \underset{2^\circ\text{Alcohol}}{R-\overset{OH}{\overset{|}{CH}}-CH_3}$$

Comparison of Hydration, Hydroboration-oxidation and Oxymercuration-reduction Methods of Alcohol Synthesis

As discussed above acid-catalyzed hydration, hydroboration-oxidation and oxymercuration-reduction, all give alcohols but not the same alcohol from the same alkene. Always hydration of alkenes gives alcohols corresponding to *Markownikoff's addition* of water to alkene. This reaction involves carbocation intermediates. Hence, if the initially formed carbocation in less stable, it will rearrange to give the more stable carbocation thereby giving rearranged alcohols derived from more stable carbocation. For example,

$$CH_3-\underset{CH_3}{\overset{CH_3}{C}}-CH{=}CH_2 \xrightarrow{\overset{\oplus}{H}} CH_3-\underset{CH_3}{\overset{CH_3}{C}}-\overset{\oplus}{C}H-CH_3 \xrightarrow[\text{shift}]{\text{1, 2-Methyl}} CH_3-\overset{\oplus}{C}(CH_3)-\overset{CH_3}{CH}-CH_3$$

3, 3-Dimethylbut-1-ene

2°- Carbocation (*less stable*)

3°-Carbocation (*more stable*)

$H_2\ddot{O}:$

$$CH_3-\underset{\overset{\oplus}{O}H_2}{\overset{CH_3}{C}}-\overset{CH_3}{CH}-CH_3 \xrightarrow{-\overset{\oplus}{H}} CH_3-\underset{OH}{\overset{CH_3}{C}}-\overset{CH_3}{C}HCH_3$$

2, 3 -Dimethylbutan-2-ol

Contrary to the above, **the hydroboration-oxidation and oxymercuration-reduction** do not proceed through carbocation intermediates and hence always give unrearranged products. *Hydroboration-oxidation* gives alcohols which seem to be formed as if there is overall *anti-Markownikoff's addition* of water.

$$CH_3-\underset{CH_3}{\overset{CH_3}{C}}-CH{=}CH_2 \xrightarrow[(ii)\ H_2O_2/\overset{\ominus}{O}H]{(i)\ (BH_3)_2} CH_3-\underset{CH_3}{\overset{CH_3}{C}}-CH_2-CH_2OH$$

3, 3-Dimethylbut-1-ene   3, 3-Dimethylbutan-1-ol

But in case of *oxymercuration-reduction* the alcohols are obtained corresponding to *Markownikoff's addition* of water to alkenes. Therefore,

$$H_3C-\underset{CH_3}{\overset{CH_3}{C}}-CH{=}CH_2 \xrightarrow[\text{THF-}H_2O]{(i)\ (CH_3COO)_2Hg} CH_3-\underset{CH_3}{\overset{CH_3}{C}}-\underset{OH}{CH}-\underset{HgOCOCH_3}{CH_2}$$

3, 3-Dimethylbut-1-ene   3, 3-Dimethyl-1-acetoxy-mercuri-butane-2-ol

$$\xrightarrow{NaBH_4} CH_3-\underset{CH_3}{\overset{CH_3}{C}}-\underset{OH}{CH}-CH_3$$

3, 3-Dimethylbutan-2-ol

A comparison of all these three methods is summarized as below:

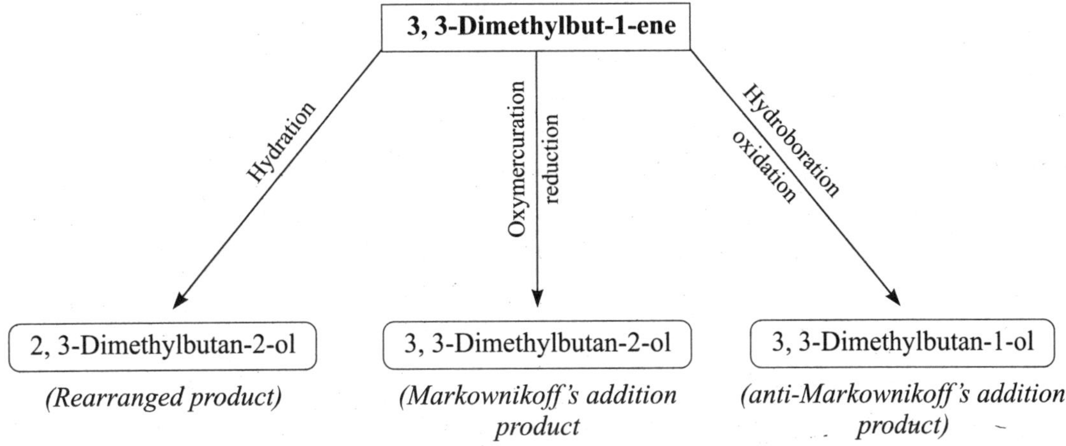

(x) Oxidation with Peracids: Epoxidation

The process of formation of *epoxide* (alkene oxides or oxiranes) by the treatment of an alkene with appropriate peracid such as perbenzoic acid (C_6H_5COOOH), peroxytrifluoroaceticacid (CF_3COOOH), etc., in inert solvents such as $CHCI_3$ is called **epoxidation**. For example,

$$\underset{\text{Ethene}}{CH_2{=}CH_2} + \underset{\text{Perbenzoic acid}}{C_6H_5COOOH} \xrightarrow{CCl_4} \underset{\substack{\text{Ethylene oxide}\\ \text{(oxirane)}}}{CH_2 \overset{O}{—} CH_2} + \underset{\text{Benzoic acid}}{C_6H_5COOH}$$

$$\underset{\text{2-Methylbut-2-ene}}{(H_3C)_2C{=}CH—CH_3} + C_6H_5COOOH \xrightarrow{CCl_4} \underset{\text{2, 2, 3-Trimethyloxirane}}{(H_3C)_2C \overset{O}{—} C(H)CH_3} + C_6H_5COOH$$

Mechanism

The peracids act like electrophiles and the reaction is believed to proceed by the following mechanism:

$$\underset{\text{Alkene}}{>C{=}C<} + \underset{\text{Peracid}}{H—O—O—\overset{O}{\overset{\|}{C}}—R} \xrightarrow{CCl_4} \underset{\text{Epoxide}}{>C \overset{O}{—} C<} + \underset{\text{Acid}}{H—O—\underset{}{C}(=O)—R}$$

Ring Opening Reactions of Epoxides

The epoxides being highly reactive, ring can be opened both by acids as well as bases to produce *trans*-1, 2-diols.

$\overset{\oplus}{H}$/ROH

$\overset{\ominus}{RO}\overset{\oplus}{Na}$/ ROH or NaOH

Epoxide
1, 2-Epoxycyclohexane

Trans -1, 2-diol

The orientation of ring opening depends upon the reaction conditions, e.g.,

***(i) In the acidic medium*:** The oxygen atom of the epoxide ring gets protonated. The nucleophilic attack of water from the side opposite to the oxygen atom gives *trans*-1, 2-diol. For example,

$\xrightarrow{\overset{\oplus}{H}}$ $\xrightarrow{:\ddot{O}H}$ $\xrightarrow{-\overset{\oplus}{H}}$

Trans-1, 2-diol

It may be noted here that in case of unsymmetrical epoxides nucleophilic attack of water (or alcohol) takes place on the more substituted carbon atom. For example,

$\xrightarrow[H_2SO_4]{CH_3OH}$

2, 2, 3-Trimethyloxirane

3-Methoxy-3-methylbutane-2-ol

Reason: This may be due to the fact that in acid-catalyzed ring opening the transition state aquires carbocation character because breaking of ring carbon-oxygen bond takes place slightly ahead to the formation of Nu—C bond. Due to this carbocation character in the transition state, nucleophilic attack is favoured at the carbon which can better support the positive charge. For example,

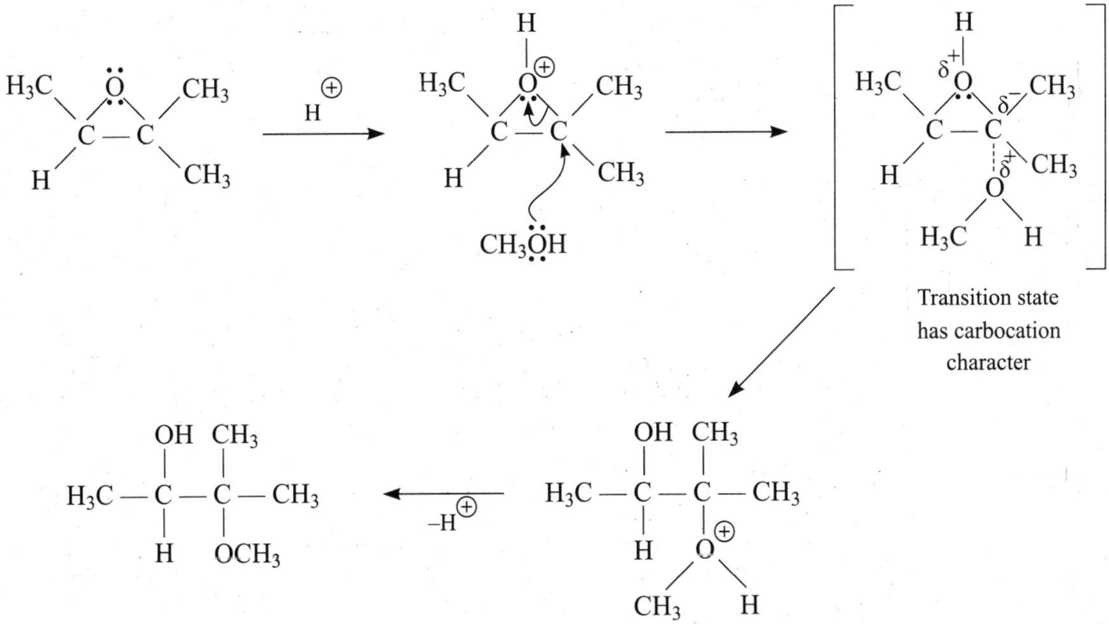

Transition state has carbocation character

(ii) ***In the basic medium***: The reaction is S_N2 type in which the nucleophile attacks the less substituted carbon atom of the oxirane ring, where the attack occurs from the backside of oxygen atom. For example,

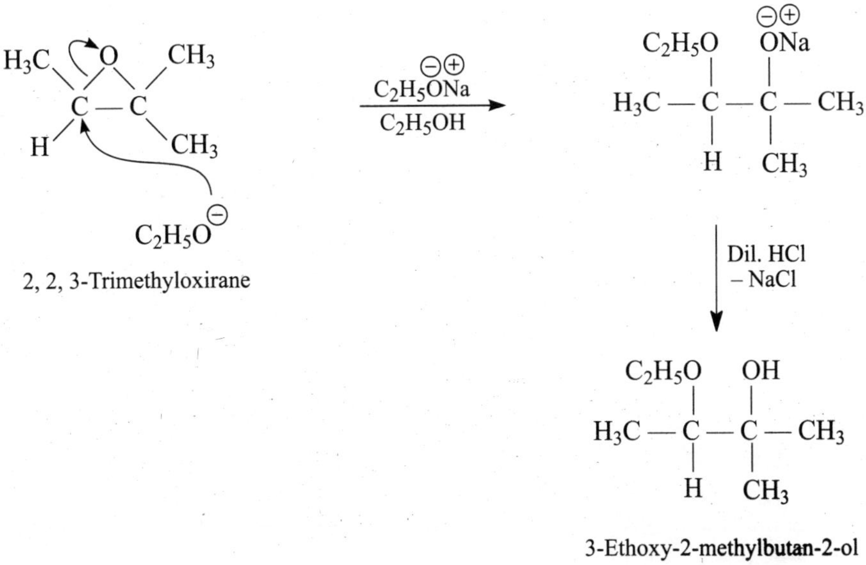

2, 2, 3-Trimethyloxirane

3-Ethoxy-2-**methylbutan**-2-ol

(*xi*) Oxidation of Alkenes by Ozone: Ozonolysis

Ozonization: Ozone can be represented as a resonance hybrid of the following contributing forms:

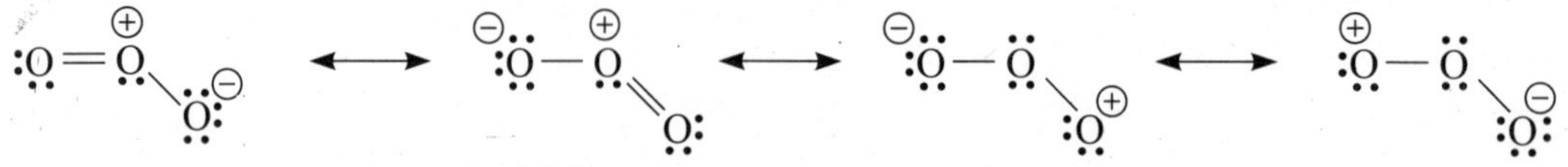

It is a system of three atoms having an electrophilic character and in the presence of a multiple bond it forms a cyclic product called ***ozonide***. These ozonides are not usually isolated and yield carbonyl compounds on reductive decomposition with zinc and acetic acid or hydrogenation over palladium catalyst.

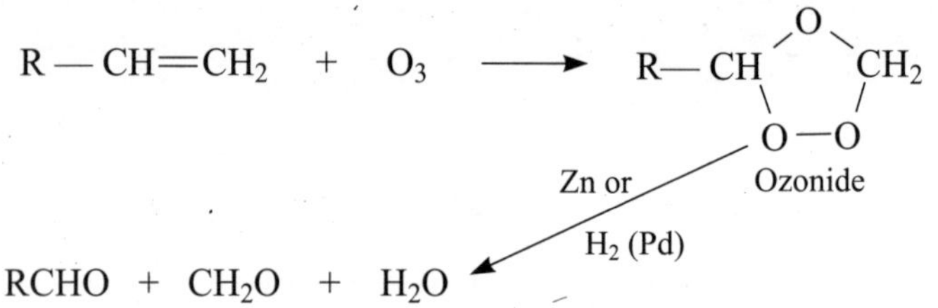

$$R-CH=CH_2 + O_3 \longrightarrow \text{Ozonide} \xrightarrow{\text{Zn or } H_2\text{ (Pd)}} RCHO + CH_2O + H_2O$$

Ozonide

The mechanism of ozonide formation is a multistep process.

Step I: A 1, 3 dipolar addition of ozone to double bond forming a molozonide.

Step II: The molozonide, being unstable, subsequently decomposes into two fragments.

Step III: Recombination of these fragments in an alternative way yields an ozonide.

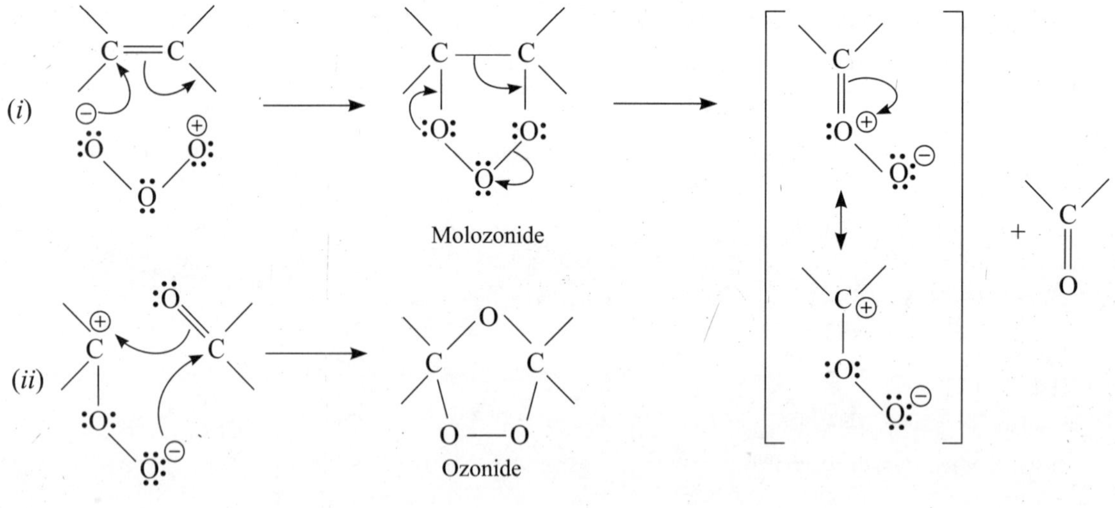

(*i*) Molozonide

(*ii*) Ozonide

Importance

The overall reaction, *i.e.*, fragmentation of the alkene into carbonyl components is termed as *ozonolysis*, and has been found to be of much use for preparative purpose and also for locating the position of a double bond in an unknown compound. This is possible as the cleavage of the double bond produces smaller carbonyl fragments in which a doubly-bonded oxygen is found attached to each of the carbon atom linked by a double bond in the parent alkene. The aldehydes and ketones thus obtained are identified and by locating the number and arrangement of carbon atoms in these products one can work back so as to arrive at the structure of the original alkene. Thus, for instance, three isomers having molecular formula C_4H_8 have been characterized as 1-butene, 2-butene and 2-methylpropene respectively on the basis of identification of the aldehydes and ketones obtained by ozonolysis.

$$\underset{\text{1-Butene}}{CH_3-CH_2-CH=CH_2} \xrightarrow[(ii)\ H_2O/Zn]{(i)\ O_3} \underset{\substack{\text{Propionaldehyde}\\ \text{(1 mole)}}}{CH_3-CH_2-CHO} + \underset{\substack{\text{Formaldehyde}\\ \text{(1 mole)}}}{CH_2O}$$

$$\underset{\text{2-Butene}}{CH_3-CH=CH-CH_3} \xrightarrow[(ii)\ H_2O/Zn]{(i)\ O_3} \underset{\text{Acetaldehyde (2 moles)}}{CH_3-CHO + CH_3-CHO}$$

$$\underset{\text{2-Methylpropene}}{(H_3C)_2C=CH_2} \xrightarrow[(ii)\ H_2O/Zn]{(i)\ O_3} \underset{\substack{\text{Acetone}\\ \text{(1 mole)}}}{(H_3C)_2C=O} + \underset{\substack{\text{Formaldehyde}\\ \text{(1 mole)}}}{CH_2O}$$

(*xii*) Oxidation

Due to the presence of carbon-carbon double bond, the alkenes are readily oxidized to give different types of products depending upon the type of oxidizing agent and/or reaction conditions. Some common oxidation reactions of alkenes are discussed below:

(a) Hydroxylation of Double Bond

Alkenes on treatment with certain oxidising agents such as cold dil. alkaline $KMnO_4$ or OsO_4 form 1, 2-diols. This reaction is called as *hydroxylation.* The reactions using these two reagents are discussed as follows:

1. *Hydroxylation using potassium permanganate*: Alkenes are readily oxidised on treatment with a cold dilute (1%) aqueous alkaline solution of $KMnO_4$. The products are *vicinal* or 1, 2-*diols* or 1, 2-*glycols* and $KMnO_4$ is itself reduced to MnO_2. For example,

$$\underset{\substack{\text{Ethylene}\\ (\textit{Colourless})}}{3H_2C=CH_2} + \underset{\text{Violet colour}}{2KMnO_4} + 4H_2O \longrightarrow 3\underset{\text{Ethylene glycol}}{\begin{array}{c}CH_2OH\\ |\\ CH_2OH\end{array}} + 2MnO_2 + 2KOH$$

This reaction forms the basis of the well known **Baeyer's Test** for detecting the presence of a carbon-carbon double bond (*i.e., test for unsaturation*). During the reaction, the violet colour of $KMnO_4$ solution is decolourized and a brown ppt. of MnO_2 is formed.

Mechanism

It is also regarded as electrophilic addition reaction and is believed to proceed with initial formation of a cyclic manganate ester which undergoes ring opening under the conditions of the reaction to form the diols.

$$CH_2{=}CH_2 + MnO_4^- \longrightarrow \text{[Cyclic manganate ester]} \xrightarrow[\text{Hydrolysis}]{H_2O/\overset{\ominus}{O}H} CH_2OH{-}CH_2OH + (HO)_2MnO_2^-$$

Ethylene Permanganate ion Cyclic manganate ester (*Not isolated*) Ethylene glycol Mn (V) species

The Mn (V) species formed above as a result of hydrolysis undergoes disproportionation to MnO_2 and MnO_4^-.

$$3\,(HO)_2MnO_2^- \xrightarrow{\text{Disproportionation}} MnO_4^- + 2MnO_2 + 2\overset{\ominus}{O}H + 2H_2O$$

Permanganate ion

This mechanism accounts for the formation of *cis*-diols (see below). Additional support for this mechanism comes from the experiments with $KMnO_4$ labelled with ^{18}O, which conclusively demonstrate that both oxygens in the diol come from the permanganate ion.

Evidence in support of the formation of a cyclic intermediate

The following observations are in agreement with the formation of cyclic manganate ester.

(*i*) ***Stereochemistry***: The addition of OH groups across the double bond is *stereospecific cis-addition.* For example,

$$\text{Cyclohexene} \xrightarrow{\text{aq. } KMnO_4,\ NaOH} \textit{cis}\text{-cyclohexane-1,2-diol}$$

(*ii*) ***Isotopic labelling experiment***: When the reaction was carried out with $KMnO_4$ labelled with ^{18}O, it resulted in the formation of the diol in which both the oxygens were found to be ^{18}O.

$$R-CH{=}CH-R + KMn^{18}O_4 \longrightarrow R-\underset{^{18}OH}{CH}-\underset{^{18}OH}{CH}-R$$

2. ***Hydroxylation using osmium tetraoxide*:** Another useful reagent for hydroxylation of alkenes is OsO_4. When an alkene is treated with OsO_4 in ether, it forms cyclic osmate ester. Pyridine is often added as a catalyst since it complexes with osmium atom in the ester. The ester so produced is isolated and then hydrolysed with aqueous Na_2SO_3 or $NaHSO_3$ to give 1, 2-glycols.

R — CH = CH — R + OsO_4 ⟶ Cyclic osmate ester (*Isolated*) $\xrightarrow[NaHSO_3]{H_2O}$ R — CH(OH) — CH(OH) — R + $[OsO_3]^{\ominus}$

Cyclic osmate ester

(*Isolated*)

Because of the formation of cyclic ester it also results in the *cis*-hydroxylations. For example,

1, 2-Dimethylcyclopentene $\xrightarrow[Pyridine]{OsO_4}$ cyclic osmate ester $\xrightarrow[NaHSO_3]{H_2O}$ *cis*-1, 2-Dimethylcyclopentane

Comparison of hydroxylation using $KMnO_4$ and OsO_4

(*i*) *$KMnO_4$ being a strong oxidising agent is not selective*, unwanted side reactions involving C—C bond cleavage take place due to which yields of hydroxylation products (diols) is often low. But under alkaline conditions, this cleavage does not take place. So, it gives better yields. However, OsO_4 is highly selective and yields of hydroxylation products are often quite high.

(*ii*) *$KMnO_4$ is quite cheap and non-toxic compared to OsO_4 which is very costly and toxic.* Therefore, OsO_4 is not used in routine laboratory preparations. However, it is used in multistep organic syntheses, where yields of the reactions are more important. Due to heavy cost a combination in catalytic amounts of OsO_4 and an oxidising agent has proved to be good.

(*b*) *Oxidation with hot $KMnO_4$*

Treatment of an alkene with hot $KMnO_4$ results is the cleavage of C = C bond which leads to the formation of carboxylic acids, ketones and carbon dioxide. The nature of the products depends upon the nature of alkenes as discussed below:

(*i*) **Terminal alkenes** invariably produce methanoic acid (formic acid) which on further oxidation gives CO_2 and H_2O. For example,

$$\underset{\text{Ethene}}{CH_2{=}CH_2} + 4[O] \xrightarrow[\text{KOH, 373–383 K}]{KMnO_4} \underset{\text{Formic acid}}{2H{-}\overset{\overset{O}{\|}}{C}{-}OH} \xrightarrow{[O]} 2CO_2 + 2H_2O$$

$$\underset{\text{Propene}}{CH_3CH{=}CH_2} + 4[O] \xrightarrow[\text{373–383 K}]{KMnO_4,\ KOH} \underset{\text{Acetic acid}}{CH_3COOH} + \underset{\text{Formic acid}}{HCOOH} \xrightarrow{[O]} CO_2 + H_2O$$

$$\underset{\text{2-Methylpropene}}{(H_3C)_2C{=}CH_2} + 3[O] \xrightarrow[\text{373–383 K}]{KMnO_4,\ KOH} \underset{\text{Acetone}}{CH_3{-}\overset{\overset{O}{\|}}{C}{-}CH_3} + HCOOH \xrightarrow{[O]} CO_2 + H_2O$$

(*ii*) **Non-terminal alkenes** produce carboxylic acids or ketones or both depending upon the nature of the alkene. For example,

$$\underset{\text{Pent-2-ene}}{CH_3{-}CH{=}CH{-}CH_2CH_3} + 4[O] \xrightarrow[\text{373–383 K}]{KMnO_4,\ KOH} CH_3COOH + \underset{\text{Propanoic acid}}{HOOCCH_2CH_3}$$

$$\underset{\text{2-Methylbut-2-ene}}{(H_3C)_2C{=}CH{-}CH_3} + 3[O] \xrightarrow[\text{373–383 K}]{KMnO_4,\ KOH} CH_3{-}\overset{\overset{O}{\|}}{C}{-}CH_3 + HOOCCH_3$$

$$\underset{\text{2, 3-Dimethylpent-2-ene}}{(H_3C)_2C{=}C(CH_3)CH_2CH_3} + 2[O] \xrightarrow[\text{373–383 K}]{KMn\Theta_4,\ KOH} \underset{\text{Acetone}}{(H_3C)_2C{=}O} + \underset{\text{Butan-2-one}}{O{=}C(CH_3)CH_2CH_3}$$

(*c*) **Combustion:** Alkenes burn in oxygen or air and get oxidized completely to CO_2 and H_2O. This process is called *combustion*. All combustion reactions are *highly exothermic*. For example,

$$CH_2{=}CH_2 + 3O_2 \xrightarrow{\Delta} 2CO_2 + 2H_2O \qquad \Delta H_c = -1411 \text{ kJ mol}^{-1}$$

(*d*) **Catalytic oxidation:** Alkenes undergo controlled oxidation when they react with oxygen in presence of silver as catalyst to give epoxides or oxiranes. For example,

$$2\,CH_2{=}CH_2 + O_2 \xrightarrow[\text{575 K}]{Ag} \underset{\text{Oxirane}}{2\,\overset{\;\;O}{CH_2{-}CH_2}}$$

$$2\,CH_3CH{=}CH_2 + O_2 \xrightarrow[\text{575 K}]{Ag} \underset{\text{1, 2-Epoxypropane}}{2\,CH_3{-}\overset{\;\;O}{CH{-}CH_2}}$$

(*xiii*) Polymerization

A reaction of great importance is the conversion of alkenes to high molecular weight compounds. The simple repeating units (alkenes) in such huge molecules are called **monomers** and the high molecular wieght compounds are called ***polymers***. The process of transformation of *monomers* into a polymers is known as ***polymerization***. For example, polyethylene (or polythene)—a highly useful industrial polymer–is manufactured by heating ethene under pressure in the absence of air.

$$\underset{\substack{\text{Ethene}\\(\textit{Monomer})}}{n\,CH_2{=}CH_2} \xrightarrow{\Delta,\ \text{high pressure}} \underset{\substack{\text{Polythene}\\(\textit{Molymer})}}{\sim\sim (CH_2)_n \sim\sim}$$

Types of polymerization: Broadly speaking, there are three types of polymerizations as given below:

I. Addition polymerization or chain growth polymerization.

II. Condensation polymerization or step growth polymerization.

III. Coordination polymerization or catalysed polymerization.

I. Addition polymerization

In this type of polymerization, there is successive addition of monomers to the growing chain having a free radical, a carbocation or a carbanion terminal. In this type of polymerization, the empirical formula of polymer is the same as that of the monomer from which it is derived.

As mentioned above in the addition or chain growth polymerization, monomers are the compounds having carbon-carbon double bonds which polymerize in the presence of catalysts as discussed below:

(*i*) Polymerization of ethene: Polymerization of ethylene (or ethene) gives rise to the following two types of polythenes (polyethylenes) depending upon the reaction conditions and catalysts used. However, both have the same repeating unit.

(*a*) Low density polyethene: It is prepared by heating ethene to 473 K at 1500 atmosphere pressure in the presence of a trace of oxygen. This polymerization involves free radical mechanism which is initiated by oxygen.

$$\underset{\text{Ethene}}{n\,CH_2{=}CH_2} \xrightarrow[\text{Traces of oxygen}]{473\text{ K, }1500\text{ atm.}} \underset{\text{Low density polythene}}{\sim\sim (CH_2-CH_2)_n \sim\sim}$$

The polythene produced by this procedure has a molecular weight of approx. 20,000 and possesses a branched structure. Due to this branching, the molecules of polythene do not pack closely and hence the density of this type of polythenes is as low as 0.92 g/cm^3. In addition its melting point is also low (384 K).

Properties

1. Low density polythene is *transparent.*
2. It has a moderate *tensile strength.*
3. It is chemically *inert* and a *poor conductor* of electric current.
4. It is highly *tough.*
5. It is highly *flexible.*

Chapter 6

Uses

Low density polythene is used for the manufacture of packing materials (in the form of thin plastic films, bags, etc.) and as insulating material for electrical wires and cables.

(b)* High density polythene:** This polymer is manufactured by a process called *coordination polymerization.* This involves heating of ethene dissolved in a hydrocarbon solvent at 333–343 K under 6–7 atmospheric pressure in presence of ***Ziegler-Natta catalyst* (For details see coordination polymerization).

$$\underset{\text{Ethene}}{n\,CH_2{=}CH_2} \xrightarrow[(CH_3CH_2)_3Al\,+\,TiCl_4]{333\text{–}343\text{ K, 6–7 atm.}} \underset{\text{High density polythene}}{\sim\!\sim (CH_2-CH_2)_n \sim\!\sim}$$

The polythene so obtained consists of linear chains of polymer molecules. Thus, these molecules pack closely and hence, the density of this type of polythene is as high as 0.97 g/cm^3. In addition, melting point is also high (403 K).

Properties

1. High density polythene is *translucent.*
2. It is chemically *inert.*
3. It has high *toughness, hardness* and *tensile strength* than the low density polythene.

Uses

It is used in the manufacture of containers (buckets, tubes, etc.) housewares, pipes, bottles, toys, etc.

***(ii)* Polymerization of propene:** Polypropene (polypropylene) is manufactured either by heating propene in presence of a trace of benzoyl peroxide as radical initiator or Ziegler-Natta catalyst.

$$\underset{\text{Propene}}{n\,CH_3CH{=}CH_2} \xrightarrow[\text{or Ziegler-Natta catalyst}]{\text{Peroxide, }\Delta} \underset{\text{Polypropene}}{-(\overset{\displaystyle CH_3}{\overset{|}{C}H}-CH_2)_n-}$$

An important characteristic of this polymerization is the stereochemical control over the way in which the monomer units enter the polymer chain. Three different stereochemical forms of polypropylene are thus produced by head to tail joining of propene molecules.

---

*Ziegler-Natta catalyst is a mixture of trialkylaluminium and titanium tetrachloride ($R_3Al + TiCl_4$). During co-ordination polymerization alkene forms a coordinated complex with titanium by donating π-electrons into the vacant d-orbital of Ti atom. The resulting π-complex then reacts stepwise with a large number of ethene molecules to give the polymer.

(*a*) **Isotactic (*Greek* meaning *'same order'*):** In which all carbon atoms having a methyl group have identical configuration as shown below:

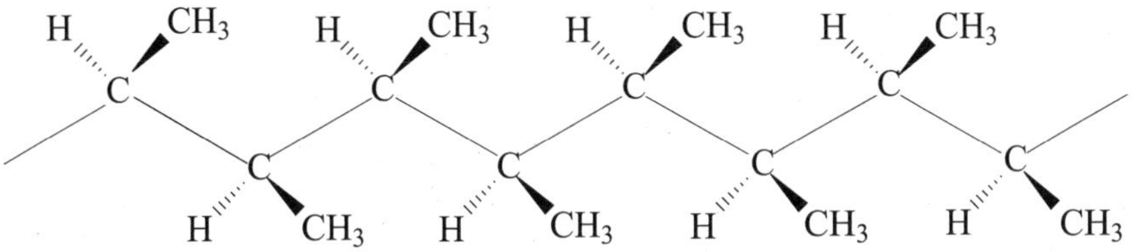

(*b*) **Syndiotactic (*Greek* meaning *'alternating order'*):** In which the configuration of carbon atoms having methyl group alternate regularly (in one CH_3 in front and in other CH_3 in back and so on) as shown below:

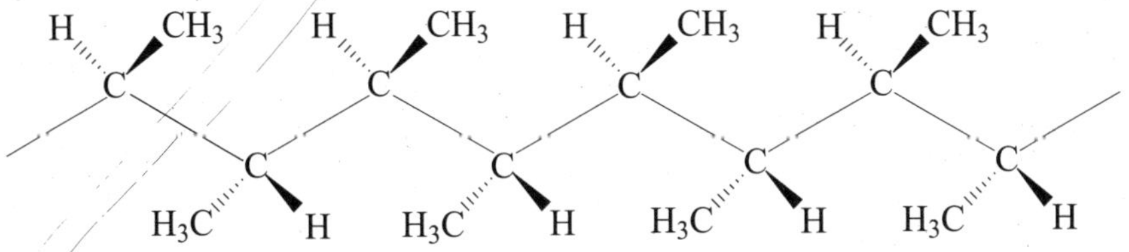

(*c*) **Atactic (*Greek* meaning *'no order'*):** In which there is random stereochemical distribution as shown below:

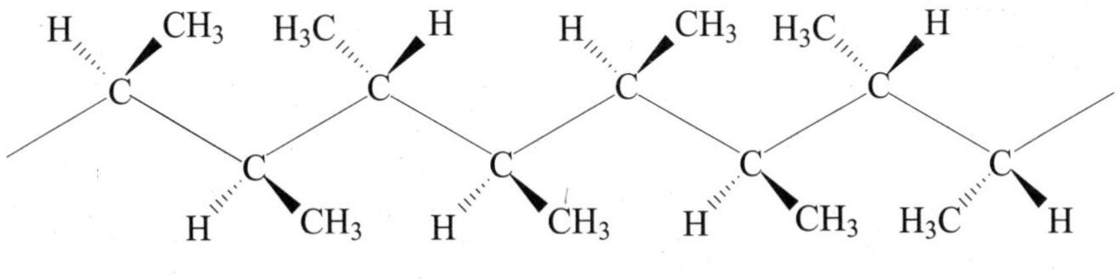

Here, it may be noted that branched atactic polymer is obtained from normal radical chain polymerization while Ziegler-Natta polymerization gives purely isotactic polypropene which is more useful than syndiotactic and atactic ones.

Uses

Although the above three stereochemical forms of polypropene have somewhat different properties, they have one thing in common that they are harder and stronger polymers than polythene. Their main uses are as follows:

1. In the manufacture of stronger pipes and bottles.
2. For making automobile mouldings, seat covers, carpets, ropes, etc.

3. For packing of foods and textiles.
4. For making liner for bags and heat shrinkable wraps for records and other articles.

***(iii)* Polymerization of vinyl chloride:** Vinyl chloride polymerises in presence of free radical initiators (peroxides) to produce a polymer called poly vinyl chloride (PVC).

$$n\,CH_2{=}CHCl \xrightarrow{\text{Peroxide}} \sim\!\sim (CH_2-\underset{\underset{Cl}{|}}{CH})_n \sim\!\sim$$

Chloroethene or Vinyl chloride → PVC

Uses

This polymer can be easily moulded and has the following uses:

1. It has good flame resistance and good electrical insulation and is used for coating wires, cables and other electrical goods.

2. When mixed with high boiling esters such as *di-n*-butylphthalate, known to possess plasticizing property, it is used for making rain coats, hand bags, plastic dolls, shower curtains, upholestery, fabrics, shoe soles and vinyl flooring. Since dibutylphthalate is slightly volatile, it evaporates slowly. As a result, soft plasticized vinyl polymer eventually loses its plasticizer property and becomes hard and brittle.

3. It is also used in making gramophone records and hose pipes.

***(iv)* Copolymerization of vinyl chloride with vinylidene chloride:** When a mixture of vinyl chloride and vinylidene chloride is polymerized the growing chain preferentially adds that monomer which does not constitute the terminal of the growing chain. In other words, this polymer chain has an alternate arrangement of the two monomer units. This polymer is called **saran**.

$$n\,CH_2{=}\underset{\underset{Cl}{|}}{CH} + n\,CH_2{=}\underset{\underset{Cl}{|}}{C}-Cl \xrightarrow{\text{Polymerization}} \sim\!\sim (CH_2-\underset{\underset{Cl}{|}}{CH}-CH_2-CCl_2)_n \sim\!\sim$$

Vinyl chloride + Vinylidene chloride → Saran

Uses

It is used for making film, for wrapping food material and as a covering for seats.

***(v)* Polymerization of vinyl cyanide:** Polymerization of vinyl cyanide or acrylonitrile under free radical conditions produces poly acrylo nitrile (PAN).

$$n\,CH_2{=}\underset{\underset{CN}{|}}{CH} \xrightarrow{\text{Peroxide}} \sim\!\sim (CH_2-\underset{\underset{CN}{|}}{CH})_n \sim\!\sim$$

Vinyl cyanide or Acrylonitrile → PAN

Uses

It is a hard, horny polymer having a high melting point. It is put to the following uses:

1. In the manufacture of **Orlon** and **Acrilan** fibres which are used for making clothes, carpets and blankets.

2. In the preparation of other polymers for improving their qualities.

***(vi)* Polymerization of styrene:** Styrene polymerizes to give polystyrene in presence of peroxides.

$$\underset{\text{Styrene}}{n\,CH_2{=}\underset{\underset{C_6H_5}{|}}{CH}} \xrightarrow{\text{Peroxide}} \underset{\text{Poly(styrene)}}{\sim\!\sim\,(CH_2-\underset{\underset{C_6H_5}{|}}{CH})_n\,\sim\!\sim}$$

Uses

Polystyrene is a transparent polymer and is put to use for making plastic toys, household wares, telephone, radio and television bodies, refrigerator linings, etc.

***(vii)* Polymerization of tetrafluoroethylene:** Polymerization of tetrafluoroethylene in presence of oxygen gives polytetrafluoroethylene (PTFE) called as teflon.

$$\underset{\text{Tetrafluoroethylene}}{n\,CF_2{=}CF_2} \xrightarrow{O_2} \underset{\text{Teflon (PTFE)}}{-\!\!\left(CF_2-CF_2\right)\!\!-_n}$$

Uses

1. Teflon is a flexible polymer which is highly resistant to the action of acids (even boiling *aqua regia)*, solvents and other chemicals. Therefore, it is used for making pipes and surgical tubes.

2. Because of its resistance to chemicals and thermal stability (up to 598 K) teflon is used for making non-stick utensils and electrical press. A thin layer of teflon is coated on the inner side of the utensils and at the base of electric press.

3. Teflon is also employed in making gaskets, pump packings, valves, seals, non-lubricated bearings, filter clothes, etc.

4. Teflon stop-cocks are used in burettes and dropping funnels due to its resistance towards chemicals.

Mechanism of Chain Growth Polymerization *i.e.*, Addition Polymerization

Mostly the polymerization of alkenes occurs by chain mechanism. Depending upon the reaction conditions and types of catalysts, they may be classified into following three categories:

(*i*) Cationic Polymerization, (*ii*) Anionic Polymerization, (*iii*) Free Radical Polymerization.

(*i*) Cationic Polymerization

It takes place in presence of strongly acidic catalysts such as H_2SO_4 or Lewis acids such as $AlCl_3$, BF_3,etc., in presence of a trace of water as inhibitor. For example, polymerization of *iso*-butylene in presence of acid catalysts produces poly (*iso*-butylene).

$$n\,CH_2{=}C(CH_3)_2 \xrightarrow{\text{Acid}} \left(CH_2 - \underset{CH_3}{\overset{CH_3}{C}} \right)_n$$

Iso-butylene Poly(*iso*-butylene)

The pure poly (*iso*-butylene) is a rubber —like substance and is used in pressure-sensitive adhesives. Polymerization of *iso*-butylene with *isoprene* (copolymerization) gives a polymer which is used as synthetic rubber.

The mechanism of cationic polymerization involves the initial addition of a proton to the alkene forming a carbocation which in the second step adds to another alkene molecule giving rise to a second carbocation. This carbocation in its turn adds to the third molecule of alkene and so on. Thus, a chain process sets up which may terminate by the elimination of a proton from the carbocation resulting in the formation of a polymer having a terminal double bond. The mechanism of polymerization is as follows:

Step I: *Chain Initiation* — Protonation of alkene to produce a carbocation (*tert*-butyl carbocation in this example).

$$(H_3C)_2C{=}CH_2 + \overset{\oplus}{H} \longrightarrow (H_3C)_2\overset{\oplus}{C} - CH_3$$

Step II: *Chain propagation* — Addition of carbocation to carbon-carbon double bond producing a larger carbocation and so on.

$$H_3C - \underset{CH_3}{\overset{CH_3}{C^{\oplus}}} + CH_2{=}C(CH_3)_2 \longrightarrow H_3C - \underset{CH_3}{\overset{CH_3}{C}} - CH_2 - \overset{\oplus}{C}(CH_3)_2 \xleftarrow{} CH_2{=}C(CH_3)_2$$

$$CH_3 - \underset{CH_3}{\overset{CH_3}{C}} - CH_2 - \underset{CH_3}{\overset{CH_3}{C}} - CH_2 - \underset{CH_3}{\overset{CH_3}{C^{\oplus}}} \longrightarrow \longrightarrow \longrightarrow (CH_3)_3C \left[CH_2 - \underset{CH_3}{\overset{CH_3}{C}} \right]_n CH_2 - \overset{\oplus}{C}(CH_3)_2$$

Step III: *Chain termination* — The growing chain terminates by elimination of a proton from polymeric terminal carbocation producing a polymer with a terminal C=C bond.

$$(CH_3)_3C\left[CH_2-\underset{CH_3}{\overset{CH_3}{C}}\right]_n CH_2-\overset{\oplus}{C}(CH_3)_2 \xrightarrow{-H^{\oplus}} (CH_3)_3C\left[CH_2-\underset{CH_3}{\overset{CH_3}{C}}\right]_n CH_2-\overset{CH_3}{C}=CH_2$$

In presence of sulphuric acid, however, *iso*-butylene forms a mixture of two dimers namely 2,4,4-trimethyl-1-pentene and 2,4,4-trimethyl-2-pentene. The mechanism of dimerization is the same as above except that chain terminates after only one molecule of *iso*-butylene has been added.

$$(H_3C)_2C=CH_2 \xrightarrow[343\text{ K}]{H_2SO_4} (H_3C)(CH_3)\overset{\oplus}{C}-CH_3 \xrightarrow{H_2C=C(CH_3)_2} H_3C-\underset{CH_3}{\overset{CH_3}{C}}-CH_2-\underset{CH_3}{\overset{CH_3}{C^{\oplus}}}$$

$$H_3C-\underset{CH_3}{\overset{CH_3}{C}}-\overset{H}{CH}-\underset{CH_3}{\overset{CH_3}{C^{\oplus}}} \xrightarrow{-H^{\oplus}} H_3C-\underset{CH_3}{\overset{CH_3}{C}}-CH=C(CH_3)_2$$

2,4,4-Trimethyl-2-pentene

OR

$$H_3C-\underset{CH_3}{\overset{CH_3}{C}}-CH_2-\underset{CH_3}{\overset{\oplus}{C}}-CH_2-H \xrightarrow{-H^{\oplus}} H_3C-\underset{CH_3}{\overset{CH_3}{C}}-CH_2-\underset{CH_3}{C}=CH_2$$

2,4,4-Trimethyl-1-pentene

This is an industrially important reaction as catalytic hydrogenation of this mixture yields ***iso-octane***, which is used to improve upon **octane number** of gasolene.

(*ii*) Anionic Polymerization

It takes place in presence of a powerful nucleophile (for instance $—NH_2^-$). The mechanism of anionic polymerization may be formulated as below:

Step I: *Chain initiation*—Addition of a strong nucleophile to carbon-carbon double bond to produce a carbanion.

$$\overset{\ominus}{Nu} + >C=C< \longrightarrow Nu-\overset{|}{\underset{|}{C}}-\overset{|}{\underset{|}{C^{\ominus}}}$$

Step II: *Chain propagation*—Addition of the above carbanion to another molecule of alkene producing a larger carbanion and so on.

$$\mathrm{Nu-\overset{|}{\underset{|}{C}}-\overset{|}{\underset{|}{C}}^{\ominus}} + \mathrm{>C=C<} \longrightarrow \mathrm{Nu-\overset{|}{\underset{|}{C}}-\overset{|}{\underset{|}{C}}-\overset{|}{\underset{|}{C}}-\overset{|}{\underset{|}{C}}^{\ominus}} \rightarrow \rightarrow \rightarrow$$

Step III: *Chain termination*—The growing chain terminates by abstraction of a proton from any proton donor.

$$\mathrm{Nu}\left(\mathrm{-\overset{|}{\underset{|}{C}}-\overset{|}{\underset{|}{C}}-}\right)_n\mathrm{\overset{|}{\underset{|}{C}}-\overset{|}{\underset{|}{C}}^{\ominus}} \xrightarrow[\text{(Any proton donor)}]{\mathrm{H}^{\oplus}} \mathrm{Nu}\left(\mathrm{-\overset{|}{\underset{|}{C}}-\overset{|}{\underset{|}{C}}-}\right)_n\mathrm{\overset{|}{\underset{|}{C}}-\overset{|}{\underset{|}{C}}-H}$$

Usually the anionic polymerization is not easy to achieve but the presence of a powerful electron attracting group on the alkene facilitates the process.

(*iii*) Free Radical Polymerization

It may be achieved at high temperatures under pressure or in the presence of a small amount of peroxides. The chain is initiated by the homolytic cleavage of the peroxide to free radicals which add to the double bond of the alkene forming another free radical in the chain propagation step. This radical adds to another alkene molecule and the sequence is repeated a number of times till the chain is terminated by coupling or disproportionation. Thus, the mechanism of free radical polymerization of ehtylene can be formulated as follows:

Step I: *Chain initiation* — Formation of free radical

$$\mathrm{R-O-O-R} \xrightarrow{h\nu \text{ or } \Delta} \mathrm{R\dot{O}}$$

Step II: *Chain propagation* — Addition of free radical to carbon-carbon double bond to form another free radical and so on to produce a polymeric radical.

$$\mathrm{R\dot{O}} + \mathrm{H_2C=CH_2} \longrightarrow \mathrm{RO-\overset{H}{\underset{H}{C}}-\overset{H}{\underset{H}{\dot{C}}}}$$

$$\mathrm{RO-\overset{H}{\underset{H}{C}}-\overset{H}{\underset{H}{\dot{C}}}} + \mathrm{H_2C=CH_2} \longrightarrow \longrightarrow \longrightarrow \mathrm{RO}\left(\mathrm{-\overset{H}{\underset{H}{C}}-\overset{H}{\underset{H}{C}}-}\right)_n\mathrm{\overset{H}{\underset{H}{C}}-\overset{H}{\underset{H}{\dot{C}}}}$$

Step III: *Chain termination* — Either by radical coupling reaction or a disproportionation reaction of polymeric radical.

$$2RO\left(\begin{matrix} H & H \\ | & | \\ C-C \\ | & | \\ H & H \end{matrix}\right)_n \begin{matrix} H & H \\ | & | \\ C-\dot{C} \\ | & | \\ H & H \end{matrix} \xrightarrow{\text{Coupling}} \left[RO{+}CH_2-CH_2{+}_n CH_2-CH_2\right]_2$$

$$\xrightarrow{\text{Disproportionation}} \left[\begin{matrix} RO{+}CH_2-CH_2{+}_n CH=CH_2 \\ + \\ RO{+}CH_2-CH_2{+}_n CH_2-CH_3 \end{matrix}\right]$$

II. Condensation polymerization (shall be discussed at relevant places)

III. Coordination polymerization

One of the great discoveries of organometallic chemistry was the catalyzed polymerization or coordination polymerization of alkenes at atmospheric pressure and ambient temperature. Large quantities of stereospecific polyethylene and poly(propylene) (over 15 million tons annually) are made by Ziegler-Natta catalysis.

The Ziegler-Natta catalyst, which is heterogeneous, is made by treating $TiCl_4$ with Al_2Et_6 in heptane to form a fibrous material that is partially alkylated (Et_2AlCl) and is used as a cocatalyst.

Titanium does not have a filled coordination sphere and acts as a Lewis acid, accepting ethylene or propylene as another legend to form π-complex which undergoes further reactions. The reaction brings the alkyl group and the alkene bound on metal in a particular stereochemical arrangement, to facilitate migration of the alkyl group to the alkene (*i.e.*, ***cis-insertion of alkene into neighbouring metal-alkyl bond***) giving a longer chain of alkyl derivative; this results in a fresh vacant site at the surface where new alkene can coordinate.

Vacant site

$$-Ti-CH_2CH_3 \xrightarrow{CH_2=CH_2} -Ti(\leftarrow H_2C=CH_2)-CH_2CH_3 \xrightarrow{\text{Insertion}} [H_2C\text{---}CH_2 \text{ / } -Ti\text{----}CH_2CH_3]$$

$$\downarrow$$

$$-Ti(CH_2CH_2CH_2CH_3) \text{ (New vacant site)} \xrightarrow{CH_2=CH_2} -Ti(CH_2CH_2CH_2CH_3)(\leftarrow CH_2=CH_2) \xrightarrow{\text{Insertion}} \cdots \rightarrow -Ti-(CH_2CH_2)_n-CH_2CH_3$$

Vacant site

The stereogularity of polypropene is assumed to arise due to the nature of sterically controlled directing sites on the $TiCl_3$ lattice. Only ***iso-tactic*** chain of poly(propylene) is obtained by coordination polymerization, *i.e.*, no branched polymer is formed.

(*xiv*) Substitution Reactions

So for we have described only those reactions of alkenes in which addition occurring across a $C=C$ double bond takes place. Besides these, however, higher alkenes undergo substitution reactions also either by replacement of allylic or vinylic hydrogen. Accordingly, there are two types of substitution reactions of alkenes. They are:

1. *Free Radical Substitution at Allylic Position.*
2. *Electrophilic Substitution at Vinylic Position.*

Both these reactions are discussed below:

1. *Free Radical Substitution at the Allylic Position*

Since higher alkenes have alkane moieties in the form of alkyl groups, it is reasonable to expect these compounds to undergo the *free radical substitution reactions* characteristic of alkanes. Such molecules have two sites of attack by the reagent and the actual site of attack will depend upon the choice of conditions. Thus, under conditions that favour free radical formation, allylic hydrogen is replaced by halogen. For instance, propene forms 3-chloropropene on treatment with chlorine at 500°C or in the presence of free radical initiator at low chlorine concentrations.

$$\underset{\text{Allylic position}}{CH_3} - CH{=}CH_2 + Cl_2 \xrightarrow[\text{or free radical initiator}]{773\text{ K}} Cl - CH_2 - CH{=}CH_2 + HCl$$

Mechanism

At high temperature or in the presence of free radical initiator, halogen molecule undergoes homolytic cleavage yielding halogen atoms which abstract a hydrogen atom from the methyl group (allylic H) to give a resonance stabilized allyl radical. The allyl radical in turn reacts with fresh molecule of halogen to form 3-halopropene and a halogen atom. Thus, the substitution proceeds in the following three steps:

Step I: Generation of free radical.

$$Cl - Cl \xrightarrow{\Delta} 2\dot{Cl}$$

Step II: Attack of chlorine atom on allylic carbon to abstract hydrogen atom and generate allyl radical which is resonance stabilized.

$$\dot{Cl} + H - CH_2 - CH{=}CH_2 \longrightarrow HCl + \left[\begin{matrix} \dot{C}H_2 - CH{=}CH_2 \\ \updownarrow \\ CH_2{=}CH - \dot{C}H_2 \end{matrix}\right] \text{Allyl radical}$$

$$\equiv \overbrace{CH_2 \overset{\bullet}{\text{---}} CH \text{---} CH_2}$$

***Step III*:** Attack of allyl radical on a new halogen molecule to generate 3-halopropene and new halogen atom.

$$Cl-Cl + \dot{C}H_2-CH=CH_2 \longrightarrow Cl-CH_2-CH=CH_2 + \dot{Cl} \quad \text{and so on.}$$

Step II and III are repeated over and over again till the chain is finally terminated.

Now the question arises as to how is it that under these conditions substitution is a preferred process whereas alkenes are known to readily undergo free radical additions?

$$CH_3-CH=CH_2 + \dot{Cl} \longrightarrow CH_3-\dot{C}H-CH_2Cl \xrightarrow{\times} CH_3-\underset{|}{\overset{Cl}{C}}H-CH_2Cl$$

1, 2-Dichloropropane
(*Addition product*)

Here, the addition product is not formed. It has been explained that addition of halogen radical does not take place. It is due to the fact that at high temperature the halogen is eliminated before the completion of the second step. Secondly, because of the low concentration of halogen, the radical generated after its addition will have to wait for an encounter with a second halogen molecule which is delayed and hence it breaks down to starting material. On the other hand, the allylic radical formed as a result of hydrogen abstraction can wait for the second halogen molecule because of its *resonance stabilization*.

In addition to free halogens, *N-bromosuccinimide (NBS), being a source of low concentration of bromine, is frequently used for allylic bromination.*

Allylic Bromination by NBS (Wohl-Ziegler Bromination)

N-Bromosuccinimide (NBS) is a reagent of choice for allylic bromination of alkenes. The reaction is catalysed by light. For example, bromination of propene by NBS.

$$CH_3-CH=CH_2 + \underset{\text{NBS}}{\begin{array}{l}CH_2-CO\\ |\qquad\quad\;\; \rangle N-Br\\ CH_2-CO\end{array}} \xrightarrow{h\nu} \underset{\text{Allyl bromide}}{BrCH_2CH=CH_2} + \underset{\text{Succinimide}}{\begin{array}{l}CH_2-CO\\ |\qquad\quad\;\; \rangle NH\\ CH_2-CO\end{array}}$$

Mechanism

The role of NBS is to provide a constant low concentration of bromine thus favouring the substitution reaction rather than the addition reaction. The mechanism of the reaction is as follows:

(*i*)
$$\begin{array}{l}CH_2-CO\\ |\qquad\quad\;\; \rangle N-Br\\ CH_2-CO\end{array} \xrightarrow{h\nu} \begin{array}{l}CH_2-CO\\ |\qquad\quad\;\; \rangle \dot{N}\\ CH_2-CO\end{array} + \dot{Br}$$

(*ii*)
$$\dot{Br} + H-CH_2-CH=CH_2 \xrightarrow{h\nu} HBr + \dot{C}H_2-CH=CH_2$$

$$(iii)\quad H-Br + \begin{array}{l}CH_2-CO\\ |\\ CH_2-CO\end{array}\!\!\!>N-Br \xrightarrow{h\nu} Br_2 + \begin{array}{l}CH_2-CO\\ |\\ CH_2-CO\end{array}\!\!\!>NH$$

$$(iv)\quad Br-Br + \dot{C}H_2-CH=CH_2 \longrightarrow BrCH_2CH=CH_2 + \dot{Br}$$

The steps (*ii*) and (*iv*) are repeated several times till the chain is finally terminated.

2. *Electrophilic Substitution at Vinylic Position*

An interesting reaction of alkenes is the electrophilic substitution at an unsaturated carbon atom (symbolized as S_E reaction). For example, treatment of an alkene with acid chloride in presence of anhydrous aluminium chloride affords an unsaturated ketone (*cf.* Friedel-Crafts acylation of aromatics, Chapter 8).

$$>C=C<^{H} + RCOCl \xrightarrow{AlCl_3} >C=C<^{COR}$$

Mechanism

The reaction takes place involving a two-step mechanism as shown below:

Step I: In the presence of aluminium chloride, the acid chloride generates the acylium ion ($RCO^{\oplus}$) which adds to the double bond as an electrophile forming a carbocation.

$$R-\overset{O}{\overset{\|}{C}}-Cl + AlCl_3 \rightleftharpoons \underset{\text{Acylium ion}}{R-\overset{\oplus}{C}=O} + \overset{\ominus}{AlCl_4}$$

$$>C=C<^{H} + O=\overset{\oplus}{C}-R \longrightarrow \underset{\text{Carbocation}}{>\overset{\oplus}{C}-C<^{H}_{COR}}$$

Step II: The carbocation, so produced, may lose a proton forming the α, β-unsaturated ketone (the S_E1 reaction, meaning **S**ubstitution **E**lectrophilic **U**nimolecular).

$$>\overset{\oplus}{C}-C<^{H}_{COR} \xrightarrow{-\overset{\oplus}{H}} >C=C<^{COR}$$

Alternatively, it may add a chloride ion forming an addition compound, the β-chloroketone, which may under the reaction conditions lose HCl forming the unsaturated ketone (addition-elimination mechanism).

$$\overset{\oplus}{\text{C}}-\underset{}{\overset{\text{H}}{\text{C}}}-\text{COR} \xrightarrow[\overset{\ominus}{(\text{AlCl}_4)}]{\overset{\ominus}{\text{Cl}}} \text{Cl}-\text{C}-\overset{\text{H}}{\text{C}}-\text{COR} \xrightarrow{-\text{HCl}} \text{C}=\text{C}-\text{COR}$$

(*xv*) Isomerisation (Not included in the syllabus of KUK)

Conversion of one isomer of an alkene into another called *isomerization* —can be achieved either by heating at high temperature (500°) or at a lower temperature (200–300°) in the presence of a suitable acid catalyst such as anhydrous aluminium chloride. The following two types of isomerizations are commonly encountered:

(*a*) Less substituted alkene may be converted into a more substituted one by the migration of the double bond as typified by the conversion of 1-pentene to 2-pentene.

$$\underset{\text{1- Pentene}}{CH_3-CH_2-CH_2-CH=CH_2} \xrightarrow[200-300°]{AlCl_3} \underset{\text{2- Pentene}}{CH_3-CH_2-CH=CH-CH_3}$$

Following mechanism can be proposed for this change:

$$\underset{\text{1- Pentene}}{CH_3-CH_2-CH_2-CH=CH_2} + AlCl_3 \rightleftharpoons CH_3-CH_2-\overset{H}{CH}-\overset{\oplus}{C}H-CH_2\overset{\ominus}{A}lCl_3$$

$$\Big\downarrow\uparrow\ -\overset{\oplus}{H}$$

$$\underset{\text{2- Pentene}}{CH_3-CH_2-CH=CH-CH_3} \underset{-AlCl_3}{\overset{+\overset{\oplus}{H}}{\rightleftharpoons}} CH_3-CH_2-CH=CH-CH_2-\overset{\ominus}{A}lCl_3$$

(*b*) A methyl group may shift from an allylic position.

$$\underset{\text{1-Butene}}{CH_3-CH_2-CH=CH_2} + AlCl_3 \rightleftharpoons CH_3-CH_2-\overset{\oplus}{C}H-CH_2-\overset{\ominus}{A}lCl_3$$

$$\Big\downarrow\uparrow\ \text{1,2-methyl shift}$$

$$\underset{\text{2-Methylpropene}}{CH_2=\underset{CH_3}{\underset{|}{C}}-CH_3} \underset{-AlCl_3}{\overset{+\overset{\oplus}{H}}{\rightleftharpoons}} CH_2=\underset{CH_3}{\underset{|}{C}}-CH_2-\overset{\ominus}{A}lCl_3 \underset{}{\overset{-\overset{\oplus}{H}}{\rightleftharpoons}} \overset{\oplus}{C}H_2-\underset{CH_3}{\underset{|}{\overset{H}{C}}}-CH_2-\overset{\ominus}{A}lCl_3$$

(*c*) Many transition metal ions and complexes promote double bond migration—that is, isomerization—in alkenes to give invariably the thermodynamically most stable isomeric mixture. Thus, 1-alkenes give (*cis* + *trans*)-2-alkenes.

Mechanism

The mechanism of isomerisation involves following steps:

Step I: *The alkene coordinates with metal.*

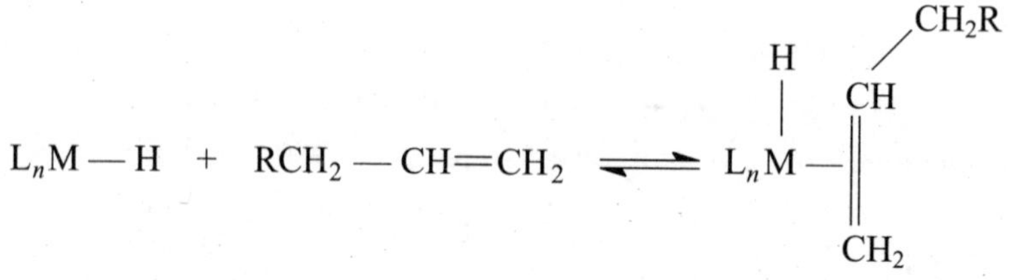

$$L_nM—H + RCH_2—CH=CH_2 \rightleftharpoons L_nM(H)—(\eta^2\text{-}CH(CH_2R)=CH_2)$$

Step II: *Insertion of alkene into M—H bond.* (Transfer of hydrogen atom from metal to alkene). With alkenes other than ethylene, there is, of course, the possibility of addition of M—H to the double bond in either the Markownikoff or the *anti*-Markownikoff direction, just as with the addition, say, of HCl. Thus, we may get either **(A)** or **(B)**.

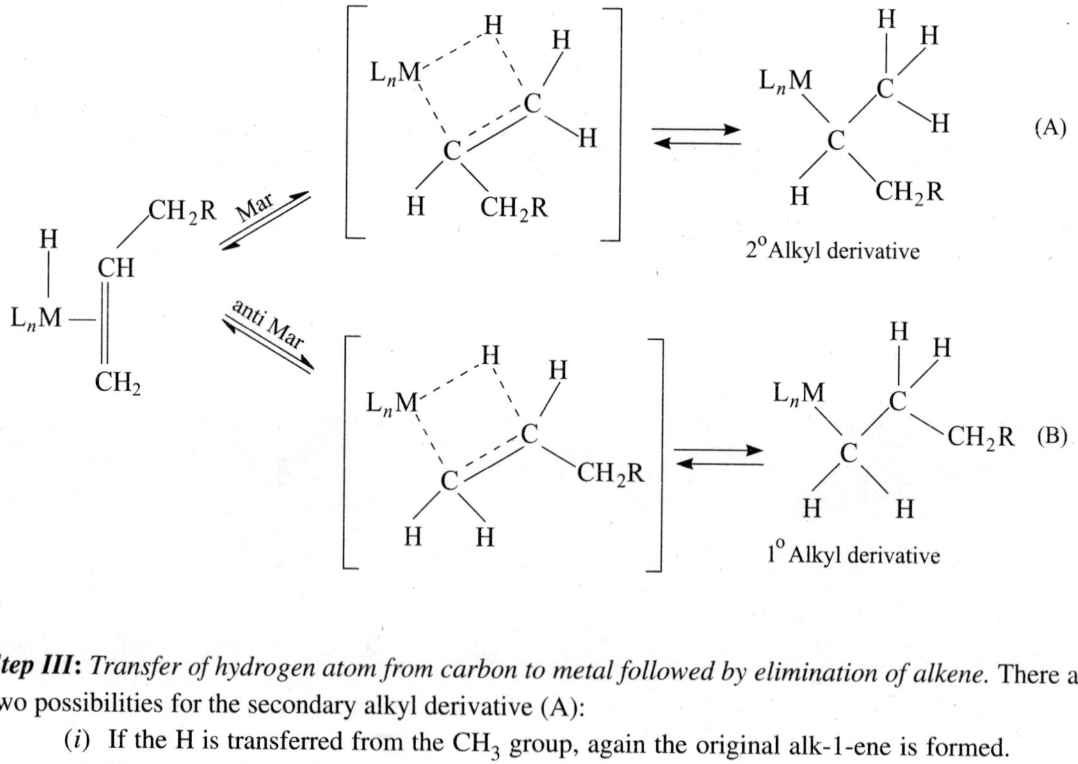

Mar: $L_nM(H)(CH_2{=}CHCH_2R) \rightleftharpoons [\text{four-centre transition state}] \rightleftharpoons L_nM—CH(CH_2R)—CH_3$ (A)

2° Alkyl derivative

anti Mar: $L_nM(H)(CH_2{=}CHCH_2R) \rightleftharpoons [\text{four-centre transition state}] \rightleftharpoons L_nM—CH_2—CH_2—CH_2R$ (B)

1° Alkyl derivative

Step III: *Transfer of hydrogen atom from carbon to metal followed by elimination of alkene.* There are two possibilities for the secondary alkyl derivative (A):

(*i*) If the H is transferred from the CH_3 group, again the original alk-1-ene is formed.

(*ii*) If it is transferred from methylene of CH_2R group, then alk-2-ene is formed.

Thus, isomerization can occur only on Markownikoff addition with the formation of either *cis* or *trans* or *both*. It should be noted that there will be no isomerization on *anti*-Markownikoff's addition as same alk-1-en is formed.

6.7 TEST FOR ALKENES

There are two tests which are used for detection of unsaturation (*i.e.*, alkenes) in a molecule.

1. Decolourisation of Bromine Solution in CCl_4

$$\underset{\text{Alkene colourless}}{>C=C<} + \underset{\text{Yellow brown}}{Br_2/CCl_4} \longrightarrow \underset{\text{Colourless}}{-\underset{Br}{\overset{|}{C}}-\overset{Br}{\underset{|}{C}}-}$$

2. Baeyer test: Decolorization of $KMnO_4$ in cold, dilute, neutral medium.

$$\underset{\text{Alkene}}{>C=C<} + \underset{\text{Purple}}{MnO_4^{\ominus}} \longrightarrow \underset{\text{Brown ppt.}}{MnO_2} + \underset{\text{Colourless}}{-\underset{OH}{\overset{|}{C}}-\underset{OH}{\overset{|}{C}}-}$$

6.8 INDUSTRIAL APPLICATIONS OF ETHYLENE AND PROPYLENE

Alkenes such as ethylene, propylene and other substituted alkenes undergo polymerization readily to form polymers which are also known as plastics. These synthetic plastics can be moulded into required shapes and possess other properties which enable them to be used for the manufacture of a wide range of articles of common use. Dyes and pigments are added to get attractive colours. Some important industrial uses of ethylene and propylene are as below:

Uses of Ethylene

(*i*) Ethylene is used for the manufacture of ethanol which is an important ingredient of alcoholic beverages and a very good solvent. It is also used for the manufacture of other chemicals such as acetaldehyde, acetic acid and ethyl acetate.

(*ii*) Ethylene oxide, which is formed by epoxidation, is used for the manufacture of ethylene glycol—an antifreeze for car radiators (***Coolant***). It is also used in the manufacture of polyester fibres.

(*iii*) It is used in the synthesis of industrial products such as polythene, vinyl chloride and styrene.

(*iv*) Vinyl chloride and styrene polymerize to form polyvinyl chloride (PVC) and polystyrene respectively. These are very useful polymers.

Uses of Propylene

(*i*) Polypropylene is used for making automobile moldings, seat covers, carpets, ropes etc.

(*ii*) Propene upon ammoxidation gives acrylontrile which polymerise to give polyacrylonitrile **(PAN)**. This polymer is mainly used for the manufacture of orlon and acrilan fibres for making clothes, blankets and carpets.

(*iii*) Propylene is also used in the production of isopropylbenzene (cumene). Cumene in turn is used in the synthesis of two very important industrial chemicals namely acetone and phenol.

A BRIEF REVIEW

- Alkenes are obtained by **dehydration of alcohols** with concentrated H_2SO_4 at 443 – 453 K. The ease of dehydration of alcohols follows the order: 3° > 2° > 1° due to the stability of carbocation intermediates in the same order.
- According to **Saytzeff rule** more highly substituted alkene is always the major product of dehydration of alcohols. Further, it is the *trans*-product which predominates *e.g.*, dehydration of butan-2-ol gives about 70% of *trans*-but-2-ene and 10% of *cis*-but-2-ene and about 20% of but-1-ene. It is the stability of transition state involved which controls the regiochemistry of elimination reaction.
- Dehydration of alcohols mostly gives alkenes obtained from rearranged carbocations *e.g.*, butan-1-ol gives the same mixture of alkenes (*i.e.*, 80% of but-2-ene + 20% of but-1-ene) as that obtained from butan-2-ol. This is due to greater stability of *sec*-butyl carbocation. In case of butan-1-ol the initially formed 1°-butyl carbocation rearranges to more stable 2°-butyl carbocation.
- Alkenes can also be obtained by **dehydrohalogenation** of alkyl halides with hot alcoholic KOH. This dehydrohalogenation of both 1° and 2° alkyl halides occurs by **E2 elimination** and in case of 3° alkyl halides by **E1 mechanism**.
- Dehydrohalogenation of 1° and 2° alkyl halides does not involve formation of carbocation intermediates. Due to this rearranged products are not formed.
- **Saytzeff rule** always holds good for dehydrohalogenation of *secondary*—alkyl halides.
- **Hoffmann elimination** (*Pyrolysis of quaternary ammonium halides*) always gives less substituted alkene as the major product. It is the acidity of the β-hydrogen with respect to nitrogen which controls the direction of elimination.
- The m. pt.'s of *trans*-alkenes are higher than those of corresponding *cis*-alkenes because symmetrial *trans*-alkenes are packed more closely and thus require higher energy to break their crystal lattice.
- The b. pt.'s of *cis*-alkenes are higher than those of corresponding *trans*-alkenes because *cis*-alkenes have higher dipole moments and hence have strong intermolecular forces of attraction.
- Alkenes undergo ***hydrogenation*** in presence of finely divided metals such as Pt, Pd or Raney Ni to form alkanes. It is an exothermic process.
- The stability of an alkene depends upon its heat of hydrogenation (*i.e.*, the heat evolved when one mole of an alkene is hydrogenated).
- The order of stability of different alkenes is : $R_2C = CR_2 > R_2C = CHR > RCH = CHR > R_2C = CH_2 > RCH = CH_2 > CH_2 = CH_2$.
- *Trans*-isomer of an alkene is always more stable than *cis*-isomer.
- **Electrophilic** addition reactions of alkenes occur in two steps. First step involves slow formation of a carbocation intermediate and the second step involves rapid nucleophilic attack on the carbocation. The first step is the rate-determining step.

- **Evidence** in favour of formation of a carbocation during electrophilic addition reactions is the formation of $CH_2Br—CH_2Cl$ and $CH_2Br—CH_2OH$ along with $CH_2Br—CH_2Br$. When ethene is bubled through an aqueous solution of Br_2 containing aq. NaCl.
- **Cyclic-halonium ion mechanism** operates during addition of halogens to alkenes to give *trans*-addition products without formation of any rearranged product. This rules out the intermediacy of any carbocation in the halogenation reaction of alkenes.
- **Markownikoff's rule** governs addition of unsymmetrical reagents to unsymmetrical alkenes during which the negative part of the reagent attacks at more substituted carbon atom of the double bond.
- According to **anti-Markownikoff's** addition or **peroxide effect** or **Kharasch effect**, the addition of HBr (but not HCl or HI) to an unsymmetrical alkene in presence of peroxides takes place contrary to Markownikoff's rule.

$$CH_3—CH=CH_2 + HBr \xrightarrow{\text{Mark. addition}} CH_3—CHBr—CH_3$$

$$CH_3—CH=CH_2 + HBr \xrightarrow[\text{Peroxide}]{\text{anti-Mark. addition}} CH_3—CH_2—CH_2Br$$

- **Hydration** of alkenes takes place when the alkene is bubled through cold conc. H_2SO_4 to give alkyl hydrogen sulphate which undergoes *hydrolysis* with boiling H_2O to give alcohols in accordance with Markownikoff's rule. Rearranged products are formed during hydration of alkenes due to the formation of carbocation intermediates.
- During **hydroboration** of alkenes with diborane, *tri-n*-alkylboranes are formed which undergo oxidation with alk. H_2O_2 to give alcohols corresponding to anti-Markownikoff's addition of H_2O to alkenes without formation of any rearranged product.
- During **oxymercuration** of alkenes with $(CH_3COO)_2Hg$ in aq. THF initially formed adducts undergo reduction with $NaBH_4$ in basic solution to give alcohols in accordance with Markownikoff's addition of water to alkenes without formation of any rearranged product.
- Addition of two hydroxyl groups across a double bond is called **hydroxylation.** It is carried out either with cold dil(1%) aq. alk. $KMnO_4$ (Baeyer's reagent) or with OsO_4 followed by hydrolysis in presence of aq. $NaHSO_3$ or Na_2SO_3. Both give *cis*-1, 2-diols.
- *OsO_4 is highly selective* whereas $KMnO_4$ is *less selective.*
- Alkenes undergo **epoxidation** on reaction with peracids which upon subsequent hydrolysis with acids or base give *trans*-1, 2-diols. The nucleophilic attack of H_2O in acidic medium occurs at the more substituted carbon but in basic medium, it occurs at the less hindered carbon atom of the epoxide.

- Cleavage of C = C double bond with **hot $KMnO_4$** gives a mixture of two acids or one acid and one ketone or two ketones depending upon the structure of alkene. However, if HCOOH is one of the products of oxidation, it is always further oxidised to $CO_2 + H_2O$.
- Alkenes undergo **ozonolysis** on passing O_3 through an alkene solution in some inert solvent (CCl_4) at a low temperature (196 K). *Ozonolysis is used to locate the position of double bond in an unknown alkene.*
- **Free radical substitution at the allylic position — Allylic halogenation** Propene undergoes allylic halogenation when treated with Cl_2 or Br_2 at 773 K to give 3-halopropene. NBS is the specific reagent for the allylic bromination.
- **Electrophilic substitution at vinylic position** occurs when an alkene is treated with an acid chloride in presence of anhydrous $AlCl_3$ to give α, β-unsaturated ketone.

SOLVED PROBLEMS

Q. 1. How will you convert *n*-hexylamine into 1-hexene?

Ans. This conversion can be accomplished *via* exhaustive methylation of amine followed by Hoffmann elimination reaction as depicted below:

$$\underset{n\text{-Hexylamine}}{CH_3(CH_2)_3CH_2CH_2NH_2} \xrightarrow[\text{(excess)}]{CH_3I} \underset{n\text{-Hexyltrimethylammonium iodide}}{CH_3(CH_2)_3CH_2CH_2-\overset{\oplus}{N}(CH_3)_3\overset{\ominus}{I}} \xrightarrow[-AgI]{Ag_2O/H_2O}$$

$$CH_3(CH_2)_3CH_2CH_2-\overset{+}{N}(CH_3)_3\overset{\ominus}{O}H \xrightarrow[\text{(Hoffmann elimination)}]{\Delta} \underset{\text{1-Hexene}}{CH_3(CH_2)_3CH{=}CH_2} + (CH_3)_3N + H_2O$$

Q. 2. Dehydration of 1-butanol and 2-butanol give the same mixture of alkenes. Explain.

Ans. Dehydration of alcohols occurs though carbocation intermediates. In case of 1-butanol, the initially formed less stable 1° carbocation, *i.e.*, *n*-butyl carbocation rearranges to the more stable 2° carbocation *i.e.*, 2° butyl carbocation by a 1,2-hydride shift. 2-Butanol, on the other hand, directly gives the more stable 2° butyl carbocation.

$$\underset{\text{1-Butanol}}{CH_3CH_2CH_2CH_2OH} \xrightarrow[-H_2O]{\overset{\oplus}{H}} \underset{\substack{n\text{-Butyl carbocation}\\ (\textit{Less stable})}}{CH_3CH_2-\underset{\mid\atop H}{CH}-\overset{\oplus}{C}H_2} \xrightarrow[\text{shift}]{\text{1, 2-Hydride}} \underset{\substack{2^\circ\text{ Butyl carbocation}\\ (\textit{More stable})}}{CH_3CH_2-\overset{\oplus}{C}H-CH_3}$$

$$\underset{\text{2-Butanol}}{CH_3CH_2-\underset{\mid\atop OH}{CH}-CH_3} \xrightarrow[-H_2O]{\overset{\oplus}{H}} 2^\circ\text{ Butyl carbocation}$$

$$2^\circ\text{ Butyl carbocation} \xrightarrow[\text{(Saytzeff rule)}]{-\overset{\oplus}{H}} \text{2-Butene (80\%)} + \text{1-Butene (20\%)}$$

Since, the same carbocation is obtained from both alcohols, therefore, both of these alcohols will give the same mixture of 2-butene (80%) and 1-butene (20%) in accordance with Saytzeff rule.

Q. 3. Dehydration of alcohols to alkenes is carried out with an acid but not with a base and the acid used is invariably H_2SO_4 and not HCl or HNO_3. Explain.

Ans. Dehydration of alcohols to alkenes requires the elimination of a molecule of H_2O. But OH is a strong base and hence is a poor leaving group. Therefore, to make it a good leaving group such as neutral H_2O molecule, some strong acid such as H_2SO_4 is required. Since, a base cannot convert OH to a good leaving group, therefore, it is never used for dehydration of alcohols.

$$\underset{\text{Bad leaving group}}{CH_3CH_2—OH} \xrightarrow{H^+} \underset{\text{Good leaving group}}{CH_3CH_2—\overset{+}{O}H_2} \xrightarrow{-H_2O} \underset{\text{Ethyl carbocation}}{CH_3—\overset{+}{C}H_2}$$

Further the carbocation intermediate thus formed can either eliminate a proton to form an alkene or may undergo nucleophilic attack by the anion of the acid left after the loss of a proton. These anions are:

$H—O—S(=O)_2—O^{\ominus}$
Bisulphate ion from H_2SO_4
(Least nucleophilic)

$O=\overset{+}{N}(O^{\ominus})_2$
Nitrate ion from HNO_3
(Less nucleophilic than HSO_4^- ion but a strong oxidising agent)

$Cl^{\ominus}$
Chloride ion from HCl
(Most nucleophilic)

Out of these anions, since Cl^- ion is the best nucleophile, therefore, it will preferentially attack the carbocation to give the substitution product rather than losing a proton to give the elimination product.

$$\underset{\text{Ethylene}}{CH_3{=}CH_2} \xleftarrow[\times]{-H^+} CH_3—\overset{+}{C}H_2 \xrightarrow[\text{(Substitution)}]{+Cl^-} \underset{\text{Ethyl chloride}}{CH_3CH_2—Cl}$$

Thus, HCI is not used for dehydration of alcohols.

Both NO_3^- ion and HSO_4^- are stabilized by resonance and hence are weaker nucleophiles than Cl^- ion. Further, HSO_4^- is better stabilized than NO_3^- ion, therefore HSO_4^- does not give substitution products but gives only elimination products. For example,

$$CH_2{=}CH_2 \xleftarrow{-H^{\oplus}} CH_3—\overset{\oplus}{C}H_2 \xrightarrow[\times]{\overset{\ominus}{HSO_4}/\ 443–453\ K} CH_3CH_2OSO_3H$$

NO_3^- ion being a weaker nucleophile than HSO_4^- does give substitution products. But the problem here is that conc. HNO_3 being a strong oxidising agent brings about the oxidation of the alcohol first to the aldehyde and then to the carboxylic acid rather than bringing about elimination to give alkenes. Thus, HNO_3 acid is also not used for dehydration of alcohols. In other words, H_2SO_4 or its acid salt *i.e.*, $KHSO_4$ are usually used for dehydration because of the following two reasons:

(*i*) H_2SO_4 is a strong acid.

(*ii*) It has non-nucleophilic anion *i.e.*, HSO_4^- anion.

Q. 4. How will you convert 2-propanol to 1-propanol and *vice-versa*?

Ans.

(*i*)
$$\underset{\text{2-Propanol}}{CH_3-CHOH-CH_3} \xrightarrow[443\text{–}453K]{\text{conc. } H_2SO_4} \underset{\text{Propene}}{CH_3-CH=CH_2} \xrightarrow[\text{(Hydroboration)}]{B_2H_6}$$

$$\underset{\text{Tri-}n\text{-propylborane}}{(CH_3-CH_2-CH_2)_3B} \xrightarrow[\text{(Oxidation)}]{\text{Alk. } H_2O_2} \underset{\text{1-Propanol}}{3CH_3CH_2CH_2OH}$$

(*ii*)
$$\underset{\text{1-Propanol}}{CH_3CH_2CH_2OH} \xrightarrow[443-453\text{ K}]{\text{conc. } H_2SO_4} \underset{\text{Propene}}{CH_3-CH=CH_2} \xrightarrow[\text{room temp.}]{\text{conc. } H_2SO_4}$$

$$\underset{\textit{iso}\text{-Propyl hydrogen sulphate}}{CH_3-\underset{\underset{OSO_2OH}{|}}{CH}-CH_3} \xrightarrow{H_2O,\text{ boil}} \underset{\text{2-Propanol}}{CH_3-\underset{\underset{OH}{|}}{CH}-CH_3} + H_2SO_4$$

Q. 5. Give the structure of the major organic product formed when 3-ethylpent-2-ene is treated with $Hg(OAc)_2/H_2O$, $NaBH_4$.

Ans.

$$\underset{\text{3-Ethylpent-2-ene}}{CH_3CH_2-\overset{\overset{CH_2CH_3}{|}}{C}=CH-CH_3} \xrightarrow[\text{(Mark addn.)}]{Hg(OAc)_2/H_2O} CH_3CH_2-\overset{\overset{CH_2CH_3}{|}}{\underset{\underset{OH}{|}}{C}}-\underset{\underset{HgOAc}{|}}{CH}-CH_3$$

$$\xrightarrow[\text{(Reduction)}]{NaBH_4} \underset{\text{3-Ethylpentan-3-ol}}{CH_3CH_2-\overset{\overset{CH_2CH_3}{|}}{\underset{\underset{OH}{|}}{C}}-CH_2CH_3}$$

UNSOLVED PROBLEMS

1. Write IUPAC names of the following

(i) $CH_3-\underset{|}{\overset{CH_3}{C}}=CHCH_3$

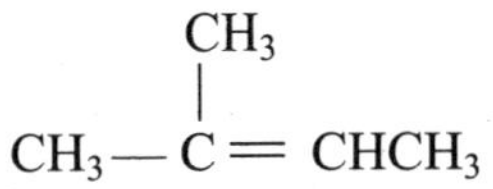

(ii) $CH_3-\underset{CH_3}{\underset{|}{C}}=CHCH_2-\underset{CH_3}{\underset{|}{C}HCH_3}$

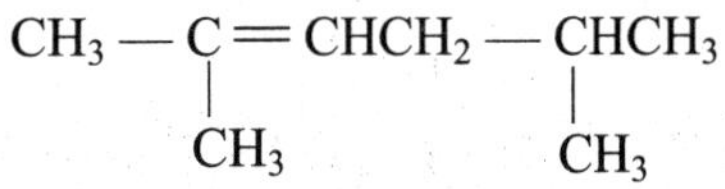

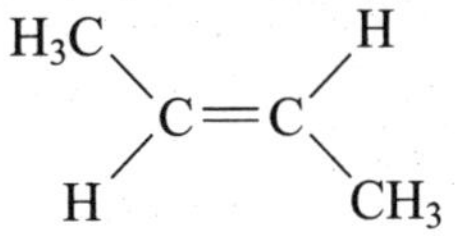

(iii) $(CH_3)_3CCH=CHC(CH_3)_3$

(iv) $H_3C(H)C=C(H)CH_3$ (H₃C and H on the left carbon; H and CH₃ on the right carbon, trans)

(v) $CH_3CH_2CH_2-\underset{CH_2}{\underset{\|}{C}}-CH_2CH_2CH_3$

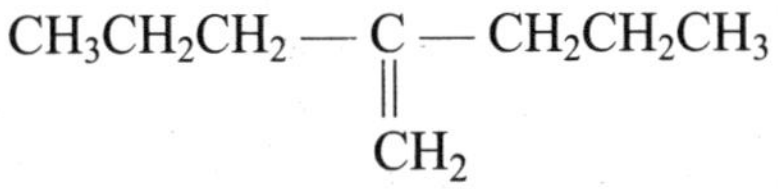

(vi) $(CH_3)_2C=C(CH_3)_2$

2. Write structural formulas for:

(i) *cis*-2-Pentene (ii) *cis*-3-Hexene

(iii) 1-Methylcyclopentene (iv) 3,5-Dimethylcyclohexene

(v) 5,5-Dimethyl-2-hexene (vi) 3-Ethylcyclohexene

(vii) *cis*-1,3-Pentadiene (viii) Phenylethene.

3. For which of the following compounds is *cis,trans* isomerism possible:

(i) 2-Heptene (ii) 1-Butene

(iii) 2-Methylpropene (iv) 1-Chloro-1-butene

(v) 1,1-Dichlorobutene (vi) 2-Methyl-2-heptene.

Write structural formulas for their geometric isomer, if any.

4. Discuss the mechanisms of the following transformations:

(i) $CH_3CH_2Br \xrightarrow{\overset{\ominus}{O}C_2H_5} CH_2=CH_2$

(ii) $(CH_3)_3COH \xrightarrow{\overset{\oplus}{H}} (CH_3)_2C=CH_2$

5. Predict the structures of alkenes expected from the dehydrohalogenation of the following alkyl halides. In case of isomeric alkenes, tell which shall be the major product:

(i) 2-Chlorobutane (ii) 1-Chlorobutane

(iii) 2-Chloro-3-methylbutane (iv) 3-Chloro-2-mehtylpentane

(v) 2-Chloro-3-methylpentane and (vi) 2-Choro-3, 3-dimethylpentane.

6. Acid catalyzed dehydration of neopentyl alcohol produces two isomeric alkenes. Predict their structures and discuss a suitable mechanistic pathway for the reaction. Which alkene do you think shall be the major product? Give reasons.

7. (*i*) Addition of an electrophile to 2, 4-hexadiene generates two intermediates. Write their structures and tell which one is more stable.

   (*ii*) Write the structures of all the products that could possibly be formed when 1-pentene is treated with bromine water containing a little NaCl. Sketch a suitable mechanistic scheme that explanis the formation of these products.

8. Which alcohol in each of the following pairs would undergo facile dehydration:

(*i*) $CH_3CH_2CH_2CH_2OH$ (A) and $CH_3CH(OH)CH_2CH_3$ (B)

(*ii*) $CH_3CH(CH_3)—CH(OH)CH_2CH_3$ (A) and $CH_3CH(CH_3)—C(CH_3)(OH)—CH_3$ (B)

(*iii*) $CH_3CH_2CH(CH_3)—C(CH_3)(OH)—CH_3$ (A) and $CH_3CH_2CH(CH_2—CH_3)—CH(OH)—CH_3$ (B)

9. Justify the statement "Dehydrohalogenation is one of the best methods for preparing alkenes."

10. Discuss briefly the following chemical reactions of alkenes

   (*i*) Addition of halogens and halogen acids. (*ii*) Oxidation reactions of alkenes.

11. On carrying out the reaction of 2-butene with HCl in the presence of ethanol, one of the products formed has the molecular formula $C_6H_{14}O$. Assign structure to this product and discuss the mechanism leading to its formation.

12. (*i*) On heating *cis*-2-butene to a temperature greater than 573 K, a mixture two isomeric 2-butene result.

   (*a*) What is the mechanism of this reaction?

   (*b*) Which butene isomer would you expect to predominate in the mixture when equilibrium is established between them?

   (*ii*) Discuss three methods by which hydration of alkenes can be achieved? Describe the merits and demerits of each.

13. (*i*) Discuss briefly free redical substiution at the allylic carbon and electrophilic substitution at the vinylic carbon with suitable examples.

(*ii*) List industrial application of ethylene and propylène.

14. Which of the following sequences would be the best for the preparation of 2-bromobutane. Discuss the mechanisms of the reactions involved in all the three cases.

(*i*) $CH_3CH{=}CH\,CH_3 \xrightarrow{Br_2}$

(*ii*) $CH_3CH_2CH{=}CH_2 \xrightarrow{HBr}$

(*iii*) $CH_3CH_2CH{=}CH_2 \xrightarrow[\text{Peroxide}]{HBr}$

15. (*i*) With suitable examples, illustrate the mechanism of hydroboration.

(*ii*) Write a feasible mechanism for the acid-catalyzed isomerization of *cis*-2-butene to *trans*-2-butene.

16. (*i*) Discuss the mechanism of ozonide formation.

(*ii*) A hydrocarbon which has the molecular formula C_6H_{12}, when subjected to ozonolysis gave equimolar amounts of ethyl methyl ketone ($CH_3CH_2COCH_3$) and acetaldehyde (CH_3CHO). Assign structure and give IUPAC name of the hydrocarbon.

17. Ozonolysis of alkenes A, B, C and D gave the following products:

(*i*) $CH_3\overset{O}{\overset{\|}{C}}\,CH_2CH_2CH_2CH_2\overset{O}{\overset{\|}{C}}{-}H$ (only)

(*ii*) $CH_3\overset{O}{\overset{\|}{C}}CH_3 + CH_3\overset{O}{\overset{\|}{C}}{-}H$

(*iii*) $CH_3CH_2\overset{O}{\overset{\|}{C}}{-}H$ (only)

(*iv*) $CH_3{-}\overset{O}{\overset{\|}{C}}{-}\overset{O}{\overset{\|}{C}}{-}H$ + 2 moles of $CH_3{-}\overset{O}{\overset{\|}{C}}{-}CH_3$

Predict the structures of alkenes A through D.

18. Complete the following equations showing the stereochemistry of the products wherever possible:

(*i*) *cis*-2-Butene $\xrightarrow{H_2/Pt}$

(*ii*) *trans*-2-Butene $\xrightarrow{Br_2/CCl_4}$

(*iii*) *cis*-2-Butene $\xrightarrow{RCOOOH}$

Discuss the mechanisms involved in the processes.

19. With suitable example, illustrate the mechanism of hydroboration-oxidation reaction.

20. Predict the products of the following reactions:

(*i*) $CH_3CH{=}CH_2 \xrightarrow{HBr}$

(*ii*) $3CH_3CH{=}CH_2 \xrightarrow{BH_3}$

(*iii*) $RCH{=}CHR \xrightarrow{KMnO_4/\overset{\ominus}{O}H}$

(*iv*) $CH_2{=}C{=}CH_2 \xrightarrow{HCl}$

(*v*) $CH_2{=}CH{-}CH{=}CH_2 \xrightarrow{Br_2}$

21. What is Hoffmann elimination? How does it differ from Saytzeff elimination? Give reasons for your answer.

22. (*i*) Predict the stereochemistry of the product formed by the addition of bromine on

(*a*) E-1, 2-dideuterioethene, (*b*) Z-1, 2-dideuterioethene, (*c*) *cis*-2-Butene.

(*ii*) Write the mechanism of dehydrohalogenation of alkyl halides. Explain the orientation and ease of formation of alkenes.

23. How many products do you expect when 3-methyl-2-butene is treated with bromine in the presence of methyl alcohol containing lithium chloride? Discuss the mechanism involved.

24. Explain why *trans*-alkenes have higher melting points but lower boiling points than their corresponding *cis*-isomers?

25. Discuss the mechanism of hydrogenation of alkenes and explain how heat of hydrogenation can be used to explain the stability of alkenes ?

26. Why do alkenes undergo electrophilic addition reactions and not electrophilic substitution reactions ? Give reasons to support your answer.

27. Discuss the mechanism of anti-Markownikoff's rule of addition of HBr to unsymmetrical alkenes.

28. (*a*) Taking propene as example, explain the synthetic utility of hydroboration-oxidation reaction.

(*b*) Discuss the mechanism of hydroboration-oxidation of propene.

29. Predict the product (s) and sketch the mechanism of the following reactions:

(*i*) $CH_3{-}CH{=}CH_2 \xrightarrow[\text{Peroxide}]{HBr} ?$ (*ii*) $CH_3{-}CH{=}CH_2 + (BH_3)_2 \xrightarrow{THF} ?$

(*iii*) $CH_3{-}\overset{\overset{CH_3}{|}}{C}{=}CH_2 + HCl \longrightarrow ?$ (*iv*) $CH_3CH{=}CHCH_3 + O_3 \longrightarrow ? \xrightarrow{Zn/H_2O} ?$

30. Explain why treatment of 3-methyl-1-butene with cold conc. H_2SO_4 followed by hydrolysis with boiling water gives 2-methyl-2-butanol as the major product but hydroboration-oxidation of the same alkene gives 3-metyl-1-butanol as the chief product?

31. Explain why addition of chlorine to propene at ordinary temperature gives 1, 2-dichloropropane but at 773 K, it gives 3-chloropropene?

32. Why peroxide effect is shown only by HBr and not by HCl or HI ?

33. How will you detect the presence of unsaturation in an organic molecule?

34. Acidic hydration of propene yields 2-propanol while the hydroboration-oxidation gives 1-propanol. Explain.

35. What is polymerization? Describe the mechanism of anionic polymerization of alkenes.

36. Write the reaction for the hydroxylation of alkene with osmium tetroxide.

37. Discuss the mechanism of oxymercuration-reduction of alkenes.

38. With suitable example, explain how oxymercuration-reduction of alkenes is a better method for preparation of alcohols than simple acid-catalysed hydration of alkenes ?

39. What happens when:

(*a*) 2,2,3-Trimethyloxirane is treated with (*i*) CH_3ONa/CH_3OH and (*ii*) CH_3OH/H_2SO_4

(*b*) 2-Methyl-2-butene is heated with HCl

(*c*) iso-butylene reacts with chlorine water and HCl.

40. Explain free radical polymerization with suitable examples.

41. Why NBS is a reagent of choice for allylic bromination of alkenes ?

SECTION IV

CHAPTER 7

Cycloalkenes, Dienes and Alkynes

Carbocyclic compounds containing one double bond in the ring are called **cycloalkenes**. They are isomeric with acyclic dienes and alkynes having the same molecular formula, (C_nH_{2n-2}). These three types of compounds differ in their properties due to difference in the number and type of multiple bonds. *Cycloalkenes have one double bond, while dienes have two double bonds and alkynes have one triple bond.* In this chapter, we will discuss the chemistry of these three classes of compounds. The chapter is divided into three sub units:

7.1 CYCLOALKENES

7.1.1 Nomenclature of Cycloalkenes

According to IUPAC system of nomenclature these compounds are named as cyloalkenes. Their names are derived by adding prefix **cyclo**-to the name of acyclic alkene having the same number of carbon atoms. The position of the double bond is indicated by numbering the ring carbon atoms in such a way that the double bond gets the least number. Following are the examples:

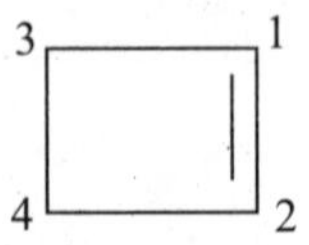

Cyclobutene 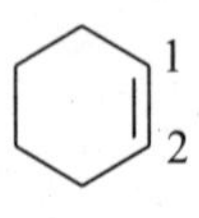   Cyclohexene 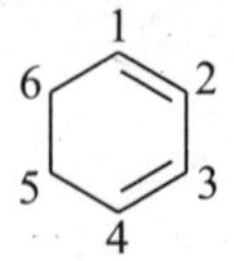   1,3-Cyclohexadiene 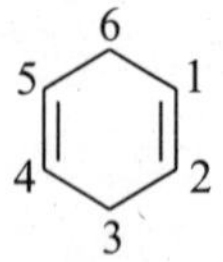   1,4-Cyclohexadiene

In case of substituted cycloalkenes, the positions of substituents are specified by using arabic numerals with the double bond getting lowest number. Numbers are then alloted to the rest of the carbons of the ring from that end which gives least number to the substituents. In di-and polysubstituted cycloalkenes, the names of the substituents are arranged in alphabetical order. Some examples are:

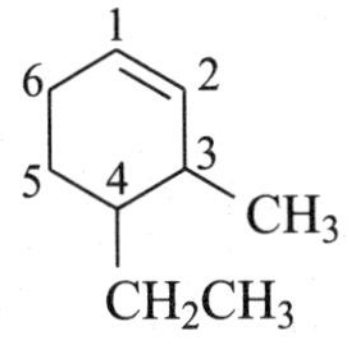

4-Ethyl-3-methylcyclohexene

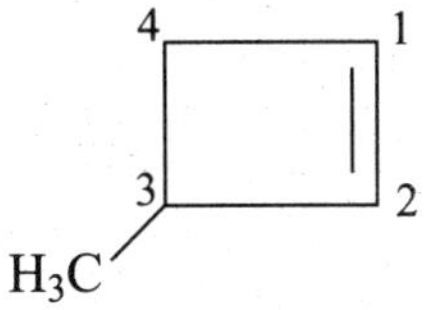

3-Methylcyclobutene

7.1.2 Methods of Formation of Cycloalkenes

Cycloalkenes are prepared by using similar methods as used for preparation of alkenes. Some important methods are described as follows:

1. Dehydration of Cycloalkanols

Cycloalkanols, which are obtained by reduction of cycloalkanones, on heating with 85% phosphoric acid at 440 – 445 K or with conc. H_2SO_4 at 413 K give cycloalkenes. For example,

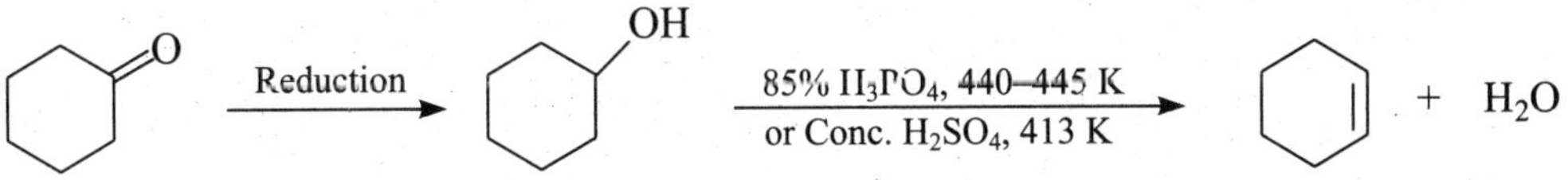

Reduction; 85% H_3PO_4, 440–445 K or Conc. H_2SO_4, 413 K; + H_2O

Cyclohexanone Cyclohexanol Cyclohexene

The dehydration of cyclohexanol is an example of β-elimination and follows acid-catalysed mechanistic path as discussed under dehydration of alcohols.

Dehydration of substituted cycloalkanols follows **Saytzeff rule** *i.e.*, more substituted cycloalkene is the major product. For example,

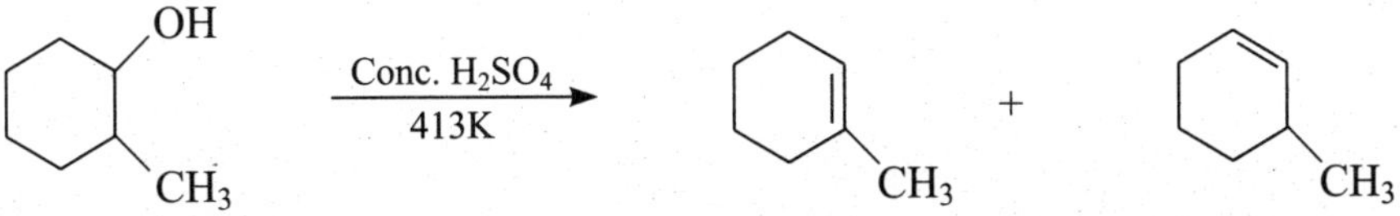

Conc. H_2SO_4, 413K

(84%) More substituted (16%) Less substituted

In this case, two β-hydrogens can be eliminated. But the one attached to the carbon having methyl group is lost to give more substituted alkene.

2. Dehydrohalogenation of Cycloalkyl Halides

Cycloalkyl halides on heating with an alcoholic potassium hydroxide solution give rise to cycloalkenes by dehydrohalogenation involving E2 mechanism.

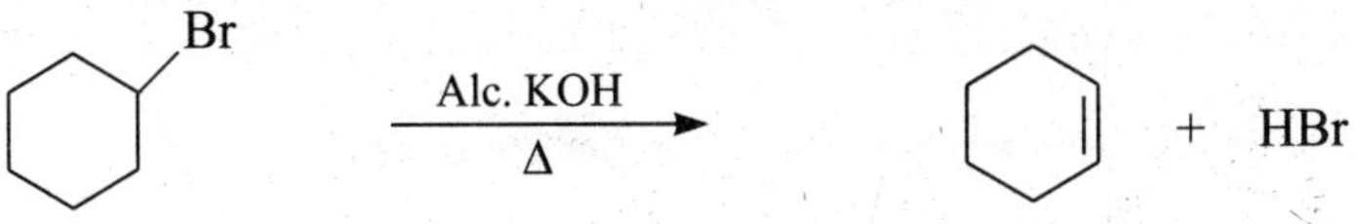

Alc. KOH, Δ; + HBr

Cyclohexyl bromide Cyclohexene

Dehydrohalogenation of substituted cycloalkyl halides follows *Saytzeff rule*. For examples,

1-Bromo-1-methylcyclohexane $\xrightarrow[\Delta]{\text{Alc. KOH}}$ 1-methylcyclohexene (94%, More substituted) + 3-methylcyclohexene (6%, Less substituted)

As with dehydrohalogenation of alkyl halides elimination in these systems is also *trans*. For example, *cis*-4-*tert*-butylcyclohexyl-*p*-toluene sulphonate (as *tert*-butyl group always occupies equatorial position the *p*-toluenesulphonyl group (Ts) has to occupy the axial position) undergoes E2 reaction in presence of a strong base to give 4-*tert*-butylcyclohexene whereas *trans*-isomer fails to undergo elimination under these conditions. This is because of the fact that in the *cis*-isomer both the leaving groups (H and OTs) are in *trans* antiparallel arrangement—a condition which is very essential for an E2 reaction.

(Axial H and OTs) *cis*-isomer $\xrightarrow{C_2H_5O^{\ominus}}$ 4-*tert*-butylcyclohexene ($C(CH_3)_3$)

trans-isomer $\xrightarrow{C_2H_5O^{\ominus}}$ No reaction

3. Dehalogenation of 1, 2-Dihalocycloalkanes

When an acetone solution of 1, 2-dibromocycloalkane is heated with zinc dust, it gives rise to cycloalkene. For example,

1, 2-Dibromocyclohexane (*cis* or *trans*) $\xrightarrow[\text{Acetone reflux}]{\text{Zinc dust}}$ Cyclohexene + $ZnBr_2$

This reaction is similar to the one used for preparation of alkenes by dehalogenation of 1, 2-dihalides.

4. Diels-Alder Reaction

In Diels-Alder reaction a 1,3-diene reacts with a dienophile alkene on heating to form a six membered ring with a double bond. For example,

1, 3-Butadiene (*Diene*) + Ethene (*Dienophile*) $\xrightarrow{\Delta}$ Cyclohexene

7.1.3 Conformations of Cycloalkenes

Since substituents attached to the sp^2-hybridized carbon atoms constituting a double bond lie in one plane, all the carbon atoms of cyclopropene and cyclobutene are essentially planar and because of bond angle distortion these molecules are highly strained. Cyclopentene is supposed to be approximately planar.

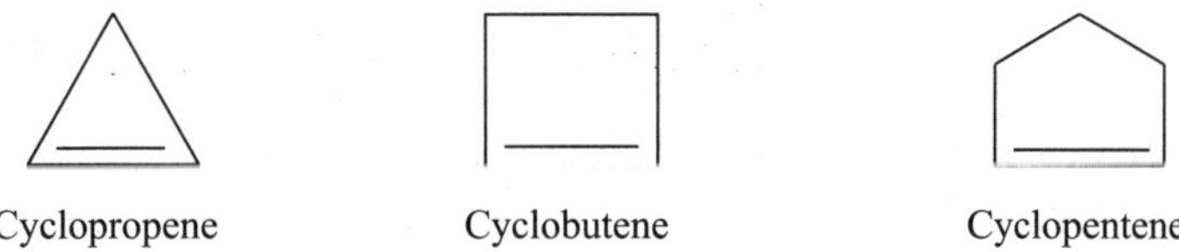

Cyclopropene Cyclobutene Cyclopentene

On the other hand, higher cycloalkenes exist in strain free puckered conformations. Diffraction studies on cyclohexene derivatives have shown that cyclohexene assumes ***'half-chair'*** conformation. Because of the presence of a double bond, four carbon atoms C_1, C_2, C_3 and C_6 lie in one plane and of the remaining two carbon atoms C_4 and C_5, one lies above and the other below this plane. Two chair forms are interconvertible. In both these forms bonds to C_4 and C_5 are normal axial (a) and normal equatorial (e) but those extending from C_3 and C_6 are *quasiaxial* (a') and *quasiequatorial* (e'). A half-boat conformation is also possible with both C_4 and C_5 above the C_1, C_2, C_3 and C_6 plane but this is unstable compared to half-chair form.

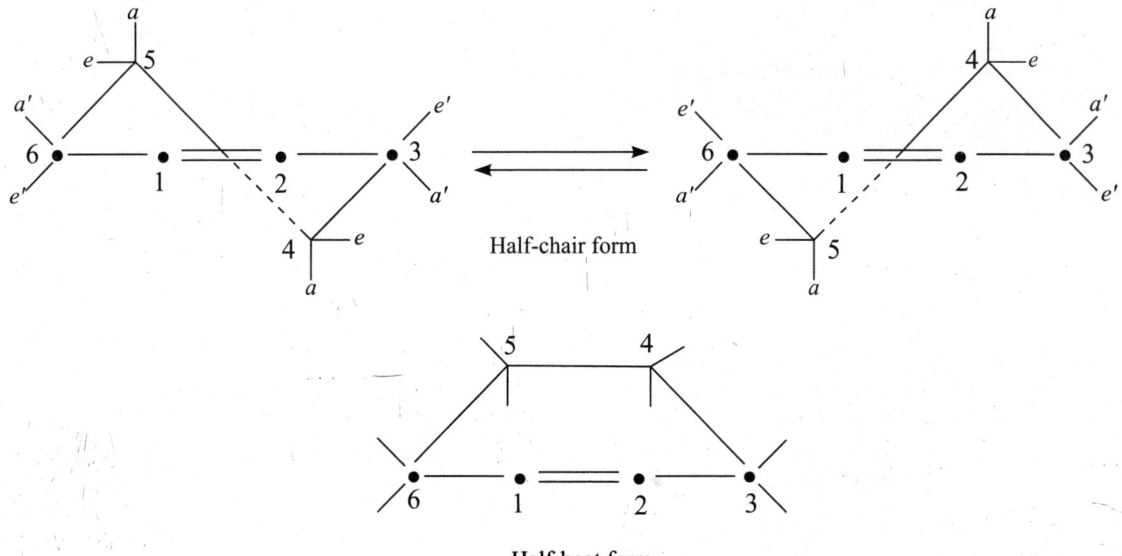

Half-chair form

Half-boat form

It must be remembered that only one geometrical isomeric form, *i.e.*, *cis* form is possible for cyclopropene to cycloheptene. This is quite understandable as the *trans* configuration of these rings would mean a lot of strain. The smallest cycloalkene capable of existing in both *cis* and *trans* forms is cyclooctene. As expected the *cis* form is more stable (heat of hydrogenation = 96.14 kJ mole^{-1}) than the *trans* form (heat of hydrogenation = –134.6 kJ mole^{-1}). Further, the *trans* cyclooctene exists in two optically active forms.

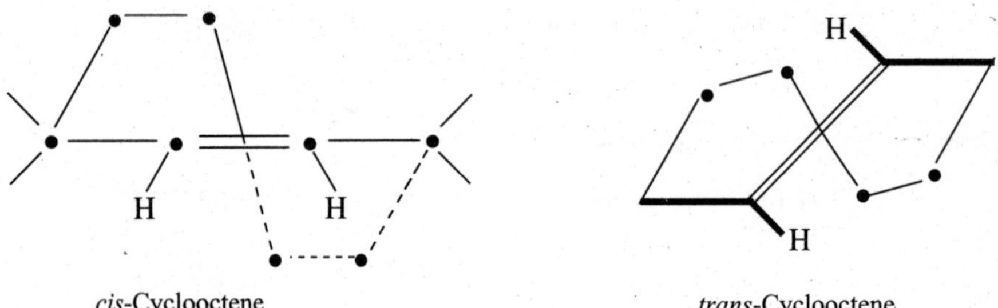

cis-Cyclooctene

trans-Cyclooctene

7.1.4 Chemical Reactions of Cycloalkenes

These compounds undergo electrophilic addition reactions which are characteristic of alkenes. The ring remains intact. Furthermore, allylic bromination is also known to take place. Some of the typical reactions are discussed below:

(*i*) Catalytic Hydrogenation

Cycloalkenes on treatment with H_2 in presence of a catalyst lead to the formation of cycloalkanes. The two hydrogen atoms are added on the same face of the double bond to give *cis*-addition products.

CH_3, CH_3 + H_2 $\xrightarrow{\text{Catalyst}}$ H, CH_3, H, CH_3

1, 2-Dimethylcyclopentene

cis-1, 2-Dimethylcyclopentane

(*ii*) Electrophilic Addition Reactions

Like alkenes cycloalkenes also undergo electrophilic addition reactions. Some examples are:

***(a) Addition of halogens*:** On treatment with halogen, they produce 1,2-dihalocycloalkanes. For example,

+ Br_2 $\longrightarrow$ Br, Br

1, 2-Dibromocyclopentane

Structures of the products of the addition reactions definitely prove that the addition across a double bond is *trans*, thereby providing evidence for the intermediacy of bridged bromonium ion. Addition of bromine to cyclopentene, for instance, yields *trans*-1, 2-dibromocyclopentane.

***(b) Addition of hypohalous acid*:** Addition of hypohalous acid to cyclohexene forms *trans*-chlorohydrin or *trans*-2-chlorocyclohexanol.

(*c*) *Addition of halogen acids*: Contrary to addition of halogens and hypohalous acids the addition of halogen acids to cycloalkenes does not involve cyclic intermediate. It rather forms a simple carbocation. In case of substituted cycloalkenes, addition occurs in accordance with **Markownikoff's rule**. For example, addition of HBr to 1-methylcyclopentene in absence of peroxides gives 1-bromo-1-methylcyclopentane.

However, addition of HBr in presence of peroxides gives ***anti-Markownikoff's addition*** product. The reaction involves free radical mechanism.

$$RO—OR \xrightarrow{\Delta} 2\,R\dot{O}$$

$$R\dot{O} + H—Br \longrightarrow ROH + \dot{Br}$$

Br CH$_3$ → Br CH$_3$ H—Br → Br CH$_3$ + $\dot{B}r$

1-Bromo-2-methyl-cyclopentane

(*d*) ***Addition of carbenes or carbenoids***: Carbenes as well as carbenoids, being electrophilic in nature, add to C = C bond to form a bicycloalkanes. For example,

1. *Addition of dichlorocarbene*

$$\text{Cyclohexene} + CH_2I_2 \xrightarrow[\text{Pentane}]{(CH_3)_3C\overset{\ominus\oplus}{OK}}$$

7,7-Dichlorobicyclo [4.1.0] heptane (60%)

2. *Addition of carbenoid*

$$\text{Cyclohexene} + CH_2I_2 \xrightarrow[\text{(Simmons-smith reaction)}]{\text{Zn-Cu/Ether}} CH_2 + ZnI_2$$

Bicyclo [4.1.0] heptane (92%)

(*e*) ***Hydration, hydroboration-oxidation, oxymercuration-reduction*** of the cycloalkenes gives alcohols. The reactions follow the same path as in the case of alkenes.

(*i*) Conc. H_2SO_4
(*ii*) Boil with H_2O
(*Hydration*)

1-Methyl-1-cyclohexanol ← 1-Methylcyclohexene

(*i*) B_2H_6/THF
(*ii*) $H_2O_2/\overset{\ominus}{O}H$
Hydroboration-oxidation

trans-2-methyl-cyclohexanol

(*i*) $Hg(OAc)_2$/THF
(*ii*) $NaBH_4/\overset{\ominus}{O}H$
Oxymercuration reduction

CH$_3$ OH

(*iii*) Oxidation Reactions

(*a*) ***Hydroxylation***: Treatment of cycloalkenes with alkaline $KMnO_4$ or OsO_4, reveals that *cis*-addition takes place in all cases. For example, treatment of cyclopentene with aq. alkaline $KMnO_4$ or OsO_4 followed by aq. $NaHSO_3$ affords *cis*-cyclopentan-1, 2-diol.

(*i*) $KMnO_4/\overset{\ominus}{O}H$ or (*ii*) OsO_4, $NaHSO_3/H_2O$

OH OH

cis-Cyclopentane-1, 2-diol

(*b*) ***Epoxidation***: Treatment of cycloalkenes with peracids gives rise to epoxides.

+ C_6H_5COOOH (Perbenzoic acid) → (H, O, H) + C_6H_5COOH

1, 2-Epoxycyclohexane

(*c*) ***Ozonolysis***: Ozonolysis of cycloalkenes results in the rupture of the ring at the olefinic centre forming dicarbonyl compounds with the same number of carbon atoms. This is an elegant method for locating the position of a double bond in an unknown cycloalkene.

CH_3 — $O_3/CHCl_3$, 196–200K → (H, O, O, O, CH_3) → (H, O, =O, CH_3)

1-Methylcyclohexene | Ozonide | 6-Oxoheptanal (A 1, 6-dicarbonyl compound)

(*iv*) Allylic Substitution

Cycloalkenes react with Br_2 at high temperature or in presence of light or with NBS in presence of light to form 3-bromocycloalkenes. Like alkenes here also substitution of allylic hydrogen takes place.

Br_2, $h\nu$ or Δ or NBS, $h\nu$ → (Br) + HBr

3-Bromocyclohexene

7.2 DIENES

Unsaturated hydrocarbons containing two double bonds are called **dienes** whereas those with more than two double bonds are called **trienes** (three double bonds), **tetraenes** (four double bonds) and in general **polyenes**. Dienes have the general formula C_nH_{2n-2} and are isomeric with alkynes. Some common examples are allenes, butadiene, etc.

Chapter 7

7.2.1 Nomenclature of Dienes

In the nomenclature of these compounds according to IUPAC system, the longest chain containing the maximum number of double bonds is chosen and the compound is assigned the parent name by replacing the suffix-*ne* of the corresponding alkene by *-diene*, *-triene*, etc.

The carbon atoms of the longest chain so selected, are numbered from one end to the other in such a manner that the doubly bonded carbon atoms get the lowest possible numbers. The position of the double bonds are indicated by the number of first carbon atoms involved in the double bonds.

The positions of the substituents are shown by placing the number of the carbon atom, to which they are attached, before the name of the substituent. The following examples are illustrative [some compounds have common names (given in brackets)]:

$$\overset{3}{H_2C}=\overset{2}{C}=\overset{1}{CH_2}$$

1, 2-Propadiene (Allene)

$$\overset{1}{H_2C}=\overset{2}{C}=\overset{3}{CH}-\overset{4}{CH_3}$$

1, 2-Butadiene

$$H_2C=CH-CH=CH_2$$

1, 3-Butadiene

$$\underset{5}{H_3C}-\underset{4}{\overset{CH_3}{\overset{|}{C}}}=\underset{3}{\overset{CH_3}{\overset{|}{C}}}-\underset{2}{\overset{CH_3}{\overset{|}{C}}}=\underset{1}{CH_2}$$

2, 3, 4-Trimethyl-1, 3-pentadiene

$$H_2C=CH-\overset{CH_3}{\overset{|}{C}}=CH_2$$

2-Methyl-1, 3-butadiene (Isoprene)

Classification of Dienes

Depending upon the relative positions of double bonds, dienes are classified into three categories: (*i*) *Dienes with isolated double bonds* (isolated dienes), (*ii*) *Dienes with alternate single and double bonds* (conjugated dienes) and (*iii*) *Dienes with cumulated double bonds* (allenes, cumulenes).

(*i*) Isolated dienes: *Dienes having two double bonds separated by more than one single bond are called* **isolated dienes**. For example,

$$\overset{1}{CH_2}=\overset{2}{CH}-\overset{3}{CH_2}-\overset{4}{CH}=\overset{5}{CH_2}$$

1, 4-Pentadiene

$$\overset{1}{CH_2}=\overset{2}{CH}-\overset{3}{CH_2}-\overset{4}{CH_2}-\overset{5}{CH}=\overset{6}{CH_2}$$

1, 5-Hexadiene

(*ii*) Conjugated dienes: *Dienes having two double bonds separated by one single bond or in other words, the dienes having alternate single and double bonds are called* **conjugated dienes**. For example,

$$\overset{1}{CH_2}=\overset{2}{CH}-\overset{3}{CH}=\overset{4}{CH_2}$$

1, 3-Butadiene

$$\overset{5}{CH_3}-\overset{4}{CH}=\overset{3}{CH}-\overset{4}{CH}=\overset{1}{CH_2}$$

1, 3-Pentadiene

(*iii*) Allenes or cumulenes: *Dienes having at least one carbon atom joined to the neighbouring two carbons by double bonds are called* **cumulenes** or **allenes**. They contain $>C=C=C<$ (group).

$$\overset{1}{CH_2}=\overset{2}{C}=\overset{3}{CH_2}$$

1, 2-Propadiene (*Allene*)

$$\overset{1}{CH_2}=\overset{2}{C}=\overset{3}{CH}-CH_2-\overset{5}{CH_3}$$

1, 2-Pentadiene

7.2.2 Structure of Allenes

Allene is a gas b. pt. 241 K. In the molecule of allene, the central carbon atom is *sp*-hybridized while the terminal carbon atoms are sp^2-hybridized. The central carbon atom forms a σ-bond with each of the terminal sp^2-hybridized carbon atoms. The remaining two π-orbitals on this carbon form two π-bonds by sideways overlapping of the π-orbitals of the terminal carbons. The molecular orbital picture that emerges is that the most stable bonding arrangement should involve two mutually perpendicular π-orbitals, since the two π-orbitals of the central carbon are directed at right angles to each other ***(orthogonal)***. The terminal groups consequently lie in mutually perpendicular planes.

As shown in Fig. 7.1, these two π-bonds are not in conjugation with each other because of the lack of planarity. The carbon-carbon bond length in allene (130.9 pm) is a little shorter than that of ethylene (133 pm) which could be attributed to hyperconjugation.

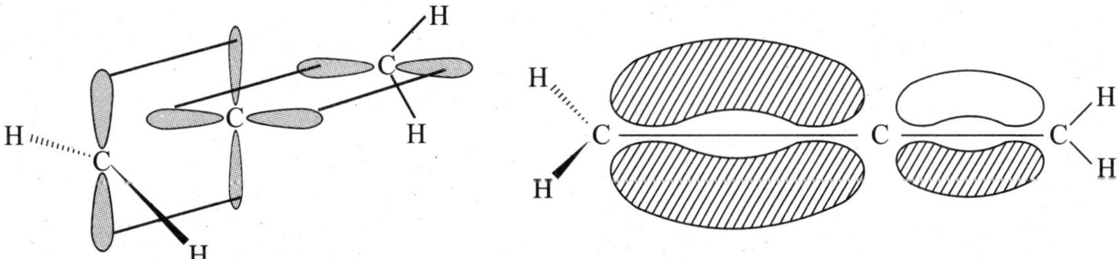

Fig. 7.1: Molecular orbital picture of allene

The following type of resonance (hyperconfugation) is possible as the σ-C—H orbital at one terminal is coplanar with the π-orbital at the other.

$$\mathrm{H_2C{=}C{=}CH_2} \longleftrightarrow \overset{\oplus}{\mathrm{H}}\quad \mathrm{H_2C{\equiv}C{-}\overset{\ominus}{C}H_2} \longleftrightarrow \text{etc.}$$

Alternatively, we may consider that the central carbon atom being *sp*-hybridized, the C — C bond distance in allenes should be intermediate between those in ethylene and acetylene, as indeed is the case.

Because of the above mentioned geometry, the allenes of the type *ab*C = C = C*ab* show molecular asymmetry and are **optically active**. *Their mirror images are non-superimposable and hence show enantiomerism.*

7.2.3 Structure of 1, 3-Butadiene—*Relative Stability of Dienes*

As we shall subsequently see butadiene-1, 3 and other conjugated dienes are more stable than dienes with isolated system of double bonds. The extra stability of conjugated dienes over isolated dienes can be explained in terms of resonance and orbital structure. We know that, resonance confers extra stability to a molecule. Butadiene can be regarded as a hybrid of the following *canonical forms*:

$$\overset{4}{H_2C}=\overset{3}{C}H-\overset{2}{C}H=\overset{1}{C}H_2 \longleftrightarrow H_2\overset{\ominus}{C}-CH=CH-\overset{\oplus}{C}H_2 \longleftrightarrow$$

(a) (b)

$$H_2\overset{\oplus}{C}-CH=CH-\overset{\ominus}{C}H_2 \longleftrightarrow H_2\dot{C}-CH=CH-\dot{C}H_2 \longleftrightarrow \text{etc.}$$

(c) (d)

A look at forms (*b*), (*c*) and (*d*) indicates that C_2—C_3 bond has some double bonds character and C_1—C_2 bond and C_3—C_4 bonds have some single bond character. This is found to be so in practice as the actual measurements of the length of C_2—C_3 bond and C_1—C_2 and C_3—C_4 bonds show that the former has a length of 146 pm (as against 154 pm for a C—C single bond) and the latter have a length of 135 pm.

The extra stability conferred by resonance is reflected by comparing the data, on the heats of hydrogenation of terminal olefins, dienes with isolated double bonds and 1,3-butadiene. The average heat of hydrogenation of terminal alkenes is 127.26 kJ mole^{-1} and that of isolated dienes is 254.14 kJ mole^{-1} (roughly double). But the heat of hydrogenation of butadiene is 239.8 kJ mole^{-1} which is less than that of a hydrocarbon with two isolated double bonds by 15.54 kJ mole^{-1}. Thus butadiene is stable to the extent of 15.54 kJ mole^{-1} because of resonance. In other words, the resonance energy of butadiene is 15.54 kJ mole^{-1}.

It has recently been pointed out that resonance contribution in case of butadiene is negligible and stability to the extent of 15.54 kJ mole^{-1} is due to hybridization of C_2—C_3 bond from C^{sp^3} — C^{sp^3} to C^{sp^2} — C^{sp^2} which as we have already seen, is more stable than the former. Similarly, the observed difference in bond lengths may also be attributed to the hybridization.

Molecular Orbital Picture of 1, 3-Butadiene

Each carbon atom in butadiene is sp^2-hybridized and forms three sigma bonds. The terminal carbons forming two sp^2-s bonds with hydrogen and sp^2-sp^2 bond with carbon, whereas the middle carbon atoms form two sp^2-sp^2 bonds with adjacent carbon and one sp^2-s bond with hydrogen. There is left, on each carbon atom, an unhybridized *p*-orbital with an electron. Such an orbital overlaps, sideways with the *p*-orbitals of the two adjacent carbon atoms. The LCAO of these four *p*-orbitals gives four molecular orbitals having wave functions ψ_1, ψ_2, ψ_3 and ψ_4 out of which ψ_1 and ψ_2 called the **bonding orbitals**, contain two electrons each while the ψ_3 and ψ_4 orbitals called **antibonding orbitals** remain vacant in the ground state of the molecule. Out of the bonding orbitals ψ_1 is of lowest energy as it has no node and energy of ψ_2 is somewhat higher because of one node. Similarly, energies of ψ_3 and ψ_4 also increase with the number of nodes. (Fig. 7.2).

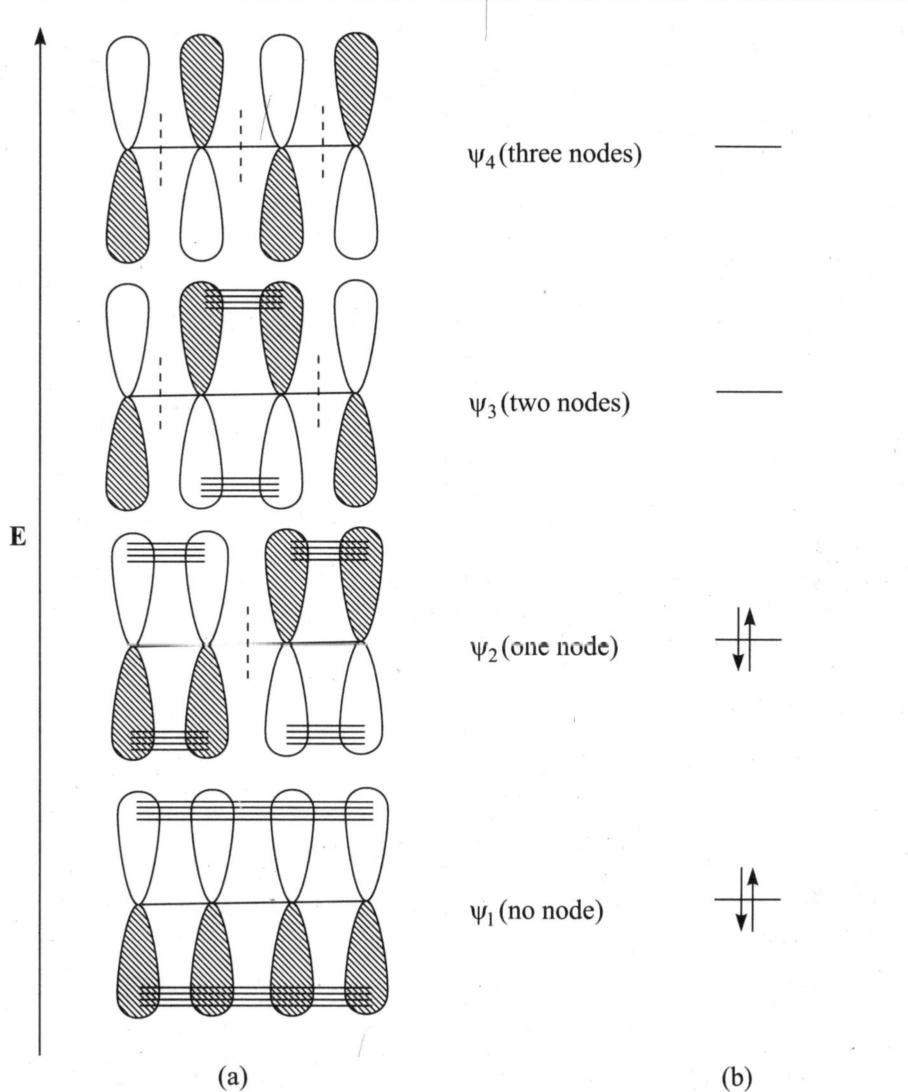

Fig. 7.2: (*a*) LCAO of 4*p*-atomic orbitals of butadiene, and (*b*) Energy levels of the ground state of butadiene

Due to rotation about C_2—C_3 'single' bond only two conformations, namely *s-trans* (transoid) and *s-cis* (cisoid) are possible. The symbol *s* refers to the fact that the isomerism is with respect to *single bond*. It has been shown that mixture of 95% *s-trans* and 5% *s-cis* form meaning thereby that the *s-trans* form is more stable than *s-cis*.

s-trans

s-cis

This is probably because of greater steric repulsions between hydrogen atoms of *s-cis* form. It may be noted however, that the energy barrier for the conversion of *s-trans* to *s-cis* form is quite low (21 kJ mol^{-1}).

7.2.4 Methods of Formation of Allenes

(*i*) Dehalogenation of 2, 3-dihaloalkenes: On treatment with zinc metal the 2, 3-dihaloalkenes yield allenes. For instance, allene is prepared from 2,3-dichloropropene.

$$\underset{\text{2, 3-Dichloropropene}}{\underset{\text{Cl}}{\underset{|}{CH_2}}-\underset{\text{Cl}}{\underset{|}{C}}=CH_2} + Zn \longrightarrow \underset{\text{Allene}}{H_2C=C=CH_2} + ZnCl_2$$

(*ii*) Dehalogenation of haloalkenes: Haloalkenes of both types (vinylic and allylic) afford allenes on treatment with a base.

$$\underset{\text{Vinylic}}{>\overset{\text{H}}{\overset{|}{C}}-\overset{\text{X}}{\overset{|}{C}}=CH_2} \xrightarrow{\overset{\ominus}{O}H} >C=C=C<\begin{smallmatrix}H\\H\end{smallmatrix} \xleftarrow{\overset{\ominus}{O}H} \underset{\text{Allylic}}{>\overset{\text{X}}{\overset{|}{C}}-\overset{\text{H}}{\overset{|}{C}}=CH_2}$$

Acetylenes are side products in the case of vinylic haloalkenes having at least one hydrogen atom attached to the terminal carbon, since such allenes are in equilibrium with the corresponding acetylenes in alkaline media.

$$>C=C=C<\begin{smallmatrix}\\H\end{smallmatrix} \underset{}{\overset{\overset{+}{H}\text{ shift}}{\rightleftharpoons}} -\underset{\text{H}}{\underset{|}{\overset{|}{C}}}-C\equiv C-$$

(*iii*) From acetylene: For reasons stated above, aliphatic alkynes having the ≡ C—H ending (called terminal alkynes) on refluxing with alcoholic KOH, undergo isomerization to allenes.

$$\underset{\text{Propyne}}{H_3C-C=CH_2} \xrightarrow[\text{150°, 6hrs}]{\text{Alc. KOH}} \underset{\text{Allene}}{H_2C=C=CH_2}$$

This rearrangement, called the **propargyl rearrangement**, is of great synthetic utility as alkynes are easier to prepare.

(*iv*) Addition to conjugated alkenynes: 1, 3-Alkenynes undergo 1, 4-addition forming allenes.

$$\underset{\text{Vinylacetylene}}{H_2C=CH-C\equiv CH} + HX \xrightarrow{\text{1,4-Addition}} XH_2C-CH=C=CH_2$$

(*v*) From dihalocyclopropanes: Dihalocyclopropanes on heating with a metal yield allenes.

$$>\overset{CX_2}{C-C}< \xrightarrow[\Delta]{M} >C=C=C< + MX_2$$

7.2.5 Chemical Reactions of Allenes

Because of a readily accessible π-electron system, allenes undergo facile additions initiated by electrophiles. The addition takes place in accordance with Markownikoff's rule which would predict that the electrophile will attack the terminal carbon atom, since there is no hydrogen attached to the central carbon. However, the attack at the central carbon atom would give a resonance stabilized allylic carbocation. In most of the reported examples, there is attack of an electrophile at the terminal carbon atom.

$$\rangle C{=}C{=}C\langle \; + \; E^{\oplus} \longrightarrow \rangle C{=}\overset{\oplus}{C}{-}\overset{E}{\overset{|}{C}}\langle$$

$$\rangle C{=}C{=}C\langle \; + \; E^{\oplus} \longrightarrow \left[\rangle \overset{\oplus}{C}{-}\overset{E}{\overset{|}{C}}{=}C\langle \longleftrightarrow \rangle C{=}\overset{E}{\overset{|}{C}}{-}\overset{\oplus}{C}\langle \right]$$

This experimental observation has been explained on the basis of orbital pictures of allene and allyl cation which reveal that the attack of electrophile at the central carbon does not give a 'true' allyl cation until the rotation around C—C bond has taken place. Consequently, this would be a relatively unstable primary carbocation and would not be expected to get formed easily, unless the attack by the nucleophile, in the second step, is delayed to permit this rotation. However, addition of the electrophile at the terminal carbon would form a secondary type of carbocation.

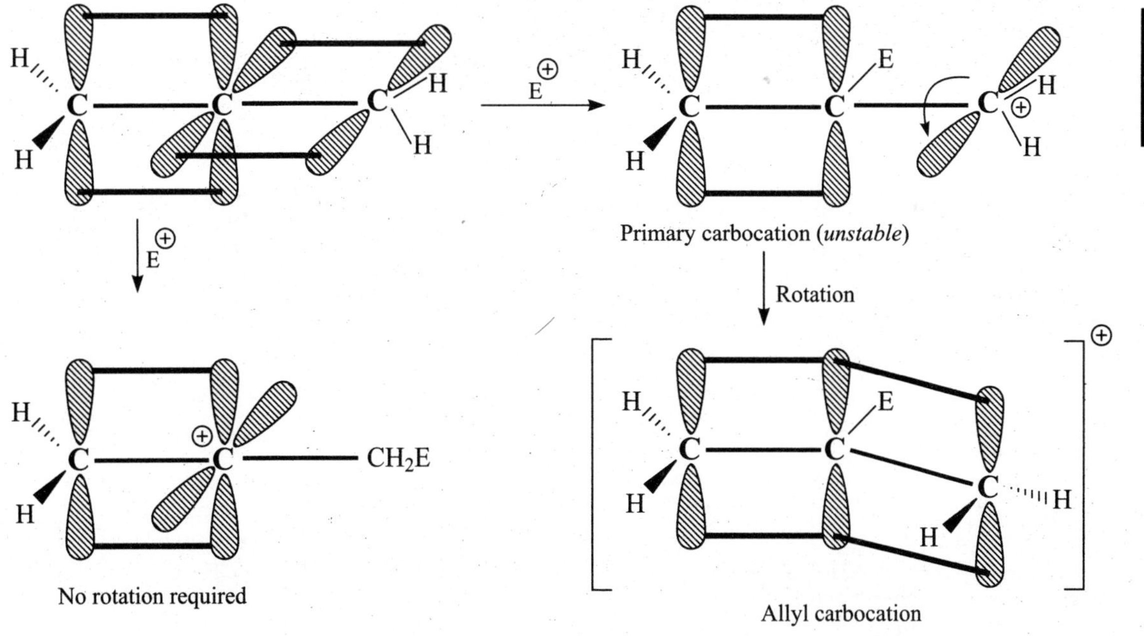

Some of the common reactions of allenes are described in the following pages:

(*i*) Addition of halogens: Allenes add halogens to form 2,3-dihalopropenes and finally the tetrahaloalkanes.

$$H_2C{=}C{=}CH_2 + Br_2 \longrightarrow \underset{Br}{\underset{|}{CH_2}}-\underset{Br}{\underset{|}{C}}{=}CH_2 \xrightarrow{Br_2} \underset{Br}{\underset{|}{CH_2}}-\overset{Br}{\overset{|}{\underset{Br}{\underset{|}{C}}}}-CH_2Br$$

1, 2, 2, 3-tetrabromopropane

The addition of second halogen molecule, however, has been found to be difficult.

(*ii*) Addition of halogen acids: Addition of molecule of hydrogen chloride to allene gives 2-chloropropene in accordance with the Markownikoff's rule.

$$H_2C{=}C{=}CH_2 + HCl \longrightarrow H_3C-\underset{Cl}{\underset{|}{C}}{=}CH_2$$

(*iii*) Hydration: On treatment with aqueous sulphuric acid, allene adds a molecule of water forming acetone.

$$CH_2{=}C{=}CH_2 + H_2O \longrightarrow H_3C-\overset{O}{\overset{\|}{C}}-CH_3$$

This is an electrophilic addition and proceeds by the acid catalyzed addition of water to the initially formed carbocation resulting in the formation of enol form of acetone which readily tautomerizes to acetone. The reaction sequence may be formulated as under:

$$H_2C{=}C{=}CH_2 + \overset{\oplus}{H} \longrightarrow H_2C{=}\overset{\oplus}{C}-CH_3 \xrightarrow{:\ddot{O}H_2} H_2C{=}\overset{\overset{\oplus}{O}H_2}{\overset{|}{C}}-CH_3$$

$$\xrightarrow{-\overset{\oplus}{H}} \left[H_2C{=}\overset{OH}{\overset{|}{C}}-CH_3\right] \longrightarrow H_3C-\overset{O}{\overset{\|}{C}}-CH_3$$

(*iv*) Addition of hydrogen: Allenes can be hydrogenated to alkanes in presence of suitable catalysts.

$$H_2C{=}C{=}CH_2 + 2H_2 \xrightarrow{Cat} H_3C-CH_2-CH_3$$

In the case of substituted allenes, the least substituted double bond is preferentially reduced giving mostly the product of *cis*-addition.

(*v*) **Ozonolysis:** Ozonolysis of allenes results in their degradation to simple carbonyl fragments whose identification often leads to the determination of the structure of the allenes.

$$\underset{R'}{\overset{R}{>}}C=C=C\underset{R'''}{\overset{R''}{<}} \xrightarrow{O_3} \underset{R'}{\overset{R}{>}}C=O + O=\overset{\overset{O}{\|}}{C} + O=C\underset{R'''}{\overset{R''}{<}}$$

This reaction is also employed for distinguishing allenes from the isomeric alkynes.

(*vi*) **Rearrangements:** In comparison to other dienes, allenes are less stable compounds. Their stability is even lesser than the corresponding isomeric alkynes. This is evident from the fact that the heat of hydrogenation of allene is greater by 8.4 kJ mole^{-1} in comparison to propyne. Thus, when treated with a basic catalyst ($NaNH_2$ or sodium), it rearranges to propyne. The various steps of this rearrangement are depicted below:

$$\overset{\ominus}{B}: + H-CH=C=CH_2 \rightleftharpoons \left[\overset{\ominus}{C}H=C=CH_2 \longleftrightarrow HC\equiv C-\overset{\ominus}{C}H_2\right]$$

$$\Big\downarrow \overset{\oplus}{+H} \text{ (From any proton donor)}$$

$$CH_3-C\equiv H$$

As we have already seen terminal alkynes also rearrange to allenes under suitable conditions. In fact, allenes are formed as intermediates in isomerization of alkynes (see Section 7.3.5).

7.2.6 Methods of Formation of 1, 3-Dienes

Butadiene is an important material for synthetic rubber whereas its homologue, isoprene, is obtained by slow distillation of natural rubber. Higher conjugated polyenes often occur in nature. For example, β-carotene —a precursor of vitamin A–which is found in carrots and green leaves, contains a system of eleven conjugated double bonds. Butadiene and isoprene are by far the most important conjugated dienes.

$$\underset{\text{1, 3-Butadiene}}{H_2C=CH-CH=CH_2} \qquad \underset{\text{Isoprene}}{H_2C=\underset{CH_3}{\underset{|}{C}}-CH=CH_2}$$

The methods usually employed for the synthesis of conjugated dienes involve elimination reactions of suitably selected molecules in order to introduce a system of two conjugated double bonds. Some of the important methods for the synthesis of butadiene are described below:

(*i*) **Pyrolysis of cycloalkenes** (*retro* **Diels-Alder reaction):** Butadiene is produced by passing vapours of cyclohexene over heated *nichrome* (Ni—Cr—Fe) alloy.

Chapter 7

$$\text{Cyclohexene} \xrightarrow[\text{Nichrome}]{\Delta} CH_2{=}CH{-}CH{=}CH_2 + CH_2{=}CH_2$$

The reaction is called ***retro* Diels-Alder reaction** as it is the reversal of the Diels-Alder reaction.

(*ii*) Dehydration of butane-1, 3-diol: Acid-catalyzed dehydration of butane-1, 3-diol gives rise to butadiene.

$$\underset{\text{Butan-1,3-diol}}{H_3C-CH(OH)-CH_2-CH_2OH} \xrightarrow[-2H_2O]{\overset{\oplus}{H},\ \Delta} H_2C{=}CH-CH{=}CH_2$$

(*iii*) Catalytic dehydrogenation: Butadiene is manufactured in petrochemical industry by the catalytic dehydrogenation of *n*-butane, 1-butene or 2-butene.

$$CH_3-CH_2-CH_2-CH_3 \xrightarrow[\Delta]{Cr_2O_3/Al_2O_3} H_2C{=}CH-CH{=}CH_2$$

$$\left.\begin{array}{c} CH_3-CH{=}CH-CH_3 \\ or \\ CH_3-CH_2-CH{=}CH_2 \end{array}\right\} \xrightarrow[\Delta]{Ca_3(PO_4)_2,\ Ni_3(PO_4)_2} H_2C{=}CH-CH{=}CH_2$$

(*iv*) Dehydrohalogenation of 1, 3-dichlorobutanes: *Isoprene* can be prepared by dehydrohalogenation of 1, 3-dichloro-3-methylbutane with alcoholic potassium hydroxide.

$$\underset{\text{1, 3-Dichloro-3-methylbutane}}{H_3C-C(Cl)(CH_3)-CH_2-CH_2-Cl} \xrightarrow[\Delta]{\overset{\ominus}{O}H} \underset{\text{Isoprene}}{H_2C{=}C(CH_3)-CH{=}CH_2}$$

Isoprene is synthesized by the following sequence of reactions utilizing acetone and acetylene as the starting materials.

$$HC\equiv CH + (H_3C)_2C{=}O \xrightarrow{NaNH_2} H-C\equiv C-\underset{CH_3}{\overset{CH_3}{C}}-OH$$

Dimethylethynylcarbinol

Partial reduction H_2 (Pd)

$$H_2C{=}CH-\overset{CH_3}{C}{=}CH_2 \xleftarrow[-H_2O]{Al_2O_3} H_2C{=}CH-\underset{CH_3}{\overset{CH_3}{C}}-OH$$

Isoprene — Dimethylethenylcarbinol

7.2.7 Chemical Reactions of 1,3-dienes

The reactions of conjugated dienes are typified by butadiene-1, 3, which readily undergoes catalytic hydrogenation and electrophilic as well free radical reactions. Butadiene-1, 3 is thermodynamically more stable than alkenes but it is more reactive than alkenes.

Some of the important reactions of butadiene-1, 3 are discussed below:

(*i*) Electrophilic Addition

Butadiene-1, 3 has been shown to undergo electrophilic addition reactions under usual conditions forming a mixture of 1, 2-and 1, 4-addition products. Quite often the 1, 4-addition product predominates.

Chapter 7

$$\overset{1}{H_2C}{=}\overset{2}{CH}-\overset{3}{CH}{=}\overset{4}{CH_2} + X-Y \longrightarrow \underset{X}{\overset{4}{H_2C}}-\underset{Y}{\overset{3}{CH}}-\overset{2}{CH}{=}\overset{1}{CH_2}$$

1, 2-Addition product

+

$$X-\overset{1}{H_2C}-\overset{2}{CH}{=}\overset{3}{CH}-\overset{4}{CH_2}-Y$$

1, 4-Addition product

(*a*) Addition of Halogens

The addition of bromine on butadiene-1, 3 for instance, yields a mixture of 3, 4-dibromo 1-butene (1, 2-addition product) and 1, 4-dibromo-2-butene (1, 4-addition product). Formation of 1, 4-adduct is quite characteristic reaction of butadiene.

$$H_2C{=}CH{-}CH{=}CH_2 \xrightarrow{Br_2} \underset{Br}{\overset{4}{H_2C}}{-}\underset{Br}{\overset{3}{CH}}{-}\overset{2}{CH}{=}\overset{1}{CH_2} \quad \text{(1, 2-addition)}$$

3, 4-Dibromo-1-butene

+

$$\underset{Br}{\overset{1}{H_2C}}{-}\overset{2}{CH}{=}\overset{3}{CH}{-}\underset{Br}{\overset{4}{CH_2}} \quad \text{(1, 4-addition)}$$

1, 4-Dibromo-2-butene

Mechanism

Like alkenes, 1, 3-dienes also undergo electrophilic addition in two steps as discussed below:

Step I**:** Addition of electrophile to carbon-carbon double bond formation of allyl cation.

The electrophile can attack either C_1 or C_2 to form carbocations I and II, respectively:

$$H_2C{=}CH{-}CH{=}CH_2 + Br{-}Br$$

Addition of C_2 | Addition of C_1

$$H_2C{=}CH{-}\underset{Br}{CH}{-}\overset{\oplus}{CH_2}$$

(I)

Non-conjugated, 1° allylic carbocation, no resonance, unstable (*not formed*)

$$\left[H_2C{=}CH{-}\overset{\oplus}{CH}{-}\underset{Br}{CH_2} \longleftrightarrow \overset{\oplus}{H_2C}{-}CH{=}CH{-}\underset{Br}{CH_2}\right] \equiv \overbrace{H_2C{\cdots}CH{\cdots}CH}^{\oplus}{-}\underset{Br}{CH_2}$$

(II)

Conjugated, 2° allylic carbocation, stabilized by resonance (*formed*)

As shown above, the addition of electrophile at C_2 produces a less stable, non-conjugated localized 1° carbocation (I), showing no resonance and not formed. On the other hand, addition of the electrophile at C_1 produces a resonance stabilized 2° allylic carbocation II.

Step II**:** Addition of nucleophile on carbocation II

Due to resonance, the allylic cation II offers two alternate sites (C_2 and C_4) for addition of the **nucleophile** ($Br^{\ominus}$). Now if $Br^{\ominus}$ adds at C_2, we get a 1,2-adduct and if it adds at C_4, we get a 1, 4-adduct.

$$\left[\overset{4}{H_2C}{=}\overset{3}{CH}-\overset{\oplus 2}{CH}-\overset{1}{CH_2Br} \longleftrightarrow \overset{4\,\oplus}{H_2C}-\overset{3}{CH}{=}\overset{2}{CH}-\overset{1}{CH_2Br} \right]$$

$\xrightarrow{Br^{\ominus}}$ At C_2: $H_2C{=}CH-C(H)(Br)-CH_2Br$ (1, 2-Adduct); At C_4: $H_2C(Br)-CH{=}CH-CH_2(Br)$ (1, 4-Adduct)

Moreover, 1, 4-adduct is more stable than 1, 2 because of the presence of non-terminal double bond (see stability of alkenes).

(*b*) Addition of Halogen Acids

Here also, we get a mixture or 1, 2- and 1, 4- adducts.

$$HC_2{=}CH-CH{=}CH_2 + HBr \longrightarrow \underset{\substack{\text{3-Bromobutene} \\ \text{(1, 2-Adduct)}}}{CH_3-\underset{Br}{\underset{|}{CH}}-CH{=}CH_2} + \underset{\substack{\text{1-Bromo-2-butene} \\ \text{(1, 4-Adduct)}}}{CH_3-CH{=}CH-CH_2Br}$$

Mechanism

The reaction takes place in two steps as discussed below:

***Step I*:** Addition of electrophile (proton in this case) to carbon-carbon double bond: formation of resonance stabilized allyl cation.

$$H-Br \rightleftharpoons \underset{\text{Electrophile}}{\overset{\oplus}{H}} + \overset{\ominus}{Br}$$

$$H_2C{=}CH-CH{=}CH_2 + \overset{\oplus}{H} \longrightarrow \left[\begin{array}{c} H_2C{=}CH-\overset{\oplus}{CH}-CH_3 \\ \updownarrow \\ \overset{\oplus}{H_2C}-CH{=}CH-CH_3 \end{array} \right] \equiv \underset{\substack{\text{Allylic cation} \\ \text{(Resonance Stabilized)}}}{\overset{\oplus}{H_2C{\overset{...}{=}}CH{\overset{...}{=}}CH}-CH_3}$$

***Step II*:** Addition of nucleophile ($Br^{\ominus}$) to allylic cation at two available sites producing a mixture of 1, 2and 1, 4-adducts

$$\left[\underset{4}{H_2C}{=}\underset{3}{CH}-\underset{2}{\overset{\oplus}{C}H}-\underset{1}{CH_3} \longleftrightarrow \underset{4}{\overset{\oplus}{H_2C}}-\underset{3}{CH}{=}\underset{2}{CH}-\underset{1}{CH_3} \right]$$

$Br^{\ominus}$

At C_2 → $H_2C{=}CH-CH(Br)-CH_3$ (1, 2-Adduct)

At C_4 → $H_2C(Br)-CH{=}CH-CH_3$ (1, 4-Adduct)

1, 2 Addition *vs* 1, 4-addition (Rate *vs* Equilibrium)

Some very interesting features of this reaction become apparent, when we consider the relative amounts of 1, 2-and 1, 4-adducts formed. Let us consider, for example, the addition of HBr to 1, 3-butadiene at different temperatures as shown below:

$$H_2C{=}CH-CH{=}CH_2 + HBr$$

At 193 K → $H_3C-CH(Br)-CH{=}CH_2$ 1, 2-Adduct (80%) + $H_3C-CH{=}CH-CH_2Br$ 1, 4-Adduct (20%)

At 313 K → $H_3C-CH(Br)-CH{=}CH_2$ 1, 2-Adduct (20%) + $H_3C-CH{=}CH-CH_2Br$ 1, 4-Adduct (80%)

As a result of detailed studies of this important reaction, the following facts have come to light:

***Fact - I*:** *At lower temperature (i.e., at 193 K) the reaction produces a mixture containing 20% of 1, 4-adduct and 80% of 1, 2-adduct. However, at higher temperature (i.e., at 313 K) the product mixture contains 80% of 1, 4-adduct and 20% of 1, 2-adduct.*

***Fact - II*:** *The relative amounts of the 1, 2- and 1, 4-adducts produced at 193 K remains the same even if the product mixture is allowed to stand at this temperature for a longer period of time.*

Fact - III: *When the product mixture obtained at 193 K is heated to 313 K, the relative amounts of the 1, 2- and 1, 4-adducts are reversed, i.e., the amount of 1, 2-adduct is reduced from 80% to 20% and that of 1, 4-adduct is increased from 20% to 80%.*

Fact - IV: *Both the adducts are quite stable at lower temperatures. However, when either of them is heated, it produces the same mixture as in III above, i.e., 20% of 1, 2-adduct and 80% of 1, 4- adduct.*

Following explanations may be offered to account for these observed facts:

1. The formation of 80% 1, 2-adduct and 20% of 1, 4- adduct at 193 K may be explained on the basis of the fact that 1, 2-*adduct is formed faster than the* 1, 4-*adduct perhaps due to lower value of* E*act for* 1, 2- *addition.*

Further, *the formation* 80% *of* 1,4- *adduct and* 20% *of*-1, 2- *adduct may be due to greater thermodynamic stability of* 1, 4-*adduct* (**Note**: The 1, 4-adduct having a non-terminal double bond should be more stable than 1, 2-adduct having a terminal double bond). Thus, at higher temperature the reaction is **thermodynamically controlled.**

2. The formation of 80% of 1, 2-adduct and 20% of 1, 4-adduct at 193 K may be governed by the *relative rates of the formation of these adducts.* In other words, in this case the *transition state for* 1, 2- *additions, being like the more stable secondary allylic cation, is of lower energy. Hence, the formation of* 1, 2- *adduct will take place at a lower temperature and at faster rate.*

3. At higher temperature (313 K) *the product which is thermodynamically more stable predominates.* The formation of this adduct is preceded by a transition state of higher energy because it resembles the less stable 1° allyl cation. Such a transition state will be favoured only at higher temperature.

4. At higher temperature the two products attain an equilibrium.

$$\text{1, 2-Adduct} \rightleftharpoons \text{1, 4-Adduct}$$

Since, at this *temperature* 1, 2-*adduct is thermodynamically less stable than* 1, 4-*adduct, the equilibrium is shifted in the direction of* 1, 4-*adduct at a faster rate*. In other words, at higher temperature the relative amounts of the two adducts depend upon their relative stabilities. Thus, at higher temperature, because thermodynamically more stable product predominates, the *reaction at higher temperature is said to be thermodynamically controlled.*

With this background, we shall examine 1, 2- and 1, 4-addition more closely with the help of energy profile diagram (Fig. 7.3).

A look at Fig. 7.3 reveals that E*act* for 1, 2-addition is less than that for 1, 4-addition. Therefore initially the 1, 2-adduct is formed at a much faster rate than the 1, 4-adduct. Further, allyl halides ionise readily to produce allyl cations. In this case, the ionisation of 1, 2-adduct as well as 1, 4-adduct, produces the same carbocation which is responsible for establishing equilibrium between the two adducts.

$$\underset{\text{1, 2-Adduct}}{H_3C-\underset{\underset{Br}{|}}{CH}-CH{=}CH_2} \xrightleftharpoons{\text{Ionisation}} H_3C-\overbrace{CH\cdots CH\cdots CH_2}^{\oplus} + Br^{\ominus} \xrightleftharpoons{\text{Ionisation}} \underset{\text{1, 4-Adduct}}{CH_3-CH{=}CH-CH_2Br}$$

As shown in Fig. 7.3 the E*act* for ionisation of 1, 4-adduct is much higher than that for 1, 2-adduct. So, 1, 4-adduct will ionise more slowly than 1, 2-adduct. But once the ionisation has taken place an equilibrium is established between formation and ionisation of both adducts. The 1, 2-adduct is formed and ionised more easily than 1, 4- adduct which is formed and ionised with difficulty. It means that the 1, 4-adduct once formed will resist ionisation and hence it will accumulate in the reaction mixture and thus become the major product at equilibrium temperature.

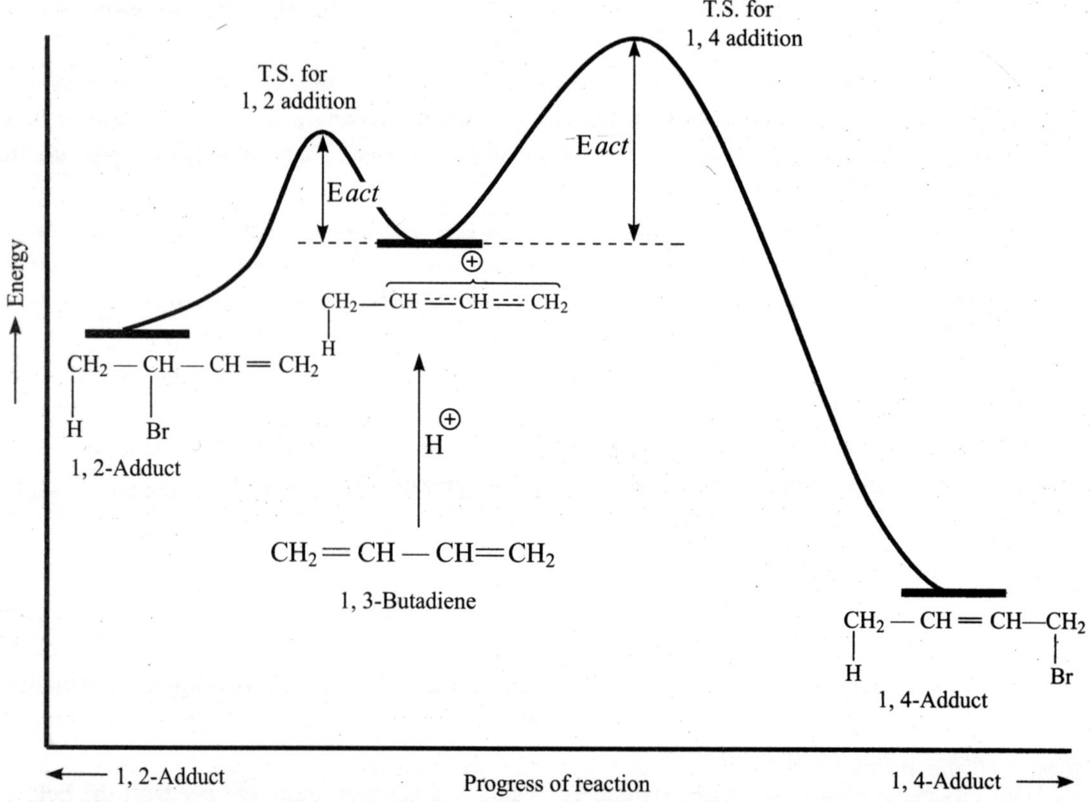

Fig. 7.3: Potential energy diagram for 1,2 *versus 1,4-addition to* 1,3-*butadiene*

(*ii*) Free Radical Addition

Conjugated dienes also undergo free radical addition reactions. They are in fact more reactive to free radical attack than alkenes resulting in the formation of a mixture of 1, 2-and 1, 4-addition products. This enhanced reactivity of the conjugated dienes toward the free radical addition, which is in direct conflict with their thermodynamic stability, can be rationalized on the basis of resonance stabilization of the radical intermediates. The following mechanism has been proposed:

$$\underset{\substack{\text{A radical initiator}\\ (\text{R = alkoxy})}}{R-R} \longrightarrow 2\dot{R}$$

$$\dot{R} + Y-X \longrightarrow R-Y + \dot{X}$$

$$H_2C=CH-CH=CH_2 + \dot{X} \longrightarrow \left[X-CH_2-\dot{C}H-CH=CH_2 \longleftrightarrow X-CH_2-CH=CH-\dot{C}H_2\right]$$

$$\xrightarrow{X-Y} \dot{X} + X-CH_2-\underset{\underset{Y}{|}}{CH}-CH=CH_2 + X-CH_2-CH=CH-CH_2-Y$$

If the transition state leading to a free radical is stabilized more than the substrate, the E*act* of the reaction becomes less, and consequently the reactant is said to be less stable than the one for which E*act* is more. Though resonance stabilizes conjugated dienes (as compared to simple alkenes) this stabilization effect is more pronounced in the transition state leading to allyl free radical. Fig. 7.4 makes this point more clear.

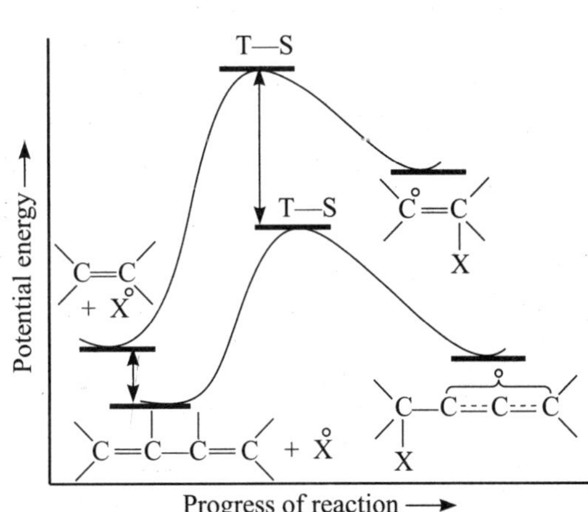

Fig. 7.4: Energy profile diagram for free radical addition to dienes and alkenes

(Note the energy gap between the alkene and diene on the one hand, and that between the respective transition states on the other. From this difference it is clear that free radical additions to dienes are more facile.)

(*a*) **Addition of $BrCCl_3$** to butadiene in presence of peroxides forms a mixture of 1, 2- and 1, 4-additon products. When a 1 : 1 mixture of alkene and butadiene is allowed to compete for 1 mole of $BrCCl_3$, it adds almost exclusively to butadiene. This a observation is in agreement with the above argument.

$$H_2C=CH-CH=CH_2 + BrCCl_3 \xrightarrow{\text{Peroxide}} \underset{\text{1, 2-Addition product}}{Cl_3C-CH_2-\underset{\overset{|}{Br}}{CH}-CH=CH_2} + \underset{\text{1, 4-Addition product}}{Cl_3C-CH_2-CH=CH-CH_2-Br}$$

Mechanism

This reaction takes place by free radical chain mechanism which occurs in the usual three steps:

***Step I*:** Chain initiation

$$(i)\quad \underset{\text{Benzoyl peroxide}}{C_6H_5CO-O-O-OCC_6H_5} \xrightarrow{\Delta} 2\,C_6H_5-CO-\dot{O} \longrightarrow \dot{C}_6H_5 + CO_2$$

$$(ii)\quad \dot{C}_6H_5 + Br-CCl_3 \longrightarrow C_6H_5Br + \dot{C}Cl_3$$

$$(iii)\quad CH_2{=}CH-CH{=}CH_2 + \dot{C}Cl_3 \longrightarrow \left[\begin{array}{c} H_2C{=}CH-\dot{C}H-CH_2-CCl_3 \\ \updownarrow \\ H_2\dot{C}-CH{=}CH-CH_2-CCl_3 \end{array}\right]$$

***Step II*:** Chain propagation

$$\left[\begin{array}{c} CH_2{=}CH-\dot{C}H-CH_2CCl_3 \\ \updownarrow \\ \dot{C}H_2-CH{=}CH-CH_2CCl_3 \end{array}\right] + Br-CCl_3 \longrightarrow \left[\begin{array}{c} CH_2{=}CH-CH(Br)-CH_2(CCl_3) \\ + \\ CH_2(Br)-CH{=}CH-CH_2(CCl_3) \end{array}\right] + \dot{C}Cl_3$$

The newly generated $\dot{C}Cl_3$ radical propagates the chain reacting with other molecules of the diene.

(*iii*) [4 + 2] Cycloaddition: The Diels-Alder Reaction

As already mentioned, the reaction is of great utility for the synthesis of six membered cycloalkenes (called adduct) and takes place by the [4 + 2] cycloaddition of a conjugated diene with an alkene, often called a dienophile. For example,

$$\underset{\substack{\text{1, 3-Butadiene}\\(Diene)}}{CH_2{=}CH-CH{=}CH_2} + \underset{\substack{\text{Ethylene}\\(Dienophile)}}{H_2C{=}CH_2} \xrightarrow{200^\circ} \underset{\substack{\text{Cyclohexene}\\(Adduct)}}{C_6H_{10}}$$

Because of the rigid structure associated with a ring system, the cyclohexene must have a *cis* double bond and this makes it obligatory for the diene to assume the *s-cis* conformation before it participates in the reaction. In agreement with this, the cyclic dienes which are essentially *s-cis* react more readily than acyclic dienes which exist preferentially in *s-trans* form:

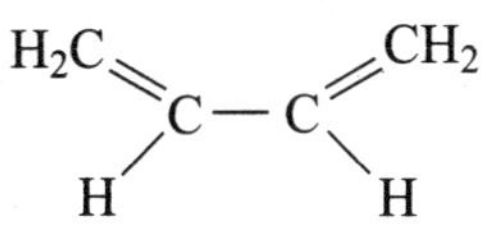

s-cis 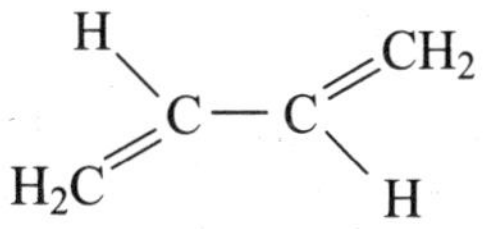 *s-trans*

Factors Affecting Rates of Cycloaddition Reactions

***(i) Presence of electron withdrawing groups in the dienophile*:** This reaction is facilitated by an electron withdrawing substituent, such as —CHO, —COOH, —COOR, —CN and —NO_2 on the dienophile, *e.g.*,

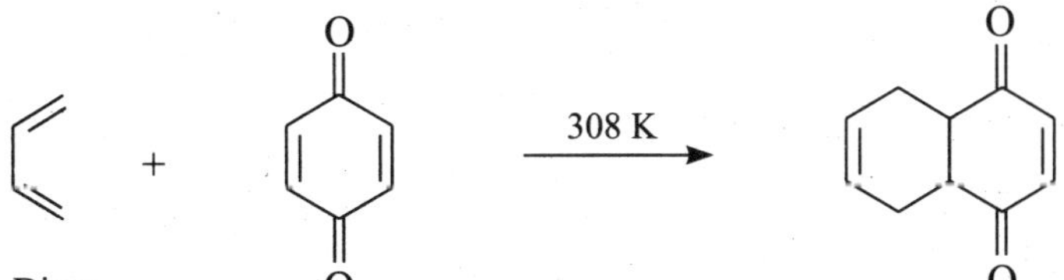

Diene + Benzoquinone —308 K→

Benzoquinone
Dienophile (Electron withdrawing carbonyl group)

323 K

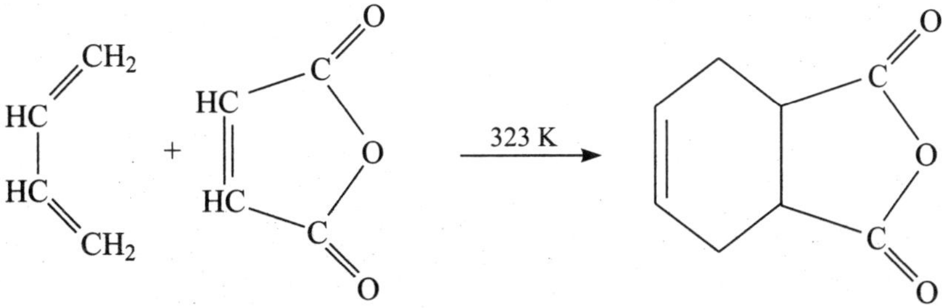

Maleic anhydride
Dienophile (Electron withdrawing carbonyl groups)

(ii) Presence of electron donating group in the diene also facilitates the addition, e.g.,

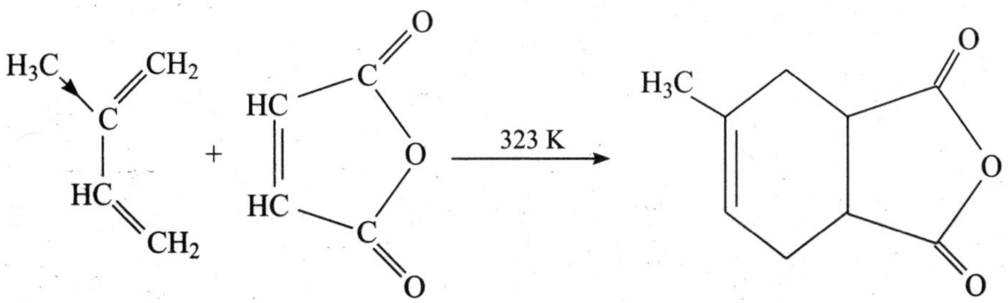

323 K

(iii) Dienophiles with triple bond also facilitate the reaction

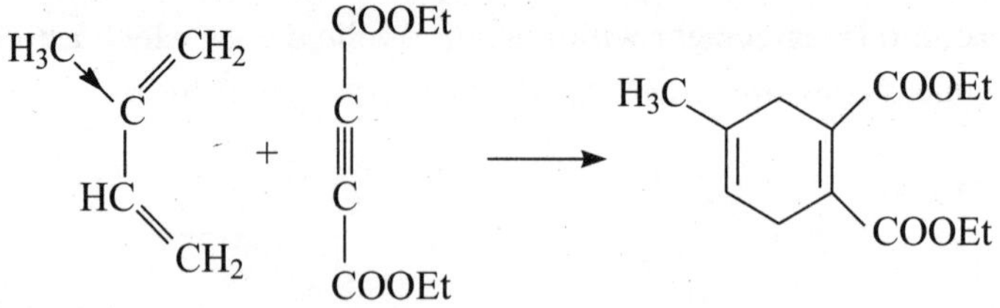

H_3C, CH_2, C, HC, CH_2 + COOEt, C≡C, COOEt → H_3C, COOEt, COOEt

Stereochemistry

An interesting feature of this reaction is that it shows extremely high degree of stereospecificity. For instance, maleic acid gives *cis*-adduct and fumaric acid gives the *trans* isomer.

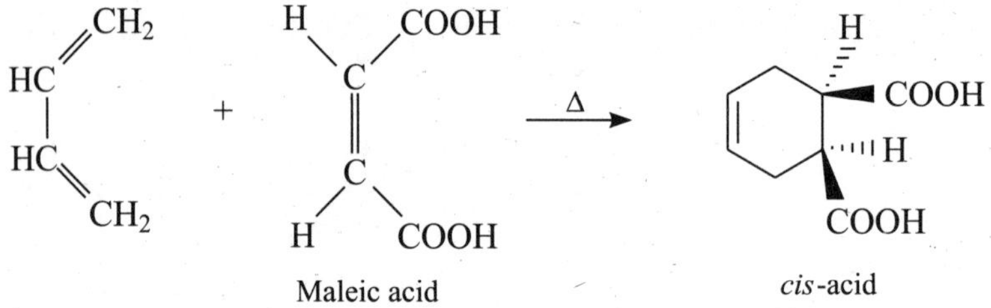

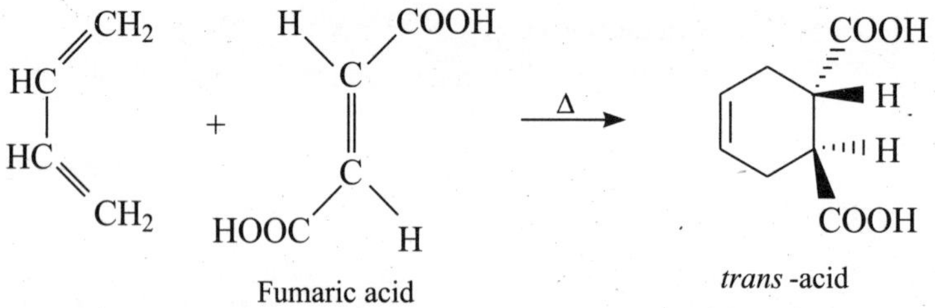

CH_2=HC–HC=CH_2 + H(COOH)C=C(COOH)H —Δ→ *cis*-acid

Maleic acid

cis-acid

CH_2=HC–HC=CH_2 + H(COOH)C=C(H)HOOC —Δ→ *trans*-acid

Fumaric acid

trans -acid

Mechanism

The Diels-Alder reaction does not involve ionic or free radical intermediates and is remarkably insensitive to the presence of solvents and catalysts. In fact, this reaction proceeds in a concerted manner involving a six membered cyclic transition state with simultaneous making of new bonds (m) and breaking of old bonds (b).

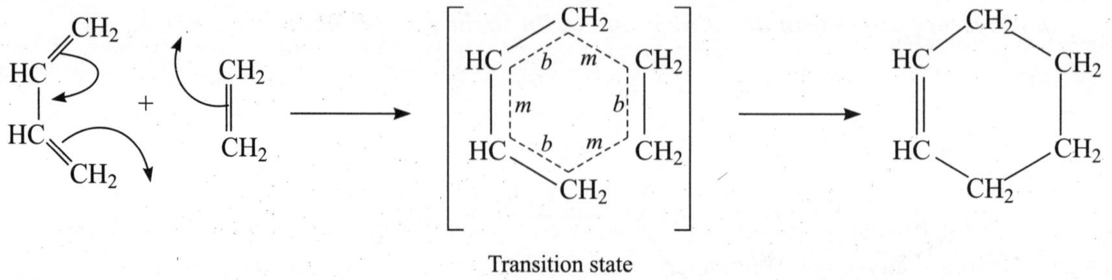

CH_2=HC–HC=CH_2 + CH_2=CH_2 → [Transition state] → cyclohexene

Transition state

The stereospecificity of this reaction is in accordance with the concerted mechanism.

Regioselectivity

Diels-Alder reactions between unsymmetrical dienes and unsymmetrical dienophiles are highly regioselective. These reactions are *ortho-para*-directing, *i.e.*, *ortho* or *para-* but never *meta*-products are obtained. For example, 1-substituted butadienes give *ortho*-products but 2-substituted butadienes give *para*-products.

CHO
CH$_3$ + CHO → *ortho*-product (CH$_3$, CHO)
CHO
meta-product
(not formed)
CH$_3$

CHO
para-product
H$_3$C
H$_3$C + CHO
meta-product
(not formed)
H$_3$C CHO

Chapter 7

(*iv*) Chemical Reduction

The enhanced chemical reactivity of conjugated dienes relative to alkenes having isolated double bonds is borne out by the fact that butadiene is easily reduced by sodium and alcohol giving the 1, 4-addition product whereas the isolated double bonds remain unaffected under these conditions. The reaction stops after reduction of one double bond, *i.e.*, the second double bond is not reduced.

$$H_2C{=}CH-CH{=}CH_2 \xrightarrow{Na\,/\,EtOH} \underset{\text{(formed)}}{\underset{\text{But-2-ene}}{H_3C-CH{=}CH-CH_3}} \xrightarrow[\times]{Na\,/\,EtOH} \underset{\text{(not formed)}}{\underset{\text{Butane}}{H_3C-CH_2-CH_2-CH_3}}$$

Mechanism

Here again, the enhanced reactivity of conjugated dienes may be rationalized by the formation of a resonance stabilized allylic carbanion intermediate whereas an isolated double bond would give a high energy dianion having localized negative charge on two adjacent carbons.

$$>C=C< \xrightarrow[\text{Metal}]{2e^{\ominus}} >\overset{\ominus}{C}-\overset{\ominus}{C}<$$

Dianion

$$2\,\dot{Na} \longrightarrow 2\,\overset{\oplus}{Na} + 2\,\overset{\ominus}{e}\,(\text{electron})$$

$$H_2C{=}CH-CH{=}CH_2 \xrightarrow{2\,e^{\ominus}} \left[H_2C{=}CH-\overset{\ominus}{CH}-\overset{\ominus}{CH_2} \longleftrightarrow H_2\overset{\ominus}{C}-CH{=}CH-\overset{\ominus}{CH_2} \right]$$

$$H_3C-CH{=}CH-CH_3 \xleftarrow{2\,EtOH} \left[\ldots\right] \equiv \overset{\ominus}{\overbrace{H_2C\,{=}{=}\,CH\,{=}{=}\,CH}}-\overset{\ominus}{CH_2}$$

(*v*) Catalytic Reduction

In contrast to chemical reduction, catalytic reduction gives a mixture of 1, 2- and 1, 4- addition products.

$$H_2C{=}CH-CH{=}CH_2 \xrightarrow{H_2/\,Ni} \underset{\substack{\text{But-1-ene} \\ (\textit{minor product})}}{H_2C{=}CH-CH_2-CH_3} + \underset{\substack{\text{But-2-ene} \\ (\textit{major product})}}{H_3C-CH{=}CH-CH_3}$$

(*vi*) Polymerization

Polymerization involving the same monomer gives **homopolymer**. Conjugated dienes behave in similar fashion but the interesting features are:

1. Their polymerization may involve 1, 2-or 1, 4-addition to the monomer.

2. The polymer obtained from conjugated dienes still contains the double bonds and hence shows geometrical isomerism.

3. In case of polymerization through 1, 4-addition of one monomer molecule to another, there may be *cis* or *trans* addition though the trans mode is preferred.

Thus, the polymerization of 1, 3-butadiene in presence of peroxide takes place with 80% by 1, 4-addition and 20% 1, 2-addition.

$$n\,\begin{matrix} HC{=}CH_2 \\ | \\ HC{=}CH_2 \end{matrix} \longrightarrow \begin{cases} \xrightarrow{\text{1, 2-Addition}} & \left(\begin{matrix} CH - CH_2 \\ | \\ CH{=}CH_2 \end{matrix}\right)_n \\ \xrightarrow{\textit{cis}\text{-1, 4-Addition}} & \left(CH_2 - C(H){=}C(H) - CH_2\right)_n \ (cis) \\ \xrightarrow{\textit{trans}\text{-1, 4-Addition}} & \left(CH_2 - C(H){=}C(H) - CH_2\right)_n \ (trans) \end{cases}$$

Copolymerization

A mixture of two or more different monomers may also undergo polymerization and the product then has two or more kinds of monomeric units in the same molecule. This process is known as *copolymerization* and such a polymer is referred to as *copolymer*.

Copolymerization of conjugated dienes is quite common. Some of the important polymers of butadiene and also of isoprene are given below:

(*a*) 1, 3-Butadiene forms a rubber like polymer called **Buna** (Bu = butadiene, Na = natrium) with metallic sodium under the influence of heat, light or peroxide.

$$n\,H_2C{=}CH - CH{=}CH_2 \xrightarrow{Na} \underset{\text{Buna}}{(CH_2 - CH{=}CHCH_2)_n}$$

Monomers are joined largely by 1, 4 and to some extent by 1, 2 linkages.

(*b*) 1, 3 Butadine (75%) and styrene (25%) undergo copolymerization in presence of sodium to form a useful synthetic rubber **Buna S** (**Bu** for butadiene, **Na** for sodium (Natarium) and **S** for styrene).

$$H_2C{=}CH - CH{=}CH_2 + H_2C{=}\underset{\displaystyle C_6H_5}{\underset{|}{CH}}$$

$$\Big\downarrow Na$$

$$\underset{\text{Buna S}}{\left(CH_2 - CH{=}CH - CH_2 - CH_2 - \underset{\displaystyle C_6H_5}{\underset{|}{CH}} - CH_2 - \underset{\displaystyle CH{=}CH_2}{\underset{|}{CH}}\right)_n}$$

The resulting polymer is treated with elemental sulphur, which results in the cross linking of the double bonds. This process called, ***vulcanization*** improves the elasticity of the material.

(*c*) **Buna N** (nitrile rubber) is another important copolymer of butadiene (75%) and acrylonitrile (25%).

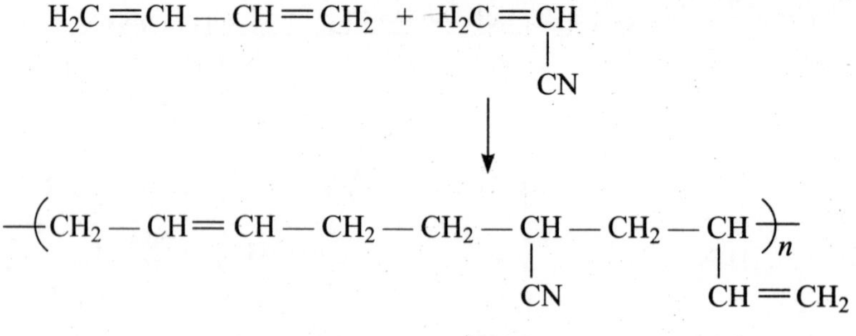

$$H_2C{=}CH{-}CH{=}CH_2 + H_2C{=}CH(CN) \longrightarrow {-}\!\!\left(CH_2{-}CH{=}CH{-}CH_2{-}CH_2{-}CH(CN){-}CH_2{-}CH(CH{=}CH_2)\right)_n$$

Buna N

It is more useful than the above materials since it is resistent to organic solvents.

(*d*) Polymerization of isoprene is perhaps the most important example of ***anionic polymerization***. Natural rubber is *cis*-polyisoprene and a similar type of polymer, called **coral rubber**, can be prepared by polymerizing isoprene on a specially made lithium dispersion.

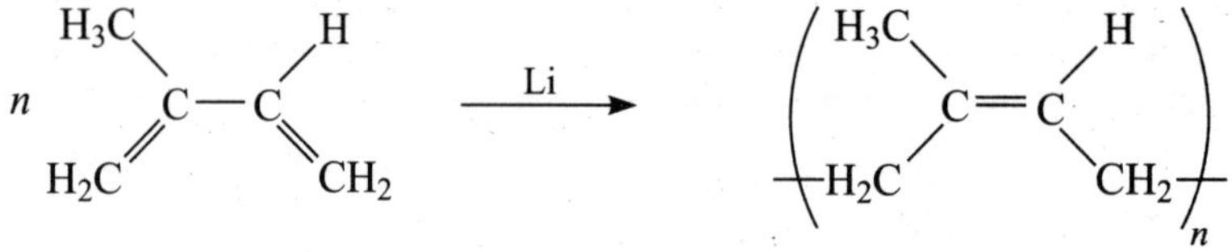

$$n\; (H_3C)(H_2C{=})C{-}C(H){=}CH_2 \xrightarrow{Li} {-}\!\!\left(H_2C{-}C(H_3C){=}C(H){-}CH_2\right)_n \quad \text{Coral rubber}$$

Mechanism

Free radical polymerization of conjugated dienes as discussed below occurs by the same mechanism as the free radial polymerization of alkenes :

(*i*) Peroxides $\longrightarrow$ Radicals (Say $\dot{R}$.)

(*ii*) $\dot{R} + CH_2{=}CH{-}CH{=}CH_2 \longrightarrow R{-}CH_2{-}CH{=}CH{-}\dot{C}H_2$

(*iii*) $R{-}CH_2{-}CH{=}CH{-}\dot{C}H_2 + CH_2{=}CH{-}CH{=}CH_2 \longrightarrow$
$R{-}CH_2{-}CH{=}CH{-}CH_2{-}CH_2{-}CH{=}CH{-}\dot{C}H_2 \longrightarrow R{-}(CH_2{-}CH{=}CH{-}CH_2)_n$
Polybutadiene

Natural and synthetic rubber: Natural rubber is polymer of isoprene (2-methyl 1, 3-butadiene) in which all the double bonds have *cis*-stereochemistry. The addition occurs by head (marked as *h*) to tail (marked as *t*) union of the isoprene units as shown below:

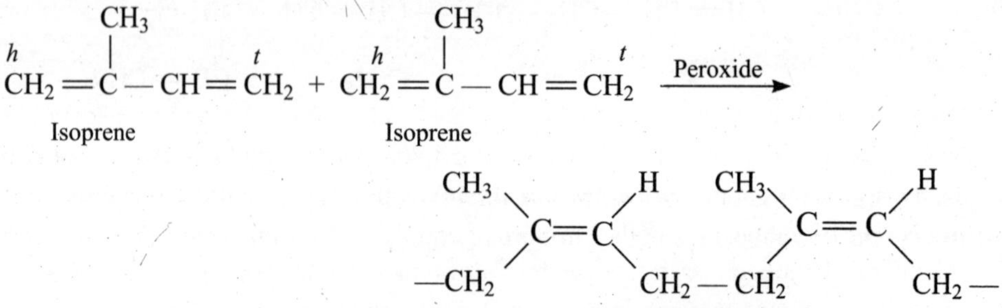

$$\overset{h}{CH_2}{=}C(CH_3){-}CH{=}\overset{t}{CH_2} + \overset{h}{CH_2}{=}C(CH_3){-}CH{=}\overset{t}{CH_2} \xrightarrow{\text{Peroxide}}$$

Isoprene   Isoprene

$$-CH_2{-}C(CH_3){=}C(H){-}CH_2{-}CH_2{-}C(CH_3){=}C(H){-}CH_2-$$

cis-Polyisoprene
(*Natural rubber*)

But the synthetic rubber (**gutta-percha**) obtained by free radical polymerization of ***isoprene,*** has all ***trans***-configuration. Here again polymerization occurs by head to tail union of isoprene units as shown below:

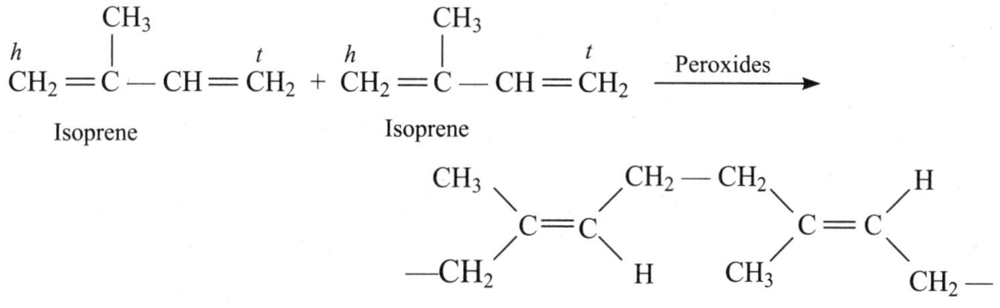

$$\underset{\text{Isoprene}}{\overset{h}{CH_2}=\overset{CH_3}{\overset{|}{C}}-CH=\overset{t}{CH_2}} + \underset{\text{Isoprene}}{\overset{h}{CH_2}=\overset{CH_3}{\overset{|}{C}}-CH=\overset{t}{CH_2}} \xrightarrow{\text{Peroxides}}$$

$$\begin{array}{c} CH_3 \\ -CH_2 \end{array}\!\!>C=C<\!\!\begin{array}{c} CH_2-CH_2 \\ H \end{array}\!\!>C=C<\!\!\begin{array}{c} H \\ CH_2- \end{array}$$

trans-Polyisoprene or Gutta-percha
(*Synthetic rubber*)

We have seen above that the free radical polymerization of dienes gives mainly all-*trans* polymers. But by using **coordination polymerization** *i.e.*, polymerization in presence of certain special catalysts such as *triethyaluminium-titanium trichloride complex* (***Ziegler-Natta catalyst***), we can control the stereochemistry of the polymers. These catalysts were discovered by Karl Ziegler and Gulio Natta in 1953 and are commonly called as **Ziegler-Natta catalysts**. Because of their discovery, these two scientists shared the **1963 Nobel Prize** in Chemistry.

By using Ziegler-Natta catalyst, isoprene can be polymerized to give a polymer which is almost identical with the natural rubber in its properties. (see **Section 6.6** for details).

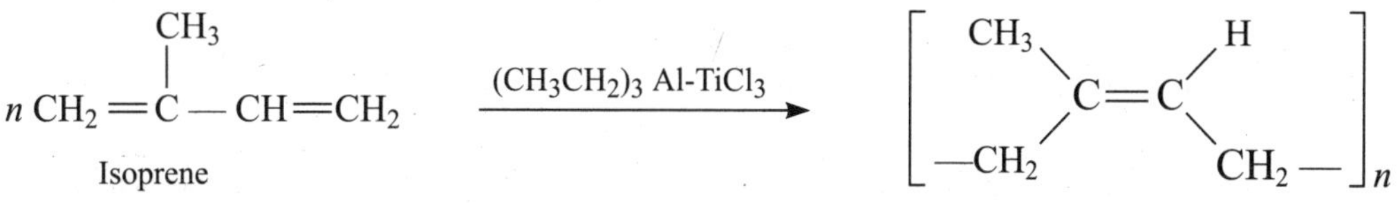

$$\underset{\text{Isoprene}}{n\,CH_2=\overset{CH_3}{\overset{|}{C}}-CH=CH_2} \xrightarrow{(CH_3CH_2)_3\,Al\text{-}TiCl_3} \left[\begin{array}{c} CH_3 \\ -CH_2 \end{array}\!\!>C=C<\!\!\begin{array}{c} H \\ CH_2- \end{array}\right]_n$$

cis-Polyisoprene
(*Natural rubber*)

Here, it may be mentioned that isoprene is regarded as the building block for a large number of plant products, for instance, **terpenes**. Further, this forms the basis of the famous **isoprene rule.**

7.2.8 Vulcanization of Rubber

Both raw natural and synthetic rubbers are soft and tacky and become even more so in hot weather. Their tensile strength and resistance to abrasion is also low and elasticity is maintained only over a low range of temperature. All these properties can be improved markedly by a process called **vulcanization**. *It consists in heating rubber with 5–8% of sulphur.*

The chemistry of vulcanization is complex but involves formation of sulphur bridges or cross-links between polymer chains through their reactive allylic positions as shown below:

Chapter 7

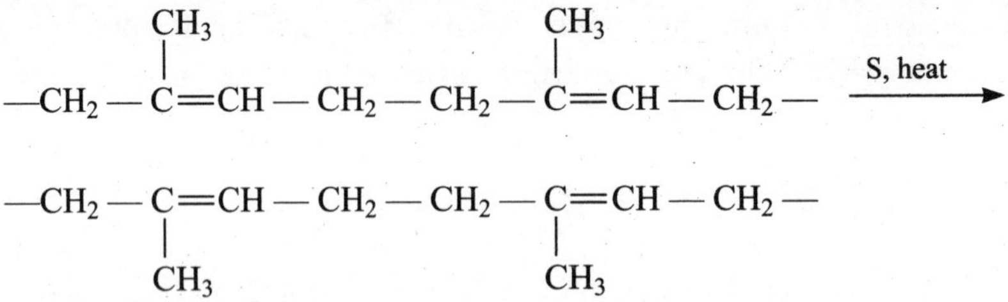

$$\begin{array}{l} -CH_2-\underset{}{\overset{CH_3}{\overset{|}{C}}}=CH-CH_2-CH_2-\overset{CH_3}{\overset{|}{C}}=CH-CH_2- \\ -CH_2-\underset{CH_3}{\underset{|}{C}}=CH-CH_2-CH_2-\underset{CH_3}{\underset{|}{C}}=CH-CH_2- \end{array} \xrightarrow{\text{S, heat}}$$

Natural or Synthetic rubber (two polymer chains)

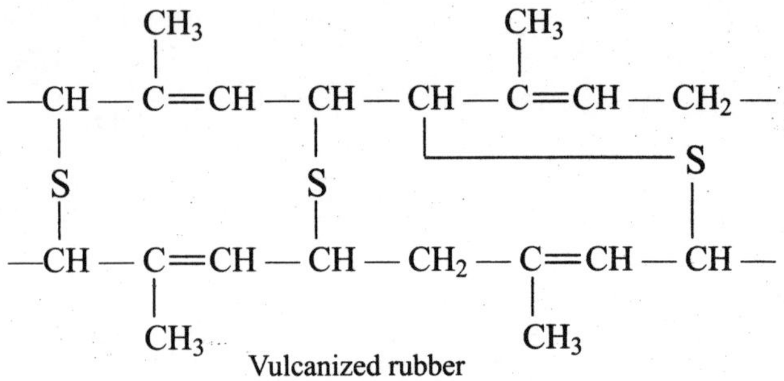

$$\begin{array}{l} -\underset{|}{CH}-\overset{CH_3}{\overset{|}{C}}=CH-\underset{|}{CH}-CH-\overset{CH_3}{\overset{|}{C}}=CH-CH_2- \\ \quad S \qquad\qquad\qquad S \qquad\qquad\qquad\qquad\quad S \\ -\overset{|}{CH}-\underset{CH_3}{\underset{|}{C}}=CH-\overset{|}{CH}-CH_2-\underset{CH_3}{\underset{|}{C}}=CH-CH- \end{array}$$

Vulcanized rubber

These cross links make the rubber harder and stronger and remove the tackiness of the raw rubber since, the individual chains can no longer slip over the other but are instead locked together in a giant single molecule.

7.3 ALKYNES OR ACETYLENES

Unsaturated hydrocarbons having a carbon-carbon triple bond are called **alkynes**. They correspond to general formula C_nH_{2n-2} *i.e.*, they have smaller proportion of hydrogen than alkenes. Though their general formula is the same as that of dienes, their chemical behaviour is quite different from them. The simplest and the most important member of the series is acetylene, $HC \equiv CH$, and due to this reason the whole series is sometimes referred to as **acetylenes** and the carbon-carbon triple bond is called **acetylenic bond**. Acetylene is an important compound from industrial point of view as it is the starting material for the manufacture of a large number of organic compounds and plastics. It is manufactured from very cheap starting materials — water and calcium carbide.

$$CaC_2 + H_2O \longrightarrow C_2H_2 + CaO$$

Some alkynes occur in nature particularly in essential oils of the plants of *compositae family.*

7.3.1 Nomenclature of Alkynes

The following two systems of nomenclature are in use:

(*i*) Common name system

(*ii*) IUPAC system.

(*i*) Common name system: In this system, these hydrocarbons are named as derivatives of acetylene from which they are regarded to be derived by replacement of one or both hydrogens by alkyl or other groups. Some alkynes are assigned special names as well. For example,

$HC \equiv CH$ — Acetylene

$H_3C - C \equiv CH$ — Methylacetylene

$H_5C_2 - C \equiv CH$ — Ethylacetylene

$H_3C - C \equiv C - CH_3$ — Dimethylacetylene

$H_5C_2 - C \equiv C - CH_3$ — Ethylmethylacetylene

***(ii)* IUPAC system:** In IUPAC system, these compounds are called **alkynes** by replacing the suffix –**ane** from the name of alkane by **–yne**. Thus

Alkynes = Alkane – **ane** + **yne**

The names of substituted alkynes are assigned according to the following rules:

1. The longest possible straight chain containing the triple bond is selected and the compound is assigned a parent name by replacing the suffix –**ane** from the name of the corresponding alkane by the suffix **–yne**.

2. The positions of the triple bond and the substituents are indicated by arabic numerals, *viz.*, 1, 2, 3, etc., which are used for numbering of the carbon atoms in the chain.

3. The carbon atoms of the parent chain are numbered in such a way that the triple bond gets the lowest number. If while numbering the carbon chain, the triple bond gets the same number from either direction, the carbon chain is numbered in such a way that the substituent gets lower number. The position of the triple bond is indicated by the number assigned to the first triply bonded carbon atom.

4. The complete name of the alkyne is then obtained by indicating the position and name of the substituent separated by a hypen before the name of the parent alkyne. The following examples illustrate both the systems. The common names are given in parentheses,

$H - C \equiv C - H$
Ethyne
(*Acetylene*)

$H_3C - C \equiv CH$
Propyne
(*Methylacetylene*)

$\overset{4}{H_3C} - \overset{3}{CH_2} - \overset{2}{C} \equiv \overset{1}{C} - H$
1-Butyne
(*Ethylacetylene*)

$\overset{4}{H_3C} - \overset{3}{C} \equiv \overset{2}{C} - \overset{1}{CH_3}$
2-Butyne
(*Dimethylacetylene*)

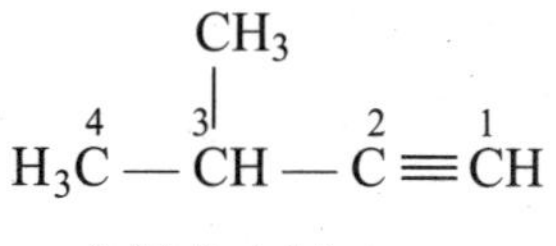

$\overset{4}{H_3C} - \overset{3}{CH}(CH_3) - \overset{2}{C} \equiv \overset{1}{CH}$
3-Methyl-1-butyne
(*Isopropylacetylene*)

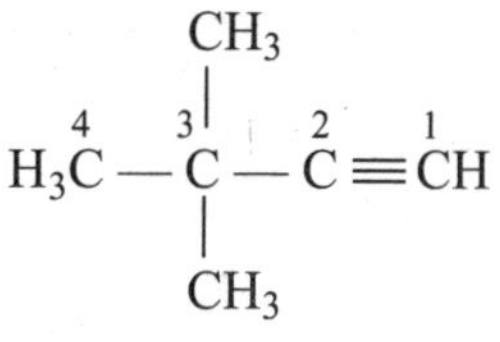

$\overset{4}{H_3C} - \overset{3}{C}(CH_3)_2 - \overset{2}{C} \equiv \overset{1}{CH}$
3-3, Dimethyl-1-butyne
(*Tert-butylacetylene*)

5. If a compound has a double and a triple bond, then such a compound is named as derivative of alkyne rather than that of alkene and while naming it, the terminal '*e*' from the suffix 'ene' is dropped. Thus, in the IUPAC system, its class name is alkenyne, *i.e.*,

Alkenyne = Alkene ***–e*** + **yne**

Chapter 7

It may be noted here that the numbering of the parent chain in such compounds is done as per lowest locant rule *i.e.*, from that end which is nearer to the double or the triple bond. If however, there is a choice in numbering, the double bond is always given lower number.

The following examples are illustrative:

$$\overset{5}{CH_3}-\overset{4}{CH_3}=\overset{3}{CH}-\overset{2}{C}\equiv\overset{1}{CH}$$

Pent-3-en-1-yne
or
3-Penten-1-yne

$$\overset{5}{HC}\equiv\overset{4}{C}-\overset{3}{CH_2}-\overset{2}{CH}=\overset{1}{CH_2}$$

Pent-1-en-4-yne
or
Penten-4-yne

$$\overset{6}{HC}\equiv\overset{5}{C}-\overset{4}{CH}=\overset{3}{CH}-\overset{2}{CH}=\overset{1}{CH_2}$$

1, 3-Hexadien-5-yne

6. In case of cyclic compounds the prefix ***cyclo*** is written before the parent name.

The smallest alicyclic ring that can accommodate the linear geometry of the triple bond is cyclooctye.

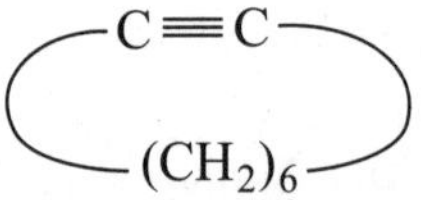

$$C\equiv C \quad (CH_2)_6$$

Cyclooctyne

Isomerism in Alkynes

Alkynes exhibit isomerism of the following four types:

1. *Chain isomerism*: The chain isomers differ from each other in their chain structure, *i.e.*, straight chain and branched chain isomers. This isomerism is called as *chain isomerism*. For example,

$$\overset{5}{H_3C}-\overset{4}{CH_2}-\overset{3}{CH_2}-\overset{2}{C}\equiv\overset{1}{CH} \quad \text{and} \quad \overset{4}{H_3C}-\overset{3}{CH}(CH_3)-\overset{2}{C}\equiv\overset{1}{CH} \quad \text{are chain isomers}$$

1-Pentyne 3-Methyl-1-butyne

2. *Position isomerism*: Alkynes which differ only in position of the triple bond show *position isomerism*. For example,

$$\overset{4}{H_3C}-\overset{3}{CH_2}-\overset{2}{C}\equiv\overset{1}{CH} \quad \text{and} \quad \overset{4}{H_3C}-\overset{3}{C}\equiv\overset{2}{C}-\overset{1}{CH_3} \quad \text{are position isomers.}$$

1-Butyne 2-Butyne

3. *Functional isomerism*: Alkynes having four or more carbon atoms are isomeric with dienes having the same molecular formula. This isomerism is called as *functional isomerism*. For example,

$\left[\begin{array}{c} CH_3-CH_2-C\equiv CH \\ \text{1-Butyne} \\ \text{or} \\ CH_3-C\equiv C-CH_3 \\ \text{2-Butyne} \end{array}\right]$ and $\left[\begin{array}{ccc} H_2C=CH-CH=CH_2 & \text{or} & H_2\overset{1}{C}=\overset{2}{C}=\overset{3}{C}H\overset{4}{C}H_3 \\ \text{1, 3-Butadiene} & & \text{1, 2-Butadiene} \end{array}\right]$

are functional isomers.

4. *Ring-chain isomerism*: Alkynes are also isomeric with cycloalkenes containing the same number of carbon atoms. This isomerism is called *ring-chain isomerism.*

(*i*) $H_3C-C\equiv CH$ (Propyne) and Cyclopropene are ring-chain isomers.

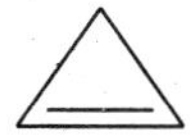

(*ii*) $H_3C-CH_2-C\equiv CH$ (1-Butyne) or $H_3C-C\equiv C-CH_3$ (2-Butyne) and Cyclobutene are ring-chain isomers.

Classification of Alkynes

The alkynes are referred to as terminal or non-terminal alkynes depending upon the position of triple bond. If it is present at the end of the carbon chain, the alkyne is called *terminal alkyne* and if it is within the chain, the alkyne is called *non-terminal alkyne.*

Terminal alkynes

$HC\equiv C-CH_3$ (Propyne) $HC\equiv C-CH_2CH_3$ (1-Butyne) $HC\equiv C-CH_2CH_2CH_3$ (1-Pentyne)

Non-terminal alkynes

$H_3C-C\equiv C-CH_3$ (2-Butyne) $H_3C-C\equiv C-CH_2CH_3$ (2-Pentyne)

7.3.2 Structure and Bonding

Alkynes have two carbon atoms linked by a triple bond. The remaining carbon atoms form C — C and C — H, σ-bonds. The C ≡ C triple bond consists of one strong σ-bond and two weak π-bonds.

The carbon atoms involved in triple bond formation are *sp*-hybridized. Therefore, the groups which are attached to the triply bonded carbon atoms are *linear.*

Due to linear nature the H — C ≡ C bond angle is 180° and the carbon-carbon bond distance

(C $\equiv$ C) in acetylene is very small (120 pm) due to smaller size of *sp*-orbitals as compared to sp^2 and sp^3 orbitals. The C $\equiv$ C bond strength in ethyne is 823 kJ mol^{-1}. It is stronger than C $=$ C bond of ethene (610 kJ mol^{-1}) and C — C bond of ethane (370 kJ mol^{-1}).

$$\mathrm{H{-}C{\equiv}C{-}H}\quad (\text{C}\equiv\text{C} = 120\ \text{pm},\ \text{C–H} = 106\ \text{pm},\ \text{angle } 180^\circ)$$

7.3.3 Methods of Formation of Alkynes

Like alkenes they can be prepared by the well-known β-elimination reactions involving the removal of the leaving groups twice. The general pattern of these reactions can be depicted as under:

$$\mathrm{R{-}C(a)(b){-}C(d)(e){-}R'} \xrightarrow{-ae} \mathrm{(R)(b)C{=}C(d)(R')} \xrightarrow{-bd} \underset{\text{Alkyne}}{\mathrm{R{-}C{\equiv}C{-}R'}}$$

Some of the most important methods of preparation of alkynes are given below:

***(i)* Dehydrohalogenation of vicinal dihalides:** On heating with a base (alcoholic KOH) the *vic*-dihalides (1, 2-dihalides, *i.e.*, compounds containing two halogen atoms on adjacent carbon atoms) yield alkynes. This reaction takes place in two stages involving the intermediate formation of vinyl halides.

$$\underset{\substack{\text{Vicinal dihalide}\\ \text{R and R}' = \text{H or alkyl}}}{\mathrm{R{-}CH(X){-}CH(X){-}R'}} \xrightarrow{\text{Base}} \underset{\text{A vinyl halide}}{\mathrm{R{-}CH{=}C(X){-}R'}} \xrightarrow{\text{Base}} \underset{\text{Alkyne}}{\mathrm{R{-}C{\equiv}C{-}R'}}$$

The mechanism of this reaction is similar to that of E2 reactions of alkyl halides. As the intermediate vinyl halides are quite unreactive, the reaction frequently stops at this stage under mild conditions. However, under more vigorous conditions, particularly in the presence of stronger base (sodium amide, etc.), alkynes are obtained. We have already seen that the *vic*-dihalides are readily prepared by the addition of halogens to alkenes. Thus, the whole sequence —the addition of halogens to alkenes followed by dehydrohalogenation of the *vic*-dihalides formed —can be regarded *as a method of converting an alkene into an alkyne*. It may also be noted that the presence of at least two hydrogens on the adjacent carbon atoms is essential for the success of this reaction.

***(ii)* Dehydrohalogenation of *gem*-dihalides:** As in (*i*) the gem-dihalides (1,1-dihalides, that is, the compounds having both the halogens linked to one carbon atom) which have two hydrogens on the adjacent carbon atoms give alkynes on heating with a strong base *via* the formation of *vic* halide as an intermediate.

$$R-CH_2-\overset{R'}{\overset{|}{C}}X_2 \xrightarrow{\text{Base}} \underset{\text{A vinyl halide}}{R-CH=\overset{X}{\overset{|}{C}}-R'} \longrightarrow R-C\equiv C-R'$$

The *gem*-dihalides are easily obtained by the action of PCl_5 on carbonyl compounds.

***(iii)* Dehalogenation of tetrahalides:** Alkynes are formed when the vapours of 1, 1, 2, 2-tetrahalides (compounds containing a $-CX_2-CX_2-$system) are passed over heated zinc.

$$R-\underset{X}{\overset{X}{C}}-\underset{X}{\overset{X}{C}}-R' \xrightarrow{Zn} (X)(R)C=C(R')(X) \xrightarrow{Zn} R-C\equiv C-R'$$

This reaction is also believed to take place in two steps and its mechanism is the same as that of E2 reaction of *vic*-dihalides with zinc. This reactions has, however, not much synthetic importance as the tetrahalides themselves are usually prepared from alkynes.

***(iv)* Action of acetylides on alkyl halides:** As we shall see later, alkynes with a terminal triple bond ($R-C\equiv C-H$) are acidic and react readily with sodium (or sodium amide) at 475 K forming salts called **acetylides**. These acetylides are good nucleophiles and react with primary alkyl halides to form higher alkynes. This reaction is an example of an S_N2 reaction and is a convenient method for the conversion of lower alkynes into higher ones. This reaction gives good yields of alkynes only with primary alkyl halides, as with secondary and tertiary alkyl halides elimination becomes a competitive side reaction resulting in comparatively poor yields of alkynes.

$$2R-C\equiv CH + 2\,Na \longrightarrow 2R-C\equiv \overset{\ominus}{C}\overset{\oplus}{Na} + H_2$$

$$R-C\equiv CH + NaNH_2 \xrightarrow{\text{liq. } NH_3} R-C\equiv \overset{\ominus}{C}\overset{\oplus}{Na} + NH_3$$

$$R-C\equiv \overset{\ominus}{C} + R'-X \xrightarrow{S_N2} R-C\equiv C-R' + \overset{\ominus}{X}$$

$$\underset{\text{Sodium acetylide}}{H-C\equiv \overset{\ominus}{C}\overset{\oplus}{Na}} + CH_3CH_2-I \xrightarrow{S_N2} \underset{\text{1-Butyne}}{H-C\equiv C-CH_2CH_3} + NaI$$

$$H-C\equiv \overset{\ominus}{C}\overset{\oplus}{Na} + CH_3CH_2CH_2-Br \xrightarrow{S_N2} \underset{\text{1-Pentyne}}{HC\equiv C-CH_2CH_2CH_3} + NaBr$$

***(v)* Action of water on calcium carbide:** Calcium carbide upon treatment with water gives rise to acetylene. This method is used to prepare acetylene in the laboratory as well as on a large scale.

$$\underset{\text{Calcium carbide}}{CaC_2} + 2H_2O \longrightarrow \underset{\text{Acetylene}}{HC{\equiv}CH} + Ca(OH)_2$$

For this purpose, calcium carbide is obtained by heating limestone (calcium carbonate) with coke in an electric furnace at 2275 K.

$$CaCO_3 \xrightarrow{2275\text{ K}} CaO + CO_2$$

$$CaO + 3C \xrightarrow{2275\text{ K}} CaC_2 + CO$$

(*vi*) **Dehalogenation of haloforms:** When chloroform or iodoform are heated with silver powder, they undergo dehalogenation to give acetylene.

$$\underset{\text{Chloroform}}{2CHCl_3} + 6Ag \xrightarrow{\Delta} \underset{\text{Ethyne}}{HC{\equiv}CH} + 6AgCl$$

$$\underset{\text{Iodoform}}{2CHI_3} + 6Ag \xrightarrow{\Delta} HC{\equiv}CH + 6AgI$$

(*vii*) **Kolbe's electrolytic reaction:** Electrolysis of a concentrated solution of sodium or potassium maleate or fumarate gives rise to acetylene.

$$\underset{\substack{\text{Potassium fumarate}\\(\textit{trans}\text{-isomer})}}{\begin{matrix}H & & COOK\\ & C & \\ & \| & \\ & C & \\ KOOC & & H\end{matrix}} \text{ or } \underset{\substack{\text{Potassium maleate}\\(\textit{cis}\text{-isomer})}}{\begin{matrix}H & & COOK\\ & C & \\ & \| & \\ & C & \\ H & & COOK\end{matrix}} + 2H_2O \xrightarrow{\text{Electrolysis}} \begin{matrix}CH\\ |||\\ CH\end{matrix} + 2CO_2 + H_2 + 2KOH$$

This reaction is called ***Kolbe's electrolytic reaction*** and is believed to occur through the following steps:

$$\begin{matrix}CHCOOK\\ \|\\ CHCOOK\end{matrix} \xrightarrow{\text{Ionization}} \begin{matrix}CHCOO^{\ominus}\\ \|\\ CHCOO^{\ominus}\end{matrix} + 2K^{\oplus}$$

$$2H_2O \underset{}{\overset{\text{Ionization}}{\rightleftharpoons}} 2\overset{\ominus}{O}H + 2H^{\oplus}$$

At cathode $$2H^{\oplus} + 2e^- \longrightarrow 2[H] \longrightarrow H_2$$

At anode $$\begin{matrix}CHCOO^{\ominus}\\ \|\\ CHCOO^{\ominus}\end{matrix} -2e^- \longrightarrow \underset{\text{Diradical }(\textit{unstable})}{\left[\begin{matrix}CHCOO^{\bullet}\\ \|\\ CHCOO^{\bullet}\end{matrix}\right]} \longrightarrow \begin{matrix}CH\\ |||\\ CH\end{matrix} + 2CO_2$$

***(viii)* Synthesis from carbon and hydrogen- Berthelot's synthesis:** Acetylene can be prepared by generating an electric spark between carbon electrodes under hydrogen atmosphere *i.e.*,

$$2C + H_2 \xrightarrow[3273\text{ K}]{\text{Electric arc}} \underset{\text{Ethyne}}{HC\equiv CH}$$

7.3.4 Physical Properties

***(i)* Physical state:** The first three alkynes (ethyne, propyne and butyne) are colourless gases,the next eight are liquids and higher ones are solids.

***(ii)* Odour:** All alkynes are odourless. However, commercial acetylene possesses garlic smell which is due to the presence of phosphine as impurity.

***(iii)* Melting and boiling points:** The alkynes have slightly higher melting and boiling points as compared to those of the corresponding alkenes and alkanes. Due to their linear structure, the molecules of alkynes can be packed more closely in the crystal lattice compared to corresponding alkenes or alkanes.

The following m.pt. /b.pt. of first member of each series are in agreement with the above explanation.

Hydrocarbon	Ethane	Ethene	Ethyne
m.pt. (K)	101	104	191
b.pt. (K)	184.5	171	198

***(iv)* Solubility:** Like alkanes and alkenes, alkynes are also non-polar.Hence, they dissolve in organic solvents such as petroleum ether,benzene, carbon tetrachloride, etc.

***(v)* Density:** Densities of alkynes fall in the range 0.69–0.77 g/cm^3. The density rises with increase in molecular size.

Chapter 7

7.3.5 Chemical Reactions of Alkynes

Due to the presence of loosely held π-electrons alkynes also undergo *electrophilic additions*, characteristic of alkenes. Although the π-electron density on a carbon-carbon triple bond is higher than that of a double bond, *alkynes are less reactive than alkenes*. It is therefore, reasonable to believe that nucleophilic addition should be a facile process. In addition, terminal alkynes behave as acids. Some of the important reactions of alkynes are:

1. Acidic character of terminal alkynes: Formation of acetylides
2. Reduction (Addition of Hydrogen)
3. Electrophilic addition
4. Nucleophilic addition
5. Oxidation
6. Polymerization
7. Isomerization

1. Formation of Acetylides: Acidic Character of Terminal Alkynes

A hydrogen atom attached to triply bonded carbon atom is called as acetylenic hydrogen and is acidic in nature. Unlike alkanes and alkenes, acetylene and other terminal alkynes are acidic in nature. Following examples are illustrative:

***(i) Formation of alkali metal acetylides*:** Ethyne as well as other alkynes having a terminal triple bond on treatment with sodium metal at 475 K or sodamide at 196 K evolve hydrogen and produce sodium salts called **sodium acetylides.**

$$2R-C\equiv C-H + 2Na \xrightarrow{475\text{ K}} 2R-C\equiv \overset{\ominus\oplus}{CNa} + H_2$$

$$R-C\equiv C-H + \overset{\oplus}{Na}\overset{\ominus}{NH_2} \xrightarrow{196\text{ K}} R-C\equiv \overset{\ominus\oplus}{CNa} + NH_3$$

Sodium acetylide

The reaction with sodium amide clearly shows that alkynes are stronger acids (pK_a = 25) than NH_3 (pK_a = 34) as a stronger acid displaces a weaker acid from its salt). On the other hand, sodium acetylides on treatment with water afford alkynes. Thus, the alkynes are weaker acids than water (pK_a = 15.7).

$$R-C\equiv C-\overset{\ominus\oplus}{C\,Na} + H_2O \longrightarrow R-C\equiv C-H + NaOH$$

***(ii) Formation of heavy metal acetylides*:** Acetylene and 1-alkynes react with some heavy metal cations such as Ag^+ and Cu^+ forming acetylides. These salts being insoluble, precipitate out from the reaction mixture. For example, formation of white precipitate on treating an alkyne with ammonical silver nitrate solution and a red precipitate with ammonical cuprous chloride solution indicates the presence of a — C ≡ CH end group.

$$RC\equiv CH + AgNO_3 + NH_4OH \longrightarrow RC\equiv CAg + H_2O + NH_4NO_3$$

White ppt

$$HC\equiv CH + AgNO_3 + NH_4OH \longrightarrow HC\equiv CAg + H_2O + NH_4NO_3$$

Acetylene

Silver acetylide (or Silver ethynide) (White ppt)

$$2RC\equiv CH + Cu_2Cl_2 + 2NH_4OH \longrightarrow 2RC\equiv CCu + 2H_2O + 2NH_4Cl$$

Red ppt

$$2HC\equiv CH + Cu_2Cl_2 + 2NH_4OH \longrightarrow 2HC\equiv CCu + 2H_2O + 2NH_4Cl$$

Cu (1) acetylide (Red ppt)

It may be noted here that unlike alkali metal acetylides, the heavy metal acetylides are not decomposed by water. However, on treatment with mineral acids they regenerate alkynes.

$$R-C\equiv CAg + HNO_3 \longrightarrow R-C\equiv C-H + AgNO_3$$

$$R-C\equiv CCu + HCl \longrightarrow R-C\equiv C-H + CuCl$$

Importance

1. These reactions can be used as methods of separation of terminal and non-terminal alkynes from a mixture. If a mixture of 1-butyne and 2-butyne, for instance, is passed through ammonical silver nitrate solution then 1-butyne, because of the presence of C ≡ CH group, forms an insoluble white precipitate while 2-butyne remains unaffected. The silver butylide is separated and treated with acid to generate 1-butyne.

$$\underset{\text{Terminal alkyne}}{CH_3CH_2-C\equiv C-H} \xrightarrow{Ag^{\oplus}} CH_3CH_2-C\equiv C-Ag \xrightarrow{H^{\oplus}} CH_3CH_2-C\equiv C-H$$

$$\underset{\text{Non-terminal alkyne}}{CH_3-C\equiv C-CH_3} \xrightarrow{Ag^{\oplus}} \text{No reaction}$$

2. These reactions are also useful in separation of alkynes from alkene and alkanes.

***(iii) Formation of Alkynyl Grignard Reagents* :** Because of acidic nature, 1-alkynes also react with Grignard reagents forming alkynyl magnesium halides or alkynyl Grignard reagents. Like other Grignard reagents, alkynyl Grignard reagents can be made to react with alkyl halides to yield higher alkynes. This is also a nucleophilic displacement reaction as the alkynyl group in these reagents has carbanion, character.

$$R-C\equiv C-H + R'MgX \xrightarrow[\text{Ether}]{\text{Dry}} \underset{\text{Alkynyl Grignard reagent}}{R-C\equiv C-MgX} + R'-H$$

$$R-C\equiv C-MgX + R''X \longrightarrow R-C\equiv C-R'' + MgX_2$$

The metal acetylides should not be allowed to dry as they have a tendency to explode.

Reasons for Acidic Character of Terminal Alkynes

We know that the strength of an acid depends upon the ease with which it can lose a proton. In other words, factors which stabilize the anion after the removal of the proton make the acid stronger. Thus, the acidic character of terminal alkynes can be easily explained on the basis of the following two parameters (factors).

1. Type of hybridization of the carbon carrying “acidic” hydrogen.
2. Stability of alkynide ion produced after removal of the hydrogen as proton.

We will now discuss how these factors govern the acid character of alkynes.

1. Hybridization: In acetylene and 1-alkynes, the hydrogen atom is bonded to an *sp*-hybridized carbon, which because of increased *s*–character, is more electronegative than that of alkenes (sp^2) and

Chapter 7

alkanes (sp^3). Due to greater electronegativity of *sp* hybridised carbon, the electrons of C—H bond are more strongly held nearer the carbon nucleus. In other words, the hydrogen atom is less tightly held by the carbon atom and thus can be easily removed in presence of a base. Hence, the alkyns behave as acids.

Further, we know that the *s*-character of the hybrid orbitals decreases in the order $sp > sp^2 > sp^3$, the decreasing order of acidic character of hydrocarbons should be $HC \equiv CH > H_2C = CH_2 > H_3C—CH_3$.

2. Stability of acetylide ions: The pair of electrons left after the removal of proton occupies an *sp*-hybrid orbital, which is closer to the nucleus as compared to an electron pair occupying π-orbital. Therefore, this pair of electrons is less available for combining with acids compared to the electron pair occupying sp^2-orbital (in the case of vinyl anion) and sp^3-orbital (in case of alkyl carbanion) which have more p-character.

In other words, the acetylide anion is the most stable of the three types of anions. The decreasing order of relative stabilities of these anions is as follows:

$$HC \equiv \overset{\ominus}{\ddot{C}} \quad > \quad H_2C = \overset{\ominus}{\ddot{C}}H \quad > \quad H_3C - \overset{\ominus}{\ddot{C}}H_2$$

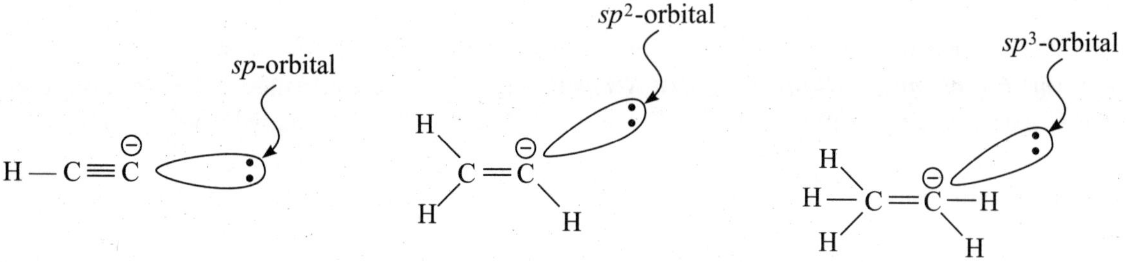

sp-orbital

$H-C \equiv \overset{\ominus}{C}$

I

Acetylide anion

Electron pair in sp-orbital, nearest to the nucleus, most stable

sp^2-orbital

H, H, C=C, H

II

Vinyl carbanion

Electron pair in sp2-orbital (more p character) less nearer to nucleus compared to I, less stable than I.

sp^3-orbital

H, H—C=C—H, H, H

III

Ethyl carbanion

Electron pair in sp3-orbital (more p-character) less nearer to nucleus compared to I and II least stable.

In agreement with the above discussion the decreasing order of acid strength of the three types of hydrocarbons is as follows (their pK_a values are given in the parenthesis).

$$\underset{(pK_a = 26)}{HC \equiv CH} \quad > \quad \underset{(pK_a = 45)}{H_2C = CH_2} \quad > \quad \underset{(pK_a = 62)}{H_3C—CH_3}$$

2. Reduction: Addition of Hydrogen

As in the case of alkenes, the alkynes can also be reduced to form alkenes and/or alkanes. The reaction can be stopped at alkene stage and the nature and type of alkene produced, *i.e.*, *cis*-or *trans*-depends upon the reaction conditions.

The following two methods are mainly used:

(*a*) Reduction with metal —ammonia.

(*b*) Catalytic reduction.

(*a*) **Reduction with Metal — Ammonia (Birch Reduction):** Alkynes on treatment with sodium (or lithium) in liquid ammonia produce *trans*-alkenes as almost exclusive products.

$$R-C\equiv C-R' \xrightarrow{\text{Na/liq. NH}_3} \underset{H}{\overset{R}{>}}C=C\underset{R'}{\overset{H}{<}}$$

Mechanism

The following mechanism may be proposed:

$$R-C\equiv C-R' \xrightarrow[\text{(Na or Li)}]{e} \underset{\text{Radical anion}}{R-\overset{\ominus}{C}=\dot{C}-R'} \xrightarrow[\overset{\ominus}{-}NH_2]{H-NH_2} \underset{trans\text{-radical}}{\underset{R}{\overset{H}{>}}C=\dot{C}-R'}$$

$$\underset{trans\text{-radical}}{} \xrightarrow{e} \underset{trans\text{-carbanion}}{\underset{R}{\overset{H}{>}}C=C^{\ominus}-R'} \xrightarrow[\overset{\ominus}{-}NH_2]{NH_3} \underset{trans\text{-alkene}}{\underset{R}{\overset{H}{>}}C=C\underset{H}{\overset{R'}{<}}}$$

The addition of first electron followed by protonation gives a radical. The second electron adds to the *trans*-radical and the resultant *trans* anion is protonated before the inversion could take place.

(*b*) **Catalytic reduction:** Catalytic reduction of alkynes to alkanes takes place in two stages. First an alkene is formed followed by an alkane.

$$R-C\equiv C-R' \xrightarrow[\text{(Ni or Pt) 573 K}]{H_2/\text{Catalyst}} R-CH=CH-R' \xrightarrow[\text{573 K}]{H_2/\text{Catalyst}} R-CH_2-CH_2-R'$$

By using calculated amount of hydrogen and choosing suitable reaction conditions the reduction can be stopped at the alkene stage.

In contrast to metal ammonia reduction, catalytic hydrogenation produces *cis*-alkenes.

$$R-C\equiv C-R' + H_2 \xrightarrow{\text{Catalyst}} \underset{cis\text{-Alkene}}{\underset{H}{\overset{R}{>}}C=C\underset{H}{\overset{R'}{<}}}$$

Thus, *cis*-or *trans*-alkenes may be produced depending upon the reaction conditions.

Alkynes on hydrogenation in presence of *Lindlar catalyst* give rise to *cis*-alkens.

$$R-C\equiv C-R' \xrightarrow{\text{Lindlar catalyst}} \underset{cis\text{-Alkene}}{(R)(H)C=C(R')(H)}$$

Lindlar catalyst: The catalyst is prepared by precipitating palladium over calcium carbonate and treating it with lead acetate and quinoline (partially poisoned catalyst which cannot reduce the double bond).

The mechanism of formation of *cis*-alkene is the same as that of *cis*-hydrogenation of alkenes.

3. Electrophilic Addition

As mentioned above the characteristic reaction of alkynes is electrophilic addition, which takes place in two stages involving the formation of olefinic intermediates.

$$-C\equiv C- + X-Y \longrightarrow -\underset{X}{C}=\underset{Y}{C}- \xrightarrow{X-Y} -\overset{X}{\underset{X}{C}}-\overset{Y}{\underset{Y}{C}}-$$

Mechanism

The mechanism of addition to alkynes resembles rather closely to that delineated for alkenes. In spite of close similarity, the reaction rates of alkynes with electrophilic reagents are slower than those of alkenes. This is evident from the fact that in some reactions a catalyst, such as Hg^{2+} ion, is needed. The reason for this is not very well understood. However, the following two explanations can be offered:

1. One explanation for this is that bridged intermediate cation formed by the initial attack of electrophile on the triple bond is less stable because it is a highly strained system.

$$-C\equiv C- + E^{\oplus} \longrightarrow -C\overset{E^{\oplus}}{\equiv}C-$$

Highly strained

2. Another explanation for the low reactivity of alkynes towards electrophiles may be attributed to *sp*-hybridization of acetylenic carbon atoms which hold the π-electrons more tightly near the nuclei. Thus, they are less easily available for combination with electrophiles.

Perhaps, both these factors –steric and electronic –play their part in diminishing the reactivity of alkynes towards electrophiles.

As we shall see, the intermediacy of cyclic cation is borne out by the fact that the olefinic intermediate is invariably *trans*. Some of the important examples of electrophilic additions are given below:

(*a*) Addition of halogens: Addition of halogens to alkyenes takes place in two stages, first forming a *trans*-1, 2-dihaloalkene and then 1, 1, 2, 2-tetrahaloalkane.

$$R-C\equiv C-R' \xrightarrow{X_2} \text{(trans) } RXC=CXR' \xrightarrow{X_2} R-CX_2-CX_2-R'$$

The order of reactivity of halogens is $Cl_2 > Br_2 > I_2$. The reaction may be limited to the formation of halogenated alkene by choosing proper reaction conditions. Ethyne, for instance, on treatment with bromine water gives only 1, 2-dibromoethene whereas with bromine alone, it forms 1, 1, 2, 2-tetrabromoethane.

$$H-C\equiv C-H + Br_2 \longrightarrow \text{(trans) } HBrC=CBrH$$

$$H-C\equiv C-H + 2Br_2 \longrightarrow CHBr_2-CHBr_2$$

Mechanism

As mentioned above, the addition of halogens to alkynes takes place in two steps

***Step I*:** Formation of trans 1, 2-dihaloalkene

$$H-C\equiv C-R' + X-X \xrightarrow[-X^{\ominus}]{\text{Slow}} RC\overset{X^{\oplus}}{=}CR' \xrightarrow[X^{\ominus}]{\text{Fast}} \text{(trans) } RXC=CXR'$$

***Step II*:** Formation of 1, 1, 2, 2-tetrahaloalkane

$$RXC=CXR' + X-X \xrightarrow[-X^{\ominus}]{\text{Slow}} RXC\overset{X^{\oplus}}{-}CXR' \xrightarrow[X^{\ominus}]{\text{Fast}} R-CX_2-CX_2-R'$$

(*b*) Addition of halogen acids: On treatment with halogen acids, alkynes first form vinyl halides followed by alkylidene halides (*gem*-dihalides). The second molecule of halogen acid adds according to Markownikoff's rules.

Mechanism

$$R-C\equiv C-R' + H-X \xrightarrow[-X^{\ominus}]{} R-\overset{\overset{H}{\oplus}}{C=C}-R' \xrightarrow{X^{\ominus}} \underset{R}{\overset{H}{>}}C=C\underset{X}{\overset{R'}{<}} \xrightarrow{HX} RCH_2-CX_2R'$$

The reaction is catalyzed by light or mercury salts. In the case of unsymmetrically substituted alkynes, the first molecule of HX adds in accordance with the Markownikoff's rule. Thus,

$$\underset{\text{Propyne}}{H_3C\rightarrow C\equiv CH} + H-Br \longrightarrow H_3C-\overset{Br}{\overset{|}{C}}=CH_2 \xrightarrow{H-Br} \underset{\text{2, 2-Dibromopropane}}{H_3C-CBr_2-CH_3}$$

Mechanism

(*i*) Kinetic studies of the addition of halogen acids to alkynes indicate that these reactions follow a rate law of **third order** which is **second order** in halogen acid.

$$\text{Rate} \propto [\text{alkyne}]\,[\text{halogen acid}]^2$$

On the basis of these studies, the following *Ad E*3 (*addition, electrophilic, termolecular*) mechanism has been proposed.

$$H-C\equiv C-H + 2\,H-X \longrightarrow H-\overset{H---X^{\delta-}}{C\equiv C}-H\ (\text{with } \overset{\delta+}{H}---X) \longrightarrow \underset{X}{\overset{R}{>}}C=C\underset{H}{\overset{H}{<}} + HX$$

(*ii*) **Free radical addition of hydrogen bromide:** In presence of peroxides, the addition of HBr (and not HCl or HI) to alkynes take place in anti-Markownikoff's manner, *e.g.*,

$$H_3C-C\equiv CH + HBr \xrightarrow{\text{Peroxide}} CH_3CH=CHBr \xrightarrow[\text{Peroxide}]{HBr} \underset{\text{1,1-Dibromopropane}}{CH_3CH_2CHBr_2}$$

(*c*) Hydration (addition of water): A molecule of water adds to the carbon-carbon triple bond in presence of sulphuric acid and mercuric sulphate to form a vinyl alcohol (C = C — OH, enol) which readily tautomerizes to the corresponding carbonyl derivative.

$$H-C\equiv C-H + H_2O \xrightarrow{\overset{\oplus}{H},\ \overset{2\oplus}{Hg}} [H_2C{=}CHOH] \rightleftharpoons H_3C-\overset{O}{\overset{\|}{C}}-H$$

Vinyl alcohol (Enol form of acetaldehyde) — Acetaldehyde (keto form)

$$R-C\equiv C-R' + H_2O \xrightarrow{\overset{\oplus}{H},\ \overset{2\oplus}{Hg}} [R-CH{=}\overset{OH}{\overset{|}{C}}-R'] \rightleftharpoons R-CH_2-\overset{O}{\overset{\|}{C}}-R$$

(Enol form of ketone) — Ketone

This is also an electrophilic addition reaction in which mercuric ions form a complex with the triple bond. The reaction then proceeds as shown below:

$$-C\equiv C- + \overset{2\oplus}{Hg} \rightleftharpoons -C\overset{\overset{2\oplus}{Hg}}{\uparrow}\!\!\equiv C- \longrightarrow -C\overset{\overset{2\oplus}{Hg}}{\overset{\oplus}{=}}C- \xrightarrow{H_2\ddot{O}:} -\overset{\overset{\oplus}{Hg}}{\overset{|}{C}}=\underset{\underset{\oplus}{O}H_2}{\underset{|}{C}}-$$

$$-\overset{H}{\overset{|}{\underset{H}{\underset{|}{C}}}}-\overset{}{\underset{O}{\underset{\|}{C}}}- \rightleftharpoons -\overset{H}{\overset{|}{C}}=\underset{OH}{\underset{|}{C}}- \xleftarrow{+\overset{\oplus}{H}} -\overset{\overset{\oplus}{Hg}}{\overset{|}{C}}=\underset{OH}{\underset{|}{C}}- \xleftarrow{-\overset{\oplus}{H}}$$

With unsymmetrical alkynes addition of water takes place in accordance with Markownikoff's rule.

$$H_3C\rightarrow C\equiv CH \xrightarrow[H_2O]{\overset{\oplus}{H},\ \overset{2\oplus}{Hg}} H_3C-\overset{OH}{\overset{|}{C}}{=}CH_2 \longrightarrow H_3C-\overset{O}{\overset{\|}{C}}-CH_3$$

In the case of unsymmetrical disubstituted alkynes, the orientation of water addition will depend on the electronic character of the substituents, electrophile adding to that end of the triple bond which is farther form the substituent having more electron donating property.

(*d*) Addition of hypohalous acids: The addition of hypohalous acids also takes place in two steps forming dihalocarbonyl derivatives.

***Step* I**

$$R-C\equiv C-R' + \overset{\delta^+}{Cl}-\overset{\delta^-}{OH} \longrightarrow R-\overset{\oplus}{C}=\underset{Cl}{\underset{|}{C}}-R' \xrightarrow{\overset{\ominus}{OH}} R-\overset{OH}{\overset{|}{C}}=\underset{Cl}{\underset{|}{C}}-R'$$

Chapter 7

***Step* II**

$$R-\underset{\underset{Cl}{|}}{\overset{\overset{OH}{|}}{C}}=C-R' \xrightarrow{HOCl} \left[R-\underset{\underset{OH}{|}}{\overset{\overset{OH}{|}}{C}}-\underset{\underset{Cl}{|}}{\overset{\overset{Cl}{|}}{C}}-R' \right] \xrightarrow{-H_2O} R-\overset{\overset{O}{\|}}{C}-CCl_2-R'$$

α, α-Dichlorocarbonyl compound

(*e*) **Addition of acetic acid:** In presence of mercuric ions, acetic acid adds to alkynes forming vinyl acetate followed by alkylidene acetate.

$$H-C\equiv C-H + CH_3COOH \xrightarrow[353\ K]{Hg^{2\oplus}} H_2C=CH-OCOCH_3$$

Vinyl acetate

$$\downarrow Hg^{2\oplus},\ CH_3COOH$$

$$H_3C-CH\begin{matrix} \diagup OCOCH_3 \\ \diagdown OCOCH_3 \end{matrix}$$

Ethylidene acetate

Importance

Vinyl acetate is an important industrial chemical as it polymerizes to PVA (polyvinyl acetate) plastics. Ethylidene acetate on heating gives acetic anhydride and acetaldehyde.

(*f*) **Ozonization:** Ozone adds to carbon-carbon triple bond forming the ozonide. The ozonides are hydrolyzed by water and the products—the 1,2-diketones undergo oxidative cleavage to acids by hydrogen peroxide formed in the reaction.

$$R-C\equiv C-R' + O_3 \longrightarrow R-\overset{\overset{O}{\diagup\ \diagdown}}{C-C}-R' \text{ (with } \underset{O-O}{|\quad|}\text{ below)}$$

Ozonide

$$\downarrow H_2O$$

$$R-COOH + R'-COOH \longleftarrow R-\underset{\underset{O}{\|}}{C}-\underset{\underset{O}{\|}}{C}-R' + H_2O_2$$

The identification of the acids formed would tell us which of the carbons were linked through triple bond in the original alkyne. Thus, this sequence can be used as an unambiguous procedure for locating the position of triple bond in an unkown alkyne.

Alkynes react with ozone in presence of some inert solvents such as CH_2Cl_2, $CHCl_3$ or CCl_4 at low temperature (196–200 K) to form ozonides. These ozonides on decomposition with Zn dust and water or H_2/Pd (reductive cleavage) give 1, 2-dicarbonyl compounds. For example,

$$HC{\equiv}CH \xrightarrow[196\text{–}200\ K]{O_3/CH_2Cl_2} \text{Ethyne ozonide} \xrightarrow[-ZnO]{Zn/H_2O} H-\underset{\underset{O}{\|}}{C}-\underset{\underset{O}{\|}}{C}-H$$

Ethyne → Ethyne ozonide → (Reductive cleavage) 1, 2-Ethanedial (*Glyoxal*)

$$CH_3-C{\equiv}CH \xrightarrow[196\text{–}200\ K]{O_3/CH_2Cl_2} \text{Propyne ozonide} \xrightarrow[-ZnO]{Zn/H_2O} CH_3-\underset{\underset{O}{\|}}{C}-\underset{\underset{O}{\|}}{C}-H$$

Propyne → Propyne ozonide → (*Reductive cleavage*) 2-Oxopropanal

$$CH_3-C{\equiv}C-CH_3 \xrightarrow[196\text{–}200K]{O_3/CH_2Cl_2} \text{But-2-yne ozonide} \xrightarrow[-ZnO]{Zn/H_2O} CH_3-\underset{\underset{O}{\|}}{C}-\underset{\underset{O}{\|}}{C}-CH_3$$

But-2-yne → But-2-yne ozonide → (*Reductive cleavage*) Butane -2, 3-dione

***(g)* Hydroboration:** Addition of borane to carbon-carbon triple bond produces vinylboranes.

$$R-C{\equiv}C-R' \xrightarrow[THF]{B_2H_6} \text{(H)(R)}C{=}C\text{(B<)(R')}$$

The vinylboranes are highly useful intermediates for the synthesis of a large variety of organic compounds. Some important examples are discussed below:

***(i)* Protonolysis:** Vinylboranes on heating with acetic acid produce *cis*-alkenes, by replacement of boron by a hydrogen atom. The conversion of alkyl or alkenylborane into the corresponding hydrocarbons on treatment with a carboxylic acid is called *protonolysis*.

$$\text{(H)(R)}C{=}C\text{(B<)(R')} \xrightarrow[373\ K]{CH_3COOH} \text{(H)(R)}C{=}C\text{(H)(R')}$$

cis-Alkene

For example: Synthesis of *cis*-2-butene from 2-butyne.

$$H_3C-C\equiv C-CH_3 \xrightarrow[\text{THF}]{B_2H_6} \underset{H_3C}{\overset{H}{}}C=C\underset{CH_3}{\overset{B-}{}} \xrightarrow[373\text{ K}]{CH_3COOH} \underset{H_3C}{\overset{H}{}}C=C\underset{CH_3}{\overset{H}{}}$$

2-Butyne — *cis*-2-Butyne

***(ii)* Oxidation:** When treated with alkaline hydrogen peroxide, the vinylboranes are oxidised to produce the corresponding carbonyl compounds (Hydroboration-oxidation). For example,

(A) *Non-terminal alkynes give rise to ketones.*

$$\left(R-\overset{H}{C}=\underset{R'}{C}\right)_3B \xrightarrow{H_2O_2/\overset{\ominus}{O}H} R-\overset{H}{C}=\overset{OH}{C}-R' \underset{}{\overset{\text{Tautomerization}}{\rightleftharpoons}} R-CH_2-\overset{O}{\overset{\|}{C}}-R'$$

Enol-form — *Keto*-form

Symmetrically substituted non-terminal alkyenes give rise to a single ketone. For example, synthesis of 2-butanone from 2-butyne.

$$H_3C-C\equiv C-CH_3 \xrightarrow[\text{THF}]{B_2H_6} CH_3-\overset{H}{C}=\overset{B<}{C}-CH_3 \xrightarrow{H_2O_2/\overset{\ominus}{O}H} H_3C-\overset{H}{C}=\overset{OH}{C}-CH_3$$

Enol-form 2-Butanone

$$\rightleftharpoons CH_3CH_2-\overset{O}{\overset{\|}{C}}-CH_3$$

Keto-form 2-Butanone

However, unsymmetrically substituted non-terminal alkynes produce a mixture of both the possible ketones in which methyl ketones predominates (boron attaches itself to less crowded carbon). For example,

$$H_3C-C\equiv C-CH_2CH_3 \xrightarrow[(ii)\ H_2O_2/\overset{\ominus}{O}H]{(i)\ B_2H_6/THF} CH_3\overset{O}{\overset{\|}{C}}CH_2CH_2CH_3 + CH_3CH_2\overset{O}{\overset{\|}{C}}CH_2CH_3$$

2-Pentanone (*major product*) — 3-Pentanone (*minor product*)

(*B*) *Terminal alkyenes produce aldehydes.* For example,

$$CH_3-C\equiv C-H \xrightarrow[THF]{B_2H_6} \left[H_3C-\overset{H}{\overset{|}{C}}=\overset{B<}{\overset{|}{C}}H\right] \xrightarrow{H_2O_2/\overset{\ominus}{O}H} CH_3-\overset{H}{\overset{|}{C}}=\overset{OH}{\overset{|}{C}}H \xrightleftharpoons[merism]{Tauto-} CH_3-CH_2-\overset{O}{\overset{||}{C}}-H$$

1-Propanal

Here, it may be added that intermediate vinylboranes still have a double bond. Therefore, there are chances of the addition of second molecule of diborane to the double bond which would create complications. To overcome this difficulty sterically hindered boranes such as disiamylborane [*bis* (1, 2-dimethylpropyl) borane] is used.

$$\begin{array}{c} H_3C-\overset{CH_3}{\overset{|}{C}H}-CH-CH_3 \\ \quad\quad\quad\quad \diagdown \\ \quad\quad\quad\quad\quad B-H \\ \quad\quad\quad\quad \diagup \\ H_3C-\underset{CH_3}{\underset{|}{C}H}-CH-CH_3 \end{array}$$

Another interesting feature of hydroboration-oxidation procedure is that, it is **complementary** to direct hydration of alkynes. Both processes lead to the formation of different products. Thus for example, direct hydration of propyne (a terminal alkyne) gives acetone (a methyl ketone) whereas hydroboration-oxidation of this compound gives rise to propanal.

$$\underset{\text{Acetone}}{CH_3-\overset{O}{\overset{||}{C}}-CH_3} \xleftarrow[Hg^{2+}]{\text{dil. } H_2SO_4} CH_3-C\equiv CH \xrightarrow[(ii)\ H_2O_2/\overset{\ominus}{O}H]{(i)\ B_2H_6/THF} \underset{\text{Propanal}}{CH_3-CH_2-CHO}$$

(***h***) **Addition of arsenic trichloride:** Acetylene, in presence of anhyd. $AlCl_3$ reacts with arsenic trichloride to give **Lewisite** which is a highly poisonous gas and was used in second world war.

$$HC\equiv CH + AsCl_3 \xrightarrow{\text{Anhyd. } AlCl_3} \underset{\text{Lewisite}}{ClHC=CHAsCl_2}$$

4. Nucleophilic Addition

Keeping in view the decreased reactivity of carbon-carbon triple bond towards electrophilic addition, it is quite reasonable to presume said that alkynes should be susceptible to nucleophilic attack. This indeed is so.

$$RC \equiv CR' + Nu-H \longrightarrow R-\overset{\displaystyle Nu}{\overset{|}{C}}=\overset{\displaystyle H}{\overset{|}{C}}-R'$$

Like electrophilic addition, the nucleophilic addition also proceeds in two steps:
***Step I*:** Addition of nucleophile to produce a vinyl-carbanion intermediate.

$$\overset{\ominus}{Y} + R-C\equiv C-R' \xrightarrow{\text{Slow}} (R)(Y)C=\overset{\ominus}{C}-R'$$

(Nucleophile)

***Step II*:** Abstraction of a proton from the solvent or the reagent.

$$(R)(Y)C=\overset{\ominus}{C}-R' + H-X \xrightarrow{\text{Fast}} (R)(Y)C=C(H)(R') + \overset{\ominus}{X}$$

Following are some of the examples of this type of addition:

***(a)* Addition of hydrogen cyanide:** Alkynes add hydrogen cyanide in presence of barium cyanide catalyst. Addition of HCN to acetylene, for instance, yields acrylonitrile.

$$H-C\equiv C-H + HCN \xrightarrow{Ba(CN)_2} \underset{\text{Acrylonitrile}}{H_2C=CH-CN}$$

Mechanism

The reaction is probably initiated by the attack of cyanide ion on a triply bonded carbon resulting in the formation of a vinyl-carbanion which in turn abstracts a proton from hydrogen cyanide.

$$\overset{\ominus}{CN} + H-C\equiv C-H \longrightarrow (CN)(H)C=\overset{\ominus}{C}-H \xrightarrow{H-CN} \underset{\text{Acrylonitrile}}{(NC)(H)C=C(H)(H)}$$

Acrylonitrile is extensively used for the manufacture of **Buna N**.

***(b)* Addition of methanol:** Methanol adds to alkynes in presence of catalytic amounts of sodium methoxide forming vinyl ethers.

$$HC\equiv CH + CH_3OH \xrightarrow{CH_3\overset{\ominus}{O}} \underset{\text{Methyl vinyl ether}}{H_2C=CHOCH_3}$$

Mechanism

This reaction is initiated by the nucleophilic attack of methoxide ion on acetylene.

$$CH_3\overset{\ominus}{O} \quad H-C\equiv C-H \longrightarrow \underset{H}{\overset{H_3CO}{>}}C=\overset{\ominus}{C}\underset{H}{<} \xrightarrow{H-OCH_3} \underset{H}{\overset{H_3CO}{>}}C=C\underset{H}{\overset{H}{<}} + \overset{\ominus}{O}CH_3$$

Methyl vinyl ether

This process of addition of compounds containing active hydrogen (methanol/base) to acetylene is also referred to as **vinylation**. The methyl vinyl ether is useful for the manufacture of polyvinyl ether plastics.

5. Oxidation

Alkynes are less reactive towards oxidising agents as compared to alkanes. Different products are obtained under different conditions. Some of the more important oxidation reactions of alkynes are discussed below:

***(i)* Aerial oxidation:** Combustion — In the presence of excess of oxygen, alkynes burn readily to produce carbon dioxide and water. This is accompanied by evolution of a large amount of heat and light. For example,

$$2HC\equiv CH + 5O_2 \longrightarrow 4\,CO_2 + 2H_2O \qquad \Delta H = -1300 \text{ kJ mol}^{-1}$$

As such acetylene burns with yellow luminous sooty flame due to less hydrogen content compared to carbon. But in excess of air, it burns with a blue flame. Due to high temperature of the order of 3000 K, this flame called as **oxyacetylene flame**, is used for cutting and welding of metals.

***(ii)* Oxidation with potassium permanganate:** Alkynes are oxidised by $KMnO_4$ giving rise to different products under different conditions, as discussed below:

***(a)* Oxidation with cold dilute $KMnO_4$:** Oxidation of alkynes with cold dilute alkaline solution of $KMnO_4$ produces 1,2-dicarbonyl compounds (1, 2-dialdehydes, 1, 2-diketones, 2-oxoacids, or 1, 2-dioic acids). The nature of the product depends upon the nature of alkyne. A noteworthy feature of this oxidation is that in terminal alkynes the $\equiv$CH unit is oxidised to $-\overset{O}{\overset{\|}{C}}-OH$ group. For example,

$$\underset{\substack{\text{Propyne}\\(\text{having}\equiv\text{CH unit})}}{CH_3-C\equiv CH} + 3[O] \xrightarrow[\substack{H_2O\\298\text{–}303\text{ K}}]{KMnO_4} \left[CH_3-\overset{O}{\overset{\|}{C}}-COOH\right] \xrightarrow{[O]} \underset{\text{Acetic acid}}{CH_3COOH} + CO_2$$

$$\underset{\substack{\text{2-Butyne}\\(\text{having}\equiv C-R\text{ unit})}}{CH_3-C\equiv C-CH_3} + 2[O] \xrightarrow[\substack{H_2O\\298\text{–}303\text{ K}}]{KMnO_4} \underset{\substack{\text{Butan-2, 3-dione}\\(\textit{Biacetyl})}}{CH_3-\overset{O}{\overset{\|}{C}}-\overset{O}{\overset{\|}{C}}-CH_3}$$

However, acetylene produces oxalic acid under these conditions. This is probably due to further oxidation of the initially formed glyoxal (1,2-dialdehyde).

$$\underset{\text{Acetylene}}{\begin{array}{c}CH\\|||\\CH\end{array}} + 2[O] \xrightarrow[\substack{H_2O\\298\text{–}303\ K}]{KMnO_4} \underset{\text{Ethan-1, 2-dial (\textit{Glyoxal})}}{\begin{array}{c}H\diagdown\\C{=}O\\|\\C{=}O\\H\diagup\end{array}} \xrightarrow{[O]} \underset{\text{Ethan-1, 2-dioic acid}}{\begin{array}{c}COOH\\|\\COOH\end{array}}$$

Note: As in the case of alkenes, the colour of $KMnO_4$ is discharged during this reaction. Hence, this reaction forms the basis of Baeyer's test for unsaturation.

***(b)* Oxidation with hot alkaline potassium permanganate:** Terminal alkynes when heated with $KMnO_4$ undergo $C \equiv C$ bond cleavage producing CO_2 and a carboxylic acid depending upon the nature of alkynes. In these cases the $\equiv CH$ moiety is oxidised to CO_2 and water and $R — C \equiv$ moiety is oxidized to RCOOH. In contrast non-terminal alkynes on oxidation produce only carboxylic acids (both moieties being $R — C \equiv$). Some examples are as follows:

(1) Oxidation of acetylene (both $\equiv CH$ units).

$$\begin{array}{c}CH\\|||\\CH\end{array} + 4[O] \xrightarrow[373\text{–}383\ K]{KMnO_4,\ KOH} \begin{array}{c}COOH\\|\\COOH\end{array} \xrightarrow{[O]} 2CO_2 + H_2O$$

(2) Oxidation of propyne (one $\equiv CH$ unit and other $\equiv CR$ unit).

$$CH_3 — C \equiv CH + 4[O] \xrightarrow[373\text{–}383\ K]{KMnO_4,\ KOH} \underset{\text{Acetic acid}}{CH_3COOH} + CO_2$$

(3) Oxidation of 2-butyne and 2-pentyne (both $\equiv CR$ units).

$$\underset{\text{2-Butyne}}{CH_3 — C \equiv C — CH_3} + 4[O] \xrightarrow[373\text{–}383\ K]{KMnO_4,\ KOH} \underset{\text{Acetic acid}}{CH_3COOH} + \underset{\text{Acetic acid}}{HOOCCH_3}$$

$$\underset{\text{2-Pentyne}}{CH_3CH_2 — C \equiv C — CH_3} + 4[O] \xrightarrow[373\text{–}383\ K]{KMnO_4,\ KOH} \underset{\text{Propanoic acid}}{CH_3CH_2COOH} + \underset{\text{Acetic acid}}{HOOCCH_3}$$

Thus on the basis of the structures of products formed during alkaline $KMnO_4$ oxidation, it is possible to determine the position of the triple bond in an alkyne molecule.

***(iii)* Oxidation with acidic potassium dichromate:** Alkynes are oxidised by acidic $K_2Cr_2O_7$ solution to give carboxylic acids. For example,

$$R-C\equiv CH + H_2O + [O] \xrightarrow[H^+]{K_2Cr_2O_7} \underset{(Unstable)}{R-\underset{OH}{\underset{|}{C}}=\underset{OH}{\underset{|}{CH}}} \longrightarrow \underset{\text{Carboxylic acid}}{RCH_2COOH}$$

$$\underset{\text{Acetylene}}{HC\equiv CH} + H_2O + [O] \xrightarrow[H^+]{K_2Cr_2O_7} \underset{\text{Acetic acid}}{CH_3COOH}$$

(*iv*) **Oxidation with selenium dioxide:** Alkynes on treatment with SeO_2 give 1,2-dicarbonyl compounds. For example,

$$R-C\equiv C-R \xrightarrow{SeO_2} \underset{\text{1, 2-Diketone}}{R-\overset{O}{\overset{\|}{C}}-\overset{O}{\overset{\|}{C}}-R}$$

$$\underset{\text{2-Butyne}}{CH_3-C\equiv C-CH_3} \xrightarrow{SeO_2} \underset{\text{Butandione}}{CH_3-\overset{O}{\overset{\|}{C}}-\overset{O}{\overset{\|}{C}}-CH_3}$$

$$CH\equiv CH \xrightarrow{SeO_2} \underset{\text{Glyoxal}}{CHO-CHO}$$

6. Polymerization

Alkynes undergo polymerization in two different ways, *i.e.*, cyclic and linear, depending upon the reaction conditions.

(*i*) **Cyclic polymerization:** At high temperatures, they form *cyclic polymers*. Acetylene and propyne, on passing through red hot iron tube, form benzene and mesitylene respectively,

$$3\,HC\equiv CH \xrightarrow[\text{Red hot Fe tube}]{773\text{ K}} \underset{\text{Benzene}}{C_6H_6}$$

$$3\,H_3C-C\equiv CH \xrightarrow[\text{Red hot Fe tube}]{773\text{ K}} \underset{\text{Mesitylene}}{C_6H_3(CH_3)_3}$$

In another mode of cyclization of acetylene, four molecules add together to form cyclooctatetraene in presence of a catalyst under pressure.

$$4\,HC{\equiv}CH \xrightarrow[\text{Ni (CN)}_2]{\text{Pressure}} \text{Cyclooctatetraene}$$

(Four HC≡CH molecules: HC≡CH, HC≡HC, CH≡CH, HC≡CH → ring: HC=CH, HC=HC, CH=CH, HC=CH)

Cyclooctatetraene

***(ii)* Linear polymerization:** Under suitable conditions alkynes form **linear polymers** also. Acetylene, for instance, on passing through a solution of cuprous chloride and ammonium chloride at 343 K forms vinylacetylene and divinylacetylene.

$$2HC{\equiv}CH \xrightarrow[NH_4Cl]{Cu_2Cl_2} \underset{\text{Vinylacetylene}}{H_2C{=}CH{-}C{\equiv}CH} \xrightarrow{C_2H_2} \underset{\text{Divinylacetylene}}{H_2C{=}CH{-}C{\equiv}C{-}CH{=}CH_2}$$

Vinylacetylene is a useful starting material for the manufacture of a highly solvent resistant synthetic rubber–**neoprene**. It adds to a molecule of hydrogen chloride in presence of aqueous solution of cupric chloride to form chloroprene (2-chloro-1, 3-butadiene) which readily polymerizes to neoprene.

$$\underset{\text{Vinylacetylene}}{H_2C{=}CH{-}C{\equiv}CH} + HCl \xrightarrow{\text{Mark. addition}} \underset{\text{Chloroprene}}{H_2C{=}\overset{\overset{Cl}{|}}{C}{-}CH{=}CH_2}$$

Acetylene polymerizes to give polyacetylene under optimum conditions. Thin films of this polymer are used as electrodes in batteries.

7. Rearrangement of Alkynes—Isomerization

Alkynes undergo rearrangement when heated with ethanolic potassium hydroxide. The triple bond generally migrates from *terminal position* to *non-terminal* positions of the chain. There is evidence to show that an allene is an intermediate. For example, each of the three compounds 1-pentyne, 1,2-pentadiene and 2-pentyne on heating with ethanolic potassium hydroxide is converted into an equilibrium mixture in which 2-pentyne predominates.

$$\underset{\substack{\text{1-Pentyne}\\2\%}}{H_3C{-}CH_2{-}CH_2{-}C_{\infty}H{\equiv}CH} \rightleftharpoons \underset{\substack{\text{1, 2-Pentadiene}\\3\%}}{H_3C{-}CH_2{-}CH{=}C{=}CH_2} \rightleftharpoons \underset{\substack{\text{2-Pentyne}\\95\%}}{H_3C{-}CH_2{-}C{\equiv}C{-}CH_3}$$

The following mechanism may be proposed for this rearrangement:

$$H_3C-CH_2-CH_2-C\equiv CH$$

1-Pentyne

$$\Big\downarrow EtOH \quad \Big\uparrow Et\overset{\ominus}{O}$$

$$\left[CH_3-CH_2-\overset{\ominus}{C}H-C\equiv CH \longleftrightarrow CH_3-CH_2-CH=C=\overset{\ominus}{C}H\right]$$

$$\Big\downarrow EtOH \quad \Big\uparrow Et\overset{\ominus}{O}$$

$$H_3C-CH_2-CH=C=CH_2$$

1, 2-Pentadiene (an allene) intermediate

$$\Big\uparrow\Big\downarrow$$

$$\left[CH_3-CH_2-C\equiv C-\overset{\ominus}{C}H_2 \longleftrightarrow H_3C-CH_2-\overset{\ominus}{C}=C=CH_2\right]$$

$$\Big\uparrow\Big\downarrow$$

$$CH_3-CH_2-C\equiv C-CH_3$$

2-Pentyne

Isomerization of 2-alkynes to 1-alkynes (and *vice-versa)* can also be carried out by heating alkyne with $NaNH_2$ in an inert solvent.

$$\underset{\text{2-Butyne}}{CH_3-C\equiv C-CH_3} \xrightarrow[\Delta]{NaNH_2 \text{ in inert solvent}} \underset{\text{1-Butyne}}{CH_3CH_2-C\equiv CH}$$

$$\underset{\text{1-Butyne}}{CH_3CH_2-C\equiv CH} \xrightarrow[\Delta]{NaNH_2 \text{ in inert solvent}} \underset{\text{2-Butyne}}{CH_3-C\equiv C-CH_3}$$

A BRIEF REVIEW

I. CYCLOALKENES

- **Cycloalkenes** are carbocyclic compounds having one double bond.
- **β-Elimination reactions** involving dehydration of cycloalkanols on heating with conc. H_2SO_4, H_3PO_4 or $KHSO_4$ or dehydrohalogenation of cycloalkyl halides on heating with alcoholic KOH solution give cycloalkenes where **Saytzeff rule** is always obeyed.

- **Diels-Alder reaction** gives cyclohexene.
- 1, 2-Disubstituted cycloalkenes on catalytic hydrogenation produce ***cis*-addition** products.
- ***Trans* –addition** products are obtained by addition of halogens or hypohalous acid to cycloalkenes. This addition involve *cyclic halonium* ion as an intermediate.
- Addition of hydrogen halides to cycloalkenes takes place through the formation of simple carbocations as intermediates. This reaction follows *Markownikoff's rule.*
- Addition of HBr in presence of peroxides follows ***anit-Markownikoff rule***. This is also called **peroxide effect or Kharasch effect.**
- *Cis*-1, 2-cycloalkanediols are obtained on *hydroxylation of cycloalkenes* with cold dilute aqueous alkaline $KMnO_4$ solution (***Baeyer's test***) or with OsO_4 followed by hydrolysis with aqueous. solution of $NaHSO_3$ or Na_2SO_3.
- **Hydration, hydroboration-oxidation and oxymercuration-reduction** of the cycloalkenes give alochols.
- **Ozonolysis** of cycloalkenes results in the rupture of the ring at the olefinic double bond producing dicarbonyl compounds.
- Cyclopropene to cycloheptene exist in the *cis*-form only. But cyclooctene shows *cis-trans* isomerism, where *cis*-isomer is more stable than *trans*-isomer. Further, *trans*-isomers being chiral molecules show *enantiomerism*.

II. DIENES

- **Dienes** contain two double bonds. Depending upon the relative positions of the two double bonds, dienes are classified as under:
  (*i*) **Isolated dienes** have two double bonds separated by more than one single bond.
  (*ii*) **Conjugated dienes** have two double bonds separated by one single bond.
  (*iii*) **Allenes** or **cumulenes** have both double bonds at the same carbon atom.
- Conjugated dienes are more stable than the isolated dienes, due to delocalization of π - electrons in the former. This delocalization makes $C_1 — C_2$ bond slightly longer (140 pm) than the normal $C = C$ bond length (134 pm) while $C_2 — C_3$ bond becomes slightly shorter (146 pm) than the normal $C — C$ bond length (154 pm).
- Conjugated dienes are obtained either by dehydration of 1, 3-alkandiols or by dehydrohalogenation of 1, 3-chloroalkanes.
- **Electrophilic addition reactions** of dienes lead to both 1, 2- and 1, 4-addition products. 1, 2-Addition products are formed at lower temperature while at room temperature 1,4-addition products are formed, *i.e.*, 1, 2-addition products are kinetically controlled and 1, 4-addition products are thermodynamically controlled.
- Dienes on catalytic hydrogenation give a mxiture of 1, 2-and 1, 4-addition products where 1, 4-*addition products predominate*.

- Dienes on chemical reduction with Na/alcohol give mainly 1, 4-addition products.
- **Diels-Alder reaction** is the thermal cycloaddition of diene with an alkene. In this reaction, the alkene is called dienophile. These reactions are favoured by the presence of electron –withdrawing groups in the dienophiles and by the presence of election donating groups in the dienes. These reactions are highly *stereospecific*.
- Diels-Alder reactions are *regioselective*.
- During free-radical polymerization of dienes 1, 4-addition predominates to give *trans*-polymer.
- Polymerization of isoprene under free radical conditions gives *trans*-polyisoprene but in presence of Ziegler-Natta catalyst the polymerization gives *cis*-polyisoprene.
- 1,3-Butadiene on copolymerization with styrene gives **Buna-S** and with acrylonitrile it gives **Buna-N** rubber.
- **Vulcanization** is heating of the raw rubber with 6–8 % sulphur. During this process, ***cross-linking*** of polymer chains occurs through their reactive allylic positions. It improves the tensile strength, resistance to abrasion and elasticity of these vulcanized rubbers.

III. ALKYNES

- **Alkynes or acetylenes** are unsaturated hydrocarbons containing a $C \equiv C$ bond and having general formula C_nH_{2n-2} where n = 2, 3, 4.....etc.
- **Acetylenes** are linear having *sp*-hybridized carbon atoms joined by a carbon-carbon triple bond.
- **Acetylene** is obtained by the action of water on CaC_2.
- **Higher alkynes** are obtained by alkylation of acetylene and other terminal alkynes.
- **Alkynes** are also obtained by the following reactions:

  (*i*) **Dehydrohalogenation of dihaloalkenes** with alcoholic KOH or with $NaNH_2$ in liquid ammonia.

  (*ii*) **Dehalogenation of tetrahalides** with Zn/CH_3OH.

  (*iii*) **Kolbe's electrolysis** of sodium maleate or fumarate.

  (*iv*) **Bertheiot's synthesis** of acetylene.

- Acetylene and other terminal alkynes are **acidic** in nature due to higher electronegativity of the *sp*-hybridized carbon atom from which the proton is lost. Such alkynes undergo following reactions:

  (*i*) On treatment with Na at 475 K or $NaNH_2$ in liquid NH_3 at 196 K corresponding acetylides are formed with the liberation of H_2 gas.

  (*ii*) On treatment with a **Tollen's reagent, white ppt.** of silver acetylide is obtained.

  (*iii*) On treatment with ammonical solution of Cu_2Cl_2, **red ppt**. of copper acetylide is obtained.

  (*iv*) On treatment with a suitable Grignard reagent, **alkynyl Grignard reagent** is formed.

- *Alkynes are less reactive than alkenes toward electrophilic addition reactions*. This is due to greater electronegativity of *sp*-hybridized carbon atom and greater stability of cylindrical electron cloud.
- Addition of halogens (but not of halogen acids) to alkynes takes place through a *cyclic halonium ion intermediate*.
- Acetylene and propyne on addition of HOCl give dichloroethanal and 1,1-dichloropropanone respectively.
- **Hydration of alkynes** with dilute H_2SO_4 in presence of Hg^{+2} gives methyl ketones. However, acetylene gives acetaldehyde.
- Addition of acetic acid in presence of Hg^{+2} to acetylene first gives vinyl acetate and then vinylidene acetate.
- **Lewisite** (a poisonous gas) is formed by addition of $AsCl_3$ in presence of $AlCl_3$ to acetylene.
- **Hydroboration oxidation** of terminal alkynes with disiamyl borane gives aldehyde but that of non-terminal alkynes with diborane gives a mixture of ketones where methyl ketone is formed preferentially.
- **Hydroboration – protonolysis** of non-terminal alkynes gives corresponding *cis*-alkene.
- **Ozonolysis** of alkynes gives ozonides which are hydrolysed to 1, 2-diketones. These diketones undergo oxidative cleavage to acids by H_2O_2 formed in *situ*.
- **Reductive ozonolysis** of alkynes gives diketones but that of acetaldehyde gives glyoxal.
- Alkynes undergo **nucleophilic addition reactions** due to higher electronegativity of the *sp*-hybridized carbon atoms.
- Nucleophilic addition of CH_3OH to acetylene at 433–473K in presence of potassium methoxide gives methyl vinyl ether.
- Nucleophilic addition of HCN to acetylene in presence of $Ba(CN)_2$ gives acrylonitrile.
- **Birch reduction** (Na in liq NH_3) of alkynes gives *trans*-alkenes.
- **Catalytic reduction** of alkynes gives *cis*-alkenes. The reduction can be stopped at alkene stage. **Lindlar's catalyst** (Pd supported over $CaCO_3$ or $BaSO_4$ and partially poisoned with S or quinoline) produces *cis*-alkenes.
- Oxidation of terminal alkynes with **Baeyer's reagent** (cold dilute $KMnO_4$ solution) produces a mixture of CO_2 and a carboxylic acid having one carbon atom less than the starting alkyne. However, acetylene gives oxalic acid.
- Oxidation of non-terminal alkynes with Baeyer's reagent produces 1, 2-diketones.
- Alkynes undergo polymerization under different conditions to give different important products. For example,

  (*i*) When passed through red hot iron tube, acetylene give benzene and propyne give mesitylene.

  (*ii*) In presence of $Ni(CN)_2$ and high pressure, acetylene gives cyclooctatetraene.

(*iii*) In presence of $CuCl/NH_4Cl$, acetylene first gives vinylacetylene and then divinylacetylene. Vinylacetylene is used for the manufacture of **neoprene**, which is a highly solvent-resistant synthetic rubber.

- Alkynes undergo isomerization on heating with $NaNH_2$ in an inert solvent or on heating with ethanolic potassium hydroxide.
- 1-Alkynes isomerize to 2-alkynes and *vice-versa*. The triple bond migrates from *terminal* position to *non-terminal* position preferentially involving an allene as an intermediate.

SOLVED PROBLEMS

Q. 1. A conjugated alkadiene having molecular formula $C_{13}H_{22}$ on ozonolysis yielded ethyl methyl ketone and cyclohexanecarbaldehyde. Identify the diene, write its structural formula and give its IUPAC name.

Ans. The structures of the products of ozonolysis are:

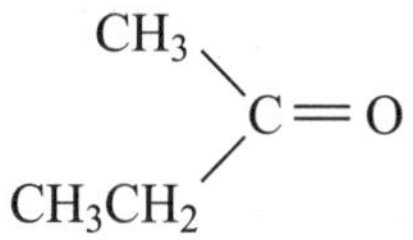

$$(CH_3)(CH_3CH_2)C{=}O \quad \text{and} \quad O{=}CH{-}C_6H_{11}$$

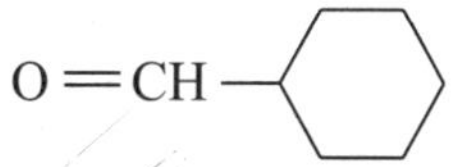

Ethyl methyl ketone (C_4H_8O) Cyclohexanecarbaldehyde ($C_7H_{12}O$)

Since, the total number of carbon atoms of the two products is 11(4 + 7), therefore, the ozonolysis must have also produced another two carbon product. Further, since the given compound ($C_{13}H_{22}$) is an alkadiene, therefore, this two carbon product must be glyoxal, $O = CH — CH = O$. The structure of the alkadiene is thus established by replacing the oxygen atoms from these three products by double bonds.

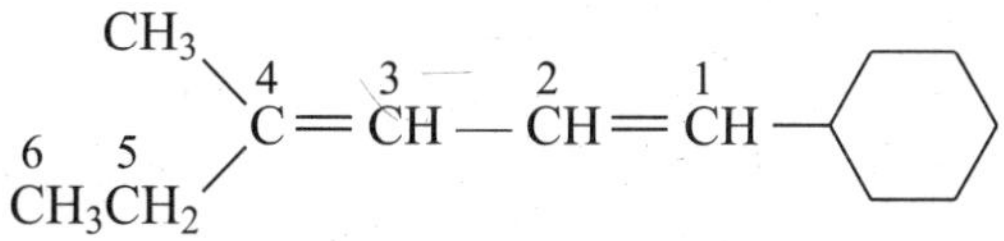

$$\overset{6}{C}H_3\overset{5}{C}H_2(CH_3)\overset{4}{C}{=}\overset{3}{C}H{-}\overset{2}{C}H{=}\overset{1}{C}H{-}C_6H_{11}$$

1-Cyclohexyl-4-methylhexa-1,3-diene

Q. 2. Write two important conformations of 1,3-butadiene. Which of the two is more stable and why?

Ans. Due to rotation about $C_2 — C_3$ single bond, an infinite number of conformations are theoretically possible but the following two conformations are very important. These are designated as *s-trans* and *s-cis* where the symbol *s* refers to the fact that the isomerism is with respect to single bond.

$$\overset{1}{CH_2}=\overset{2}{C}H-\overset{3}{C}H=\overset{4}{CH_2}$$

s-trans and *s-cis*

Steric hindrance

1, 3-Butadinene at ordinary temperatures contains, 95% of *s-trans* and only 5% of *s-cis* form thereby suggesting that *s-trans* form is more stable than the *s-cis* form probably due to steric repulsions between the H-atoms of the terminal carbon atoms which are very close to each other in the *s-cis* form. It may be noted here that the energy barrier between the *s-trans* and *s-cis* form is quite low, *i.e.*, 21 kJ mol^{-1}.

Q. 3. How will you convert acetylene into (*i*) 2-pentanone and (*ii*) 1, 3-butadiene?

Ans.

(*i*) $HC \equiv CH \xrightarrow[196\ K]{NaNH_2 \text{ in liq. } NH_3} \underset{\text{Sod. acetylide}}{HC \equiv \bar{C}\overset{+}{Na}} \xrightarrow[-NaI]{CH_3CH_2I} \underset{\text{1-Butyne}}{CH_3CH_2C \equiv CH} \xrightarrow[196\ K]{NaNH_2 \text{ in liq. } NH_3}$

$\underset{\text{Sod. butynide}}{CH_3CH_2C \equiv \bar{C}\ \overset{+}{Na}} \xrightarrow[-NaI]{CH_3-I} \underset{\text{2-Pentyne}}{CH_3-C \equiv C-CH_2CH_3} \xrightarrow[333\ K]{\text{Dil. } H_2SO_4,\ HgSO_4}$ 2-Pentanone

$$CH_3-\overset{\overset{O}{\|}}{C}-CH_2CH_2CH_3$$

(*ii*) $HC \equiv CH \xrightarrow[-CH_3OH]{CH_3\bar{O}\ \overset{+}{Na}} HC \equiv \bar{C}\overset{+}{Na} \xrightarrow{HCHO} \underset{\text{Prop-2-yn-1-ol}}{HC \equiv C-CH_2OH} \xrightarrow{CH_2O/CH_3ONa}$

$\underset{\text{But-2-yne-1, 4-diol}}{HOCH_2-C \equiv C-CH_2OH} \xrightarrow{H_2/Ni} HOCH_2-CH_2-CH_2-CH_2OH \xrightarrow[-2H_2O]{Al_2O_3,\ \Delta}$

$$\underset{\text{1, 3-Butadiene}}{CH_2=CH-CH=CH_2}$$

Q. 4. Acetylene reacts with dil. H_2SO_4 in presence of mercury salts to give acetaldehyde but with dil. HCl under similar conditions, it gives vinyl chloride. Explain.

Ans. First of all mercury ions form a complex (I) with acetylene. Since H_2O is more nucleophilic then SO_4^{2-} ion, it attacks the complex to form first vinyl alcohol which then tautomerises to give acetaldehyde.

$$\underset{\text{Acetylene}}{H-C\equiv C-H} + Hg^{2+} \longrightarrow \underset{\text{Complex (I)}}{H-C\equiv C-H \;(\text{with } Hg^{2+})} \xrightarrow{H_2O} \underset{\text{Vinyl alcohol}}{\left[H_2C=C(H)OH \right]}$$

$$\xrightarrow{\text{Tautomerises}} \underset{\text{Acetaldehyde}}{CH_3-\overset{O}{\overset{\|}{C}}-H}$$

In case of. HCl, since Cl^- -ion is more nucleophilic than H_2O, it reacts with complex (I) to form vinyl chloride.

$$\underset{\text{(I)}}{H-C\equiv C-H \;(\text{with } Hg^{2+})} \xrightarrow{Cl^-} \underset{H}{\overset{Cl}{>}}C=C\underset{H}{\overset{Hg^+}{<}} \xrightarrow{H^+} \underset{H}{\overset{Cl}{>}}\overset{\oplus}{C}-C\underset{H}{\overset{Hg^+}{<}}H \xrightarrow{-Hg^{2+}} \underset{\text{Vinyl chloride}}{\underset{H}{\overset{Cl}{>}}C=CH_2}$$

Q. 5. Why are conjugated dienes more reactive than alkenes or alkynes towards electrophilic addition reactions?

Ans. The relative reactivity of alkenes, alkynes and conjugated dienes depends upon the stability of the carbocations they generate on addition of an electrophile. Thus,

$$\underset{\text{Ethene}}{CH_2=CH_2} + \overset{+}{Br} \longrightarrow \underset{\text{I}}{\overset{\oplus}{C}H_2-CH_2-Br}$$

$$\underset{\text{Ethyne}}{CH\equiv CH} + \overset{+}{Br} \longrightarrow \underset{\text{II}}{\overset{\oplus}{C}H=CH-Br}$$

$$\underset{\text{Buta-1, 3-diene}}{CH_2=CH-CH=CH_2} + \overset{+}{Br} \longrightarrow \underset{\text{III}}{Br-CH_2-\overset{\oplus}{C}H-CH=CH_2} \longleftrightarrow Br-CH_2-CH=CH-\overset{\oplus}{C}H_2$$

Out of the three carbocations (I, II and III), the carbocation (III) obtained from buta-1, 3-diene is the most stable, since it is stabilised by resonance. Out of carbocations (I and II), carbocation (II) obtained from alkyne is less stable since the +ve charge is located on a more electronegative sp^2-hybridised carbon. Thus, the stability of carbocations follows the order: III > I > II. Therefore, the reactivity decreases in the same order, *i.e.*, conjugated diene > alkene > alkyne .

Q. 6. An organic compound (A), C_4H_6 absorbs two molecules of hydrogen on hydrogenation. It reacts with ammonical $AgNO_3$ solution to form a white precipitate. On treatment with dil. H_2SO_4 in presence of $HgSO_4$, it forms a ketone (B),C_4H_8O. Propose the structure of (A) and (B) and write the reactions involved.

Ans. (*i*) Since the compound (A), C_4H_6 on hydrogenation absorbs two molecules of H_2, therefore (A) can be either a cumulene, conjugated diene or an alkyene, *i.e.*,

$$\underset{\text{1, 2-Butadiene}}{CH_2{=}C{=}CHCH_3} \quad \underset{\text{1, 3-Butadiene}}{CH_2{=}CH{-}CH{=}CH_2} \quad \underset{\text{1-Butyne}}{CH_3CH_2{-}C{\equiv}CH} \text{ or } \underset{\text{2-Butyne}}{CH_3C{\equiv}CCH_3}$$

(*ii*) Since compound (A) forms white precipitate with $AgNO_3$ solution, therefore, (A) must be a terminal alkyne, *i.e.*, 1-butyne.

$$\underset{n\text{-Butane}}{CH_3CH_2CH_2CH_3} \xleftarrow{2H_2,\ Ni} \underset{\text{1-Butyne (A)}}{CH_3CH_2{-}C{\equiv}CH} \xrightarrow{AgNO_3/NH_4OH} \underset{\text{White ppt.}}{CH_3CH_2CH{\equiv}CAg\downarrow}$$

(*iii*) Since compound (A) on treatment with dil. H_2SO_4 in presence of $HgSO_4$, gives ketone (B) C_4H_8O, therefore, (B) must be 2-butanone.

$$\underset{\text{1-Butyne}}{CH_3CH_2{-}C{\equiv}CH} + H_2O \xrightarrow{HgSO_4/H_2SO_4} \underset{\text{2-Butanone}}{CH_3CH_2{-}\overset{\overset{O}{\|}}{C}{-}CH_3}$$

Q. 7. 3-Methyl-1-butyne cannot be prepared by alkylation of ethyne. Suggest a suitable method for its preparation.

Ans. Since 2° alkyl halides prefer to undergo elimination rather than substitution, therefore, 3-methyl-1-butyne cannot be prepared by alkylation of sod. acetylide with isopropyl bromide.

$$HC{\equiv}CH \xrightarrow[\text{196–200 K}]{NaNH_2,\ \text{liq. } NH_3} HC{\equiv}C^-Na^+ \xrightarrow[\times]{(CH_3)_2\,CH{-}Br} (CH_3)_2CH{-}C{\equiv}CH + NaBr$$

$$HC{\equiv}C^-Na^+ \xrightarrow{(CH_3)_2\,CH{-}Br} CH_3CH=CH_2 + HC{\equiv}CH + NaBr$$

It can, however, be prepared as follows:

$$HC{\equiv}CH \xrightarrow[(ii)\ CH_3CH_2Br]{(i)\ NaNH_2,\ liq.\ NH_3} CH_3CH_2-C{\equiv}CH \xrightarrow[-2C_4H_{10}]{n\text{-BuLi (2 equiv.)}}$$

$$\underset{\text{Dianion}}{CH_3-\overset{\ominus}{\ddot{C}}H-C{\equiv}C\!:^{\ominus}} \xrightarrow{CH_3-I\ (1\ equiv.)} CH_3-\overset{CH_3}{\overset{|}{C}H}-C{\equiv}C\!:^{\ominus} \xrightarrow{H_2O} CH_3-\overset{CH_3}{\overset{|}{C}H}-C{\equiv}CH$$

Since *n*-BuLi is a very strong base, it not only abstracts more acidic acetylenic hydrogen but also less acidic propargylic hydrogen (hydrogen on the carbon atom next to the triple bond) to form a dianion. When this dianion is treated with one equiv. of CH_3I, alkylation occcurs exclusively at the more reactive propargylic position, and not at the terminal alkyne position leading to the formation of 3-methyl-1-butyne.

Q. 8. How will you distinguish between buta-1, 3-diene and but –1-yne?

Ans. But -1-yne is a terminal alkyne and hence can be distinguished from buta-1, 3-diene by Tollen's reagent which forms a white ppt. with but-1-yne and not with buta-1, 3-diene.

$$\underset{\text{But-1-yne}}{CH_3-CH_2-C{\equiv}CH} + \underset{\text{Tollen's reagent}}{[Ag(NH_3)_2]^+\ OH^-} \longrightarrow \underset{\text{Silver but-1-ynide (white ppt.)}}{CH_3CH_2-C{\equiv}C\,Ag} + 2NH_3 + H_2O$$

$$\underset{\text{Buta-1, 3-diene}}{CH_2{=}CH-CH{=}CH_2} + [Ag(NH_3)_2]^+\ OH^- \longrightarrow \text{No reaction}$$

Q. 9. (*a*) Write the structures of all possible carbocation intermediates in the addition of one equivalent of HI to 2, 4-hexadiene.

(*b*) Which carbocation would you expect to be formed at a faster rate?

Ans.

(*a*)

$$CH_3CH{=}CH-CH{=}CHCH_3$$

$$\Big\downarrow \overset{\oplus}{H}$$

$$\underset{\text{(I)}}{[CH_3CH_2-\overset{\oplus}{C}H-CH{=}CHCH_3]} \text{ or } \underset{\text{(II)}}{[CH_3\overset{\oplus}{C}H-CH_2-CH{=}CHCH_3]}$$

(Addition of $\overset{\oplus}{H}$ to the other double bond gives identical intermediate).

(*b*) The first carbocation (I) would be formed at a faster rate because it is resonance stabilized, allylic carbocation.

UNSOLVED PROBLEMS

1. What are dienes? How are they classified? Give examples.

2. Draw the structures of all the isomeric alkynes having molecular formula C_5H_{10} and give their common and IUPAC names.

3. Discuss the methods of formation of conjugated dienes.

4. Discuss briefly the methods of preparation of alkynes.

5. What happens when 2, 3-dichloropropene is heated with Zn dust?

6. Discuss the reactions of conjugated dienes.

7. Why are conjugated dienes more stable than isolated dienes?

8. Draw molecular orbital picture of alkenes and 1,3-butadiene.

9. Explain why 1,3-butadienes undergoes both 1,2-and 1,4-addition reactions?

10. Discuss briefly electrophilic addition reactions of alkynes and their mechanism.

11. Write notes on the following:

(*i*) Isomerization reactions of alkynes.

(*ii*) Oxidation and reduction reactions of alkynes.

12. Discuss the polymerization of alkynes with special reference to acetylene.

13. Alkynes participate in both electrophilic and nucleophilic addition reactions. Discuss.

14. (*i*) What are alkynes? Why are they called acetylenes?

(*ii*) Why are alkynes linear in nature?

15. (*i*) Discuss the orbital structure of acetylene.

(*ii*) How does the orbital picture of triple bond show the absence of geometrical isomerism?

16. Write the IUPAC names of the following:

(*i*) $HC = C — CH_2 — CH = CH_2$

(*ii*) $CH_3CH = CH — C \equiv C — C \equiv CH$

(*iii*) $(CH_3)_2CH — C \equiv C — CH(CH_3)_2$

(*iv*) $HC \equiv C — CH_2CH_2Br$

17. Discuss briefly the preparation of higher alkynes from acetylene.

18. Discuss the preparation of alkynes from 1, 2-dihaloalkanes. Why is sodium in liquid ammonia perferred over alcoholic potassium hydroxide in these reactions?

19. Discuss the physical properties of dienes and alkynes.

20. (*i*) How will you explain acidic nature of C — H bond in acetylene?

(*ii*) Do all alkynes show acidic nature? Explain.

21. What is the reason for the low reactivity of alkynes towards electrophilic addition reactions?

22. How will you account for the fact that alkynes undergo nucleophilic addition reactions but alkenes do not?

23. What happens when acetylene is passed through dilute sulphuric acid containing mercuric sulphate at 333 K? Write the mechanism of the reaction.

24. Predict the products of the following reactions:

(*i*) 1, 3-Butadiene + Maleic anhydride $\longrightarrow$ (*ii*) 1, 3-Butadiene $\xrightarrow{\text{Peroxide}}$

(*iii*) $HC{\equiv}CH + C_2H_5OH \xrightarrow{\text{KOH}}$ (*iv*) $CH_3{-}C{\equiv}C{-}H + H_2O \xrightarrow[H_2SO_4]{HgSO_4}$

(*v*) $HC{\equiv}CH + H_2O \xrightarrow[H_2SO_4]{HgSO_4}$ (*vi*) $HC{\equiv}CH \xrightarrow{\text{Iron tube, 773K}}$

(*vii*) $HC{\equiv}CH \xrightarrow{CuCl / NH_4Cl}$ (*viii*) $HC{\equiv}CH \xrightarrow{Zn / CH_3COOH}$

25. Give the mechanism of the following reaction:

$$Ph_2\underset{\substack{|\\OH}}{C}{-}C{\equiv}CH \xrightarrow{H^+} Ph_2C{=}CHCHO$$

26. Write a short note on Diels-Alder reaction.

27. Give the reaction with conditions for preparing the following from ethyne:

(*i*) Acetaldehyde (*ii*) Cyclooctatetraene (*iii*) Oxalic acid (*iv*) Ag_2C_2

28. What happens when (Give chemical equations).

(*i*) 1, 3-Butadiene is treated with $BrCCl_3$ in presence of peroxide.

(*ii*) Ethyne reacts with ammonical silver nitrate solution.

(*iii*) Ethyne reacts with ammonical cuprous chloride solution.

(*iv*) Ethyne is treated with (*a*) CH_3COOH (*b*) HCN and (*c*) HCl.

29. Starting from acetylene, how will you prepare:

(*i*) Ethylidene chloride (*ii*) Vinyl acetate (*iii*) Ethylene (*iv*) Ag_2C_2 (*v*) Benzene

(*vi*) Methyl vinyl ether ?

30. Compare the reduction of alkynes with Lindlar' catalyst and with alkali metals in liquid ammonia.

31. Discuss briefly hydroboration-oxidation of alkynes. In what respect, does it differ from hydration reaction ?

32. Discuss the mechanism of isomerization of 1-alkynes to 2-alkynes by heating with alcoholic potassium hydroxide solution.

33. Write the products of the following reactions:

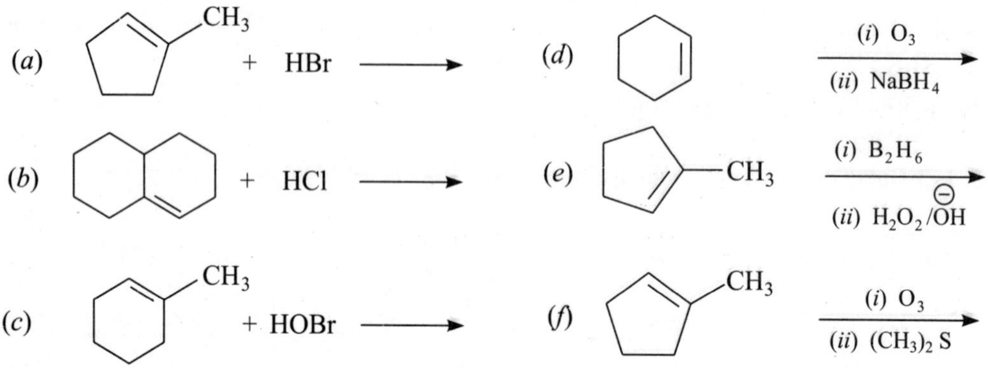

(*a*) 1-methylcyclopentene + HBr ⟶

(*b*) [bicyclic alkene] + HCl ⟶

(*c*) 1-methylcyclohexene + HOBr ⟶

(*d*) cyclohexene $\xrightarrow[(ii)\ NaBH_4]{(i)\ O_3}$

(*e*) 1-methylcyclopentene (CH_3) $\xrightarrow[(ii)\ H_2O_2/\overset{\ominus}{O}H]{(i)\ B_2H_6}$

(*f*) 1-methylcyclopentene (CH_3) $\xrightarrow[(ii)\ (CH_3)_2S]{(i)\ O_3}$

34. Write the structure of alkenes that would yield the following alcohol (I) by hydroboration-oxidation reaction.

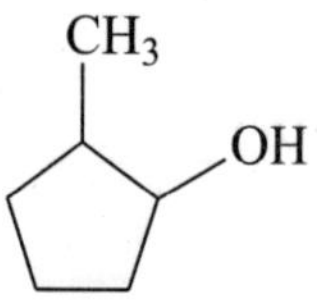

2-methylcyclopentanol (CH_3, OH)

I

35. (*a*) Predict the isomer of 1, 2-dimethylcyclohexane that is produced when 1, 2-dimethylcyclohexene is treated with hydrogen in presence of finely divided platinum.

(*b*) What isomer would you expect to obtain from the reactions of cyclohexene and deuterium in presence of finely divided platinum ?

36. Write chemical equations for the following:

(*i*) Propyne reacts with alkaline $KMnO_4$ at 298 K.

(*ii*) 1,3-Butadiene reacts with HBr at 313 K.

(*iii*) Ozonolysis of cyclohexene.

37. Explain the following:

(*a*) Acetylene is more acidic than ethene.

(*b*) 1-Alkynes are acidic in nature.

(*c*) Stereoselective reduction of alkynes.

38. How will you distinguish between the following pairs:

(*a*) Acetylene and ethylene (*b*) Propyne and propene.

(*c*) 1-Heptyne and 2-Heptyne (*d*) Methylacetylene and dimethylacetylene

39. When allene reacts with lithium amide, $LiNH_2$ in liquid ammonia and the reaction mixture is treated with D_2O, $CH_3—C\equiv C—D$ is produced. Provide a mechanism for this conversion.

40. An unknown compound 'A' molecular formula C_5H_8 takes up two moles of chlorine, decolourises a solution of bromine in CCl_4 and yields no precipitate with ammonical silver nitrate solution. Treatment of A with water in presence of H_2SO_4 and $HgSO_4$ catalyst yields an equimolar mixture of 2-pentanone and 3-pentanone. Deduce the structure of A.

41. What are the products of ozonolysis of acetylene? Suggest an appropriate mechanism for their formation.

42. Do alkynes exhibit geometrical isomerism? Give reasons for your answer.

43. Write the product for each of the following reactions:

(*a*) $CH_3-C\equiv CH \xrightarrow[(ii)\ H_2O/H^{\oplus}]{(i)\ O_3}$

(*b*) $C_4H_9C\equiv CC_4H_9 \xrightarrow{Na/NH_3}$

(*c*) $CH_3CH_2C\equiv CH \xrightarrow[(ii)\ H_2O_2/OH^{\ominus}]{(i)\ BH_3}$

(*d*) $C_6H_5\underset{\large Br}{\underset{|}{C}}H-\underset{\large Br}{\underset{|}{C}}HC_6H_5 \xrightarrow[\Delta]{alc,\ KOH}$

(*e*) $CH_3CH_2CH_2C\equiv CH \xrightarrow{Li,\ NH_3}$

44. A hydrocarbon has the formula corresponding to that of an alkene, C_nH_{2n-2} but it adds only one molecule of bromine per mole of the hydrocarbon. What type of structure does the hydrocarbon have?

45. An organic compounds (A), C_6H_{10}, on reduction first gives (B), C_6H_{12}, and finally (C) C_6H_{14}. (A) on ozonolysis followed by reductive hydrolysis gives two aldehydes (D) C_2H_4O and (E) $C_2H_2O_2$. Oxidation of (B) with acidified $KMnO_4$ gives the acid (F) $C_4H_8O_2$. Determine the structures of the compounds (A) to (F) with proper reasoning.

46. 1, 3-Butadiene is treated with one equivalent of Br_2 at 123 K. Two structural isomers are obtained: 46% A and 54% B. When the reaction is carried out at 333 K, the product mixture contains 90% A. What are the structures of A and B.

47. Explain the following:

(*i*) Mechanism of Diels-Alder reaction.

(*ii*) Why Diels-Alder reaction is referred to the [4+2] cyclo-addition reaction.

(*iii*) Mechanism of free radical addition to conjugated dienes.

(*iv*) Mechanism of the conversion of 1, 3-butadiene into polybutadiene.

(*v*) 1, 3-Pentadiene is more stable than 1, 4-pentadiene.

48. (*a*) Explain why does 1, 3-butadiene undergo both electrophilic and nucleophilic addition reactions.

(*b*) Predict the product of the addition of one equivalent of bromine to the following dienes:

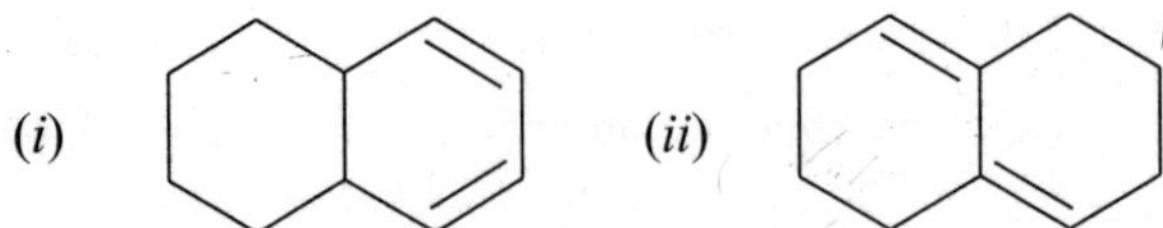

49. (*a*) Discuss the configurations of cycloalkenes by taking suitable example.

(*b*) Write chemical equations for the following:

(*i*) 2-Butyne reacts with diborane followed by alkaline H_2O_2.

(*ii*) 1, 3-Butadiene reacts with acrolein at 373 K.

(*c*) Write structures of the products expected from the following two reactions indicating the most stable conformations:

(*i*) $—C(CH_3)_3$ $\xrightarrow[\text{dark}]{\text{HBr}}$

(*ii*) CH_3 $\xrightarrow{\text{HOBr}}$

(*iii*) CH_3 $\xrightarrow[2.\ H_2O_2/\overset{\ominus}{O}H]{1.\ B_2H_6}$

50. Discuss the chemical reactions of cycloalkenes with particular emphasis on the mechanistic course. In what way the conformation of cyclopentene differs from that of cyclohexene?

51. What is the state of hybridization of carbons in butadiene? Draw the molecular orbital picture of butadiene and show the relative energies of various molecular orbitals. Which of the two conformations of butadiene is more stable?

52. (*i*) With suitable examples, discuss the mechanism of electrophilic addition to alkynes.

(*ii*) Write structures of the products of following reactions indicating the stereochemistry wherever possible:

(*a*) $CH_3C \equiv CH \xrightarrow{\text{HF}}$

(*b*) $C_6H_5C \equiv CH \xrightarrow{\text{NaNH}_2}$

(*c*) $HC \equiv CCH_2CH_3 \xrightarrow[(ii)\ CH_3CHBrCH_3]{(i)\ \overset{\ominus}{N}H_2}$

(*d*) $RC \equiv CH \xrightarrow[H_2O]{Hg^{2+},\ H^+}$

(*e*) $CH_3CH_2CH_2C \equiv CH \xrightarrow{\text{ethanolic KOH, }\Delta}$

CHAPTER 8

Arenes-I
Benzene and Aromaticity

During the nineteenth century, a large number of organic compounds, isolated from balsams, resins etc., were termed as ***'aromatic'*** since many of them had *fragrant odour* (*aroma* = fragrant odour). These compounds differ widely from the aliphatic compounds in their chemical behaviour and all of them have been shown to be closely related to the parent monocyclic hydrocarbon, ***benzene*** (C_6H_6). This hydrocarbon though expected to be highly unsaturated, shows a high degree of stability. Benzene and all the aromatic compounds show a tendency to undergo *substitution* rather than *addition* reactions. In all such reactions the parent structure called benzene nucleus is retained. Thus, aromatic compounds are also termed as ***benzenoid*** compounds. The term aromatic, however, also includes another class of compounds possessing aromatic character but having structures that differ considerably from that of benzene. Such compounds are termed as ***non-benzenoid aromatics*** and include compounds such as azulene, tropolone, ferrocene, etc.

8.1 NOMENCLATURE OF BENZENE DERIVATIVES

IUPAC System

Systematic nomenclature of benzene derivatives is based upon the fact that all the benzenoid compounds are named as *substituted benzenes*. The IUPAC names of various substituted benzenes are discussed below:

(*i*) Monosubstituted Benzene Derivatives

(*a*) The name of a monosubstituted benzene derivative is written by prefixing the name of the substituent before the root name benzene. For instance,

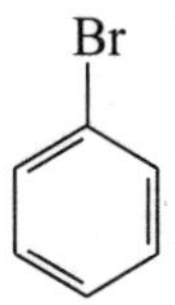

Br — Bromobenzene

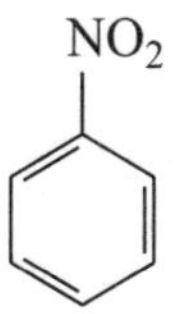

NO_2 — Nitrobenzene

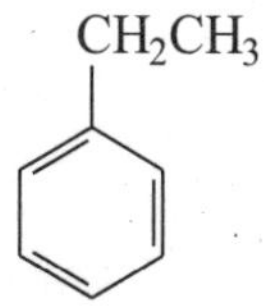

CH_2CH_3 — Ethylbenzene

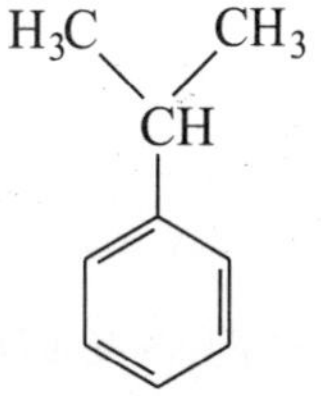

H_3C CH_3 CH — Isopropylbenzene

(*b*) Some compounds still retain their common names. Names of certain compounds based upon the history of their discovery are still in use. For instance, the name *aniline* is derived from the word *anil* (meaning *indigo*) as this compound was first obtained by the distillation of indigo. The common names of some of the important compounds are:

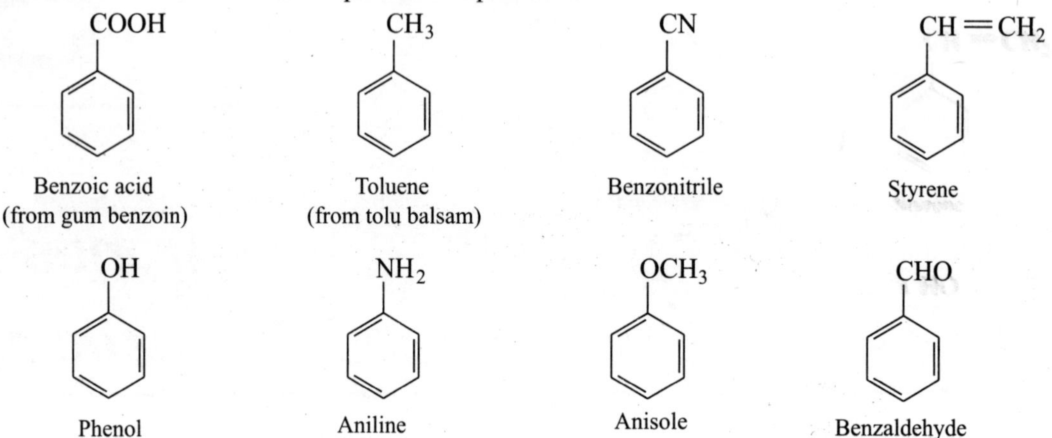

Benzoic acid (from gum benzoin) | Toluene (from tolu balsam) | Benzonitrile | Styrene

Phenol | Aniline | Anisole | Benzaldehyde

(*ii*) Disubstituted Benzene Derivatives

(*a*) *Disubstituted benzene derivatives having similar substituents*

In the case of disubstituted benzene derivatives, three *positional isomers* are possible. These are called *ortho* (or 1,2-) , *meta* (or 1,3-) and *para* (or 1,4-) isomers, respectively, signifying the relative positions of the substituents. For example,

ortho-Isomer | *meta*-Isomer | *para*-Isomer

In compounds having two similar substituents the prefix *di* is used before the name of the substituent and the relative positions of the substituents are indicated by putting the prefixes *o*-(for *ortho*), *m*-(for *meta*) and *p*-(for *para*) before the names of the substituents to designate 1,2-, 1,3- and 1,4-positions, respectively. Sometimes, numbers are also used for designating the positions of the substituents. Thus,

o-Dibromobenzene or 1,2-Dibromobenzene | *m*-Dinitrobenzene or 1,3-Dinitrobenzene | *p*-Dichlorobenzene or 1,4-Dichlorobenzene

(*b*) *Disubstituted benzene derivatives having dissimilar substituents*

(*i*) In case of disubstituted benzene derivatives having two *dissimilar substituents*, if one of the substituent gives a common name to the molecule, it is generally named as a derivative of that molecule (the root name). For instance,

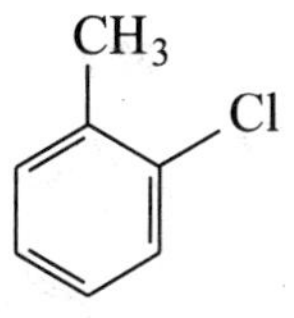

CH_3 Cl OH COOH NH_2 NO_2

o-Chlorotoluene
(Root name, toluene)

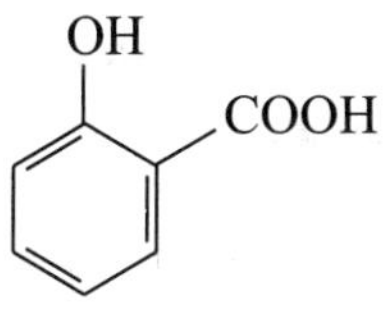

o- Hydroxybenzoic acid
(Salicyclic acid)
(Root name, benzoic acid)

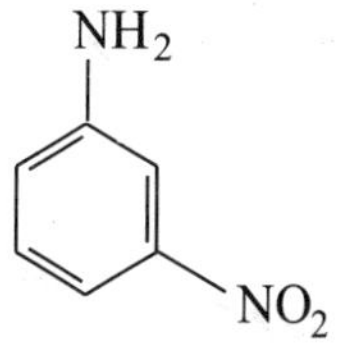

m-Nitroaniline
(Root name, aniline)

(*ii*) In other cases, the substituents are arranged in *alphabetical order.*

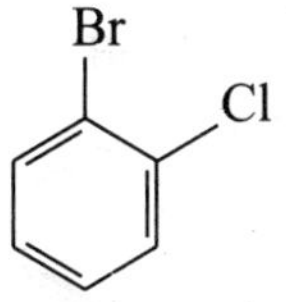

Br Cl NO_2 F $CH_2—CH_3$ NO_2

m-Bromochlorobenzene
or
1-*Bromo*-3-chlorobenzene

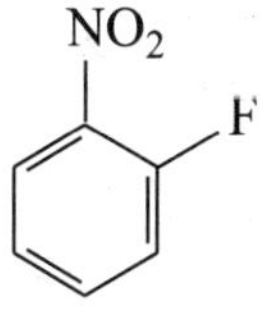

o-Fluoronitrobenzene
or
1-*Fluoro*-2-nitrobenzene

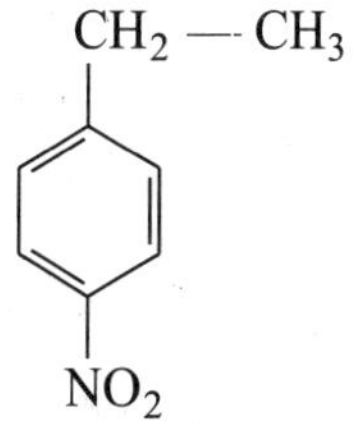

p-Ethylnitrobenzene
or
1-*Ethyl*-4-nitrobenzene

(*iii*) Polysubstituted Benzene Derivatives

(*a*) *For polysubstituted benzene derivatives having similar substituents,* the carbon atoms of the ring are numbered in such a way that the substituents get the least possible numbers. The positions of these substituents are indicated by the numbers.

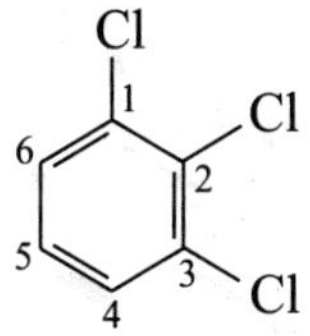

Cl 1 Cl 2 Cl 3 4 5 6 CH_3 1 2 3 CH_3 4 5 H_3C 6 H_3C 6 CH_3 1 2 CH_3 3 4 H_3C 5

1,2,3-Trichlorobenzene

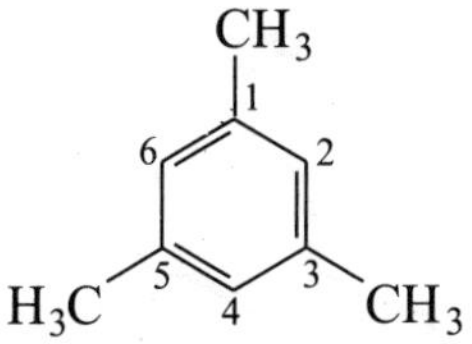

1,3,5-Trimethylbenzene
(common name, mesitylene)

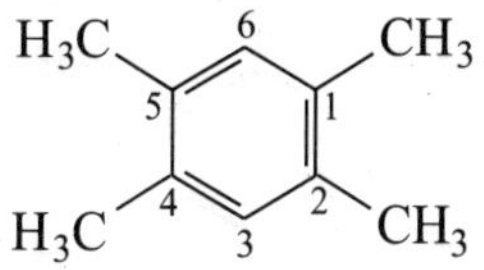

1,2,4,5-Tetramethylbenzene
(common name, durene)

(*b*) *If the substituents are different*, the numbering of ring carbon atoms commences with the substituent coming last in the alphabetical order.

Chapter 8

2-Bromo-4-chloro-6-iodonitrobenzene

(*c*) If one of the substituents gives a special name to the monosubstituted derivative then the numbering of the ring carbon atoms is commenced with that substituent.

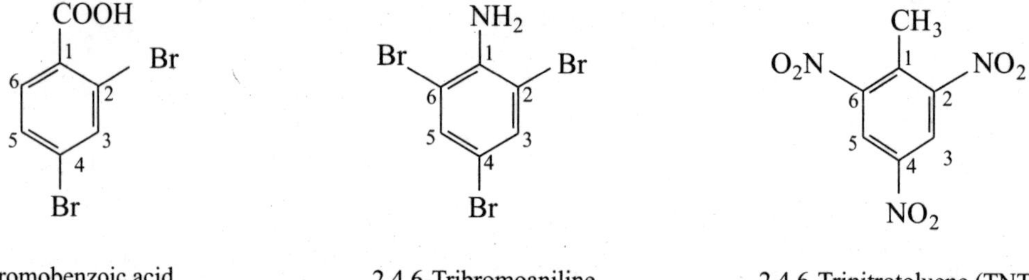

2,4-Dibromobenzoic acid (Root name benzoic acid)

2,4,6-Tribromoaniline (Root name aniline)

2,4,6-Trinitrotoluene (TNT) (Root name toluene)

8.2 THE ARYL GROUP

The univalent group obtained by the removal of one hydrogen atom from benzene nucleus is called ***phenyl*** (C_6H_5–, Ph–, or ϕ) group, and the univalent group derived from any other aromatic hydrocarbon in this manner is known as ***aryl group*** and is symbolized as **Ar**. Removal of two hydrogen atoms from the benzene nucleus gives rise to bivalent ***phenylene*** group. Some of the common aryl groups are:

$C_6H_5 \equiv$

Phenyl   *o*-Tolyl   *m*-Tolyl   *p*-Tolyl

o-Phenylene   *m*-Phenylene   *p*-Phenylene   Benzo   Benzal   Benzyl

8.3 AROMATIC NUCLEUS AND SIDE CHAIN

The basic carbon skeleton of the benzene ring or any other aromatic compound is called the *aromatic nucleus*. An alkyl group attached to the ring by the replacement of a hydrogen atom is termed as the *side chain*. Thus, in toluene, methyl group constitutes the side chain.

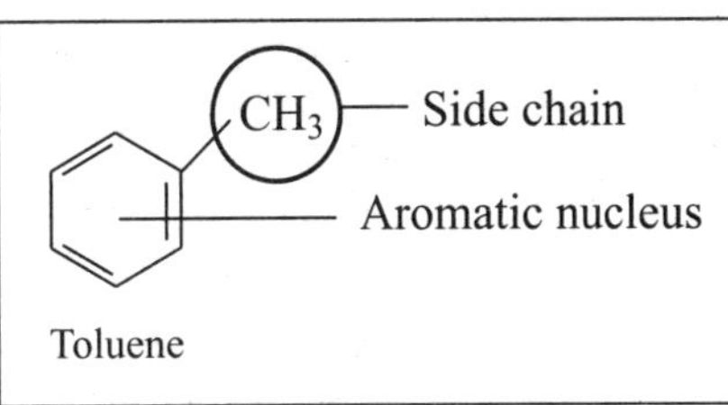

8.4 STRUCTURE OF BENZENE: MOLECULAR FORMULA AND KEKULÈ STRUCTURE

From elemental analysis and molecular weight determination, the molecular formula of benzene has been found to be C_6H_6. Since, the carbon to hydrogen ratio in this compound is much less than the corresponding alkane (C_6H_{14}), it is expected to be highly unsaturated. In other words, it is expected to undergo addition reactions like alkenes and alkynes. Following reactions show the presence of three double bonds and a six-membered carbocyclic ring:

(*i*) It adds six atoms of halogens forming benzene hexahalides under suitable conditions.

$$C_6H_6 + 3X_2 \longrightarrow C_6H_6X_6$$

(*ii*) It forms a triozonide on ozonolysis.

$$C_6H_6 + 3O_2 \longrightarrow \underset{\text{Triozonide}}{C_6H_6O_9}$$

(*iii*) It can be hydrogenated catalytically, adding a maximum of six hydrogen atoms to yield cyclohexane.

$$C_6H_6 + 3H_2 \xrightarrow{\text{Catalyst}} \underset{\text{Cyclohexane}}{C_6H_{12}}$$

Further studies have shown that the three double bonds of benzene are different from the ordinary double bonds and have a special type of arrangement which is responsible for **'unusual'** stability of this molecule. The high degree of stability of the benzene molecule is revealed from the following observations:

(i) Benzene remains unaffected on treatment with potassium permanganate under usual conditions, i.e., benzene resists oxidation like saturated compounds.

(ii) Halogen acids fail to add to the molecule of benzene.

(iii) In the absence of sunlight and in the presence of some suitable catalyst, it undergoes electrophilic substitution rather than addition reactions. Some of the important reactions of benzene are nitration, sulphonation, halogenation and Friedel-Crafts reactions. Further, it forms only one monosubstitution and three disubstitution products.

* Kekulè got this intuition from a dream. He was writing his textbook and the work did not progress. He started thinking and at last he turned his chair to fire and dozed. He dreamed that a few snakes were dancing mockingly before him. Then all of a sudden he saw that one of the snakes caught hold of the tail of the other. The dance went on and finally he saw that one snake had seized hold of its own tail and the form whirled mockingly before his eyes. Then as if with a flash of lightening, he woke up and got the idea that carbon atoms can form long chains and that they could also unite in such a fashion so as to form rings. This also led him to propose two structures for benzene having alternating double bonds.

Chapter 8

8.4.1 Kekulès Structure

To accommodate all the observations F.A. Kekulè, in 1865, proposed formula I based upon intuition*. This is known as Kekulè formula and it suggests the presence of a six-membered symmetrical ring having alternate single and double bonds.

Objections to Kekulè structure. There are three objections.

(i) Position of double bonds

Kekulè structure accounts for the observations regarding addition reactions, and the fact that it forms only one monosubstitution product. Structure (I) could explain nearly all the observed properties of benzene. But a serious drawback of this structure was that it would predict four (II to V) rather than the observed three isomeric disubstitution products for $C_6H_4X_2$.

A closer scrutiny of these formulas reveals that structures II and III represent two *o*-disubstituted structural isomers which differ in the position of substituents relative to double bonds. This would make the existence of four isomeric disubstituted benzene derivatives possible. But there are only three isomers found so far. To overcome this objection, Kekulè himself suggested that structure I of benzene is in dynamic equilibrium with another structure VI which differs from the former in the position of the double bonds.

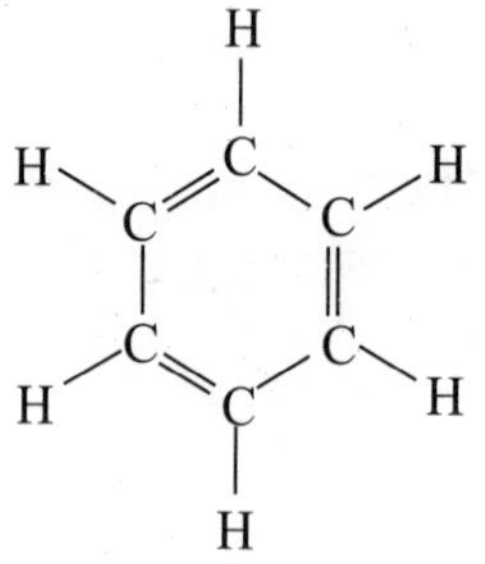

I 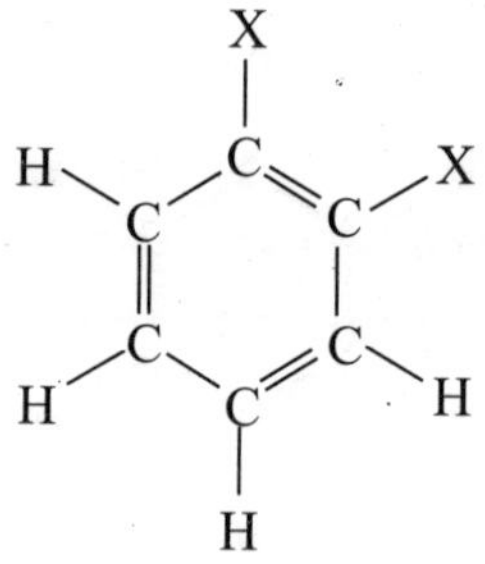 II

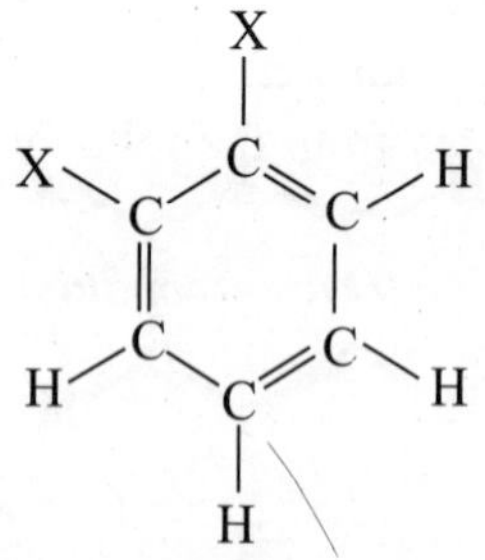

III 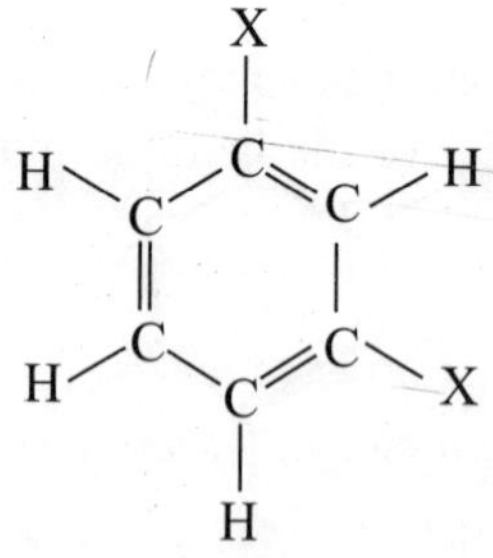 IV 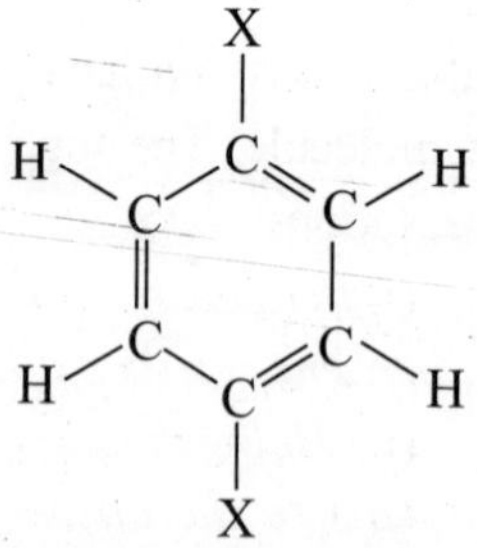 V

As the forms I and VI are indistinguishable, the forms II and III should also not be separable. Kekulè got the idea of **'oscillation'** of double bonds *i.e.*, position of single and double bonds are not fixed but interchange compared to arrangement seen in the dream in which Kekulè saw the ring form of the snakes **"whirling mockingly"** before his eyes.

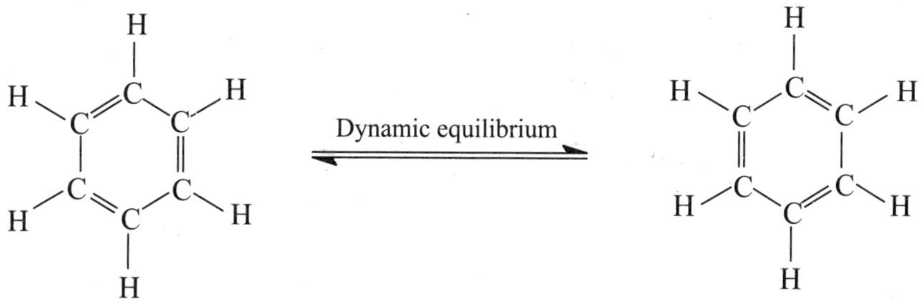

Dynamic equilibrium

I VI

OR

Kekulè had predicted our present concept of electron delocalization based purely on intuition without knowing anything about it. In fact even now the two "Kekulè forms" of benzene are written though in a different context.

Alternative Structure of Benzene

A number of structures, as shown in Fig 8.1,were proposed from time to time to account for all the observed properties of benzene. However, none of them could satisfy all the observed properties of benzene and were consequently rejected.

Structures VII to XI were rejected because they should all undergo addition reactions when treated with halogens or hydrogen halides at room temperature or should be oxidised with common reagents like $KMnO_4$, HNO_3, etc., at moderate temperature.

Dewar suggested that benzene should have structure XII but it was rejected on the basis of the fact that it was not symmetrical. It is worthwhile to mention here that this molecule now known as **Dewar benzene** has actually been synthesized in 1863 by two American workers, Van Tamelen and Pappas, and has been found to be widely different from benzene.

Claus proposed his famous **diagonal structure** XIII in 1868. But it was also rejected on the ground that it did not account for the number of isomers possible for mono and di-substituted benzenes.

Armstrong and Baeyer proposed the **centric formula** XIV also in 1868. The main feature of this formula is that fourth valency of each carbon is directed towards the centre of a hexagon. These valencies balance each other and are in equilibrium. When benzene is reduced to dihydrobenzene, this equilibrium is disturbed and the resulting product has typical olefinic bonds. However, no satisfactory explanation for the behaviour of polynuclear hydrocarbons could be given on the basis of this formula, so this was also rejected.

$$CH_3-C\equiv C-C\equiv C-CH_3 \qquad CH_2=CH-C\equiv C-CH=CH_2$$

VII VIII

$$HC\equiv C-C\equiv C-CH_2-CH_3$$

IX

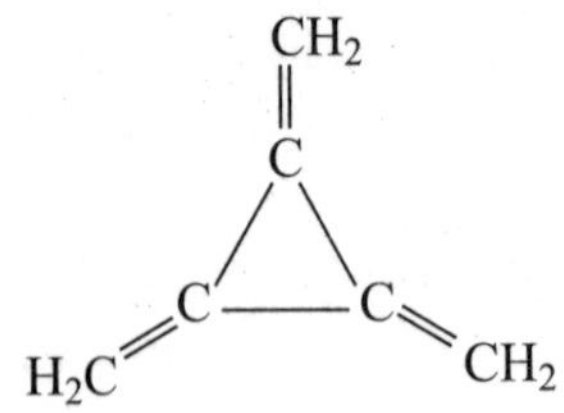

X

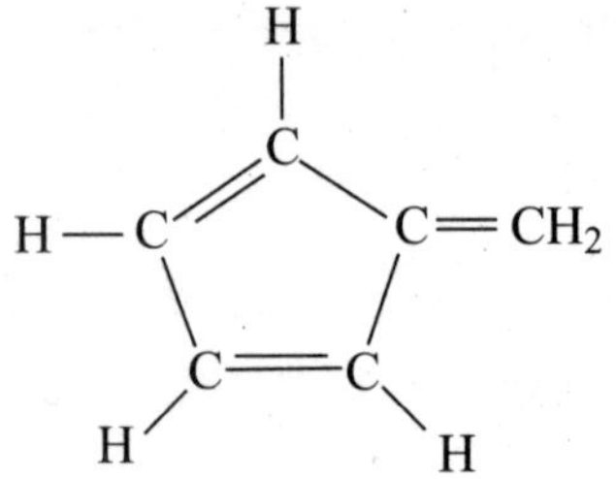

XI

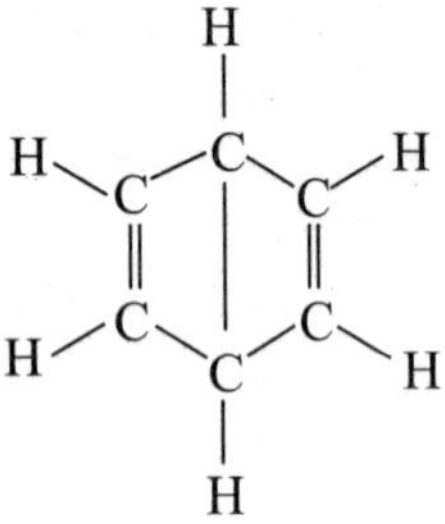

XII
Dewar's Formula

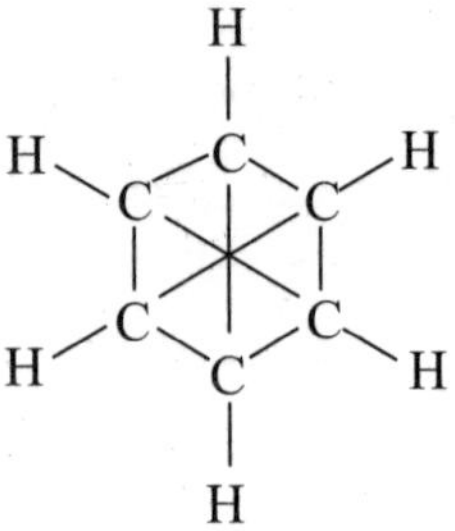

XIII
Claus Diagonal Formula

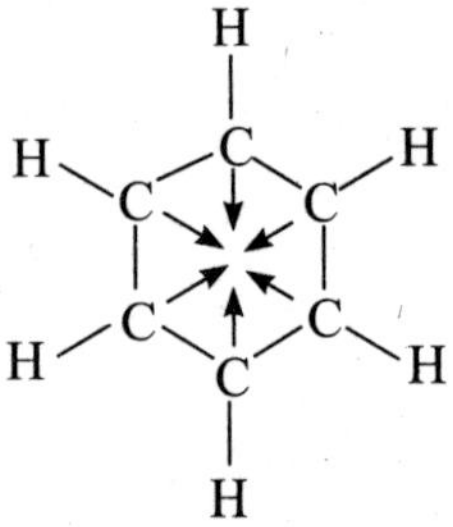

XIV
Armstrong and
Baeyers Centric Formula

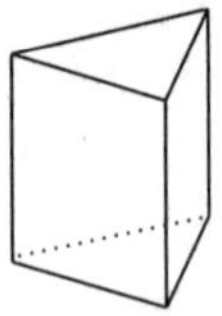

XV
Ladenburg Prism Formula

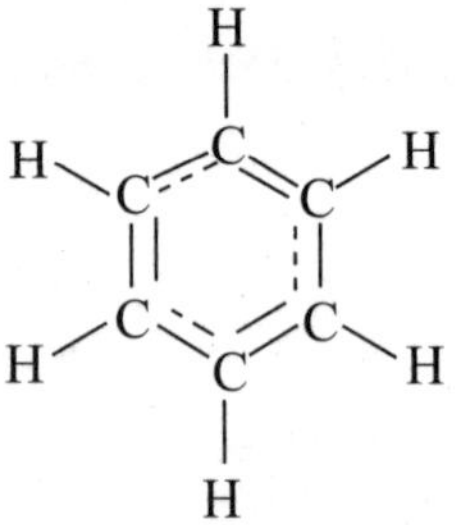

XVI
Thiele's Partial Valency Formula

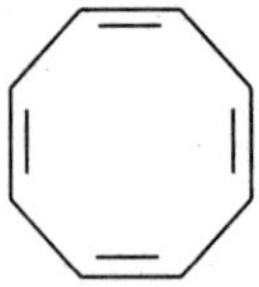

XVII
Cyclooctatetraene

Fig. 8.1: Various alternate formulas for benzene

Ladenburg proposed a **prism structure** XV in order to account for unusual stability of benzene. However, if this structure is correct then a trisubstituted benzene, $C_6H_3X_3$ is expected to be optically active. In fact such a compound does not posses any optical activity.

Thiele proposed his familiar **partial valency formula** XVI for benzene. This structure was able to account for the observed behaviour. However, if Thiels formula is correct then compounds such as cyclooctatetraene (XVII) should also exhibit aromatic character as it has four conjugated double bonds. But this was also not found to be true. So, it was also rejected by Willstater.

Willstater further found that the introduction of one or two double bonds in cyclohexane ring produced normal unsaturated compounds. But when third double bond was introduced, there was a complete change in the properties, the molecule became quite stable.

This led to the revival of Kekulè dynamic equilibrium or oscillation formulas again.

Evidence in Support of Kekulè Structure

Levine and Cole (1932) observed that ozonolysis of *o*-xylene produces methylglyoxal, biacetyl and glyoxal. Now, if *o*-xylene had the structure XVIII, ozonolysis would give two moles of monomethylglyoxal and one mole of glyoxal per mole of *o-xylene* as shown below:

On the other hand, if it were XIX, ozonolysis would give one mole of biacetyl and two moles of glyoxal per mole of *o*-xylene.

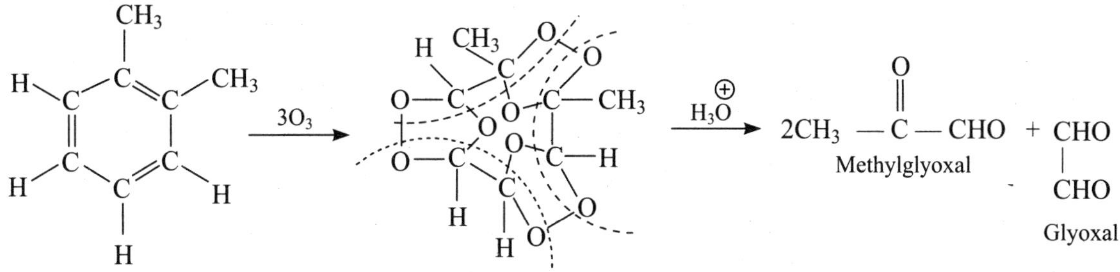

$$\text{XVIII} \xrightarrow{3O_3} \text{Triozonide} \xrightarrow{H_3O^{\oplus}} 2CH_3-\overset{\overset{O}{\|}}{C}-CHO + \underset{CHO}{\overset{CHO}{|}}$$

XVIII — Triozonide — Methylglyoxal — Glyoxal

Now in actual experiment the methylglyoxal, biacetyl and glyoxal are obtained in the molar ratios of 2 : 1 : 3 from 2 moles of *o*-xylene.

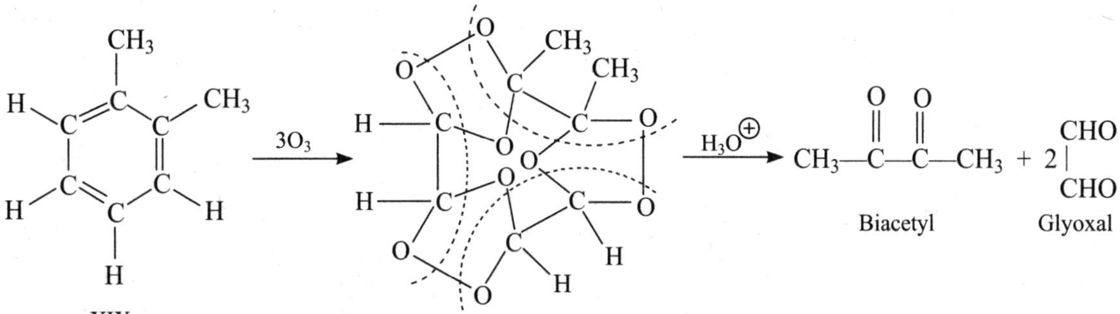

$$\text{XIX} \xrightarrow{3O_3} \text{Triozonide} \xrightarrow{H_3O^{\oplus}} CH_3-\overset{\overset{O}{\|}}{C}-\overset{\overset{O}{\|}}{C}-CH_3 + 2\underset{CHO}{\overset{CHO}{|}}$$

XIX — Triozonide — Biacetyl — Glyoxal

$$2o\text{-Xylene} \xrightarrow{6O_3} CH_3-\overset{\overset{O}{\|}}{C}-CHO + CH_3-\overset{\overset{O}{\|}}{C}-\overset{\overset{O}{\|}}{C}-CH_3 + \underset{CHO}{\overset{CHO}{|}}$$

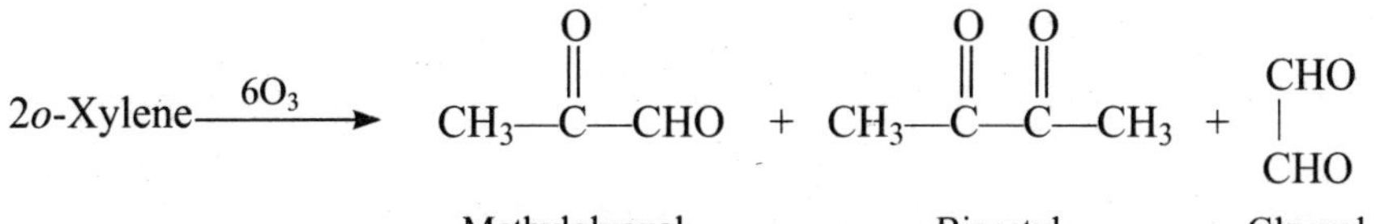

Methylglyoxal (2 moles) — Biacetyl (1 mole) — Glyoxal (3 moles)

On the basis of this observation, it may be concluded that *o*-xylene consists of a dynamic equilibrium mixture of XVIII and XIX.

Chapter 8

(ii) Benzene ring shows unusual stability

Though the molecule of benzene contains three double bonds, it undergoes electrophilic substitution reactions, instead of addition reactions. It may be interesting to note that in all these substitution reactions, the 'characteristic' structure of benzene in retained. Thus, it may be said that benzene resists addition, simply because it would destroy the benzene ring system.

The unusual stability of the benzene ring is reflected in its *heats of hydrogenation and combustion which are much lower than expected.* This is illustrated as under:

The heat of hydrogenation (–ΔH°) of cyclohexene (localised double bond) as determined experimentally is –119.66 kJ mol^{-1}

$$\text{Cyclohexene} + H_2 \longrightarrow \text{Cyclohexane} \qquad \Delta H = -119.66 \text{ kJ mol}^{-1}$$

Form this value, we will expect that the heat of hydrogenation of hypothetical compound 'Cyclohexatriene', with three 'localized' double bonds will be three times this value.

$$\text{Cyclohexatriene} + 3H_2 \longrightarrow \text{Cyclohexane} \qquad \Delta H^\circ = 3 \times (-119.66) \text{ kJ mol}^{-1} = -358.98 \text{ kJ mol}^{-1}$$

Hypothetical Cyclohexatriene
(Isolated double bonds)

But the experimental value of heat of hydrogenation of actual benzene molecule is –208 kJ mol^{-1}, which is much less than the value calculated for hypothetical 'cyclohexatriene'.

$$\text{Benzene} + 3H_2 \longrightarrow \text{Cyclohexane} \qquad \Delta H^\circ = -208 \text{ kJ mol}^{-1}$$

Benzene

Because both benzene and hypothetical cyclohexatriene produce same product cyclohexane on hydrogenation, the difference in heats of hydrogenation of the two may be attributed to the difference in their relative stabilities.

A look at the above values of ΔH° reveals that benzene is more stable than cyclohexatriene by 150.98 kJ mol^{-1}(358.98–208). This difference in energy is known as **resonance energy** of benzene. It is the energy lost due to complete delocalization of π-electrons and measures the additional stability of the aromatic system compared to a system with localized π-electrons.

(iii) All the carbon-carbon bond lengths of benzene are equal

Kekulè formula (I) predicts that benzene should have three single (154 pm) and three double (134 pm) bonds. But the actual measurements have revealed that all the carbon-carbon bond lengths of benzene are equal (139 pm) which are somewhat intermediate between the two bond lengths.

8.4.2 Resonance Structure

To account for limitations of Kekulè structures pointed out above it has been conceived, on the basis of quantum theory, that benzene is a resonance hybrid of two Kekulè and three Dewar formulas, with the first two making major contributions. It may be mentioned that benzene molecule is shown as a regular hexagon, each corner of which represent a **methine** (CH) group. By calculations, it has been shown that the two Kekulè forms differing only in the arrangement of electrons make major contributions towards resonance.

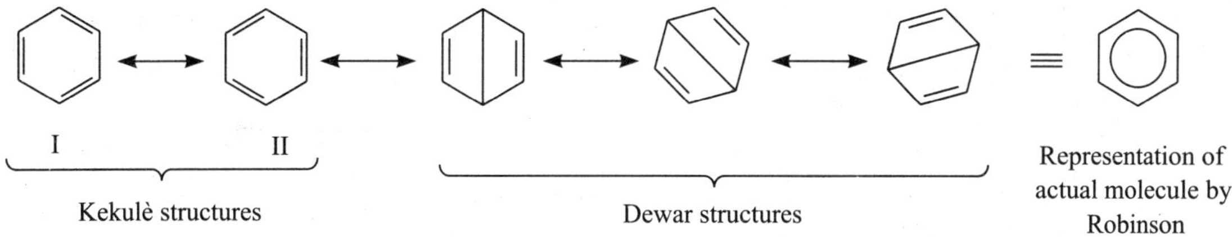

Since, the two Kekulè structures have exactly the same stability, they make equal contribution to the hybrid. Therefore stabilization due to resonance should be large. The **resonance energy** of benzene (150.98 kJ mol^{-1}) is in agreement with this and it is this high resonance energy which is responsible for 'aromatic properties'. Further, *resonance theory also explains the equivalence of all the six carbon-carbon bond lengths.* With the realization of the equivalence of C—C bond lengths, it becomes understandable that there is no difference between the two Kekulè structures (I and II) which are canonical forms of the resonance hybrid *i.e.*, the actual structure of benzene is neither represented by I and nor by II but is a hybrid of these two.

Actual molecule (**resonance hybrid**) is always more stable than any of the canonical forms. Thus, *the actual molecule of benzene is much more stable compared to either of the two Kekulè structures*. The magnitude of stability conferred on a molecule as a result of resonance is expressed in terms of **resonance energy** or **delocalization energy**. *It is defined as the difference in internal energy of the resonance hybrid and that of the most stable canonical form.* In case of benzene, this resonance energy has been determined to be 150.98 kJ mol^{-1}.

Chapter 8

8.4.3 Molecular Orbital Picture

According to molecular orbital theory, in benzene, each carbon atom is sp^2-hybridized. Two of these sp^2-hybrid orbitals are used for end-on overlaping with sp^2-hybrid orbital of each one of the two adjacent carbon atoms. The third sp^2-orbital overlaps with 1s orbital of hydrogen. Thus, each carbon atom forms three σ -bonds in all. The whole of this σ-skeleton is planar and the molecule is symmetrical with each carbon lying at the corner of a regular hexagon. All the bond angles (C—C—C and C—C—H) are 120° [Fig. 8.2 (*a*)]. There are still six unhybridized 2*p*-orbitals —one on each carbon atom which are parallel to each other and are at right angles to the σ-skeleton. Any two *p*-orbitals on adjacent carbon atoms being closer enough to each other can overlap sideways to form a π-bond. Thus, this type of orbital overlap can take place in two ways to form three localised π-bonds in each case [Fig. 8.2 (*a*) and (*b*)]. Each of these 2*p*-orbitals overlap both of its neighbours equally.

In other words, the electrons in the *p*-orbital are not localized to a particular pair of carbon atoms. In fact they are delocalized over whole of the molecule. As each *p*-orbital consists of two lobes, one lying above and the other lying below the plane of the carbon skeleton, this delocalization results in two **doughnut shaped electron clouds**; one above and the other below the plane of the ring [Fig. 8.2(*b*)]. In other words, the σ-skeleton of the benzene ring is sandwiched between these doughnut shaped electron clouds. The picture that emerges out is shown in Fig. 8.2 (*b*). This structure of benzene gets additional support from X-ray diffraction studies.

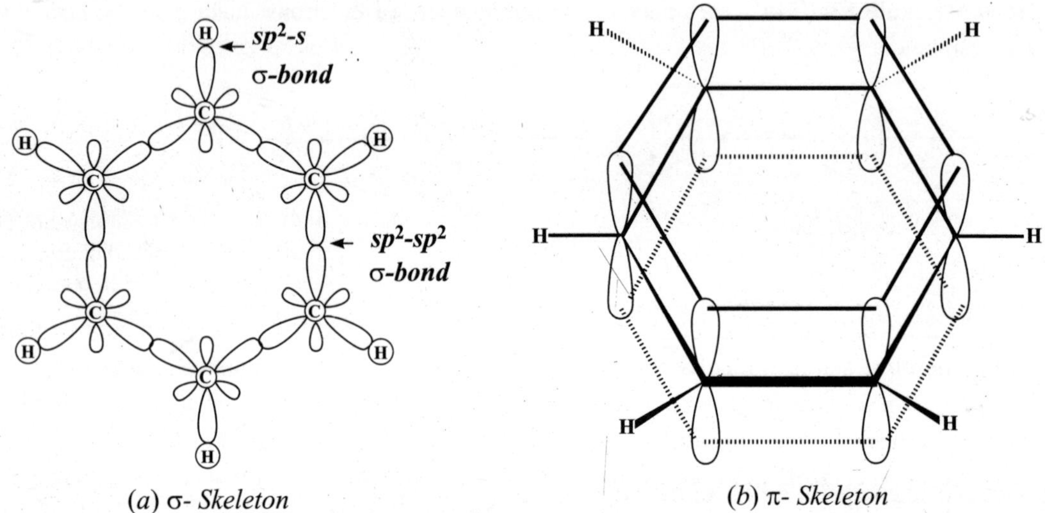

(*a*) σ- *Skeleton* (*b*) π- *Skeleton*

Fig. 8.2: σ and π skeleton of benzene

8.4.4 Representation of Benzene Structure

Finally, it may be noted that for the sake of convenience the benzene ring is represented by drawing the structure as a regular hexagon with a circle in the centre as proposed by **Robinson**. The circle is drawn to give an idea of electron delocalization. However, the conventional Kekulè structure is still being widely used and we shall be using both the representations in this book.

8.5 AROMATICITY: THE HÜCKEL RULE

A most significant feature of benzene and other compounds containing this ring system is that they are *more stable* in comparison to the corresponding acyclic polyenes. The stability of the ring compounds is due to the fact that they possess large resonance energies and are called *aromatics* or *aromatic compounds*. The extra stability of such compounds is due to a special property called as ***aromatic character*** or ***aromaticity***. Since benzene is a cyclic conjugated polyene, earlier chemists thought that the lower and higher vinylogues of benzene, for instance, cyclobutadiene and cyclooctatetraene can also be prepared and they should also exhibit aromaticity. Attempts were made to synthesize these types of compounds and it was found that the former compound is highly unstable while the later though prepared does not possess comparable stability and behaves like an ordinary polyene with isolated double bonds. Evidently aromaticity is related to some characteristic feature of benzene and related systems generally having some 'magic' number of π-electrons in a regular planar, monocyclic completely conjugated polyene.

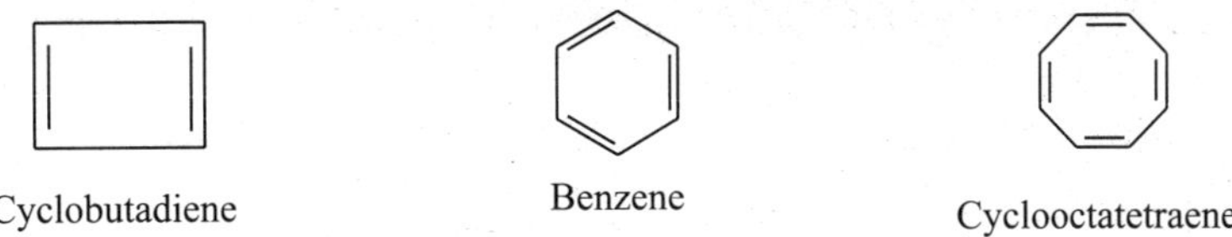

Hückel rule: Cyclic, planar, completely conjugated polyenes having an even number of CH units represented as neutral monocyclic polyenes are also referred to as annulenes. They are also shown as $(CH=CH)_n$ where $n = 2, 3, 4$ etc. They are named by writing the number of carbon atoms in the ring in square bracket and writing this as a prefix before the root name annulene. Thus, benzene is [6] annulene, square planar cyclobutadiene is [4] annulene, etc.

It was Hückel, who in 1931 formulated a simple rule for predicting whether or not a given annulene would be aromatic. This rule commonly referred to as Hückel's rule, states that *a monocyclic, planar and completely conjugated polyene called as annulene will be aromatic, if and only if, it contains* $(4q + 2)$ *π-electrons, where* $q = 0, 1, 2, 3,$ ....etc. *On the other hand, if such a system contains* $4q\pi$ *-electrons, it will not be aromatic, i.e., it will either be antiaromatic or non-aromatic.*

8.5.1 Theoretical Basis of Hückel Rule

From quantum mechanical point of view, an aromatic system is the one that possesses a closed shell of π-electrons corresponding to a rare gas configuration. As shown by Hückel, the solution of Schrodinger wave equation for π-electron system of any regular planar, cyclic conjugated polyene, reveals that energies of the molecular orbitals in such systems have a very simple characteristic pattern. In these systems, there is always one orbital of lowest energy which is followed by degenerate pairs of orbitals (*i.e.*, orbitals having same energy) in the order of increasing energy and finally there is one orbital of highest energy (Fig. 8.3). Filling of electrons in these orbitals takes place in the following manner: first of all two electrons are filled in the lowest energy molecular orbital and then rest of the electrons go to the degenerate pairs of orbitals strictly according to *Hund's rule*.

Thus, in an annulene $(CH=CH)_n$, if n is an *even number*, the lowest orbital will always be filled by two electrons and other orbital being filled in the manner described above. But there will always be two singly occupied degenerate orbitals in such systems. Because of this arrangement of electrons, such systems are highly unstable and are termed as *antiaromatic*. For instance, cyclobutadiene ($n = 2$; $4q\pi$-electrons) would be antiaromatic because out of 4π-electrons, two would fill the lowest orbital and the remaining two would go to the first pair of degenerate orbitals, each being singly occupied. As expected, cyclobutadiene would be an antiaromatic compound.

On the other hand, when n is an *odd number*, out of $4q + 2\pi$-electrons, two would fill the lowest orbital, then each of the degenerate pairs of orbitals would be occupied by 4 electrons. *Because of this closed shell filling of orbitals with electrons, such compounds are aromatic.* For instance, benzene (n =3) having 6π-electrons, $(4q+2)$ is aromatic. This set of 6π-electrons of benzene is also called an *aromatic sextet.*

Chapter 8

This explains the stability of $(4q + 2)$ *vis-à-vis* $(4q)$ π-electron systems which are antiaromatic as they possess higher energy than the hypothetical structures having independent double bonds without mutual overlap, that is without delocalization. The Hückel's molecular orbital π-energy diagrams of some of the important systems such as cyclopropenyl cation ($4q + 2$; $q = 0$), cyclobutadiene ($4q$; $q = 1$), benzene ($4q + 2$; $q = 1$), cyclooctatetraene ($4q$; $q = 2$) and cyclodecapentaene ($4q + 2$; $q = 2$) are shown in Fig. 8.3 assuming that they are regular polygen (cyclooctatetraene is, however, a non-planar compound).

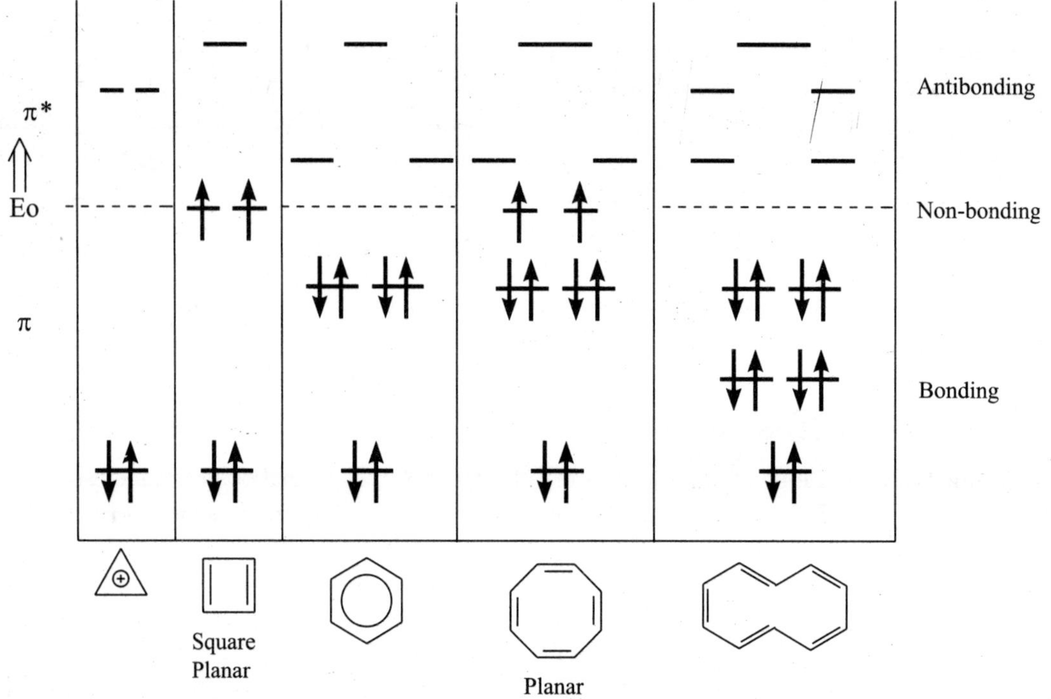

Fig. 8.3: HMO π-energy level diagram of some annulenes

8.5.2 Experimental Basis of Hückel Rule

Apart from theoretical considerations, the phenomenon of aromaticity can also be confirmed on the basis of experimental data. In addition to high degree of stability and their tendency to participate in substitution rather than addition reactions another characteristic feature that distinguishes aromatic compounds from others is their nmr* spectral characteristics. Under the influence of applied magnetic field, the π-electrons in these compounds begin to circulate in a plane at right angles to the direction of the applied field. This electron circulation generates an induced magnetic field which opposes the applied field in the region of π-electron cloud (inner side of the molecule) and reinforces the latter in the peripheral region. Thus, the protons lying in the former region are shielded (showing upfield shift of the resonance signal) while those lying in the latter region are deshielded (showing downfield shift of the resonance signals) [Fig. 8.4].

* NMR spectroscopy is not included in the syllabus.

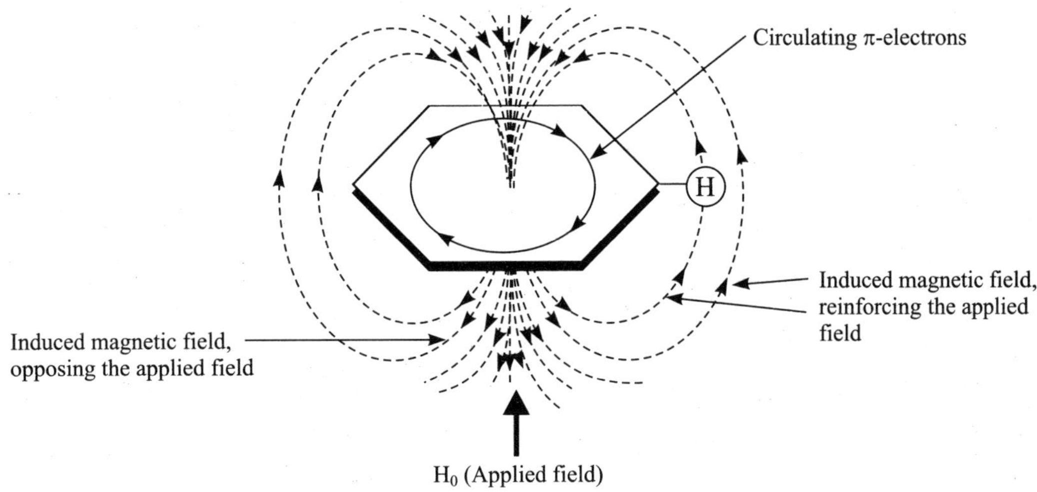

Fig. 8.4: Induced π-electron ring current in benzene which generates a local induced field

Benzene provides an illustrative example which shows a six proton singlet at δ 7.27. Thus, NMR spectral study leads to an excellent spectroscopic test of aromaticity and this has been successfully applied to the other members of the annulene series. The annulenes having $(4q + 2)$ π-electrons show similar behaviour in their low temperature* nmr spectra which exhibit downfield signals due to external protons and the internal protons show upfield signals as expected. On the other hand, annulenes with $4q$ π-electrons show lowfield and upfield signals due to internal and external protons respectively. Such annulenes are *antiaromatic*. Thus, for instance [16] annulene shows in its NMR spectrum at room temperature all the protons located at δ 6.7 but at –110°, it shows a four proton signal at much lower field strength and a 12 proton multiplet at higher field strength.

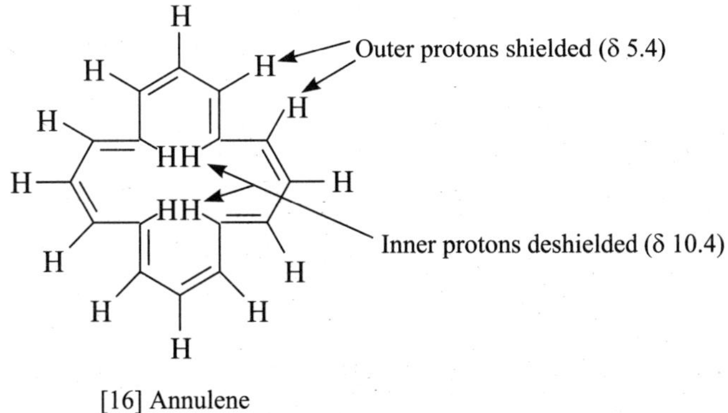

[16] Annulene

8.5.3 Criteria for (4q+2) Rule or Hückel Rule or Aromaticity

An aromatic compound should fulfill the following criteria:

(*i*) *It should be a planar, conjugated, cyclic polyene.*

* These large conjugated rings are conformationally mobile, there being averaging of the chemical shifts of internal and external protons and consequently, it has been necessary to examine the annulenes at low temperature to determine whether or not, they are aromatic.

Chapter 8

(ii) *The total no. of π-electrons should be (4q + 2) where q = 0, 1, 2, 3.......*

(iii) *It should have cyclic cloud of delocalized electrons above and below the plane of molecule.*

(iv) *It should posses high resonance energy, i.e., it should be highly stable. Such compound should undergo electrophilic substitution in preference to addition reactions.*

(v) *It should show appreciable deshielding of peripheral protons in its NMR spectrum.*

It has experimentally been shown that conjugated rings with 2, 6, 10 and 14 π-electrons are aromatic, while those with 4, 8, 12, 16, and 20 π-electrons surely are not.

8.5.4 Applications of Hückel Rule

It should be clear that aromatic compounds need not necessarily have a benzene ring. All the cyclic planar, completely conjugated polyenes containing $(4q + 2)$ π-electrons should be aromatic. Some examples are discussed below:

(*i*) Systems having 2π-electrons: Aromatic duet ($q = 0$)

Examples of aromaticity in 2π-electron system are found in XX (as perchlorate), XXI (as perchlorate) and XXII inspite of the bond angle of 60°. The dication (XXIII) (as fluoroborate) is the latest addition to this class of stable systems.

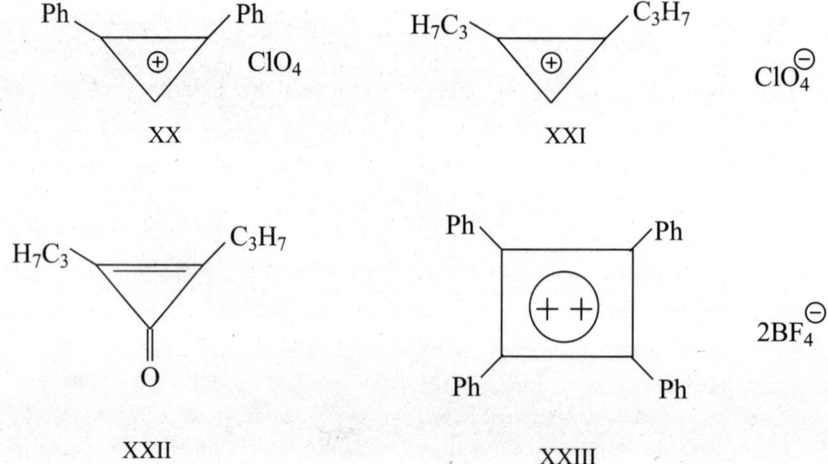

(*ii*) Systems having 6 π-electrons: Aromatic sextet ($q = 1$)

Benzene is the best known example of the aromatic sextet. Cyclopentadienyl anion (XXIV) and cycloheptatrienyl cation (XXV) are also well known. Further illustrations are provided by tropone (XXVI) which is stable since, the electronegative oxygen atom attracts the carbonyl π-electrons leaving the seven-membered ring with the aromatic sextet as evident by high value for its dipole moment (4.17 D). Indeed, tropolone (XXVII), a derivative of tropone, is a natural product. Azulene (XXVIII) is stable and may be regarded as a combination of (XXIV) and (XXV). The ylids (XXIX) and (XXX) are further examples of aromatic sextet. Metallocenes, of which ferrocene (XXXI) is the prime example, owe their stability to the aromatic sextet. Hückel's rule also applies to heterocyclic systems such as furan, thiophene and pyrrole, each of which contains a group of six annular electrons including the lone pair from oxygen, sulphur and nitrogen respectively. Pyridine and pyrrillium cation offer two other examples of aromatic sextet. The stable inorganic compound (XXXII) and (XXXIII) as well as sydnones (XXXIV) though apparently having π-electron sextet are not considered aromatic. However, sydnones (XXXIV) are stable aromatic compounds.

XXIV XXV XXVI XXVII XXVIII XXIX XXX XXXI XXXII

XXXIII XXXIV

(*iii*) Systems having ten π-electrons ($q = 2$) and other annulenes

[10] Annulene (XXXV) is very unstable apparently because of the non-bonded interaction of the hydrogens inside the planar ring, but stable derivatives (XXXVI; X = CH_2, O, NH) of 10-annulene are known in which the hydrogen interference is absent. These substances undergo aromatic substitution and sustain a ring current. Cyclooctatrienyl dianion (XXXVII) and cyclononatetraenyl anion (XXXVIII) have been prepared as the potassium and the lithium salts, respectively. One of the spectacular examples of Hückel's rule is the discovery that the *trans*-dimethyldihydropyrene derivative (XXXIX) displays aromatic stability of a much higher order than 14-annulene (XL). This is explained by the absence of non-bonded interactions of the hydrogens inside the planar ring. The compound (XXXIX) undergoes bromination, nitration and Friedel-Crafts reaction under mild conditions; [18] and [22]-annulenes (XLI) and (XLII), are aromatic if we go by their NMR spectra. The π-electrons in aromatic compounds, due to the induced ring current, produce characteristic shielding effects on the inner protons and deshielding effects on the outer protons. These annulenes show similar NMR characteristics.

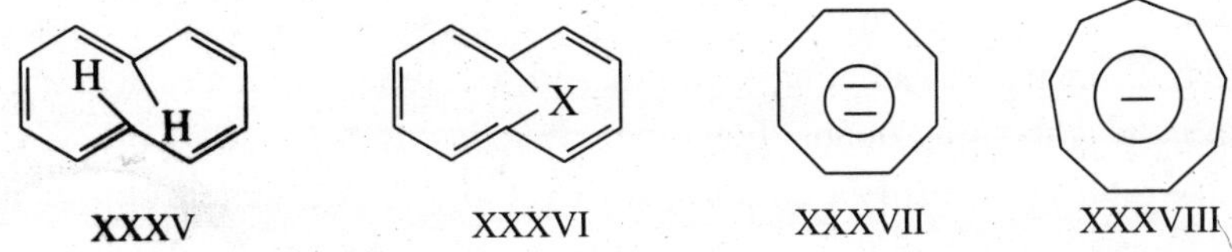

XXXV XXXVI XXXVII XXXVIII

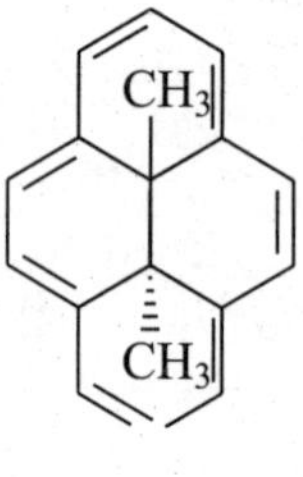

XXXIX

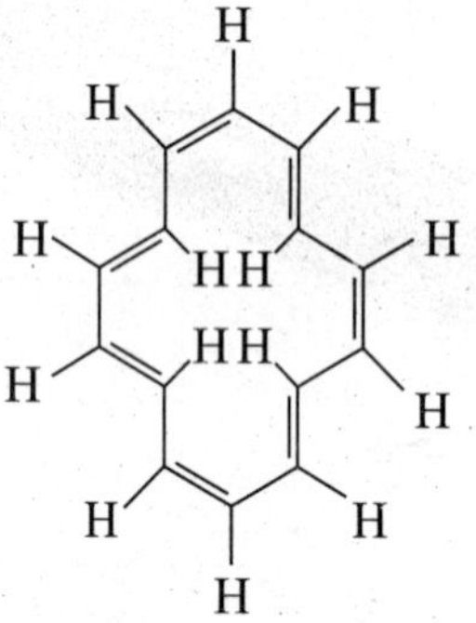

[14] - Annulene
XL

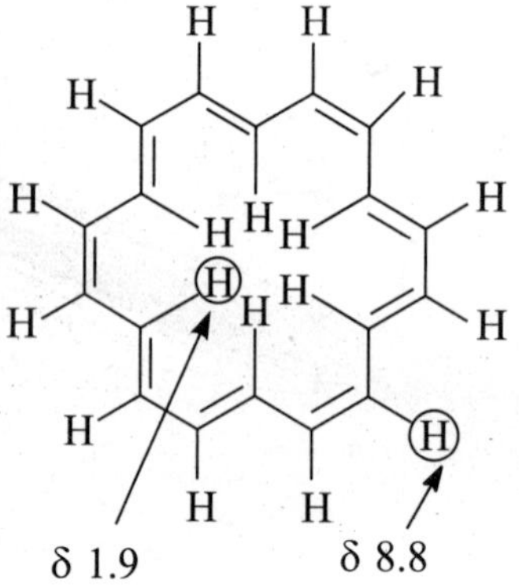

δ 1.9 δ 8.8

[18] - Annulene
XLI

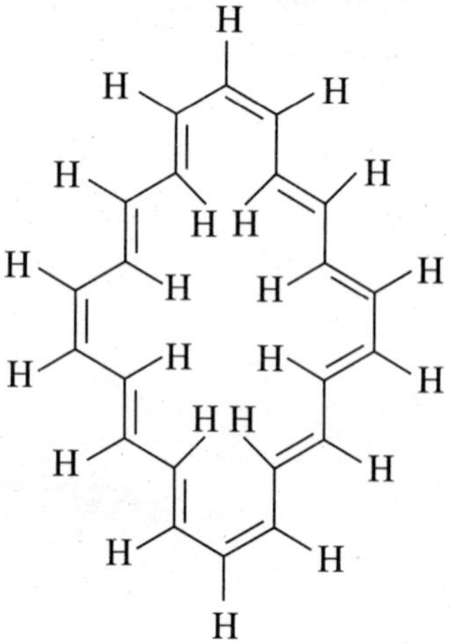

[22] - Annulene
XLII

Thus, Hückel's rule seems to hold good up to an extent, but it fails to explain that aromaticity disappears even in $(4q + 2)$ π-electron systems above a certain critical ring size. It must be noted, however, that $(4q + 2)$ annulenes which are not aromatic still remain relatively more stable than their neighbouring $(4q)$ annulenes in the vinylogous series. Medium size annulenes are aromatic only on the basis of magnetic measurements. Chemically they show no aromatic stability; rather they are very reactive and polymerize readily like polyenes. The decrease in aromaticity may be explained first by increasing tendency of σ-bonds with increasing ring size to favour alternating structures with localized bonds, and secondly by the non-bonded interaction among the hydrogens inside the planar rings of the medium size annulenes,

Hückel's rule has been extended to polycyclic ring systems as well. Aromatic stability, however, decreases in the order, benzene, naphthalene, anthracene, tetracene, pentacene, hexacene and heptacenes with 6, 10, 14, 18, 22, 26 and 30 carbons atoms. The higher members of this series (acenes) show extreme reactivity and are very unstable. Further, the rule does not account for the difference in aromatic stability of the above hydrocarbons and the angularly fused isomeric hydrocarbons which are more stable. Again, it is applied only to half of the most stable aromatic hydrocarbons having $6q$ π-electrons, such as biphenyl, triphenyl, triphenylene dibenzpyrene, etc., which may be regarded as derivatives of polyphenyls. Hückel's rule is, therefore not suitable for predicting aromatic stability in polycyclic hydrocarbons.

(*iv*) Systems with 4 *q* π-electrons: Cyclobutadiene and cyclooctatetraene

According to Hückel's rule both *square* cyclobutadiene and *planar* cyclooctatetraene with $4q$ π-electrons should be antiaromatic. Cyclooctatetraene, however can escape, as it does its antiaromaticity by abandoning planar geometry. In fact, it has a tub shaped structure in which each double bond is almost perpendicular to its neighbouring double bond thereby reducing the π-overlap to a very great extent. This explains its relative stability compared to cyclobutadiene which enjoys no such option. Cyclobutadiene, however, can acquire some stability by assuming a rectangular form though even in this form, it may be unstable due to ring strain (Fig. 8.5).

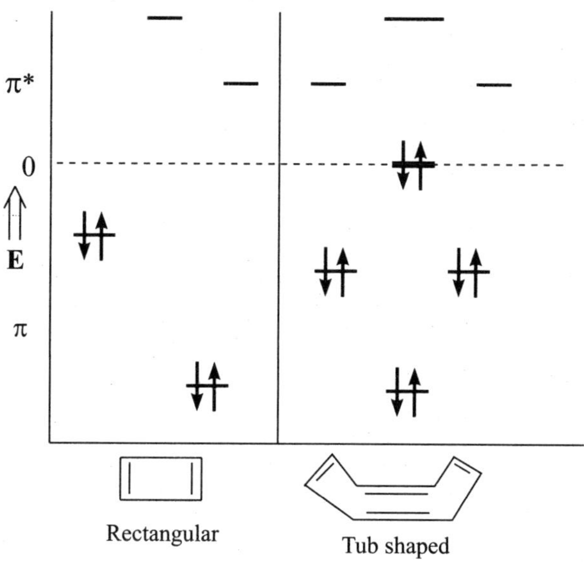

Fig. 8.5: HMO π-energy level diagram of rectangular cyclobutadiene and tub-shaped cyclooctatetraene

(*v*) Heterocyclic compounds

Furan, thiophene, pyrrole, pyridine, etc., all contain a cyclic system of 6 π-electrons which are completely delocalized. Hence, they are aromatic. Whereas in pyridine, the heteroatom contributes one π-electron, in furan, thiophene and pyrrole each heteroatom contributes two π-electrons towards the aromatic system.

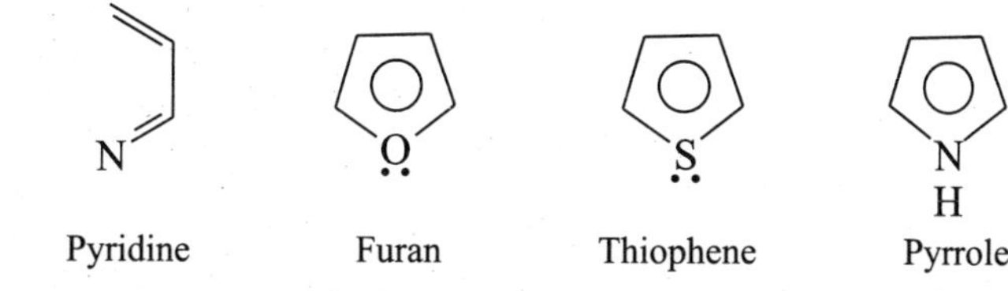

(*vi*) Polycyclic compounds

Aromaticity of some polycyclic compounds can also be explained on the basis of Hückel's rule. For example, naphthalene with 10 π-electrons ($q = 2$), anthracene and phenanthene each with 14 π-electrons ($q = 3$) are all **aromatic**.

Naphthalene   Anthracene   Phenanthrene

Special case of pyrenes

There are 16 π-electrons in pyrene but it has aromatic properties. Thus, it can be said that only the peripheral electrons are involved in the formation of a closed loop of 14 π-electrons. The two π-electrons of the internal C_{15}—C_{16} double bond do not participate. This is confirmed on the basis of the fact that when this double bond is replaced by two methyl substituents, the resulting 15, 16-dihydrodimethylpyrene still shows aromatic character.

Pyrene
Aromatic (16 π-electrons)
(Peripheral π-electrons = 14)

Trans -15,16-Dimethyldihydropyrene
Aromatic (14π-electrons)

8.6 AROMATIC, ANTI-AROMATIC AND NON-AROMATIC COMPOUNDS

Cyclic conjugated polyenes are divided into the following three categories depending upon the extent of delocalization.

(*i*) **Aromatic:** *If the delocalization energy (DE) of cyclic, planar, completely conjugated polyene is higher than that of its acyclic conjugated analogue, it is said to be* ***aromatic****.*

For examples

1. The delocalization energy of benzene is much greater than that of its acyclic analogue 1,3, 5-hexatriene, so benzene is aromatic.

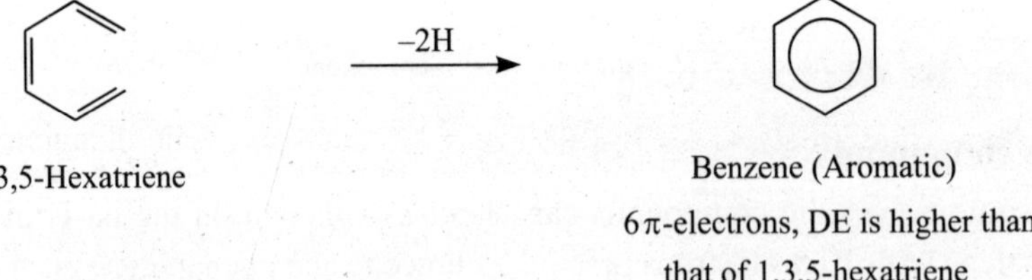

1,3,5-Hexatriene

Benzene (Aromatic)
6 π-electrons, DE is higher than that of 1,3,5-hexatriene

2. The delocalization energy of cyclopropenyl cation is much greater than that of its acyclic analogue allyl cation. So, cyclopropenyl cation is aromatic.

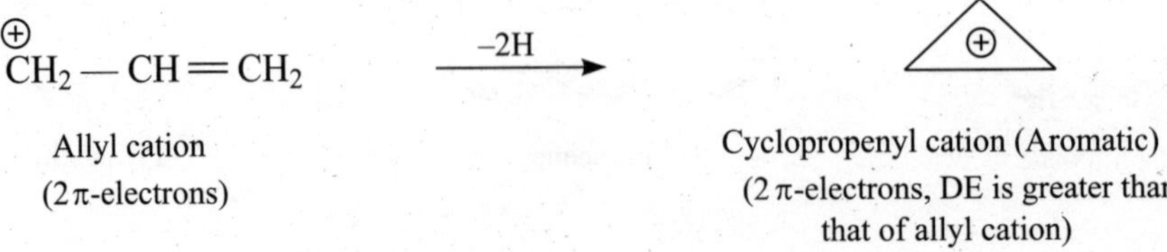

Allyl cation
(2 π-electrons)

Cyclopropenyl cation (Aromatic)
(2 π-electrons, DE is greater than that of allyl cation)

(*ii*) **Antiaromatic:** *If the delocalization energy of the cyclic, planar, completely conjugated polyene is less than that of its acyclic conjugated analogue, it is said to be* ***antiaromatic***. For example, the delocalization energy of cyclobutadiene (square planar) is much lower than that of its acyclic analogue 1,3-butadiene. So, cyclobutadiene is antiaromatic.

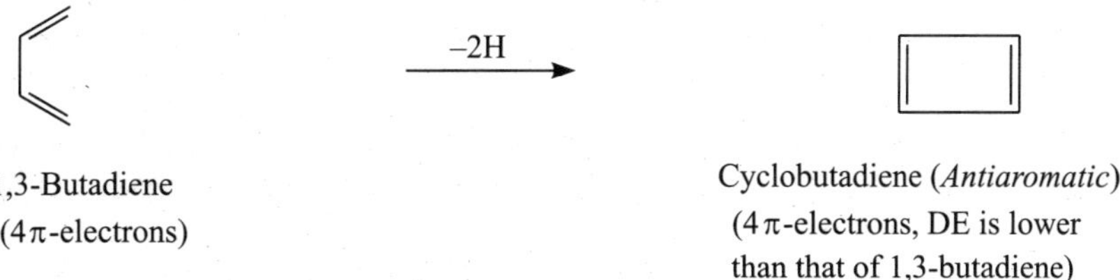

1,3-Butadiene (4π-electrons)   Cyclobutadiene (*Antiaromatic*) (4π-electrons, DE is lower than that of 1,3-butadiene)

(*iii*) **Non-aromatic:** *If the delocalization energy of a cyclic, completely conjugated polyene is equal to that of its acyclic analogue, it is said to be* ***non-aromatic***. For example, the delocalization energy of cyclooctatetraene is equal to that of 1,3,5,7-octatetraene, so cyclooctatetraene is non-aromatic.

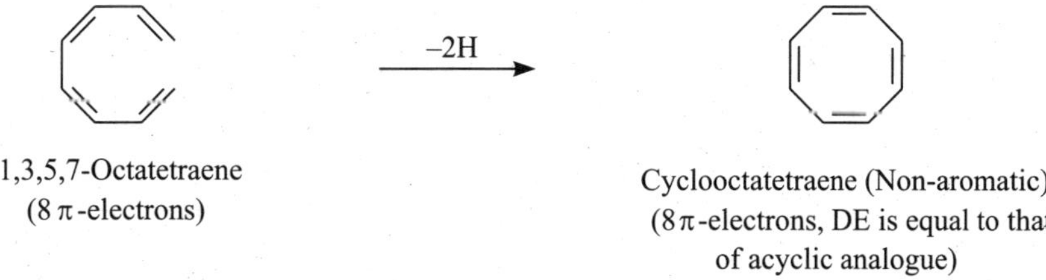

1,3,5,7-Octatetraene (8 π-electrons)   Cyclooctatetraene (Non-aromatic) (8π-electrons, DE is equal to that of acyclic analogue)

Here, it may be mentioned that the term delocalization is used to measure the stability of the cyclic planar conjugated polyene in comparison to the corresponding cyclic or acyclic conjugated polyene. In fact, it has nothing to do with either chemical reactivity or thermodynamic stability of the molecule which is measured in terms of resonance energy. Thus, it may be said that a molecule having a large delocalization energy may still be reactive.

Some examples of antiaromatic and non-aromatic compounds are discussed below:

8.6.1 Antiaromatic Compounds

We have seen above that greater delocalization of π-electrons in cyclic planar completely conjugated polyene compared to its acyclic analogue makes the compound **aromatic**. However, this is not universally true. Molecular orbital calculations have shown that in cyclic, planar completely conjugated polyenes having $4q$ π-electrons delocalization leads to destabilization. Such systems which upon delocalization of $4q$ π-electrons are destabilized are called **antiaromatic**. This property is called antiaromaticity. Some examples are:

1. Cyclopropenyl Anion

Molecular orbital calculations have revealed that cyclopropenyl anion has high energy than any one of the contributing forms. Hence, it is **antiaromatic**.

2. Cyclopentadienyl Cation

Similar to cyclopropenyl anion, cyclopentadienyl cation possesses higher energy than the contributing forms. So, it is antiaromatic.

3. Cyclobutadiene

It is well studied example of an antiaromatic compound. Cyclobutadiene is so unstable that all attempts to synthesize it, have failed. However, when generated at 196 K, it undergoes dimerization to give different poroducts as shown below:

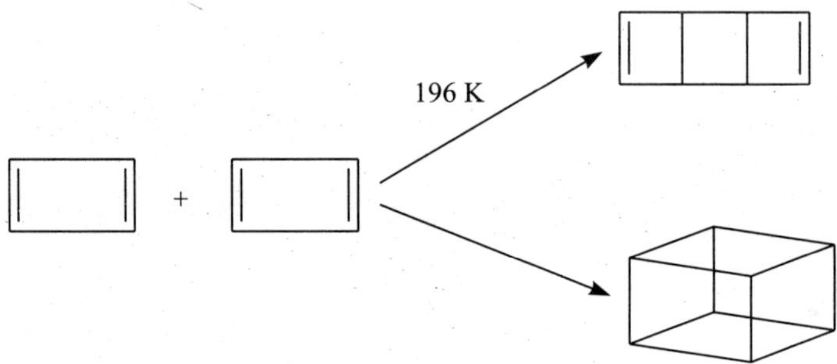

The high reactivity of cyclobutadiene is not merely due to angle strain since cyclopropene derivatives are well known compounds despite having still more angle strain.

On the basis of the above observations, it is clear that possibility of existence of many resonance structures of any compound does not necessarily confer aromaticity on it. In fact, the main criteria for stability or instability is the number of π-electrons. Thus, the instability of compounds having $4q$ π-electrons is due to the reason that two of their electrons are present either in non-bonding or anti-bonding molecular orbitals in the ground state of the molecule.

8.6.2 Non-aromatic Compounds

If a cyclic planar completely conjugated polyene is as stable as its analogues (cyclic or acyclic polyene) with localized bonds, it is said to be **non-aromatic** and this phenomenon is called **non-aromaticity**.

As discussed above cyclic planar completely conjugated polyenes containing $4q$ π-electrons are destabilized by the delocalization and hence are **antiaromatic**. However, if the structure permits such a molecule may assume a non-planar geometry which prevents the delocalization of π-electrons. Such compounds now have localized π-electrons and hence have some life time. For example, cyclooctatetraene escapes antiaromaticity by assuming a tub shaped geometry.

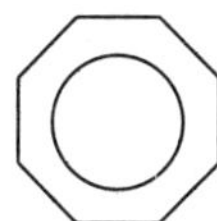

Planar Cyclooctatetraene
(8π-electrons, antiaromatic)

Tub shaped Cyclooctatetraene
(8π-electrons, non- aromatic)

Planar Cycloheptatrienyl anion
(8π-electrons, antiaromatic)

Since cyclooctatetraene is not planar, it is neither aromatic nor antiaromatic but is non-aromatic. This is due to the reason that in the tub shaped geometry, all the π-orbitals are no longer parallel and hence destabilization due to delocalization is prevented. As a result, 8 π-electrons are localized in four distinct π-bonds. In other words, it behaves as a typical alkene and undergoes electrophilic addition (Br_2, HCl, etc.) and oxidation reaction ($KMnO_4$ etc.).

Similarly, planar cycloheptatrienyl anion is supposed to be antiaromatic, but in reality, like cyclooctatetraene, it has non-planar geometry. As a result, delocalization is avoided and hence it behaves as a non-aromatic compound.

A BRIEF REVIEW

- **Benzenoid compounds** are aromatic compounds containing one or more benzene rings *e.g.*, benzene, phenol, toluene, aniline, anthracene, naphthalene, etc.
- **Non-benzenoid compounds** are aromatic compounds having no benzene ring *e.g.*, tropolone, cyclopentadienyl anion, tropylium ion, azulenes, etc.
- Benzene is highly unsaturated but it behaves like a saturated molecule which undergoes **substitution reactions**, such as halogenation, sulphonation, nitration and Friedel Crafts reactions. Benzene does not undergo **addition reactions** which are characteristic reactions of unsaturated compounds. It also does not undergo oxidation.
- Kekulè structure of benzene is represented by a hexagon having three localized double bonds. This structure *fails to explain the unusual stability of the benzene ring and the equivalence of all carbon-carbon bond lengths.*
- **Resonance hybrid** theory explains the unusual stability of benzene ring and equivalence of all C—C bond lengths in terms of delocalization of π-electrons.
- **Structure of benzene** can be explained on the basis of M.O. theory. According to this theory each carbon atom is sp^2-hybridized where two of the sp^2-hybridized orbitals are used for C—C bond formation with two adjacent carbon atoms and the third sp^2 hybrid orbital is used for C—H bond formation. Each carbon atom has one unhybridized π-orbital containing one electron which overlap laterally to form π-bonds leading to formation of two continuous doughnut shaped electron clouds, one lying above and the other below the plane of carbon atoms.
- **Hückel rule:** Monocyclic planar completely conjugated polyenes containing $(4q + 2)$ π-electrons and having a single cyclic π-electron cloud above and below the ring are called **aromatic**.

The non-aromatic molecule does not satisfy one or more of the above conditions.

- **Aromatic system:** Delocalization energy of cyclic conjugated polyene system is *greater* than that of its acyclic analogue.
- **Antiaromatic system:** Delocalization energy of cyclic conjugated polyene system is *lower* than that of the corresponding acyclic analogues.
- **Non-aromatic system:** Delocalization energy of cyclic conjugated polyene is *equal* to that of the corresponding acyclic analogue.

SOLVED PROBLEMS

Q.1 An aqueous solution of tropylium bromide (C_7H_7Br) on treatment with silver nitrate gives a precipitate of silver bromide. How do you account for this observation?

Ans. In aqueous solution, tropylium bromide readily undergoes ionization to form tropylium cation and bromide anion which reacts with $AgNO_3$ solution to form a precipitate of AgBr. The driving force for this ionization is the aromatic character of tropylium ion due to high resonance energy.

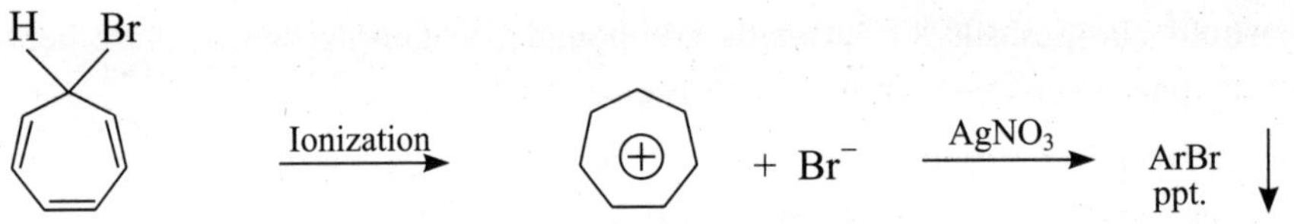

H Br

Ionization → ⊕ + Br^- $\xrightarrow{AgNO_3}$ ArBr ppt. ↓

Tropylium bromide   Tropylium cation (*aromatic*)

Q.2. Though benzene is an unsaturated hydrocarbon yet it fails to respond to Baeyer's test. Explain.

Ans. The six π-electrons of the benzene ring are completely delocalized over all the six carbon atoms of the ring. As a result of this delocalization, benzene has extra stability and hence does not undergo addition reactions because such reactions would lead to the destruction of the delocalized π-electron cloud thereby decreasing the stability of the benzene ring.

Q.3. Cycloheptatriene (A), though a cyclic nearly planar triene, containing six π-electrons, does not exhibit usual stability of aromatic compounds whereas tropolone (B), a derivative of (A),behaves like a typical phenol. Explain.

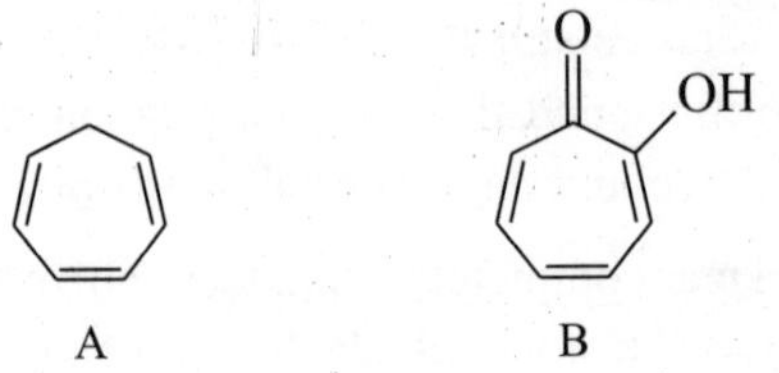

O OH

A   B

Ans. Although cycloheptatriene (A) has six π-electrons, these are not completely delocalized over all the seven carbon atoms of the ring since it has one sp^3-hybrized carbon atom. As a result, it is not aromatic and hence does not have the usual stability of aromatic compounds.

On the other hand in tropolone (B), the keto group withdraws the π-electrons of the C=O towards itself thereby creating positive charge on the ring carbon atom. As a result, the six π-electrons of the tropolone ring are completely delocalized over the seven carbon atoms of the ring (C) thereby making tropolone an aromatic compound.

O OH ⟷ $O^{\ominus}$ OH ⊕

B C

Since (C) is an aromatic compound containing an OH group, tropolone (B) behaves like a phenol.

Q.4. Cyclooctatetraene reacts with two equivalents of potassium to yield an unusually stable compound with the molecular formula, 2 K^+, $C_8H_8^{2-}$. Write its structure and account for its unusual stability.

Ans. Cyclooctatetraene has 8 π-electrons which on treatment with two moles of potassium gains two electrons, thereby raising the number of π-electrons to 10.

In other words, cyclooctatetraenyl dianion having 10 π-electrons is aromatic and hence quite stable.

+ 2 K ⟶ + 2 K^+

Cyclooctatetraene Cyclooctatetraenyl dianion (10 *π-electrons, aromatic*)

Q.5. 3-Chlorocyclopropene reacts with $SbCl_5$ to form a salt but chlorocyclopropane does not. Explain.

Ans. Due to aromatic character of cyclopropenyl cation, 3-chlorocyclopropene reacts with $SbCl_5$ to form the corresponding salt.

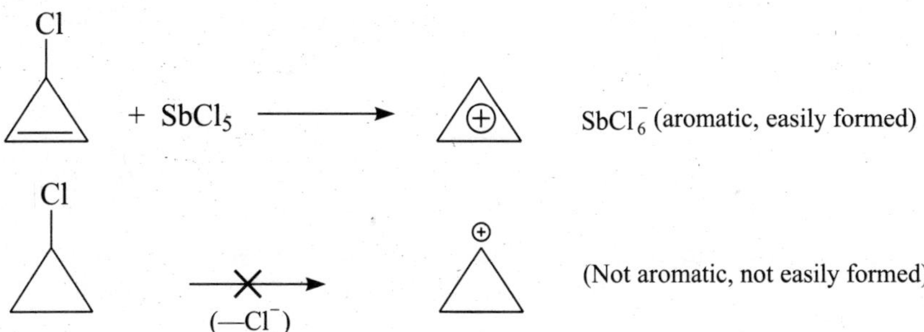

Since, cyclopropyl cation is not aromatic, it is not easily formed from chlorocyclopropane.

Q.6. Cyclopentadienone is less stable than cyclopropenone and cycloheptatrienone. Explain.

Ans. The less stability of cyclopentadienone is due to the antiaromatic character of the ring in which one of the contributing forms has 4π-electrons.

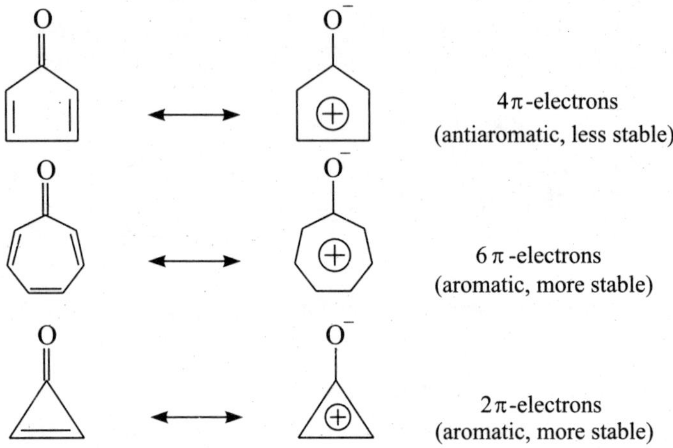

Q.7. Comment on the aromaticity of cyclooctatetraene.

Ans. Cyclooctatetraene is not an aromatic compound since, it does not contain $(4q + 2)$ π-electrons. Since, it contain 8π-electrons, therefore, it must be antiaromatic if planar. However, to avoid destabilization due to resonance, it assumes *tub shaped structure* where all the 4π-bonds are isolated (orthogonal to each other) and behaves as a typical alkene (non-aromatic).

Q.8. The following compound contains a keto group but does not undergo nucleophilic addition reactions. Explain.

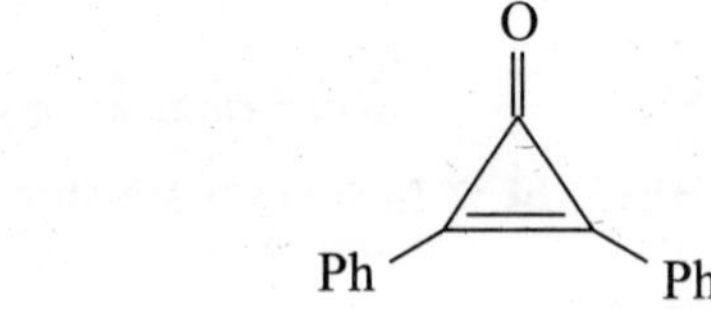

Ans.

A B

The carbon-oxygen bond in compound (A) does not have a true carbonyl character due to its aromatic nature and hence it does not undergo nucleophilic addition reactions.

UNSOLVED PROBLEMS

1. (*i*) The following names are inconsistent with the prevalent rules. Rename these compounds in accordance with the accepted conventions.

(*a*) 1-Hydroxy-2-nitrobenzene (*b*) 2,4,6-Trimethylbenzene

(*c*) 2-Ethyl-1-methylbenzene (*d*) 2-Phenyl-3-butanol.

(*ii*) Write structures and names for all the possible isomers of benzene derivatives having the following molecular formulas:

(*a*) C_8H_{10} (*b*) C_9H_{12}

2. How does Kekulè structure account for all the observed properties of benzene?

3. Carbon-carbon single bond and carbon-carbon double bond lengths in non-aromatic hydrocarbons are 154 pm and 134 pm respectively. However, carbon-carbon bond length in benzene is uniformly 139 pm. Account for this observation.

4. Following alternate structures were proposed for benzene:

(*i*) $HC \equiv C - CH_2 - CH_2 - C \equiv CH$ (*ii*) $HC \equiv C - C \equiv C - CH_2CH_3$

(*iii*) $HC \equiv C - CH_2 - C \equiv C - CH_3$ (*iv*)

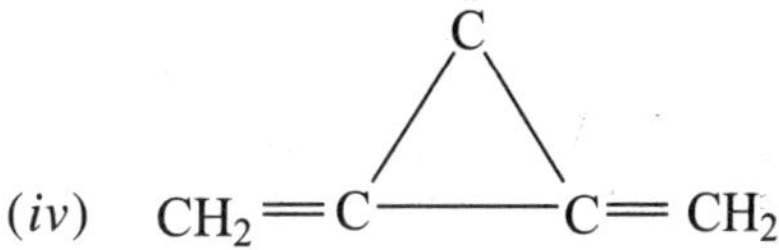

Indicate for each one of them how many (*i*) mono, and (*ii*) disubstitution products are possible? Which structure fits in the isomer number observed for benzene?

5. Since Kekulè formula was not able to explain all the observed properties of benzene, many other formulas were proposed. One of them, the prism formula (prismane) was proposed by Ladenburg. How many mono and disubstituted products are possible by the replacement of one and two hydrogen atoms of prismane? Write their structures and pick up chiral and achiral structures from them.

6. Define the term aromaticity. How is it related with Hückel rule? Account for the aromaticity of cyclopentadienyl anion and cycloheptatrienyl cation.

7. Write the resonance structures for the cyclopentadienyl anion. Can the equivalence of the five hydrogen atoms in this compound be explained by resonance?

8. (*i*) An aqueous solution of tropylium bromide (C_7H_7Br) on treatment with silver nitrate gives precipitate of silver bromide. How do you account for this observation?

(*ii*) Cycloheptatrienyl cation has a low π-electron energy than its open chain counterpart. What conclusion do you draw from this?

9. (*i*) The cyclopentadienyl cation is an antiaromatic compound, whereas cyclopropenyl cation is an aromatic compound. What do these observations mean in terms of π-electron energies of these systems?

(*ii*) Compound (A) easily undergoes keto-enol tautomerism while compound (B) does not why?

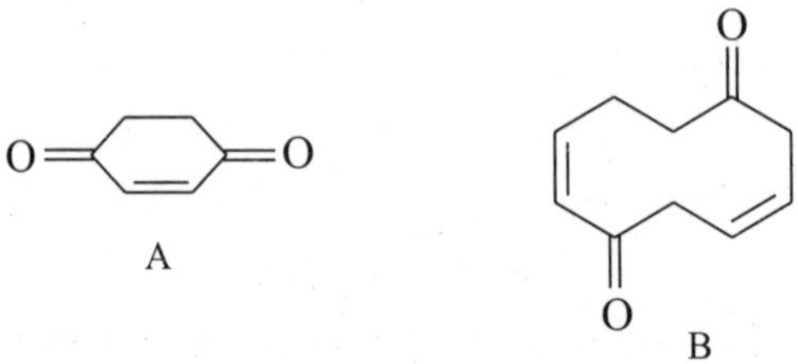

10. Comment upon the statement. "Delocalization alone is not a sufficient criterion for a molecule to be aromatic."

11. (*i*) Draw a labelled diagram of molecular orbital picture of benzene and discuss.

(*ii*) Justify the statement "Cyclohexatriene is different from benzene."

12. Explain

(*i*) All the carbon-carbon bond lengths are equal in benzene.

(*ii*) Although benzene contains three double bonds, it resists addition reactions.

(*iii*) Cyclopentadiene is acidic while cyclopropene is not.

(*iv*) Solvolysis of cyclopentadienyl iodide is 1000 times slower than that of cyclopropenyl iodide.

13. What are annulenes? Discuss their aromatic, non-aromatic and antiaromatic character.

14. Which one of the following species is aromatic?

(*i*) [10] Annulene (*ii*) [20] Annulene

(*iii*) [8] Annulene (*iv*) C_9H_{10} (*v*) $C_9H_9^+$ (*vi*) $C_9H_9^-$

15. When 3-chlorocyclopropene, H Cl is treated with antimony pentachloride, it gives a stable salt (B) $\overset{\oplus}{C_3H_3}\overset{\ominus}{SbCl_6}$ which is highly polar. Further, when A is treated with $AgBF_4$, it gives a white precipitate of AgCl and a crystalline salt (C). Both B and C are much stable than the corresponding allylic substances. Identify B and C giving reasons for their unusual stability.

16. Compound (A) reacts with water much more slowly than the resonance stabilized allyl cation or *tert*-butyl cation. Similarly, compound (B) though having a strained ring system is extremely stable and does not participate in many nucleophilic additions.

(*i*) Explain why A is highly stable?

(*ii*) Draw a resonance structure of B which accounts for these observations.

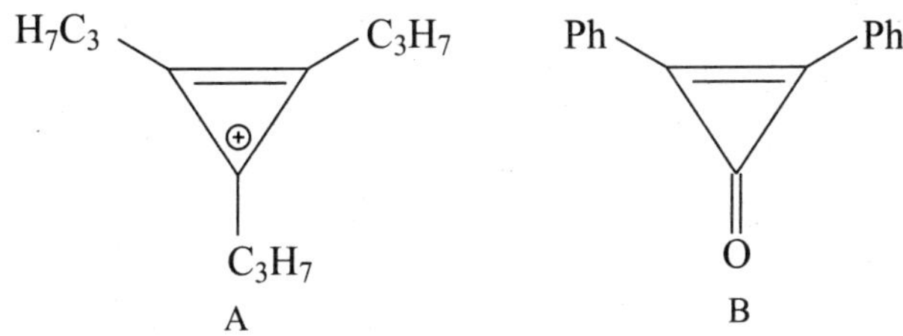

17. *p*-Aminobenzaldehyde behaves like an amide is evident by the fact that it is not easily attacked by the nucleophile at the carbonyl carbon. Explain.

18. What are non-benzenoid aromatic compounds? Give three examples along with structures.

19. Cyclopropenyl cation is much more stable than allyl cation while cyclopropenyl anion is less stable than allyl anion. Explain.

20. Explain why cyclopropenyl anion is antiaromatic though a number of resonance structures can be written for it.

CHAPTER 9

Arenes-II

We have already seen that resonance confers extra stability to the aromatic rings and such systems undergo substitution rather than addition reactions. Further, these substitutions are brought about by the reagents, which are mainly electrophilic in character. A look at the structure of benzene reveals that the σ-skeleton of benzene is planar and is sandwiched between two lobes of π-electron cloud, one above and the other below this plane. This delocalization of electrons in the aromatic system is responsible for holding the carbons together with a greater strength than in the case of alkenes. But still they are loosely held in comparison to *s* electrons. Hence these π-electrons are easily available to an electrophile — which is a positive ion or a dipole or induced dipole. The group that departs must, therefore, leave without its electron pair.

Thus, the typical reactions of the benzene ring are *electrophilic substitution reactions* just as the typical reactions of alkenes are *electrophilic addition reactions.*

Further, these reactions are characteristic of benzene ring wherever it is present and also of many other aromatic systems (benzenoid or non-benzenoid). They are very important synthetically as well as mechanistically. We shall now discuss the mechanism of these reactions.

9.1 AROMATIC ELECTROPHILIC SUBSTITUTION: GENERAL PATTERN OF THE REACTION

As the name implies, in electrophilic aromatic substitution, a covalent bond between the aromatic compound and the electrophilic reagent (E^+) is formed for which the necessary electron pair is supplied by the former. The reaction can be visualized as taking place through following two different pathways:

(*i*) Concerted process and (*ii*) Two-step process

(*i*) Concerted Process

In a concerted process one-step formation of new covalent bond and fission of old C—H bond are simultaneous. Thus, the reaction would take place *via a transition state* (I) which resembles the transition state of an S_N2 process.

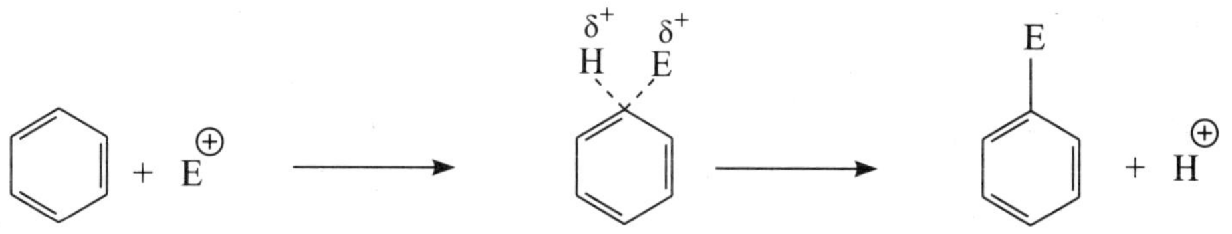

Transition state I

Substituted product

(*ii*) Two-Step Process

In a two step process the **first step** may involve the addition of the electrophile to the aromatic nucleus forming a resonance stabilized carbocation intermediate (II) called as cyclohexadienyl cation. Intermediates of this type which correspond to the energy minimum on the reaction coordinate diagram were first described by **Wheland** and are sometimes referred to a *Wheland intermediates* also. The Wheland intermediate may lose a proton in the **second step** producing the substituted benzene derivative. These steps are shown below and either of these steps may be rate-determining.

Step I

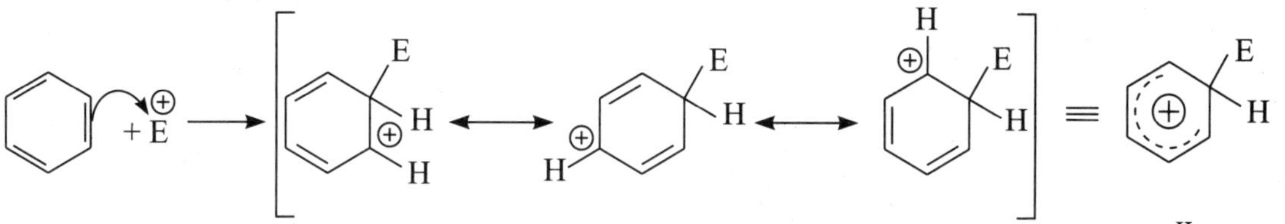

Resonance stabilised carbocation

II

Wheland intermediate

Step II

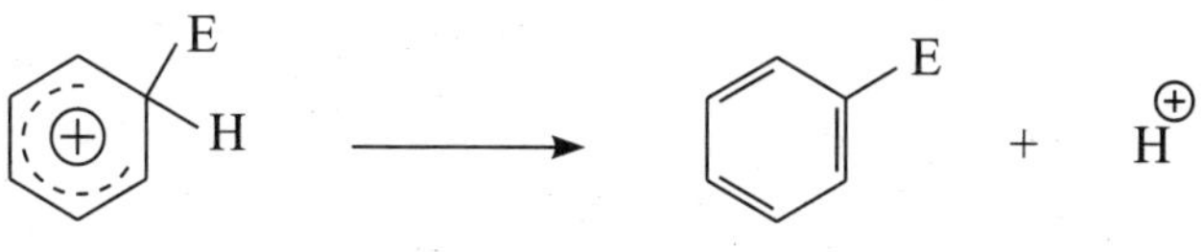

Substituted product

Now the question arises: *which of the two processes represents the correct mechanism*? On the basis of the following observations it is now certain that it is the second mechanism that operates in aromatic electrophilic substitution.

(*a*) **Isolation of intermediates:** The intermediates of aromatic substitution have not only been detected but have also been isolated and characterized in several cases. Following two examples are illustrative:

(*i*) When benzotrifluoride is treated with nitryl fluoride in presence of boron trifluoride at low temperatures (193 K), it gives a yellow crystalline complex which has been isolated and identified. On warming to 223 K, it yields *m*-nitrobenzotrifluoride, BF_3 and HF.

CF_3-benzene + FNO_2 + BF_3 $\xrightarrow{193\text{ K}}$ [arenium ion, H, NO_2] $BF_4^{\ominus}$ $\xrightarrow{223\text{ K}}$ m-$NO_2C_6H_4CF_3$ + BF_3 + HF

Yellow crystalline solid (*isolated*)

m-Nitrobenzotrifluoride

(***ii***) A carbocation intermediate has also been isolated by the treatment of mesitylene with ethyl fluoride and BF_4 at 193 K. The normal substitution product was obtained on warming this intermediate.

Mesitylene + CH_3CH_2F + BF_3 $\xrightarrow{193\text{ K}}$ [arenium ion with H, CH_2CH_3] $BF_4^{\ominus}$ $\longrightarrow$ ethylmesitylene + BF_3 + HF

(***b***) **Isotope effect:** In general aromatic substitutions do not exhibit the isotopic effect. For example, the rates of nitration of *deuterio* and *tritio*-benzenes are the same as that in benzene. Thus, it is reasonable to conclude that proton elimination is not the rate-determining step of the reaction and, therefore, the one step mechanism is inconsistent with the experimental observations.

Substitution *versus* Addition Reaction

The tendency of aromatic compounds to undergo substitution rather than addition is associated with aromaticity (the stability conferred upon a cyclic planar conjugated polyene having $(4q + 2)$ π-electrons). The addition product will no longer retain this special feature and will, therefore, be far less stable than the starting aromatic compound. Hence, it will not be formed. Completely delocalized π-orbitals are regained by the expulsion of a proton from the *Wheland intermediate* (II) thus regaining aromatic stability.

Addition product (*Less stable*) $\xleftarrow[\times]{Nu^{\ominus}}$ II $\xrightarrow{-H^{\oplus}}$ Substitution product (*Aromatic*)

Thus, *benzene and other aromatic compounds do not undergo electrophilic addition reactions since the second step is endothermic at least by* 150.48 *kJ mol*$^{-1}$ (equal to the resonance energy of benzene).

On the other hand, *alkenes prefer to undergo electrophilic addition reaction since no such effects are involved.*

It is worthwhile to mention here that although the first step involving the initial attack of an electrophile on the double bond to produce a carbocation as intermediate is the same for both alkenes as well as benzene. *But the rate of addition to benzene is much slower than that of alkenes.* The reason being that the carbocation intermediate from benzene is much less stable than benzene itself because it has 150.48 kJ mol^{-1} higher energy (resonance energy) than benzene. In other words, E*act for the formation of carbocation intermediate from benzene is much higher than that of alkenes.*

9.2 ROLE OF σ-AND π-COMPLEXES IN AROMATIC ELECTROPHILIC SUBSTITUTIONS

The carbocations formed as intermediates in aromatic substitutions are commonly called as σ-complexes. In these complexes there is actual bonding of the reagent to one of the ring carbon atom and since the π-orbital now involves only five atoms, there is loss of aromaticity.

Experimental evidence, however, suggests that the formation of a σ-complex follows the initial formation of another complex known as π-complex. In contrast, to the σ-complex, the π-complex does not involve actual bonding. Actually the electrophile is held near the π-electron cloud of the aromatic ring.

The following two experiments clearly support the formation of σ-and π-complexes:

(*i*) First Experiment

In the presence of aluminium chloride, benzene forms a 1:1 complex with HCl. The complex is coloured and its solution conducts electricity, signifying its ionic character. Moreover, when DCl is taken in place of HCl, the benzene protons are exchanged with deuterium.This can be explained by assuming that the deuterium cation attaches itself to the ring carbon and thus becomes equivalent to the proton on the same carbon. Either of them can be lost on decomposition of the complex to yield the products. Structure III is in agreement with these observations. Thus under these conditions the reaction gives rise to a σ-complex.

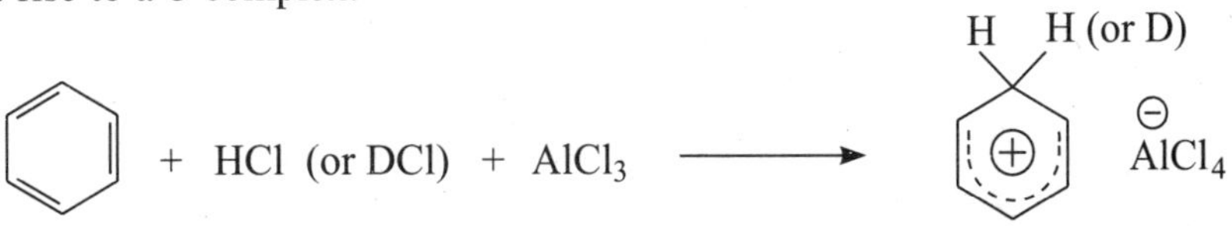

H H (or D)

$$+ \ HCl\ (or\ DCl) + AlCl_3 \longrightarrow \quad \overset{\ominus}{AlCl_4}$$

III

(*ii*) Second Experiment

In the absence of $AlCl_3$, however, benzene still forms a 1 : 1 complex with HCl. But, in contrast to the σ-complex, the solution of this complex is colourless and does not conduct electricity. Finally, when HCl is substituted for DCl, the recovered benzene is not found to contain any deuterium. This type of complex is called a π-complex. It is formed by the loose association of a proton (structure IV) with the π-electron cloud of the benzene ring as shown below:

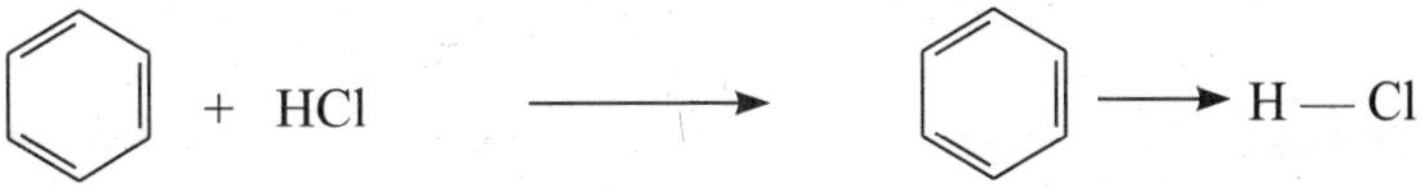

$$+ \ HCl \longrightarrow \quad \longrightarrow H—Cl$$

IV

The Complete Picture

As mentioned above, the initial step of electrophilic aromatic substitution involves the attack of the electrophile on the π-electron cloud of the ring leading to the formation of a π-complex. Now in the π-complex the electrophile removes a pair of electrons from the ring and enters into covalent bond formation with any one of the ring carbon to yield the σ-complex. This σ-complex loses a proton which before departing from the substrate again forms a π-complex. Thus, a complete picture of the mechanism of electrophilic substitution would be as follows:

+ $\overset{\oplus}{E}$ → $\overset{\oplus}{E}$ → E, H → $\overset{\oplus}{H}$ → E + $\overset{\oplus}{H}$

π-Complex σ-Complex π-Complex

The question now arises: *which of the two complexes σ- or π- is involved in the rate-determining step of this reaction*? In case the π-complex is formed in the rate-determining step, the energy profile of substitution would be as shown in Fig. 9.1(*a*); with σ-complex formation being the rate-determining step, the energy profile of the reaction would be as shown in Fig. 9.1(*b*).

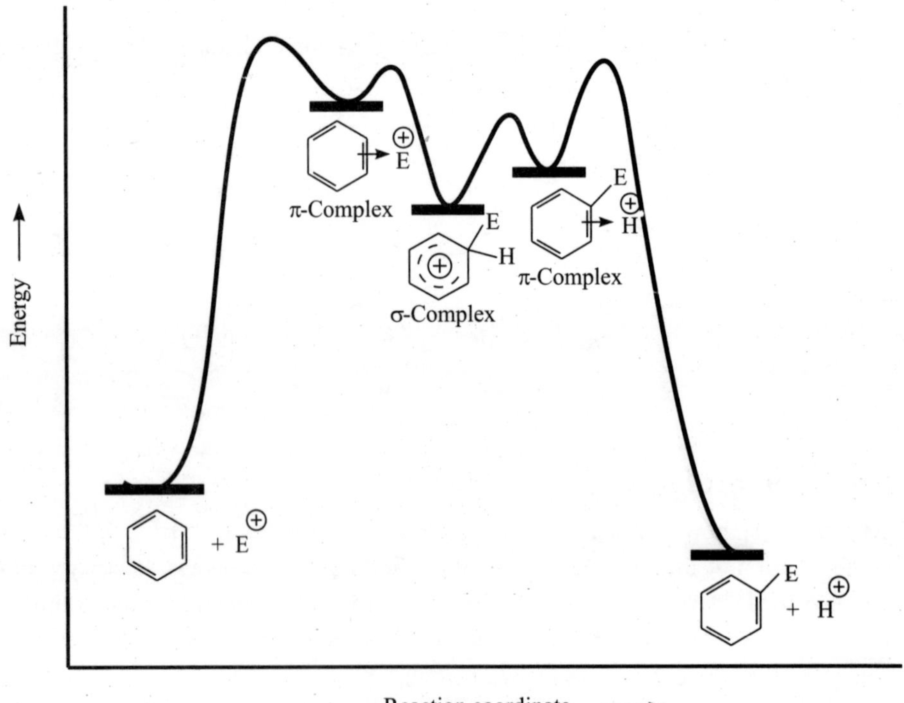

Fig. 9.1(a): Energy profile diagram of aromatic substitution having formation of π-complex as rate-determining step

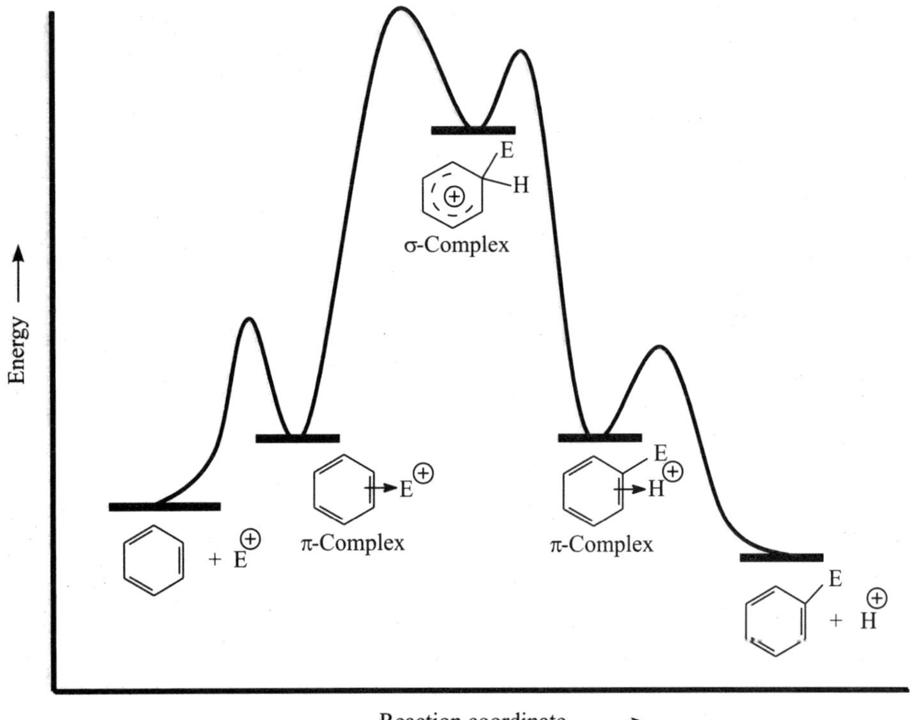

Fig. 9.1(*b*): Energy profile diagram of aromatic substitution having formation of σ-complex as rate-determining step

This problem has been solved by comparing the relative rates of substitution with the relative stabilities of the σ- and π-complexes. **Bromination studies provide a good example**. The rate of bromination of a number of alkylbenzenes has been compared with the relative stabilities of σ as well as π-complexes. It has been shown that the rate of bromination parallels the stabilities of the corresponding σ-complexes more closely than those of π-complexes. Thus it can be said that in this reaction the first step is rapid, reversible formation of a π-complex and a high energy transition state is involved in the conversion of the π-complex to the σ-complex (*c.f.* Fig. 9.1(*b*)).

+ Br — Br $\xrightleftharpoons{\text{Fast}}$ → Br — Br $\xrightarrow{\text{Slow}}$ H, Br $\overset{\ominus}{\text{Br}}$

Olah, on the other hand, has provided evidence that it is the *formation of π-complex which constitutes the rate-determining step in nitration of a series of aromatic hydrocarbons* using nitronium fluoroborate (NO_2BF_4) in tetramethyl sulphone solvent [Fig. 9.1(*a*)]. The initially formed π-complex rapidly collapses to a σ-complex subsequently yielding the product.

However, available data suggest that in most of the aromatic substitutions the high energy transition state strongly resembles the σ-complex and thus the more probable course for the reaction involves the initial reversible, rapid formation of a π-complex followed by the rate-determining formation of a σ-complex.

9.3 ELECTROPHILIC SUBSTITUTION REACTIONS OF BENZENE

We are now in a position to discuss in detail the mechanisms of some common electrophilic substitution reactions of benzene.

9.3.1 Nitration

Nitration is usually brought about by treating an aromatic compound with a mixture of nitric acid and sulphuric acid which is called the *nitrating mixture.*

$$C_6H_6 \xrightarrow{HNO_3/H_2SO_4} C_6H_5NO_2 + H_2O$$

Mechanism

This reaction involves the following steps:

***Step I*: Generation of the electrophile**

The reaction involves electrophilic attack by the ***nitronium ion***, NO_2^+, which is produced under the conditions of the reaction as shown by the following equilibria:

$$H_2SO_4 + HNO_3 \rightleftharpoons \overset{\ominus}{HSO_4} + \overset{\oplus}{H_2O}—NO_2$$

$$\overset{\oplus}{H_2O}—NO_2 \rightleftharpoons HO_2 + \overset{\oplus}{NO_2}$$

Nitronium ion

$$H_2SO_4 + H_2O \rightleftharpoons \overset{\ominus}{HSO_4} + \overset{\oplus}{H_3O}$$

$$\text{overall} \quad 2H_2SO_4 + HNO_3 \rightleftharpoons \overset{\oplus}{H_3O} + 2\overset{\ominus}{HSO_4} + \overset{\oplus}{NO_2}$$

Evidence

The existence of nitronium ion in the acid mixture is evident from the following studies:

*(i) The best evidence for its existence comes from cryoscopic studies. The **vant Hoff** factor for the reaction is 4.0 which is in complete agreement with the formation of nitronium ion. The results of these studies are consistent with the establishment of the above equilibria.*

(ii) Further evidence for the participation of nitronium ion comes from the fact that some other species known to produce nitronium ions, such as $\overset{\oplus}{NO_2}\overset{\ominus}{BF_4}$, $\overset{\oplus}{NO_2}\overset{\ominus}{NO_3}$ and $\overset{\oplus}{NO_2}\overset{\ominus}{ClO_4}$ also nitrate aromatic hydrocarbons.

The relative reactivities of different aromatics towards these nitronium salts are the same as their relative reactivities towards the nitric acid-sulphuric acid mixtures.

***Step II*: Formation of σ-complex or carbocation as intermediate**

The isolation of intermediates and the absence of any isotope effect support the view that addition of the nitronium ion to the aromatic nucleus is the rate-determining step.

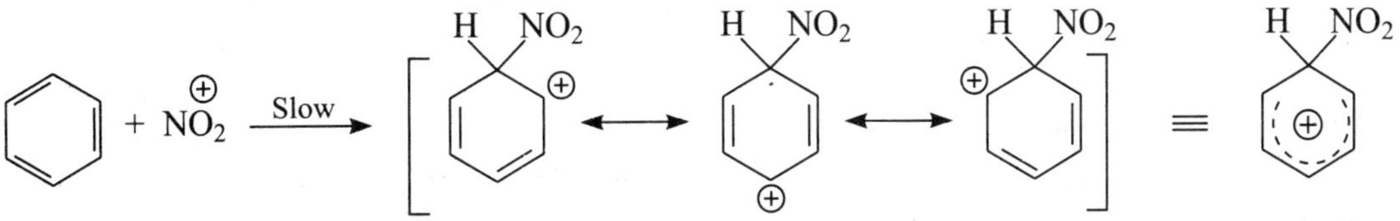

Resonance stabilised carbocation or σ-complex

σ-complex
Wheland intermediate

***Step III*: Loss of a proton from σ-complex**

In this final step the σ-complex or the carbocation intermediate loses a proton to the base $HSO_4^{\ominus}$ to give nitrobenzene.

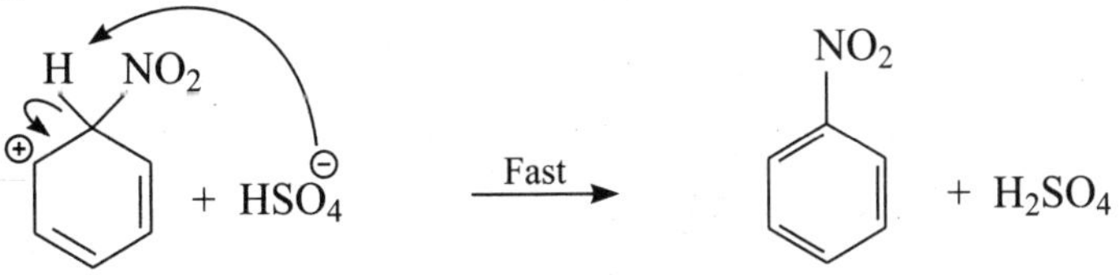

Nitrobenzene

Evidence

Another evidence in support of this mechanism comes from studies of *isotope effect*. It has been found that the rate of nitration of hexadeuterobenzene (C_6D_6) and benzene (C_6H_6) are the same. Hence, the reaction does not show any isotope effect. Had it been the case then the rate of reaction in case of C_6D_6 should have been slower than in the case of C_6H_6. Therefore, the loss of a proton is not the *rate-determining* step.

9.3.2 Halogenation

Halogenation is generally brought about by treating benzene with halogen in presence of metal halides known as *halogen carriers* at low temperature and in dark. It can take place chosing any one condition.

(i) Halogenation with free halogen

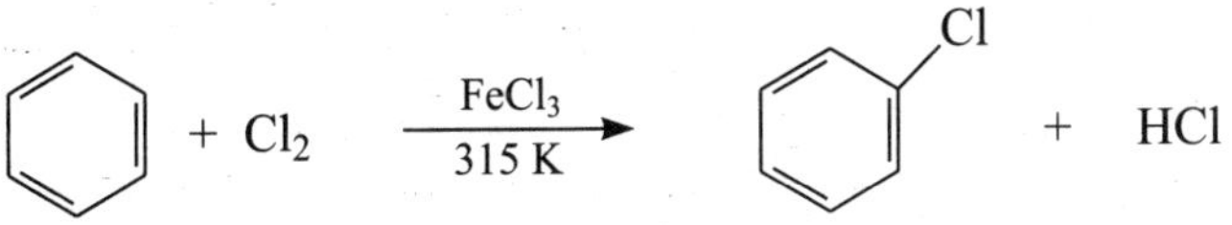

$$C_6H_6 + Cl_2 \xrightarrow[315\ K]{FeCl_3} C_6H_5Cl + HCl$$

Mechanism: Halogenation takes place in the following steps:

***Step I*: Generation of an electrophile**

The main function of the catalyst (metal halide) is to partially or completely polarize the halogen-halogen bond.

Chapter 9

$$FeCl_3 + Cl_2 \rightleftharpoons \overset{\delta^+}{Cl}-\overset{\delta^-}{Cl} \ldots\ldots FeCl_3 \quad \text{or} \quad \overset{\oplus}{Cl} + \overset{\ominus}{FeCl_4}$$

***Step II*: Formation of σ-complex or carbocation as intermediate**

The positive end of the chlorine dipole (or the free chloronium ion, Cl^+, itself) attacks the aromatic ring resulting in the formation of a σ-complex.

$$C_6H_6 + Cl_2 \underset{}{\overset{\text{Fast}}{\rightleftharpoons}} C_6H_6 \rightarrow Cl-Cl$$

π-complex

$$C_6H_6 \rightarrow Cl-Cl + FeCl_3 \rightleftharpoons C_6H_6 \rightarrow \overset{\delta^+}{Cl}-\overset{\delta^-}{Cl} \ldots\ldots FeCl_3$$

Slow

Cl H, Cl H, Cl H ⟷ resonance structures ≡ Cl H σ-complex $\overset{\ominus}{FeCl_4}$

σ-complex

***Step III*: Loss of a proton**

The carbocation intermediate or s-complex loses a proton to the base, $FeCl_4^-$, to give chlorobenzene.

$$\text{Cl H (σ-complex)} + \overset{\ominus}{FeCl_4} \xrightarrow{\text{Fast}} C_6H_5Cl + HCl + FeCl_3$$

This step is fast and thus does not affect the rate of reaction.

(ii) Halogenation with interhalogen compounds

Aromatic compounds undergo halogenation on treatment with some interhalogen compounds in presence of Lewis acid catalysts. More electropositive atom acts as an electrophile. For example, with iodine monochloride (ICl), benzene gives iodobenzene (and not chlorobenzene).

$$I-Cl + AlCl_3 \longrightarrow \overset{\oplus}{I} + \overset{\ominus}{AlCl_4}$$

$$C_6H_6 + \overset{\oplus}{I} \xrightarrow{\text{Slow}} \text{σ-complex (H, I)} \xrightarrow[\text{Fast}]{-H^{\oplus}} C_6H_5I$$

σ-complex Iodobenzene

(iii) *Halogenation with hypohalous acids*

Halogenation can also be affected by using hypochlorous or hypobromous acids in presence of strong acids. The function of strong acid catalyst is to protonate the hypohalous acids thus making them powerful electrophilic species.

$$H-\ddot{\underset{\cdot\cdot}{O}}-Cl \;+\; \overset{\oplus}{H} \xrightarrow{\text{Fast}} H_2\overset{\oplus}{O}-Cl$$

$$C_6H_6 \;+\; Cl-\overset{\oplus}{O}H_2 \xrightarrow{\text{Fast}} [C_6H_6Cl]^{\oplus} \xrightarrow[\text{Fast}]{-\overset{\oplus}{H}} C_6H_5Cl$$

Instead of a protonated hypohalous acid, the attacking species can also be a positive halogen ion, without affecting the course of the reaction.

$$H_2\overset{\oplus}{O}Cl \rightleftharpoons H_2O + \overset{\oplus}{Cl}$$

9.3.3 Sulphonation

Sulphonation is brought about by treating benzene or other aromatic compounds with conc. H_2SO_4 or oleum $H_2S_2O_7$ ($H_2SO_4 + SO_3$).

$$C_6H_6 \xrightleftharpoons{H_2SO_4 \text{ or } (H_2SO_4 + SO_3)} C_6H_5SO_3H + H_2SO_4$$

Benzenesulphonic acid

Mechanism

Overall reaction takes place in the following steps:

***Step I*: Generation of the electrophile**

There is a controversy regarding the possible sulphonating species. Probably it is the free sulphur trioxide produced by the following "acid-base" equilibrium between two sulphuric acid molecules.

$$H_2SO_4 + H_2SO_4 \rightleftharpoons SO_3 + H_3\overset{\oplus}{O} + HS\overset{\ominus}{O_4}$$

Chapter 9

***Step II*: Formation of resonance stabilized σ-complex (carbocation)**

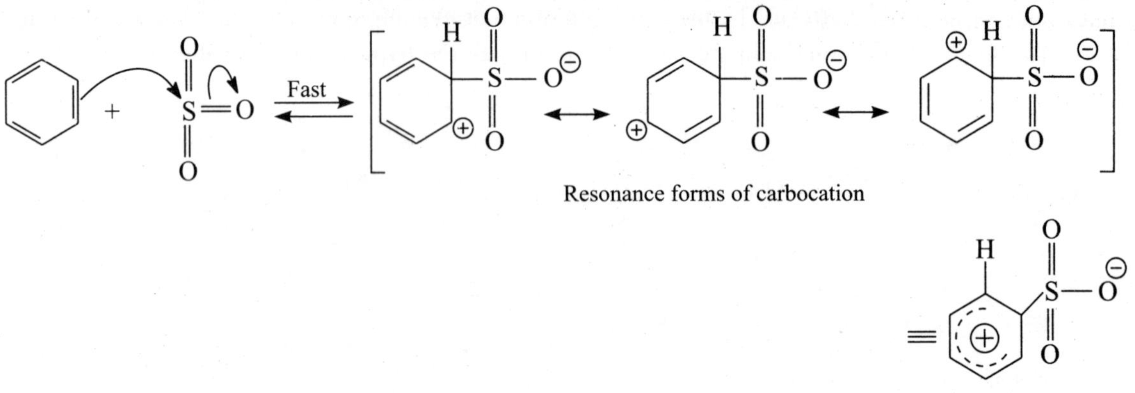

Fast

Resonance forms of carbocation

σ-Complex

Although SO_3 is not positively charged, it has electron deficient sulphur atom. Hence it behaves as an electrophile.

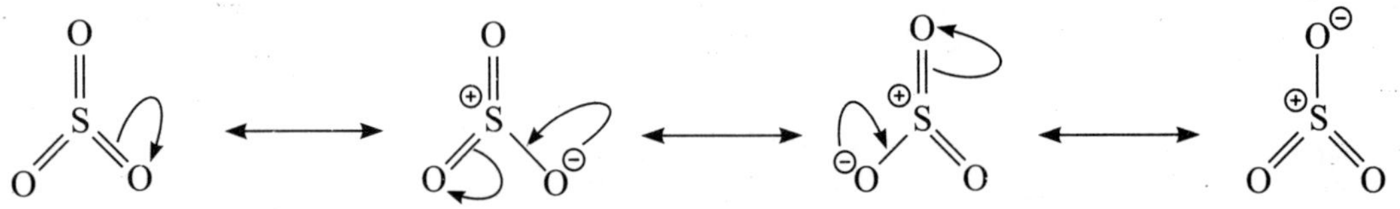

***Step III*: Loss of a proton from the resonance stabilized σ-complex**

Slow

+ HSO_4^- ⇌ $C_6H_5SO_3^-$ + H_2SO_4

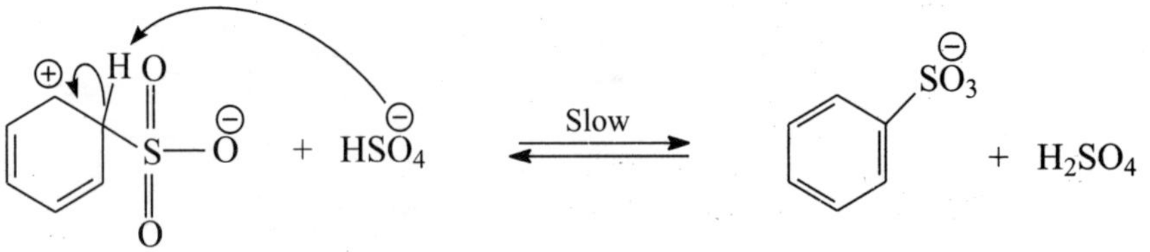

***Step IV*: Gain of proton from H_3O^+ formed in step I to produce unionized benzenesulphoric acid which is highly dissociated**

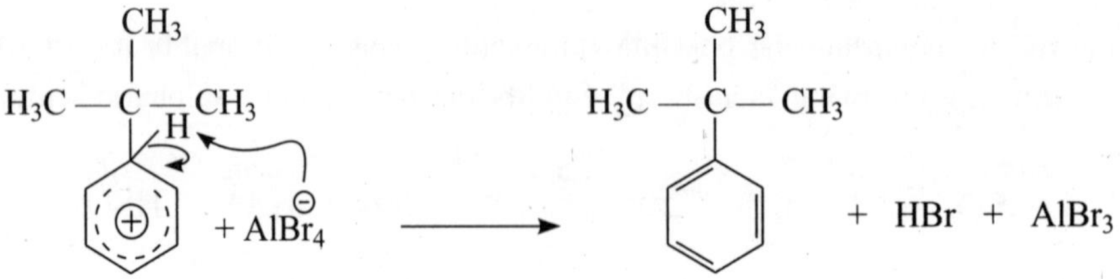

+ $AlBr_4^-$ → + HBr + $AlBr_3$

Difference from other electrophilic substitutions

Here it may be worthwhile to mention that sulphonation differs from other electrophilic substitution reactions in the following two aspects:

(*i*) *It shows positive isotope effect* (*ii*) *It is reversible*

(i) It shows positive isotope effect

It has been observed that hexadeuterated or hexatritiated benzene reacts more slowly than benzene, it means that the C—H bond cleavage (removal of proton in step III) is the rate-determining step.

(ii) It is reversible

Sulphonic acids are known to undergo desulphonation when heated in presence of dilute mineral acid and superheated steam.

$$C_6H_5SO_3H + \underset{\text{(aq. acid)}}{H_3O^{\oplus}} \xrightarrow[\text{Steam}]{\text{Pressure}} \underset{\text{Benzene}}{C_6H_6} + H_2SO_4$$

9.3.4 Mercuration

This reaction is carried out by heating benzene with mercuric acetate. The product is phenylmercuric acetate.

$$C_6H_6 + (CH_3COO)_2\,Hg \xrightarrow{363\text{–}430\ K} \underset{\text{Phenylmercuric acetate}}{C_6H_5HgOCOCH_3} + CH_3COOH$$

It provides another example of an electrophilic aromatic substitution exhibiting a *positive isotope effect*. It has been observed that benzene reacts six times faster than perdeuteriobenzene under similar conditions thereby demonstrating that like sulphonation this reaction also involves breaking of a C—H bond in the rate-determining step. An inspection of the energy profile diagram shown in Fig. 9.2 makes it clear.

Mechanism

Like other electrophilic aromatic substitutions, this reaction also takes place in three steps as follows:

***Step I*: Generation of the electrophile**

The mercurating species is a positively charged acetomercuric ion $\overset{\oplus}{H}gOCOCH_3$ formed by the initial ionization of mercuric acetate.

$$Hg(O-\overset{\overset{O}{\|}}{C}-CH_3)_2 \xrightleftharpoons{\text{Ionization}} CH_3CO\overset{\ominus}{O} + \overset{\oplus}{H}gO\overset{\overset{O}{\|}}{C}CH_3$$

***Step II*: Formation of a resonance stabilized σ-complex**

$$C_6H_6 + \overset{\oplus}{H}gOCOCH_3 \xrightleftharpoons{\text{Fast}} [C_6H_6(HgOOCCH_3)]^{\oplus} \quad (\text{ring carbon bearing } CH_3COOHg \text{ and } H)$$

***Step III*: Loss of a proton from σ-complex**

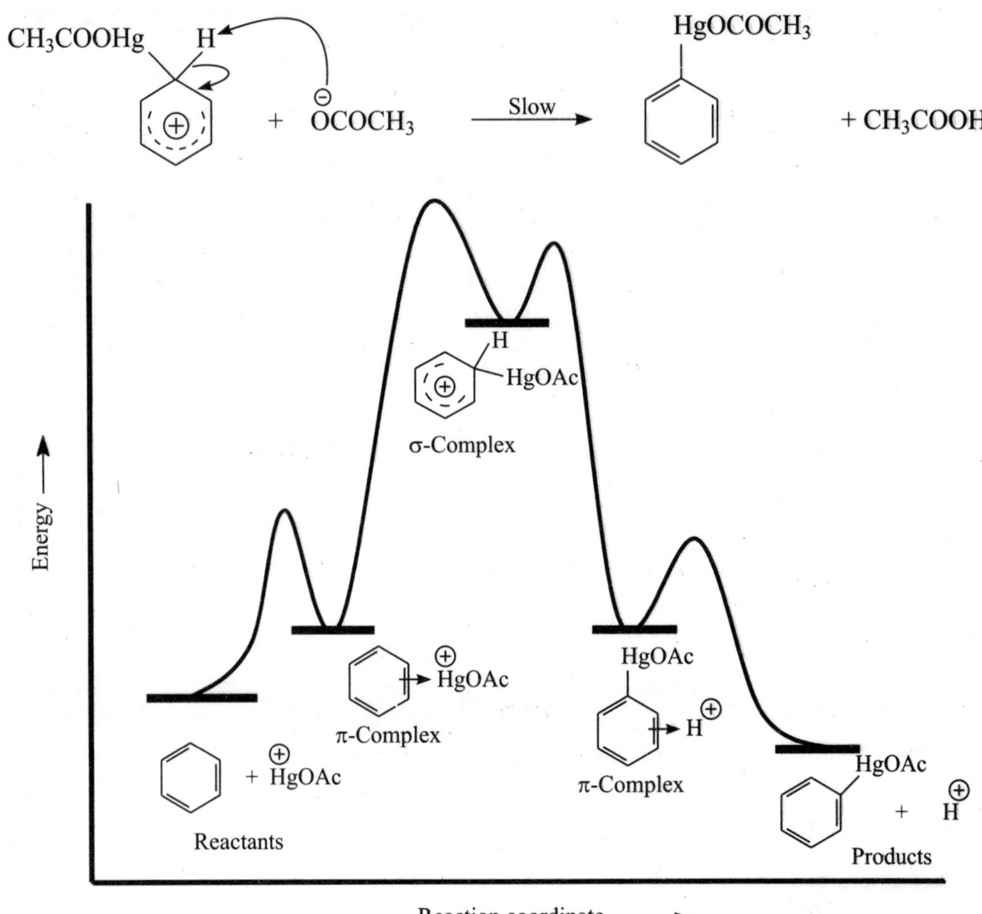

Fig. 9.2: Energy profile diagram of mercuration of benzene

9.3.5 Friedel-Crafts Reaction

In fact the name Friedel-Crafts Reaction, is used to describe two different processes taking place by the same mechanism. These processes are called

(*i*) Friedel-Crafts Alkylation and (*ii*) Friedel-Crafts Acylation

(*i*) Friedel-Crafts Alkylation

Aromatic compounds react with alkyl halides in presence of Lewis acid catalysts to yield alkylated products.

$$C_6H_6 + \underset{\text{Alkyl halide}}{RX} \xrightarrow{\text{Anhydrous } AlCl_3} \underset{\text{Alkylbenzene}}{C_6H_5R} + HX$$

This reaction is known as the ***Friedel-Crafts alkylation reaction.***

Mechanism: It also takes place by stepwise mechanism as discussed below:

***Step I*: Generation of an electrophile:** In this step an alkyl halide reacts with anhydrous $AlCl_3$ (or any other Lewis acid such as BF_3, $GaCl_3$, etc.) to produce a free or incipient alkyl carbocation which acts as an electrophile.

In case of alkylation with **tertiary alkyl halides**, the electrophilic species is a free carbocation. For example,

$$(CH_3)_3C{-}Br + AlBr_3 \longrightarrow (CH_3)_3C^{\oplus} + AlBr_4^{\ominus}$$

tert-Butyl bromide → *tert*-Butyl carbocation

***Step II*: Formation of resonance stabilized σ-complex:** The alkyl carbocation is attracted towards the aromatic ring resulting in the formation of a resonance stabilized carbocation (σ-complex).

$$C_6H_6 + (CH_3)_3C^{\oplus} \longrightarrow [C_6H_6C(CH_3)_3]^{\oplus}$$

σ-Complex intermediate

The intermediate sigma complexes have been isolated and characterized.

***Step III*: Loss of a proton from carbocation (σ-complex)**

$AlBr_4^{\ominus}$ abstracts a proton from σ-complex to produce alkylbenzene.

$$[C_6H_6C(CH_3)_3]^{\oplus} + AlBr_4^{\ominus} \longrightarrow C_6H_5C(CH_3)_3 + HBr + AlBr_3$$

However, with **primary and secondary alkyl halides** it appears that instead of free carbocations the electrophilic species is an alkyl halide-Lewis acid complex with a positively polarized carbon.

$$R{-}\ddot{X}: + AlX_3 \longrightarrow \overset{\delta^+}{R}\text{.........}X\text{.............}\overset{\delta^-}{AlX_3}$$

The principal **evidence against** the participation of free carbocations in these reactions is based on the following two observations:

1. *These reactions are first-order with respect to the aromatic compound, the Lewis acid and the alkyl halide.*

2. *When toluene is methylated using methyl bromide and methyl iodide, the isomeric xylenes are formed in different ratios.*

		o-xylene	*m*-xylene	*p*-xylene
Toluene	CH_3Br →	54%	17%	29%
	CH_3I →	49%	11%	40%

Free carbocation (CH_3^+ in either case) should have reacted identically with toluene irrespective of their source of generation.

Nature of Electrophile

On the basis of above observations it may be concluded that alkylation *via* polarized alkyl halides takes place in the case of primary halides. The reason for this is the high energy required to generate a primary carbocation. Secondary alkyl halides are on the border line and alkylation may take place through either mechanism. With *tertiary* alkyl halides the alkylation involves a free carbocation.

Other alkylating agents: Aromatic compounds can be alkylated with carbocations obtained from other sources such as alkenes or alcohols as shown below:

$$(CH_3)_2C{=}CH_2 \text{ (Alkene)} \xrightarrow{H^{\oplus}} H_3C{-}\overset{\oplus}{C}(CH_3)_2 \text{ (Carbocation)} \xleftarrow[-H_2O]{H^{\oplus}} (CH_3)_3C{-}OH \text{ (Alcohol)}$$

$$H_3C{-}\overset{\oplus}{C}(CH_3)_2 \xrightarrow{C_6H_6} (CH_3)_3C{-}C_6H_5$$

Limitations of Friedel-Crafts Alkylation

Though quite versatile, this reaction has the following two main limitations:

1. *Rearrangement of carbocations*: Friedel-Crafts alkylation is often accompanied by rearrangements in the alkylating groups. As an example, the alkylation of benzene with *n*-propyl chloride leads to the formation of a mixture of *n*-and *iso*-propylbenzenes. The formation of the rearranged product can be rationalized on the basis of the relative carbocation stability although free carbocations may not be present in the reaction.

$$CH_3-CH_2-CH_2-Cl + AlCl_3 \longrightarrow CH_3-\underset{|}{CH}(H)-\overset{\delta^+}{CH_2}\text{.........}Cl\text{.........}\overset{\delta^-}{AlCl_3}$$

1, 2-Hydride shift

$$H_3C-\overset{\delta^+}{CH}(CH_3)-Cl\text{..........}\overset{\delta^-}{AlCl_3} \xrightarrow{C_6H_6} C_6H_5CH(CH_3)_2 + C_6H_5CH_2CH_2CH_3$$

2. *Formation of polysubstituted products*: An experimental limitation of the Friedel-Crafts alkylations is the formation of polysubstituted products. The reason for this is that the introduction of one alkyl substituent activates the ring towards a second substitution. This can be avoided, to some extent, using an excess of the hydrocarbon substrate.

Other limitations of this reaction are as follows:

(*a*) Aryl halides cannot be used in place of alkyl halides.

(*b*) Substrates like nitrobenzene cannot be alkylated because of the deactivating influence of nitro group on the ring. ***In fact nitrobenzene is used as a solvent in many Friedel-Crafts reactions.***

(*c*) Aniline and other aromatic amines fail to undergo Friedel-Crafts reaction, probably because of the formation of a complex between the base and Lewis acid. Since, this complex has a positive charge on nitrogen, the ring is deactivated.

$$C_6H_5\ddot{N}H_2 + AlCl_3 \longrightarrow C_6H_5\overset{\oplus}{N}H_2\overset{\ominus}{AlCl_3}$$

However, it may be added here that in spite of these limitations Friedel-Crafts reaction is widely applicable in organic synthesis.

The important aspects of Friedel-Crafts alkylation can be summarized in the following chart using benzene, aluminium chloride and an alkyl halide.

$$R-X \xrightarrow{AlCl_3} \begin{cases} \xrightarrow{R = \textit{tertiary}} \overset{\oplus}{R} + X-\overset{\ominus}{AlCl_3} \xrightarrow{C_6H_6} C_6H_5R \text{ and / or } C_6H_5R' \\ \xrightarrow{R = \textit{primary or secondary alkyl}} \overset{\delta^+}{R}\text{------}X\text{----}\overset{\delta^-}{AlX_3} \xrightarrow{C_6H_6} C_6H_5R \text{ and / or } C_6H_5R' \end{cases}$$

R′ = rearranged alkyl group

(ii) Friedel-Crafts Acylation

Benzene and other aromatic compounds react with acid chlorides or acid anhydrides in presence of anhydrous $AlCl_3$ (or other Lewis acid catalysts) to provide acylated products (ketones). For example,

$$C_6H_6 + CH_3COCl \xrightarrow{AlCl_3} C_6H_5COCH_3 + HCl$$

$$C_6H_6 + C_6H_5COCl \xrightarrow{AlCl_3} C_6H_5COC_6H_5 + HCl$$

$$C_6H_6 + (CH_3CO)_2O \xrightarrow{AlCl_3} C_6H_5COCH_3 + CH_3COOH$$

Mechanism

Like Friedel-Craft alkylation, the acylation reaction also takes place by a stepwise mechanism. Various steps involved in the reaction are as follows:

***Step I*: Generation of electrophile**

Acetylium ion (RCO^+), which is the electrophile in the reaction, is generated by the reaction of acyl chloride or an acid anhydride with the Lewis acid ($AlCl_3$, etc.).

$$\underset{\text{Acetyl chloride}}{CH_3-\overset{O}{\overset{\|}{C}}-Cl} + AlCl_3 \longrightarrow \underset{\text{Acetylium ion (electrophile)}}{\left[CH_3-\overset{\oplus}{C}=\ddot{O}: \longleftrightarrow CH_3-C\equiv\overset{\oplus}{O}:\right]} + \overset{\ominus}{AlCl_4}$$

$$\underset{\text{Acetic anhydride}}{CH_3-\overset{O}{\overset{\|}{C}}-O-\overset{O}{\overset{\|}{C}}-CH_3} + AlCl_3 \longrightarrow \underset{\text{Acetylium ion (electrophile)}}{\left[CH_3-\overset{\oplus}{C}=\ddot{O}: \longleftrightarrow CH_3-C\equiv\overset{\oplus}{O}:\right]} + (CH_3\overset{O}{\overset{\|}{C}}O)\overset{\ominus}{AlCl_3}$$

***Step II*: Formation of resonance stabilized σ-complex**

$$C_6H_6 + CH_3-\overset{\oplus}{C}=\ddot{O}: \longrightarrow [C_6H_6(H)(COCH_3)]^{\oplus}$$

***Step III*: Loss of a proton from σ-complex**

$AlCl_4^-$ or $(CH_3COO)AlCl_3^-$ abstract a proton from σ-complex to produce acylbenzene.

H X⊖ COCH₃ —Fast→ COCH₃ + HX

$\left[\overset{\ominus}{X} = \overset{\ominus}{Cl} \text{ in case of } \overset{\ominus}{AlCl_4} \text{ and } CH_3CO\overset{\ominus}{O} \text{ in case of } (CH_3COO)\overset{\ominus}{AlCl_3} \right]$

Olah prepared a number of acetylium salts such as $CH_3\overset{\oplus}{C}O\overset{\ominus}{B}F_4$ and $CH_3\overset{\oplus}{C}O\overset{\ominus}{S}bCl_6$, which have been used for aromatic acylation reactions. The structures of these salts have been established by NMR spectroscopy.

Unlike alkylation, acylation reactions are **not** accompanied by rearrangement within the acyl group and also there is **no polysubstitution** as the aromatic ring is deactivated after the introduction of the first acyl group.

***Note*:** The ketone produced in acylation reaction forms a complex with $AlCl_3$. Therefore, this reaction is carried out using more than one equivalent of $AlCl_3$. After the reaction is over the complex is broken by adding water to the reaction mixture to liberate the product from the complex by hydrolysing it.

$C_6H_5-\overset{O}{\overset{\|}{C}}-CH_3 + AlCl_3 \rightleftharpoons C_6H_5-\overset{\overset{\oplus}{O}-\overset{\ominus}{AlCl_3}}{\overset{\|}{C}}-CH_3 \xrightarrow{3H_2O} C_6H_5-\overset{O}{\overset{\|}{C}}-CH_3 + Al(OH)_3 + 3HCl$

9.4 ORIENTATION EFFECTS OF SUBSTITUENTS

When an electrophilic reagent attacks a substituted aromatic ring, the group already attached to the ring determines how in this reaction the attack occurs and at what position of the ring it occurs. The ability of a group already present in the ring to orient or direct the incoming group is called *directive influence of the group*. In other words, the group attached to the ring not only affects the *reactivity* but also determines the *orientation*, of substitution. On the basis of these two characteristics, various substituents have been divided into two categories.

(*i*) Groups such as —NR_2, —NH_2, —OH, —$\overset{\ominus}{O}$, — NHCOR, — OR, —OCOR and alkyls lead to *o*, *p*-orientation with **activation**,

(*ii*) Groups such as —$\overset{\oplus}{N}R_3$, —CX_3, —C ≡ O, —C ≡ N, —SO_2OH and —NO_2 lead to *meta*-orientation with **deactivation**.

Thus, for instance, nitration of toluene is relatively easier (taking place at low temperature) and gives *o*-and *p*-nitrotoluenes as major products. On the other hand, nitration of nitrotoluenes is somewhat difficult (taking place at 363 K) and yields *m*-dinitrobenzene as the major product. From this example it is clear that although the incoming group is the same, *i.e.*, NO_2 group but the position occupied by it is different. In case of toluene, the substituent methyl group directs the incoming nitro group to *ortho* and *para*-positions but in case of nitrobenzene, the substituent nitro group *directs* the incoming group to *meta*-position.

Toluene $\xrightarrow{HNO_3/H_2SO_4}$ *o*-Nitrotoluene (58%) + *p*-Nitrotoluene (38%) + *m*-Nitrotoluene (04%)

Nitrobenzene $\xrightarrow{HNO_3/H_2SO_4}$ *m*-Dinitrobenzene (93%) + *o*-Dinitrobenzene (6.4%) + *p*-Nitrobenzene (0.6%)

9.5 THEORY OF REACTIVITY AND ORIENTATION

9.5.1 Theory of Reactivity

The rate of electrophilic substitution, like any other reaction, depends on the difference between the energy of the transition state and the ground state of the reactants, *i.e.*, *energy of activation* (*Eact*). Since, the rate-determining step of electrophilic aromatic substitution is usually the formation of a σ-complex, a slightly modified form of the reaction coordinate diagram for an aromatic electrophilic substitution on benzene ignoring the π-complex formation can be constructed as shown in Fig. 9.3.

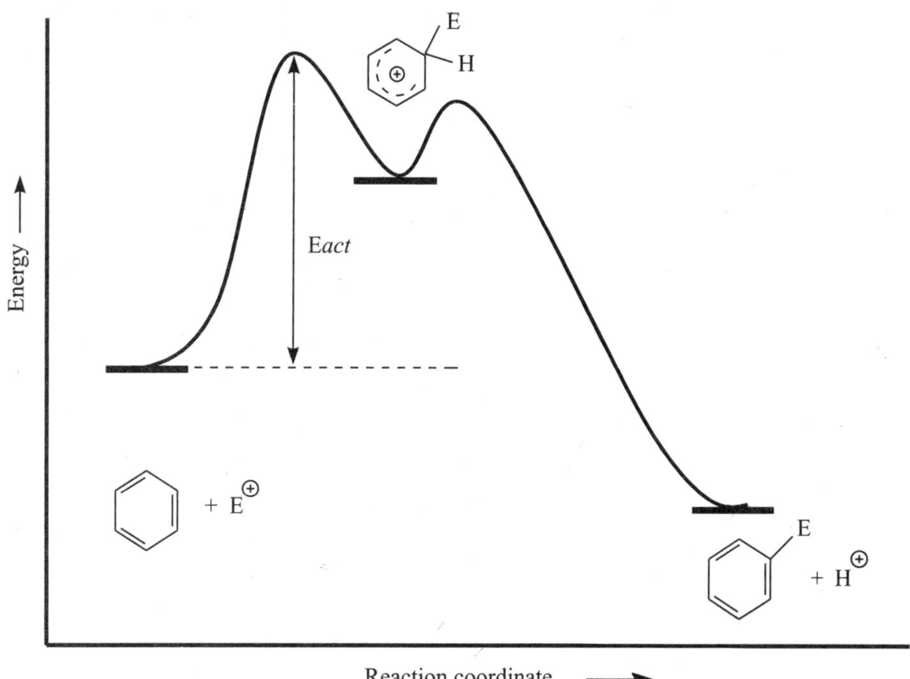

Fig. 9.3: Energy profile for a typical electrophilic aromatic substitution ignoring the formation of π-complex

Similar energy profiles can be constructed for substituted aromatic rings. The rate of substitution at *o*-, *p*- or *m*-positions then depends on the height of the energy barrier between the reactant and the transition state (*Eact*). If any effect due to the substituent lowers *Eact.* The rate of substitution increases and the favoured orientation of attack is the one with the lowest *Eact*. Since it is difficult to assess *the energy of transition state directly, the energy of the σ-complex is often assessed as the transition state closely resembles it in energy and geometry. Thus any effect of the substituent which stabilizes the σ-complex will enhance the rate of substitution.* The substituents of category (*i*) are electron-donating (either by inductive or by resonance effect) and exert an electrostatic effect which stabilizes the positively charged σ-complex and thus increases the rate of the reaction. On the other hand, substituents of category (*ii*) are electron-withdrawing which destabilize the σ-complex, *i.e.*, the positive charge is intensified on the carbocation due to which the rate of the reaction decreases. It is worthwhile to mention here that the release or withdrawal of electrons may occur due to inductive effect alone or due to a combination of both inductive and resonance effects. The two effects may reinforce or may oppose each other.

9.5.2 Theory of Orientation

The orientation effect of the substituents can be explained by writing all the possible resonance forms of the charged intermediate for each of the three possible reaction courses. Following examples illustrate these points:

Various substituents which orient an incoming group to a particular position may be classified as under:

(*a*) ***Ortho-para*** **directing groups** with electron releasing inductive effect **(+I effect).**

In toluene where the methyl group is electron-donating (by inductive effect) the following resonance forms can be written for attack of electrophile at *para-meta-*and *ortho-*positions.

Attack of electrophile at para-position

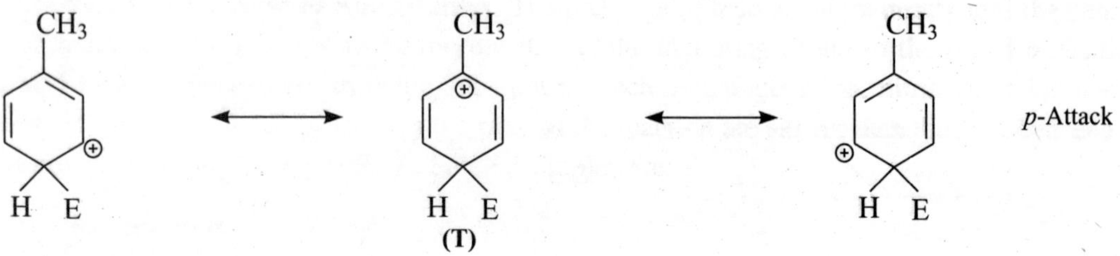

CH_3 H E ⟷ CH_3 H E (T) ⟷ CH_3 H E *p*-Attack

Attack of electrophile at meta-position

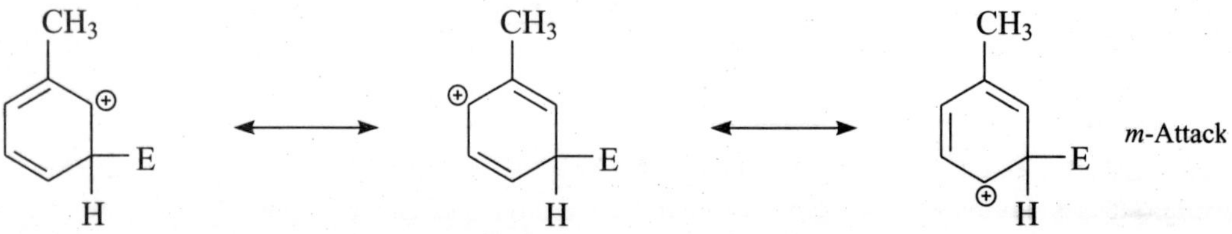

CH_3 E H ⟷ CH_3 E H ⟷ CH_3 E H *m*-Attack

Attack of electrophile at ortho-position

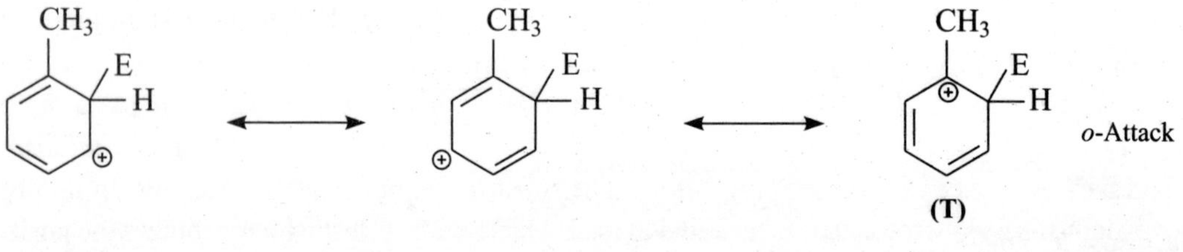

CH_3 E H ⟷ CH_3 E H ⟷ CH_3 E H (T) *o*-Attack

In case of *ortho-*and *para-*attacks, there in one structure (marked **T**) in each case where the positive charge is located on the carbon atom to which the methyl group is attached, whereas no such structure can be written when the electrophile attacks at the *meta-*position. As structures marked **T** are *tertiary carbocations, they are more stable* than rest of structures which are all *secondary carbocations.* Thus in this case substitution at *o-*and *p-*is preferred to *m-*substitution.

In other words, the energy of *o-*and *p-*substituted Wheland intermediates (σ-complexes) is considerably lower than that of *m-*substituted intermediates. However, it may be noted that its energy is lower than the corresponding intermediate in benzene. This can be easily understood by an inspection of the energy profile, depicting substitution in benzene and *o-*, *m-* and *p-*substitution in toluene (Fig. 9.4).

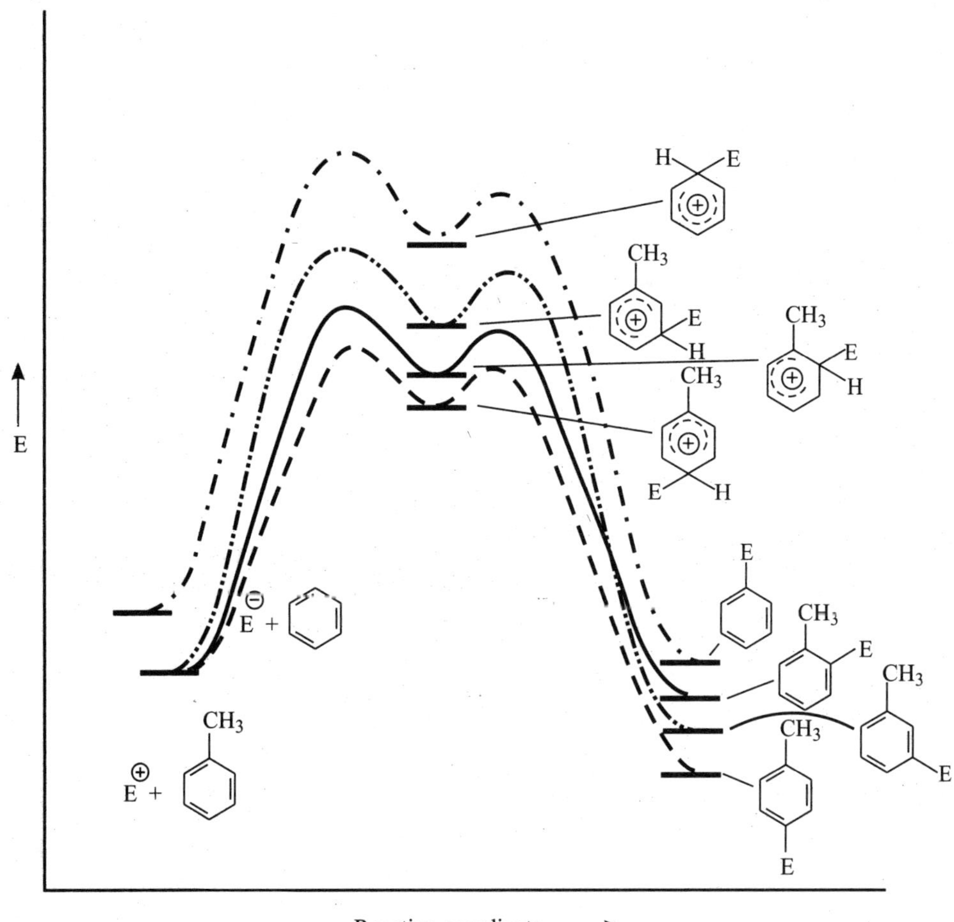

Fig. 9.4: Energy profile for electrophilic substitution in benzene and toluene

Thus electrophilic substitution in toluene occurs faster than in benzene and substituition will occur mainly at *ortho*-and *para* - positions.

(*b*) *Ortho-para* directing groups with Electron-withdrawing inductive effect **(–I effect)** and electron-releasing resonance or mesomeric effect (+ R or + M-effect.)

Groups such as $—\ddot{\underset{\cdot\cdot}{O}}H$, $—\ddot{\underset{\cdot\cdot}{O}}R$, $—\ddot{N}H_2$, $—\ddot{N}HR$, $—\ddot{N}R_2$, etc., have lone pair(s) of electrons on nitrogen and oxygen atoms and are, therefore, powerful electron donors. This effect is due to strong interaction of their lone pair of electrons with the π -electrons of benzene ring.

The orientation effect of these substituents can be understood by taking $—\ddot{N}H_2$ group as an example.

As in the case of toluene, the following resonance contributing forms can be written for σ-complex formed as a result of attack of an electrophile on *para*-, *meta*- and *ortho*-positions.

Attack of electrophile at para-position

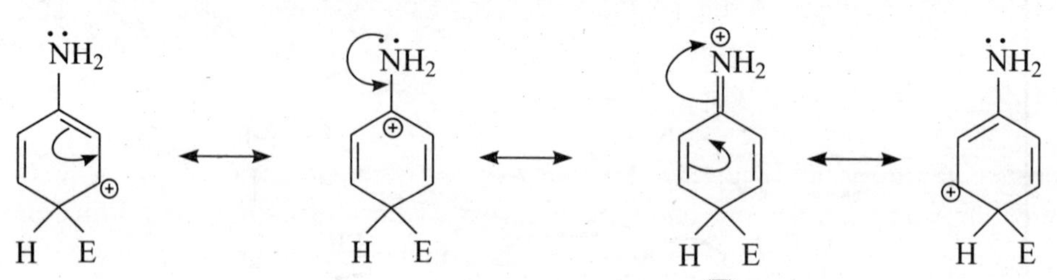

$\ddot{N}H_2$ $\ddot{N}H_2$ $\overset{\oplus}{N}H_2$ $\ddot{N}H_2$

H E ↔ H E ↔ H E ↔ H E

(T)

Attack of electrophile at meta-position

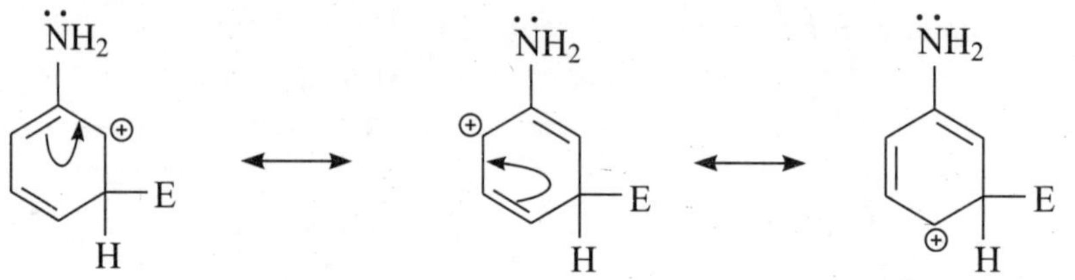

$\ddot{N}H_2$ $\ddot{N}H_2$ $\ddot{N}H_2$

E ↔ E ↔ E

H H H

Attack of electrophile at ortho-position

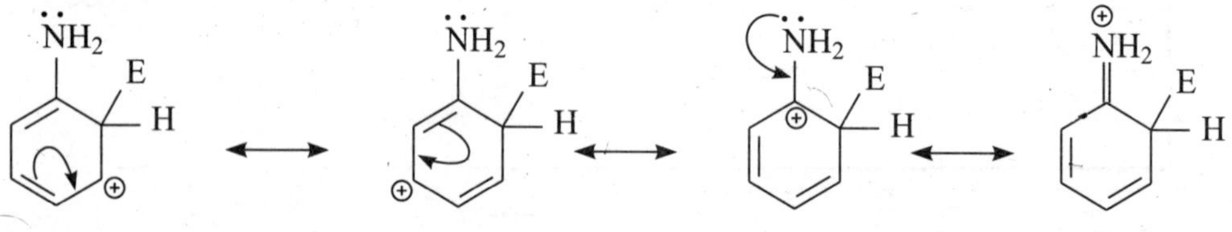

$\ddot{N}H_2$ $\ddot{N}H_2$ $\ddot{N}H_2$ $\overset{\oplus}{N}H_2$

E H ↔ E H ↔ E H ↔ E H

(T)

The cyclohexadienyl cations resulting from *ortho*-and *para*-attack of electrophile are more stable than the one obtained by *meta*-attack of electrophile.

It may be due to the following two reasons:

(*i*) The cations produced as a result of *o*-and *p*-attack are hybrids of four contributing forms each as against three such forms of a similar intermediate resulting from *m*-attack. We know that greater is the number of contributing forms of equal energy, stabler is the species. Hence, cations derived from *o*-and *p*-attack are more stable than the one derived from *m*-attack.

(*ii*) The fourth resonating structure marked **T** of the cations derived from *o*-and *p*-attack is derived by delocalization of nitrogen lone pair into the π-system of the ring. This is an especially stable resonance contributing form because in this form nitrogen as well as all carbon atoms (except H) have octet of electrons.

Thus all groups which donate electrons by resonance are *o*-and *p*-directing groups.

(c) *Meta*-directing groups: Electron-withdrawing groups such as —NO_2, —CN, —COOH, —CHO, —SO_3H, etc. deactivate all ring positions relative to benzene due to their electron-withdrawing inductive (–I) and resonance (–R) or mesomeric (–M) effects. In fact they are more deactivating for *o*- and *p*- substitutions than *m*-substitution. Hence in the presence of such substituents further substitution is slower and takes place only at *m*-position.

The reason for this is not far to seek if we compare the resonance forms of σ-complexes resulting from electrophilic attack on benzene and on *o-m*-and *p*-positions of nitrobenzene which provides an example of aromatic substrate having a substituent belonging to category (*ii*). The σ-complex resulting from the addition of electrophile to benzene has only one positive charge in all its contributing forms, while the σ-complex obtained from similar addition of electrophile to nitrobenzene has two positive charges—one on the ring carbon and the other on the nitrogen atom of the nitro group. This results in the electrostatic repulsions rendering the transition states (leading to the intermediates) less stable in nitrobenzene as compared to benzene. This deactivating influence is more pronounced for *ortho*-and *para*-positions than for *meta*-position. Hence, substitution is slower and takes place at *meta*-position by **'default'.** This is best illustrated by considering the following contributing forms for the σ-complexes resulting from an electrophilic attack at *ortho-meta*-and *para*-positions of nitrobenzene:

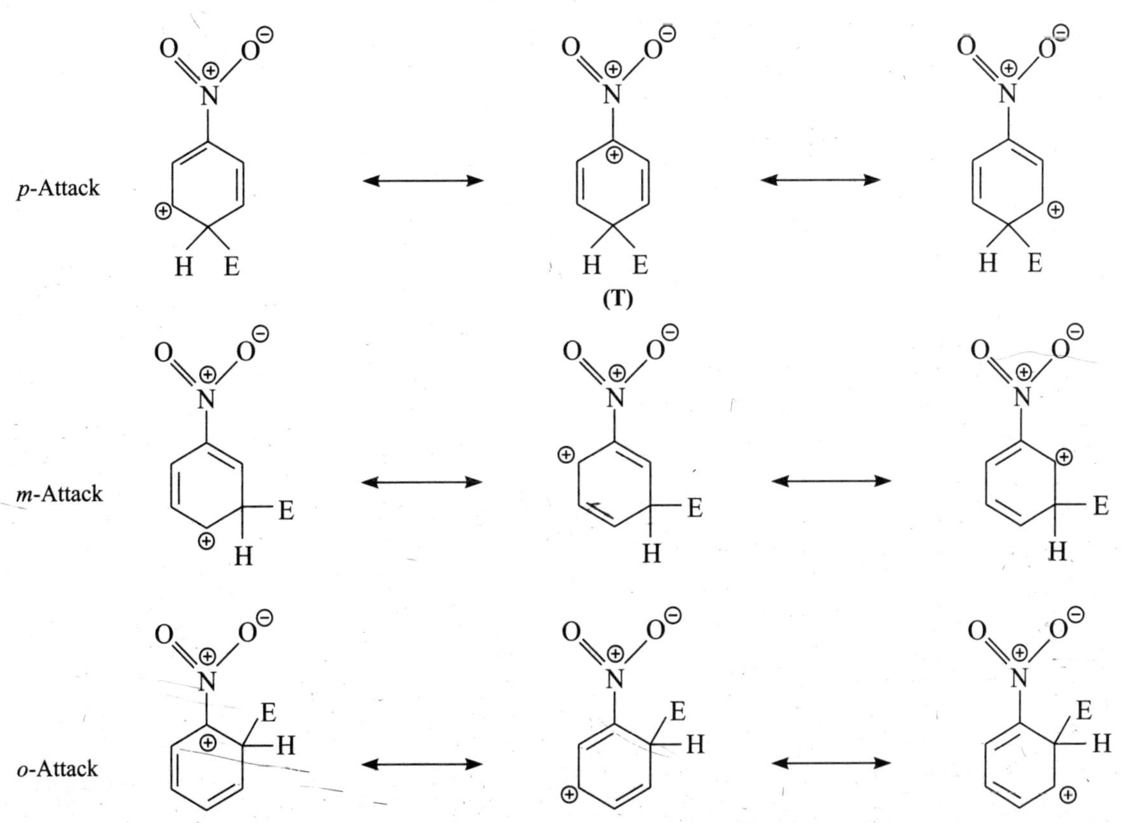

Chapter 9

Since attack at *ortho* and *para* positions places a positive charge adjecent to nitro group already having a positively charged nitrogen atom, structures marked T are strongly *destabilized* because of greater electrostatic repulsions. The intermediate from *meta*-attack has no such form with similar charges on adjacent atoms. Therefore, its transition state is the most stable and attack at *meta* position is preferred. The situation can best be illustrated by an energy profile depicting electrophilic substitution of nitrobenzene and benzene (Fig. 9.5).

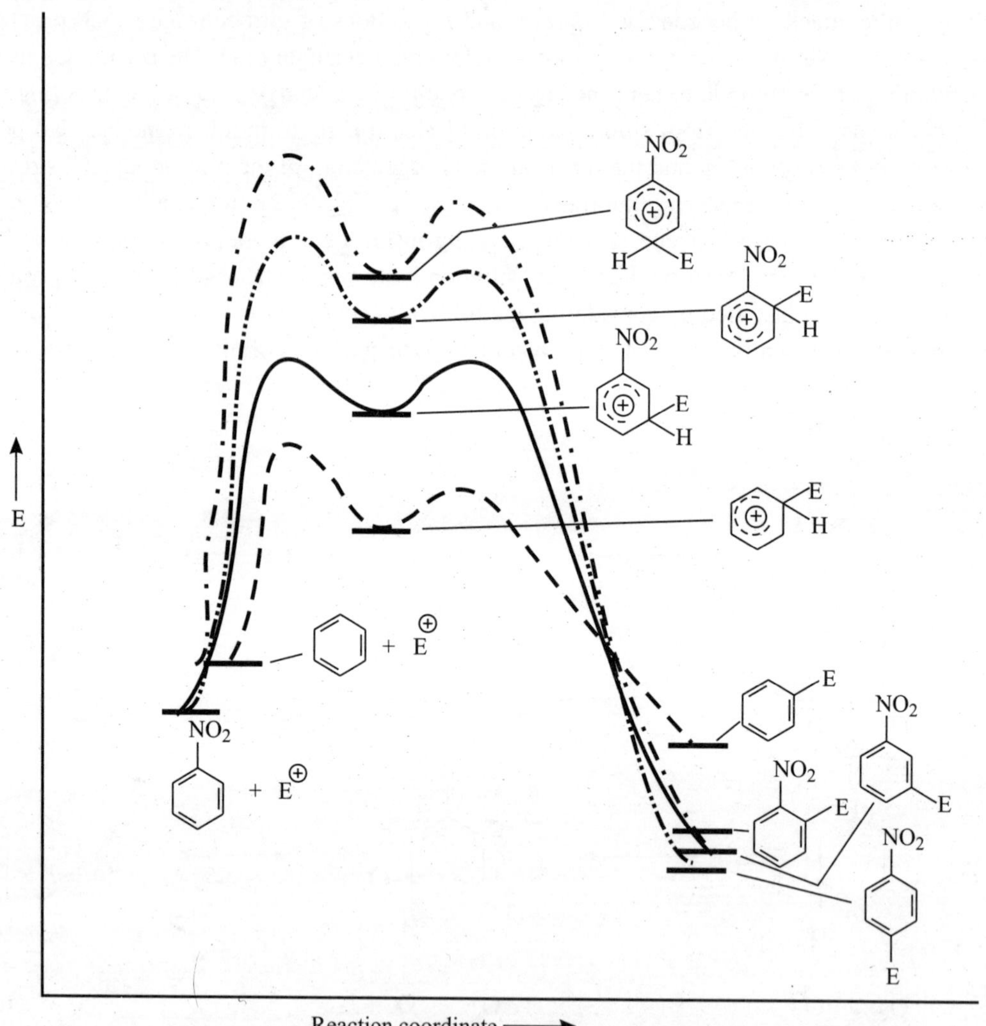

Fig. 9.5: Energy profile for electrophilic substitution in benzene and nitrobenzene

(d) The case of halogens **:** Halobenzenes are less reactive than benzene but yield *ortho*- and *para*-substitution products. This apparent anomalous phenomenon can be explained by a consideration of the inductive and resonance effects of the halogens. Halogens withdraw electrons through inductive effect (–I) thereby deactivating the ring towards electrophilic substitution. On the other hand, halogens are capable of sharing their pairs of electrons with the aromatic ring (+ R) and can accommodate a positive charge. Writing various resonating structures of the positively charged intermediate for *o*-, *p*- and *m*-attacks makes it clear that the attack at *o*- and *p*-positions places the positive charge on the carbon atom adjacent to halogens.

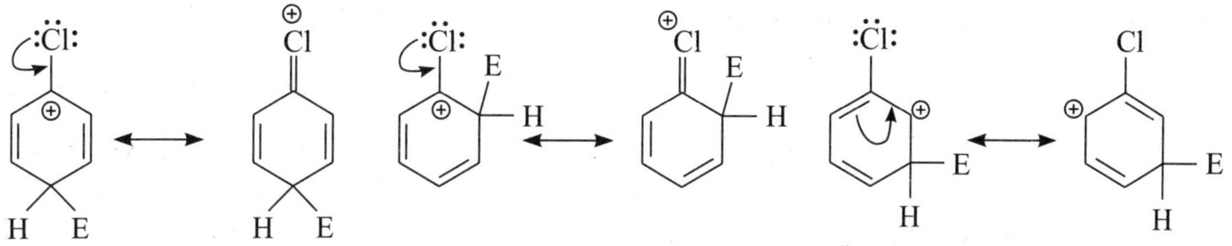

p-attack *o*-attack *m*-attack

Although *o*-and *p*-positions are activated to some extent because of this resonance effect, *the inductive effect overshadows it and causes net electron-withdrawl leading to deactivation at all the positions.* But this deactivation is less for *o*- and *p*-positions than for the *m*-position. Reactivity is thus controlled by the stronger inductive effect whereas the weaker resonance effect controls the orientation.

Here it may be noted that the activating or deactivating effect of any group is strongest at *o*- and *p*-positions than at *m*-position.

9.6 THE *ORTHO-PARA* RATIO

Since in a given monosubstituted benzene there are two *ortho* and one *para* positions which are available for further substitution, it is expected that if the substituent is *ortho-para*-directing, the *ortho* to *para* ratio should be 2 : 1. But usually in *o*-, *p*-directing systems, the *ortho*-substitution occurs to lesser extent than expected.

The above difference may be due to many factors, *i.e.*, polar effects of the substituent, and effects of solvent, temperature, sizes of the substituents and the entering groups (steric factors). Some factors are discussed below:

It has conclusively been shown that the last two factors play dominant role in determining the *ortho* to *para* ratios.

Thus nitration of various alkylbenzenes has revealed that *ortho* to *para* ratio decreases gradually (from 1.5 to 0.25) as one goes from toluene to *t*-butylbenzene. This is in accordance with the order: toluene (1.5) > ethylbenzene (0.95) > *iso*-propylbenzene (0.5) > *t*-butylbenzene (0.25). This decrease in *o/p* ratio has been explained on the basis of **steric effects** of the substituents already present. In the case of *o*-substitution the steric interaction between the electrophile and the group already present on the ring raises the energy of the transition state, thus making it less stable.

Probably, this is the reason why the proportion of *o*-isomer in the *o*-, *p*-mixture increases if the reaction is carried out at higher temperatures. For instance, nitration of acetanilide at 0°C gives *p*-nitroacetanilide almost exclusively. However, the amount of *o*-isomer increases with the rise in reaction temperature. The role of steric effects in determining the *o/p* ratio has elegantly been demonstrated in the halogenation of *t*-butylbenzene. Thus, when *t*-butylbenzene is chlorinated, there is appreciable substitution at *o*-position but its bromination at *o*-position is almost negligible. This is consistent with the fact that bromine molecule is bulkier than chlorine.

A classic example of the steric factors dominating in electrophilic substitution (bromination, nitration, Friedel-Crafts reaction) is *p*-cymene. From electronic considerations the substitution should be expected to occur at the position *ortho* to isopropyl group, but invariably, it takes place *ortho* to the methyl group.

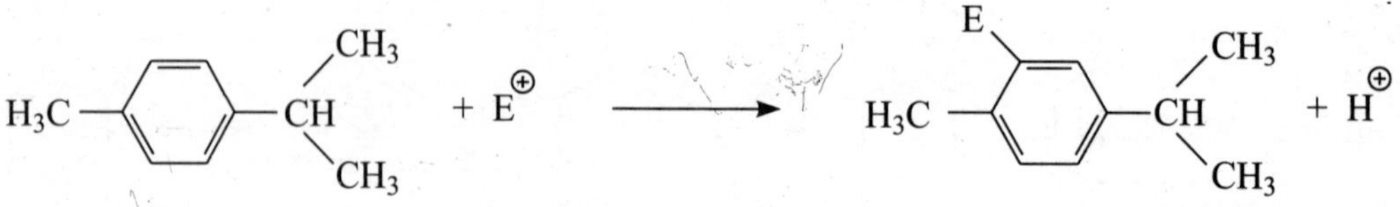

$$H_3C-C_6H_4-CH(CH_3)_2 + E^{\oplus} \longrightarrow H_3C-C_6H_3(E)-CH(CH_3)_2 + H^{\oplus}$$

p-Cymene

However, a quite different situation arises when the substituent already present has a pair of non-bonding electrons. Electronic factors now dominate and result in increase in the amount of *p*-isomer. This is due to the fact that one of the resonance forms of the σ-complex assumes *p*-quinonoid structure in the case of *p*-attack, while that obtained from *ortho*-attack assumes an *ortho*-quinonoid structure. As the *para* quinonoid structures are more stable than *ortho*-quinonoids, it can be assumed that the form (T′) would be more stable than the form (T) and hence substitution is proferred at *para*-position.

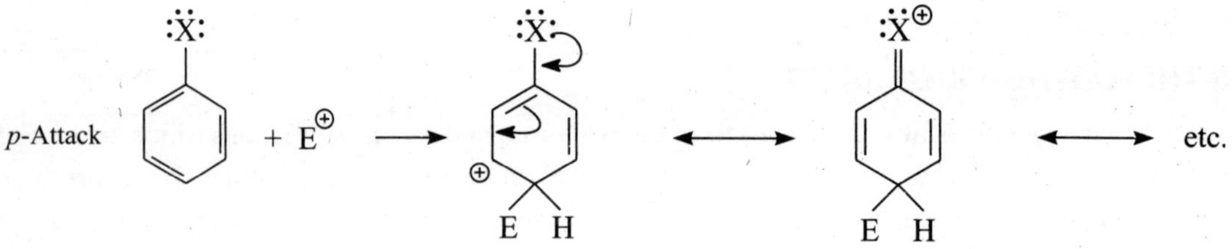

p-Attack + $E^{\oplus}$ ⟶ ⟷ ⟷ etc.

E H E H

(T′)
para-Quinonoid
(more stable)

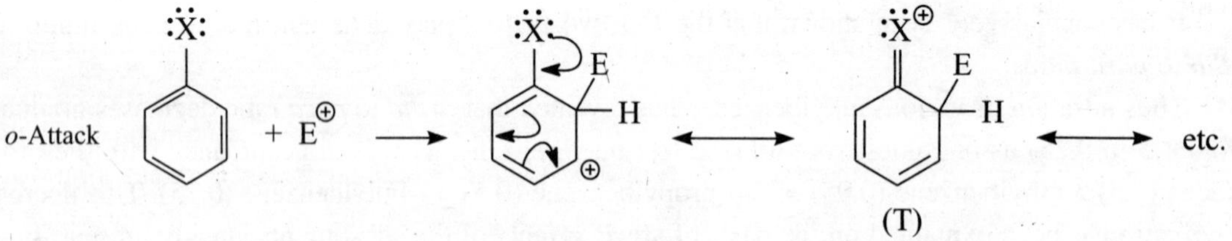

o-Attack + $E^{\oplus}$ ⟶ ⟷ ⟷ etc.

(T)
ortho-Quinonoid
(less stable)

But sometimes because of the *chemical interaction between the substituent and the electrophile the ortho to para ratio becomes abnormally high.* For instance, nitration of anisole with the conventional nitrating mixture (HNO_3/H_2SO_4) where the nitrating agent is $NO_2^{\oplus}$, gives *ortho* and *para* nitroanisoles in the ratio 31 : 67 but when the same reaction is carried out with N_2O_5 (HNO_3–AC_2O) this ratio becomes 71 : 28. The mechanism of the reaction with N_2O_5 is as shown below:

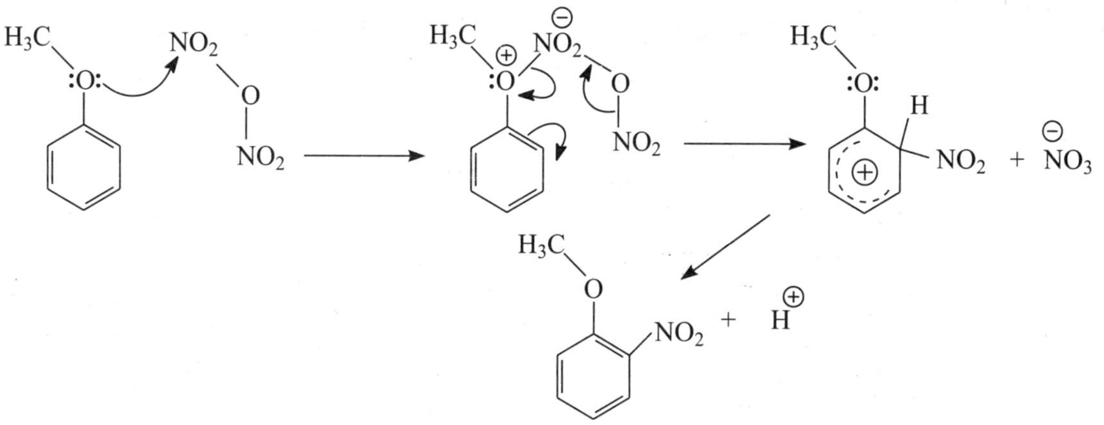

$$H_3C-\ddot{O}-C_6H_5 + O_2N-O-NO_2 \longrightarrow \text{[intermediate]} \longrightarrow \text{[σ-complex]} + NO_3^{\ominus} \longrightarrow o\text{-}O_2N-C_6H_4-OCH_3 + H^{\oplus}$$

9.7 FURTHER SUBSTITUTION IN DISUBSTITUTED BENZENES: COMPETITION BETWEEN SUBSTITUENTS FOR THIRD GROUP

When two or more groups are present in the ring, the orientation and reactivity effects as described above still hold good. Both the groups can affect further substitution as discussed below:

(*a*) Usually the effects of these groups are additive provided they are suitably situated. Thus, for instance, the position to be occupied by the new group (indicated by means of arrows) is in accordance with this prediction.

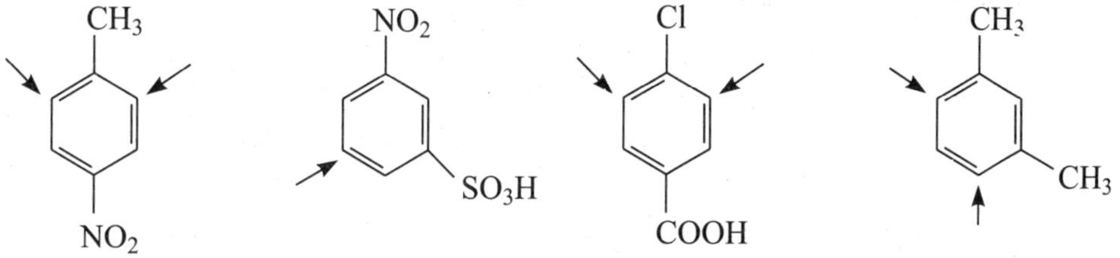

CH$_3$ / NO$_2$; NO$_2$ / SO$_3$H; Cl / COOH; CH$_3$ / CH$_3$

(*b*) However, when the two groups are located at such positions where they oppose each other, predictions regarding the position taken by the electrophile may become somewhat difficult. For instance, when two groups belonging to the same category are present in competing positions, it is hard to predict which will control the orientation. Thus all four isomers are possible in the case of *o*-methoxyacetanilide, however, the amount of isomer having the incoming group *ortho* to anilide group would be less due to steric effect. In some cases it is possible to predict the product composition by invoking the resonance and inductive effects of the substituents already present in the molecule. But in practice mixtures having all possible isomers in nearly all proportions are usually obtained. In many cases, it is hard to exaplain the experimental results. For instance, in the case of *ortho*-bromofluorobenzene, substitution takes place *para* to fluorine while in the case of *para*-bromofluorobenzene electrophile enters *ortho* to bromine.

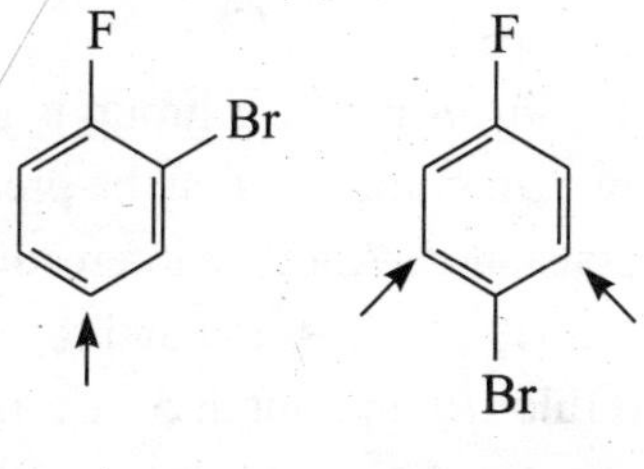

F / Br; F / Br

However, it is possible to make the following generalizations which are based on the experimental observations as well as theoretical considerations:

(*i*) If there is a competition between an *ortho-para*-directing group and *meta*-directing group the former controls the site of substitution. Even among the *ortho-para*-directing groups, the one with stronger tendency to release electrons will control the orientation.

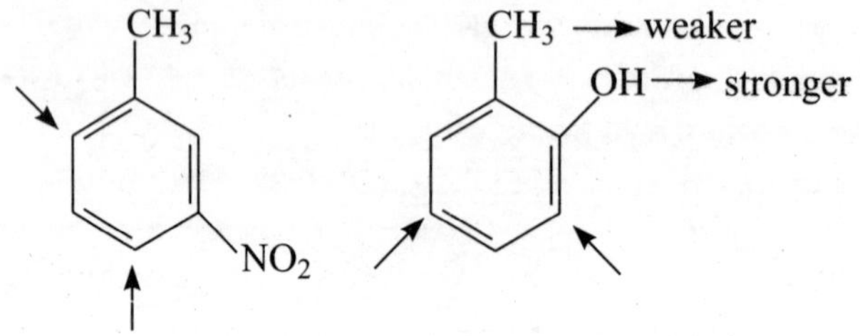

Thus various groups can be arranged in the decreasing activation of the ring:

$$—OH, —NR_2, \bar{O}, —NH_2 > —OR > —O—CO—R > —NHCOCH_3 > C_6H_5 > CH_3 >$$

meta-directing groups.

(*ii*) When two groups are present *meta* to each other, seldom does the substitutent enter the position between these groups. This is because of greater setric hindrance at this position due to bulk of the substituents and the electrophile.

(*iii*) If a *meta*-directing group is situated *meta* to an *ortho para*, directing group, the third group enters *ortho* to the *meta*-directing group, rather than *para* to it. For instance 3-chloronitrobenzene on chlorination yields mainly 3,6-dichloronitrobenzene with a little amount of 2,3-dichloronitrobenzene. The latter compound is formed in violation of rule (*ii*) but 3, 4-dichloronitrobenzene is not formed at all.

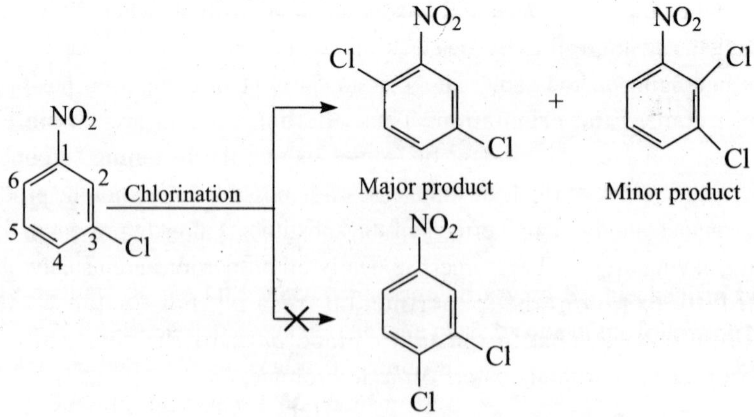

This special orienting effect of the *meta*-directing group is known as ***ortho* effect**. Although no suitable explanation is available for this observation, it may be presumed that the *meta*-directing group possibly gives some intramolecular assistance. Another interesting example of this is dinitration of 4-bromotoluene. The first nitro group enters at 2-position and the second enters at 3-position (*ortho* to first NO_2 group though in violation to rule (*ii*) and not at 5-or 6-position.

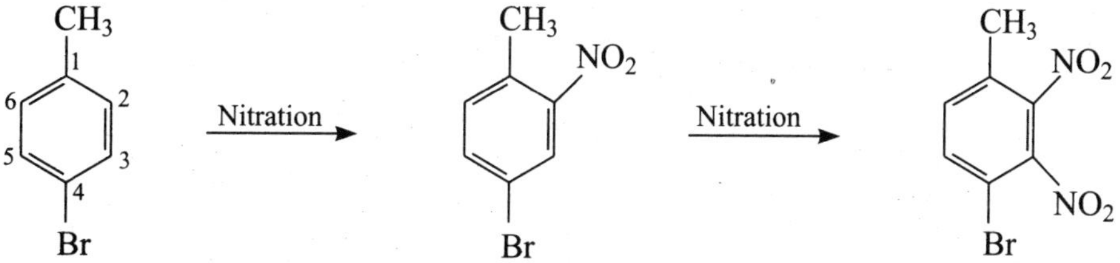

CH3 (1), Br (4), positions 6, 2, 5, 3 — Nitration → CH3, NO2, Br — Nitration → CH3, NO2, NO2, Br

(*iv*) In case where both the substituents are *o/p* directing, the one which exerts stronger influence controls the orientation of the third group. Generally speaking, resonance effects take predominance over inductive effects and if both the groups operate via resonance effects the one which is more capable of sharing electron pair exerts stronger influence. The following examples illustrate this rule:

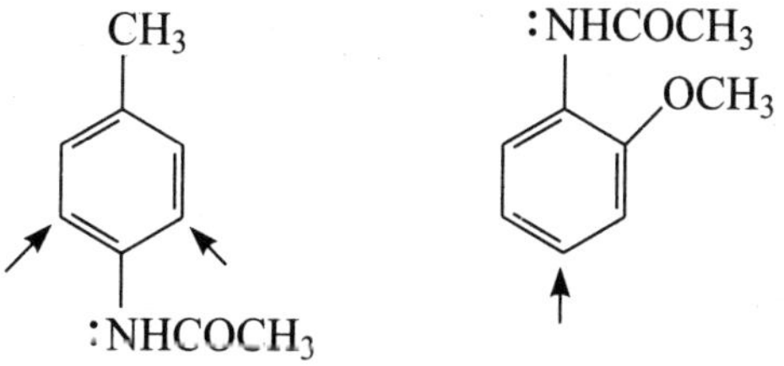

CH_3, $:NHCOCH_3$ — $:NHCOCH_3$, OCH_3

(*v*) When both the groups already present on the ring are *meta* directing, further substitution in the ring becomes more difficult. However, if the third group is actually introduced, its orientation is governed by the less powerful of the two *meta*-directing groups. The decreasing order of *m*-directing effect of some groups is as follows:

$—^{\oplus}NR_3 > —^{\oplus}NH_3 > —NO_2 > —CN > —SO_3H > —COOH > —CONH_2 > —COOR > —CHO$

(*vi*) If the two *ortho/para* directing groups are *meta* to each other, little substitution occurs at the position which is *ortho* to both (marked by arrow ⟶)

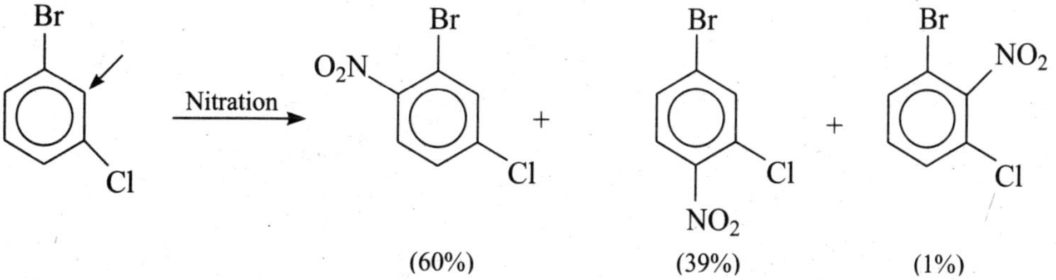

Br, Cl — Nitration → Br, O_2N, Cl (60%) + Br, Cl, NO_2 (39%) + Br, NO_2, Cl (1%)

9.8 SIDE CHAIN REACTIONS OF BENZENE DERIVATIVES

Two important reactions involving side chains only are as follows: (*i*) Halogenation of the side chain and (*ii*) Oxidation of the side chain.

(*i*) Halogenation of the Side Chain

The alkyl groups constituting the side chain are *o, p*-directing. So, when any alkylbenzene, *e.g.*, toluene is treated with halogen (chlorine or bromine) in presence of Fe or anhyd. Fed_3 or anhyd. $AlCl_3$ (Lewis acid) at low temperature, it gives rise to a mixture of *o*- and *p*-chlorotoluenes. For example,

$$\text{Toluene} + Cl_2 \xrightarrow[\text{303 K}]{\text{Anhydrous } AlCl_3} \text{o-Chlorotoluene} + \text{p-Chlorotoluene}$$

This reaction in which a H atom of the benzene ring is replaced by halogen atom is called **nuclear halogenation.**

However, when toluene is treated with halogen at high temperatures, in sunlight or in presence of free radical initiators it undergoes free radical substitution of side chain hydrogen by chlorine (*i.e., side chain halogenation*). For instance, toluene forms successively benzyl chloride, benzal chloride and benzotrichloride on passing chlorine in boiling toluene or on exposure to light.

$$C_6H_5CH_3 + Cl_2 \xrightarrow{\Delta \text{ or } h\nu} C_6H_5CH_2Cl \xrightarrow[\Delta \text{ or } h\nu]{Cl_2} C_6H_5CHCl_2 \xrightarrow[\Delta \text{ or } h\nu]{Cl_2} C_6H_5CCl_3$$

Toluene Benzyl chloride Benzal chloride Benzotrichloride

Instead of Cl_2, sulphuryl chloride (SO_2Cl_2) at 475 K in presence of light and traces of peroxide can also be used for side chain chlorination of arenes. For example,

$$C_6H_5CH_3 + SO_2Cl_2 \xrightarrow[\text{Peroxide}]{475 \text{ K}, h\nu} C_6H_5CH_2Cl + SO_2\uparrow + HCl\uparrow$$

This reaction, in which one or more hydrogen atoms of the side chain are replaced by halogen, is called the **side chain halogenation.**

Mechanism

This reaction takes place *via* a free radical chain mechanism involving a benzyl radical as shown below:

***Step I*: Chain initiation**

$$Cl—Cl \xrightarrow{\Delta \text{ or } h\nu} 2\dot{Cl}$$

***Step II*: Chain propagation**

(*i*) $C_6H_5CH_2—H + \dot{Cl} \longrightarrow C_6H_5\dot{C}H_2 + HCl$

Benzyl radical

(ii) $C_6H_5\dot{C}H_2 + Cl—Cl \longrightarrow C_6H_5CH_2Cl + \dot{C}l$

The reactions (*i*) and (*ii*) are repeated again and again till the side chain halogenation of toluene is complete.

In presence of excess of Cl_2, the chain reaction continues until all the three hydrogen atoms are displaced.

$C_6H_5CHCl–H + \dot{C}l \longrightarrow C_6H_5\dot{C}HCl + HCl$

$C_6H_5\dot{C}HCl + Cl—Cl \longrightarrow C_6H_5CCl_2–H + \dot{C}l$

$C_6H_5CCl_2–H + \dot{C}l \xrightarrow{-HCl} C_6H_5\dot{C}Cl_2$

$C_6H_5\dot{C}Cl_2 + Cl—Cl \longrightarrow C_6H_5CCl_3 + \dot{C}l$ and so on.

***Step III*: Chain termination**

$\dot{C}l + \dot{C}l \longrightarrow Cl_2$

$C_6H_5\dot{C}H_2 + \dot{C}l \longrightarrow C_6H_5CH_2Cl$

$C_6H_5\dot{C}H_2 + H_2\dot{C}C_6H_5 \longrightarrow C_6H_5CH_2CH_2C_6H_5$

1,2-Diphenylethane

Side Chain Halogenation with Side Chain Longer than Methyl

With side chain longer than methyl, it is expected that the free radical substitution should take place on all the carbon atoms with equal probability. In the case of ethylbenzene, for example, the probability of hydrogen abstraction from methyl and methylene carbon atoms is 3 : 2, respectively. But it is interesting to note that hydrogen abstraction from benzylic carbon (carbon atom directly linked to benzene ring) takes place preferentially. Thus, chlorination of ethylbenzenes under free radical conditions gives a mixture of α- and β-chloroethylbenzenes with the predominance of the former isomer (91%).

$$C_6H_5CH_2-CH_3 \xrightarrow[\Delta \text{ or } h\nu]{Cl_2} C_6H_5CH_2-CH_2Cl \;(\beta\text{-Chloroethylbenzene } (9\%)) + C_6H_5CH(Cl)-CH_3 \;(\alpha\text{-Chloroethylbenzene } (91\%))$$

This reaction in which a benzylic hydrogen is preferentially replaced by a halogen is called **benzylic halogenation**.

A comparison of bond dissociation energies of benzylic C—H bond in toluene, (355 kJ mol^{-1}) and CH_3 — H bond in methane (434 kJ mol^{-1}) reveals that 78 kJ mol^{-1} (433–355) less energy is needed for the *homolytic fission* of the former bond. In other words, *benzyl radical is more stable than methyl radical.* The greater stability of benzyl radical is ascribed to delocalization of the electron over the ring.

$$C_6H_5CH_2-H \longrightarrow C_6H_5\dot{C}H_2 \longleftrightarrow \text{(radical at ortho position)} \longleftrightarrow \text{(radical at para position)} \longleftrightarrow \text{(radical at other ortho position)}$$

Bromination of ethylbenzene differs from chlorination in the orientation. Unlike chlorination (which gives a mixture of products) bromination gives the α-bromoethylbenzene as the only product.

$$C_6H_5CH_2-CH_3 + Br_2 \xrightarrow{\Delta \text{ or } h\nu} C_6H_5CH(Br)-CH_3$$

(The path to β-Bromoethylbenzene, $C_6H_5CH_2-CH_2Br$, is crossed out.)

β-Bromoethylbenzene

α-Bromoethylbenzene
(Sole product)

Thus more reactive chlorine atom is less selective than the less reactive bromine atom which is more selective. This can be rationalized by consideration of the fact that during attack by highly reactive chlorine atom the transition state (T.S.) is reached a little earlier than the fission of the C—H bond, so that it (T.S.) has little free radical character. Hence, the factors that stabilize the free radicals have little role to play in stabilizing the T-S. The chlorine atom can, therefore, attack any one of the two carbon atoms of the chain. On the other hand, when the attacking reagent is less reactive (bromine atom) the C—H bond fission takes place slightly ahead of the formation of the T-S, and the substrate acquires a considerable free radical character. The factors that stabilize the benzylic radical will also stabilize the incipient benzylic radical in the T.S. and it is thus reasonable to expect that bromine will be more selective in its attack on a carbon.

(*ii*) Oxidation of the Side Chain

We have earlier seen that the oxidizing agents such as potassium permanganate or potassium dichromate have little or no effect on alkanes and benzene. In contrast, an alkyl side chain is quite susceptible to oxidation. The oxidation of the side chain is usually carried out with potassium permanganate, potassium dichromate or even with dilute nitric acid. It has been shown that the whole of the side chain, irrespective of its length, is completely oxidized to a carboxylic group. Thus, the number and position of the carboxylic groups produced in such a reaction indicates the number and position of the side chains attached to the aromatic rings. Some examples are as follows:

(*a*) Oxidation with Hot $KMnO_4$

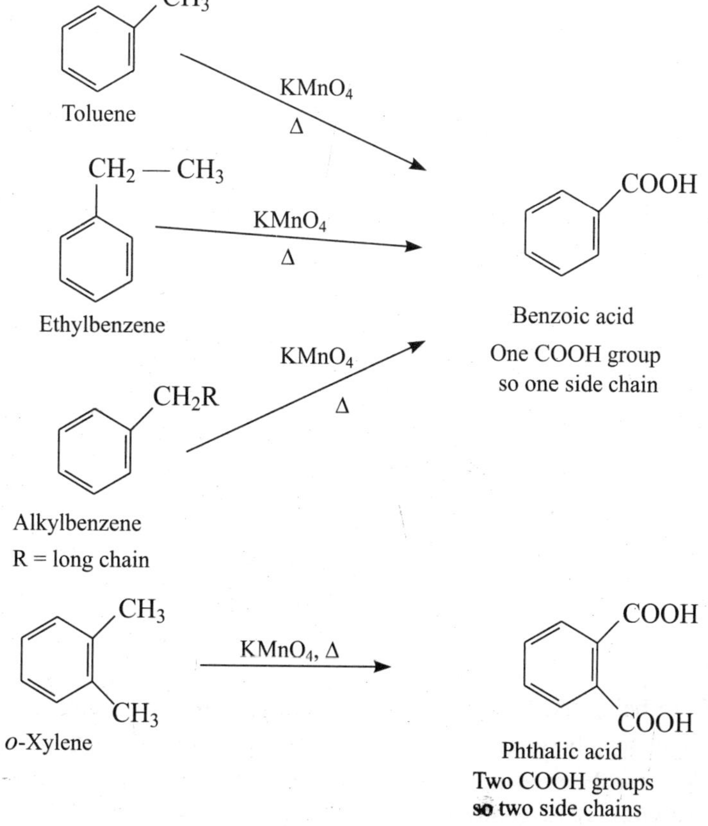

Chapter 9

$$\text{m-Xylene (1,3-dimethylbenzene)} \xrightarrow{KMnO_4,\ \Delta} \text{Iso-phthalic acid (benzene-1,3-dicarboxylic acid)}$$

m-Xylene — *Iso*-phthalic acid

$$H_3C-C_6H_4-CH_3 \xrightarrow{KMnO_4,\ \Delta} HOOC-C_6H_4-COOH$$

p-Xylene — *Ter*-phthalic acid

Mechanism

The mechanism of this oxidation of side chain reaction is quite uncertain. However, it seems that the oxidising agent abstracts the benzylic hydrogen to form benzyl radical as intermediate. This is in agreement with the fact that alkylbenzenes which do not have a benzylic hydrogen and do not undergo oxidation easily. For example,

$$C_6H_5-C(CH_3)_3 \xrightarrow[373\text{–}383\ K]{KMnO_4,\ KOH} \text{No reaction}$$

Here it may be interesting to note that in comparison to toluene, benzyl chloride is easily oxidised to give benzaldehyde and benzoic acid under different conditions.

$$C_6H_5CH_2Cl \xrightarrow[HNO_3]{[O]} C_6H_5COOH \text{ (Benzoic acid)}$$

$$C_6H_5CH_2Cl \xrightarrow[Pb(NO_3)_2]{[O]} C_6H_5CHO \text{ (Benzaldehyde)}$$

Benzyl chloride

Uses: Apart from being used as a procedure for determining the number and position of side chains and thus distinguishing between the isomeric alkylbenzenes, this reactions can also be used as a valuable synthetic route for the preparation of substituted carboxylic acids. *e.g.*, *o*-chlorobenzoic acid can be obtained by oxidation of *o*-chlorotoluene.

$$\text{o-Chlorotoluene} \xrightarrow{KMnO_4, \Delta} \text{o-Chlorobenzoic acid}$$

o-Chlorotoluene   o-Chlorobenzoic acid

(*b*) Oxidation with Chromyl Chloride: *Etard Reaction*

When toluene is treated with chromyl chloride in CCl_4 or CS_2 solution, it is converted into benzaldehyde.

$$C_6H_5CH_3 \xrightarrow[CS_2]{CrO_2Cl_2} C_6H_5CHO$$

Toluene   Benzaldehyde

In a similar manner *o*-xylene gives *o*-tolualdehyde.

$$\text{o-Xylene} \xrightarrow[CS_2]{CrO_2Cl_2} \text{o-Tolualdehyde}$$

o-Xylene   *o*-Tolualdehyde

This reaction is called **Etard reaction.**

9.9 REDUCTION OF BENZENE (BIRCH REDUCTION)

The benzene molecule is highly resistant to reduction. However, when a mixture of benzene and hydrogen is passed over a nickel catalyst at 473 K complete reduction of benzene produces cyclohexane. In all, three moles of hydrogen are consumed per mole of benzene.

$$C_6H_6 + 3H_2 \xrightarrow[CS_2]{Ni,\ 473\ K} C_6H_{12}$$

Cyclohexane

Benzene can be reduced partially by **Birch reduction** as given below:

Birch Reduction (*Partial Reduction*)

Partial reduction of aromatic rings with alkali metals (Li, Na, K) and liquid ammonia in presence of an alcohol results in the formation of a non-conjugated cyclic diene. This reaction is called **Birch Reduction** (after the name of Australian chemist **Arthur J. Birch** who demonstrated the usefulness of this reaction in 1940). The simplext example of Birch reduction is the partial reduction of benzene into 1,4-cyclohexadiene.

Benzene + 2Na + 2C_2H_5OH $\xrightarrow{NH_3}$ 1,4-Cyclohexadiene + C_2H_5ONa

Mechanism

The mechanism of Birch reduction resembles that of the metal ammonia reduction of alkynes to form *trans*-alkenes. The reaction takes place in four steps as discussed below:

***Step I*: Transfer of one electron from sodium metal (the reducing agent) to the π-system of the ring** resulting in the formation of cyclohexadienyl radical anion (1,4).

Benzene + $\dot{e}$ (from $\dot{Na}$) ⟶ Cyclohexadienyl radical anion (1,4) + $Na^{\oplus}$

It may be noted here that the radical anion (1, 2) is not formed due to electron-electron repulsions:

Benzene + $\dot{e}$ (from $\dot{Na}$) ⟶ (✕) Cyclohexadienyl radical anion (1, 2)

Electron-electron repulsion

***Step II*: Abstraction of a proton, by the cyclohexadienyl radical anion (1, 4), from ethanol giving rise to cyclohexadienyl radical.**

Cyclohexadienyl radical anion + H—OC_2H_5 ⟶ Cyclohexadienyl radical + $C_2H_5O^{\ominus}$

***Step III*: Transfer of one electron from sodium to cyclohexadienyl radical to produce cyclohexadienyl anion.**

Cyclohexadienyl radical + $\dot{e}$ (from Na) ⟶ Cyclohexadienyl anion + $\overset{\oplus}{Na}$

***Step IV*: Abstraction of a proton, by the cyclohexadienyl anion, from ethanol to give 1, 4-cyclohexadiene**

Cyclohexadienyl anion + $H-OC_2H_5$ ⟶ 1,4-Cyclohexadiene + $C_2H_5\overset{\ominus}{O}$

The Rate and Regiochemistry

Both rate and regiochemistry of Birch reduction are influenced by electronic and steric effects of the substituents. These effects are discussed below:

Electronic Effects of the Substituents

***(a) Effect of electron-releasing groups*:** Electron-releasing substituents such as alkyl or alkoxy slow down the rate of the reaction. Also in the reduction product these substituents are present on vinylic carbon (sp^2 hybridized carbon). For example, the reduction of anisole.

Anisole $\xrightarrow{Na/liq.\ NH_3/C_2H_5OH}$ 1-methoxy-1,4-cyclohexadiene

Reason

(*i*) Slow rate: The rate lowering effect of methoxy group can be explained on the basis of the fact that addition of an electron to the ring is opposed by electron-donating group.

***(ii)* Substituent at vinylic carbon in the product:** The orientation of the addition of electron is such that in the resulting radical anion the negative charge does not develop on that carbon to which the electron-donating group is attached. Thus,

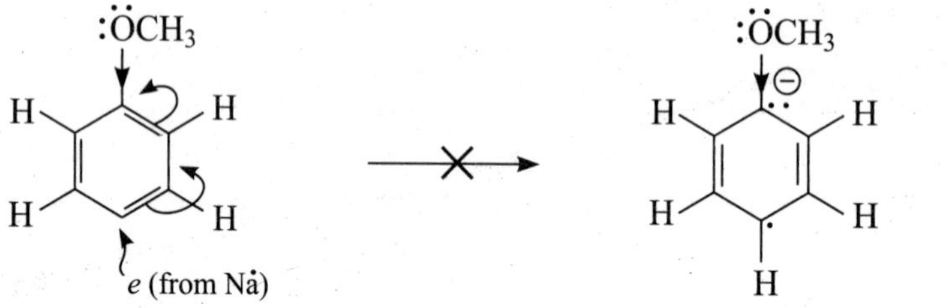

$:\ddot{O}CH_3$ H H H H *e* (from Nȧ) → (crossed) $:\ddot{O}CH_3$ ⊖ H H H H H

Unfavourable, not formed
(Negative charge on carbon to which electron-donating-OCH_3 group is attached).

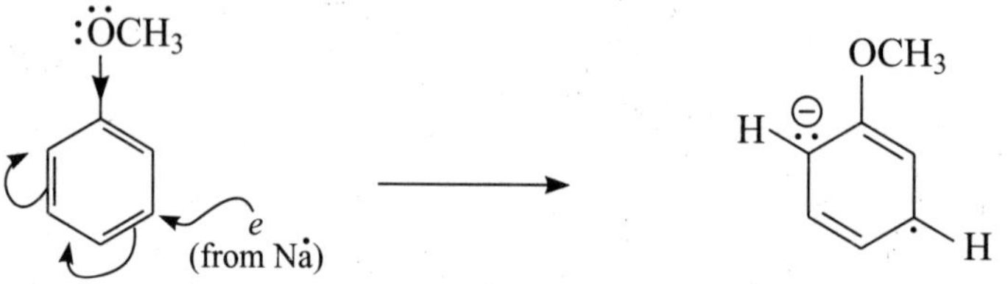

$:\ddot{O}CH_3$ *e* (from Nȧ) → OCH_3 H ⊖ H

Favourable, formed
(Negative charge on carbon which does not bear OCH_3 group)

***(b)* Effect of electron-withdrawing groups:** Electron-withdrawing substituents such as —COOH or $CONH_2$ enhance the rate of the reaction. Also the substituent is found on sp^3 carbon atom in the reduction product. For example, the reduction of benzoic acid.

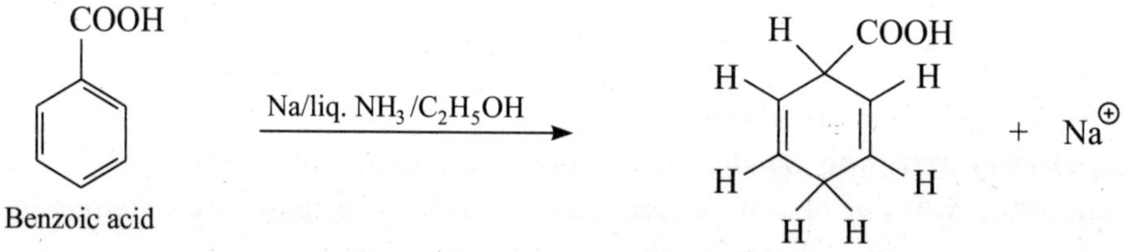

COOH
Benzoic acid

$\xrightarrow{Na/liq.\ NH_3/C_2H_5OH}$

H COOH H H H H H H + $Na^{\oplus}$

Cyclohexa-2,5-dien-1-carboxylic acid

Reason

***(i)* Fast rate:** The rate enhancing effect of carboxylic group can be explained on the basis of the fact that addition of an electron to the ring is favoured by electron-withdrawing group.

***(ii)* Substituent at sp^3 carbon in the product:** Orientation of addition of an electron is such that in the resulting anion the negative charge develops on that carbon to which an electron-withdrawing group is attached.

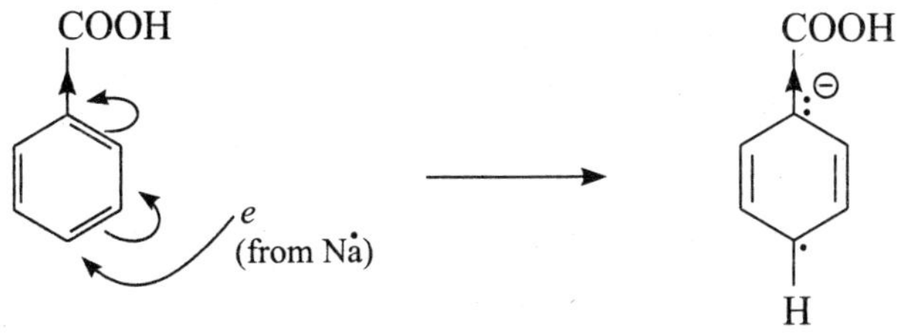

COOH

(from Na)

COOH

H

Favourable (Negative charge on carbon to which an electron-withdrawing carboxylic group is attached)

Uses: Besides formation of non-conjugated cyclic dienes Birch reduction finds use in the synthesis of a variety of compounds. For example, the synthesis of cyclohex-2-en-1-one form anisole as shown below:

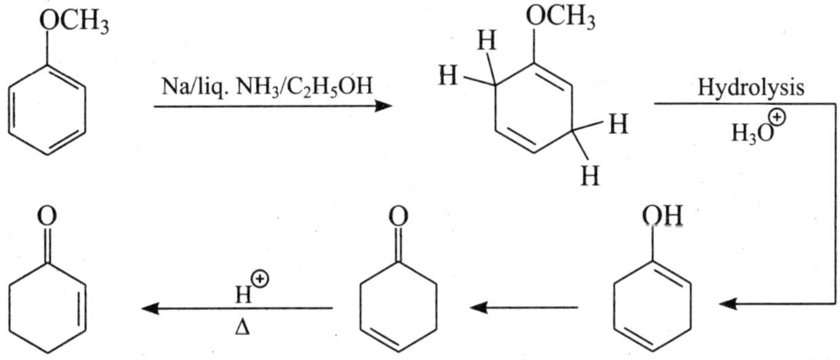

OCH_3

Na/liq. NH_3/C_2H_5OH

OCH_3

Hydrolysis $H_3O^{\oplus}$

OH

O

$H^{\oplus}$ Δ

O

(Double bond migration to produce conjugated ketone from non-conjugated ketone.)

9.10 ARENES

The term **arene** is used to designate aromatic hydrocarbons which are derived by replacement of one or more hydrogen atoms of the benzene ring by hydrocarbon substituents such as alkyl, alkenyl, alkynyl or aryl groups. They can be divided into the following two categories:

(*i*) Arenes containing one benzene ring.

(*ii*) Arenes containing more than one benzene ring either linked directly or through one or more carbon atoms.

Some of the more common examples are given below:

(*i*) **Arenes containing one benzene ring:** These are further divided into following three sub-classes:

(*a*) **Alkylbenzenes**

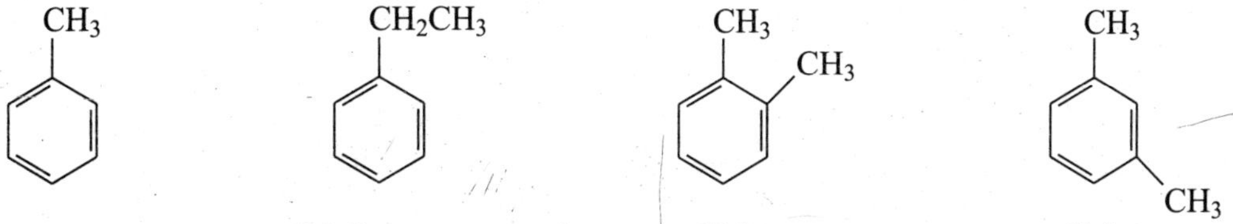

CH_3 — Toluene (Methylbenzene)

CH_2CH_3 — Ethylbenzene

CH_3, CH_3 — *o*-Xylene

CH_3, CH_3 — *m*-Xylene

Chapter 9

(b) Alkenylbenzenes

$CH{=}CH_2$

Styrene or Ethenylbenzene

$CH{=}CH{-}CH_3$

1-Phenylpropene

(c) Alkynylbenzenes

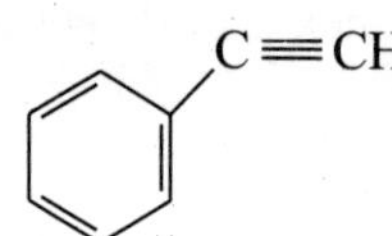

Phenylacetylene or Ethynylbenzene

$C{\equiv}C{-}CH_3$

1-Phenylpropyne

(ii) Arenes containing two or more benzene rings

Biphenyl

Triphenyl

CH_2

Diphenylmethane

CH

Triphenylmethane (Tritane)

9.10.1 Nomenclature of Arenes

Some arenes bearing an alkyl substituent on the benzene ring are known by common names with their parent name as *benzene*. Arenes are thus referred to as *alkylbenzenes*. Nomenclature of some common alkylbenzenes is given below. For a hydrocarbon with a common name the systematic name is written in the bracket.

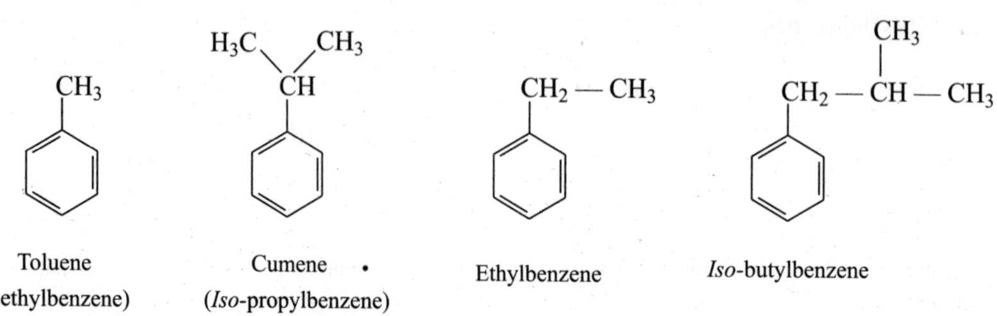

Toluene (Methylbenzene)

Cumene (*Iso*-propylbenzene)

Ethylbenzene

Iso-butylbenzene

Among the dialkylbenzenes the simplest ones are three isomeric dimethylbenzenes which are commonly named as xylenes.

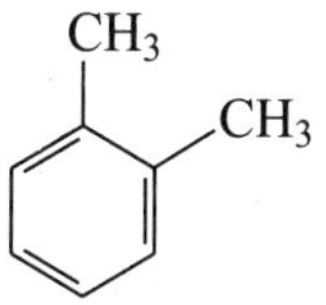

o-Xylene
(*o*-Dimethylbenzene)

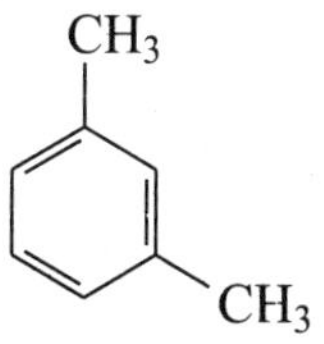

m-Xylene
(*m*-Dimethylbenzene)

p-Xylene
(*p*-Dimethylbenzene)

Other dialkylbenzenes are named by anyone of the following procedures:

(i) If a part of the compound is derived from an arene having a common name, then it is named as the alkyl derivative of that arene. e.g.,

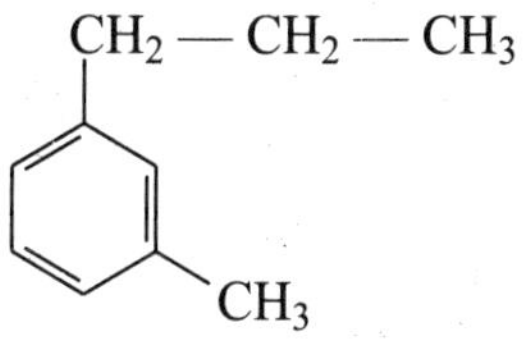

m-*n*-Propyltoluene

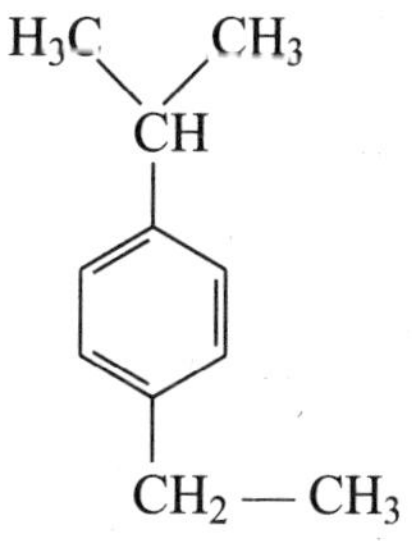

p-Ethylcumene

(ii) In the systematic nomenclature, the names of the alkyl substituents are prefixed before the parent name benzene in the alphabetical order. If the positions of the substituents are designated by numbers, the numbering is commenced with the alkyl substituent coming first in this order and the whole ring is numbered in such a manner that the substituents get the minimum numbers. e.g.,

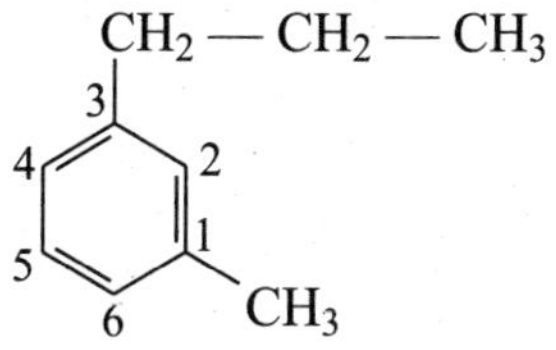

m-Methyl-*n*-propylbenzene
or
1-Methyl-3-*n*-propylbenzene

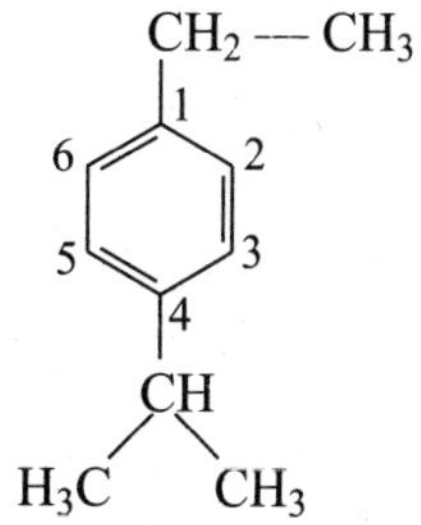

p-Ethyl-*iso*-propylbenzene
or
1-Ethyl-4-*iso*-propylbenzene

Arenes having long alkyl chains are also named as **phenylalkanes**, the position of the phenyl group being designated by the number of carbon atom to which the ring is attached.

$$\overset{}{\underset{1}{CH_3}}-\underset{2}{CH}-\underset{3}{CH_2}-\underset{4}{CH}-\underset{5}{CH_3}$$

(with CH_3 on C-2 and a phenyl ring on C-4)

2-Methyl-4-phenylpentane

The simplest alkenyl derivative of benzene is vinylbenzene (ethenylbenzene) and is known by its common name, *styrene*. Other members are named either as *phenylalkenes* or *alkenylbenzenes*. Alkynyl derivatives are described as substituted acetylenes or alkynes.

$C_6H_5-CH=CH_2$ — Styrene (Vinylbenzene, ethenylbenzene or phenylethylene)

$C_6H_5-\overset{3}{CH_2}-\overset{2}{CH}=\overset{1}{CH_2}$ — Allylbenzene (3-Phenylpropene)

$C_6H_5-C\equiv CH$ — Phenylacetylene (Ethynylbenzene)

Arenes containing two benzene (or substituted benzene) rings joined directly with each other are named as *biphenyls* or *arylbenzenes*. The position of substituents on one of the rings is indicated by numbering its carbon atoms as 1, 2, 3, 4, 5 and 6 commencing with carbon atom at the junction; the carbon atoms of the other ring are numbered as 1′, 2′, 3′, 4′, 5′ and 6′commencing with the corresponding carbon atom of the ring.

Biphenyl (Phenylbenzene) — ring positions 1–6 and 1′–6′

3, 3′-Dimethylbiphenyl — H_3C at position 3 and CH_3 at position 3′

Arenes containing two benzene rings joined through a carbon chain consisting of one or more carbon are named as *polyphenylalkanes*.

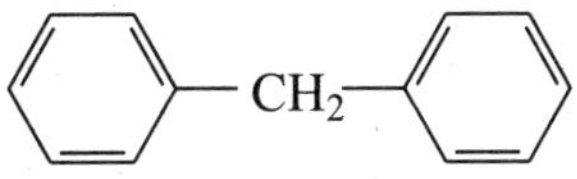

Diphenylmethane

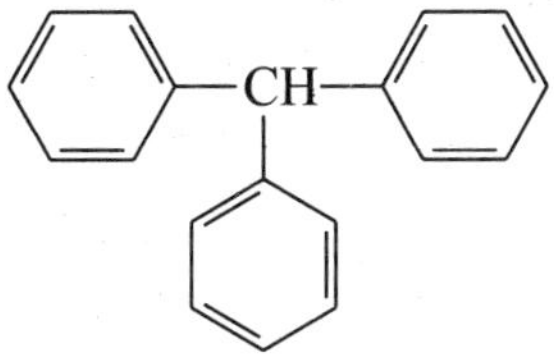

Triphenylmethane (Tritane)

9.10.2 Alkylbenzenes

Methods of Formation of Alkylbenzenes

(*i*) **Friedel-Crafts alkylation:** (Already discussed Sec. 9.3.5).

(*ii*) **Würtz-Fittig reaction:** Fittig modified Würtz reaction and found that cross coupling takes place between an alkyl and aryl halide when they are treated with metallic sodium in ethereal solution to give an alkylbenzene mixed with diaryl and dialkyl derivatives.

$$C_6H_5X + 2Na + RX \longrightarrow C_6H_5R + 2NaX$$

The synthetic utility of the reaction depends upon the fact that three products are easily separable. This method has an advantage over Friedel-Crafts reaction as there is no isomerization in the alkyl group during the course of reaction. When an aryl halide, such as bromobenzene, is treated with metallic sodium, it gives biphenyl (Fittig reaction).

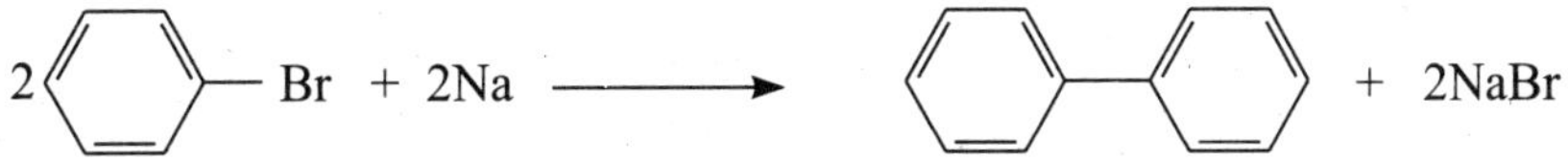

$$2C_6H_5Br + 2Na \longrightarrow C_6H_5-C_6H_5 + 2NaBr$$

As in the case of Würtz reaction, two mechanisms, the free radical and ionic, have been proposed. However, available evidence favours the ionic mechanism involving the intermediate formation of organometallics and can be represented in the following steps:

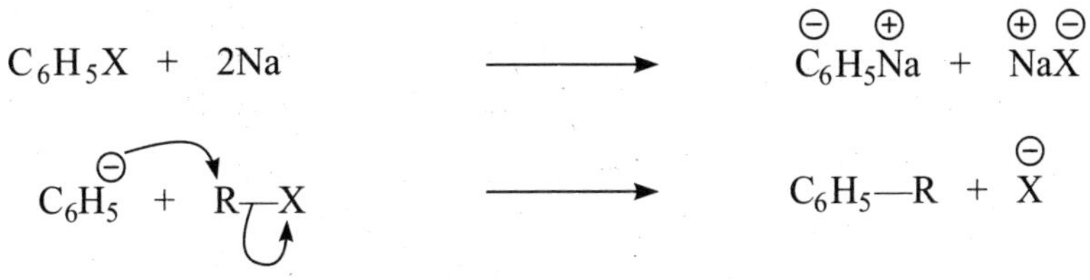

$$C_6H_5X + 2Na \longrightarrow \overset{\ominus}{C_6H_5}\overset{\oplus}{Na} + \overset{\oplus}{Na}\overset{\ominus}{X}$$

$$\overset{\ominus}{C_6H_5} + R-X \longrightarrow C_6H_5-R + \overset{\ominus}{X}$$

Other products can similarly be rationalized by this mechanism.

$$R{-}X + 2Na \longrightarrow \overset{\ominus\oplus}{RNa} + \overset{\oplus\ominus}{NaX}$$

$$\overset{\ominus}{R} + R{-}X \longrightarrow R{-}R + \overset{\ominus}{X}$$

$$\overset{\ominus}{C_6H_5} + C_6H_5{-}X \longrightarrow C_6H_5{-}C_6H_5 + \overset{\ominus}{X}$$

Biphenyl

(*iii*) **Reduction of alkyl aryl ketones:** The alkyl aryl ketones, synthesized by Friedel-Crafts acylation, can be reduced to alkylbenzenes by any one of the procedures available for converting $>C=O$ to $>CH_2$ group.

The ***Clemmensen reduction*** consists in refluxing the ketone with zinc amalgam and concentrated hydrochloric acid.

$$C_6H_5{-}CO{-}R \xrightarrow{Zn{-}Hg/HCl} C_6H_5{-}CH_2{-}R$$

The ketones can also be subjected to ***Wolff-Kishner reduction***, which involves heating of the hydrazone with sodium ethoxide in a high boiling alcohol such as diethylene glycol.

$$C_6H_5{-}CO{-}R \xrightarrow{NH_2-NH_2} C_6H_5{-}C(=N{-}NH_2){-}R \xrightarrow[473\ K]{NaOEt/ROH} C_6H_5{-}CH_2{-}R + N_2$$

Friedel-Crafts acylation followed by reduction is a better method for synthesizing alkylbenzenes from benzene as compared to direct alkylation as it does not involve polyalkylation or rearrangement in the alkyl side chain. For example, synthesis of *n*-propylbenzene from benzene.

$$C_6H_6 + CH_2CH_2COCl \xrightarrow{AlCl_3} C_6H_5COCH_2CH_3 \xrightarrow[\text{Wolff-Kishner reduction}]{\text{Clemmensen reduction or}} C_6H_5CH_2CH_2CH_3$$

Propiophenone

(*iv*) **Grignard reaction:** Arylmagnesium halides on treatment with alkyl halides form alkylbenzenes in good yields.

$$C_6H_5MgBr + R-X \longrightarrow C_6H_5R + MgXBr$$

Grignard reaction is also useful for preparing alkylbenzenes with branched side chain. This can be affected by treating an alkyl aryl ketone with a suitably chosen alkylmagnesium halide followed by dehydration of the tertiary alcohol and catalytic reduction of the alkene so produced. *Iso*-propylbenzene, for instance, can be synthesized from acetophenone, according to the reaction sequence outlined below:

$$\underset{\text{Acetophenone}}{C_6H_5COCH_3} \xrightarrow[2.\ H^{\oplus}]{1.\ CH_3MgI} C_6H_5C(CH_3)_2OH \xrightarrow[-H_2O]{\Delta} \underset{\alpha\text{-Methylstyrene}}{C_6H_5C(CH_3)=CH_2} \xrightarrow{H_2\text{-Pd}} \underset{\text{Isopropylbenzene (Cumene)}}{C_6H_5CH(CH_3)_2}$$

(*v*) Decarboxylation of aromatic acids: Occasionally, this method is used for the synthesis of aromatic hydrocarbons. This involves heating of aromatic carboxylic acids with soda lime. For instance, toluene from toluic acids.

$$\underset{\text{Toluic acid}}{CH_3C_6H_4COOH} \xrightarrow{\text{Soda lime}} \underset{\text{Toluene}}{C_6H_5CH_3}$$

(*vi*) From phenols: Phenols, for instance, *o*-cresol, on heating with zinc dust produce toluene.

$$\underset{o\text{-Cresol}}{CH_3C_6H_4OH} \xrightarrow{\text{Zn, }\Delta} \underset{\text{Toluene}}{C_6H_5CH_3}$$

Physical Properties of Alkylbenzenes

The alkylbenzenes are colourless liquids possessing a strong and rather pleasant odour. Some of these compounds are, however, quite toxic and continuous inhaling of their vapours may cause serious health hazards. They are volatile and highly inflammable compounds burning with a sooty flame. Because of their low polarity they are insoluble in water and more soluble in non-polar solvents. Their physical properties are listed in Table 9.1 which reveals that the boiling points show a regular gradation and there is an increase of about 20–30°C in the boiling point for each additional carbon atom.

Table 9.1: Physical Properties of Alkylbenzenes

Compound	M.P. °C	B.P. °C
Benzene*	5.5	80
Toluene	–95	111
Xylenes		
ortho-	–25	144
meta-	–48	139
para-	13	138
Ethylbenzene	–95	136
1,2,3-Trimethylbenzene (Hemimellitene)	–25	176
1,2,4-Trimethylbenzene(Pseudocumene)	–44	169
1,3,5-Trimethylbenzene (Mesitylene)	–45	165
n-Propylbenzene	–99	159
Isopropylbenzene	–96	152
1,2,3,4-Tetramethylbenzene (Prehnitene)	–6.5	205
1,2,3,5-Tetramethylbenzene (Isodurene)	–24	197
1,2,4,5-Tetramethylbenzene (Durene)	80	195
n-Butylbenzene	–81	183
sec-Butylbenzene	–83	173
t-Butylbenzene	–58	169
1-Methyl-4-isopropylbenzene (*p*-cymene)	–70	177
Pentamethylbenzene	53	231
Hexamethylbenzene	165	261

**Physical constants given here for the sake of comparison.*

A perusal of the table further reveals that there is little correlation between the melting points and molecular weights. In fact the melting point is highly dependent on the symmetry of the molecule. The more symmetrical a compound, the better it fits in the crystal lattice, meaning thereby that the intermolecular forces in the crystals of symmetrical molecules are greater. Thus benzene, being a symmetrical molecule melts at about 100°C higher than toluene. Dependence of melting point on symmetry is further revealed by a comparison of melting points of *o*-, *m*-, and *p*-xylenes. The *p*-isomer, because of its symmetry, melts at a higher temperature than any of the *o*- or *m*-isomer. The lower solubility of symmetrical *p*-isomer in water can also be explained on the basis of this fact.

Chemical Reactions of Alkylbenzenes

The characteristic chemical reactions of alkylbenzenes are of the following three types:

(*i*) *Electrophilic aromatic substitution reactions in which the alkyl group is o- and p- directing with activation.*

(*ii*) *Rearrangement of alkyl group and disproportionation reactions*

(*iii*) *Free radical substitution of the alkyl side chain, i.e., side chain reactions.*

These types of reactions are discussed below:

(*i*) Electrophilic Aromatic Substitution

As mentioned earlier, the alkyl substituent activates the ring for further substitution and directs the incoming group to *ortho*-and *para* positions. Some of the examples are:

(*a*) *Nitration*

(*i*) Nitration with a mixture of conc. HNO_3 and conc. H_2SO_4 gives a mixture of *ortho* and *para*- isomers mainly.

CH_3 (Toluene) $\xrightarrow{\text{conc. } HNO_3 \text{ + conc. } H_2SO_4}$ CH_3, NO_2 (*o*-Nitrotoluene (58%)) + CH_3, NO_2 (*p*-Nitrotoluene (38%)) + *m*-isomer (4%)

(*ii*) Nitration with fuming nitric acid gives a low yield of the explosive 2,4,6-trinitrotoluene (TNT).

CH_3 (benzene ring) $\xrightarrow{\text{Fuming } HNO_3}$ CH_3, O_2N, NO_2, NO_2 (TNT) + $3H_2O$

(b) Sulphonation

With fuming H_2SO_4 toluene gives *o*-and *p*-toluenesulphonic acid.

$$\underset{\text{Toluene}}{C_6H_5CH_3} \xrightarrow[313\ K]{\text{conc. } H_2SO_4\ (SO_3)} \underset{o\text{-Toluenesulphonic acid}}{o\text{-}CH_3C_6H_4SO_3H} + \underset{p\text{-Toluenesulphonic acid}}{H_3C\text{-}C_6H_4\text{-}SO_3H}$$

(c) Nuclear Halogenation

Toluene on treatment with halogen in presence of halogen carrier in dark affords *o*- and *p*-chloro derivatives.

$$\underset{\text{Toluene}}{C_6H_5CH_3} \xrightarrow{Cl_2/FeCl_3} \underset{o\text{-Chlorotoluene}}{o\text{-}CH_3C_6H_4Cl} + \underset{p\text{-Chlorotoluene}}{H_3C\text{-}C_6H_4\text{-}Cl}$$

(d) Friedel-Crafts Alkylation

It occurs readily and the nature of products depends upon the reaction temperature. Thus Friedel-Crafts alkylation of toluene at 273 K gives a mixture of *o*-and *p*-xylenes while at 353 K it produces *m*-xylene.

$$\underset{m\text{-Xylene}}{m\text{-}C_6H_4(CH_3)_2} \xleftarrow[353\ K]{CH_3Cl/AlCl_3} \underset{\text{Toluene}}{C_6H_5CH_3} \xrightarrow[273\ K]{CH_3Cl/AlCl_3} \underset{o\text{-Xylene}}{o\text{-}C_6H_4(CH_3)_2} + \underset{p\text{-Xylene}}{p\text{-}C_6H_4(CH_3)_2}$$

(e) Friedel-Crafts Acylation

Treatment of toluene with an acid chloride or acid anhydride in presence of anhyd. $AlCl_3$ gives *o*-and *p*-acyltoluenes.

$$C_6H_5CH_3 + RCOCl \text{ or } (RCOO)_2O \xrightarrow{AlCl_3} o\text{-Isomer} + p\text{-Isomer}$$

CH3 COR o-Isomer; CH3 COR p-Isomer

(*ii*) Rearrangement of Alkyl Groups and Disproportionation Reactions

On treatment with halogen acids in presence of a Lewis acid, the alkylbenzenes undergo a characteristic chemical reaction which involves movement of alkyl group around benzene ring or from one ring to another. The former types of reactions are called **isomerizations** and latter ones are called **disproportionations**. It may be mentioned that both the reactions may occur simultaneously. With methyl group the disproportionation takes place to a lesser extent while with other alkyl groups such as *t*-butyl, isopropyl and ethyl the diproportionation is an easier process and the ease of intermolecular alkyl transfer parallels the carbocation stability.

(*a*) **Isomerization:** Heating of any of the isomeric xylenes with hydrogen fluoride—boron trifluoride affords a mixture of the three xylenes in the fixed ratio ($o : m : p = 18 : 60 : 22$). *m*-Xylene is formed in larger amounts as it is a thermodynamically controlled product.

$$C_6H_4(CH_3)_2 \xrightarrow[\Delta]{HF - BF_3} \underset{18}{o\text{-Xylene}} + \underset{60}{m\text{-Xylene}} + \underset{22}{p\text{-Xylene}}$$

o, m or *p*-Xylene

The following mechanism may be proposed for this reaction:

$$p\text{-Xylene } 22\% + H^{\oplus} \underset{-H^{\oplus}}{\overset{+H^{\oplus}}{\rightleftharpoons}} \text{arenium ion} \quad \xrightleftharpoons{\text{Me shift}} \quad \text{arenium ion} \underset{-H^{\oplus}}{\overset{+H^{\oplus}}{\rightleftharpoons}} m\text{-Xylene } 60\% + H^{\oplus}$$

$$\text{arenium ion} \xrightleftharpoons{\text{1, 2-Methyl shift}} \text{arenium ion} \underset{+H^{\oplus}}{\overset{-H^{\oplus}}{\rightleftharpoons}} o\text{-Xylene } 18\%$$

p-Xylene 22%

o-Xylene 18%

m-Xylene 60%

***(b)* Disproportionation:** Under almost similar conditions if the alkyl group is secondary or tertiary it undergoes facile migration from one substrate to another. Such a transformation may be envisaged as taking place through free carbocations.

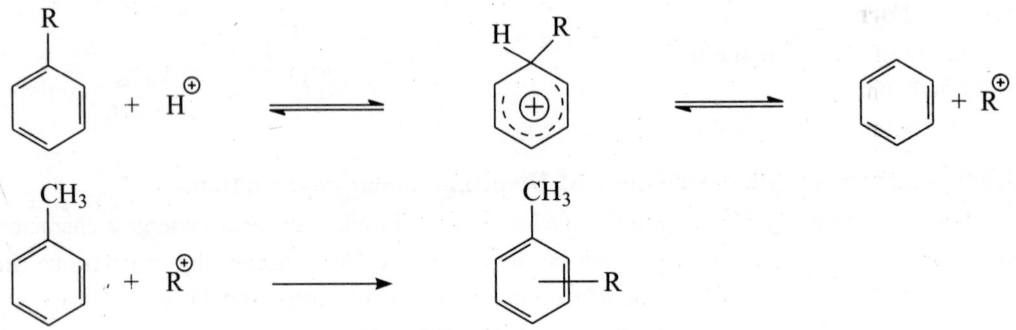

A mixture of *o*-, *m*- and *p*-isomers

According to an alternative mechanism, which appears to be more plansible, the alkyl group never gets free from the σ-complex of the parent hydrocarbon substrate and is transferred to the other aromatic substrate in a concerted manner.

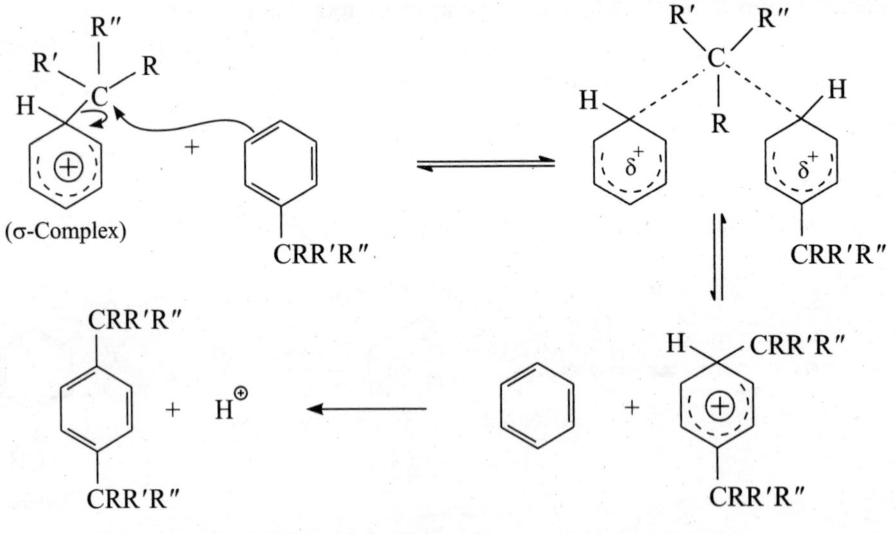

(σ-Complex)

The above mechanism is in agreement with the observed facts that:

(i) There is no rearrangement of the migrating alkyl substituent, and

(ii) Neopentylbenzene fails to undergo the disproportionation reaction under the above conditions because of the fact that steric hindrance at the neopentyl moiety blocks the backside attack of the benzene ring at the α-carbon atom.

It may be noted that the conditions used in these reactions are more or less similar to those of Friedel-Crafts alkylation reaction which is of little synthetic utility because of poly-alkylations and accompanying rearrangements, etc.

***(iii)* Side chain reactions:** The side chain halogenation and oxidation reactions of alkylbenzenes have already been discussed in Section 9.8.

9.10.3 Alkynylbenzenes

Methods of Formation

The chemistry of alkynylbenzenes is a combination of acetylenic side chain and benzene ring. The simplest compound, phenylacetylene may be prepared by the decarboxylation of phenylpropiolic acid and by dehydrobromination of β-bromostyrene with alcoholic potassium hydroxide.

$$C_6H_5-C{\equiv}C-COOH \xrightarrow{\Delta} C_6H_5-C{\equiv}CH$$

Phenylpropiolic acid → Phenylacetylene

$$C_6H_5-CH{=}CHBr \xrightarrow[\Delta]{\text{Alc. KOH}} C_6H_5-C{\equiv}CH$$

β -Bromostyrene → Phenylacetylene

Properties: Phenylacetylene is a colourless liquid *b.pt.* 415 K.

Reactions

***(i)* Acidic character**: Like acetylene, it also shows some acidic character and forms metallic salts, *e.g.*,

$$C_6H_5-C{\equiv}CH + NaNH_2 \longrightarrow C_6H_5-C{\equiv}\overset{\ominus\oplus}{C}Na + NH_3$$

Phenylacetylene → Sodium phenylacetylide

Sodium phenylacetylide is used for preparation of higher alkynylarenes.

***(ii)* Reduction**: It produes styrene upon reduction with Zn dust/CH_3COOH.

$$C_6H_5-C{\equiv}CH + 2\,[H] \xrightarrow{Zn\,/\,CH_3COOH} C_6H_5-CH{=}CH_2$$

Styrene

Chapter 9

(iii) Hydration: Phenylacetylene on treatment with mercuric perchlorate-perchloric acid mixture in dioxan gives acetophenone.

$$C_6H_5-C\equiv CH + H_2O \xrightarrow[\text{Dioxan}]{Hg(ClO_4)_2-HClO_4} C_6H_5-\overset{O}{\overset{\|}{C}}-CH_3$$

Acetophenone

9.10.4 Biphenyl or Diphenyl

Biphenyl occurs in coal-tar in small amounts. It is the simplest and most important example of an aromatic hydrocarbon in which two rings are directly linked to each other.

Methods of Formation

It can be synthesized by using the following well-known methods:

(*i*) **Fittig reaction:** It is an extension of Würtz reaction and consists in heating an ethereal solution of bromobenzene with metallic sodium. The mechanism of the reaction is similar to that of Würtz reaction.

$$C_6H_5-Br + 2Na + Br-C_6H_5 \longrightarrow C_6H_5-C_6H_5 + 2NaBr$$

Bromobenzene Biphenyl

A number of substituted triphenyls are also obtained by this reaction.

(*ii*) **Ullmann biaryl synthesis:** Ullmann discovered that iodobenzene, on heating with copper or bronze in a sealed tube affords biphenyl.

$$C_6H_5-I + 2Cu + I-C_6H_5 \xrightarrow[\text{Sealed tube}]{\Delta} C_6H_5-C_6H_5 + 2CuI$$

Iodobenzene Biphenyl

Chloro- or bromobenzene fail to undergo this coupling reaction unless some strong electron-withdrawing group (*e.g.*, NO_2) is present in the *ortho* or *para*-position. The reaction is quite vigorous with such substances and solvents such as dimethylformamide are used to moderate the vigour of the reaction.

$$o\text{-}O_2N-C_6H_4-Cl + 2Cu + Cl-C_6H_4-NO_2\text{-}o \longrightarrow o\text{-}O_2N-C_6H_4-C_6H_4-NO_2\text{-}o + 2CuCl$$

Mechanism of this reaction is uncertain. Most probably, organocopper compounds are the intermediates in the reaction (*c.f.* Würtz reaction).

***(iii)* Grignard reaction:** Phenylmagnesium bromide reacts with bromobenzene in the presence of a little cobaltous, or nickelous chloride.

$$C_6H_5MgBr + BrC_6H_5 \xrightarrow{CoCl_2} C_6H_5-C_6H_5 + MgBr_2$$

The reaction between bromobenzene and phenylmagnesium bromide does not take place without the addition of small amount of catalyst. On the basis of this evidence a free radical chain mechanism may be proposed which is initiated by the formation of phenylcobaltous chloride by a simple displacement reaction. The latter then undergoes facile homolytic cleavage under the reaction conditions to give $^{\bullet}CoCl$ and phenyl radicals which subsequently couple to form biphenyl.

$$C_6H_5MgBr + CoCl_2 \rightleftharpoons C_6H_5CoCl + MgBrCl$$

$$C_6H_5CoCl \longrightarrow C_6H_5^{\bullet} + \dot{C}oCl$$

$$C_6H_5^{\bullet} + {}^{\bullet}C_6H_5 \xrightarrow{\text{Coupling}} C_6H_5-C_6H_5$$

Also, formation of small quantities of benzene and triphenyl is in accordance with a free radical chain mechanism.

$$C_6H_5Br + \dot{C}oCl \longrightarrow C_6H_5^{\bullet} + CoClBr$$

$$C_6H_5^{\bullet} \xrightarrow{\text{Solvent etc.}} C_6H_6$$

$$C_6H_5^{\bullet} + C_6H_5-C_6H_5 \longrightarrow C_6H_5-C_6H_4-C_6H_5 + \dot{H}$$

***(iv)* From benzidine:** Diazotization of benzidine followed by reduction with H_3PO_2 gives biphenyl.

$$H_2N-C_6H_4-C_6H_4-NH_2 \xrightarrow[\substack{273\text{–}278 \\ (Diazotization)}]{NaNO_2/HCl} ClN_2-C_6H_4-C_6H_4-N_2Cl \xrightarrow[\text{Reduction}]{H_3PO_2} C_6H_5-C_6H_5$$

Benzidine Biphenyl

(v)* From aniline:** Aniline upon diazotisation gives benzendiazonium salt which upon heating in presence of Cu in ethanol gives rise to biphenyls. This is a special case of ***Gattermann reaction.

$$2\,C_6H_5-\overset{\oplus}{N}\equiv N\,H\overset{\ominus}{SO_4} \xrightarrow[C_2H_5OH]{Cu,\ \Delta} C_6H_5-C_6H_5 + 2N_2$$

***(vi)* Industrial preparation:** Biphenyl can be obtained by pyrolytic coupling of benzene in vapour phase (1273–1373 K) in presence of a little moisture and Fe_3O_4. This reaction is called *intermolecular dehydrogenation* and may take place *via* free radical coupling.

$$C_6H_5-H \xrightarrow[Fe_3O_4]{\Delta} C_6H_5^{\bullet} + H^{\bullet}$$

$$2\,C_6H_5^{\bullet} \longrightarrow C_6H_5-C_6H_5$$

$$H^{\bullet} + H^{\bullet} \longrightarrow H_2$$

Physical Properties

Biphenyl is a colourless solid, *m. pt.* 344 K, having a characteristic smell.

Reactions

***(a)* *Electrophilic aromatic substitution*:** It undergoes the usual electrophilic substitution reactions. In a broader sense, the molecule can be regarded as a benzene derivative having a phenyl substituent. Since aryl groups are electron-withdrawing (–I) they might seem to be deactivating and *m*-orienting. But it has been experimentally shown that the presence of one benzene ring activates the other for electrophilic substitution and directs the incoming group to the *o*-and *p*-positions. As discussed in the earlier section, the orientation effect of one benzene ring on the other can best be explained by writing all the possible resonance forms for the charged intermediates for each of the three possible reaction courses.

ortho Attack

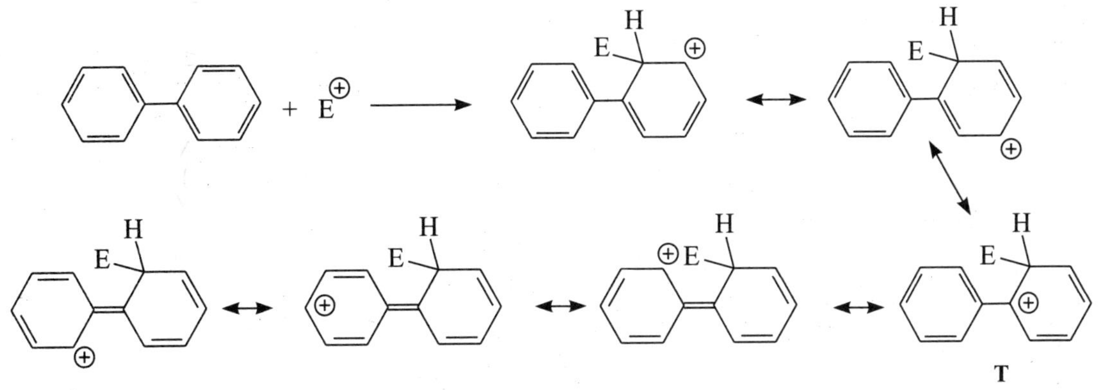

T

para Attack

etc.

T

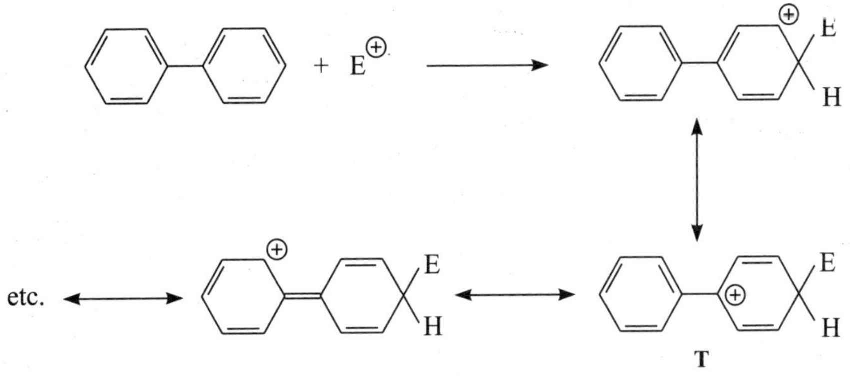

meta Attack

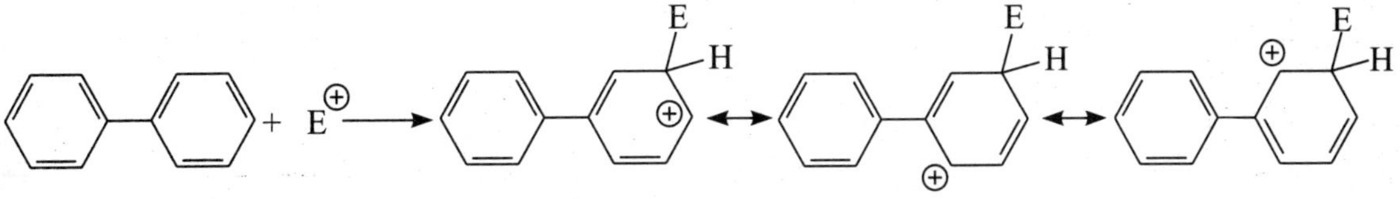

In the case of *ortho* and *para* attack there is one structure (marked **T**) in each case where the positive charge is located on the carbon atom to which the phenyl group is attached, whereas no such structure is possible for the charged intermediate formed by the *meta* attack. The structures marked **T** are more stable, firstly because they are *tertairy carbocations* whereas all other forms are *secondary carbocations*, and secondly because of the *p*-π-conjugation, there is delocalization of positive charge from this carbon to the second ring.

It has been shown that monosubstitution in the biphenyl results in the formation of *para* isomer as the major product. For instance, nitration of biphenyl yields mainly 4-nitrobiphenyl together with a small amount of 2-nitrobiphenyl.

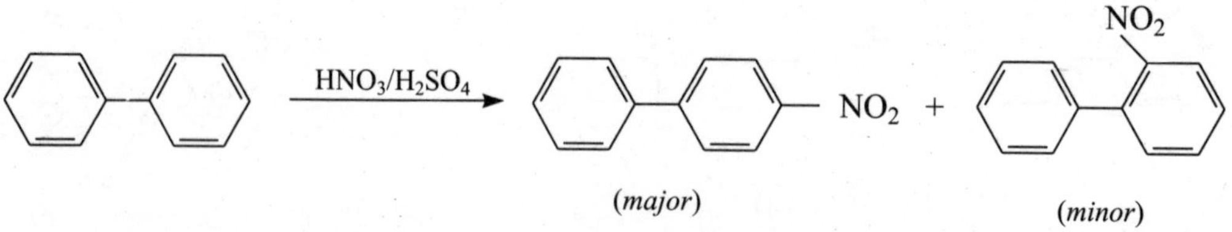

HNO_3/H_2SO_4 → NO_2 + NO_2

(*major*) (*minor*)

Similar product composition has been found in sulphonation and halogenation reactions.

Another special feature of the biphenyl ring system is the behaviour towards second substitution in a monosubstituted biphenyl. Whether the substituent present is activating or deactivating, the second substituent invariably enters the unsubstituted ring in the *ortho*-and *para*-position. This is again due to the fact that attack at the *para*-position (or *ortho*-position) places positive charge, in one of the resonance forms of the intermediate, on the carbon atom directly linked to the other benzene ring (**T**). Such a form is not possible in the case of the intermediate formed as a result of *meta* attack. Thus due to the reasons already discussed, the *para* and *ortho* substitution is preferred over *meta* substitution, even though the substituted ring may contain a deactivating group.

para Attack

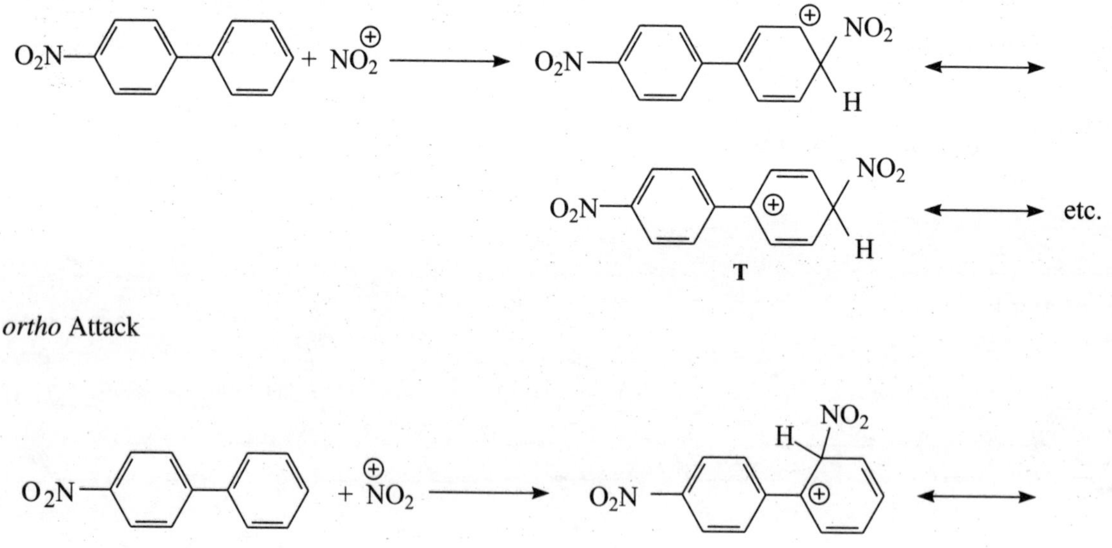

O_2N– + $\overset{\oplus}{NO_2}$ → O_2N– $\oplus$ NO_2 H ⟷

O_2N– $\oplus$ NO_2 H ⟷ etc.

T

ortho Attack

O_2N– + $\overset{\oplus}{NO_2}$ → O_2N– H NO_2 $\oplus$ ⟷

T

O_2N– H NO_2 $\oplus$ ⟷ etc.

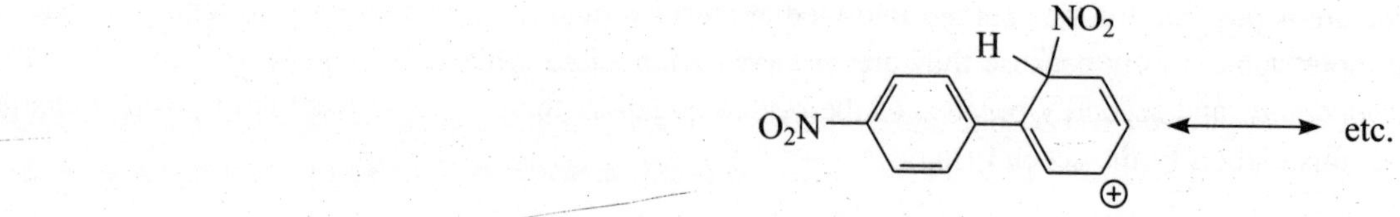

meta Attack

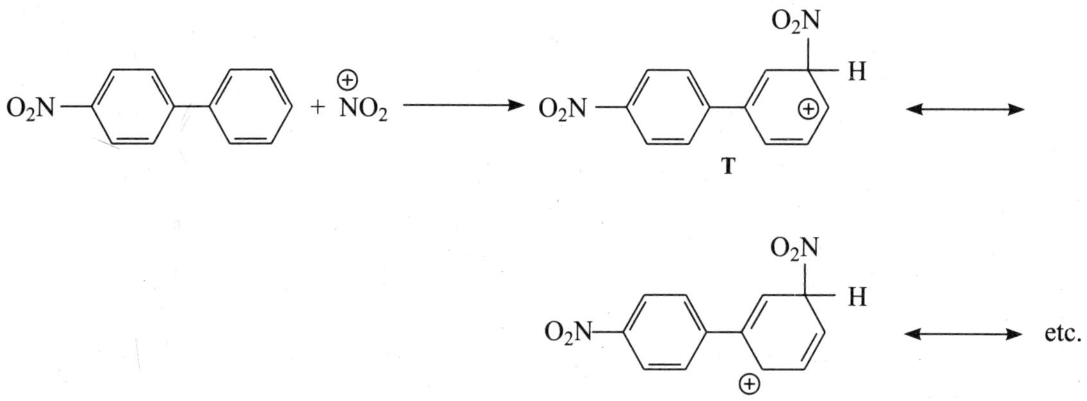

$$O_2N-C_6H_4-C_6H_5 + \overset{\oplus}{N}O_2 \longrightarrow \text{T} \longleftrightarrow \longleftrightarrow \text{etc.}$$

(b) *Oxidation*: Oxidation of biphenyl with chromic acid affords benzoic acid in small amounts.

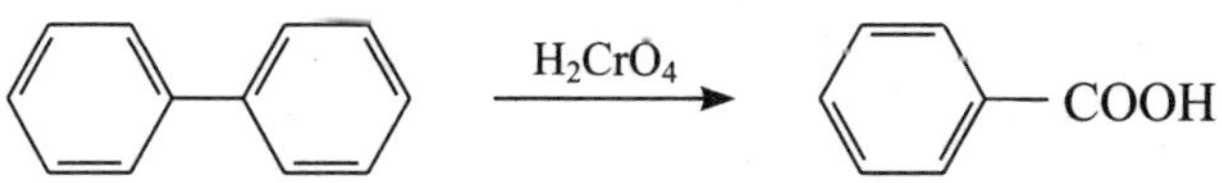

$$C_6H_5-C_6H_5 \xrightarrow{H_2CrO_4} C_6H_5-COOH$$

Stereochemistry: Stereochemistry of 2,2′,6,6′-tetrasubstituted biphenyl derivatives has already been discussed in Chapter-3.

Uses: Biphenyl finds extensive use in industry. It is used as heat-transfer fluid in the form of mixture with diphenyl ether. Its chloro derivatives are used as transformer oils and plasticizers. Benzidine (4,4′-diaminobiphenyl) is a useful dye intermediate.

A BRIEF REVIEW

- The typical reactions of arenes are **electrophilic substitution reactions** (*i.e.*, halogenation, nitration, sulphonation, mercuration, Friedel-Crafts alkylation and acylation reactions).
- During electrophilic aromatic substitution, first a π-complex is formed between the aromatic compound and the electrophile. This π-complex then gets converted into a σ-complex through a high energy T. S. In some cases, these σ-complexes have actually been isolated and characterized.
- The σ-complex finally eliminates a proton to form the substitution product. In majority of the electrophilic substitution reactions, this step does not show isotope effect thereby showing that cleavage of C—H bond is not involved in the rate-determining step of the reaction. Therefore, formation of σ-complex constitutes the rate-determining step of the reaction.
- Unlike nitration and halogenation, mercuration and sulphonation show **positive isotope effect** thereby showing that in these cases, C—H bond cleavage is the rate-determining step of the reaction.

- Sulphonation is a reversible process and arene sulphonic acids can be **desulphonated** by heating with steam under pressure.
- The substituents such as OH, OR, OCOR, NH_2, NHR, NR_2, NHCOR, OH, R, Ph and halogens which direct the incoming group to *o*-and *p*-positions are called ***o, p*-directing groups.** All these substiuents except halogens tend to increase the electron density on the benzene ring and hence are **activating groups** and as such they increase the rates of electrophilic substitution reactions.
- Substituent such as NR_3^+, CX_3, NO_2, CN, CHO, COR, COOH, SO_3H, etc. which direct the incoming groups to *meta*-positions are called ***m*-directing groups**. All these substitutents including halogens which tend to decrease the electron density are called **deactivating groups** and as such they decrease the rates or electrophilic substitution reactions.
- When any alkylbenzene is treated with a halogen in presence of Fe or anhyd. $AlCl_3$ (Lewis acid) at low temperature, a mixture of *o*-and *p*-alkylbenzene is obtained. This is called **nuclear halogenation**.
- When an alkylbenzene is treated with a halogen at high temperatures in sunlight or UV light and in presence of a radical initiator, side **chain halogenation** occurs.

  (*i*) *If excess of Cl_2 is passed through boiling toluene, in presence of light and a trace of benzoyl peroxide, all the three H-atoms of the CH_3 group are successively replaced to give first benzyl chloride, then benzal dichloride and finally benzotrichloride.*

  (*ii*) *When the side chain is larger than methyl group, halogenation preferentially occurs at the benzylic position. For example, bromination of ethylbenzene gives only* 1-*bromo*-1-*phenylethane without even a trace of* 2-*bromo*-1-*phenylethane. This is called* ***benzylic halogenation***.

  (*iii*) *Benzylic bromination can also be carried out by* ***NBS***.

  (*iv*) *Side chain halogenation occurs by a free-radical mechanism while nuclear halogenation occurs by ionic mechanism.*
- Oxidation of alkylbenzene with hot alkaline $KMnO_4$, $K_2Cr_2O_7$ / H_2SO_4 or with dil. HNO_3 eventually gives benzoic acid irrespective of the length of the carbon chain. If there are two or more alkyl groups, each alkyl group is oxidised to —COOH. Oxidation of toluene with CrO_2Cl_2 in CCl_4 or CS_2 gives benzaldehyde. This reaction is called **Etard reaction.**
- **Birch reduction** or *dissolving metal reduction* is a partial reduction of aromatic compounds by electron transfer from dissolving metals such as Li or Na or K in liquid ammonia in presence of a weak proton donor such as ethyl, isopropyl or tert-butyl alcohol. During this reduction, 1,4-addition of hydrogen occurs to benzene to form 3,6-dihydrocyclohexa-1,4-diene. If an electron donating group is present in the benzene ring, then addition of hydrogen atoms occurs at 3,6-positions with respect to the substituent to give 1-substituted 1,4-cyclohexadiene. Thus, Birch reduction of anisole followed by acid hydrolysis gives cyclohex-2-en-1-one.

  If, on the other hand, an electron-withdrawing group is present in the benzene ring, then addition of hydrogen atoms occurs at 1,4-positions with respect to the substituent to give 1-substituted 2,5-cyclohexadiene.

- Alkylbenzenes can be prepared by Friedel-Crafts alkylation, Wurtz-Fittig reaction of alkyl aryl ketones, Grignard reaction with alkyl halides, decarboxylation of aromatic acids with soda lime and deoxygenation of phenols.

  The most characteristic reactions of alkylbenzenes are (*i*) *electrophilic substitution reactions,* (*ii*) *free radical side chain halogenation and* (*iii*) *side chain oxidation reactions.*

- Friedel-Crafts reaction of toluene with CH_3Cl in presence of anhyd. $AlCl_3$ at 273 K gives a mixture of *o*-xylene and *p*-xylene. However, when this reaction is carried out at 353 K, only *m*-xylene is the major product. The formation of *m*-xylene from *o*-and *p*-xylene involves 1, 2-methyl migration.

- Phenylacetylene is prepared by dehydrobromination of β-bromostyrene with alcoholic KOH or by decarboxylation of phenylpropiolic acid. It gives the usual reactions of acetylene. Being weakly acidic, it reacts with $NaNH_2$ to give sodium salt. Upon reduction with Zn/CH_3COOH, it gives styrene and upon hydration in presence of $Hg(ClO_4)_2$—$HClO_4$ mixture, it gives acetophenone.

- **Biphenyl** is obtained by Fitting reaction, Ullmann biaryl synthesis, Grignard reaction and by heating benzene diazonium hydrogen sulphate with Cu in C_2H_5OH.

  It is more reactive than benzene and substitution preferentially occurs at *p* and *p'*-positions. For example, nitration of biphenyl first gives 4-nitrobiphenyl and then 4, 4'-dinitrobiphenyl.

SOLVED PROBLEMS

Q.1. Predict the product, of the following reaction and give mechanism.

$$C_6H_6 + F—CHO \xrightarrow{BF_3}$$

Ans. BF_3, being a Lewis acid, polarises the F — CHO in the following manner:

$$BF_3 + F—CHO \longrightarrow BF_3 ......\overset{\delta^-}{F}............\overset{\delta^+}{CHO}$$

The positive ends of this complex attack the benzene ring forming benzaldehyde.

$$C_6H_6 + \overset{\delta^+}{OHC}......\overset{\delta^-}{F}.........BF_3 \xrightarrow[-BF_4^-]{Slow} C_6H_6^{\oplus}(H)(CHO) \xrightarrow{BF_4^-} C_6H_5CHO + BF_3 + HF$$

Benzaldehyde

Q.2. It has been shown that chlorination of activated substrates can be affected with hypochlorous acid in presence of a proton catalyst. Suggest a suitable mechanistic scheme showing the role of proton in this reaction.

Ans. The role of proton is to generate chloronium ion needed for the reaction.

$$Cl{-}\ddot{\underset{..}{O}}{-}H + \overset{\oplus}{H} \longrightarrow Cl{-}\overset{\oplus}{\underset{..}{O}}\begin{smallmatrix}H\\H\end{smallmatrix} \longrightarrow \overset{\oplus}{Cl} + H_2O$$

Phenol (Activated aromatic substrate) + $\overset{\oplus}{Cl}$ ⟶ [arenium ion intermediate with H and Cl at the para carbon] ⟶ *p*-Chlorophenol + $\overset{\oplus}{H}$

Q.3. Explain why, compared to anilline ($C_6H_5NH_2$), acetanilide (C_6H_5 $NHCOCH_3$) is somewhat deactivated towards electrophilc aromatic substitution?

Ans. Because of resonance with the CH_3CO group, lone pair of electrons on nitrogen in acetanilide is less easily available for donation to the ring as compared to aniline.

$$C_6H_5{-}\ddot{N}H{-}C(=\ddot{O}:){-}CH_3 \longleftrightarrow C_6H_5{-}\overset{\oplus}{N}H{=}C(-\ddot{O}:^{\ominus}){-}CH_3$$

Consequently, as compared to aniline, acetanilide is somewhat deactivated towords electrophilc aromatic substitution.

Q.4. In organic synthesis, why is Friedel-Crafts acylation preferred over Friedel-Crafts alkylation?

Ans. Friedel-Crafts acylation is preferred over Friedel-Crafts alkylation because of the following two reasons:

(*i*) *During Friedel-Crafts alkylation, an alkyl group is introduced in the benzene ring. Since alkyl groups are electron-donating, therefore, the products, i.e., alkylbenzene is more reactive than benzene. In other words, further substitution will occur faster in the alkylbenzene than in benzene to form disubstituted and then polysubstituted benzene. As a result, of this polysubstitution, a mixture of products is obtained which will be difficult to separate. In contrast, in Friedel-Crafts acylation, an acyl group is introduced in the benzene ring. Since, an acyl group is electron-withdrawing, therefore, the product of acylation, i.e., acylbenzene is less reactive than benzene towards further substitution and the reaction stops after the first acylation. In other words, only a single product is obtained.*

(*ii*) Since carbocations or incipient carbocations are involved as intermediates in Friedel-Crafts alkylation reactions, therefore, if possible, they will rearrange to more stable carbocations before alkylating the ring. In other words, rearranged products are often obtained in Friedel-Crafts alkylation reactions. In contrast in Friedel-Crafts acylation reactions, no such rearranged products are possible.

Q.5. Write the mechanisms for the reactions of (*i*) *tert*-butyl bromide, (*ii*) *iso*-butyl bromide, (*iii*) *sec*-butyl bromide and (*iv*) *n*-butyl bromide with benzene in the preence of anhydrous $AlCl_3$.

Ans. With all the butyl bromides, the same product, *i.e.*, *tert*-butylbenzene is obtained since, the initially formed incipient carbocations rearrange to the more stable *tert*-butyl carbocation as shown below:

$$CH_3CH_2\underset{\underset{H}{|}}{CH}-\overset{\delta+}{CH_2}\cdots\cdots\overset{\delta-}{Br}\cdots\cdots AlCl_3 \xrightarrow[\text{shift}]{\text{1, 2-Hydride}} \underset{\underset{CH_3}{|}}{CH_2}-\overset{\overset{CH_3}{|\delta+}}{CH}\cdots\cdots\overset{\delta-}{Br}\cdots\cdots AlCl_3 \xrightarrow[\text{shift}]{\text{1, 2-Methyl}}$$

n-Butyl bromide -$AlCl_3$, complex   2°-Butyl bromide -$AlCl_3$, complex

$$AlCl_3\cdots\cdots\overset{\delta-}{Br}\cdots\cdots\overset{\delta+}{CH_2}-\underset{\underset{H}{|}}{\overset{\overset{CH_3}{|}}{C}}-CH_3 \xrightarrow[\text{shift}]{\text{1, 2-Hydride}} CH_3-\underset{\underset{CH_3}{|}}{\overset{\overset{CH_3}{|}}{C^{\oplus}}} + AlCl_3Br^-$$

Iso-Butyl bromide -$AlCl_3$, complex   *tert*-Butyl carbocation

Q.6. The *o*/*p*-ratio of sulphonation of toluene-4-D is higher than that of toluene. What information does it convey regarding the rate-determining step of sulphonation?

Ans. Since *o*/*p*-ratio of sulphonation of toluene-4-D is higher than that of toluene, it implies that C—D bond cleavage is involved in the rate-determining step of the reaction.

Q.7. Using Friedel-Crafts reactions, sketch the following transformation.

$$C_6H_6 \longrightarrow CH_3-\overset{\overset{CH_3}{|}}{CH}-CH_2-C_6H_4-CH_2-CH_3$$

Ans. Since a substituted alkyl group, *i.e.*, *iso*-butyl group is to be introduced in the ring, therefore, Friedel-Crafts acylation is preferred over Friedel-Crafts alkylation. We know that *iso*-butyl carbocation will rearrange to the more stable *tert*-butyl carbocation before alkylating the benzene ring. Therefore, acylation of benzene is carried out with *iso*-butyryl chloride. The complete sequence of reactions is as under:

$$C_6H_6 \longrightarrow \underset{\textit{Iso-Butyryl chloride}}{CH_3-CH(CH_3)-COCl} \xrightarrow[\text{Friedel-Crafts acylation}]{\text{Anhyd. AlCl}_3} \underset{\textit{Iso-Butyrophenone}}{CH_3-CH(CH_3)-CO-C_6H_5} \xrightarrow[\text{Clemmensen reduction}]{\text{Zn/Hg–HCl}}$$

$$\underset{\textit{Iso-Butylbenzene}}{CH_3-CH(CH_3)-CH_2-C_6H_5} \xrightarrow[\text{Friedel-Crafts acylation}]{CH_3COCl/AlCl_3} CH_3-CH(CH_3)-CH_2-C_6H_4-\overset{O}{\overset{\|}{C}}-CH_3$$

$$\xrightarrow[\text{Clemmensen reduction}]{\text{Zn/Hg–HCl}} CH_3-CH(CH_3)-CH_2-C_6H_4-CH_2-CH_3$$

Q.8. In Friedel-Crafts reactions, nitrobenzene is often used as a solvent. Why does it not complete with aromatic substrates for this reaction?

Ans. (*i*) Due to strong –I and –R-effects of the NO_2 group the electron density on the benzene ring decreases to such an extent that electrophiles fail to react and hence Friedel-Crafts reaction does occur.

(*ii*) Nitrobenzene being a highly polar solvent, dissolves $AlCl_3$ to form a homogeneous medium, which increases the rate of the reaction.

Q.9. Out of toluene and nitrobenzene, which will be nitrated more easily?

Ans. Toluene, because of +I-effect and hyperconjugation effect of the CH_3 group, the electron-density in toluene is much higher than that in nitrobenzene.

Q.10. Pick out as *ortho-para* or *meta*-directors:

—OR, —R, —NO_2, —NH_2, —CHO, —COOH.

Ans. *o, p*-directors : —OR, —R, —NH_2. *m*-directors : NO_2, —CHO, —COOH.

Q.11. Predict the major product when *o*-acetamidoanisole is subjected to bromination.

Ans. (Benzene ring with $NHCOCH_3$ at C-1, OCH_3 at C-2, Br at C-4) Since $NHCOCH_3$ is a better electron-donating group than OCH_3.

Q.12. A mixture of three isomeric xylenes obtained from coal-tar distillation is subjected to sulphonation. Which isomer will preferentially react?

Ans.

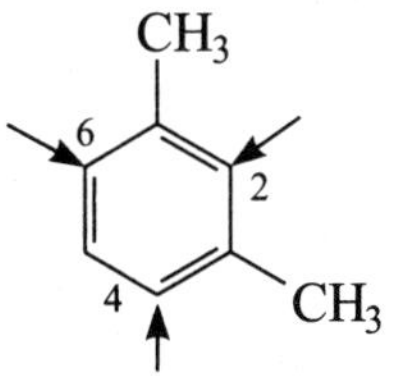

CH$_3$ 6 2 4 CH$_3$

Only in *m*-xylene, both the CH_3 groups increase the electron density at *o, p*-positions while in other isomers only one of the CH_3 group does so. Therefore, *m*-xylene is most reactive.

Q.13. Arrange the following compounds in increasing order of reactivity towards electrophilic substitution reactions: benzene, aniline, acetophenone, acetanilide.

Ans. Reactivity increase in the order:

Acetophenone < benzene < acetanilide < aniline, since electron-density increase in the same order.

Q.14. Mesitylene reacts with HF/SbF$_5$ in liquid SO$_2$ to form a coloured crystalline complex. Suggest a suitable structure of the product.

Ans. A σ-complex is formed due to electrophilic attack of H^+ on the ring carbon atom.

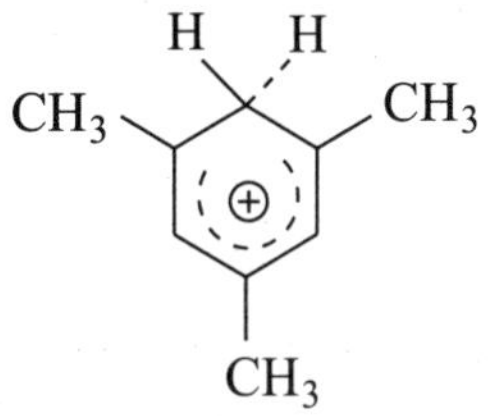

H H CH$_3$ CH$_3$ ⊕ CH$_3$

Q.15. Write the structure of the reduction product formed when toluene is subjected to Birch reduction.

Ans. Toluene on Birch reduction gives 1-methyl-1, 4-cyclohexadiene.

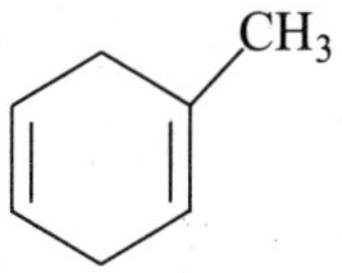

CH$_3$

Q.16. How will you convert anisole into cyclohex-2-en-1-one?

Ans. Birch reduction of anisole followed by acid hydrolysis gives cyclohexan-2-en-1-one.

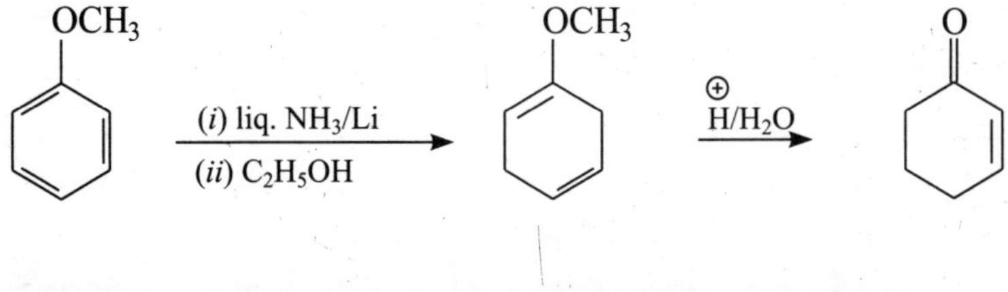

OCH$_3$ —(*i*) liq. NH$_3$/Li, (*ii*) C$_2$H$_5$OH→ OCH$_3$ —$\overset{\oplus}{H}$/H$_2$O→ O

UNSOLVED PROBLEMS

1. What is the role of Lewis acid catalyst in electrophilic reactions of benzene?

2. What are σ-and π-complexes ? Illustrate their role in aromatic electrophilic substitution reactions with the help of an energy profile diagram.

3. Write a note on sigma (σ) and pi (π) complexes.

4. It is well known that aromatic electrophilic substitution is a two-step process. Describe evidence for stepwise mechanism of electrophilic substitution in benzene.

5. Give the mechanism of Friedel-Crafts alkylation of benzene. Briefly discuss its limitations.

6. Following reactions have been designated as electrophilic substitution reactions. Name the electrophile in each case and show with the help of chemical equations how each one of them is produced?

(*i*) Sulphonation (*ii*) Nitration

(*iii*) Friedel-Crafts alkylation of *iso*-butylene in presence of an acid.

(*iv*) Bromination with bromine in presence of $FeBr_3$.

7. What are electrophiles? Can a proton act as an electrophile? Predict the product obtained from toluene upon treatment with D_2SO_4.

8. The following experiments are proposed to be carried out in order to decide whether aromatic sulphonation exhibits isotope effect. What do you expect to observe in each case, if this is so?

(*i*) Measurement of relatives rates of sulphonation of benzene and benzene-D_6 .

(*ii*) Determination of *o*/*p*-ratio of the sulphonation products in case of toluene and toluene-4-D.

(*iii*) Sulphonation of a 1 : 1 mixture of benzene and benzene-D_6 with one mole of sulphonating agents.

9. Discuss the mechanism of the following reaction.

$$C_6H_6 + (CH_3CO)_2O \xrightarrow{\text{Anhydrous } AlCl_3} C_6H_5COCH_3 + CH_3COOH$$

10. Give the mechanism of sulphonation of benzene. In what respect does it differ from nitration?

11. Explain the mechanism of nitration of benzene.

12. Explain the mechanism of halogenation of benzene.

13. Give the mechanism of reaction of benzene with $CH_3CH_2CH_2Cl$ in presence of anhydrous $AlCl_3$.

14. Discuss the mechanism of acylation of benzene. Give evidence in support of the mechanism.

15. How will you convert benzene into acetophenone? Give the mechanism of the reaction involved.

16. What is the essential difference between Friedel-Crafts alkylation and Friedel-Crafts acylation?

17. Polysubstitution products are often obtained in Friedel-Crafts alkylation when the alkylation agent is present in excess. However, no polysubstitution is observed in Friedel-Crafts acylation reaction. How do you explain this observation?

18. Write the important resonance contributing forms for σ-complexes for the *meta*-chlorination of aniline and then show why does this reaction take place at *ortho*-and *para*-positions?

19. Giving mechanism, complete the following reactions:

(*i*) $C_6H_6 + CH_3COCl \xrightarrow{\text{Anhyd. AlCl}_3}$ (*ii*) $C_6H_5NO_2 + HNO_3 \xrightarrow{H_2SO_4}$

(*iii*) $C_6H_6 + H_2SO_4 \xrightarrow{\text{Heat}}$ (*iv*) $C_6H_6 + CH_3CH{=}CH_2 \xrightarrow{H_2SO_4}$

(*v*) $C_6H_6 + ICl \xrightarrow{FeX_3}$ (*vi*) $C_6H_6 + (CH_3)_3COH \xrightarrow{H_2SO_4}$

(*vii*) $C_6H_6 + (CH_3)_3CCl \xrightarrow{\text{Anhyd. AlCl}_3}$ (*viii*) $C_6H_6 + (CH_3)_2CHCH_2Cl \xrightarrow{\text{Anhyd. AlCl}_3}$

(*ix*) $C_6H_6 + F - CHO \xrightarrow{BF_3}$

20. (*i*) Will the presence of a trifluoromethyl group activate or deactivate the benzene ring for further electrophilic aromatic substitution? Is it *meta* or *ortho, para*-directing group?

(*ii*) How do you explain the *ortho, para*-directing ability of alkyl group in electrophilic aromatic substitution.

21. Draw σ-complex for the electrophilic nitration of styrene ($C_6H_5CH = CH_2$) at *o-m*-and *p*-positions. On the basis of stability of these complexes, predict whether $—CH = CH_2$ group is *o*- or *p*-directing.

22. Compare the relative reactivity of nitration of benzene, toluene and nitrobenzene towards electrophilic substitution with the help of energy profile diagram.

23. Pick out the substituents which are (*i*) *o/p*-directing with activation, (*ii*) *o, p*-directing with deactivation, and (*iii*) and *m*-directing with deactivation

$$—Cl, —NH_2, —OH, —OCH_3, —NO_2, —CHO, —COOH$$

Illustrate your answer with the help of energy profile diagrams.

24. Arrange the following in order of decreasing reactivity towards electrophilic substitution reactions:

(*i*) Benzene (*ii*) Toluene (*iii*) Chlorobenzene (*iv*) Phenol

(*v*) Nitrobenzene.

25. From each of the following sets, select the most reactive and least reactive substrate towards ring nitration:

(*i*) Benzene, toluene, nitrobenzene and bromobenzene.

(*ii*) *m*-Dinitrobenzene, *m*-nitrotoluene and toluene.

(*iii*) Benzene, mesitylene, *m*-xylene and toluene.

(*iv*) Chlorobenzene, 2, 4-dinitrochlorobenzene and *p*-chloronitrobenzene.

26. How do you account for the observation that in *o, p*-directing systems, usually *ortho*-substitution occurs to a lesser extent than expected? Are there exceptions to this general rule?

27. In the presence of two groups—an *ortho/para* and the other *meta*-directing systems, what will be the position for the third substituent on the phenyl ring?

28. Explain the following:

(*i*) It is easier to introduce a nitro group in *ortho* or *para* position in phenol than in toluene.

(*ii*) $—NH_2$ group when attached to aromatic ring directs the incoming group to *ortho* and *para* position.

(*iii*) Chlorination of nitrobenzene gives *m*-chloronitrobenzene while nitration of chlorobenzene gives *ortho*-and *para*-chloronitrobenzenes.

(*iv*) Alkoxy group acts as activating and *ortho, para*-directing even though oxygen is more electronegative than carbon.

(*v*) In case of *o*-and *p*-disubstitution, the *p*-isomer generally predominates.

29. Justify the following statements:

(*a*) Nitration of bromobenzene gives *ortho*-and *para*-bromonitrobenzene while bromination of nitrobenzene gives *meta*-bromonitrobenzene.

(*b*) It is easier to nitrate toluene than benzene.

(*c*) Chlorine atom in chlorobenzene has deactivating influence for electrophilic substitution, yet it directs the incoming groups to *ortho* and *para*-positions.

(*d*) Nitration of benzene takes place more readily than that of nitrobenzene.

(*e*) Amino group in aniline is *o/p* directing but in aniline hydrochloride it is *meta*-directing.

30. Comment upon the statement: In case of halogens, reactivity is determined by stronger inductive effects while orientation is largely controlled by resonance.

31. Predict the major product in each of the following reactions:

(*i*) Monochlorination of *p*-cymene.

(*ii*) Monobromination of *o*-cresol.

(*iii*) Mononitration of *m*-bromochlorobenzene.

(*iv*) Mononitration of benzotrichloride.

(*v*) Mononitration of *m*-bromochlorobenzene.

32. Methyl group attached to benzene ring has no lone pairs of electrons but is still *ortho, para*-directing in nature. Explain why?

33. Explain *m*-directing nature of nitro group.

34. Justify that the carboxyl group attached to the benzene ring is *meta*-directing in nature for electrophilic substitutions.

35. Explain the following:

(*i*) Friedel-Crafts reaction between benzene and *n*-propyl chloride gives a considerable quantity of *iso*-propylbenzene.

(*ii*) More than one equivalent of $AlCl_3$ is used in Friedel-Crafts acylation to obtain a good yield.

(*iii*) Nitrobenzene is used as a solvent in Friedel-Crafts reactions, and it does not compete with the substrates in this reaction.

(*iv*) In synthetic organic chemistry, Friedel-Crafts acylation is often preferred to Friedel-Crafts alkylation.

36. (*i*) Explain why electron-withdrawing groups on benzene direct the incoming electrophile to *meta*-position?

(*ii*) Write the important resonance contribution forms for σ-complexes for the *ortho, para* and *meta*-chlorination of aniline and then tell why does this reaction take place at *ortho* and *para*-positions?

37. How will you account for the fact that —OH group attached to benzene is *ortho* and *para*-directing while —CHO group is *meta*-directing in nature?

38. Giving mechanism, complete the following reactions:

(*i*) Toluene $\xrightarrow{\text{Conc.HnO}_3\text{/ Conc.H}_2\text{SO}_4}$

(*ii*) Toluene $\xrightarrow{\text{Br}_2\text{, light}}$ (*iii*) Toluene $\xrightarrow{\text{CH}_3\text{Cl/ALCl}_3\text{353 K}}$

(*iv*) $C_6H_5CH_2CH_2CH_3$ $\xrightarrow{\text{Cl}_2/h\nu}$ (*v*) $C_6H_5CH_2CH_2CH_3$ $\xrightarrow[\Delta]{\text{KMnO}_4\text{, KOH}}$

39. What happens when:

(*i*) *o*-Xylene is heated with hot alk. $KMnO_4$ solution.

(*ii*) Chlorine is passed through boiling toluene.

(*iii*) Toluene is subjected to oxidation by chromyl chloride.

(*iv*) Toluene is treated with methyl chloride at 273 K.

40. (*i*) Why does side chain halogenation in alkylbenzenes take place preferentially at position *alpha* to the aromatic ring?

(*ii*) How many products do you expect to be formed by addition of 1mole of bromine to $C_6H_5-CH=CH-CH=CH_2$? Write their structures and state which is most stable.

41. What methods would you adopt to differentiate between the three isomeric xylenes?

42. Treatment of one of the diethylbenzenes with bromine in the presence of $AlBr_3$ results in the formation of three isomeric monobromo derivatives of which one is the major product. Suggest structure of the diethylbenzene and the products obtained. Indicate the major product.

43. Sketch the mechanism for each of the following transformations:

(*i*) $C_6H_5C(CH_3)_3 + Br_2 \xrightarrow{AlCl_3} C_6H_5Br + (CH_3)_2C=CH_2$

(*ii*) $C_6H_5-CH_3 + BrCCl_3 \xrightarrow{h\nu} C_6H_5CH_2Br + CHCl_3$

44. Sketch the following transformations:

(*i*) Acetophenone ⟶ Ethylbenzene

(*ii*) Propylbenzene ⟶ 1-Phenylpropene

(*iii*) Toluene ⟶ $C_6H_5CH_2OCH_3$

(*iv*) Styrene ⟶ Ethylbenzene

(*v*) Benzyl alcohol ⟶ 1, 2-Diphenylethane

45. Commencing with benzene, show with the help of equations how would you obtain the following:

(*i*) Phenylacetylene

(*ii*) 1-Phenylpropyne

(*iii*) 2-Phenylpropene

(*iv*) *p*-Dinitrobenzene

(*v*) *p*-Bromodiphenylmethane.

46. Sketch the following transformations:

(*i*) $C_6H_5C \equiv CCOOH \longrightarrow C_6H_5COCH_3$

(*ii*) $C_6H_5CH=CHCOOH \longrightarrow C_6H_5CH=CH_2$

(*iii*) $p\text{-}CH_3-C_6H_4COOH \longrightarrow C_6H_5CH_3$

(*iv*) $C_6H_6 \longrightarrow C_6H_5CH=CH_2$

47. Writing all the possible resonance forms of the charged intermediate, discuss the mechanism of aromatic electrophilic substitution in biphenyl.

48. (*a*) Write the structure of the major monosubstituted products formed when Br^+ attacks the following molecule:

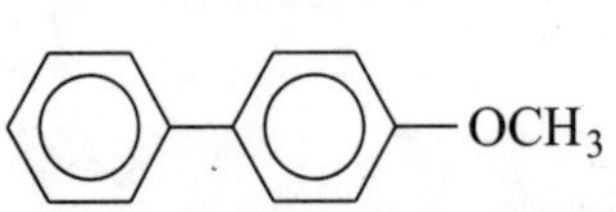

OCH_3

(*b*) Predict the product of mononitration of the following:

(*i*) NO_2

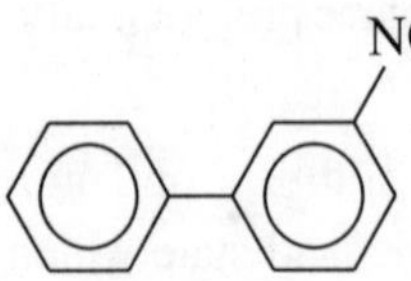

(*ii*) NO_2

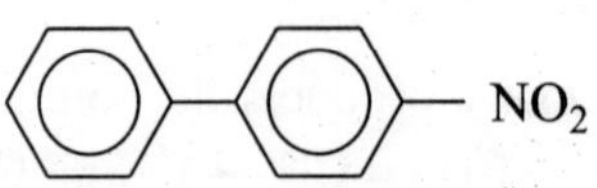

(*iii*) CH_2 COOH

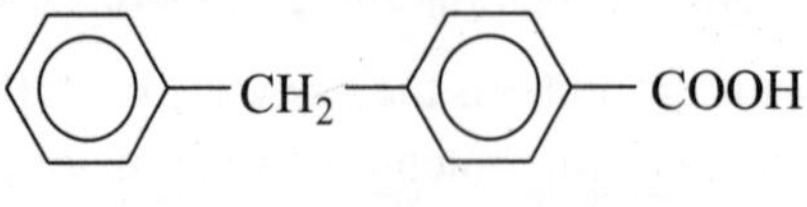

49. Explain why nitration and halogenation of biphenyl goes with activation of the *ortho* and *para*-positions and deactivation of the *meta*-positions? Suggest a reason why biphenyl is more reactive than 2, 2′-dimethylbiphenyl in nitration.

50. Phenyl group is known to exert negative inductive effect. But each phenyl ring in biphenyl is more reactive than benzene towards electrophilic substitution. Explain.

51. Write short notes on:

(*i*) Rearrangement of methyl groups in xylenes.
(*ii*) Fittig reaction
(*iii*) Free radical halogenation of alkyl side chain.
(*iv*) Ullmann biaryl synthesis.

52. In each of the following pairs, which compound will be more reactive towards electrophilic aromatic substitutions? Give reason:

(*i*) Phenol and toluene,
(*ii*) Toluene and nitrobenzene,
(*iii*) Chlorobenzene and *m*-dichlorobenzene,
(*iv*) Benzene and toluene,
(*v*) Phenol and chlorobenzene,
(*vi*) *m*-Dinitrobenzene and *m*-nitrotoluene, and
(*vii*) Benzene and acetanilide.

53. (*i*) It has been shown that chlorination of activated aromatic substrates can be affected with hypochlorous acid in presence of a proton catalyst. Suggest a suitable mechanistic scheme showing the role of proton in this reaction.

(*ii*) Compared to aniline ($C_6H_5NH_2$), acetanilide ($C_6H_5NHCOCH_3$) is somewhat deactivated towards electrophilic aromatic substitution. What explanation do you offer for this observation?

54. (*i*) Why are the cyanoanilinium ions ($C_6H_4(CN)\overset{\oplus}{N}H_3$) stronger acids than the anilinium ions ($C_6H_5\overset{\oplus}{N}H_3$)?

(*ii*) Rank the three cyanoanilinium ions (*o*-, *p*-and *m*-) in order of decreasing acidity. Give reasons.

55. Three isomeric xylenes, obtained as a mixture from coal-tar distillation, are subjected to the following treatment. Tell which isomer will preferentially react in each case.

(*i*) Sulphonation of the mixture.
(*ii*) Treatment of the mixture of respective sulphonic acids with dilute aqueous acid.
(*iii*) Treatment of the mixture with DF in presence of BF_3.

56. Commencing with propylbenzene, how will you synthesize 1-(*p*-chlorophenyl) propene (*p*-$ClC_6H_4CH=CHCH_3$)? Discuss the mechanism of the reactions involved in each step. Which reaction would you carry out first? Why?

57. Enumerate various methods employed for the synthesis of styrene. How would you convert styrene into (*i*) $C_6H_5CH_2Br$, and (*ii*) $C_6H_5CH(Br)CH_3$. Discuss the mechanisms of all the reactions?

58. Predict the major products in each of the following reactions:

(*i*) Friedel-Crafts reaction of chlorobenzene, (*ii*) Bromination of acetanilide,

(*iii*) Bromination of *m*-hydroxybenzaldehyde, (*iv*) Nitration of benzoic acid,

(*v*) Nitration of *p*-cresol, and (*vi*) Nitration of *p*-nitrotoluene.

59. Aniline reacts instantly with bromine to yield 2,4,6-tribromoaniline. However, if aniline is first acetylated, the monobromination product is prepared without any danger of over bromination. Account for the effect of acetylation upon the rate of bromination.

60. Account for the relative ease of bromination *para* to the methoxy group that is shown in the following series:

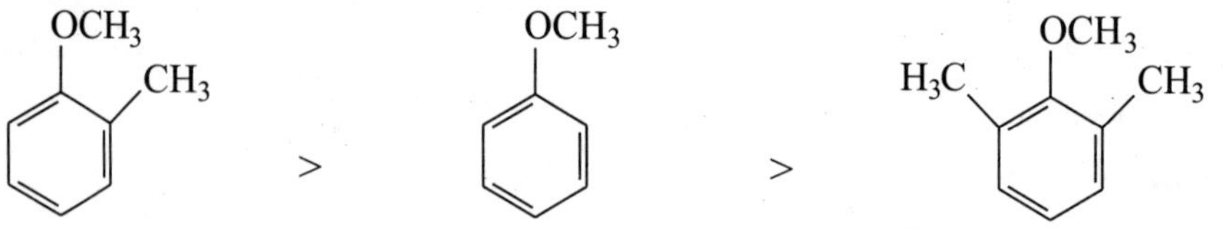

2-methylanisole (OCH_3, CH_3) > anisole (OCH_3) > 2,6-dimethylanisole (H_3C, OCH_3, CH_3)

61. Explain why Friedel-Crafts reaction cannot be carried out with (*a*) nitrobenzene, (*b*) aniline and (*c*) phenol?

62. (*i*) Suggest the structure of a compound, C_7H_6BrCl, that gives *m*-chlorobenzoic acid on oxidation.

(*ii*) Decarboxylation of the six isomeric diaminobenzoic acids ($C_6H_2(NH_2)_2COOH$) yielded the corresponding diaminobenzenes ($C_6H_4(NH_2)_2$). Three of the isomeric acids gave the same diamine (A), two acids gave another diamine (B) whereas diamine (C) was obtained by only one acid. Give names and structures for products A, B and C.

63. (*i*) Predict the products of each of the following reactions:

(*a*) $C_6H_6 \xrightarrow[H_2SO_4]{CH_2=CH_2}$ (*b*) $C_6H_6 \xrightarrow[AlCl_3]{CH_3CH_2Cl}$

(*c*) $C_6H_6 \xrightarrow[H_2SO_4]{CH_3CH_2OH}$

(*ii*) Which of the following pathways would produce *tert*-butylbenzene?

(*a*) $C_6H_6 \xrightarrow[H_2SO_4]{(CH_3)_3COH}$ (*b*) $C_6H_6 \xrightarrow[H_2SO_4]{(CH_3)_2C=CH_2}$

(*c*) $C_6H_6 \xrightarrow[AlCl_3]{(CH_3)_3CCl}$ (*d*) $C_6H_6 \xrightarrow[H_2SO_4]{CH_3CH(CH_3)CH_2OH}$

64. (*i*) An aromatic hydrocarbon of the formula, C_9H_{12}, upon oxidation gives a dibasic acid, $C_8H_6O_4$. Nitration of this acid yields only one mononitro derivative. Suggest the structure for the hydrocarbon.

(*ii*) Strong oxidation of an aromatic hydrocarbon ($C_{11}H_{12}$) produces benzoic acid as the only product without evolving CO_2. Write its structure.

65. Arrange the following in decreasing order of acidity (strongest acid first):

Benzene, propyne, ethane.

Suggest reasons for this order.

CHAPTER 10

Alkyl and Aryl Halides

Organic halogen compounds are the hydrocarbon derivatives obtained by replacing one or more atoms of hydrogen by the corresponding number of halogen atoms. Halogen containing organic compounds have a wide spectrum of uses covering various fields such as medicine, health, agriculture, industry, etc. Some of the important example are given below:

$O_2N-C_6H_4-CH(OH)-CH(NHCOCHCl_2)(CH_2OH)$

Chloromycetin (Chloramphenicol)

A naturally occurring antibiotic (secreted by *Streptomyces venezuelae*) used for treatment of typhoid.

HO—(3,5-diiodophenyl)—O—(3,5-diiodophenyl)—CH_2—CH(NH_2)—COOH

Thyroxine (hormone)

A natural product. Its deficiency causes gout.

(Quinolin-4-yl)—HN—CH(Cl)—$(CH_2)_3$—N$(C_2H_5)_2$

Chloroquine

(A synthetic antimalarial)

$F_3C—CHClBr$

Haloethane

(A synthetic anaesthetic used in surgery)

DDT
(A synthetic insecticide
used in agriculture)

BHC (Benzene hexachloride)
(A synthetic antitermite
used in agriculture)

In addition, a number of halogen derivatives are versatile intermediates in the synthesis of a large number of organic compounds.

Classification

The halogen derivatives of hydrocarbons are classified into two categories which are further subdivided into various subcategories as shown in **chart I.**

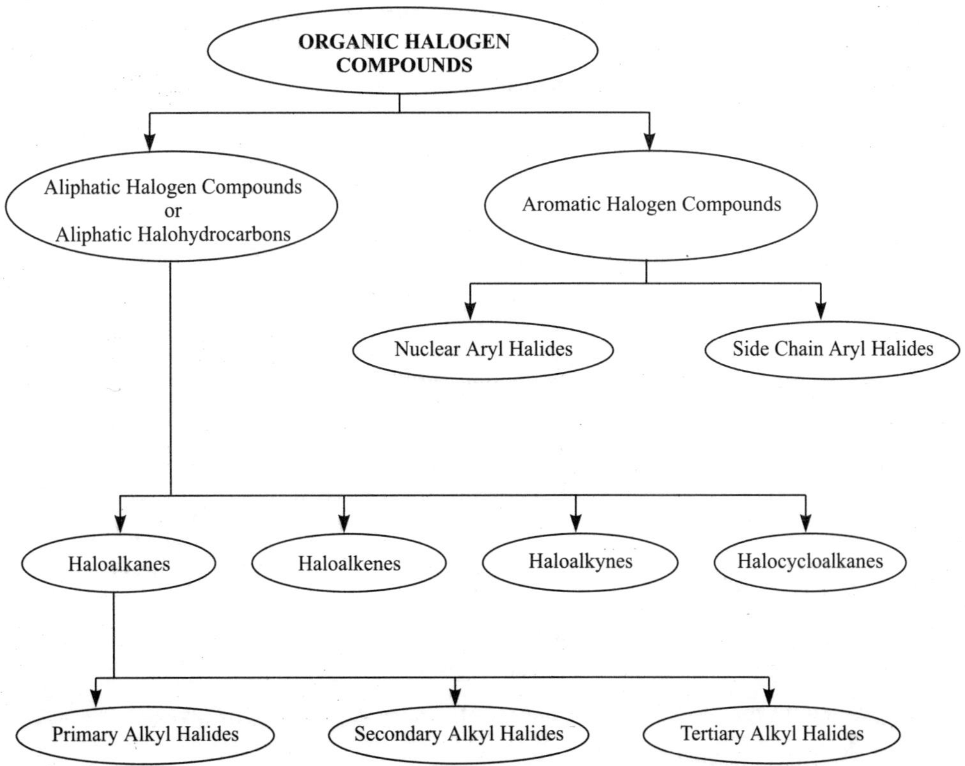

Chart I : Classification of halogen derivatives of hydrocarbons

10.1 ALIPHATIC HALOHYDROCARBONS

Nomenclature

As in case of other types of compounds, halogen derivatives of aliphatic hydrocarbons can also be named according to the following two systems follows: 1. Common or trivial name system and 2. IUPAC system.

1. Common or Trivial Name System

In this system of nomenclature the aliphatic halogen compounds are named as alkyl, alkenyl, alkynyl or cycloalkyl derivatives of hydrogen halides. Thus their names are written by naming the alkyl group followed by the name of the halogen as halide. The full name is written as two words, *e.g.*, CH_3Cl is named methyl chloride.

2. IUPAC System

The class names are derived by prefixing halo before the name of alkane, alkene, alkyne or cycloalkane. The full name is written as one word, *e.g.*, CH_3Cl is named as chloromethane. Like hydrocarbons the IUPAC names of this class of compounds are derived as follows:

(*i*) *The longest continuous straight chain of carbon atoms bearing the halogen atom and other substituents is chosen and the compound is named as the derivative of that hydrocarbon.*

(*ii*) *The carbon atoms of the longest continuous chain so chosen are numbered starting from one end to the other to indicate the position of the halogen atom and other substituents. The end chosen for numbering of the carbon atoms is such that the sum of all the numbers used to designate the positions of substituents is minimum.*

(*iii*) *If some other functional groups are also present, then these get priority in numbering the carbon atoms of the chain.*

Following examples (with common names in the parenthesis) are illustrative:

(*a*) Haloalkanes (Alkyl halides): They are halogen derivatives of alkanes and are represented by the general formula $C_nH_{2n+1}X$.

CH_3CH_2Cl

Chloroethane
(Ethyl chloride)

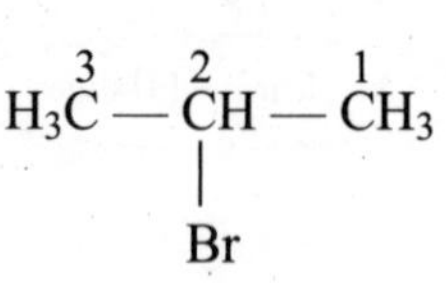

$\overset{3}{H_3C}—\overset{2}{CH}(Br)—\overset{1}{CH_3}$

2-Bromopropane
(Isopropyl bromide)

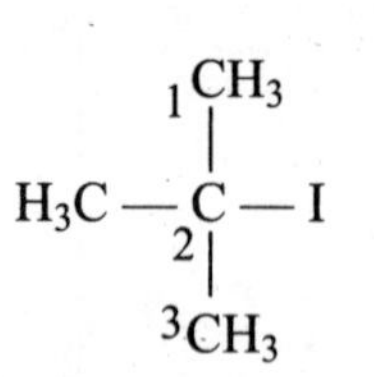

$H_3C—\overset{2}{C}(\overset{1}{CH_3})(\overset{3}{CH_3})—I$

2-Iodo-2-methylpropane
(*tert*-Butyl iodide)

$Cl—\overset{2}{CH_2}—\overset{1}{CH_2}—OH$

2-Chloroethanol
(Ethylene chlorohydrin)

$C_6H_5CH_2Cl$

Phenylchloromethane
(Benzyl chloride)

(*b*) Haloalkenes (Alkenyl halides): They are halogen derivatives of alkenes and have general formula $C_nH_{2n-1}X$.

$$H_2C{=}CHCl$$

Chloroethene
(Vinyl chloride)

$$I-\overset{3}{C}H_2-\overset{2}{C}H{=}\overset{1}{C}H_2$$

3-Iodoprop-1-ene
(Allyl iodide)

$$\overset{4}{C}H_3-\overset{3}{C}H{=}\overset{2}{C}H-\overset{1}{C}H_2Br$$

1-Bromobut-2-ene
(Crotyl bromide)

(*c*) **Haloalkynes (Alkynyl halides):** They are halogen derivatives of alkynes and have general formula $C_nH_{2n-3}X$.

$$H-C{\equiv}C-Br$$

Bromoethyne

$$HC{\equiv}C-CH_2Cl$$

3-Chloropropyne
(Propargyl chloride)

$$H_3C-C{\equiv}C-I$$

1-Iodoprop-1-yne

(*d*) **Halocycloalkanes (Cycloalkyl halides):** They are halogen derivatives of cycloalkanes and have general formula $C_nH_{2n-1}X$.

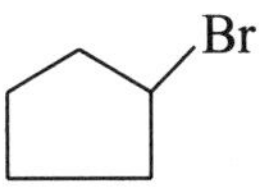

Br

Bromocyclopentane
(Cyclopentyl bromide)

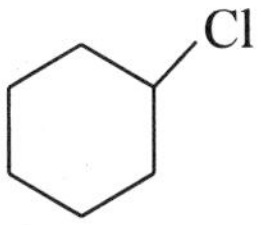

Cl

Chlorocyclohexane
(Cyclohexyl chloride)

10.2 ALKYL HALIDES

The compounds derived by replacement of one hydrogen atom by one halogen atom are called alkyl halides or monohaloalkanes or simply haloalkanes. These are generally written as R—X where R is any alkyl group and X is a halogen atom. Their general formula is $C_nH_{2n+1}X$, where n = 1, 2, 3, ... etc.

10.2.1 Classification of Alkyl Halides

Alkyl halides are divided into following three categories:

(*i*) Primary, (1°), (*ii*) Secondary, (2°), and (*iii*) Tertiary (3°) alkyl halides depending upon the halogen atom being attached to a primary, secondary and tertiary carbon atom, respectively.
For example,

(*i*) ***Primary*** **alkyl halides**

$$CH_3\overset{1°}{C}H_2-Cl$$

Chloroethane
(Ethyl chloride)

$$CH_3CH_2\overset{1°}{C}H_2-I$$

1-Iodopropane
(*n*-propyl iodide)

$$CH_3-\underset{}{\overset{CH_3}{\overset{|}{C}H}}-\overset{1°}{C}H_2-Cl$$

1-Chloro-2-methylpropane
(*iso*-butyl chloride)

$$CH_3-\underset{\underset{CH_3}{|}}{\overset{CH_3}{\overset{|}{C}}}-\overset{1°}{C}H_2-Cl$$

1-Chloro-2, 2-dimethylpropane
(*neo*-Pentyl chloride)

(*ii*) ***Secondary*** **alkyl halides**

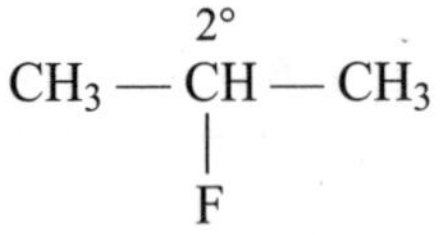

$$\underset{\displaystyle \text{F}}{\overset{2°}{\text{CH}_3-\text{CH}-\text{CH}_3}}$$

2-Fluoropropane
(*sec*-Propyl or *iso*-Propyl fluoride)

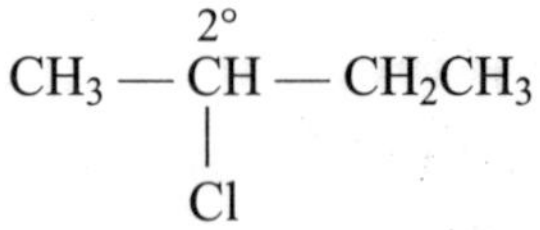

$$\text{CH}_3-\underset{\text{Cl}}{\overset{2°}{\text{CH}}}-\text{CH}_2\text{CH}_3$$

2-Chlorobutane
(*sec*-Butyl chloride)

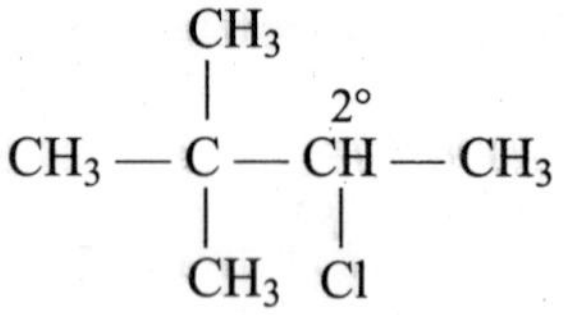

$$\text{CH}_3-\underset{\text{CH}_3}{\overset{\text{CH}_3}{\text{C}}}-\underset{\text{Cl}}{\overset{2°}{\text{CH}}}-\text{CH}_3$$

2-Chloro-3,3-dimethylbutane

(*iii*) ***Tertiary*** **alkyl halides**

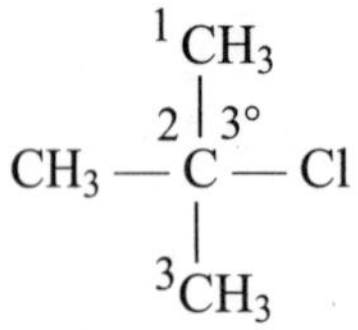

$$\text{CH}_3-\underset{{}^3\text{CH}_3}{\overset{{}^1\text{CH}_3}{\overset{2\;3°}{\text{C}}}}-\text{Cl}$$

2-Chloro-2-methylpropane
(*tert*-Butyl chloride)

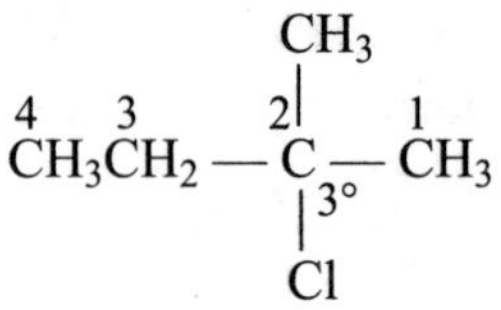

$$\overset{4}{\text{CH}_3}\overset{3}{\text{CH}_2}-\underset{\text{Cl}}{\overset{\text{CH}_3}{\overset{2}{\underset{3°}{\text{C}}}}}-\overset{1}{\text{CH}_3}$$

2-Chloro-2-methylbutane
(*tert*-Pentyl chloride)

10.2.2 Methods of Formation of Alkyl Halides

Usually the following general methods are employed for the preparation of alkyl halides:

1. From alcohols: Alcohols on treatment with a variety of reagents, are converted into alkyl halides by replacement of the —OH group by halogen. This transformation is one of the most convenient procedures used for the preparation of alkyl halides. Following are some of the more conveniently used methods:

(*i*) ***By the action of hydrogen halides*****:** Treatment of alcohols with hot halogen acids results in the formation of corresponding alkyl halides.

$$\underset{\text{Alcohol}}{\text{R}-\text{OH}} + \underset{\text{Hydrogen acid}}{\text{HX}} \xrightarrow{\Delta} \underset{\text{Alkyl halide}}{\text{R—X}} + \text{H}_2\text{O}$$

Mechanism

The reaction takes place in a step-wise mode. Two different mechanisms can operate depending upon the nature of alcohols. Thus,

(*a*) ***For primary and secondary alcohols*****:** The reaction takes place in two steps as follows:

Step I**: Protonation of alcohol to give the corresponding alkyloxonium ion.**

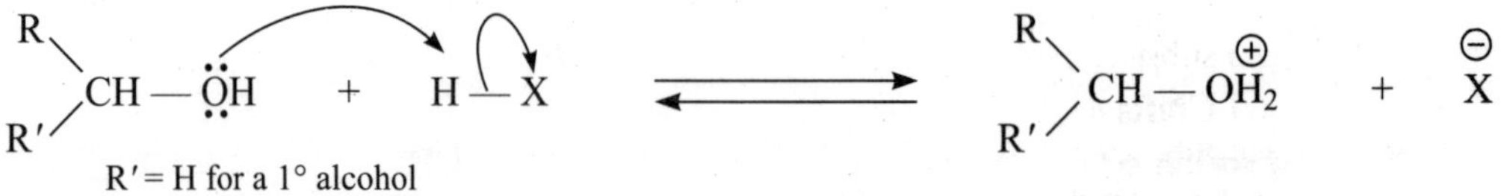

$$\text{RR}'\text{CH}-\ddot{\text{O}}\text{H} + \text{H}-\text{X} \rightleftharpoons \underset{\text{Alkyloxonium ion}}{\text{RR}'\text{CH}-\overset{\oplus}{\text{O}}\text{H}_2} + \overset{\ominus}{\text{X}}$$

R′ = H for a 1° alcohol
R′ = alkyl for a 2° alcohol

***Step II*: Nucleophilic attack of halide ion on alkyloxonium ion to produce haloalkane.**

$$\overset{\ominus}{X} + \underset{R'}{\overset{R}{>}}CH - \overset{\oplus}{O}H_2 \xrightarrow{\text{Slow}} \underset{R'}{\overset{R}{>}}CH - X + H_2O$$

The reaction is an example of an S_N2 displacement.

(*b*) *For tertiary alcohols*: The reaction takes place in three steps as follows:

***Step I*: Protonation of alcohol to give the corresponding alkyloxonium ion.**

$$R_3CD - \ddot{O}H + H - X \underset{}{\overset{\text{Fast}}{\rightleftharpoons}} R_3CD - \overset{\oplus}{O}H_2 + \overset{\ominus}{X}$$

3°-Alcohol 3°-Alkyloxonium ion

***Step II*: Dissociation of 3°-alkyloxonium ion to produce a stable 3° carbocation.**

$$R_3C - \overset{\oplus}{O}H_2 \rightleftharpoons R_3\overset{\oplus}{C} + H_2O$$

3°-Alkyloxonium ion 3° -Carbocation

***Step III*: Addition of nucleophilic halide ion to the carbocation.**

$$R_3\overset{\oplus}{C} + \overset{\ominus}{X} \rightleftharpoons R_3C - X$$

3°-Carbocation Halide ion 3°-Alkyl halide

This reaction is an example of an S_N1 displacement.

Reactivity of Hydrogen Halides and Alcohols

(*i*) *For a given type of alcohol the order of reactivity of the hydrogen halides is the same as their acid strength, i.e.*, the order is HI > HBr > HCl >> HF.

In these reactions HI is seldom used (as it can reduce the product alkyl iodide to hydrocarbon). Hydrogen Fluoride is also not useful.

(*ii*) *For a given hydrogen halide, the order of reactivity of alcohols is*

R_3COH	>	R_2CHOH	>	RCH_2OH	>	CH_3OH
3° Alcohol		2°Alcohol		1° Alcohol		Methanol

With this background, we will now discuss the methods of preparation of alkyl chlorides, alkyl bromides and alkyl iodides separately.

I. Alkyl Chlorides

(*a*) ***Primary*** **and** ***secondary*** **alkyl chlorides** are prepared by passing dry HCl gas through a primary or secondary alcohol. The reaction of these alcohols with HCl is slow (being S_N2) as the Cl^- ion is a poor nucleophile. The rate of this reaction is increased by using a Lewis acid such as anhydrous $ZnCl_2$ as a catalyst.

$$\underset{\text{Ethanol}}{CH_3CH_2OH} + HCl \xrightarrow{\text{Anhydrous } ZnCl_2} \underset{\text{Chloroethane}}{CH_3CH_2Cl} + H_2O$$

$$\underset{\text{2-Propanol}}{CH_3-\overset{\overset{OH}{|}}{CH}-CH_3} + HCl \xrightarrow{\text{Anhydrous } ZnCl_2} \underset{\text{2-Chloropropane}}{CH_3-\overset{\overset{Cl}{|}}{CH}-CH_3} + H_2O$$

Mechanism

$ZnCl_2$ being a Lewis acid complexes strongly with the non-bonding electrons of oxygen. Formation of this complex makes the C—O bond weak and creates a better leaving group. Various steps involved in this reaction are as follows:

***Step I*: Formation of Alcohol—$ZnCl_2$ Complex.**

$$R-\underset{\underset{H}{|}}{\ddot{O}:} + ZnCl_2 \longrightarrow \underset{\text{Alcohol-Lewis acid complex}}{R-\underset{\underset{H}{|}}{\overset{\oplus}{O}}-\overset{\ominus}{Z}nCl_2}$$

***Step II*: Cleavage of C—O bond to give a carbocation.**

$$\underset{\text{Alcohol-Lewis acid complex}}{R-\underset{\underset{H}{|}}{\overset{\oplus}{O}}-\overset{\ominus}{Z}nCl_2} \longrightarrow \underset{\text{Carbocation}}{\overset{\oplus}{R}} + \overset{\ominus}{Zn}(OH)Cl_2$$

***Step III*: Abstraction of chloride ion from HCl by the carbocation.**

$$H-Cl + \overset{\oplus}{R} \longrightarrow \underset{\text{Alkyl halide}}{R-Cl} + \overset{\oplus}{H}$$

***Step IV*: Regeneration of the catalyst ($ZnCl_2$).**

$$\overset{\oplus}{H} + H\ddot{O}-\overset{\ominus}{Z}nCl_2 \longrightarrow H_2\overset{\oplus}{O}-\overset{\ominus}{Z}nCl_2 \longrightarrow H_2O + ZnCl_2$$

If the carbocation produced is capable of undergoing rearrangement, we get rearranged alkyl halide. For example,

$$H_3C-\underset{CH_3}{\overset{CH_3}{C}}-CH_2-\ddot{O}H + \overset{\oplus}{H} \longrightarrow H_3C-\underset{CH_3}{\overset{CH_3}{C}}-CH_2-\overset{\oplus}{O}H_2$$

$$\xrightarrow{-H_2O} H_3C-\underset{CH_3}{\overset{CH_3}{C}}-\overset{\oplus}{C}H_2 \xrightarrow[\text{shift}]{\text{1, 2-Methyl}} CH_3-\underset{CH_3}{\overset{\oplus}{C}}-CH_2CH_3 \xrightarrow{\overset{\ominus}{Cl}} H_3C-\underset{CH_3}{\overset{Cl}{C}}-CH_2CH_3$$

1° Carbocation (*less stable*) → 3° Carbocation (*more stable*) → Rearranged alkyl chloride

Pyridine and trimethylamine can also catalyse the reaction between alcohols and HCl without rearrangement.

***(b) Tertiary* alkyl chlorides:** Tertiary alcohols, however, react readily with conc. HCl even in absence of $ZnCl_2$. For example,

$$H_3C-\underset{CH_3}{\overset{CH_3}{C}}-OH + HCl \xrightarrow[\text{temperature}]{\text{Room}} H_3C-\underset{CH_3}{\overset{CH_3}{C}}-Cl + H_2O$$

2-Methylpropan-2-ol → 2-Chloro-2-methylpropane

The above reaction forms the basis of the **Lucas Test** for distinction between 1°, 2° and 3° alcohols.

II. Alkyl Bromides

Alcohols on refluxing with constant boiling HBr (48%) in presence of a little conc. H_2SO_4 give rise to alkyl bromides. *e.g.*,

$$\underset{\text{Ethanol}}{CH_3CH_2-OH} + HBr \xrightarrow{H_2SO_4,\ \text{reflux}} \underset{\text{Bromoethane}}{CH_3CH_2Br} + H_2O$$

Chapter 10

This reaction can also be carried out by heating the alcohol with KBr and conc.H_2SO_4. In this case HBr is produced *in situ* as shown below:

$$\underset{\text{Ethanol}}{CH_3CH_2—OH} + KBr \xrightarrow{H_2SO_4,\ \Delta} \underset{\text{Bromoethane}}{CH_3CH_2—Br} + KHSO_4 + H_2O$$

$$KBr + H_2SO_4 \xrightarrow{\Delta} KHSO_4 + HBr$$

III. Alkyl Iodides

When a mixture of an alcohol and a constant boiling HI (57%) is refluxed, it produces alkyl iodide.

$$\underset{\text{Ethanol}}{CH_3—CH_2—OH} + HI \xrightarrow{\text{Reflux}} \underset{\text{Iodoethane}}{CH_3CH_2—I} + H_2O$$

As in case of preparation of alkyl bromides, this reaction can also be carried out by heating a mixture of alcohol, KI and 95% H_3PO_4. Here also HI is formed *in situ.*

$$CH_3CH_2OH + KI + H_3PO_4 \xrightarrow{\Delta} CH_3CH_2I + KH_2PO_4 + H_2O$$

$$KI + H_3PO_4 \xrightarrow{\Delta} KH_2PO_4 + HI$$

***(ii)* By the action of phosphorus halides:** Treatment of alcohols with phosphorus halides gives alkyl halides in very good yields (> 85%).

I. Alkyl Chlorides

Alcohols are converted into corresponding alkyl chlorides on treatment with PCl_3 or PCl_5.

$$\underset{\text{Alcohol}}{R—OH} + \underset{\text{Phosphorus pentachloride}}{PCl_5} \longrightarrow \underset{\text{Chloroalkane}}{R—Cl} + POCl_3 + HCl$$

$$\underset{\text{Alcohol}}{3R—OH} + \underset{\text{Phosphorus trichloride}}{PCl_3} \longrightarrow \underset{\text{Chloroalkane}}{3R—Cl} + \underset{\text{Phosphoric acid}}{H_3PO_3}$$

For examples,

$$\underset{\text{Ethanol}}{CH_3CH_2—OH + PCl_5} \longrightarrow \underset{\text{Chloroethane}}{CH_3CH_2—Cl + POCl_3} + HCl$$

$$\underset{\text{Butan-2-ol}}{3CH_3CH_2—CHOH—CH_3} + PCl_3 \longrightarrow \underset{\text{2-Chloroethane}}{3CH_3CH_2—CHCl—CH_3} + H_3PO_3$$

II. Alkyl Bromides and Alkyl Iodides

These compounds are formed by the treatment of alcohols with PBr_3 or PI_3, respectively. We know that PBr_3 and PI_3 are not very stable compounds. Hence, they are prepared *in situ* by the action of red phosphorus on Br_2 and I_2, respectively. For example,

A.

$$P_4 + 6Br_2 \longrightarrow 4\,PBr_3$$

$$\underset{\text{Ethanol}}{3CH_3CH_2OH} + PBr_3 \longrightarrow \underset{\text{Bromoethane}}{3CH_3CH_2Br} + H_3PO_3$$

B.

$$P_4 + 6I_2 \longrightarrow 4PI_3$$

$$\underset{\text{Propan-1-ol}}{3CH_3CH_2CH_2OH} + PI_3 \longrightarrow \underset{\text{1-Iodopropane}}{3CH_3CH_2CH_2I} + H_3PO_3$$

***(iii)* By the action of thionyl chloride:** Treatment of alcohols with thionyl chloride ($SOCl_2$) in presence of pyridine results in the formation of alkyl chlorides.

$$\underset{\text{Alcohol}}{R-OH} + SOCl_2 \xrightarrow{\text{Pyridine}} \underset{\text{Chloroalkane}}{R-Cl} + SO_2\uparrow + HCl\uparrow$$

For example,

$$\underset{\text{Ethanol}}{CH_3CH_2-OH} + SOCl_2 \longrightarrow \underset{\text{Chloroethane}}{CH_3CH_2-Cl} + SO_2\uparrow + HCl\uparrow$$

Mechanism: Following mechanism may be proposed:

$$\underset{\text{Alcohol}}{R-\ddot{O}(H)\!:} + Cl-\overset{O}{\overset{\|}{S}}-Cl \longrightarrow R-\overset{\oplus}{O}(H)-\overset{O}{\overset{\|}{S}}-Cl + \overset{\ominus}{Cl}$$

(pyridine, $\ddot{N}$, abstracts H)

$$\underset{\text{Chloroalkane}}{R-Cl} \xleftarrow[-SO_2,\ -\overset{\ominus}{Cl}]{\overset{\ominus}{Cl}} \underset{\text{Chlorosulphite}}{R-O-\overset{O}{\overset{\|}{S}}-Cl} + \underset{\text{Pyridinium ion}}{C_5H_5\overset{\oplus}{N}H}$$

The role of pyridine is that of an **'acid scavenger'.**

This method is better than other ones because the two by-products HCl and SO_2 being gases escape the reaction mixture leaving behind the alkyl chlorides in almost pure state.

2. From hydrocarbons: Alkyl halides can be prepared from alkenes and alkanes as discussed below:

***(a) From alkenes*:** Electrophilic addition of hydrogen halides to C = C produces alkyl halides. The order of reactivity is: HI > HBr > HCl > HF.

$$\underset{\text{Ethene}}{CH_2{=}CH_2} + HI \longrightarrow \underset{\text{Iodoethane}}{CH_3CH_2I}$$

$$\underset{\text{Propene}}{CH_3CH{=}CH_2} + HBr \xrightarrow[\text{Absence of peroxide}]{\text{Mark. addition}} \underset{\text{2-Bromopropane}}{CH_3-\underset{\displaystyle Br}{\underset{|}{CH}}-CH_3}$$

$$\underset{\text{Propyne}}{CH_3C{\equiv}CH} \xrightarrow[\text{Mark addition}]{HCl} CH_3\underset{\displaystyle Cl}{\underset{|}{C}}{=}CH_2 \xrightarrow[\text{Mark addition}]{HCl} \underset{\text{2, 2-Dichloropropane}}{CH_3-CCl_2-CH_3}$$

$$\underset{\text{Propene}}{CH_3CH{=}CH_2} + HBr \xrightarrow[\text{Anti-Mark. addition}]{\text{Peroxide}} \underset{\text{1-Bromopropane}}{CH_3-CH_2-CH_2Br}$$

Peroxide effect or *Kharasch effect* has already been discussed in details under chemical reactions of alkenes.

***(b) From alkanes*:** Alkanes on treatment with Cl_2 or Br_2 at high temperature or in presence of light usually produce complex mixture containing mono-, di- and polyhaloalkanes. For example, chlorination of methane gives four products, *i.e.*,

$$CH_4 \xrightarrow[h\nu \text{ or } 520\text{–}670\text{ K}]{Cl_2} \underset{\text{Chloromethane}}{CH_3Cl} + \underset{\text{Dichloromethane}}{CH_2Cl_2} + \underset{\text{Trichloromethane}}{CHCl_3} + \underset{\text{Tetrachloromethane}}{CCl_4}$$

It may be mentioned here that polyhalogenation can be avoided by controlling the amount of halogen. Yet this method is not a good preparatory method since, the boiling points of the haloalkanes produced are so close that they cannot be separated easily. However, in industry mixture can be separated using long fractionating columns.

In the case of propane and above, even monohalogenation gives rise to a mixture of all the possible isomers because all types of hydrogens. *e.g.*, 1°, 2° and 3° are susceptible to be substituted by the halogen atom. For example,

$$\underset{\text{Propane}}{CH_3CH_2CH_3} \xrightarrow[298\text{ K}]{Cl_2,\ h\nu} \underset{\text{1-Chloropropane (45\%)}}{CH_3CH_2CH_2Cl} + \underset{\text{2-Chloropropane (55\%)}}{CH_3-CHCl-CH_3}$$

$$CH_3CH_2CH_2CH_3 \xrightarrow[298\ K]{Cl_2,\ h\nu} CH_3CH_2CH_2CH_2Cl + CH_3-CHCl-CH_2CH_3$$

n-Butane — 1-Chlorobutane (28%) — 2-Chlorobutane (72%)

(Formation of relative amounts of these products has already been discussed under chemical reaction of alkanes).

As such **iodination** is not a good method because of the reversibility of the reaction.

$$CH_4 + I_2 \rightleftharpoons CH_3\ I + HI$$

However in presence of an oxidizing agent, such as HIO_3, HNO_3, HgO, etc., the reaction takes place in the forward direction. This is because of the fact that HI formed during the reaction is oxidised to give iodine.

$$5HI + HIO_3 \longrightarrow 3I_2 + 3H_2O$$

Fluorination being an explosive reaction, is not good for preparation of alkyl fluorides. However, this may be achieved by diluting F_2 with an inert gas such as N_2, Ar, etc. Alkyl fluorides can also be prepared indirectly by heating suitable alkyl chloride or bromide with some fluoride salts such as AsF_3, SbF_3, AgF, Hg_2F_2, etc. For example,

$$CH_3Br + AgF \longrightarrow CH_3F + AgBr$$

Bromoethane — Fluoromethane

$$2CH_3CH_2Cl + Hg_2F_2 \longrightarrow 2CH_3CH_2F + Hg_2Cl_2$$

Chloroethane — Fluoromethane

This reaction is called as **Swarts reaction**.

Further when alkyl halides contain two or more halogen atoms at the same carbon on treatment with SbF_5 in presence of CoF_3 they produce corresponding alkyl fluorides.

$$3CH_3-CCl_2-CH_3 + 2CoF_3 \longrightarrow 3CH_3-CF_2-CH_3 + 2CoCl_3$$

2, 2-Dichloropropane — 2, 2-Difluoropropane

SbF_3, when used in place of CoF_3, gives similar products.

3. Formation of alkyl iodides from other alkyl halides: This is a good procedure for preparing alkyl iodides which are otherwise not easily obtained. The method consists of treating alkyl chloride or alkyl bromide with a solution of KI in acetone or methanol.

$$\underset{(X=Cl\text{ or }Br)}{R-X} + KI \longrightarrow R-I + KX$$

This is not unexpected as the iodide ion being a very good nucleophile displaces other halogens from alkyl halides.

4. Hunsdiecker reaction: Dry silver salt of a carboxylic acid upon refluxing with bromine in carbon tetrachloride, affords the corresponding alkyl bromide with the liberation of carbon dioxide.

$$\underset{\text{Silver salt of carboxylic acid}}{R{-}COOAg} + Br_2 \xrightarrow[\text{reflux}]{CCl_4} \underset{\text{Alkyl bromide}}{RBr} + CO_2 + AgBr$$

The synthetic utility of this procedure is limited as *only the primary bromides are obtained in preparative yields whereas the yields of secondary and tertiary bromides are quite low.*

Mechanism

The reaction is believed to take place by a free radical chain mechanism being initiated by the formation of acyl hypobromite, which loses CO_2 forming an alkyl radical. This alkyl radical reacts with another molecule of acyl hypobromite giving rise to acyloxy radical and alkyl bromide. Alternatively, the alkyl or acyloxy radical may react with molecular bromine forming alkyl bromide and bromine atoms.

$$R{-}\overset{O}{\overset{\|}{C}}{-}\overset{\ominus}{O}\ \overset{\oplus}{Ag} + Br{-}Br \longrightarrow \underset{\text{Acyl hypobromite}}{R{-}\overset{O}{\overset{\|}{C}}{-}O{-}Br} + AgBr$$

$$R{-}\overset{O}{\overset{\|}{C}}{-}O{-}Br \xrightarrow[-\dot{B}r]{} R{-}\overset{O}{\overset{\|}{C}}{-}\dot{O} \longrightarrow \dot{R} + CO_2$$

$$\dot{R} + R{-}\overset{O}{\overset{\|}{C}}{-}O{-}Br \longrightarrow R{-}Br + R{-}\overset{O}{\overset{\|}{C}}{-}\dot{O} \longrightarrow \longrightarrow \longrightarrow \text{etc.}$$

or

$$\dot{R} + Br{-}Br \longrightarrow R{-}Br + \dot{B}r$$

Evidence: The above mechanism is supported by the following observations:

(*i*) Acyl hypobromites have been isolated and converted into alkyl halides under the conditions of reactions.

(*ii*) Silver salts having chiral carbon atom at the α-position (with respect to carboxyl) give racemic products.

The yields of alkyl bromides can be considerably improved by adding bromine to fatty acid in presence of mercuric oxide in carbon tetrachloride at reflux temperature.

$$2R{-}COOH + 2Br_2 + HgO \longrightarrow 2R{-}Br + CO_2 + HgBr_2 + H_2O$$

Chloroalkanes can also be prepared by this method using Cl_2 instead of Br_2 but yields are low. In case of I_2, silver salts give esters instead of iodoalkanes.

$$2RCOOAg + I_2 \xrightarrow[\text{Reflux}]{CCl_4} RCOOR + CO_2 + 2AgI$$

This reaction is called **Birnbaum-Simonini reaction.**

10.2.3 Physical Properties

Some important physical constants of alkyl halides are listed in Table 10.1. Some physical properties are discussed below:

1. Physical state: Methyl chloride, methyl bromide, ethyl chloride and some chlorofluoromethanes and chlorofluoroethanes are gases. Other halides are either liquids or solids.

2. Colour: Pure alkyl halides are colourless. The bromides and especially iodides become coloured when exposed to light.

3. Smell: A number of halogen compounds, which are volatile, have sweet smell.

4. Melting points and boiling points: (*i*) The melting points and boiling points of haloalkanes are much higher than those of parent alkanes. The reason for this is that their molecular masses are much higher than those of corresponding alkanes. Due to high mass number the magnitude of *Van der Waals attractions* is quite large.

(*ii*) The boiling points of alkyl halides decrease in the order RI > RBr > RCl > RF. This is again due to increase in the magnitude of *Van der Waals forces of attraction* because of increased molecular masses from F to I.

Compound	CH_3I	CH_3Br	CH_3Cl	CH_3F	CH_4
b.pt.(K)	249	277	249	195	111

(*iii*) With halogen atom being same, the boiling points of alkyl halides increase with increase in the size of alkyl group.

Haloalkane	CH_3Cl	CH_3CH_2Cl	$CH_3(CH_2)_2Cl$	$CH_3(CH_2)_3Cl$
b.pt.(K)	249	285.5	320	351.5

(*iv*) In case of isomeric halides, branching decreases the boiling points. This is due to the reason that branching reduces the surface area of alkyl halides thereby lowering the magnitude of the *Van der Waals forces of attraction*. Thus,

Haloalkane	$CH_3(CH_2)_3Cl$	$(CH_3)_2CHCH_2Cl$	$CH_3CH_2CHClCH_3$	$(H_3C)_3CCl$
b.pt. (K)	351.5	342	341	324

Table 10.1: Physical Properties of Alkyl Halides

Name	Formula	M.Pt. K	B.Pt. K	Density g/mL
Methyl fluoride	CH_3F	—	195	—
Methyl chloride	CH_3Cl	176	249	0.920
Methyl bromide	CH_3Br	180	277	1.732
Methyl iodide	CH_3I	209	249	1.279
Ethyl fluoride	C_2H_5F	130	241	1.306
Ethyl chloride	C_2H_5Cl	134	285.5	0.92
Ethyl bromide	C_2H_5Br	154	311	1.45
Ethyl iodide	C_2H_5I	—	345	1.93
n-Propyl fluoride	*n*-C_3H_7F	—	270	—
iso-Propyl fluoride	$CH_3—CH(F)—CH_3$	140	263	1.324
n-Propyl chloride	*n*-C_3H_7Cl	150	220	0.89
iso-Propyl chloride	$CH_3—CH(Cl)—CH_3$	156	208	0.86
n-Propyl bromide	*n*-$C_3H_7—Br$	163	244	1.353
iso-Propyl bromide	$CH_3—CH(Br)—CH_3$	184	232.5	1.310
n-Propyl iodide	*n*-C_3H_7I	172	375	1.747
iso-Propyl iodide	$CH_3—CH(I)—CH_3$	182	362.4	1.703
n-Butyl chloride	$CH_3CH_2CH_2CH_2Cl$	150	351.5	0.90
iso-Butyl chloride	$CH_3—CH(CH_3)—CH_2Cl$	148	342	0.859
sec-Butyl chloride	$CH_3—CH(Cl)—CH_2—CH_3$	142	341	0.87
t-Butyl chloride	$(CH_3)_3CCl$	245	324	0.85
Cyclohexyl chloride	$C_6H_{11}Cl$	—	416	1.000
Vinyl chloride	$CH_2=CH—Cl$	—	259	—
Alkyl chloride	$CH_2=CH—CH_2Cl$	—	318	0.938
Propargyl chloride	$HC\equiv C—CH_2Cl$	—	338	—
Benzyl chloride	$C_6H_5—CH_2Cl$	—	452	1.102

(*v*) Generally, with the increase in number of halogen atoms, the boiling points of chloro, bromo and iodo compounds increase, for ecample,

Haloalkane	CH_3Cl	CH_2Cl_2	$CHCl_3$	CCl_4
b.pt (K)	249	313	334	350

5. Density: Alkyl fluorides and chlorides are lighter than water whereas alkyl bromides and iodides are heavier. Their relative densities follow the order:

$$RI > RBr > RCl > RF$$

Methyl iodide has the highest density. However, as the size of the alkyl group increases, the density of alkyl halides decreases. Their densities also decrease with decreasing number of halogen atoms involved in alkyl halides.

Compound	CCl_4	$CHCl_3$	CH_2Cl_2	CH_3Cl
No. of Halogen atoms	4	3	2	1
Density (g/mL)	1.595	1.489	1.336	0.920

6. Solubility: Alkyl halides are polar (dipole moment 2.05 – 2.15 D) but they are insoluble in water. This may be explained on the basis of the fact that they cannot form hydrogen bonds with water molecule. As expected they are quite soluble in less polar organic solvents such as petroleum ether, benzene, ether, chloroform carbon tetrachloride, etc. It is due to similar type of intermolecular forces in both. This is in accordance with the saying **"Like dissolves like"**.

7. Stability: Stability of alkyl halides having the same alkyl group decreases in the order: alkyl fluorides > alkyl chlorides > alkyl bromides > alkyl iodides. This is in close agreement with decreasing bond strength which is in the order:

$$C—F > C—Cl > C—Br > C—I$$

Due to weak C—I bond, alkyl iodides on standing become violent or brown. This is due to their decomposition in presence of light to give iodine

$$\underset{\text{Colourless}}{2\,R—I} \longrightarrow \underbrace{R—R + I_2}_{\text{Coloured (brown or violet)}}$$

Also, the most commonly used solvents in industry are organic chlorides rather than bromides and iodides. This is due to higher stability and easy availability of organic chlorides.

8. Inflammability: Alkyl halides are less inflammable than the corresponding hydrocarbons. The inflammability, further decreases with increasing halogen content. Interestingly, carbon tetrachloride is used as a fire extinguisher under the name ***'pyrene'***.

Organic halogen compounds are good solvents for fats and oils and also do not catch fire easily. Polychloro compounds such as tri-and tetra-chloroethylenes are widely used as solvents for dry cleaning.

10.2.4 Chemical Reactions

Due to the presence of polar C—X bond alkyl halides are highly reactive. They undergo a wide variety of reactions such as nucleophilic substitution, elimination, halogenation, formation of Grignard's reagents, etc. Some important reactions are discussed below:

Nucleophilic Substitution

The most widely studied reaction of alkyl halides is a nucleophilic displacement involving the attack of a nucleophile, ($\overset{\ominus}{\text{Nu}}$:), along with a pair of non-bonding electrons, on the carbon atom bearing the halogen (called the leaving group, $\overset{\ominus}{\text{L}}$:). In this process the leaving group departs along with the bonding pair of electrons and the electrons for the formation of the new bond are furnished by the nucleophile. The general picture of this reaction may be depicted in the following equation:

$$\overset{\ominus}{\text{Nu}}: + \text{R}'\text{—}\underset{\text{R}}{\overset{\text{R}''}{\text{C}}}\text{—L} \longrightarrow \text{Nu—}\underset{\text{R}'}{\overset{\text{R}''}{\text{C}}}\text{—R} + \overset{\ominus}{\text{L}}:$$

Example: The simplest example of a nucleophilic substitution is the conversion of methyl chloride into methyl alcohol on treatment with aqueous NaOH.

$$CH_3\text{—Br} + \underset{\text{Nucleophile}}{\overset{\ominus}{\text{OH}}} \longrightarrow CH_3\text{—OH} + \underset{\text{Leaving group}}{\overset{\ominus}{\text{Br}}}$$

Though the nucleophile must possess a pair of non-bonding electrons, it may either be a negatively charged or a neutral species. Likewise, the substrate may be neutral or positively charged. Four possibilities may thus be visualized.

(*i*) $\text{H}\overset{\ominus}{\text{O}} + CH_3\text{—Br} \longrightarrow \text{HO—}CH_3 + \overset{\ominus}{\text{Br}}$

(*ii*) $H_3\ddot{N} + CH_3\text{—Br} \longrightarrow [H_3\overset{\oplus}{N}\text{—}CH_3]\,\overset{\ominus}{\text{Br}}$

(*iii*) $\overset{\ominus}{\text{Cl}} + CH_3\text{—}\overset{\oplus}{OH_2} \longrightarrow \text{Cl—}CH_3 + H_2O$

(*iv*) $H_2\ddot{S}: + \text{R—}\overset{\oplus}{N}(CH_3)_3 \longrightarrow \text{R—}\overset{\oplus}{SH_2} + N(CH_3)_3$

$$\text{R—}\overset{\oplus}{SH_2} \longrightarrow \text{R—SH} + \overset{\oplus}{\text{H}}$$

Mechanism and Stereochemistry

This process has many mechanistic implications and has been extensively investigated both *kinetically and stereochemically*. It has been shown that the reaction may proceed by any one of the following two mechanistic schemes:

(A) One-step process. (B) Two-step process.

A. One-step process (S_N2 Reaction)

One of the two probabilities involves the attack of the nucleophile with simultaneous departure of the leaving group. This path has been referred to as *one-step* or *direct* displacement or **S_N2 process**. Ingold has used the short hand notation **S_N2** for this (**S**ubstitution **N**ucleophilic **B**imolecular) reaction.

B. Two-step process (S_N1 Reaction)

This reaction may take place by a *two-step mechanism* in which the C—L bond is broken in the *first step forming a carbocation intermediate which then combines with the nucleophile in the subsequent step giving rise to the product.*

In the following pages, we shall discuss the mechanism and stereochemical consequences of both these processes.

A. One-step process: As mentioned above a typical example of this process is the hydrolysis of methyl chloride to methyl alcohol using aqueous sodium hydroxide, in which the hydroxide ion displaces the bromide ion.

$$\overset{\ominus}{HO} + CH_3—Br \longrightarrow HO—CH_3 + \overset{\ominus}{Br}$$

Kinetic studies have revealed that the rate of such a reaction depends upon the concentration of the alkyl halide as well as the hydroxide ion, both the reactants being involved in the rate determining step. Therefore, it is a **second order reaction.**

$$\text{Rate} = k\,[CH_3Br]\,[\overset{\ominus}{OH}]$$

Thus, if the concentration of any of the reactant is doubled then the rate of the reaction is also doubled and when the concentration of any of them is halved, the rate of the reaction is also halved.

It has been demonstrated that in such a process the hydroxide ion (nucleophile) attacks the carbon atom from the side opposite to that of the bromine (leaving group). The bond breaking and bond making are simultaneous processes. In other words, the bond between carbon and oxygen starts forming and at the same time the bond between carbon and bromine starts breaking. Such a reaction would, therefore, involve a *transition state* in which the carbon atom appears to be pentavalent, *i.e.*, joined to three substitutents by full covalent bonds and to the nucleophile and the leaving group by partial bonds. The three substituents acquire a coplanar geometry.

$$\overset{\ominus}{HO} + H_3C—Br \longrightarrow \left[\overset{\delta^-}{HO}\text{---}\underset{H}{\overset{H\ \ H}{C}}\text{---}\overset{\delta^-}{Br}\right] \longrightarrow HO—CH_3 + \overset{\ominus}{Br}$$

Transition state

It may noted here that in the transition state the hydroxide ion has a diminished negative charge since it begins to share its electrons with carbon atom and at the same time bromine develops partial

Chapter 10

negative charge as it is tending to depart along with the binding electrons. When the C—O bond is completely formed, the C—Br bond is completely broken. The energy necessary to break the C—Br bond is more than compensated by the energy released during the making of C—O bond. The energy profile of such a reaction is shown in Fig. 10.1.

***Steric orientation of an S_N2 reaction*:** We have seen that in S_N2 reaction of methyl bromide the nucleophile attacks from that side of the carbon atom which is opposite to that of the leaving group. But as the carbon atom of methyl bromide is not chiral, one cannot distinguish this process from the one in which the hydroxide ion attacks from the same side (*front side*) as that of the departing bromine atom because the products I and II obtained by two processes are identical.

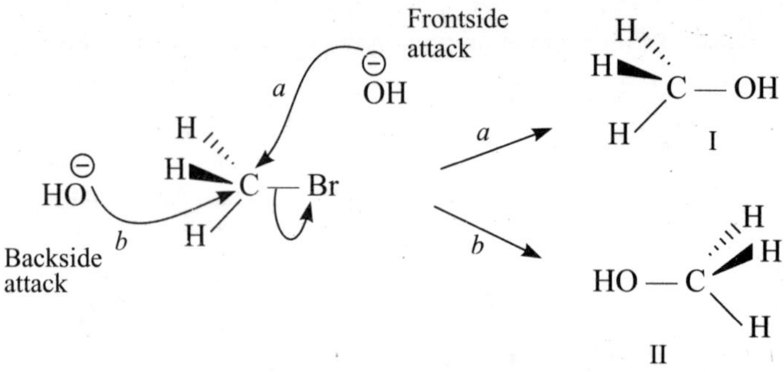

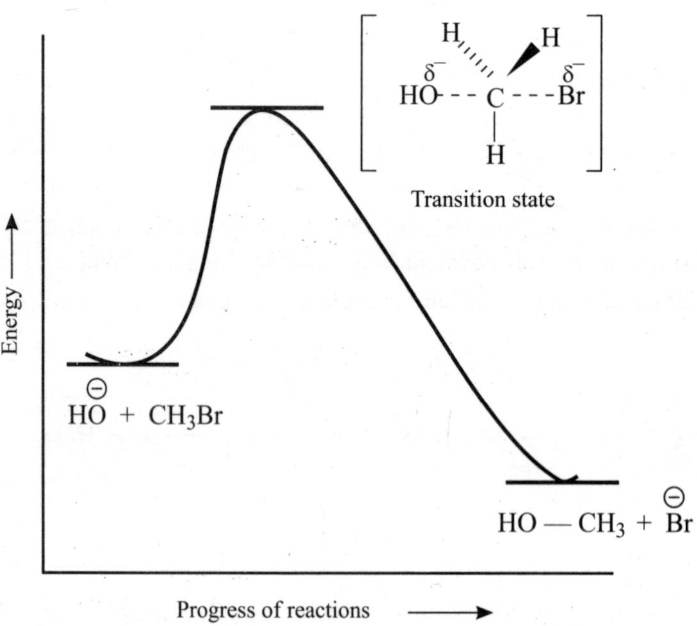

Fig. 10.1: Energy profile for an S_N2 reaction

However, by choosing appropriately substituted substrates it can be demonstrated that the results of the frontside and backside attacks are just opposite to each other. A lot of work has been done in this direction. In cyclic compounds, for instance, the two types of attacks would produce two distinct isomeric products. Thus, the S_N2 reaction of *cis*-3 methylcyclopentyl bromide and hydroxide ion can give rise to both *cis*-and *trans*-alcohols by front and backside attacks, respectively. But in actual experiment there is an exclusive formation of the *trans*-product. This proves exclusive backside attack in the S_N2 process.

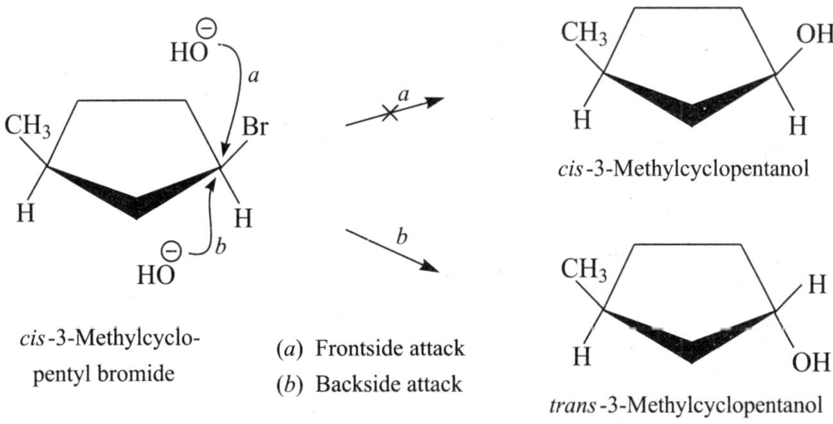

Backside attack in S_N2 reactions in acyclic systems has been proved beyond doubt by choosing compounds having chiral carbon atom involved in substitution. Since such compounds are optically active, it is possible to investigate the steric course of the reaction. The two possible products will no longer be identical in such cases. They would actually be non-superimposable mirror images of each other and may be distinguished by their absolute configuration. In actual practice the hydrolysis of optically active 2-bromooctane with hydroxide gives optically active 2-octanol with inverted configuration.

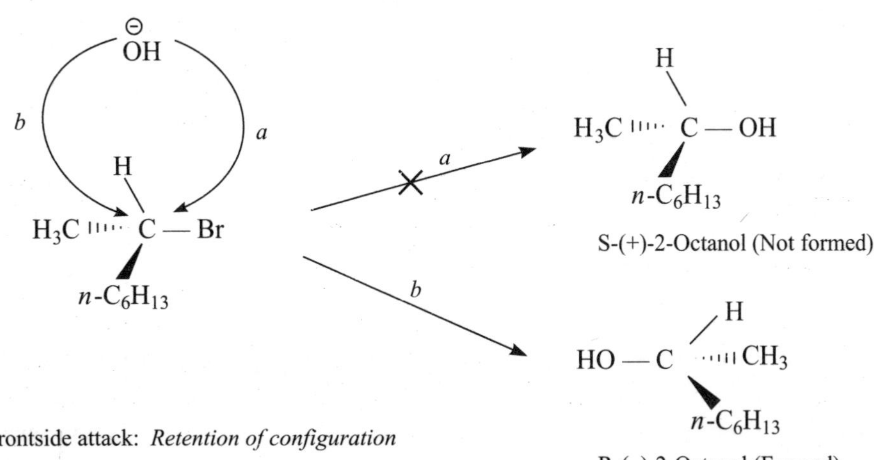

Thus, it can be said that an S_N2 *reaction proceeds with complete inversion of configuration usually referred to as* ***Walden inversion.***

We can now give an accurate description of an S_N2 reaction as involving backside attack of the nucleophile which prompts the tetrahedral arrangement of the substrate to assume a transition state forcing the three atoms or groups attached to the carbon to a planar configuration perpendicular to the line of attack. After the departure of the leaving group the tetrahedral arrangement is regained but with the inverted configuration. This process is quite similar to turning inside-out of an umbrella in strong winds.

Still one problem remains to be solved that is relating the configuration of the reactant with that of the product. We know that there is no distinct relation between the sign of rotation and the configuration of different compounds. They may be having same sign of rotation but would be different in configuration.

Evidence

Following experiments have clearly proved that all S_N2 reactions involve inversion of configuration:

(*i*) A clear-cut evidence for ***Walden inversion*** was provided by the conversion of an optically active alcohol into its enantiomer by a series of chemical transformations as shown below:

$$C_6H_5CH_2\text{—}C(CH_3)(H)\text{—}OH \xrightarrow[\text{Tosyl chloride (TsCl)}]{A} C_6H_5CH_2\text{—}C(CH_3)(H)\text{—}OTs$$

$[\alpha]_D = +33°$

$$\xrightarrow[]{B\ \ \overset{\ominus}{\text{AcO}}\overset{\oplus}{\text{K}}} AcO\text{—}C(CH_3)(H)\text{—}C_2H_6CH_5$$

$$AcO\text{—}C(CH_3)(H)\text{—}C_2H_6CH_5 \xrightarrow[\text{Hydrolysis}]{C} HO\text{—}C(CH_3)(H)\text{—}C_2H_6CH_5$$

$[\alpha]_D = -33°$

A look at the sign and magnitude of rotation of the two alcohols reveals that in any one of the three steps in the above sequence, inversion of configuration has taken place. This can happen only in that step which involves C—O bond fission. Now reactions A (tosylation) and C (hydrolysis) do not involve the cleavage of C—O bond. The configuration would, therefore, remain unaffected during these reactions. The cleavage of C—O bond could, therefore, take place only during the step B which is an S_N2 reaction also.

(*ii*) Further proof for inversion of configuration in an S_N2 process has been provided by **Hughes** and **Ingold**. They treated optically active 2-iodooctane with sodium iodide, having radioactive iodide, in dry acetone. This resulted in an exchange of radioactive iodine between 2-iodooctane and inorganic iodide.

$$C_6H_{13}\,CH(I)\,CH_3 \;+\; \overset{\ominus}{I}{}^{*} \longrightarrow C_6H_{13}CH\,(I^{*})\,CH_3 \;+\; \overset{\ominus}{I}$$

The reaction was found to be first-order with respect to alkyl iodide and first order with respect to inorganic iodide, *i.e.*, it showed second order kinetics (S_N2 character).

$$\text{Rate} = k\,[C_6H_{13}CH\,(I)\,CH_3]\,[\overset{\ominus}{I}{}^{*}]$$

Further, with the progress of iodine exchange reaction there was loss of optical activity. The rate of loss of optical activity was twice the rate of iodine exchange. In other words, when one molecule exchanged iodine its configuration was inverted which compensated for the optical activity of the second molecule. Thus, it becomes very clear that during iodine exchange, $\overset{\ominus}{I}{}^{*}$ has attacked from backside passing through the transition state II. On the other hand, if the attack had taken place from the frontside it should have passed through the transition state I thereby forming the product with retention of configuration, *i.e.*, without any loss of optical activity.

To account for the above observation there is no other alternative but to infer that the nucleophile attacks from the backside and every individual act of S_N2 displacement passess through the transition state II and inverts configuration.

$H_{13}C_6$, H, H_3C — C—I + $\overset{\ominus}{I}{}^{*}$ ⟶✕ $[H_{13}C_6,\ H,\ H_3C \text{—C} \cdots I^{*\,\delta^-} \cdots I^{\delta^-}]$ ⟶ C_6H_{13}, H, CH_3 — C — I^{*} + $\overset{\ominus}{I}$

Frontside attack

I

$\overset{\ominus}{I}{}^{*}$ + $H_{13}C_6$, H, H_3C — C — I ⟶ $[\overset{\delta^-}{I^{*}} \text{- - - -} C(H)(C_6H_{13})(CH_3) \text{- - - -} I^{\delta^-}]$ ⟶ I^{*}—C(C_6H_{13})(CH_3)—H + $\overset{\ominus}{I}$

Backside attack

II

B. Two-step process: Carbocation Mechanism (S_N1 reaction): In contrast to the hydrolysis of methyl bromide and other examples discussed above, the typical example of a *two-step process* is the hydrolysis of *tertiary butyl bromide*.

$$H\overset{\ominus}{O} + (CH_3)_3CBr \longrightarrow (CH_3)_3COH + \overset{\ominus}{Br}$$

Kinetic studies have shown that the rate of such a reaction depends only on the concentration of *t*-butyl bromide and is not affected at all by the concentration of the hydroxide ion even though they are consumed during the process.

$$\text{Rate} = k\,[(CH_3)_3\,CBr]$$

On the basis of kinetics of this reaction, the following two-step mechanism may be proposed:

Step I*: *Slow ionization of *tert*-butyl bromide producing bromide ion and *t*-butylcarbocation. Thus this step is the rate-determining step.

$$(H_3C)_3C-Br \xrightarrow{\text{Slow}} \left[(CH_3)_3\overset{\oplus}{C}\right] + Br^{\ominus}$$

t-Butyl carbocation

***Step II*:** A *fast* addition of nucleophile to the carbocation which having only a sextet of electrons, shows a strong tendency to react with any substance possessing an unshared electron pair.

$$\left[(H_3C)_3\overset{\oplus}{C}\right] + \overset{\ominus}{O}H \xrightarrow{\text{Fast}} HO-C(CH_3)_3 + (H_3C)_3C-OH$$

The energy profile diagram of this reactions is shown in Fig. 10.2.

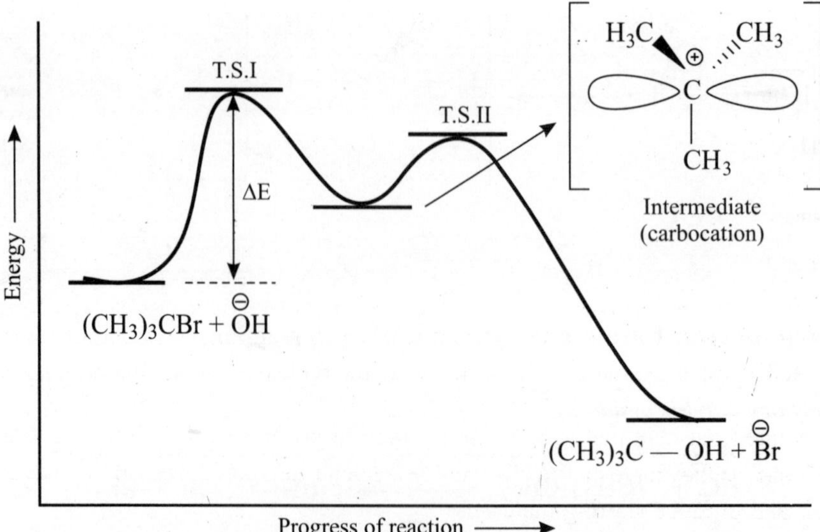

Fig. 10.2: Energy profile for an S_N1 reaction

Solvents play a very important role in S_N1 reactions as the ionization is facilitated by the solvation of carbocation which is defined as a process of the formation of ion-dipole bonds between the ions produced and solvent molecules.

Evidence for Carbocation Intermediates

Following experiments have clearly proved that S_N1 reaction involves the formation of carbocations as intermediates:

When benzhydryl chloride is treated with tetramethylammonium fluoride in liquid sulphur dioxide, it results in the formation of benzhydryl fluoride by a nucleophilic displacement of $\overset{\ominus}{Cl}$ by $\overset{\ominus}{F}$.

$$(CH_3)_4\overset{\oplus\ominus}{NF} + \underset{\text{Benzhydryl chloride}}{C_6H_5-\underset{}{\overset{Cl}{\overset{|}{C}H}}-C_6H_5} \xrightarrow{\text{liq. } SO_2} \underset{\text{Benzhydryl fluoride}}{C_6H_5-\overset{F}{\overset{|}{C}H}-C_6H_5} + (CH_3)_4\overset{\oplus}{N}\,\overset{\ominus}{Cl}$$

The rate of the reaction is dependent only on the concentration of benzhydryl chloride and is independent of that of fluoride.

$$\text{Rate} = k\,[C_6H_5\,CH\,(Cl)\,C_6H_5]$$

Further, as the reaction progresses the rate shows a gradual fall. In addition the rate of the reaction can also be slowed down by the addition of chloride ions from outside.

These observation can be interpreted if we assume the formation of a carbocation in the rate-determining step. This carbocation is attacked by the available nucleophiles (fluoride and chloride) with equal ease. As the concentration of the chloride ions increases, either by the gradual ionization of the substrate or by adding a chloride salt from outside, they start competing with fluoride ions. This results in the regeneration of the starting material, thereby retarding the rate of the reaction.

$$C_6H_5-\underset{\underset{Cl}{|}}{CH}-C_6H_5 \xrightarrow{\text{Slow}} C_6H_5-\overset{\oplus}{C}H-C_6H_5 + \overset{\ominus}{Cl}$$

$$C_6H_5-\overset{\oplus}{C}H-C_6H_5 \xrightarrow{\overset{\ominus}{F}} C_6H_5-\underset{\underset{F}{|}}{CH}-C_6H_5$$

$$C_6H_5-\overset{\oplus}{C}H-C_6H_5 + \overset{\ominus}{Cl} \longrightarrow C_6H_5-\underset{\underset{Cl}{|}}{CH}-C_6H_5$$

It may be pointed out here that such a rate retardation is not observed in the reactions involving unstable carbocations, as they have no time to wait for the departing group present in small concentration to combine again and produce starting materials.

Steric Orientation of an S_N1 Reaction

It has been mentioned above that an S_N1 process involves the initial ionization of the halide to a carbocation in the rate-determining step with the consequent rehybridization of the carbon from sp^3 to sp^2 state. The three substituents in the resulting carbocation are attached to the central carbon through three sp^2-hybrid orbitals. All the groups thus lie in one plane and make an angle of 120° with each other. There remains an empty *p*-orbital at right angles to the plane of the sp^2 orbitals. It can, therefore, be assumed that if a carbocation, involving a chiral carbon atom, is formed either lobe of the empty *p*-orbital is equally available to the nucleophile to yield one or the other optical isomer with equal ease. We would thus expect equal number of molecules with retention and inversion of configuration in the product mixture, which would be racemic. In other words, it can be said that an S_N1 *reaction involves racemization of the optically active substrate.* The above process may be illustrated by the hydrolysis of 1-phenylethyl bromide.

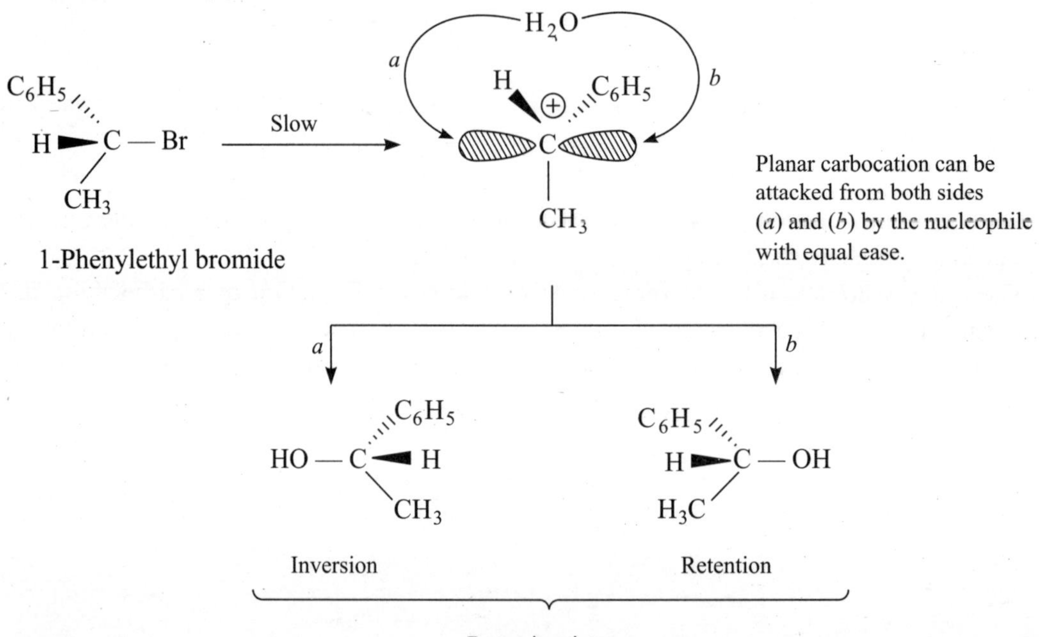

However, it has been observed that complete racemization in such cases is rather uncommon. Usually there is some excess of molecules with inverted configuration over those with the same configuration. It means that the product is not racemic and there is always some **net inversion**.

The above observation could be explained by considering the fact that ionization does not lead to the formation of a free carbocation. Actually, an ion pair is formed in which the leaving group does not depart completely from the front side and thus shields this side from attack by the nucleophile. If the nucleophile is the solvent molecule, it results in the formation of *unsymmetrically solvated carbocation* **(V)** which may form product with inverted configuration. However, when the carbocation is long lived, the leaving group is replaced by another solvent molecule leading to a *symmetrically solvated* *ion*(VI). Being symmetrical, this intermediate would now collapse into product by forming covalent bond either with solvent molecule or nucleophile on either side with equal ease giving equal amounts of products with inverted and retained configurations, *i.e.*, a **racemic mixture**. It can, therefore, be concluded that the *extent of racemization depends upon the lifetime (i.e., stability) of the carbocation.* The more stable carbocation gives more racemization because of symmetrical solvation, whereas the less stable carbocation forms product with net inversion of configuration due to unsymmetrical solvation.

lone pair

V

VI

Inversion

Inversion

Retention

Racemization

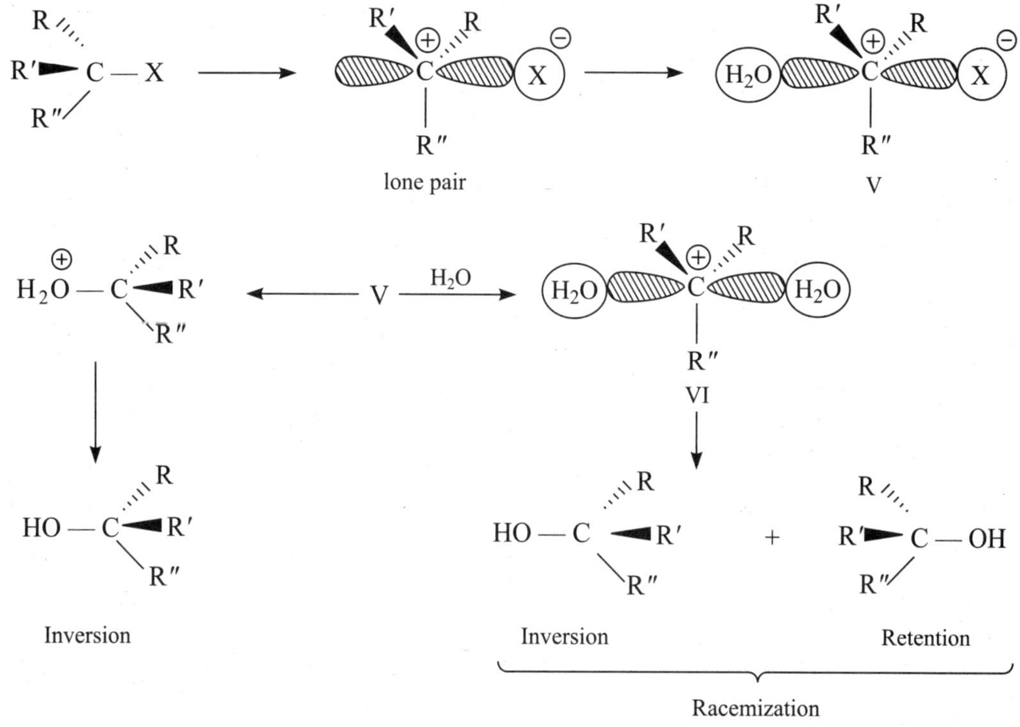

10.2.5 Factors Affecting Reactivity of Alkyl Halides in Substitution Reactions

We have seen that the hydrolysis of methyl bromide takes place by S_N2 mechanism whereas *t*-butyl bromide undergoes hydrolysis by S_N1 mechanism. Other alkyl halides react either by S_N1 or S_N2 mechanism depending on the structure and the experimental conditions employed. How can we predict which of the two mechanism is operating with a given substrate under the given experimental conditions? It has been found that at least the following four factors influence the reactivity of alkyl halides in substitution reactions.

(*a*) Structure of the alkyl group (*b*) Nature of the nucleophile
(*c*) Nature of the leaving group (*d*) Nature of the solvent.

We will now discuss the effect of these factors on the reactivity of S_N1 and S_N2 reactions.

(*a*) Structure of the alkyl Group

(*i*) S_N2 *Reactions*

In S_N2 *reactions* the order of reactivity of various alkyl halides is : CH_3X > primary > secondary > tertiary. The electron withdrawal or electron release do not affect the rate of the S_N2 reaction. Usually primary and secondary alkyl halides undergo S_N2 displacement and tertiary alkyl halides undergo S_N1 reactions. However, in 1° and 2° alkyl halides *steric hindrance* around the carbon site undergoing displacement does not affect the rate. As the hydrogen atoms of the methyl halides are gradually replaced by bulkier alkyl groups, the nucleophile finds it more and more difficult to push in among the three carbon substituents in order to reach the transition state. Moreover, if the substituents are bulky the transition state will be heavily crowded and the non-bonding interactions between various substituents would raise its energy.

Rlative rates of S_N2 reactions of some typical alkyl bromides are illustrated in Table 10.2.

Table 10.2: Relative Rates of S_N2 Reactions of Alkyl Halides

$$R—Br + \overset{\ominus}{I} \xrightarrow{S_N2} R—I + \overset{\ominus}{Br}$$

Alkyl halides	**Relative rate of displacement**
CH_3—Br	150
CH_3CH_2—Br	1
$CH_3CH(Br)\,CH_3$	0.008
$(CH_3)_3C$—Br	0.001
$(CH_3)_3CCH_2$—Br	0.00001

Thus, *neo*-pentyl bromide although a primary alkyl halide, is very unreactive in S_N2 reactions because of the *steric crowding* in the transition state by the methyl groups of the β-carbon.

(*ii*) S_N1 *Reactions*

In S_N1 *reactions*, the sequence of reactivity of alkyl halides is, benzyl > alkyl > tertiary > secondary > primary > CH_3—X. This is so pronounced that it is almost impossible to avoid S_N1 reactions with tertiary halides. The relative rates of various alkyl bromids in S_N1 reactions are listed in Table 10.3.

Table 10.3: Relative Rates of S_N1 Reaction of Alkyl Bromides

$$R—Br + H_2O \xrightarrow{S_N1} R—OH + HBr$$

Alkyl halides	**Relative rates of displacement**
CH_3—Br	1.0
CH_3CH_2—Br	1.0
$CH_3CH(Br)CH_3$	12.0
$(CH_3)_3CBr$	1.2×10^6

Two structural features control the reactivity of alkyl halides:

1. Stability of Carbocation Intermediate

Since, carbocations are involved in the S_N1 reactions, *the factors favouring the formation and stability of these ions would increase the reactivity in S_N1 reactions.* The alkyl groups are electron releasing and so when attached to a carbocation, they tend to release electrons to that carbon and thus reduce the positive charge by acquiring a little positive charge themselves. The carbocation acquires stability because of the dispersal of the positive charge over alkyl groups. Thus, the order of reactivity for S_N1 reactions is the same as that of carbocation stability.

Another factor that controls the reactivity of alkyl halides is purely steric. If the alkyl groups attached to the carbon atom undergoing substitutions are large, they facilitate the attainment of coplanarity of the carbocation, because in the process the bond angles are expanded from 109.5° to 120°. This reduces non-bonding interactions amongst the alkyl groups. This rate enhancing influence of the bulk of alkyl group is termed as **steric acceleration**.

So far we have discussed the effect of structure on reactivity of the saturated alkyl halides. Now we shall consider the reactivity of unsaturated halides. Unsaturation at α-carbon atom results in inertness of halides towards nucleophilic displacement. For instance, vinyl, alkynyl and aryl halides do not undergo nucleophilic displacement easily. Most of the leaving groups have non-bonding electron pairs and *inertness towards S_N1 reaction* may be explained on the basis of resonance stabilization of the substrate resulting in the strengthening of the bond between carbon and the leaving group. As a result the leaving group displays little tendency to depart. Further, as the leaving groups are attached to *sp* or sp^2-hybridized carbon atoms which are more electronegative than sp^3-hybridized carbon atoms, they have a greater hold on the bonding pair of electrons. For this reason they are reluctant to lose the leaving group in a direct displacement process (S_N2). In contrast to vinylic systems the allylic and benzylic halides readily undergo S_N1 reactions because of the resonance stabilization of the carbocation intermediates involved.

$$R''-C(R')=C(R)-\ddot{X}: \longrightarrow R''-\overset{\ominus}{\ddot{C}}(R')-C(R)=\overset{\oplus}{X}:$$

$$CH_2=CH-CH_2-X \rightleftharpoons \left[CH_2=CH-\overset{\oplus}{C}H_2 \longleftrightarrow \overset{\oplus}{C}H_2-CH=CH_2\right] + \overset{\ominus}{X}$$

$$\equiv \left[CH_2 \overset{...}{=} CH \overset{...}{=} CH_2\right]^{\oplus}$$

$$C_6H_5CH_2X \rightleftharpoons \left[C_6H_5\overset{\oplus}{C}H_2 \longleftrightarrow \text{(ortho } \oplus\text{)} \longleftrightarrow \text{(para } \oplus\text{)}\right] + \overset{\ominus}{X}$$

$$\equiv \left[C_6H_5 \cdots CH_2\right]^{\oplus} + \overset{\ominus}{X}$$

An interesting feature of allyl halides is that their nucleophilic substitution reactions are often accompanied by rearrangements. Mostly they react in a stepwise mode producing the allyl cation which is actually a resonance stabilized carbocation. The allyl cation can be attacked by the nucleophile at either of the two allylic positions giving rise to two products which differ only in the positions of substituents. With simple allylic substrate the addition of nucleophile would yield the same product as both the sites are chemically equivalent.

$$CH_2{=}CH-CH_2-X \underset{}{\xrightarrow{\text{Slow}}} \overset{\oplus}{CH_2 \cdots CH \cdots CH_2} \xrightarrow{\overset{\ominus}{Y}} \left[\begin{array}{c} Y-CH_2-CH{=}CH_2 \\ + \\ H_2C{=}CH-CH_2-Y \end{array}\right]$$

Same product

However, with substituted allylic compounds, the possible products are different as the two sites of the allylic carbocation are no more chemically equivalent.

$$R-CH{=}CH-CH_2-X \xrightarrow{\overset{\ominus}{Y}} R-\underset{}{\overset{Y}{\overset{|}{CH}}}-CH{=}CH_2 + R-CH{=}CH-CH_2-Y$$

Rearranged product (S_N1') Normal product (S_N1)

The reaction in which the nucleophile attaches itself to the carbon atom from which the leaving group has departed is a usual S_N1 reaction, whereas that in which the nucleophile adds to the other carbon atom is called S_N1' (Substitution Nucleophilic Unimolecular with rearrangement), or *allylic rearrangement.*

Now if allyl cation is produced as an intermediate, the two products should be produced in the same relative ratio no matter which of the two isomeric allylic substrates are substituted.

This has been shown to be true as the hydrolysis of 3,3- and 1,2-dimethylallyl chlorides yields a mixture of 3,3- and 1,2-dimethylallyl alcohols in the same ratio, no matter which substrate is used.

$$\left.\begin{array}{c} (H_3C)_2C{=}CH-CH_2-Cl \\ \text{or} \\ (H_3C)_2\underset{\underset{Cl}{|}}{C}-CH{=}CH_2 \end{array}\right\} \xrightarrow{H_2O} \begin{array}{c} (CH_3)_2C{=}CH-CH_2OH \\ \text{Minor product} \\ + \\ (CH_3)_2\underset{\underset{OH}{|}}{C}-CH{=}CH_2 \\ \text{Major product} \end{array}$$

Some allylic substrates are known to undergo displacement reactions by the bimolecular mechanism. Here again a mixture of products is obtained because of nuleophilic attack at both the allylic positions. The reaction resulting in the formation of a rearranged product is called S_N2' **(Substitution**

Nucleophilic Bimolecular with rearrangement) in contrast to the normal S_N2 reaction. One of the earliest known example of an S_N2' process has been the reaction of 1-ethylallyl chloride with carbanion derived from diethylmalonate.

$$\overset{\ominus}{Y} + \underset{R}{\underset{|}{CH}}=CH-\underset{R'}{\underset{|}{CH}}-X \longrightarrow Y-\underset{R}{\underset{|}{CH}}-CH=CH-R' + \overset{\ominus}{X}$$

$$H_3C-CH_2-\underset{Cl}{\underset{|}{CH}}-CH=CH_2 + \overset{\ominus}{C}H(COOEt)_2 \longrightarrow$$

$$CH_3-CH_2-CH=CH-CH_2-CH(COOEt)_2$$

The following points are worth mentioning in connection with the S_N2' mechanism.

(*i*) *The reaction exhibits second order kinetics, i.e., it is first order with respect to the substrate and first order with respect to nucleophile.*

(*ii*) The reaction gives the 'abnormal' (*i.e., rearranged*) product in isolable amount.

(*iii*) *It is well established that neither the substrate nor the normal S_N2' product undergoes rearrangement under the reaction conditions.*

The preponderance of S_N2 or S_N2' product depends upon the structure of the substrate. With substitutents at the 1-position, bimolecular substitution takes place predominantly by S_N2' process. For instance, 1-*t*-butylallyl chloride and 1,1-dimethylallyl chloride on treatment with nucleophilies yield rearranged products almost exclusively.

$$Et\overset{\ominus}{O} + H_2C=CH-\underset{C(CH_3)_3}{\underset{|}{CH}}-Cl \longrightarrow EtO-CH_2-CH=CHC(CH_3)_3 + \overset{\ominus}{Cl}$$

$$Ph\overset{\ominus}{S} + H_2C=CH-\overset{CH_3}{\underset{CH_3}{\overset{|}{\underset{|}{C}}}}-Cl \longrightarrow Ph-S-CH_2-CH=C(CH_3)_2 + \overset{\ominus}{Cl}$$

The factors that are responsible for predominance of S_N2' product over S_N2 product are purely steric. Bulk of the substituent at 1-position prevents the attack at this carbon. However, if such a substituent is present at 3- or 2-position it inhibits S_N2' reaction thereby resulting in the formation of the normal S_N2 product, again due to steric reasons.

(*b*) Nature of the Nucleophile

(*i*) S_N2 *Reactions*

Since, the transition state of an S_N2 reaction involves the nucleophile, the nature of the nucleophilic reagent plays an important role in determining the rate of the reaction. It means that the rate of the reaction will depend upon the reactivity of the nucleophile which is called **nucleophilicity**. *The nucleophilicity of a reagent may be defined as its ability to donate an electron pair to carbon.* Two features of nucleophiles control their nucleophilicity.

1. Basic Nature
2. Steric Factors.

We will discuss these features one by one.

1. Basic Nature: All the nucleophiles may be regarded as bases. However, their nucleophilicity does not always parallel their base strength. For example, the hydroxide ion is a strong base and is also a good nucleophile but iodide ion is a very good nucleophile though it is a very weak base.

In a series of nucleophiles belonging to the same family, the nucleophilicity increases with the increase in size of the atom bearing unshared electron pair. Thus among halide ions the order of nucleophilicity is: $\overset{\ominus}{I} > \overset{\ominus}{Br} > \overset{\ominus}{Cl} > \overset{\ominus}{F}$. The following two explanations have been put forward to explain this:

(*i*) *Polarizability*: The electrons in the larger atoms are less tightly bound by the nucleus and consequently they are more readily available to be donated to the carbon atom undergoing displacement in an S_N2 process. In other words, *because of their bigger size the atoms become more polarizable and hence better nucleophiles.*

(*ii*) *Solvation*: The nucleophile is surrounded by the solvent molecules through hydrogen bonding, *i.e.*, it is solvated and before attacking a carbon atom it must break away from some of these bonds. *It is well known that solvation energies of smaller molecules are always greater than those of larger ones, and so latter are good nucleophiles.* This has been confirmed by the observation that in non-hydroxylic solvents such as dry acetone, iodide ion is less nucleophilic than bromide ion. Thus basicity, polarizability and solvation are all involved in determining the nucleophilicity of a given reagent.

2. Steric Factors: Steric factors in the attacking nucleophile also play an important role and may sometimes assume significant importance in an S_N2 reaction. Thus, quinuclidine reacts with isopropyl bromide hundred times faster than triethylamine. It is not surprising if we consider that the substituents in quinuclidine are tied back by the ring system, whereas freely rotating ethyl groups of the triethylamine partly cover the pair of electrons on nitrogen.

N (quinuclidine) + $(H_3C)_2CH$—Br $\xrightarrow{\text{Fast}}$ $\overset{\oplus}{N}$—$CH(CH_3)_2$ + $\overset{\ominus}{Br}$

$$\text{(triethylamine, lone pair on N)} + \text{H}_3\text{C(H}_3\text{C)(H)C—Br} \xrightarrow{\text{Slow}} (C_2H_5)_3\overset{\oplus}{N}CH(CH_3)_2\overset{\ominus}{Br}$$

(ii)* S_N1 *Reactions

Since in S_N1 reactions the nucleophile reacts with the carbocation in a fast step, the nucleophilicity of the reagent does not play any significant role. However, if more than one nucleophile is present the product derived from the stronger one will predominate.

(*c*) Nature of the Leaving Group

The nature of the leaving group at least partly determines the reactivity of the substrate. Since, the leaving group breaks away as a base, it is easier to displace weaker bases as compared to stronger bases. Thus we can say that the less basic the substituent, the more easily it is displaced by the solvent in S_N1 reactions or by an attacking nucleophilic reagent in S_N2 reactions. For instance, alcohols are resistent to nucleophilic displacement in non-acidic media. This is because of the fact that the base strength of $\overset{\ominus}{O}H$ group makes it a poor leaving group. However, in presence of acids, the oxygen of the —OH group gets protonated and the potential leaving group (H_2O) which is a very weak base is, displaced readily. Consequently, alcohols undergo nucleophilic substitution only in acidic medium.

$$R—OH + \overset{\oplus}{H} \rightleftharpoons R—\overset{\oplus}{O}H_2$$

$$R—\overset{\oplus}{O}H_2 \longrightarrow \overset{\oplus}{R} \xrightarrow{\overset{\ominus}{Br}} R—Br \qquad (S_N1)$$

$$\text{or} \qquad \overset{\ominus}{Br} + R—\overset{\oplus}{O}H_2 \longrightarrow Br—R + H_2O \qquad (S_N2)$$

Other strong bases which are poor leaving groups are: $\overset{\ominus}{H}$, $R\overset{\ominus}{O}$, $H_2\overset{\ominus}{N}$, $R_2\overset{\ominus}{N}$. In case of halides ease of displacement follows the order: $\overset{\ominus}{I} > \overset{\ominus}{Br} > \overset{\ominus}{Cl} > \overset{\ominus}{F}$. As expected this is the reverse of the carbon-halogen bond strength.

Phenoxide ion or 2,4-dinitrophenoxide ion are good leaving groups since, they are weak bases. The weakly basic character of these groups is due to dispersal of negative charge involving resonance with the benzene ring.

(*d*) Nature of the Solvent: The Solvation of Ions

Solvents play a dominant and sometimes decisive role in deciding the rate and mechanism of nucleophilic substitution reactions. Ions are stabilized in good ionizing* (polar) solvents by solvation. Cations are

*The ionizing power of a solvent is reflected in its dielectric constant. Higher the dielectric constant, greater the ionizing power. Thus water with a dielectric constant of 81 is more ionizing than methanol having a dielectric constant of only 33.

solvated by coordination whereas anions are solvated through hydrogen bonding. Reactions that produce ions from neutral substrates would, therefore, be facilitated in polar solvents whereas reverse would be true for ions reacting to form neutral products.

$H_2O----\overset{\oplus}{R}----OH_2$ $\qquad$ $H_2O\text{-}H----\overset{\ominus}{X}----H\text{-}OH$

We shall now see the effect of solvent polarity on the reactivity of S_N2 and S_N1 reactions.

(i) S_N2 ***Reactions***: In S_N2 displacements, the most common case is reaction of an anion with a neutral molecule, hence a *good ionizing solvent slows down the reaction* because it stabilizes one of the reactant (charged nucleophile) more effectively than the transition state.

Therefore, polar protic solvents such as water, alcohols, acids, etc. would solvate these ions and affect the rate of the reaction.

(ii) S_N1 ***Reactions***: In contrast to S_N2 reactions, the S_N1 reactions begin with the initial ionization of a neutral molecule and thus it is facilitated by good ionizing solvents such as water, formic acid, methanol, etc. Thus, S_N1 *reactions are favoured by polar solvents. Higher the polarity of the solvent, more will be its ionizing power and hence faster the rate of* S_N1 *reaction.*

S_N1 *vs* S_N2 Mechanism

S_N1 Mechanism (Substitution Nucleophilic Unimolecular)	S_N2 Mechanism (Substitution Nucleophilic Bimolecular)
1. Takes place in two steps.	1. Takes place in one step.
2. Follows first order kinetics.	2. Follows second order kinetics.
3. Involves carbocations as intermediates.	3. Does not involve formation of carbocations as intermediates.
4. Takes place by both backside and frontside attack of the nucleophile. But the backside attack predominates.	4. Takes place by exclusive backside attack of the nucleophile.
5. Generally involves racemization with some net inversion. (partial racemization is observed).	5. Involves complete inversion of configuration.
6. Generally gives rearranged products.	6. Does not give rearranged products.
7. In these reactions elimination is an important side reaction.	7. In these reactions elimination is not an important side reaction.
8. The rates of these reactions are greatly effected by electronic factors.	8. The rates of these reactions are effected by steric and not electronic factors.
9. In these reactions the order of reactivity of alkyl halides is: 3° > 2° > 1° > methyl halide.	9. In these reactions the order of reactivity of alkyl halides is: methyl halides > 1° > 2° > 3°.
10. Weak nucleophilies such as water, alcohols, etc. favour these reactions.	10. Strong nucleophiles such as alkoxide ions favour these reactions.
11. These reaction are generally favoured by low concentration of the nucleophile.	11. There reactions are generally favoured by high concentration of the nucleophile.

***Solvents can also alter the mechanistic course*:** Sometimes the nature of the solvent plays such a dominant role that it is possible to alter the mechanistic course of the reaction by choosing an appropriate solvent. Thus methyl bromide which normally undergoes S_N2 reaction in aqueous alkaline medium can be forced to hydrolyze by S_N1 mechanism using formic acid containing a small amount of water.

$$CH_3OH \xleftarrow[S_N2]{\overset{\ominus}{O}H,\ EtOH} CH_3Br \xrightarrow[S_N1]{HCOOH,\ H_2O} CH_3OH$$

S_N1 Mechanism *vs* S_N2 Mechanism

Uses of Nucleophilic Substitution Reactions in Organic Synthesis

Some examples of the nucleophilic substitutions, which have made this reaction synthetically important, are listed below:

(*i*) Formation of alkyl cyanides: Alkyl cyanides are formed on heating alkyl halides with sodium or potassium cyanide. A small amount of alkyl isocyanides (alkyl carbylamines) is also formed in this reaction.

$$\underset{\text{Haloalkane}}{R-X} + \overset{\oplus\ominus}{KCN}\ (\text{alc.}) \xrightarrow{\Delta} \underset{\text{Alkanenitrile}}{R-C{\equiv}N} + \overset{\oplus\ominus}{KX}$$

For example,

$$\underset{\text{Iodomethane}}{CH_3-I} + \overset{\oplus\ominus}{KCN}\ (\text{alc.}) \xrightarrow{\Delta} \underset{\text{Ethanenitrile}}{CH_3-C{\equiv}N} + \overset{\oplus\ominus}{KI}$$

$$\underset{\text{Bromoethane}}{CH_3CH_2-Br} + \overset{\oplus\ominus}{KCN}\ (\text{alc.}) \xrightarrow{\Delta} \underset{\text{Propanenitrile}}{CH_3CH_2-C{\equiv}N} + \overset{\oplus\ominus}{KBr}$$

Alkyl cyanides are very useful compounds in organic synthesis. Some examples are given below:

(*a*) Conversion into amides by partial hydrolysis with conc. HCl or alkaline hydrogen peroxide.

$$\underset{\substack{\text{Ethanenitrile}\\ \text{(Acetonitrile)}}}{CH_3-C{\equiv}N} + H_2O \xrightarrow[\text{or } H_2O_2-NaOH]{\text{conc. HCl}} \underset{\substack{\text{Ethanamide}\\ \text{(Acetamide)}}}{CH_3-CONH_2}$$

(*b*) Conversion into carboxylic acids by complete hydrolysis with dil. acid or dil.alkali.

$$CH_3-C{\equiv}N + 2H_2O \xrightarrow[\text{or dil. alkali}]{\text{dil. acid}} \underset{\substack{\text{Ethanoic acid}\\ \text{(Acetic acid)}}}{CH_3-COOH}$$

Chapter 10

(*c*) Conversion into primary amine by reduction **(Mendius reduction)**.

$$CH_3-C\equiv N + 4[H] \xrightarrow[\text{(Mendius reduction)}]{Na/C_2H_5OH} \underset{\substack{\text{Ethanamine}\\ \text{(Ethylamine)}}}{CH_3CH_2-NH_2}$$

Note: The conversion of alkyl halides into alkyl cyanides is a convenient method for **ascending the homologous series,** *i.e.*, increasing the length of carbon by one atom at a time.

(*ii*) Formation of alkyl isocyanides (carbylamines): In contrast to KCN or NaCN, when the aqueous ethanolic solution of alkyl halide is treated with silver cyanide, **alkyl isocyanide** (**alkyl isonitrile** or **alkylcarbylamine**) is formed as the major product along with a minor amount of the corresponding alkyl cyanide.

$$\underset{\text{Haloalkane}}{R-X} + Ag-CN \xrightarrow[\Delta]{C_2H_5OH/H_2O} \underset{\text{Alkyl isocyanide}}{R-N \rightleftarrows C} \quad \text{or} \quad \underset{\text{or Alkylcarbylamine}}{R-\overset{\oplus}{N}\equiv\overset{\ominus}{C}\colon + AgX}$$

For example,

$$CH_3-I + Ag-CN \xrightarrow[\Delta]{C_2H_5OH/H_2O} \underset{\text{Methyl isocyanide}}{CH_3-N \rightleftarrows C} \quad \text{or} \quad \underset{\text{or Methylcarbylamine}}{CH_3-\overset{\oplus}{N}\equiv\overset{\ominus}{C}\colon + AgX}$$

Reduction of alkyl isocyanides with Na/C_2H_5OH produces secondary amines having $N—CH_3$ group.

$$\underset{\text{Methyl isocyanide}}{CH_3-N \rightleftarrows C} + 4\,[H] \xrightarrow{Na/C_2H_5OH} \underset{\text{Dimethylamine}}{CH_3-NH-CH_3}$$

Formation of alkyl cyanides in the presence of alkali metal cyanides and alkyl isocyanides in the presence of silver cyanide: Ambident Nucleophiles: Cyanide has a lone pair of electrons each on carbon and nitrogen atom. So, both these atoms can act as nucleophilic centres. Such nucleophiles which can offer two nucleophilic sites are called **ambident nucleophiles**. Thus cyanide is an ambident nucleophile.

$$:\overset{\ominus}{C}\equiv N:$$

Cyanide ion, ambident nucleophile

(*lone pair of electrons on both carbon and nitrogen*)

Now, nucleophilic substitution in which carbon acts as a nucleophile produces alkyl cyanides and that in which nitrogen acts as nucleophile produces alkyl isocyanides.

Alkali metal cyanides (producing predominant amount of alkyl cyanides) are ionic in nature and give cyanide ion which has a negatively charged carbon atom and an uncharged nitrogen atom as nucleophiles.

$$KCN \longrightarrow \overset{\oplus}{K} + \overset{\ominus}{CN}$$

Since, negatively charged carbon is a stronger nucleophile than uncharged nitrogen, cyanide will produce alkyl cyanide as the major product.

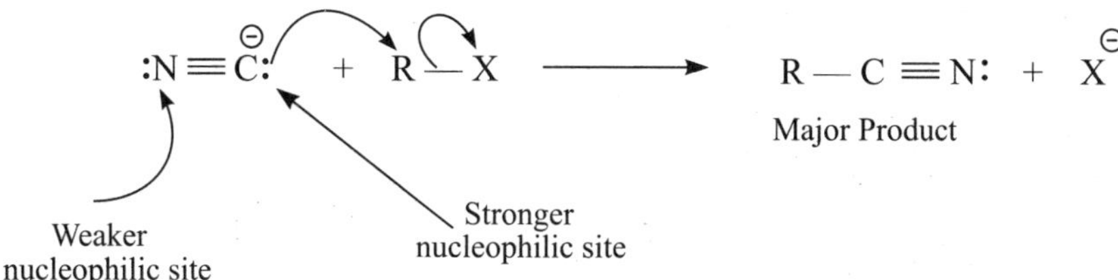

Contrary to alkali metal cyanides, the silver cyanide is predominantly covalent in nature, *i.e.*, it cannot give all free $\overset{\oplus}{Ag}$ and $\overset{\ominus}{CN}$. However, it contains traces of $\overset{\oplus}{Ag}$ which abstracts the halogen from alkyl halides producing a carbocation. The carbocation so formed reacts with silver cyanide (covalent) which can offer only nitrogen as nucleophilic site to produce alkyl isocyanides. Various steps involved in the process are given below:

***Step I*:** Abstraction of halogen from alkyl halides.

$$Ag—C{\equiv}N \rightleftharpoons \underset{\text{(Traces)}}{\overset{\oplus}{Ag}} + :\overset{\ominus}{C}{\equiv}N:$$

$$R—\ddot{X}: + \overset{\oplus}{Ag} \longrightarrow \left[\overset{\delta^+}{R}\text{----}X\text{----}\overset{\delta^+}{Ag}\right] \longrightarrow \overset{\oplus}{R} + AgX$$

***Step II*:** Nucleophilic addition of C ≡ N to carbocation involving nitrogen.

$$\underset{\substack{\text{Only one nucleophilic site}\\ \text{(N atom)}}}{Ag—C{\equiv}N:} + \overset{\oplus}{R} \longrightarrow Ag—C{\equiv}\overset{\oplus}{N}—R \longrightarrow \overset{\oplus}{Ag} + \underset{\substack{\text{Alkyl}\\ \text{isocyanide}}}{:\overset{\ominus}{C}{\equiv}\overset{\oplus}{N}—R}$$

***(iii)* Formation of alkyl nitrites:** Alkyl halides react with alkali metal nitrites giving rise to **alkyl nitrites**.

$$\underset{\text{Haloalkane}}{R—X} + \underset{\text{Potassium nitrite}}{K—O—N{=}O} \xrightarrow{\Delta} \underset{\text{Alkyl nitrite}}{R—O—N{=}O} + \overset{\oplus\ominus}{KX}$$

For example,

$$\underset{\text{Bromoethane}}{CH_3CH_2—Br} + K—O—N{=}O \xrightarrow{\Delta} \underset{\text{Ethyl nitrite}}{CH_3CH_2—O—N{=}O} + \overset{\oplus\ominus}{KBr}$$

Chapter 10

***(iv)* Formation of nitroalkanes:** Treatment of alkyl halides with alcoholic silver nitrite, however, produces nitroalkanes.

$$\underset{\text{Haloalkane}}{R-X} + \underset{\text{Silver nitrite}}{Ag-O-N=O} \xrightarrow[\Delta]{C_2H_5OH/H_2O} \underset{\text{Nitroalkane}}{R-\overset{\oplus}{N}(=O)-O^{\ominus}} + AgX$$

For example,

$$\underset{\text{Bromoethane}}{CH_3CH_2-Br} + Ag-O-N=O \xrightarrow[\Delta]{C_2H_5OH/H_2O} \underset{\text{Nitroethane}}{CH_3CH_2-\overset{\oplus}{N}(=O)-O^{\ominus}} + AgBr$$

Reason for formation of alkyl nitrites in the presence of alkali metal nitrite and nitroalkane in the presence of silver nitrite: Like cyanide ion, the nitrite ion is also **ambident nucleophile** and in this case also both oxygen and nitrogen atoms can act as nucleophilic centres. *Nucleophilic substitutions in which oxygen acts as a nucleophile produce alkyl nitrites and those in which nitrogen acts as a nucleophile produce nitroalkanes.*

Alkali metal nitrites are ionic in nature and produce nitrite ion which has a negatively charged oxygen atom and uncharged nitrogen atom as nucleophiles. Since negatively charged oxygen is a stronger nucleophile than uncharged nitrogen, these nitrites will produce alkyl nitrites as the major products.

$$O=\ddot{N}-\ddot{O}:^{\ominus} + R-X \longrightarrow R-O-N=O + X^{\ominus}$$

Weaker nucleophilic site (N)   Stronger nucleophilic site (O)

Silver nitrite, on the other hand, is predominately covalent in nature and hence does not have negatively charged oxygen atom. Instead oxygen and nitrogen atoms both are uncharged having lone pairs of electrons. Now since nitrogen is less electronegative than oxygen, its lone pair of electrons are more readily available for covalent bond formation. Hence, nitrogen acts as nucleophile resulting in the formation of nitroalkanes,

$$Ag-O-\ddot{N}=O + R-X \longrightarrow R-\overset{\oplus}{N}(=O)-O^{\ominus} + AgX$$

(v) Formation of ethers: Treatment of alkyl halides with sodium or potassium alkoxides results in the formation of ethers.

$$\overset{\oplus\ \ominus}{\text{Na OR}'} + \text{R—X} \xrightarrow{\Delta} \text{R—O—R}' + \overset{\oplus\ \ominus}{\text{Na X}}$$

Sodium alkoxide — Alkyl halide — Ether

For example,

$$\overset{\oplus\ \ominus}{\text{Na OCH}_2\text{CH}_3} + \text{CH}_3\text{CH}_2\text{—Br} \xrightarrow{\Delta} \text{CH}_3\text{CH}_2\text{OCH}_2\text{CH}_3 + \overset{\oplus\ \ominus}{\text{NaBr}}$$

Sodium ethoxide — Bromoethane — Ethoxyethane

$$\overset{\oplus\ \ominus}{\text{Na OCH}_3} + \text{CH}_3\text{CH}_2\text{Br} \xrightarrow{\Delta} \text{CH}_3\text{CH}_2\text{OCH}_3 + \overset{\oplus\ \ominus}{\text{NaBr}}$$

Sodium methoxide — Methoxyethane

This reaction is called ***Williamson's synthesis of ethers.***

Alkyl halides can give symmetrical ethers on heating with dry Ag_2O also. *e.g.*,

$$2\text{CH}_3\text{CH}_2\text{Br} + \text{Ag}_2\text{O} \xrightarrow{\Delta} \text{CH}_3\text{CH}_2\text{OCH}_2\text{CH}_3 + 2\text{AgBr}$$

Bromoethane — Ethoxyethane

(vi) Formation of alcohols: Hydrolysis of haloalkanes on treatment with aqueous alkalies or silver oxide in boiling water (AgOH) results in the formation of alcohols.

$$\overset{\oplus\ \ominus}{\text{K OH}} + \text{R—X} \xrightarrow{\Delta} \text{R—OH} + \overset{\oplus\ \ominus}{\text{KX}}$$

Haloalkane — Alcohol

For example

$$\text{CH}_3\text{CH}_2\text{Br} + \text{KOH(aq.)} \longrightarrow \text{CH}_3\text{CH}_2\text{OH} + \overset{\oplus\ \ominus}{\text{KBr}}$$

Bromoethane — Ethanol

$$\text{CH}_3\text{CH}_2\text{Cl} + \overset{\oplus\ominus}{\text{AgOH}} \xrightarrow{\Delta} \text{CH}_3\text{CH}_2\text{OH} + \text{AgCl}\downarrow$$

Chloroethane — Moist silver oxide — Ethanol

(vii) Formation of mercaptans (thioalcohols): Alkyl mercaptans are obtained by heating alkyl halides with aqueous ethanolic solution of sodium or potassium hydrosulphide.

$$\text{R—X} + \overset{\oplus\ \ominus}{\text{NaSH}} \xrightarrow[\Delta]{\text{C}_2\text{H}_5\text{OH / H}_2\text{O}} \text{R—SH} + \overset{\oplus\ \ominus}{\text{NaX}}$$

Haloalkane — Sodium hydrosulphide — Thioalcohol

$$\underset{\text{Bromoethane}}{CH_3CH_2Br} + \overset{\oplus\ \ominus}{NaSH} \xrightarrow[\Delta]{C_2H_5OH\,/\,H_2O} \underset{\text{Ethanethiol}}{CH_3CH_2SH} + \overset{\oplus\ \ominus}{NaBr}$$

(*viii*) Formation of alkyl sulphides (Thioethers): Two methods can be used for converting alkyl halides into thioethers. They are as follows:

(*a*) By heating a mixture of alkyl halides and aqueous alcoholic sodium or potassium alkyl sulphides.

$$\underset{\text{Haloalkane}}{R-X} + \underset{\text{Sodium mercaptide}}{\overset{\oplus\ \ominus}{NaSR'}} \xrightarrow[\Delta]{C_2H_5OH\,/\,H_2O} \underset{\text{Thioether}}{R-S-R'} + \overset{\oplus\ \ominus}{NaX}$$

For example,

$$CH_3CH_2Br + \overset{\oplus\ \ \ominus}{NaSCH_3} \xrightarrow[\Delta]{C_2H_5OH\,/\,H_2O} \underset{\text{Ethylmethylthioether}}{CH_3CH_2-S-CH_3} + \overset{\oplus\ \ominus}{NaBr}$$

(*b*) By heating alkyl halide and aqueous alcoholic sodium or potassium sulphide.

$$2\underset{\text{Haloalkane}}{R-X} + \underset{\text{Sodium sulphide}}{Na_2S} \xrightarrow[\Delta]{C_2H_5OH\,/\,H_2O} \underset{\text{Thioether}}{R-S-R} + 2NaX$$

$$\underset{\text{Bromoethane}}{2CH_3CH_2Br} + Na_2S \xrightarrow[\Delta]{C_2H_5OH\,/\,H_2O} \underset{\text{Diethylthioether}}{CH_3CH_2-S-CH_2CH_3} + 2NaBr$$

(*ix*) Formation of esters: Esters can be prepared by heating silver* salt of a carboxylic acid with alkyl halides.

$$\underset{\text{Silver salt of a carboxylic acid}}{R'-\overset{\overset{\displaystyle O}{\|}}{C}-O-Ag} + \underset{\text{Haloalkane}}{R-X} \xrightarrow[\Delta]{C_2H_5OH} \underset{\text{Ester}}{R'-\overset{\overset{\displaystyle O}{\|}}{C}-O-R} + AgX$$

For example,

$$\underset{\text{Silver acetate}}{CH_3-\overset{\overset{\displaystyle O}{\|}}{C}-O-Ag} + \underset{\text{Bromoethane}}{CH_3CH_2Br} \longrightarrow \underset{\text{Ethyl ethanoate}}{CH_3-\overset{\overset{\displaystyle O}{\|}}{C}-OCH_2CH_3} + AgBr$$

---

*$\overset{\oplus}{Ag}$ is known to abstract halogen from a substrate producing carbocation.

(*x*) Formation of alkylmalonic and alkyl acetoacetic esters: Carbanions derived from malonic and acetoacetic esters on treatment with alkyl halides form the corresponding alkyl derivatives.

$$R-Br + \overset{\ominus}{C}H(COOEt)_2 \longrightarrow R-CH(COOEt)_2$$

$$R-Br + CH_3CO\overset{\ominus}{C}H-COOEt \longrightarrow CH_3-CO-\underset{}{\overset{R}{\overset{|}{C}H}}-COOEt$$

(*xi*) Formation of amines: Alkyl halides on heating with an ethanolic solution of ammonia in a sealed tube give rise to a mixture of 1°, 2°, and 3° ***amines*** along with the formation of ***quaternary ammonium salts.***

$$R-X \xrightarrow[C_2H_5OH,\,\Delta]{NH_3} R-NH_2 \xrightarrow[C_2H_5OH,\,\Delta]{R-X} R-NH-R \xrightarrow[C_2H_5OH,\,\Delta]{R-X} NR_3 \xrightarrow[C_2H_5OH,\,\Delta]{R-X} R_4\overset{\oplus}{N}X$$

For example,

$$\underset{\text{Bromoethane}}{CH_3CH_2Br} + NH_3 \xrightarrow[\Delta]{C_2H_5OH} \underset{\text{Ethanamine (1°)}}{CH_3CH_2-NH_2} + HBr$$

$$\underset{\text{Ethanamine}}{CH_3CH_2NH_2} + \underset{\text{Bromoethane}}{CH_3CH_2Br} \xrightarrow[\Delta]{C_2H_5OH} \underset{\text{N-Ethylethanamine (2°)}}{(CH_3CH_2)_2NH} + HBr$$

$$\underset{\text{N-Ethylethanamine}}{(CH_3CH_2)_2NH} + CH_3CH_2Br \xrightarrow[\Delta]{C_2H_5OH} \underset{\text{N, N-Diethylethanamine (3°)}}{(CH_3CH_2)_3N} + HBr$$

$$\underset{\text{N, N-Tryethylethanamine}}{(CH_3CH_2)_3N} + CH_3CH_2Br \xrightarrow[\Delta]{C_2H_5OH} \underset{\text{Tetraethylammonium bromide}}{(CH_3CH_2)_4\overset{\oplus}{N}\overset{\ominus}{Br}}$$

This reaction is called ***Hoffmann ammonolysis*** of alkyl halides.

(*xii*) Formation of higher acetylenes: Higher acetylenes are obtained by treating alkyl halides with salts of terminal alkynes.

$$\underset{\text{Haloalkane}}{R-X} + \underset{\text{Soidum alkynide}}{\overset{\oplus}{Na}\overset{\ominus}{C}\equiv CR'} \longrightarrow \underset{\text{Higher alkyne}}{R-C\equiv C-R'} + \overset{\oplus}{Na}\overset{\ominus}{X}$$

For example,

$$CH_3CH_2Br + \overset{\oplus}{Na}\overset{\ominus}{C}\equiv CCH_3 \longrightarrow \underset{\text{2-Pentyne}}{CH_3CH_2-C\equiv C-CH_3} + \overset{\oplus}{Na}\overset{\ominus}{Br}$$

As discussed in Chapter 7, salts of terminal alkynes are obtained by the action of sodamide (or sodium in liquid ammonia at 196 K) on terminal alkynes.

$$R-C\equiv CH + NaNH_2 \xrightarrow[192\,K]{liq.\ NH_3} R-C\equiv \overset{\ominus\oplus}{C}Na + NH_3$$

(*xiii*) Formation of alkyl iodides: Alkyl chlorides or bromides upon treatment with sodium or potassium iodide in acetone produce alkyl iodides by a halogen exchange reaction.

$$\underset{\text{Chloroethane}}{CH_3CH_2-Cl} + \overset{\oplus\ \ominus}{K\,I} \xrightarrow{\text{Acetone}} \underset{\text{Iodoethane}}{CH_3CH_2-I} + \overset{\oplus\ \ominus}{KCl}$$

This reaction called **Finkelstein reaction**, is commonly used for the preparation of alkyl iodides from alkyl chlorides or bromides.

2. Elimination Reactions

As discussed in Chapter 2, when an alkyl halide is heated with a concentrated alcoholic solution of KOH, it produces an *alkene* by elimination of a molecule of hydrogen halide (*dehydrohalogention*).

Since in this reaction, the hydrogen atom is lost from β-carbon, therefore, this reaction is classified as β-elimination reaction.

$$\underset{\text{Bromoethane}}{H-\overset{H}{\underset{H}{\overset{|}{\underset{|}{C^{\beta}}}}}-\overset{H}{\underset{H}{\overset{|}{\underset{|}{C^{\alpha}}}}}-Br} + \overset{\oplus\ominus}{KOH}\,(\text{alcoholic}) \xrightarrow{\Delta} \underset{\text{Ethene}}{H_2C=CH_2} + \overset{\oplus\ominus}{KBr} + H_2O$$

Similarly, 1-bromopentane on dehydrohalogenation gives 1-pentene.

$$CH_3CH_2CH_2-\overset{H}{\underset{H}{\overset{|}{\underset{|}{C}}}}-\overset{H}{\underset{H}{\overset{|}{\underset{|}{C}}}}-Br + \overset{\oplus\ominus}{KOH}(\text{alcoholic}) \xrightarrow{\Delta} \underset{\text{1-Pentene}}{CH_3CH_2CH_2CH=CH_2} + \overset{\oplus\ominus}{K\,Br} + H_2O$$

Details of β-elimination during dehydrohalogenation reaction have already been discussed in Chapter 2.

3. Formation of Alkanes

Alkyl halides can be reduced to produce alkanes in the presence of a variety of reagents. Some of them are discussed below:

(*i*) Catalytic reduction: When a mixture of vapours of an alkyl halide and hydrogen is passed overheated metal catalyst such as Ni/Pd or Pt, it produces corresponding alkanes.

$$CH_3CH_2Br + H_2 \xrightarrow{Ni} CH_3-CH_3 + HBr$$

***(ii)* Reduction with HI in presence of red phosphorus:** Alkyl iodides can be readily reduced to alkanes on treatment with HI in presence of red phosphorus at 420 K.

$$CH_3CH_2—I + HI \xrightarrow[420\ K]{Red\ P} CH_3—CH_3 + I_2$$

***(iii)* Reduction with dissolved metals:** Alkyl halides on treatment with Zn/CH_3COOH or Zn/NaOH or a Zn—Cu couple in alcohol produce alkanes.

$$CH_3CH_2Br \xrightarrow[or\ (Zn—Cu + Alcohol)]{(Zn + CH_3COOH)\ or\ (Zn + NaOH)} CH_3—CH_3 + HBr$$

***(iv)* Reduction with $LiAlH_4$:** Primary and secondary alkyl halides on treatment with lithium aluminium hydride give rise to alkanes.

$$\overset{\ominus}{H} + R—X \longrightarrow R—H + \overset{\ominus}{X}$$

(from $LiAlH_4$)

e.g., $LiAlH_4 + 4CH_3CH_2Br \longrightarrow 4CH_3CH_3 + AlBr_3 + LiBr$

However, *tertiary* alkyl halides undergo dehydrohalogenation producing alkenes. In contrast, sodium borohydride ($NaBH_4$) reduces secondary and tertiary alkyl halides without effecting primary halides. Triphenyltin hydride (Ph_3SnH) reduces all the three types of alkyl halides.

4. Reaction with Metals

Alkyl halides (chlorides, bromides and iodides) react with some metals resulting in the formation of products having a formal carbon metal bond. Such compounds are called as **organometallic compounds.**

Some of the important reactions of alkyl halides and metals are discussed as under:

***(i)* Formation of Grignard Reagents:** Alkyl halides upon treatment with magnesium turnings in dry ether result in the formation of alkyl magnesium halides commonly known as **Grignard reagents.**

$$\underset{\text{Alkyl halide}}{R—X} + Mg \xrightarrow{\text{Dry ether}} \underset{\text{Grignard reagent}}{R—Mg—X}$$

For example,

$$CH_3CH_2Br + Mg \xrightarrow{\text{Dry ether}} CH_3CH_2—Mg—Br$$

Grignard reagents are extremely useful class of compounds which are used in the synthesis of a wide variety of organic compounds. (Refer Chapter 11 for details).

***(ii)* Formation of higher alkanes: Würtz reaction:** When an ethereal solution of an alkyl halide is treated with metallic sodium, it produces alkane having double the number of carbon atoms present in the alkyl halides.

$$2R—X + Na \xrightarrow{\text{Dry ether}} R—R + 2\overset{\oplus}{Na}\overset{\ominus}{X}$$

For example,

$$2\,CH_3CH_2Br + Na \xrightarrow{\text{Dry ether}} \underset{\text{Butane}}{CH_3CH_2-CH_2CH_3} + 2\overset{\oplus}{Na}\overset{\ominus}{Br}$$

This reaction is called **Würtz reaction** (Refer Chapter 4 for details).

***(iii)* Formation of other organometallic compounds:** Alkyl halides also react with other active metals such as lithium, zinc, mercury, lead, etc. in presence of dry ether to form the corresponding organometallic compounds. For example,

$$CH_3CH_2Br + 2Li \xrightarrow{\text{Dry ether}} \underset{\text{Ethyllithium}}{CH_3CH_2\,Li} + LiBr$$

$$4CH_3CH_2Br + \underset{\text{Sodium lead alloy}}{4Pb/Na} \xrightarrow{\text{Dry ether}} \underset{\text{Tetraethylead (TEL)}}{(CH_3CH_2)_4\,Pb} + 4Na\,Br + 3Pb$$

Tetraethyllead (TEL) is used as an *antiknocking agent* in gasoline.

5. Formation of Polyhalogen Compounds

Alkyl halides when treated with halogens in presence of light, heat or a catalyst, result in the formation of *di-*, *tri-*, *tetra-*or *poly-*substituted alkanes. For example,

$$\underset{\text{Chloromethane}}{CH_3Cl} + Cl_2 \xrightarrow[\text{or 520–670 K}]{\text{light}} \underset{\text{Dichloromethane}}{CH_2Cl_2} + HCl$$

$$\downarrow Cl_2,\ \text{light or 520–670 K}$$

$$\underset{\text{Tetrachloromethane}}{CCl_4} + HCl \xleftarrow[\text{or 520–670 K}]{Cl_2,\ \text{light}} \underset{\text{Trichloromethane}}{CHCl_3} + HCl$$

6. Isomerization

Alkyl halides on heating or treatment with a Lewis acid such as anhydrous $AlCl_3$ as a catalyst at lower temperature, isomerize in such a way that halogen migrates from one position to the other without any change in the carbon skeleton of the molecule. For example,

$$\underset{\text{1-Bromopropane}}{CH_3-CH_2-CH_2-Br} \xrightarrow[\text{or Anhyd. AlCl}_3]{573\text{ K}} \underset{\text{2-Bromopropane}}{CH_3-\underset{}{\overset{\overset{\displaystyle Br}{|}}{C}H}-CH_3}$$

$$CH_3-CH_2-CH_2-CH_2-Br \xrightarrow[\text{or Anhyd. } AlCl_3]{573\text{ K}} CH_3-CH_2-\underset{}{\overset{Br}{\overset{|}{C}}}H-CH_3$$

1-Bromobutane — 2-Bromobutane

10.3 POLYHALOGEN COMPOUNDS

Organic compounds containing more than one halogen atom are called **polyhalogen compounds.** They are classified as *di, tri, tetra,* etc. depending upon whether they contain two, three, four, etc. halogen atoms per molecule. For example,

CH_3CHBr_2	$BrCH_2—CH_2Br$
1,1-Dibromoethane	*1,2-Dibromoethane*
(*Ethylidene dibromide*)	(*Ethyene dibromide*)
$CHCl_3$	CCl_4
Trichloromethane	*Tetrachloromethane*
(*Chloroform*)	(*Carbon tetrachloride*)

Out of the polyhalogen compounds derived from methane, methylene chloride, chloroform and carbon tetrachloride are commonly used as commercial solvents. Chloroform had earlier been used as a general anaesthetic in surgery. However, because of its adverse effect on the heart and the formation of a toxic side product phosgene, $COCl_2$, it is not used now-a-days. Carbon tetrachloride is used as a fire extinguisher under the trade name **pyrene** apart from being a solvent.

In this section, we shall discuss the chemistry of chloroform and carbon tetrachloride with a special emphasis on the methods of their preparation and characteristic reactions.

10.3.1 Chloroform ($CHCl_3$)

Methods of Formation: Chloroform is prepared by the following methods:

1. From acetone or ethanol: Treatment of acetone or ethanol with sodium hypochlorite (NaOCl, *i.e.*, $NaOH/Cl_2$) gives rise to chloroform.

$$\underset{\text{Acetone}}{H_3C-CO-CH_3} + 3Cl_2 + 4NaOH \longrightarrow \underset{\text{Chloroform}}{CHCl_3} + 3NaCl + \underset{\text{Sodium acetate}}{CH_3COONa} + 3H_2O$$

$$\underset{\text{Ethanol}}{CH_3CH_2OH} + 4Cl_2 + 6NaOH \longrightarrow \underset{\text{Chloroform}}{CHCl_3} + 5NaCl + \underset{\text{Sodium formate}}{HCOONa}$$

The above conversions are **multistep processes**. The sequence of reactions involved is as follows:

For Acetone

Step I: Formation of sodium hypochlorite.

$$2NaOH + Cl_2 \longrightarrow NaOCl + NaCl$$

Step II: Base-catalyzed chlorination of acetone to produce trichloroacetone.

$$H_3C-\overset{\overset{O}{\|}}{C}-CH_3 + 3NaOCl \xrightarrow{H\overset{\ominus}{O}} H_3C-\overset{\overset{O}{\|}}{C}-CCl_3 + 3NaOH$$

Step III: Conversion of trichloroacetone into chloroform by the action of NaOH.

$$H_3C-\overset{\overset{O}{\|}}{C}-CCl_3 + NaOH \longrightarrow H_3C-CO\overset{\ominus}{O}\overset{\oplus}{Na} + CHCl_3$$

The mechanism of the last step is as follows:

(*a*) *Nucleophilic addition of* $H\overset{\ominus}{O}$ *to carbonyl group of trichloroacetone to give a tetrahedral intermediate.*

$$H_3C-\overset{\overset{O}{\|}}{C}-CCl_3 + \overset{\ominus}{O}H \longrightarrow H_3C-\underset{\underset{OH}{|}}{\overset{\overset{\overset{\ominus}{O}}{|}}{C}}-CCl_3$$

(*b*) *Cleavage of the intermediate to give chloroform and sodium acetate.*

$$H_3C-\underset{\underset{OH}{|}}{\overset{\overset{\overset{\ominus}{O}}{|}}{C}}-CCl_3 \longrightarrow H_3C-\overset{\overset{O}{\|}}{C}-OH + :\overset{\ominus}{C}Cl_3 \underset{+\overset{\oplus}{H}}{\xrightarrow{-\overset{\oplus}{H}}} H_3C-\overset{\overset{O}{\|}}{C}-\overset{\ominus}{O} + CHCl_3$$

For Ethanol

Step I: Oxidation of ethanol by chlorine to produce acetaldehyde (ethanal).

$$CH_3CH_2OH + Cl_2 \longrightarrow CH_3CHO + 2HCl$$

Step II: Base-catalyzed chlorination of acetaldehyde to produce trichloroacetaldehyde or chloral.

$$H_3C-CHO + 3NaOCl \xrightarrow{H\overset{\ominus}{O}} Cl_3C-CHO + 3NaOH$$

Trichloroacetaldehyde
(*Chloral*)

***Step III*: Conversion of trichloroacetaldehyde (or chloral) into chloroform by the action of NaOH.**

$$Cl_3C—CHO + NaOH \longrightarrow CHCl_3 + HCO\overset{\ominus}{O}\ \overset{\oplus}{Na}$$

Note: This conversion is characteristic of compounds containing **CH_3CO—**group and is known as **haloform reaction.**

Both **industrial** as well as laboratory preparation of chloroform involve heating ethyl alcohol or acetone with bleaching powder. The latter acts as a source of chlorine as well as a mild alkali.

$$CaOCl_2 + H_2O \longrightarrow Ca(OH)_2 + Cl_2$$

With **ethyl alcohol** the reaction takes place in three steps:

***Step I* : Oxidation of ethyl alcohol by chlorine to give acetaldehyde.**

$$\underset{\text{Ethanol}}{CH_3CH_2OH} + Cl_2 \longrightarrow CH_3CHO + 2HCl$$

***Step II*: Base-catalyzed chlorination of acetaldehyde to produce trichloroacetaldehyde (chloral).**

$$CH_3CHO + 3Cl_2 \longrightarrow \underset{\text{(Chloral)}}{CCl_3\,CHO} + 3HCl$$

***Step III*: Conversion of chloral into chloroform and calcium formate by the action of $Ca(OH)_2$.**

$$\underset{\text{Chloral}}{2CCl_3\,CHO} + Ca(OH)_2 \longrightarrow \underset{\text{Chloroform}}{2CHCl_3} + (HCOO)_2\,Ca$$

With **acetone** the conversion takes place in two steps:

***Step I*: Chlorination of acetone to produce trichloroacetone.**

$$\underset{\text{Acetone}}{CH_3COCH_3} + 3Cl_2 \longrightarrow \underset{\text{Trichloroacetone}}{CCl_3COCH_3} + 3HCl$$

***Step II*: Conversion of trichloroacetone into chloroform and calcium acetate by the action of $Ca(OH)_2$.**

$$\underset{\text{Trichloroacetone}}{2CCl_3\,COCH_3} + Ca(OH)_2 \longrightarrow \underset{\text{Chloroform}}{2CHCl_3} + (CH_3COO)_2\,Ca$$

2. From carbon tetrachloride: On the industrial scale, chloroform is manufactured by the reduction of carbon tetrachloride.

$$CCl_4 + 2[H] \longrightarrow CHCl_3 + HCl$$

3. From chloral hydrate: Pure chloroform is obtained by treating pure chloral hydrate with sodium hydroxide.

$$\underset{\text{Chloral hydrate}}{CCl_3\,CH(OH)_2} + NaOH \longrightarrow CHCl_3 + HCO\overset{\ominus}{O}\overset{\oplus}{Na} + H_2O$$

Chapter 10

4. From methane: Chlorination of methane is also an industrial method for preparation of chloroform.

$$CH_4 \xrightarrow[h\nu]{\frac{1}{2}Cl_2} CH_3Cl \xrightarrow[h\nu]{\frac{1}{2}Cl_2} CH_2Cl_2 \xrightarrow[h\nu]{\frac{1}{2}Cl_2} CHCl_3$$

The mixture thus formed is separated by fractional distillation.

Physical Properties

Chloroform is a sickly sweet-smelling colourless liquid with a burning taste. It is heavier than water heaving b.pt 334 K and freezing point 210 K. It is highly soluble in alcohols, ethers, etc. but sparingly soluble in water. Chloroform is a very good solvent for oils, fats, halogens, etc.

Temporary unconsciousness is produced on inhalation of vaporous of chloroform. Due to this property it was used as an **anaesthetic agent in surgery.**

Chemical Properties

1. Oxidation: It undergoes auto-oxidation in presence of light to form carbonyl chloride **(phosgene)** which is a highly toxic substance. It is therefore, stored in amber coloured bottles and 1% ethanol is added to it, which converts the phosgene into diethyl carbonate.

$$2CHCl_3 + O_2 \longrightarrow \underset{\text{Phosgene}}{2COCl_2} + 2HCl$$

$$COCl_2 + 2C_2H_5OH \longrightarrow \underset{\text{Diethyl carbonate}}{O{=}C(OC_2H_5)_2}$$

2. Nitration: Chloroform can be nitrated with nitric acid to form **chloropicrin** (nitrochloroform) which is a useful insecticide and a lachrymetry substance.

$$CHCl_3 + HONO_2 \longrightarrow O_2NCCl_3 + H_2O$$

3. Carbylamine reaction: When a drop of chloroform, dissolved in ethanol is warmed with a primary amine and alcoholic potassium hydroxide, it forms isocyanides (carbylamines) which are easily detected due their foul odour.

$$R{-}NH_2 + CHCl_3 + 3KOH \longrightarrow R{-}\overset{\oplus}{N}{\equiv}\overset{\ominus}{C}{:} + 3KCl + 3H_2O$$

This reaction forms the basis of **identification of primary amines**, and is known as **carbylamine reaction**.

Mechanism: The reaction is a multistep process which takes place in the following stages:

***Step I*: Conversion of chloroform into dichlorocarbene.**

$$CHCl_3 + \overset{\ominus}{OH} \longrightarrow \underset{\text{Dichlorocarbene}}{:CCl_2} + H_2O + \overset{\ominus}{Cl}$$

(This stage is a two step process, the mechanism for which, is discussed under next reaction, *i.e.*, hydrolysis of chloroform).

Step II : Interaction of 1° amine with : CCl_2 to produce a **nitrogenylide*** which subsequently eliminates two molecules of HCl to produce the carbylamine. The mechanistic scheme for this transformation may be written as follows:

$$R-\underset{H}{\overset{H}{N}}: + :CCl_2 \longrightarrow R-\underset{H}{\overset{H}{\overset{\oplus}{N}}}-\overset{\ominus}{C}:(Cl)_2$$

Nitrogenylide

$-H^{\oplus}$, $+H^{\oplus}$ Proton transfer from N to C

$$R-\underset{H}{\ddot{N}}-\underset{Cl}{\overset{Cl}{C}}-H \xrightarrow{-\overset{\ominus}{Cl}} R-\underset{H}{\overset{\oplus}{N}}=C\begin{smallmatrix}Cl\\H\end{smallmatrix}$$

$$\downarrow -\overset{\ominus}{H}$$

$$R-\ddot{N}=C\begin{smallmatrix}Cl\\H\end{smallmatrix} \xrightarrow{-\overset{\ominus}{Cl}} R-\overset{\oplus}{N}\equiv C-H \xrightarrow{\overset{\ominus}{O}H} R-\overset{\oplus}{N}\equiv\overset{\ominus}{C}:$$

Alkyl isonitrile
(*carbylamine*)

4. Hydrolysis: On warning with aqueous sodium hydroxide, it is hydrolysed to produce carbon monoxide and sodium formate.

$$CHCl_3 + 4\overset{\ominus}{O}H \longrightarrow 3\overset{\ominus}{Cl} + HCO\overset{\ominus}{O} + 2H_2O$$

$$CHCl_3 + 3\overset{\ominus}{O}H \longrightarrow 3\overset{\ominus}{Cl} + CO + 2H_2O$$

Mechanism: The mechanism of hydrolysis of chloroform is different from that of other alkyl halides. The reaction involves two stages:

(*i*) Formation of Dichlorocarbene (*ii*) Conversion of Dichlorocarbene into products.

Stage (*i*): Formation of Dichlorocarbene

It takes place in two steps:

* Species having opposite charged on C and a hetero atom directly linked to each other.

***Step I* : Formation of trichloromethyl anion.**

Due to the presence of three strong electron-withdrawing chlorine atoms, the hydrogen atom of chloroform is highly acidic. Thus it does not look surprising that the base abstracts the proton rather than displacing a chlorine atom by an S_N2 process. The removal of proton would result in the formation of trichloromethyl carbanion.

$$\overset{\ominus}{OH} + \underset{\text{Acidic}}{H-CCl_3} \underset{}{\overset{\text{Fast}}{\rightleftharpoons}} H_2O + \underset{\text{Trichloromethyl carbanion}}{\overset{\ominus}{C}Cl_3}$$

***Step II*: Conversion of trichloromethyl carbanion into dichlorocarbene.**

The trichloromethyl carbanion may eliminate chloride anion to give *dichlorocarbene* (:CCl_2)— a bivalent carbon species having two non-bonding electrons on the carbon atom.

$$:\overset{\ominus}{C}Cl_3 \xrightarrow{\text{Slow}} :CCl_2 + \overset{\ominus}{Cl}$$

Evidence

The above mechanism gets support from the following observations:

(*i*) This is a second order reaction, being first order with respect to base and first order with respect to chloroform.

(*ii*) When water enriched with D_2O is used and the reaction is stopped before completion, as expected, recovered unreacted chloroform becomes enriched in deuterium. Note that this observation gives credibility to the fact that proton abstraction takes place in a fast reversible step.

$$:\overset{\ominus}{C}Cl_3 \overset{D_2O}{\rightleftharpoons} CDCl_3 + \overset{\ominus}{O}D$$

(*iii*) The rate of the reaction is considerably slowed down by the addition of chloride ions meaning thereby that the removal of chloride is the slow step.

Since, in this stage the overall rate of reaction is limited to slower second step which depends upon the concentration of the conjugate base ($:\overset{\ominus}{C}Cl_3$) of chloroform, this mechanism is also called as **1,1-E1cB mechanism**.

Stage (*ii*): Conversion of dichlorocarbene into products.

The following probable mechanistic pathway may be proposed for the further hydrolysis of dichlorocarbene to either carbon monoxide (CO) or sodium formate (HCOONa).

Both CO and HCOONa are formed through a common intermediate- *the highly unstable 'formyl chloride'*, which is formed as follows:

The reaction is initiated by the addition of the hydroxide ion to highly electrophilic $:CCl_2$ producing an anion **(X)** which picks up a proton from water to give dichlorohydrin **(Y)**. Base catalysed elimination of HCl from **(Y)** produces the intermediate **(Z)**.

$$\overset{\ominus}{O}H + :CCl_2 \longrightarrow \underset{\text{X}}{H-O-\overset{\ominus}{C}Cl_2:} \xrightarrow[H_2O]{+\overset{\oplus}{H}} \underset{\text{Y}}{H-O-CHCl_2} \xrightarrow[-H_2O]{-\overset{\ominus}{Cl}} \underset{\text{Z}}{\left[H-\overset{O}{\overset{\|}{C}}-Cl\right]}$$

Z (Formyl chloride; unstable)

The intermediate **Z** is now converted into the products through two different mechanistic courses.

(*a*) ***Formation of carbon monoxide*:** Base catalysed elimination of HCl from Z produces carbon monoxide as shown below:

$$\overset{\ominus}{O}H + \underset{(Z)}{H-\overset{:\ddot{O}}{\overset{\|}{C}}-Cl} \longrightarrow \underset{\text{Carbon monoxide}}{:C{=}\ddot{O}:} + H_2O + \overset{\ominus}{Cl}$$

(*b*) ***Formation of formate*:** Nucleophilic addition of $\overset{\ominus}{O}H$ to Z followed by base catalyzed elimination of HCl produces formate as shown below:

$$\underset{(Z)}{H-\overset{O}{\overset{\|}{C}}-Cl} + \overset{\ominus}{O}H \longrightarrow H-\overset{O^{\ominus}}{\overset{|}{C}}(OH)-Cl \xrightarrow{\overset{\ominus}{O}H} \underset{\text{(Formate)}}{H-\overset{O^{\ominus}}{\overset{|}{C}}{=}O} + H_2O + \overset{\ominus}{Cl}$$

Electrophilic Character of Carbenes: Being electron-deficients the carbenes are electrophilic in nature. They undergo two principal reactions: *cycloaddition* with the alkenes forming cyclopropane derivatives and *insertion* into a C—H bond. Dichlorocarbene and other carbenes including the parent compound carbene (or methylene, $:CH_2$) are quite versatile for effecting the synthesis of cyclopropanes. This reaction is called as **cycloaddition**. The reaction has been extensively used for the determination of the intermediacy of carbenes (*trapping experiment*).

$$\text{Cyclohexene} + :CCl_2 \longrightarrow \text{7,7-dichlorobicyclo[4.1.0]heptane (C with Cl, Cl)}$$

5. Acidic nature: As discussed above due to presence of three strong electron-withdrawing chlorine atoms, the hydrogen of chloroform is acidic in nature and can be abstracted by a base.

$$\overset{\ominus}{O}H + H-CCl_3 \rightleftharpoons H_2O + :\overset{\ominus}{C}Cl_3$$

The trichloromethyl anion left after the removal of the proton from $CHCl_3$ is stabilized by a special type of resonance involving expansion of the valence shell of chlorine due to the presence of *d*-orbitals.

$$:\ddot{C}l-\overset{\ominus}{C}:(Cl)_2 \longleftrightarrow :\ddot{C}l-C(Cl)=\overset{\ominus}{Cl}: \longleftrightarrow :\ddot{C}l-C(Cl)=\overset{\ominus}{Cl}: \longleftrightarrow \overset{\ominus}{Cl}=C(Cl)_2$$

This type of resonance is called ***dπ-pπ resonance or d-orbital resonance.***

6. Reduction: Chloroform produces methylene chloride upon reduction with zinc and hydrochloric acid.

$$CHCl_3 + 2[H] \xrightarrow{Zn + HCl} CH_2Cl_2 + HCl$$

With zinc dust and water it can be reduced to methane.

$$CHCl_3 + 6[H] \xrightarrow{Zn + H_2O} CH_4 + 3HCl$$

7. Chlorination: It reacts with chlorine in presence of sunlight to produce carbon tetrachloride.

$$CHCl_3 + Cl_2 \xrightarrow{Sunlight} CCl_4 + HCl$$

8. Dehalogenation: Chloroform undergoes dehalogenation on warming with silver power to produce acetylene.

$$2CHCl_3 + 6Ag \longrightarrow HC \equiv CH + 6AgCl$$

9. Condensation with ketones: In presence of alkali, chloroform adds to ketones. Thus with acetone it forms **chloretone** which is a well known **hypnotic (sleep inducing drug)**.

$$\underset{\text{Acetone}}{(H_3C)_2C=O} + \underset{\text{Chloroform}}{H-CCl_3} \xrightarrow{KOH} \underset{\text{Chloretone}}{(H_3C)_2C(OH)CCl_3}$$

The reaction involves the following steps:

***Step I*:** Formation of trichloromethyl carbanion.

$$\overset{\ominus}{O}H + H-CCl_3 \longrightarrow H_2O + \overset{\ominus}{C}Cl_3$$

***Step II*:** Nucleophilic addition of $:\overset{\ominus}{C}lC_3$ to acetone to give chloretone.

$$(H_3C)_2C{=}O + \overset{\ominus}{C}Cl_3 \longrightarrow H_3C-C(CCl_3)(CH_3)-\overset{\ominus}{O} \xrightarrow[H_2O]{+\overset{\oplus}{H}} H_3C-C(CCl_3)(CH_3)-OH$$

Chloretone

10. Reimer-Tiemann reaction: Chloroform, when treated with phenol and alcoholic KOH at 340 K, produces salicylaldehyde as the major product.

$$C_6H_5OH + CHCl_3 \xrightarrow[-KCl,\ -H_2O]{KOH,\ 340\ K} o\text{-}(\overset{\ominus}{O}\overset{\oplus}{K})C_6H_4CHCl_2 \xrightarrow[-2KCl]{2KOH} [o\text{-}(\overset{\ominus}{O}\overset{\oplus}{K})C_6H_4C(OH)(OH)H]$$

Phenol; Unstable

$$\longrightarrow o\text{-}(\overset{\ominus}{O}\overset{\oplus}{K})C_6H_4CHO \xrightarrow[-KCl]{Dil.\ HCl} o\text{-}(OH)C_6H_4CHO$$

Salicylaldehyde

p-Hydroxybenzaldehyde is also obtained in this reaction though in very small amounts.

This reaction is called **Reimer-Tiemann reaction** and takes place through dichlorocarbene as intermediate.

Uses of chloroform: Chloroform is a very useful compound both in laboratory as well as in industry. Some of the uses are as follows:

1. It is used as a very good solvent for resins, rubbers, oils and fats.
2. It is used as a preservative for anatomical specimens.

3. It is used as a laboratory reagent for testing primary amines.
4. It is used in medicines.
5. In olden days it was being used as anaesthetic agent in surgery and as a flavouring agent in cough syrups. But because of its toxicity its use has been abandoned.

10.3.2 Carbon Tetrachloride (CCl_4)

Methods of formation: It is prepared using the following methods:

***(i)* From Methane:** Methane upon treatment with chlorine in presence of partly reduced cupric chloride catalyst produces carbon tetrachloride.

$$CH_4 + 4Cl_2 \xrightarrow{CuCl_2} CCl_4 + 4HCl$$

***(ii)* From Carbon Disulphide:** Carbon disulphide reacts with chlorine in presence of a Lewis acid catalyst ($FeCl_3$ or $AlCl_3$) to produce carbon tetrachloride.

$$CS_2 + 3Cl_2 \xrightarrow{AlCl_3} \underset{\text{Carbon tetrachloride}}{CCl_4} + \underset{\text{Sulphur monochloride}}{S_2Cl_2}$$

Sulphur monochloride obtained above is removed from the reaction mixture by fractional distillation and treated with fresh carbon disulphide to yield more carbon tetrachloride.

$$2S_2Cl_2 + CS_2 \longrightarrow CCl_4 + 6S$$

Physical Properties

Carbon tetrachloride is a colourless, volatile, non-inflammable liquid having b.pt. 350 K. It has sickly smell. It is insoluble in water but readity soluble in common organic solvents such as ether, alcohol, etc.

Chemical Properties

Some important chemical reactions of carbon tetrachloride are given below:

1. Stability: It is quite stable to heat (up to 773 K). Its vapours do not catch fire. This property has made CCl_4 a useful fire extinguisher under the name **pyrene**. At the high temperature of fire it is hydrolysed to some extent by water vapours forming phosgene which is a highly poisonous gas.

$$CCl_4 + \underset{\text{(Steam)}}{H_2O} \longrightarrow \underset{\text{Phosgene}}{COCl_2} + 2HCl$$

Therefore, after using it to extinguish fire, the room should be well ventilated .

2. Reduction: Reduction of CCl_4 with moist iron filing, produces chloroform.

$$CCl_4 + 2[H] \xrightarrow{Fe/H_2O} CHCl_3 + HCl$$

3. Reaction with hydrogen fluoride: When HF is passed into CCl_4 in presence of antimony pentachloride, dichlorodifluoromethane (Freon) is obtained.

$$CCl_4 + 2HF \xrightarrow{SbCl_5} \underset{\text{Freon}}{CCl_2F_2} + 2HCl$$

Freon is used as a refregerant (cooling agent) in refrigerators and air conditioners.

4. Reactions with antimony trifluoride: Also on treatment with antimony trifluoride in presence of antimony pentachloride as catalyst it gives rise to freon.

$$3CCl_4 + 2SbF_3 \xrightarrow{SbCl_5} 3CCl_2F_2 + 2SbCl_3$$

5. Hydrolysis: On boiling with alcoholic KOH solution, it gets hydrolysed to yield potassium carbonate.

$$CCl_4 + 4KOH \xrightarrow[-4KCl]{\Delta} \underset{\text{Unstable}}{[C(OH)_4]} \xrightarrow{-H_2O} \underset{\text{Carbonic acid}}{(HO)_2C{=}O}$$

$$\underset{\text{Carbonic acid}}{(HO)_2C{=}O} + 2KOH \longrightarrow \underset{\text{Potassium carbonate}}{K_2CO_3} + 2H_2O$$

6. Reimer-Tiemann reaction: In contrast to chloroform, treatment of CCl_4 with phenol in presence of KOH at 340 K produces salicylic acid.

$$C_6H_5OH + CCl_4 \xrightarrow[-KCl,\ -H_2O]{KOH,\ 340K} o\text{-}CCl_3C_6H_4O^{\ominus}K^{\oplus} \xrightarrow[-3KCl]{3KOH} \underset{\text{Unstable}}{[o\text{-}CH(OH)_3C_6H_4O^{\ominus}K^{\oplus}]} \xrightarrow[-3KCl]{3KOH}$$

$$o\text{-}KOOCC_6H_4O^{\ominus}K^{\oplus} \xrightarrow[-2KCl]{\text{dil. HCl}} \underset{\text{Salicylic acid}}{o\text{-}HOOCC_6H_4OH}$$

Here also small amount of *p*-hydroxybenzoic acid is produced.

Uses

Carbon tetrachloride finds immense use in laboratory as well as in industry. Some of the important uses of this compound are as follows:

1. It is used as an industrial solvent for resins, lacquers, oils, fats, etc.
2. It is also used as a solvent in dry cleaning.
3. It is used as a fire extinguisher under the name **'pyrene'.**
4. It is used for industrial preparation of chloroform.
5. It is used as a medicine for hookworms.

10.4 ARYL HALIDES

In contrast to alkyl halides, the aryl halides are the compounds having halogen atom directly attached to the aromatic ring. They are represented by the general formula Ar—X where Ar may be phenyl, substituted phenyl or any other aryl ring. These compounds are characterized by their relative inertness towards most of the common nucleophiles.

10.4.1 Nomenclature

Systematic nomenclature includes naming these compounds as substitution products of the parent hydrocarbons. But some compounds are best known by their common names based upon the aryl radical present in them. Following examples will illustrate both the systems. The common names are written in parenthesis.

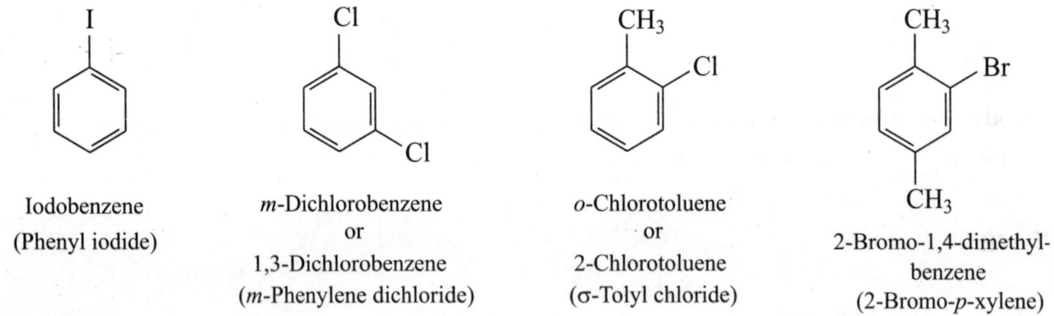

Iodobenzene (Phenyl iodide)

m-Dichlorobenzene or 1,3-Dichlorobenzene (*m*-Phenylene dichloride)

o-Chlorotoluene or 2-Chlorotoluene (σ-Tolyl chloride)

2-Bromo-1,4-dimethyl-benzene (2-Bromo-*p*-xylene)

10.4.2 Methods of Formation

In general, the methods used for the preparation of alkyl halides are not applicable for the synthesis of aryl halides. Ethanol can be converted into ethyl chloride by reagents such as hydrogen chloride, thionyl chloride and phosphorus *tri-or penta*chloride, but phenol cannot be converted into chlorobenzene by these reagents. Similarly, it can also be said that the methods used for the preparation of aryl halides do not work for the preparation of alkyl halides. Some of the commonly used methods for the formation of aryl halides and aralkyl halides are listed below:

1. Halogenation

(a) Formation aryl halides by halogenation of the ring: Benzene or its derivatives can be converted into halobenzenes by direct halogenation with chlorine or bromine in presence of a metal halide catalyst (*halogen carrier*).

$$C_6H_6 + X_2 \xrightarrow{AlX_3} C_6H_5X$$

Aryl halide

(b) Formation of aralkyl halides by halogenation of side chain: The most convenient method for the preparation of this class of compounds involves free radical halogenation of a suitable hydrocarbon. For example, phenylchloromethane (benzyl chloride) is formed by passing Cl_2 through boiling toluene in the presence of sunlight and absence of halogen carrier.

$$C_6H_5CH_3 + Cl_2 \xrightarrow[\text{Sunlight}]{383\text{ K}} C_6H_5CH_2Cl + HCl$$

Phenylchloromethane
(Benzyl chloride)

However, if an excess of Cl_2 is used, the initially formed benzyl chloride reacts to produce benzal chloride followed by benzotrichloride.

$$C_6H_5CH_2Cl + Cl_2 \xrightarrow[\text{Sunlight, –HCl}]{383\text{ K}} C_6H_5CHCl_2 \xrightarrow[\text{Sunlight, –HCl}]{Cl_2,\ 383\text{ K}} C_6H_5CCl_3$$

Benzal chloride | Benzotrichloride

As discussed earlier when the side chain has more than one carbon atom, side chain halogenation occurs mainly at benzylic carbon (*i.e.*, carbon atom next to the benzene ring). This type of halogenation is also called **benzylic halogenation**.

$$C_6H_5CH_2—CH_3 + Cl \xrightarrow[273\text{ K, –HCl}]{Cl_2,\ h\nu} C_6H_5CHClCH_3 \xrightarrow[273\text{ K, –HCl}]{Cl_2,\ h\nu} C_6H_5CCl_2CH_3$$

Ethylbenzene | 1-Chloro-1-phenylethane | 1,1-Dichloro-1-phenyl ethane

Benzylic bromination can also take place with Br_2 in presence of high temperature or light. For example, N-Bromosucciniamide (NBS) is a specific reagent for benzylic bromination.

$$C_6H_5CH_2CH_2CH_3 \xrightarrow[CCl_4,\ 333\ K]{NBS,\ (C_6H_5CO_3)_2} C_6H_5CH(Br)—CH_2CH_3$$

1-Bromo-1-phenylpropane (99%)

2. From Aniline *via* Diazonium Salts

(*a*) Sandmeyer reaction: The arenediazonium halides, obtained by treating aromatic amines with nitrous acid, upon decomposition by warming with cuprous chloride or bromide in presence of excess of halogen acids yield aryl halides. This is known as **Sandmeyer reaction.**

$$C_6H_5NH_2 + HONO + HX \xrightarrow{273-278\ K} C_6H_5\overset{\oplus}{N}{\equiv}N\overset{\ominus}{X} + 2H_2O$$

Benzenediazonium salt

$$C_6H_5\overset{\oplus}{N}{\equiv}N\overset{\ominus}{X} \xrightarrow[HX]{Cu_2X_2} C_6H_5X + N_2$$

(X = Cl, Br) Aryl halide

In Sandmeyer reaction, it is the halogen attached to copper which goes to the benzene ring. This reaction is used for preparation of chloroarenes and bromoarenes.

Fluoroarenes or Aryl Fluorides: Aryl fluorides can also be obtained using this procedure by introducing the following modification. The aromatic amine is first diazotized in the usual manner and then the solution of diazonium salt is treated with fluoroboric acid or its salt. This causes the precipitation of the sparingly soluble arenediazonium fluoroborate. The solid so obtained is carefully heated after drying to yield the aryl fluoride. This reaction is called **Balz-Schiemann reaction**.

$$C_6H_5\overset{\oplus}{N_2}\overset{\ominus}{BF_4} \xrightarrow{\Delta} C_6H_5F + N_2 + BF_3$$

Fluorobenzene

Iodoarenes or Aryl Iodides: Aryl iodides are, however, obtained by treating the diazonium salt with aqueous potassium iodide.

$C_6H_5N^{\oplus}{\equiv}NX^{\ominus}$ $\xrightarrow{KI}$ C_6H_5I (Iodobenzene) + N_2 + KX

It may be pointed out that the above sequence of reactions constitutes important method for introducing a halogen atom into a particular position of the aromatic ring. The substrate chosen is a suitably substituted aromatic amine. For instance,

(*a*) o-toluidine (CH_3, NH_2) $\xrightarrow[0°]{HCl/\ HNO}$ (CH_3, $N_2^{\oplus}Cl^{\ominus}$) $\xrightarrow[HCl]{Cu_2Cl_2}$ (CH_3, Cl)

(*b*) Benzene $\xrightarrow[60°]{HNO_3/H_2SO_4}$ (NO_2) $\xrightarrow[H_2SO_4,\ 90°]{HNO_3\ (fuming)}$ (NO_2, NO_2) $\xrightarrow{Sn/HCl\ \ Reduction}$ (NH_2, NH_2) $\xrightarrow[0°]{HNO_2/HCl}$ ($N_2^{\oplus}Cl^{\ominus}$, $N_2^{\oplus}Cl^{\ominus}$) $\xrightarrow[HCl]{Cu_2Cl_2}$ (Cl, Cl)

(*b*) Gattermann reaction: It is a modification of the Sandmeyer reaction. In this reaction the arenediazonium halide (particularly chloride or bromide) is treated with a mixture of freshly prepared copper powder in the presence of corresponding halogen acid (HCl or HBr) in place of cuprous halide dissolved in the corresponding halogen acid (CuCl/HCl or CuBr/HBr) which are used in Sandmeyer reaction.

Benzenediazonium chloride ($C_6H_5N^{\oplus}{\equiv}NCl^{\ominus}$) $\xrightarrow[\Delta]{Cu/HCl}$ C_6H_5Cl (Chlorobenzene) + N_2

Benzenediazonium chloride $\xrightarrow[\Delta]{Cu/HBr}$ C_6H_5Br (Bromobenzene) + N_2

3. Hunsdiecker Reaction

Aryl halides are obtained by heating the silver salts of aromatic acids with bromine.

$$C_6H_5COOAg + Br_2 \longrightarrow C_6H_5Br + AgBr + CO_2$$

Bromobenzene

10.4.3 Physical Properties

(1) Aryl halides are heavier than water.

(2) Though polar in character they are insoluble in water. This may be attributed to their being incapable of forming hydrogen bonds.

(3) The melting and boiling points of aryl halides are listed in Table 10.4. A glance at the Table reveals the following:

(*a*) That simple aryl halides such as chloro-and bromobenzene have boiling points comparable to alkyl halides having the same number of carbon atoms.

(*b*) For a group of aryl halides having the same aryl substrate but different halogen atoms the melting and boiling points increase with an increase in the molecular weight. For instance, the melting point of iodobenzene is the highest and that of fluorobenzene is the lowest.

(4) The physical constants of dihaloarenes show another interesting feature, *i.e.*, like arenes their boiling points are almost constant, but melting points and water solubilities of *para*-isomers are the highest. These observations can be explained on the basis of the fact that the *para* isomers, being symmetrical, are better packed in the crystal lattice and form larger molecular aggregates.

Table 10.4: Physical Properties of Aryl Halides

S.No.	Name	Formula	M. Pt. K	B. Pt. K
1.	Fluorobenzene	C_6H_5F	231	358
2.	Chlorobenzene	C_6H_5Cl	229	405
3.	Bromobenzene	C_6H_5Br	242	429
4.	Iodobenzene	C_6H_5I	243	461
5.	Fluorotoluenes	$CH_3C_6H_4F$		
	(*i*) *ortho*-isomer		—	388
	(*ii*) *meta*-isomer		163	389
	(*iii*) *para*-isomer		—	389
6.	Chlorotoluenes	$CH_3C_6H_4Cl$		
	(*i*) *ortho*-isomer		238	463
	(*ii*) *meta*-isomer		224	467
	(*iii*) *para*-isomer		281	467

(*Contd...*)

7.	Bromotoluenes	$CH_3C_6H_4Br$		
	(*i*) *ortho*-isomer		247	487
	(*ii*) *meta*-isomer		233	488
	(*iii*) *para*-isomer		301	489
8.	Iodotoluenes	$CH_3C_6H_4I$		
	(*i*) *ortho*-isomer		—	479
	(*ii*) *meta*-isomer		—	483
	(*iii*) *para*-isomer		—	484
9.	Difluorobenzenes	FC_6H_4F		
	(*i*) *ortho*-isomer		239	365
	(*ii*) *meta*-isomer		214	356
	(*iii*) *para*-isomer		259	362
10.	Dichlorobenzenes	ClC_6H_4Cl		
	(*i*) *ortho*-isomer		256	453
	(*ii*) *meta*-isomer		249	447
	(*iii*) *para*-isomer		324	448
11.	Dibromobenzenes	BrC_6H_4Br		
	(*i*) *ortho*-isomer		279	494
	(*ii*) *meta*-isomer		266	491
	(*iii*) *para*-isomer		360	491
12.	Diiodobenzenes	IC_6H_4I		
	(*i*) *ortho*-isomer		300	559
	(*ii*) *meta*-isomer		308	559
	(*iii*) *para*-isomer		402	558

10.4.4 Chemical Reactions

1. Formation of Organometallic Compounds: Treatment of an ethereal solution of aryl halides with magnesium metal results in the formation of arylmagnesium halides **(Grignard reagent)**. For instance,

$$C_6H_5Br + Mg \xrightarrow{\text{Ether}} \underset{\text{Phenylmagnesium bromide}}{C_6H_5MgBr}$$

Chlorobenzene and iodobenzene are relatively less reactive. Aryllithium compounds are obtained by treating an ethereal solution of aryl halides with lithium metal.

$$\underset{(X=Cl,Br)}{C_6H_5X} + 2Li \xrightarrow{\text{Ether}} C_6H_5Li + LiX$$

The aryllithium compounds may also be obtained by the halogen-metal exchange when aryl halides are treated with *n*-butyllithium.

$$Br-C_6H_4-Cl + nC_4H_9Li \longrightarrow Li-C_6H_4-Cl + nC_4H_9Br$$

2. Aromatic Electrophilic Substitution: The presence of halogen atom deactivates the benzene ring for electrophilic aromatic substitution but directs the incoming group to the *o*-and *p*-positions. Thus, aryl halides undergo electrophilic substitution such as halogenation, nitration, sulphonation, Friedel-Crafts reaction. In all these cases the rates of the reaction are slower than those of benzene and second group enters at *o*-or *p*-positions. Some examples are:

(i) Halogenation

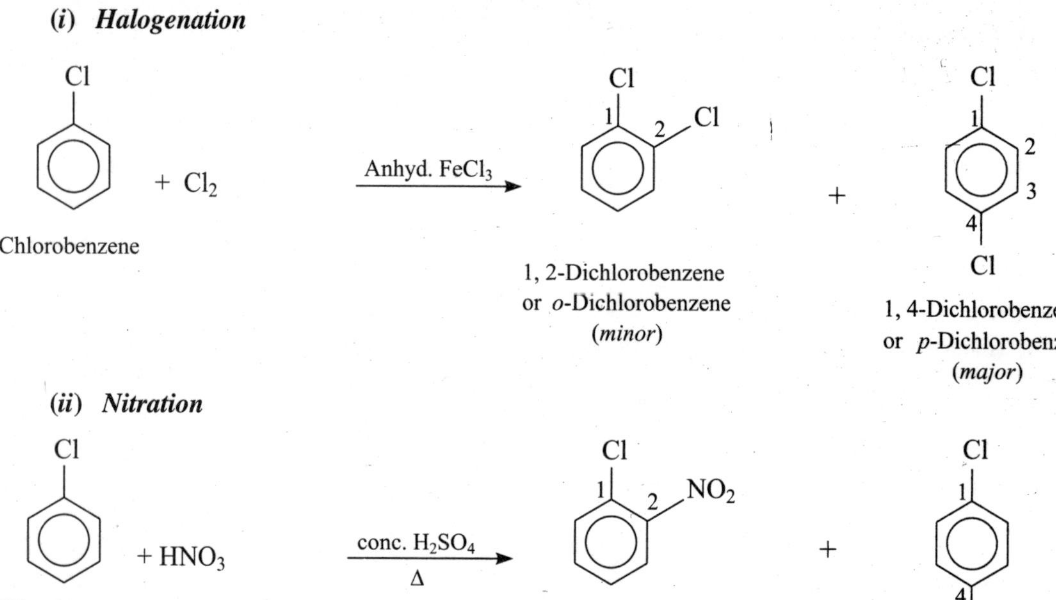

4-Nitrochlorobenzene or *p*-Nitrochlorobenzene (*major*)

(iii) Sulphonation

$$C_6H_5Cl + H_2SO_4\ (conc.) \xrightarrow{\Delta} \text{2-}ClC_6H_4SO_3H + \text{4-}ClC_6H_4SO_3H$$

Chlorobenzene

2-Chlorobenzene-sulphonic acid (*minor*)

4-Chlorobenzene-sulphonic acid (*major*)

(iv) ***Frieded-Crafts Alkylation***

Chlorobenzene + $CH_3—Cl$ $\xrightarrow{\text{Anhyd. } AlCl_3}$ *o*-Chlorotoluene (*minor*) + *p*-Chlorotoluene (*major*)

(v) ***Frieded-Crafts Acylation***

Chlorobenzene + $CH_3—\overset{O}{\overset{\|}{C}}—Cl$ $\xrightarrow{\text{Anhyd. } AlCl_3}$ *o*-Chloroacetophenone (*minor*) + *p*-Chloroacetophenone (*major*)

3. Würtz-Fittig Reaction: Action of metallic sodium on an ethereal solution of a 1 : 1 mixture of an alkyl halide and an aryl halide results in the formation of alkylbenzene as a major product.

Aryl halide + 2Na + $X—CH_2CH_3$ (Alkyl halide (Ethyl halide)) $\longrightarrow$ Ethylbenzene + 2NaX

This reaction, known as **Würtz-Fittig reaction,** has already been discussed in Chapter 9.

4. Ullmann Biaryl Synthesis: Iodobenzene on heating with copper power in a sealed tube affords biphenyl.

Iodobenzene + Cu + Iodobenzene $\xrightarrow[\text{Sealed tube}]{\Delta}$ Biphenyl + 2CuI

This reaction has already been discussed in Chapter 9. It may be mentioned, however, that aryl chlorides and bromides react only when an electron-withdrawing group is present *ortho* or *para* to the halogen.

$$2\ C_6H_4(NO_2)Cl + 2Cu \xrightarrow{\Delta} (NO_2)C_6H_4{-}C_6H_4(NO_2) + 2CuCl$$

5. Reaction with NaOH (Dow's Process): Haloarenes do not react with sodium hydroxide under the conditions used for a similar reaction of alkyl halides. However, when they are treated with NaOH under drastic conditions (623 K, 300 atom) they form sodium phenoxide which upon acidification gives phenol. This reaction known as **Dow's process**, is used for industrial production of phenol. This is also an example of elimination-addition mechanism involving benzyne as an intermediate.

Chlorobenzene $\xrightarrow[623\ K,\ 300\ atm.]{NaOH\ (6–8\%)}$ [carbanion] → [Benzyne] $\xrightarrow{\overset{\ominus}{O}H}$ → $\xrightarrow[-\overset{\ominus}{O}H]{H_2O}$ Phenol $\xrightarrow[-H_2O]{NaOH}$ Sodium phenoxide ($\overset{\ominus}{O}\overset{\oplus}{Na}$)

Support for the above mechanism comes from the observation that diphenyl ether and *p*-phenylphenol are obtained as side product.

6. Reaction with Ammonia: When haloarenes are heated with aqueous ammonia solution in presence of cuprous oxide as catalyst at 475 K and under a pressure of 60 atmospheres, the halogen atom is replaced by amino group. For example,

$$2\ C_6H_5Cl + 2NH_3 + Cu_2O \xrightarrow[60\ atm.]{475\ K} 2\ C_6H_5NH_2 + 2CuCl + H_2O$$

Chlorobenzene → Aniline

This reaction also occurs through benzyne mechanism (*Section 10.6*).

7. Reaction with Cuprous Cyanide : Aryl bromides, on heating with cuprous cyanide at 475 K in presence of pyridine or dimethylformamide (DMF) are converted into aryl cyanides by replacement of bromine by a —CN group.

$$C_6H_5Br + CuCN \xrightarrow[\text{Pyridine or DMF}]{475\text{ K}} C_6H_5CN + CuBr$$

Bromobenzene → Benzonitrile

Like alkyl nitriles, benzonitrile is also used in the synthesis of a number of compounds as shown below:

$$C_6H_5COOH \xleftarrow[\text{Complete hydrolysis}]{\text{dil. HCl or NaOH}} C_6H_5CN \xrightarrow[\text{or } Na/C_2H_5OH]{\substack{(\textit{Reduction})\\ LiAlH_4 \text{ or } Ni/H_2}} C_6H_5CH_2NH_2$$

Benzoic acid ← Benzonitrile → Benzylamine

$$C_6H_5CN \xrightarrow[\text{(Partial hydrolysis)}]{\text{conc. HCl or Alkaline } H_2O_2} C_6H_5CONH_2$$

Benzonitrile → Benzamide

8. Reduction: Aryl halides are reduced to corresponding arenes on treatment with Ni-Al alloy in presence of alkali.

$$C_6H_5Cl + 2[H] \xrightarrow[\Delta]{Ni–Al / NaOH} C_6H_6 + HCl$$

10.5 REACTIONS OF ARALKYL HALIDES (SIDE CHAIN SUBSTITUTED HALOGEN COMPOUNDS)

As in the case of alkyl halides, in these compounds also the halogen is attached to sp^3-hybridized carbon atom. Thus, their reactivity is comparable to that of alkyl halides. Further, benzyl halides are even more reactive than alkyl halides. This is due to the fact that such halides readily ionise to form resonance stabilized benzyl cation.

$CH_2—Cl$ Ionisation, $-Cl^{\ominus}$ → [Resonance structures of $\overset{\oplus}{C}H_2$ benzyl cation]

Benzyl chloride — Resonance stabilized benzyl cation

We will now discuss some of the more important reactions of benzyl halides.

1. Nucleophilic Substitution Reactions: Benzyl halides undergo a variety of nucleophilic substitution reactions. For example,

***(a) Formation of alcohols*:** On heating with **aqueous alkali**, benzyl chloride forms benzyl alcohol.

$$C_6H_5CH_2Cl + KOH\,(aq) \xrightarrow{\Delta} C_6H_5CH_2OH + KCl$$

Benzyl chloride — Benzyl alcohol

It may be pointed out here that di-and trihalides are hydrolysed to form the corresponding aldehydes and carboxylic acids, respectively.

$$C_6H_5CHCl_2 + KOH\,(aq.) \xrightarrow{-2KCl} [C_6H_5CH(OH)_2] \xrightarrow{-H_2O} C_6H_5CHO$$

Benzal chloride — Unstable — Benzaldehyde

The hydrolysis of benzyl chloride to benzaldehyde can also be accomplished in aqueous acids.

$$C_6H_5CCl_3 + KOH\,(aq.) \xrightarrow{-3KCl} [C_6H_5C(OH)_3] \xrightarrow{-H_2O} C_6H_5COOH$$

Benzotrichloride — Unstable — Benzoic acid

***(b) Formation of amines*:** Like alkyl halides, benzyl halides also react with alcoholic ammonia to form benzylamine (*Hoffmann Ammonolysis*).

$$C_6H_5CH_2Cl + 2\,NH_3\,(alc.) \longrightarrow C_6H_5CH_2NH_2 + NH_4Cl$$

Benzyl chloride → Benzylamine

(*c*) ***Formation of cyanides and isocyanides***: On treatment with NaCN and AgCN, benzyl chloride gives rise to benzyl cyanide and benzyl isocyanide, respectively.

$$C_6H_5CH_2Cl \xrightarrow[-NaCl]{NaCN\,(alc.),\,\Delta} C_6H_5CH_2CN \text{ (Benzyl cyanide)}$$

$$C_6H_5CH_2Cl \xrightarrow[-AgCl]{AgCN} C_6H_5CH_2-\overset{\oplus}{N}\equiv\overset{\ominus}{C}: \text{ (Benzyl isocyanide)}$$

Benzyl cyanide in turn can be converted into benzylamine by reduction and to phenylacetic acid by hydrolysis.

$$\underset{\text{Phenylacetic acid}}{C_6H_5CH_2COOH} \xleftarrow[-NH_3]{\text{dil. acid or alkali}} \underset{\text{Benzyl cyanide}}{C_6H_5CH_2CN} \xrightarrow[Na/C_2H_5OH]{\textit{(Mendius Reduction)}} \underset{\text{Benzylamine}}{C_6H_5CH_2NH_2}$$

2. Formation of Grignard Reagents: Treatment of an ethereal solution of benzyl bromide with magnesium gives rise to benzylmagnesium bromide.

$$\underset{\text{Benzyl bromide}}{C_6H_5CH_2Br} + Mg \xrightarrow{\text{Dry ether}} \underset{\text{Benzylmagnesium bromide}}{C_6H_5CH_2MgBr}$$

3. Friedel-Craft's Reaction: Benzyl chloride on treatment with benzene in presence of anhydrous aluminium chloride gives rise to diphenylmethane.

$$C_6H_5CH_2Cl + C_6H_6 \xrightarrow{\text{Anhyd. } AlCl_3} C_6H_5CH_2C_6H_5 + HCl$$

Benzyl chloride Diphenylmethane

4. Würtz-Fittig reaction: When benzyl chloride (two molecules) is heated with sodium in ether, it produces 1,2-diphenylethane.

$$C_6H_5CH_2Cl + 2Na + ClH_2CC_6H_5 \xrightarrow[-2NaCl]{\text{Dry ether, } \Delta} C_6H_5CH_2-CH_2C_6H_5$$

Benzyl chloride 1, 2 Diphenylethane

On treatment with alkyl halides under similar conditions, benzyl chloride produces higher alkylbenzenes.

$$C_6H_5CH_2Cl + 2Na + ClCH_3 \xrightarrow[-2NaCl]{\text{Dry ether, } \Delta} C_6H_5CH_2CH_3$$

Benzyl chloride Chloromethane Ethylbenzene

5. Reduction: Reduction with Zn-Cu couple and alcohol gives toluene.

$$C_6H_5CH_2Cl \xrightarrow{Zn\text{-}Cu/C_2H_5OH} C_6H_5CH_3 + HCl$$

Toluene

6. Halogenation of the side chain: When benzyl chloride is treated with chlorine in presence of sunlight, it produces benzal chloride followed by benzotrichloride, depending upon the amount of chlorine available,

$$C_6H_5CH_2Cl \xrightarrow[-HCl]{Cl_2,\ h\nu} C_6H_5CHCl_2 \xrightarrow[-HCl]{Cl_2,\ h\nu} C_6H_5CCl_3$$

Benzyl chloride Benzal chloride Benzotrichloride

7. Oxidation: Benzyl chloride is readily oxidised to benzaldehyde by heating with aqueous lead nitrate in an atmosphere of CO_2 .

CH_2Cl $\xrightarrow[-PbCl_2]{Pb(NO_3)_2}$ (H–C(H)–O–N⊕(=O)–O⊖ attached to benzene) $\longrightarrow$ CHO (benzene) + HNO_2

Benzaldehyde

Strong oxidising agents such as conc. HNO_3, $KMnO_4$/KOH, $K_2Cr_2O_7$/H_2SO_4 convert benzyl chloride into benzoic acid.

8. Electrophilic Substitution Reactions of Benzene Ring: Benzyl chloride undergoes the usual electrophilic substitution reactions such as halogenation, nitration and sulphonation producing a mixture of *ortho* and *para*-isomers. Some examples are listed below:

CH_2Cl (benzene) $\xrightarrow[(Halogenation)]{Cl_2/FeCl_3}$ CH_2Cl Cl (ortho) + CH_2Cl … Cl (para)

CH_2Cl (benzene) $\xrightarrow[(Sulphonation)]{Conc.\ H_2SO_4}$ CH_2Cl SO_3H (ortho) + CH_2Cl … SO_3H (para)

CH_2Cl (benzene) $\xrightarrow[(Nitration)]{HNO_3/H_2SO_4}$ CH_2Cl NO_2 (ortho) + CH_2Cl … NO_2 (para)

10.6 MECHANISM OF AROMATIC NUCLEOPHILIC SUBSTITUTION

Aromatic nucleophilic substitution may be described as a process in which a nucleophile ($Nu^{\ominus}$) forms a bond with a ring carbon atom with the corresponding displacement of the leaving group along with its bonding electrons.

$$\text{C}_6\text{H}_5\text{X} + \overset{\ominus}{\text{Nu}} \longrightarrow \text{C}_6\text{H}_5\text{Nu} + \overset{\ominus}{\text{X}}$$

Since, the potential leaving group is usually a halogen atom having non-bonding pair of electrons which are in conjugation with the ring, its displacement is a rather difficult process. However, appropriately substituted aryl halides do undergo nucleophilic displacement by a bimolecular mechanism abbreviated as **S_NAr2**. For instance, 2,4-dinitrochlorobenzene undergoes facile transformation to the corresponding methyl ether upon treatment with potassium methoxide.

$$\text{2,4-}(NO_2)_2C_6H_3Cl + \overset{\ominus}{O}CH_3 \longrightarrow \text{2,4-}(NO_2)_2C_6H_3OCH_3 + \overset{\ominus}{Cl}$$

In the case of simple aryl halides such a reaction takes place under rather drastic conditions (already discussed).

$$C_6H_5Cl \xrightarrow{\overset{\ominus}{O}H/573K} C_6H_5OH$$

$$C_6H_5Cl \xrightarrow[CuO]{NH_3/573\ K} C_6H_5NH_2$$

$$C_6H_5Cl \xrightarrow[\Delta]{CuCN/DMF} C_6H_5CN$$

Mechanism

The above transformations do not take place by a straightforward S_N mechanism but through some alternate mechanism. In fact these displacements can take place by one of the following two mechanisms:

1. *Addition-Elimination or Bimolecular Mechanism.*
2. *Elimination-Addition or Benzyne Mechanism.*

1. Addition-Elimination or Bimolecular Mechanism

The reactions taking place by this mechanism need relatively milder nucleophiles or bases and are characteristic substitutions of aryl halides having strong electron-withdrawing substituents such as nitro, etc. at *o*-and *p*-positions with respect to the leaving group.

It has been shown that like other S_N2 reactions these substitution reactions also follow second order kinetics (*i.e.*, they are bimolecular in nature). Further, they have been shown to take place by a two-step *addition-elimination mechanism* as discussed below:

***Step* I: Addition of nucleophile to produce cyclohexadienylide anion**

This step involves the addition of the nucleophile to the carbon carrying the halogen atom to produce an intermediate called ***cyclohexadienylide anion***, which is stabilized by resonance.

In many cases this step is rate-determining. The reason for this is that there is rehybridization of carbon undergoing substitution from sp^2 to sp^3 thereby destroying the aromaticity of the ring. This raises the ***Eact*** for this step (Fig. 10.3).

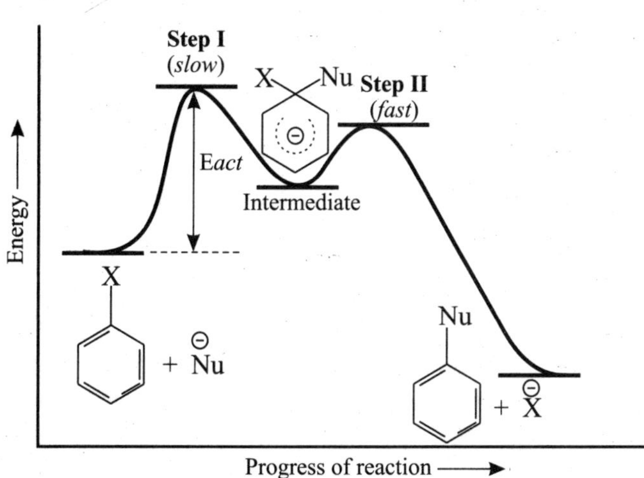

Fig. 10.3: Energy profile diagram for addition-elimination nucleophilic aromatic subsitution

***Step* II: Elimination of halide ion**

This step involves the fast elimination of the halide anion from the intermediate. The driving force for the elimination of halide ion is that in doing so, the ring regains its aromatic character.

It may be noted here that the formation of cylohexadienylide ion is not possible unless some group, capable of accommodating the negative charge, is present at *ortho* or *para*-position. Such groups help in dispersing the negative charge, thus stabilizing the intermediate as shown below:

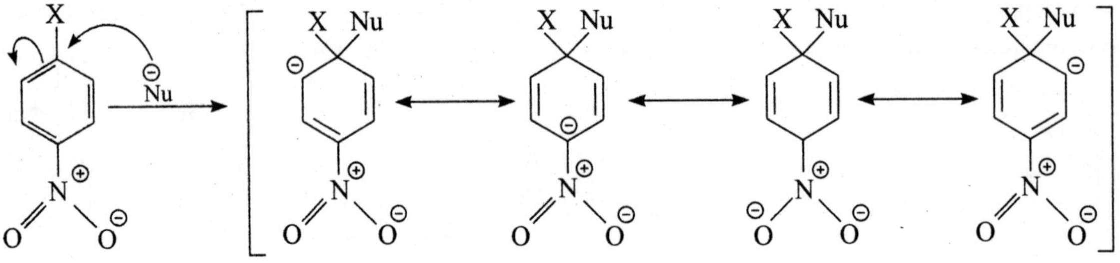

Since, delocalization of the negative charge *via* structures such as (**T**) is not possible with *meta* nitro derivatives, they do not undergo such a reaction.

Evidence in support of addition-elimination mechanism

The rate of this reaction depends upon first step which involves the formation of carbanion. In contrast to the reactivity of halogens in alkyl halides there is negligible difference in reactivities of aryl iodides, bromides, chlorides and fluorides* in these reactions.

Thus it can be said that the reactivity of various aryl halides is independent of the strength of the carbon-halogen bond. This can be possible only if the cleavage of carbon-halogen bond is not involved in the rate-determining step.

2. Elimination-Addition or Benzyne Mechanism

In contrast to bimolecular nucleophilic substitution mechanism, the nucleophilic aromatic substitution may also take place by an alternate **elimination-addition mechanism** when the substrate (aryl halide) is treated with strongly basic nucleophiles or with usual nucleophiles under drastic conditions. For example,

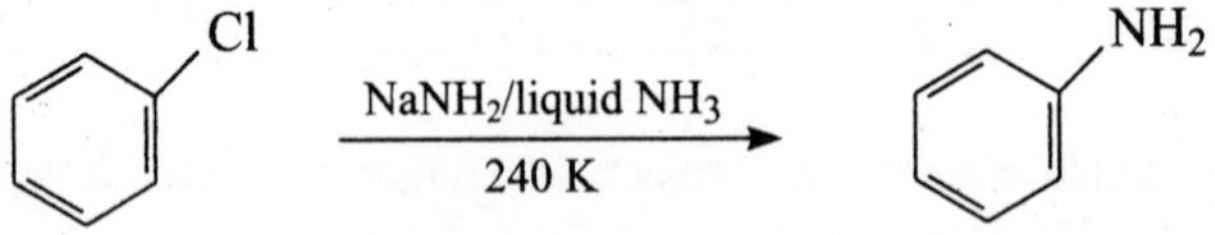

$$C_6H_5Cl \xrightarrow[240\ K]{NaNH_2/\text{liquid } NH_3} C_6H_5NH_2$$

The most important requirement for this mechanism to operate is the presence of an *ortho*-hydrogen.

*Aryl fluorides react faster than chlorides, bromides and iodides, probably due to the strong electron with drawing effect of fluorine atom, which favours the first step, *i.e.*, nucleophilic attack at the ring.

Mechanism

As discussed in Chapter 8, the reaction takes place in two stages as follows:

Stage I: Two-step Elimination of HX from o-positions.

Stage II: Two-step Addition of Nucleophilie.

Stage I: Two-step elimination of HX

***Step I*: Abstraction of a proton from *o*-position by a strongly basic amide anion (nucleophile) to produce a carbanion.**

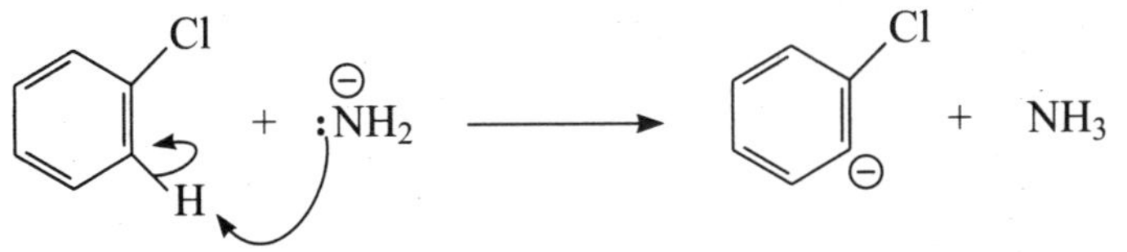

***Step II*: Loss of chloride ion from carbanion to produce benzyne.**

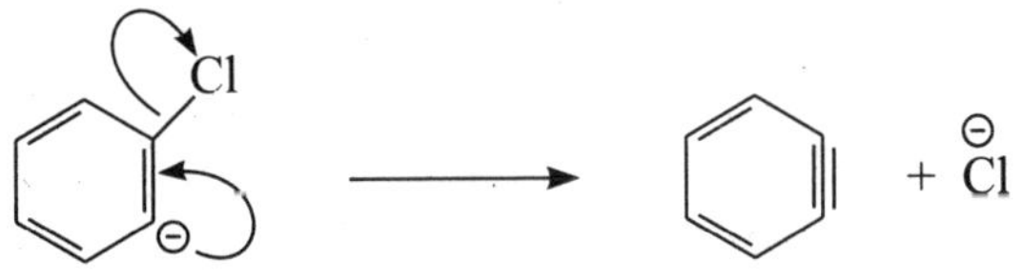

Benzyne

Stage II: Two-step addition of nucleophile

***Step I*: Addition of amide ion to benzyne to produce another carbanion.**

***Step II*: Abstraction of a proton by carbanion from ammonia.**

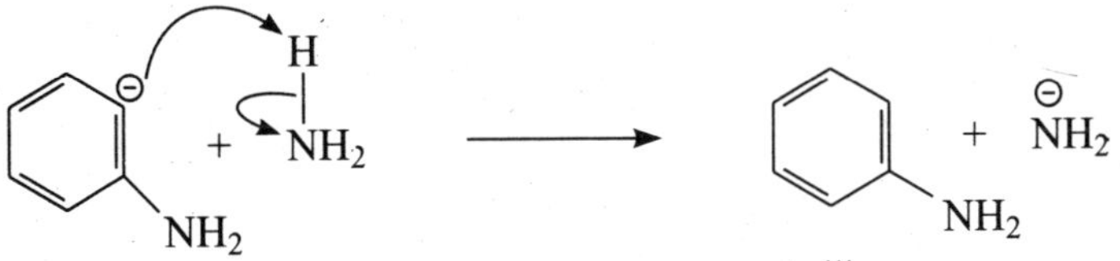

Aniline

Evidence in Support of Benzyne Mechanism

This mechanism is supported by the following observations:

1. Tracer technique: The most convincing evidence for the intermediacy of benzyne in these types of substitution reactions has been provided by the fact that [1-^{14}C] -chlorobenzene on treatment with potassium amide in liquid ammonia at 240 K yields a mixture containing almost equal amounts of [1-^{14}C]-aniline and [2-^{14}C]-aniline. The formation of these two products in equal amounts can only be explained by assuming a symmetrical intermediate which can be attacked by ammonia at either of the two positions with equal probability (except for small isotope effect).

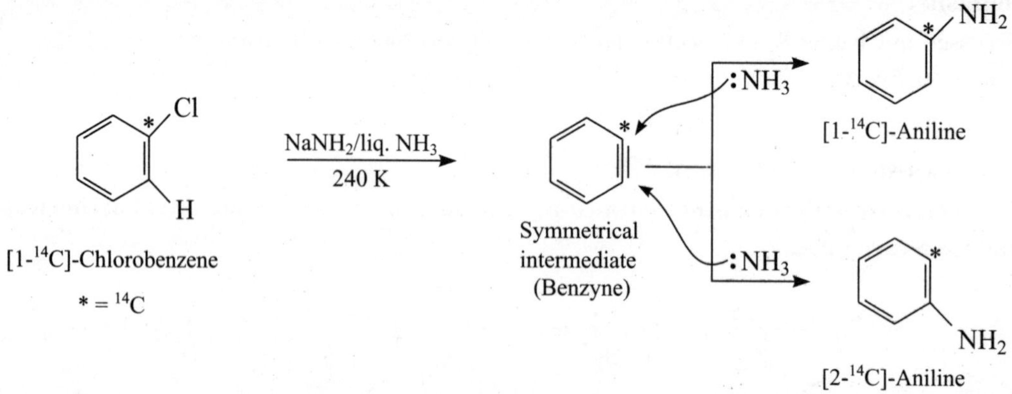

[1-^{14}C]-Chlorobenzene

* = ^{14}C

NaNH$_2$/liq. NH$_3$, 240 K

Symmetrical intermediate (Benzyne)

:NH$_3$

[1-^{14}C]-Aniline

[2-^{14}C]-Aniline

As we have seen above that at least in 50% molecules the incoming nucleophile goes to a position *ortho* to that of the leaving halide. Thus, this reaction may also be regarded as proceeding through a vicinal rearrangement. Hence, nucleophilic displacement proceeding through benzyne mechanism has also been referred to as ***cine-substitution.***

2. Condition of *ortho*-hydrogen atom: Further confirmation of the benzyne mechanism comes from the observation that *halobenzenes devoid of ortho-hydrogen fail to undergo the reaction* under the given conditions. For instance, bromodurene and bromomesitylene do not form the corresponding amines under these conditions.

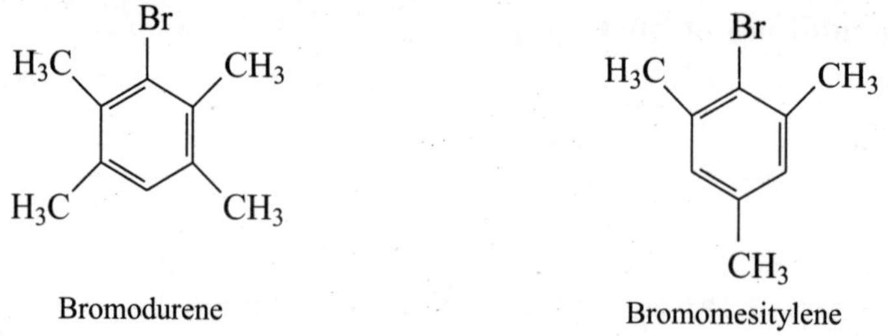

Bromodurene

Bromomesitylene

3. Isotope effect: *Ortho*-dideuteriobromobenzene reacts at a much slower rate than bromobenzene.

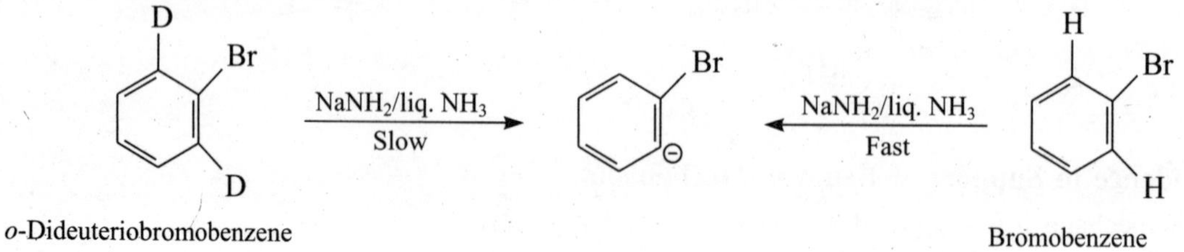

o-Dideuteriobromobenzene —NaNH$_2$/liq. NH$_3$, Slow→ (aryl anion) ←NaNH$_2$/liq. NH$_3$, Fast— Bromobenzene

The observed isotope effect clearly proves that in this reaction abstraction of *ortho*-hydrogen takes place in the rate-determining step.

4. Formation of *m*-anisidine from both *o*-and *m*-bromoanisoles: Both *ortho* and *meta*-bromoanisoles upon treatment with NaNH$_2$ in liquid ammonia at 240 K give the same product, *i.e*, *meta*-anisidine, which can be explained on the basis of the fact that same **benzyne intermediate** is formed in both the cases as shown below:

OCH_3 Br

o-Bromoanisole

OCH_3 Br

m-Bromoanisole

$\xrightarrow[240\ K]{NaNH_2/\text{liquid } NH_3}$ OCH_3 Same Benzyne Intermediate $\xrightarrow[\text{liq. } NH_3]{\overset{\ominus}{N}H_2}$ OCH_3 NH_2 *m*-Anisidine

5. Formation of *meta* and *para*-toluidines from *para*-chlorotoluene: *p*-Chlorotoluene on treatment with $NaNH_2$ in liquid ammonia at 240 K produces a mixture of *m*-and *p*-toluidines in which *m*-isomer predominates. This can be explained on the basis of the fact that the attack of the nucleophile ($\overset{\ominus}{N}H_2$) in the second step can occur on the benzyne intermediate at two different carbon atoms to give two different isomers as shown below:

CH_3 Cl *p*-Chlorotoluene $\xrightarrow[240\ K]{NaNH_2/\text{liquid } NH_3}$ CH_3 (benzyne) $\overset{\ominus}{N}H_2$ (a, b)

a (Attack at *m*-position) → CH_3, NH_2 *m*-Toluidine (68%)

b (Attack at *p*-position) → CH_3, NH_2 *p*-Toluidine (32%)

Attack at *p*-position is less favoured due to more deactivating effect of electron-donating methyl group on the ring at *ortho*-and *para*-positions as compared to that at the *meta*-position.

6. Trapping Experiment: Direct evidence for the intermediacy of benzyne in these reactions has been provided by *'trapping experiments'* which yield benzyne adducts in addition to the expected substitution product(s). The most commonly used trapping agent is anthracene with which it forms triptycene.

Benzyne + Anthracene → Triptycene

10.7 STRUCTURE OF BENZYNE

The structure of benzyne has not been fully settled. The following possibilities have been considered:

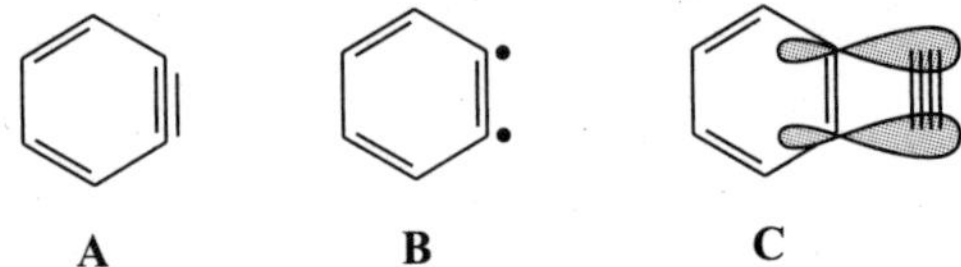

Structure A appears untenable as it has a formal triple bond and because of the hexagonal geometry associated with benzyne, it would be a highly strained system incapable of existence. Structure B representing a diradical species is also rejected since benzyne does not show any resemblance, either chemical or spectroscopic, with free radicals. The structure C, which looks more plausible depicts the new bond as being formed by a **"slight sideways overlap"** of the sp^2-hybrid orbitals. Such a bond is also unstable. Therefore, benzynes are extremely reactive species. However, benzynes are usually represented by triple bond structures due to simplicity in writing such a structure.

or

Finally, it may be pointed out that the term benzyne is somewhat misleading as it does not have an alkyne character and, therefore, the name 1, 2-dehydrobenzene is more appropriate.

Von Richter Reaction

In addition to the two types of nucleophilic displacements discussed above, there is yet another reaction in which the nucleophile enters the ring *ortho* to the position occupied by the leaving group. For instance, when ethanolic solution of a halogenonitrobenzene is heated (423 K) with potassium cyanide, nitro (and not bromine) is knocked off and the—CN group enters the ring *ortho* to the position occupied by the leaving nitro group. In presence of aqueous ethanol the corresponding acid is obtained.

NO_2 / Br $\xrightarrow{KCN,\ C_2H_5OH,\ \Delta}$ Br / CN

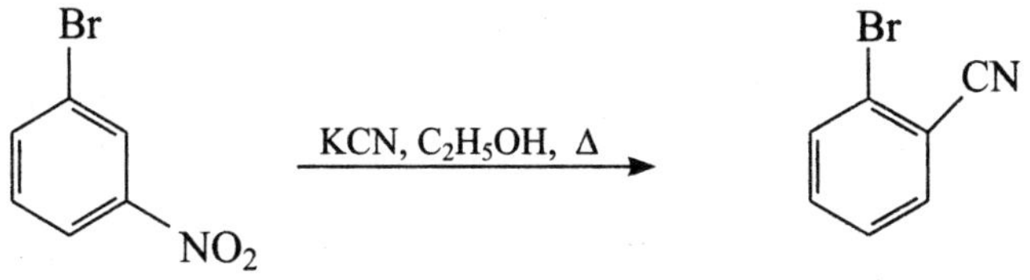

Br, NO$_2$ → (KCN, C_2H_5OH, Δ) → Br, CN

The following mechanistic scheme seems to be quite plausible :

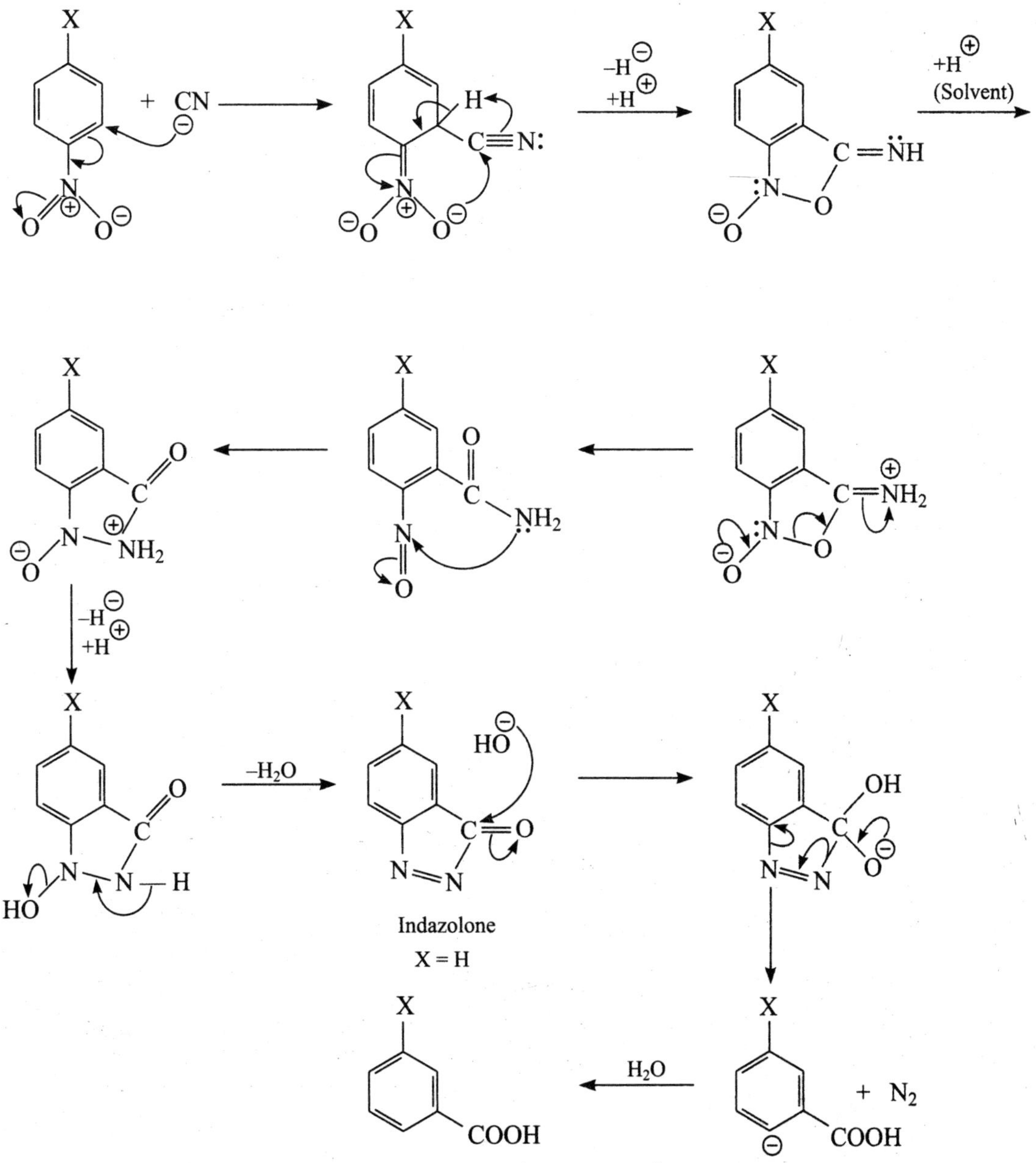

X, +CN$^{\ominus}$ → H, C≡N: → ($-H^{\ominus}$, $+H^{\oplus}$) → C=NH, :N–O → ($+H^{\oplus}$ (Solvent)) → C=NH$_2^{\oplus}$ → O, C, NH$_2$, N=O → C=O, N, NH$_2^{\oplus}$ → ($-H^{\ominus}$, $+H^{\oplus}$) → C=O, N–N–H, HO → ($-H_2O$) → HO$^{\ominus}$, C=O, N=N → OH, C, O$^{\ominus}$, N=N → COOH + N_2 → (H_2O) → COOH

Indazolone

X = H

10.8 REACTIVITIES IN NUCLEOPHILIC AROMATIC SUBSTITUTION

The rate of nucleophilic aromatic substitution depends upon the rate of formation of the carbanion. It means that greater the stability of the carbanion, greater is the ease of its formation and hence higher is the reactivity of the aryl halide. This will be more clear if we compare the reactivity of chlorobenzene, *p*-nitrochlorobenzene and *p*-chloroanisole which during the nucleophilic attack of $\overset{\ominus}{O}H$ results in the formation of the following carbanions I, II, and III, respectively.

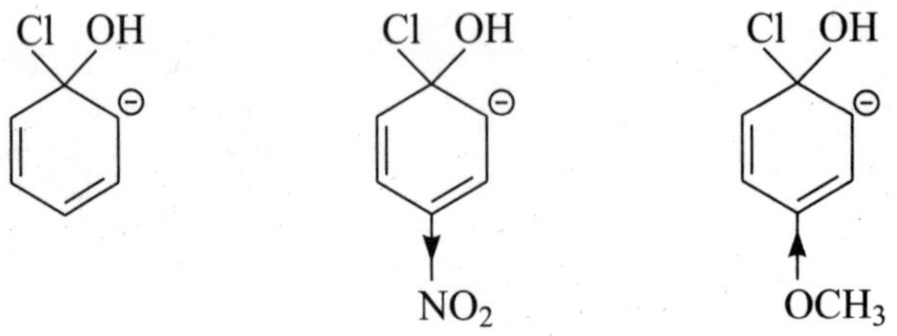

Cl OH ⊖ (I)   Cl OH ⊖ NO_2 (II)   Cl OH ⊖ OCH_3 (III)

The electron-withdrawing—NO_2 group in carbanion (II) disperses the negative charge both by—R and—I-effect, therefore, carbanion (II) is more stable than cabanion (I). On the other hand, electron-donating—OCH_3 group in carbanion (III) tends to intensify the negative charge relative to carbanion (I). As a result carbanion (III) is less stable than carbanion(I).

Therefore, the relative stabilities of the three carbanions follows the order: II > I > III.

In other words, *p*-nitrochlorobenzene is more reactive than chlorobenzene which, in turn, is more reactive than *p*-chloroanisole. In general *the presence of electron-withdrawing groups increases the reactivity while that of electron releasing group decreases the reactivity of aryl halides towards nucleophilic aromatic substitution reactions.*

Let us now explain why the presence of electron-withdrawing groups at only *o* and *p*-position with respect to the halogen atom activates the aryl halides towards nucleophilic attack by the $\overset{\ominus}{O}H$ ion on *o*, *p* and *m*-chloronitrobenzenes.

ortho-isomer: Cl, NO_2 $\xrightarrow[403\ K]{\overset{\ominus}{O}H}$ Cl OH, $\overset{\oplus}{N}$, :Ö, ⊖:Ö: ⟷ Cl OH, $\overset{\oplus}{N}$, ⊖:Ö:, ⊖:Ö:

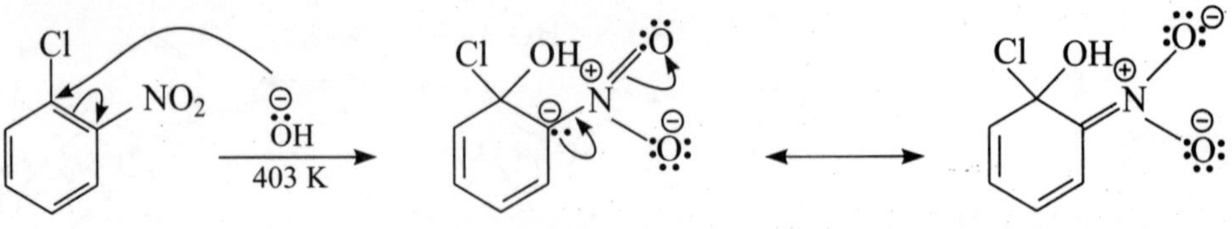

para-isomer: Cl, NO_2 $\xrightarrow[403\ K]{\overset{\ominus}{O}H}$ Cl OH, ⊖, NO_2 ⟷ Cl OH, ⊖, $\overset{\oplus}{N}$, ⊖:Ö, Ö: ⟷ Cl OH, $\overset{\oplus}{N}$, ⊖:Ö, ⊖Ö:

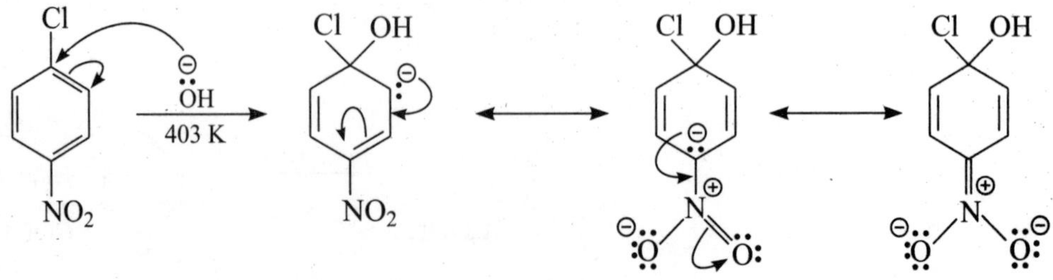

meta-isomer [Cl, NO_2 benzene] $\xrightarrow{\overset{\ominus}{O}H}$ ✗ [Cl, OH, NO_2 carbanion] No stabilization of charge by nitro group

The carbanions resulting from *o*-nitrochlorobenzene and *p*-nitrochlorobenzene are stabilized by nitro group but that resulting from *m*-nitrochlorobenzene is not stabilized by the nitro group. In other words, carbanions resulting from *o*-nitrochlorobenzene and *p*-nitrochlorobenzene are more stable than that resulting from *m*-nitrochlorobenzene. Thus, *presence of electron-withdrawing groups at ortho and para-positions (but not at meta) activates the aryl halides towards nucleophilic aromatic substitution reaction. Further, greater the number of such groups at o and p-positions more will be the reactivity of the corresponding aryl halide.*

For example,

Chlorobenzene $\xrightarrow[(ii)\ \text{dil. HCl}]{(i)\ 6\text{–}8\%\ \text{NaOH, 623 K, 300 atm}}$ Phenol

p-Nitrochlorobenzene $\xrightarrow[(ii)\ \text{dil. HCl}]{(i)\ 15\%\ \text{NaOH, 433 K}}$ *p*-Nitrophenol

2,4-Dinitrochlorobenzene $\xrightarrow[(ii)\ \text{dil. HCl}]{(i)\ \text{Aq. } Na_2CO_3\text{, 403 K}}$ 2,4-Dinitrophenol

Cl, O_2N, NO_2, NO_2 $\xrightarrow{H_2O,\ 333\ K}$ OH, NO_2, NO_2, NO_2

2, 4, 6 Trinitrochlorobenzene (*Picryl chloride*) → 2,4,6-Trinitrophenol (*Picric acid*)

10.9 RELATIVE REACTIVITIES OF ALKYL HALIDES *vs* ALLYL, VINYL AND ARYL HALIDES

The reactivity of alkyl halides towards nucleophilic substitution is more compared to vinyl and aryl halides. Further, like alkyl halides the halogen atom in allyl halides is also attached to sp^3-hybridised carbon. Hence allyl halides are expected to be at least as reactive as alkyl halides. It has actually been shown that *allyl halides are much more reactive than 1° alkyl halides*. The only difference in their behaviour is that 1° alkyl halides always react by S_N2 mechanism whereas allyl halides react both by S_N1 and S_N2 mechanisms.

10.9.1 Relative Reactivity of Alkyl Halides *vs* Allyl Halides

(*i*) Reactivity in S_N1 reactions: We know that an S_N1 reaction involves the formation of a carbocation as intermediate. Hence, the rate of such a reaction will depend upon the stability of the intermediate carbocation. It means that *stable is the carbocation, faster is the reaction.*

The bond dissociation energy of allyl chloride (724 kJ mol $^{-1}$) is comparable to that of isopropyl chloride (711 kJ mol $^{-1}$) and less than that of *n*-propyl chloride (780 kJ mol $^{-1}$). It means that allyl cation although a 1° carbocation is as stable as a 2° alkyl cation. Thus, *allyl halides, though primary in nature, are more reactive than 1° alkyl halides and almost as reactive as 2° alkyl halides. This is due to the fact that they readily ionize to give resonance stabilized allyl cation*. For example, solvolysis of allyl iodide in water proceeds much faster than solvolysis of *n*-propyl iodide.

$$\underset{\text{Allyl iodide}}{CH_2{=}CH{-}CH_2{-}I} \xrightarrow{\text{Ionization}} \left[CH_2{=}CH{-}\overset{\oplus}{C}H_2 \longleftrightarrow \overset{\oplus}{C}H_2{-}CH{=}CH_2\right] \equiv \overset{\oplus}{CH_2 \text{- - -} CH \text{- - -} CH_2}$$

Resonance stabilized carbocation

***(ii)* Reactivity in S_N2 reactions:** As the S_N2 reactions do not involve carbocation intermediates, the reactivity of alkyl halides mainly depends upon steric factors. The size of allyl group is approximately the same as that of an unsubstituted 1° alkyl group. Therefore, it can be said that both allyl halides and1° alkyl halides do not provide any steric hindrance to the attack of the nucleophile. In other words, the relative reactivities of unsubstituted 1° alkyl halides and allyl halides are almost comparable. An example of S_N2' (**S**ubstitution **N**ucleophilic, **B**imolecular with rearrangement) is provided by the reaction of 3-chloro-1-pentene with carbanion derived from diethyl malonate.

$$\underset{\text{3-Chloro-1-pentene}}{CH_3CH_2—\underset{\underset{Cl}{|}}{CH}—CH{=}CH_2} + :\overset{\ominus}{C}H(COOC_2H_5)_2 \xrightarrow{S_N2'} \underset{\text{Diethyl (2-pentenyl)-malonate}}{CH_3CH_2—CH{=}CH—CH_2—\underset{\underset{COOC_2H_5}{|}}{\overset{\overset{COOC_2H_5}{|}}{CH}}}$$

However, when there is no steric hindrance, allyl halides react much faster than simple alkyl halides not only in S_N1 reactions but also in S_N2. For example, allyl bromide is about 100 times more reactive than propyl bromide towards S_N2 reactions. This may be explained on the basis of the fact that in incipient allyl cation the double bond stabilizes the transition state, for S_N2 process, by conjugation with *p*-orbital at the carbon atom under attack.

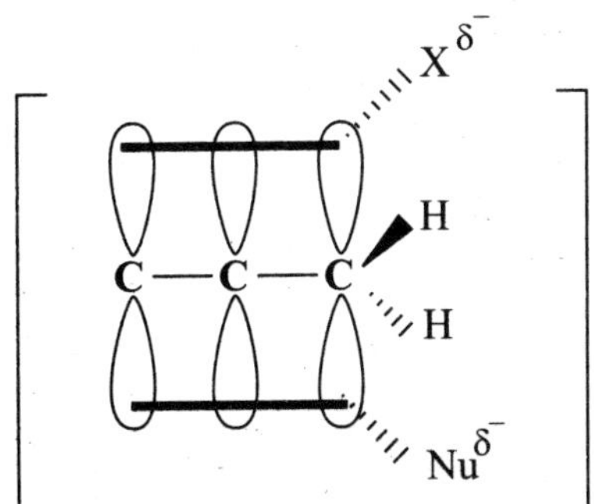

Lower energy T.S. for S_N2 displacement of an allyl halide

$X^{\delta-}$
H
$H_3C — CH_2 — C$
H
$Nu^{\delta-}$

Higher energy T.S. for S_N2 displacement of propyl halide

10.9.2 Relative Reactivity of Alkyl Halides *vs* Vinyl and Aryl Halides

Vinyl halides and aryl halides have been found to be much less reactive than alkyl halides towards nucleophilic substitution reactions. This can be explained as follows:

1. Resonance effect: In vinyl halides the lone pair of electrons on the halogen atom (leaving group) is involved in resonance with the olefinic double bond. The molecule can be represented as a resonance hybrid of the following contributing forms:

$$CH_2{=}CH—\ddot{\underset{..}{Cl}}: \longleftrightarrow :\overset{\ominus}{C}H_2—CH{=}\overset{\oplus}{\underset{..}{Cl}}:$$

It clearly shows that C—Cl bond in vinyl chloride acquires some double bond character and hence requires higher energy for the loss of chloride ion during a nucleophilic attack. Therefore, vinyl halides are less reactive than alkyl halides in which there will be no such resonance.

Similar to vinyl halides, aryl halides can also be represented as resonance hybrids of the following structures where the lone pair of electrons of halogen atom is involved in resonance with the ring double bond.

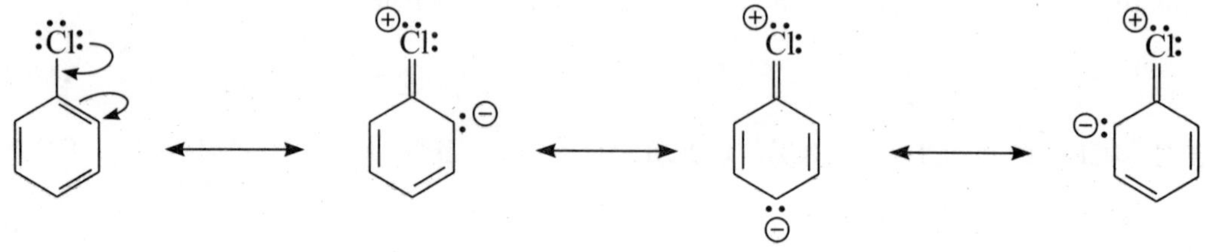

Due to above resonance the C—Cl bond tends to acquire some double bond character which makes it strong enough. Due to this the halogen atom shows little or no tendency to be displaced and hence aryl halides are less reactive than alkyl halides.

2. Hybridization: The type of hybridization of the carbon atom bearing the halogen atom, provides another explanation for less reactivity of vinyl and aryl halides as compared to alkyl halides. The carbon atom is sp^2-hybridised in case of vinyl and aryl halides while in case of alkyl halides it is sp^3- hybridised. Due to larger *s*-character, sp^2-hybridised orbitals have smaller size than sp^3-orbitals and hence C—Cl bond is shorter in vinyl and aryl halides. In other words, C—Cl bond is stronger in vinyl and aryl halides which makes them less reactive than alkyl halides.

$CH_2{=}CH{-}Cl$ (sp^2-hybridised) Cl (sp^2-hybridised) $CH_3{-}CH_2{-}Cl$ (sp^3-hybridised)

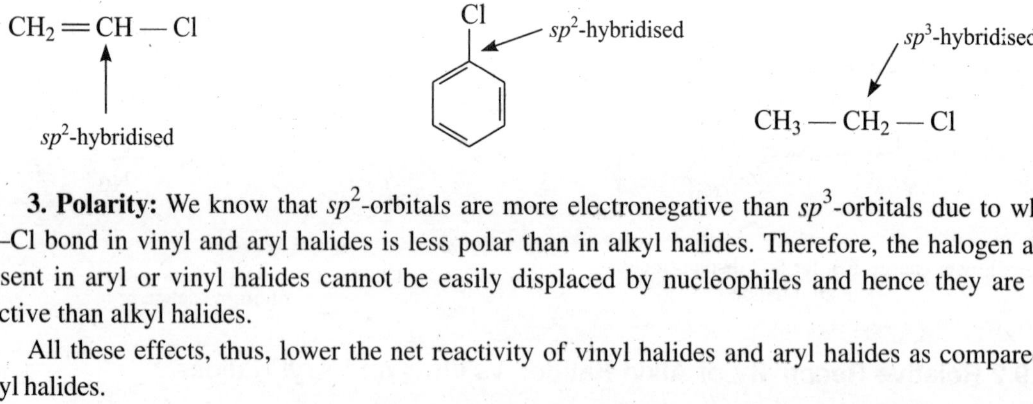

3. Polarity: We know that sp^2-orbitals are more electronegative than sp^3-orbitals due to which C—Cl bond in vinyl and aryl halides is less polar than in alkyl halides. Therefore, the halogen atom present in aryl or vinyl halides cannot be easily displaced by nucleophiles and hence they are less reactive than alkyl halides.

All these effects, thus, lower the net reactivity of vinyl halides and aryl halides as compared to alkyl halides.

10.9.3 Relative Reactivities of Nuclear and Side Chain Substituted Halogen Derivatives

It has been found that the side chain substituted halogen derivatives (aralkyl halides) are much more reactive than aryl halides. At least two factors are responsible for this difference.

1. Resonance effect: As discussed earlier, due to resonance the C—Cl bond in chlorobenzene acquires some double bond character due to which it is inert towards S_N1 reactions, *i.e.*, the halogen atom shows little or no tendency to be displaced by the nucleophile.

In aralkyl halides there is an sp^3-hybridised carbon between the aromatic ring and the leaving group, due to this no resonance can occur and hence the C—X bond retains its single bond character. Hence, it can be said that the C—X bond in aralkyl halides is weaker than that in aryl halides. So, the former are more reactive.

2. Hybridization: As discussed earlier, in aryl halides the leaving group is attached to sp^2-hybridised carbon atom whereas in aralkyl halides it is attached to an sp^3-hybridised carbon. Due to more *s*-character, the sp^2-carbon is more electronegative and smaller in size, *i.e.*, it has greater hold on bonding electrons than sp^3-carbon. Hence, the sp^2-carbon is reluctant to lose the leaving halogen atom. Thus we may conclude that halides such as benzyl chloride are much more reactive than aryl halides such as chlorobenzene.

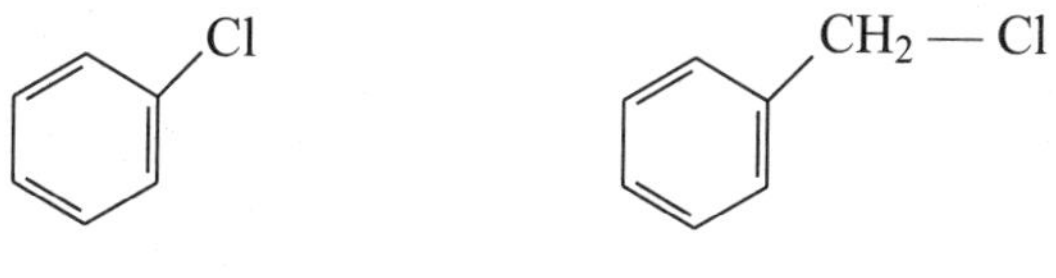

Cl | CH_2 — Cl

Chlorobenzene (less reactive) | Benzyl chloride (more reactive)

10.9.4 Relative Reactivity of Alkyl Halides *vs* Benzyl Halides

Similar to allyl halides, the benzyl halides are more reactive than alkyl halides especially in S_N1 reactions. This is due to resonance stabilization of the benzyl cation which is formed as the intermediate. This stability imparted to carbocations by resonance is mainly responsible for the easy ionization of benzyl halides. For example, solvolysis of benzyl chloride proceeds much faster than that of propyl chloride.

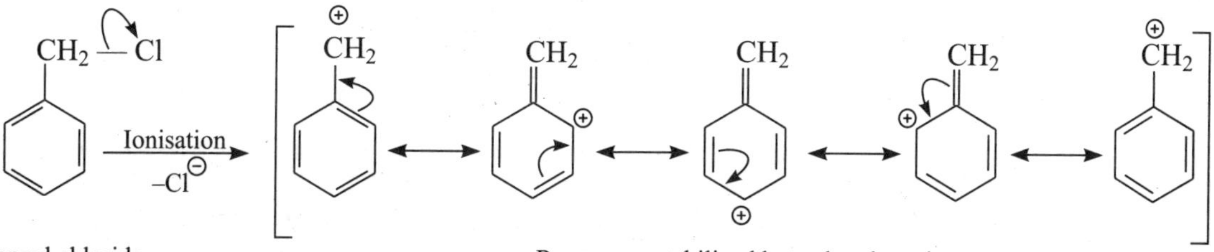

CH_2—Cl $\xrightarrow[-Cl^{\ominus}]{\text{Ionisation}}$ [$\overset{\oplus}{C}H_2$ ⟷ CH_2 ⟷ CH_2 ⟷ CH_2 ⟷ $\overset{\oplus}{C}H_2$]

Benzyl chloride | Resonance stabilized benzyl carbocation

10.10 SOME INDUSTRIAL HALOGEN COMPOUNDS

10.10.1 DDT [2,2-Bis (*p*-chlorophenyl)-1,1,1-trichloroethane]

The short hand notation DDT, for the compound has been derived from the commercial name of the compound, *p,p'-dichloro* diphenyl trichloroethane. In fact this name is a misnomer, the actual name is 2, 2-bis (4-chlorophenyl)-1,1-1-trichloroethane.

Preparation

Chlorobenzene on heating with trichloroacetaldehyde (chloral) in presence of catalytic amount of concentrated sulfuric acid produces DDT.

$$2\ C_6H_5Cl + Cl_3C-CHO \xrightarrow[\Delta]{H_2SO_4} (p\text{-}ClC_6H_4)_2CH-CCl_3$$

Chlorobenzene Chloral DDT

DDT is a solid having m.pt. 382 – 383 K.

Uses: It is specifically toxic for various insects including housefly and mosquito. It is widely used in agriculture for sugarcane and fodder crops to protect them from pests. It is very effective to kill *Anopheles mosquitoes* which spread malaria. DDT has been used as insecticide since 1946, both as preventive measure against malaria and control of agricultural pests. However, it was discovered that DDT is not readily metabolized and therefore cannot be destroyed by environmental destructive forces. Such compounds are said to be **non-biodegradable compounds**. *They get accumulated and remain for years together in soil, plants and animals, and consequently cause unforeseen ecological effects.*

Therefore, its use has been banned in many western countries. But, in India, it is still being used because of the non-availability of other economically cheap insecticides. However, the use of DDT has declined in recent years. This is due to its detrimental effects upon the environment and also due to public being more health conscious.

10.10.2 BHC [Benzene Hexachloride]

Its common name is *Gammexene* but IUPAC name is 1,2,3,4,5,6-hexachlorocyclohexane.

Preparation

Photochemical chlorination of benzene gives addition rather than substitution product. Thus, in presence of sunlight or UV light, chlorine adds to benzene to give benzene hexachloride.

$$C_6H_6 + 3Cl_2 \xrightarrow[\text{or UV light}]{\text{Sunlight}} C_6H_6Cl_6$$

BHC (γ-isomer)

This reaction involves free radical mechanism, which is initiated by chlorine forming chlorine atoms in presence of light.

$$Cl-Cl \xrightarrow{h\nu} \dot{Cl} + \dot{Cl}$$

The chlorine atom reacts with a benzene molecule to form the free radical (I) which reacts with a molecule of chlorine to give a dichloride (II) and also another chlorine atom which propagates the chain reaction to give finally the hexachloride (III).

$$C_6H_6 + \dot{Cl} \longrightarrow \text{I} \xrightarrow{Cl_2} \text{II} \xrightarrow{Cl_2} \longrightarrow \longrightarrow \text{III}$$

I II III

BHC

There are nine stereoisomers of BHC. Though the separation of these isomers is difficult, careful fractional crystallization using different organic solvents may result in their separation. Actually, four isomers α, β, γ and δ have been obtained in the pure state.

The γ-isomer, called **gammexene**, is thc most powerful insecticide of all the isomers. This isomer, in actual practice, is not separated but being sold as mixture along with other ones under the name BHC or gammexene or lindane or 666.

Further, the γ-isomer is odourless. The offensive smell of BHC is due to the presence of other isomers.

Uses: BHC is extensively used as a pesticide in agriculture for killing termites in the soil.

10.10.3 Freons

Freons are the chlorofluoro compounds of methane and ethane.

These are classified further on the basis of number of fluorine atoms present in them. For example, CF_4 (Freon-14), CF_3Cl (Freon-13), CF_2Cl_2 (Freon-12), $CFCl_3$ (Freon-11).

Out of these freons, CF_2Cl_2 (Freon-12) is most widely used refrigerant.

It is manufactured as follows from CCl_4 by **Swarts reaction**.

$$3CCl_4 + SbF_3 \xrightarrow{SbCl_5} 3CCl_2F_2 + 2SbCl_3$$

It can also be obtained by passing HF through CCl_4 in presence of $SbCl_3$ as catalyst.

$$CCl_4 + 2HF \xrightarrow{SbCl_3} CCl_2F_2 + 2HCl$$

Freon is an odourless, non-corrosive, non-toxic gas. It is quite stable even at high temperatures and pressures. It has low b.pt., low specific heat and can be easily liquified under pressure. Due to this property it is widely used as a refrigerant in refrigerators and air conditioners. It is also used as a propellant in aerosols and foams (*i.e.*, hair sprays, deodorants, shaving forms, cleansers, insecticides, etc.).

Recently it has been shown that freon is responsible for depletion of ozone layer in the upper atmosphere **(ozone hole formation)**. Ozone protect us from UV radiations coming from the sun. Therefore, its use as propellants in aerosols has been banned in USA.

A BRIEF REVIEW

I. Methods of formation of alkyl halides

(*i*) Alcohol + Halogen acid $\longrightarrow$ Alkyl halide

The order of reactivity of alcohols is : 3° > 2° > 1°.

The order of reactivity of halogen acids is: HI > HBr > HCl.

(*ii*) Alcohol + PCl_5 or $PCl_3 \longrightarrow$ Alkyl halide

(*iii*) Alcohol + $SOCl_2 \longrightarrow$ Alkyl halide (pure)

(*iv*) Alkene + HX $\longrightarrow$ Alkyl halide

The order of reactivity of HX is: HI > HBr > HCl > HF. In case of HBr in presence of peroxide, there will be anti-Markonikoff's addition of HBr to alkene.

(*v*) Alkane + $Cl_2 \xrightarrow{h\nu \text{ or } 520\text{ K}}$ Mixture of mono, di, tri and tetra-haloalkanes.

Ease of abstraction of hydrogens: 3° > 2° > 1° but their relative rates of abstraction respectively are:

(*a*) 5 : 3.8 : 1 by Cl_2 at 298 K and

(*b*) 1600 : 82 : 1 by Br_2 at 400 K

(*vi*) **Finkelstein reaction** (formation of alkyl iodides)

$$\text{R—Cl or R—Br} + \text{NaI} \xrightarrow{\text{Acetone or Methanol}} \text{R—I} + \text{NaCl or NaBr}$$

(*vii*) **Swarts reaction** (Formation of alkyl fluorides)

$$\text{R—Cl} \xrightarrow{AsF_3 \text{ or } SbF_3 \text{ or } AgF \text{ or } Hg_2F_2} \text{R—F}$$

(*viii*) **Hundsdiecker reaction** (formation of alkyl bromides)

$$\text{R—COOAg or Ar—COOAg} + Br_2 \xrightarrow{CCl_4 \text{ Reflux}} \text{R—Br or Ar—Br} + CO_2 + AgBr$$

II. (A) Physical Properties of Alkyl Halides

(*i*) Boiling points decrease in the order: RI > RBr > RCl > RF. For isomeric alkyl halides, the boiling points decrease with branching.

(*ii*) Melting points of *p*-isomers are highest due to their symmetry.

(*iii*) Density decreases in the order: RI > RBr > RCl > RF.

(*iv*) Stability decreases in the order: RF > RCl > RBr > RI.

(B) Chemical Properties of Alkyl Halides

- Alkyl halides undergo S_N1 and S_N2 types of nucleophilic substitution reactions.
- S_N1 reactions take place in two steps. First step involves formation of carbocation intermediates. These reactions are favoured by low concentration of nucleophiles and by high polarity of solvents. Optically active alkyl halides produce racemic mixture during S_N1 reactions. The order of reactivity of alkyl halides is 3° > 2° > 1°, which is the order of stability of carbocation intermediate formed in the first step. The rate of these reactions is first order with respect to concentration of alkyl halide, *i.e.*, Rate = k[RX].
- S_N2 reactions take place in one step. These reactions are favoured by high concentration of nucleophile and by low polarity of solvents. Optically active alkyl halides produce optically active substitution products during S_N2 reactions. The order of reactivity of alkyl halides is 1° > 2° > 3°. The rate of these reactions is second order which depends on the concentration of both the alkyl halide and the nucleophile, *i.e.*, Rate = $k\,[RX]\,[\overset{\ominus}{Nu}]$.
- For the same alkyl group, the order of reactivity of different alkyl halides is RI > RBr > RCl > RF.
- The order of nucleophilicity of halide ions is $\overset{\ominus}{I} > \overset{\ominus}{Br} > \overset{\ominus}{Cl} > \overset{\ominus}{F}$.
- **Formation of alcohols** (Hydrolysis of alkyl halides)

$$\text{Alkyl halide} \xrightarrow[\text{Boiling water}]{\text{KOH or } Ag_2O} \text{Alcohol}$$

- **Formation of alkene** (Dehydrohalogenation of alkyl halides)

$$\text{Alkyl halide} \xrightarrow{\text{Alc. KOH, } \Delta} \text{Alkene}$$

- **Williamson's ether synthesis**

$$R{-}X \xrightarrow{\text{NaOR}' \text{ or KOR}'} R{-}O{-}R'$$

Symmetrical Ethers can also be obtained on heating alkyl halide with dry Ag_2O.

$$R{-}X \xrightarrow{Ag_2O,\ \Delta} R{-}O{-}R$$

- Haloalkanes on reaction with alcoholic solution of KCN give cyanides as the major products while with AgCN they give isocyanides as the major product.
- Haloalkanes on reaction with KNO_2 give alkyl nitrites while with $AgNO_2$ they give nitroalkanes.
- **Formation of esters**

$$R{-}X + R'COOAg \xrightarrow{D} R'COOR$$

- **Hoffmann ammonolysis of alkyl halides**

$$\underset{}{R{-}X} \xrightarrow{NH_3} \underset{1^\circ\text{ amine}}{RNH_2} \xrightarrow{R{-}X} \underset{2^\circ\text{ amine}}{R_2NH} \xrightarrow{R{-}X} \underset{3^\circ\text{ amine}}{R_3N} \xrightarrow{R{-}X} \underset{\text{Quaternary ammonium salt}}{R_4\overset{\oplus\ominus}{N}X}$$

- **Formation of Grignard reagents**

$$R—X + Mg \xrightarrow{\text{Dry ether}} R—Mg—X$$

- **Formation of higher alkynes**

$$R—X + R'C{\equiv}\overset{\ominus\oplus}{C\,Na} \longrightarrow R'C{\equiv}CR + \overset{\oplus\ \ominus}{Na\,X}$$

- **Würtz reaction**

$$R—X \xrightarrow{\text{Na in dry ether}} R—R$$

- Reduction of alkyl halides with HI and red P at 420 K gives alkanes.
- Alkyl halides undergo isomerization on heating in presence of anhydrous $AlCl_3$.

Chloroform

- Ethyl alcohol or acetone on distillation with a suspension of bleaching powder or on treatment with cold alkaline NaOCl (NaOH + Cl_2) gives chloroform.
- Chloroform undergoes **auto oxidation** in presence of air and light to give highly poisonous phosgene ($COCl_2$) gas. To prevent this oxidation chloroform is stored in amber coloured bottles which are filled to brims to exclude any air and 1% alcohol is added which converts phosgene to non-toxic diethyl carbonate.
- **Carbylamine test for 1° amines:** Chloroform on heating with an aliphatic or aromatic primary amine in presence of alcoholic KOH gives corresponding isocyanide or carbylamine having unbearable (unpleasant) smell.
- Chloroform on heating with silver powder gives acetylene.
- The addition reaction of chloroform with acetone in presence of KOH gives **chloretone** which is used as an hypnotic.
- Chloroform on nitration with conc.HNO_3 gives **chloropicrin** which is used as an insecticide as well as in tear gas.
- **Reimer-Tiemann reaction:** Chloroform gives salicylaldehyde on reaction with phenol and alcoholic KOH at 340 K.
- Chloroform is used as an **anaesthetic agent**, as a solvent and as a preservative for anatomical specimens.

Carbon Tetrachloride

- Chlorination of methane in presence of $CuCl_2$ or chlorination of carbon disulphide in presence of $FeCl_3$ or $AlCl_3$ gives CCl_4.
- Freon is obtained by flourination of CCl_4 with SbF_3 in presence of $SbCl_5$.

$$CCl_4 + SbF_3 \xrightarrow{SbCl_5} CF_2Cl_2 \text{ (Freon)}$$

Freon is used mainly as a refrigerant.

- CCl_4 is used as a fire extinguisher under the name **pyrene**.

III. Methods of formation of aryl halides

(i) **Nuclear halogenation** takes place at low temperatures in presence of Lewis acids or halogen carriers such as $AlCl_3$, $FeCl_3$, etc.

$$\text{Benzene} + Cl_2 \xrightarrow[-HCl]{\text{Anhyd. } AlCl_3} \text{Chlorobenzene}$$

$$\text{Benzene} + Br_2 \xrightarrow[-HCl]{\text{Anhyd. } AlBr_3} \text{Bromobenzene}$$

Iodobenzene is formed in presence of an oxidising agent such as HNO_3, HIO_3, etc.

$$\text{Benzene} + I_2 \xrightarrow{HNO_3} \text{Iodobenzene}$$

(ii) **Side chain halogenation** takes place at high temperature in presence of light and in absence of halogen carriers.

$$\underset{\text{Toluene}}{C_6H_5CH_3} \xrightarrow[h\nu]{Cl_2,\ 383\ K} \underset{\text{Benzyl chloride}}{C_6H_5CH_2Cl} \xrightarrow[h\nu]{Cl_2,\ 383\ K} \underset{\text{Benzal chloride}}{C_6H_5CHCl_2} \xrightarrow[h\nu]{Cl_2,\ 383\ K} \underset{\text{Benzotrichloride}}{C_6H_5CCl_3}$$

Benzylic hydrogen is preferentially substituted during side chain halogenation as shown below:

$$C_6H_5CH_2-CH_3 \xrightarrow[h\nu]{Cl_2,\ 383\ K} \underset{\text{1-Chloro-1-phenylethane}}{C_6H_5CH(Cl)-CH_3}$$

(iii) ***Sandmeyer reaction***

$$C_6H_5Br \xleftarrow[-N_2]{CuBr/HBr} C_6H_5N_2Cl \xrightarrow[-N_2]{CuCl/HCl} C_6H_5Cl$$

(iv) **Gattermann reaction**

$$C_6H_5Br \xleftarrow[-N_2]{Cu/HBr} C_6H_5N_2Cl \xrightarrow[-N_2]{Cu/HCl} C_6H_5Cl$$

(v) **Balz-Schiemann reaction** *(formation of fluoroarenes)*

$$C_6H_5N_2Cl \xrightarrow{HBF_4} C_6H_5N_2BF_4 \xrightarrow{\Delta} C_6H_5-F + N_2 + BF_3$$

(*vi*) **Formation of iodoarenes**

$$C_6H_5N_2Cl \xrightarrow{KI, \Delta} C_6H_5—I + N_2$$

IV. Properties of Aryl Halides

- **Nucleophilic substitution reactions**

  Aryl halides undergo nucleophilic substitution reactions under drastic conditions. The lower reactivity of aryl halides, as compared to that of alkyl halides, is due to **resonance**, ***sp*^{2}*-hybridization*** and **less polarity of C—X bond.**

- **Dow's process** (***formation of phenol***)

$$C_6H_5Cl \xrightarrow[623\ K]{10\%\ NaOH,\ 300\ atm.} C_6H_5ONa \xrightarrow{HCl} C_6H_5OH$$

- **Formation of aniline**

$$2\,C_6H_5Cl + 2NH_3 + Cu_2O \xrightarrow[60\ atm.]{475\ K} 2\,C_6H_5NH_2 + 2CuCl + H_2O$$

- Nucleophilic substitution of halogen is facilitated by the presence of electron-withdrawing groups such as $—NO_2$, $—SO_3H$, —CN, —COOH, —COR, etc. at *ortho* and *para*-positions (but not at *meta*) to halogen atom. Further, higher the number of such groups more is the reactivity of haloarene.

- **Benzyne intermediacy:** Chlorobenzene when treated with $NaNH_2$ in liquid ammonia at 240 K forms aniline through the formation of benzyne intermediate.

- **Fittig reaction**

$$2\,C_6H_5—Cl \xrightarrow{Na} C_6H_5—C_6H_5 \text{ (Biphenyl)}$$

- **Würtz-Fittig reaction**

$$C_6H_5Cl + R—X \xrightarrow{Na} C_6H_5—R$$

- **Ullmann Biaryl Synthesis**

$$C_6H_5—I \xrightarrow[\text{Sealed tube}]{Cu, \Delta} C_6H_5—C_6H_5$$

- **Grignard reagent formation**

$$C_6H_5X + Mg \xrightarrow{\text{Dry ether}} C_6H_5Mg—X$$

- **Electrophilic substitution reactions:** Aryl halides undergo usual electrophilic substitution reactions such as halogenation, nitration, sulphonation and Friedel-Crafts alkylation and acylation but with less reactivity as compared to that of benzene. This is due to the fact that halogens are somewhat deactivating with *o*-, *p*-directing nature. In all the electrophilic substitution reactions *o*-and *p*-products are formed where *p*-product always predominates.

SOLVED PROBLEMS

Q. 1. Give a suitable IUPAC name for each of the following compounds:

(*i*) 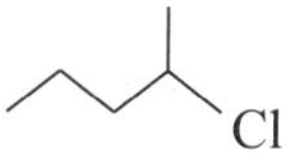 Cl (*ii*) 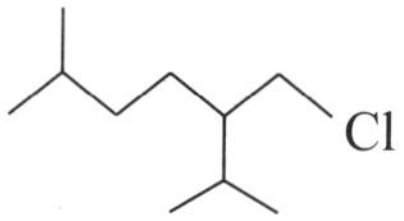 Cl (*iii*) 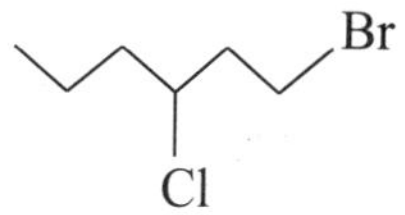 Br Cl

(*iv*) $(CH_3)_2C{=}CHCl$ (*v*) H_3C–$C_6H_3(Br)$–CH_3

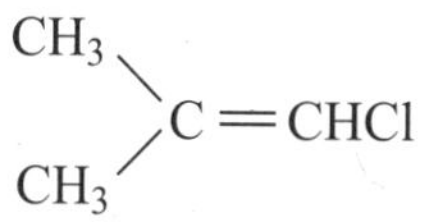

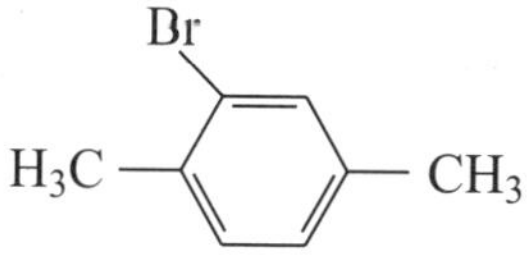

Ans.

(*i*) 2-Chloropentane (*ii*) 1-Chloro-2-isopropyl-5-methylhexane

(*iii*) 1-Bromo-3-chlorohexane (*iv*) 1-Chloro-2-methylpropene

(*v*) 2-Bromo-1, 4-dimethylbenzene

Q. 2. Classify each of the following alky halides as 1°, 2°, 3°.

(*i*) $CH_3-CH(CH_3)CH_2Cl$ (*ii*) cyclohexane with CH_2CH_3 and Cl (*iii*) $CH_3CHCl(CH_2CH_2CH_3)$

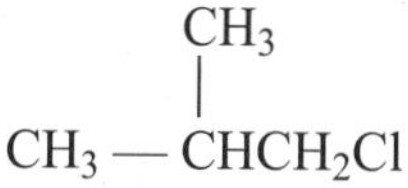

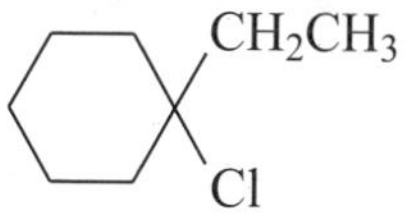

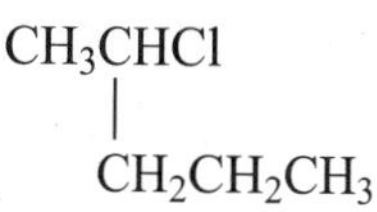

Ans. (*i*) 1° (*ii*) 3° (*iii*) 2°

Q. 3. Würtz reaction fails in case of *tert*-alkyl halides. Explain.

Ans. *Tert*-alkyl halides prefer to undergo dehydrohalogenation in presence of a strong base such as Na metal instead of undergoing Würtz reaction as shown below:

$$\underset{tert\text{-Butyl bromide}}{(CH_3)_3C{-}Br} + 2Na \longrightarrow \underset{tert\text{-Butylsodium}}{(CH_3)_3C^- Na^+} + NaBr$$

Chapter 10

$$(CH_3)_3\overset{-}{C}\overset{+}{Na} + H-CH_2-\underset{CH_3}{\overset{CH_3}{\underset{|}{\overset{|}{C}}}}-Br \xrightarrow[-NaBr]{} \underset{\text{Isobutane}}{(CH_3)_3CH} + \underset{\text{Isobutylene}}{CH_2=\overset{CH_3}{\overset{|}{C}}-CH_3}$$

Thus, only 1° and 2° alkyl halides undergo Würtz reaction while 3° alkyl halides prefer to undergo dehydrohalogenation to form alkenes.

Q. 4. Explain why the treatment of alkyl halides with aqueous KOH leads to the formation of alcohols but alkenes are the products in presence of alcoholic KOH?

Ans. In aqueous solution, KOH is almost completely ionized to give OH^- ions which being a strong nucleophile brings about a substitution reaction on alkyl halides to form alcohols. Further in the aqueous solution, OH^- ions are highly solvated (hydrated). This solvation reduces the basic character of OH^- ions which, therefore, fail to abstract a hydrogen from the β-carbon of the alkyl halide to form an alkene.

In contrast, an alcoholic solution of KOH contains alkoxide (RO^-) ions which being much stronger base than OH^- ions, preferentially eliminates a molecule of HX from an alkyl halide to from alkenes.

Q. 5. Predict the order of reactivity of the following compounds in S_N1 and S_N2 reactions.

(*a*) The four isomeric bromobutanes.

(*b*) $C_6H_5CH_2Br$, $(C_6H_5)_2CH—Br$, $C_6H_5CH(CH_3)Br$, $(C_6H_5)_2C(CH_3)Br$

Ans. (*a*) In S_N1 reactions, the order of reactivity depends upon the stability of the intermediate carbocations, therefore, $(CH_3)_3CBr$ which gives a 3° carbocation, *i.e.*, $(CH_3)_3C^+$ is the most reactive. $CH_3CH_2CH(Br)CH_3$ which gives a 2° carbocation, *i.e.*, $CH_3CH_2C^+HCH_3$ is less reactive than $(CH_3)_3CBr$. Out of the remaining two 1° alkyl halides, the carbocation $(CH_3)_2CHCH_2^+$ is more stable than the carbocation $CH_3CH_2CH_2CH_2^+$ due to greater + I effect of $(CH_3)_2CHCH_2$ group as compared to $CH_3CH_2CH_2CH_2$ group and hence the alkyl halide,$(CH_3)_2CHCH_2Br$ is more reactive than $CH_3CH_2CH_2CH_2Br$. Thus, the overall increasing reactivity of the four isomeric bromobutanes towards S_N1 reaction follows the order:

$$CH_3CH_2CH_2CH_2Br < (CH_3)_2CHCH_2Br < CH_3CH_2CH(Br)CH_3 < (CH_3)_3CBr$$

The reactivity in S_N2 reactions, however, follows the reverse order since, the steric hindrance around the electrophilic carbon (*i.e.*, α-carbon) increases in that order, *i.e.*,

$CH_3CH_2CH_2CH_2Br > (CH_3)_2CHCH_2Br > CH_3CH_2CH(Br)CH_3 > (CH_3)_3CBr$.

(*b*) Since, the reactivity in S_N1 reactions increases as the stability of the intermediate carbocation increase, therefore, $(C_6H_5)_2C(CH_3)Br$ which gives 3° carbocation, *i.e.*, $(C_6H_5)_2C^+ (CH_3)$ is the most reactive. Of the two 2° bromides, the carbocation intermediate derived from $(C_6H_5)_2CHBr$ *i.e.*, $(C_6H_5)_2CH^+$ is more stable as compared to carbocation $(C_6H_5C^+HCH_3)$ obtained from $C_6H_5CH(Br)CH_3$ because it is stabilized by two phenyl groups due to resonance. The fourth alkyl bromide, *i.e.*, $C_6H_5CH_2Br$ which gives the 1° carbocation, *i.e.*, $C_6H_5CH_2^+$ is, however, the least reactive. Thus, the overall reactivity of these alkyl bromides towards S_N1 reaction follows the order:

$(C_6H_5)_2C(CH_3)Br > (C_6H_5)_2CHBr > C_6H_5CH(CH_3)Br > C_6H_5CH_2Br$.

In S_N2 reactions, it is the steric hindrance, which determines the reactivity. Since a phenyl group is much bulkier than a methyl group, therefore, $(C_6H_5)_2C(CH_3)Br$ is the least reactive followed by $(C_6H_5)_2CHBr$ and $C_6H_5CH(CH_3)Br$ is the most reactive. Thus, the overall reactivity of these bromides towards S_N2 reactions follows the order:

$C_6H_5CH_2Br > C_6H_5CH(CH_3)Br > (C_6H_5)_2CHBr > (C_6H_5)_2C(CH_3)Br$.

Q. 6. Explain why CHF_3 is less acidic than $CHCl_3$?

Ans. Due to stronger –I effect of F than Cl, CHF_3 should be more acidic than $CHCl_3$. But actually reverse is true. This is due to the reason that $:CCl_3^-$ left after the removal of a proton from $CHCl_3$ is stabilized by resonance due to the presence of *d*-orbitals but $:CF_3^-$ left after the removal of a proton from CHF_3 is not stabilized by resonance due to the absence of *d*-orbitals in F.

Q. 7. Explain why does nucleophilic substitution reaction of chlorobenzene take place through benzyne mechanism and that of p-nitrochlorobenzene proceeds via addition-elimination mechanism? Explain.

Ans. In *p*-nitrochlorobenzene, the carbanion **(I)** resulting due to initial addition of the nucleophile is stabilized by electron withdrawing—NO_2 group as well as by the benzene ring (extended conjugation) as shown below:

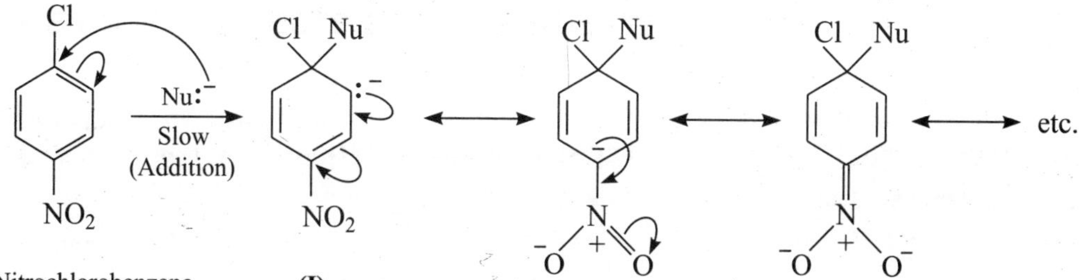

Cl, NO_2 $\xrightarrow[\text{(Addition)}]{\text{Nu:}^- \text{ Slow}}$ Cl Nu, NO_2 $\longleftrightarrow$ Cl Nu, $N^+(O^-)O$ $\longleftrightarrow$ Cl Nu, $=N^+(O^-)O^-$ $\longleftrightarrow$ etc.

p-Nitrochlorobenzene **(I)**

In the second step, the carbanion **(I)** loses a chloride ion to complete the substitution.

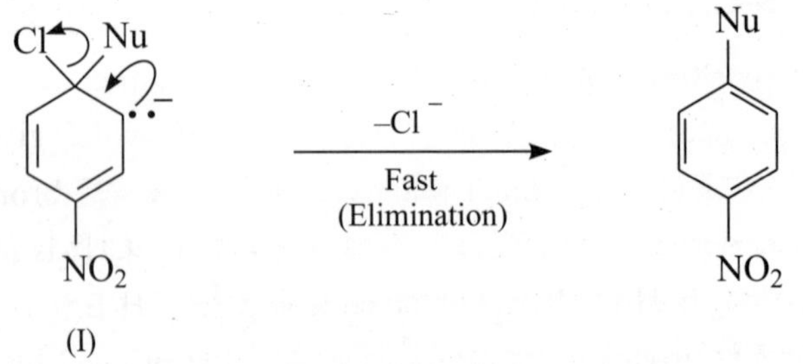

Cl Nu

$-Cl^-$

Fast
(Elimination)

Nu

NO_2

(I)

NO_2

This two-step bimolecular aromatic nucleophilic substitution mechanism is also called **addition-elimination mechanism**.

In contrast, the carbanion **(II)** resulting due to nucleophilic attack on chlorobenzene is stabilized only by the benzene ring. Moreover, due to electrostatic repulsion, the nucleophile cannot approach the benzene ring easily.

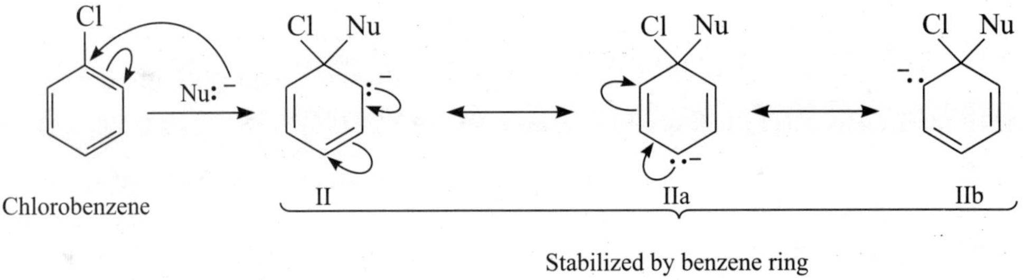

Cl

Nu:$^-$

Cl Nu

Cl Nu

Cl Nu

Chlorobenzene

II

IIa

IIb

Stabilized by benzene ring

Since stabilization due to NO_2 group is much greater than due to benzene ring, therefore, carbanion **(I)** is easily formed and hence *p*-nitrochlorobenzene undergoes substitution *via* addition-elimination mechanism. In contrast, the carbanion **(II)** is only slightly stabilized and hence is not easily formed. Therefore, chlorobenzene undergoes nucleophilic substitution by an alternate mechanism involving benzyne intermediate or elimination-addition mechanism as shown below:

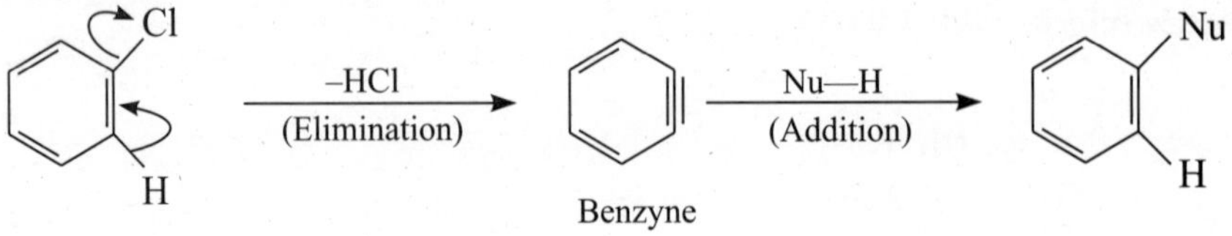

Cl

H

–HCl
(Elimination)

Benzyne

Nu—H
(Addition)

Nu

H

Q. 8. Which compound would undergo aromatic nitration more readily $C_6H_5CH_3$ or $C_6H_5CCl_3$?

Ans. While the CH_3 group is electron releasing and activates the ring, the CCl_3 group is strongly electron withdrawing because of the influence of the electronegative chlorines. Therefore, $C_6H_5CH_3$ has an activated ring and undergoes nitration more readily.

Q. 9. Show how would you synthesize 1, 3, 5-tribromobenzene from aniline (remember that bromine is an *o, p*-director in electrophilic substitution reactions).

Ans. Aniline undergoes tribromination readily. Therefore, the following reaction sequence will yield the desired product:

$$C_6H_5NH_2 \xrightarrow[-3HBr]{3Br_2} \text{2,4,6-}Br_3C_6H_2NH_2 \xrightarrow[0°C]{NaNO_2,\ HCl}$$

$$\text{2,4,6-}Br_3C_6H_2\overset{\oplus}{N_2}\overset{\ominus}{Cl} \xrightarrow{H_3PO_2} \text{1,3,5-}Br_3C_6H_3$$

Q. 10. Give the structures of the substitution products (if any) of the following reactions:

(*a*) $CH_3CH_2CH_2CH_2I + \overset{\ominus}{C}N \longrightarrow$

(*b*) $C_6H_5Cl + \overset{\ominus}{O}H \longrightarrow$

(*c*) $Cl-C_6H_4-CH_2Cl + \overset{\ominus}{O}H \longrightarrow$

Ans. (*a*) $CH_3CH_2CH_2CH_2CN$ (*b*) No reaction (*c*) $Cl-C_6H_4-CH_2OH$.

Q. 11. Write the product formed when each of the following compounds react with N-bromosuccinimide in CCl_4 in presence of light.

(*a*) [bicyclic structure: octahydronaphthalene with central double bond]

(*b*) $C_6H_5-CHCH_3$ with $CH(CH_3)_2$ on the CH

(*c*) [cyclohexene with CH_3, CH_3 on one carbon]

(*d*) $C_6H_5-CH_2CH_3$

Ans.

(*a*) (*b*) CH_3 C Br $CH(CH_3)_2$ Br

(*c*) Br CH_3 CH_3 (*d*) Br CH CH_3

Q. 12. Three isomers (A), (B) and (C), having molecular formula C_8H_9Cl yield the following compounds on oxidation:

(A) gives benzoic acid; (B) gives phthalic acid and (C) gives *p*-chlorobenzoic acid.

(A) is optically active. Both (B) and (C) but not (A) give a white precipitate when warmed with an alcoholic silver nitrate solution. Deduce the structure of (A), (B) and (C).

Ans.

(A) is 1-Chloro-1-phenylethane

(B) is *o*-Methylbenzyl chloride

(C) is *p*-Chloroethylbenzene.

Q. 13. How will you convert toluene into *m*-toluidine?

Ans.

CH_3 $\xrightarrow{Br_2/AlCl_3}$ CH_3, Br (Major isomer) $\xrightarrow[\text{liq. } NH_3]{NaNH_2 \text{ in}}$ CH_3, NH_2 *m*-Toluidine (68%) + CH_3, NH_2 *p*-Toluidine (32%)

Q. 14. How will you prepare HC $(OC_2H_5)_3$ from chloroform?

Ans. When $CHCl_3$ is heated with sodium ethoxide, it forms ethyl orthoformate as shown below:

$$C_2H_5O^- + CHCl_3 \rightleftharpoons :CCl_3^- + C_2H_5OH$$

$$:CCl_3^- \xrightarrow{\text{Slow}} :CCl_2 + Cl^-$$

Dichlorocarbene

$$:CCl_2 + 2C_2H_5O^- \longrightarrow :C(OC_2H_5)_2 + 2Cl^-$$

$$:C(OC_2H_5)_2 + C_2H_5OH \longrightarrow \underset{\text{Trimethyl formate}}{HC(OC_2H_5)_3}$$

Q. 15. Arrange the following halides in the decreasing order of S_N2 reactivity: $CH_3CH_2CH_2Cl$(I), $CH_2 = CHCHClCH_3$ (II) and $CH_3CH_2CHClCH_3$ (III).

Ans. II > III > I.

Q. 16. Out of the various possible isomers of M.F. C_7H_7Cl containing a benzene ring, suggest the structure with the weakest C—Cl bond.

Ans. Benzyl chloride.

Q. 17. What effect should the following resonance of vinyl chloride have on its dipole moment?

$$CH_2 = CH—\ddot{\underset{..}{Cl}}: \longrightarrow \bar{:}\,CH_2—CH=\overset{+}{\underset{..}{Cl}}:$$

Ans. Decrease the dipole moment.

Q. 18. Which alkyl halide has the highest density and why?

Ans. CH_32I; because of its smallest carbon content and heaviest halogen.

Q. 19. A hydrocarbon (C_5H_{10}) does not react with chlorine in dark but gives a single monochloro compound C_5H_9Cl in bright sunlight. Identity the hydrocarbon.

Ans. Cyclopentane.

UNSOLVED PROBLEMS

1. (*i*) Arrange the following compounds in order of reactivity towards sodium iodide in acetone (most reactive first):

A: $CH_3—C(CH_3)_2—CH_2Cl$

B: $CH_3CH_2CH_2CH_2CH_2Cl$

C: $CH_3CH_2CH(CH_3)—CH_2Cl$

(*ii*) Arrange the following compounds in order of reactivity towards aqueous formic acid (most reactive first):

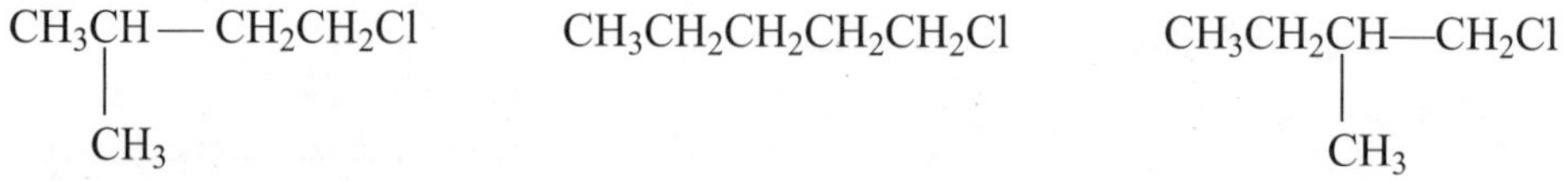

$$\underset{\text{A}}{CH_3\underset{|\;CH_3}{CH}-CH_2CH_2Cl} \qquad \underset{\text{B}}{CH_3CH_2CH_2CH_2CH_2Cl} \qquad \underset{\text{C}}{CH_3CH_2\underset{|\;CH_3}{CH}-CH_2Cl}$$

2. Which of the two alkyl chlorides in each set will undergo faster hydrolsis in an S_N1 process? Suggest reason.

(*i*) $$\underset{\text{A}}{(CH_3)_3-\overset{CH_3}{\underset{Cl}{C}}-C(CH_3)_3} \qquad \underset{\text{B}}{CH_3-\overset{CH_3}{\underset{Cl}{C}}-CH_3}$$

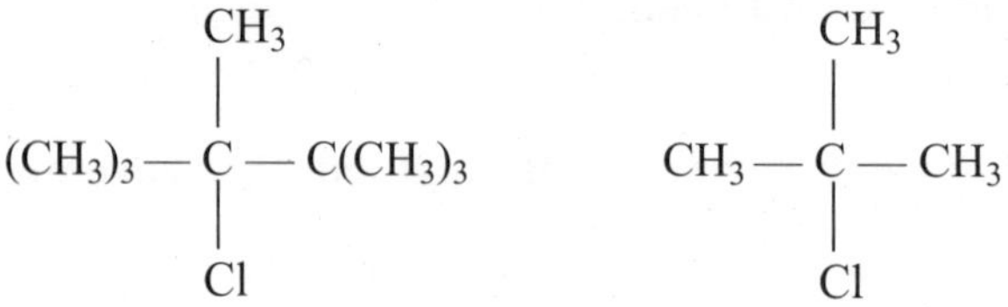

(*ii*) [bicyclic structure]—Cl (A) [bicyclic structure]—Cl (B)

3. (*i*) Draw energy profile diagrams for S_N1 and S_N2 reactions.

(*ii*) When methyl bromide is hydrolyzed using hydroxide ions, methanol and bromide ions are produced. What will be the rate of reaction if the concentration of methyl bromide is tripled and that of hydroxide ions doubled?

4. With suitable examples illustrate El, E2 and E1cB reactions. Outline conditions for each of these mechanisms to operate.

5. (*i*) Draw a comparison of S_N1 and S_N2 reactions.

(*ii*) Haloalkanes are polar in nature but are still insoluble in water. Explain.

(*iii*) Explain the rearrangement of carbocation during S_N1 reaction.

6. Briefly account for the stereochemistry of the following reaction sequences.

(*i*) $$\underset{\text{(R)-1-chloro-1-phenylethane}}{C_6H_5-\overset{Cl}{\overset{|}{C}H}-CH_3} \xrightarrow[\text{Acetone}]{NaN_3} C_6H_5\underset{CH_3}{\underset{|}{C}H}-N_3 \xrightarrow[2.\ H_3\overset{\oplus}{O}]{1.\ LAH} C_6H_5\underset{CH_3}{\underset{|}{C}H}NH_2$$

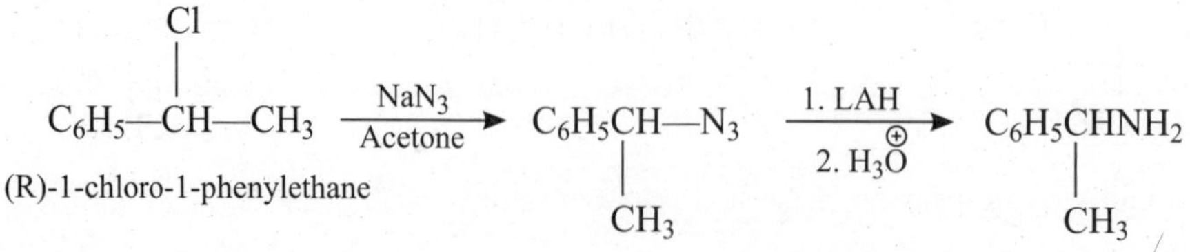

(*ii*)

$$\underset{\text{(R)-1-phenyl-2-propanol}}{C_6H_5CH_2\overset{\overset{\displaystyle OH}{|}}{C}HCH_3} \xrightarrow{p\text{-}CH_3C_6H_4SO_2Cl} C_6H_5CH_2\overset{\overset{\displaystyle OSO_2C_6H_4CH_3(p)}{|}}{C}HCH_3$$

$$\downarrow \text{Pot. acetate, acetone}$$

$$\underset{\text{(S)-1-phenyl-2-propanol}}{C_6H_5CH_2\overset{\overset{\displaystyle OH}{|}}{C}HCH_3} \xleftarrow{\overset{\ominus}{OH}} C_6H_5CH_2\overset{\overset{\displaystyle OCOCH_3}{|}}{C}HCH_3$$

7. (*i*) How do you account for the racemization of (+)-2-iodobutane when it is allowed to react with iodide ion?

   (*ii*) Why does quinuclidine react faster than triethylamine with isopropyl iodide in an S_N2 reaction?

   (*iii*) Predict the effect that increasing solvent polarity will have on the rate of an S_N1 reaction.

8. Write a reasonable mechanism for the reaction of 1,3-dibromopropane with magnesium to yield cyclopropane.

9. Outline the synthesis of the following compounds from suitable nucleophilies and an appropriate halide.

   (*i*) $CH_3C \equiv CCH_2CH_2CH_3$ (*ii*) $CH_3CH_2CH_2OCH_3$

   (*iii*) $(CH_3)_2CH\text{—}\underset{\underset{\displaystyle NH_2}{|}}{C}H\text{—}COOH$ (*iv*) $(CH_3)_2CHCH_2CH_2OCOCH_2CH(CH_3)_2$

10. Mustard gas ($ClCH_2CH_2SCH_2CH_2Cl$) gets easily hydrolyzed in aqueous NaOH. The rate of hydrolysis is independent of hydroxide ions, and is decreased by chloride ion. Suggest a mechanism consistent with this data.

11. (*i*) Arrange the following nucleophiles in order of increasing nucleophilicity:

 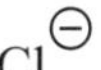 

$\overset{\ominus}{I}$ $Cl^{\ominus}$ $Br^{\ominus}$

   (*ii*) Explain the fact that allyl chloride ($CH_2 = CH\text{—}CH_2Cl$) undergoes substitution reaction by S_N1 mechanism whereas *n*-propyl chloride ($CH_3CH_2CH_2Cl$) reacts by S_N2 mechanism.

   (*iii*) How do you explain the inertness of chlorine in chlorobenzene and *vinyl* chloride in substitution reactions?

   (*iv*) Explain why do primary alkyl halides react mainly by S_N2 mechanism while tertiary alkyl halides react by S_N1 mechanism.

12. Which substrate in each of the following pair will react more rapidly with hydroxide under S_N2 reaction conditions:

(*i*) CH_3CH_2I and CH_3CH_2Cl (*ii*) *p*-$BrC_6H_4CH_3$ and $C_6H_5CH_2Br$

(*iii*) CH_3Cl and CH_2Cl_2 (*iv*) cyclohexyl–Cl and 2-chloro-1,1-dimethylcyclohexane (ring bearing Cl, and adjacent carbon bearing CH_3, CH_3)

13. (*i*) It is proposed to carry out a number of nucleophilic displacement reactions of ethyl chloride. Tell which of the following reagents will yield the product (s):

(*a*) C_6H_5ONa (*b*) CH_3SNa (*c*) KI (*d*) KCl

(*ii*) In the reaction, $CH_3I + CH_3\overset{\ominus}{O} \longrightarrow CH_3OCH_3 + \overset{\ominus}{I}$, rate is $k\,[CH_3I]\,[\overset{\ominus}{O}CH_3]$. What shall be the effect of the following on the rate of reaction?

(*a*) the concentration of both the reactants is doubled,

(*b*) the concentration of CH_3O^- is increased four times and that of CH_3I is tripled,

(*c*) the concentration of CH_3I is halved.

14. What is solvolysis? In general, it has been said that S_N1 reactions of optically active substrates are accompanied by racemization. But in actual practice, it has been found that there is always some net inversion. How do you account for this observation?

15. The kinetic rate law for the reaction

$$\overset{\ominus}{F} + C_6H_5\underset{}{\overset{Cl}{\overset{|}{C}}}HC_6H_5 \xrightarrow{\text{liq. } SO_2} C_6H_5\underset{\underset{F}{|}}{C}HC_6H_5 + \overset{\ominus}{Cl}$$

is rate $= k\,[C_6H_5\underset{\underset{Cl}{|}}{C}H—C_6H_5]$

It has been observed that:

(*i*) the rate of the reaction shows a gradual fall with the passage of time.

(*ii*) the rate of the reaction also slows down by the addition of chloride ions from outside.

Write a mechanistic scheme for the above reaction and explain these observations.

16. How do the products differ when ethyl bromide reacts separately with

(*i*) Aqueous KOH and alcoholic KOH, (*ii*) KCN and AgCN, (*iii*) KNO_2 and $AgNO_2$? Illustrate your answer with proper reasoning.

17. (*i*) Why are the nucleophilic substitution reactions of allylic halides often accompanied by rearrangements?

(*ii*) On treatment with diethylamine, substances A and B yield same product, C. Explain.

$$\underset{\textbf{A}}{C_6H_5\underset{\underset{Cl}{|}}{C}HCH{=}CH_2} \xrightarrow{(C_2H_5)_2NH} \underset{\textbf{C}}{C_6H_5CH{=}CHCH_2\overset{\oplus}{N}H(C_2H_5)_2\overset{\ominus}{Cl}}$$

$$\underset{\textbf{B}}{C_6H_5CH{=}CHCH_2Cl}$$

18. How is chloroform obtained from acetone? Discuss the mechanism of the reaction.

19. How is carbon tetrachloride formed from CS_2 and how it is converted into freon?

20. Both benzyl bromide and isopropyl bromide yield the corresponding alcohols, benzyl alcohol and isopropyl alcohol, on heating with water. The rate of reaction of either alkyl bromide depends upon the conc. of halide only. However, when both the alkyl halides are treated separately under identical conditions, the former reacts at a faster rate than the latter. What inference do you draw from these observations?

21. (*i*) Predict the product (s) of S_N2 reaction of the following compounds with hydroxide ions:

(*a*) H, H, H_3C, Cl   (*b*) H_3C, CH_3, H, Cl

(*ii*) What product (s) would be formed in an S_N1 reaction of these compounds?

22. Sketch a suitable mechanistic scheme for each of the following transformations:

(*i*) cyclopropyl$-CH_2Cl \xrightarrow{H_2O}$ cyclopropyl$-CH_2OH$ + cyclobutyl$-OH$ + $CH_2{=}CH{-}CH_2CH_2OH$

Chapter 10

(*ii*) 1-(chloromethyl)-1-methylcyclopentane (CH_3, CH_2Cl) $\xrightarrow{\text{Ag. AgNO}_3}$ 1-methylcyclohexanol (CH_3, OH)

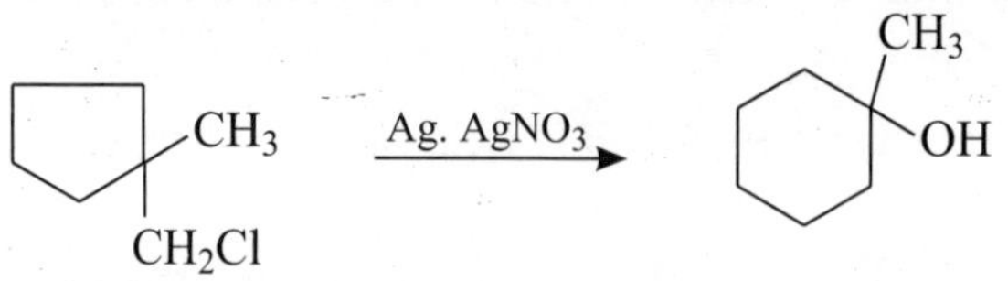

23. Predict the order of reactivity of the following halides with (*a*) sodium iodide in acetone, and (*b*) aqueous alcoholic silver nitrate:

CH_3Br	CH_3CH_2Br	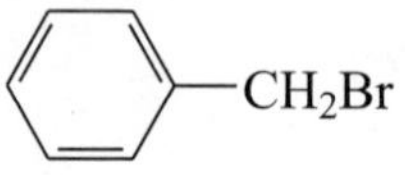 C_6H_5—CH_2Br	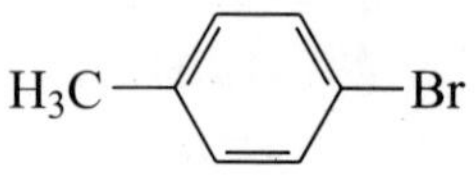 H_3C—C_6H_4—Br
A	B	C	D

24. How would you synthesize (*i*) *m*-dichlorobenzene from benzene and inorganic reagents and (*ii*) $CH_3C\equiv C—CD_3$ from $CH_3CH=CH_2$ and CD_3Cl?

25. How do you account for the formation of dichlorocarbene in the alkaline hydrolysis of chloroform? Comment upon the subsequent conversion of dichlorocarbene to the observed products.

26. (*i*) What are the evidences for the intermediacy of carbocations in S_N1 reactions?

(*ii*) Comment upon the nucleophilic substitution reactions of allyli halides.

27. There can be two mechanisms for nucleophilic aromatic substitution, namely addition-elimination mechanism and elimination-addition (benzyne) mechanism. How will you establish as to which of the two mechanisms is operating in a particular case?

28. Write the structure of the nucleophilic substitution product (s) in each of the following:

(*i*) 2-bromo-1-methylenecyclohexane (CH_2, Br) $\xrightarrow{\overset{\ominus}{\text{CN}}}$

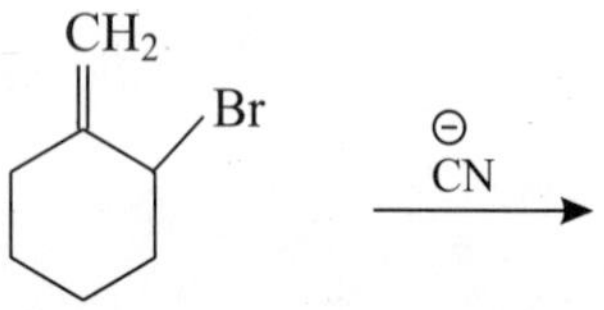

(*ii*) $CH_3CH_2CH=CH—CH=CH—CH_2Br \xrightarrow{HS^{\ominus}}$

(*iii*) $CH_2=CH—C(CH_3)_2—Cl \xrightarrow{PhS^{\ominus}}$

(*iv*) $CH_2=CH—CH(C_6H_5)—Cl \xrightarrow{\overset{\ominus}{C}H(COOEt)_2}$

29. Offer a suitable explanation for each of the following observations:

(*i*) Chloromethyl ether ($ClCH_2OCH_3$) undergoes S_N1 reaction.

(*ii*) (S)-4-Bromo-2-pentene forms a racemic product on treatment with sodium iodide.

(*iii*) Vinyl halides are inert towards nucleophilic displacement reactions.

(*iv*) Bromomesitylene and bromodurene do not form the corresponding amine upon treatment with KNH_2.

(*v*) Allyl halides are more reactive than alkyl halides towards nucleophilic substitution reactions.

(*vi*) Aryl halides are less reactive than alkyl halides towards nucleophilic substitution reactions

(*vii*) Benzyl halides are more reactive than aryl halides towards nucleophilic substitution reactions.

(*viii*) 2,4,6-Trinitrochlorobenzene readily reacts with ammonia whereas chlorobenzene fails to do so.

(*ix*) S_N2 reactions proceeds with inversion of configuration.

30. When the three isomeric chlorotoluenes are treated with KNH_2 in liquid NH_3, they yield different toluidines as shown below:

(*i*) *o*-Chlorotoluene ⟶ *o*-toluidine + *m*-toluidine (no *p*-isomer).

(*ii*) *m*-Chlorotoluene ⟶ *o*-toluidine + *m*-toluidine + *p*-toluidine.

(*iii*) *p*-Chlorotoluene ⟶ *m*-toluidine + *p*-toluidine (no *o*-isomer) + *p*-toluidine.

Explain these observations.

31. Design an experiment to show the intermediacy of cyclohexadienylide intermediate in bimolecular aromatic nucleophilic displacement reactions.

32. Give the elimination-addition mechanism of conversion of chlorobenzene into aniline.

33. What happens when chlorobenzene reacts with sodamide in liquid ammonia? Discuss mechanism of this reaction and give evidence in favour of benzyne intermediacy.

34. Write a chemical equation for each of the following:

(*i*) Chlorobenzene reacts with acetyl chloride in presence of anhydrous $AlCl_3$.

(*ii*) Carbon tetrachloride is reduced with moist iron fillings.

(*iii*) Chlorobenzene reacts with NH_3/Cu_2O.

(*iv*) Chloroform is heated with aniline in presence of alcoholic KOH.

(*v*) Chloroform is heated with phenol in presence of KOH at 340 K.

(*vi*) Chloroform is heated with silver powder.

35. How will you convert benzyl chloride into

(*i*) phenylacetic acid (*ii*) benzaldehyde?

36. Name the organic halides used as insecticides, pesticides and weedicides. Give the synthesis of DDT and BHC.

37. Discuss the stereochemistry of S_N1 and S_N2 reactions.

38. In E2 elimination, *trans*-1, 2-dichloroethylene undergoes transformation to chloroacetylene twenty times faster than the corresponding *cis*-isomer. Explain.

39. Chlorobenzene does not give precipitate with ammoniacal $AgNO_3$ but *tert*-butyl chloride gives. Explain.

40. Explain why

(*i*) *o*-Chlorotoluene when treated with sodamide in liq. NH_3 yields a mixture of *o*-toluidine and *m*-toluidine in almost equal proportions ?

(*ii*) *m*-Bromoanisole and *o*-bromoanisole yield the same product, *m*-anisidine, when treated with $NaNH_2$/liq. NH_3?

41. How can you distinguish the following pairs of compounds?

(*i*) Benzyl chloride and *o*-chlorotoluene.

(*ii*) Chlorobenzene and benzyl chloride.

(*iii*) Ethyl chloride and vinyl chloride.

(*iv*) Alkyl chloride and allyl chloride.

42. What are vicinal and geminal halides ? Give one method for preparation of each.

43. How are side chain substituted aryl halides prepared? Discuss the mechanism of the reaction.

44. Discuss the preparation of alkyl halides and aryl halides.

45. Explain why alkyl fluorides and iodides are not usually prepared directly from alkenes ? Give alternate method of their preparation.

CHAPTER 11

Organometallic Compounds

Organic derivatives of a metal having at least one carbon-metal bond are called **organometallic compounds**.The nature of such a bond depends largely upon the difference in electronegativity of the metal and the carbon.

$$\text{C—M} \longleftrightarrow \overset{\ominus}{\text{C}}\!: \overset{\oplus}{\text{M}} \equiv \overset{\delta^-}{\text{C}}—\overset{\delta^+}{\text{M}}$$

If the bond is described only by the structure on the right it would be purely ionic, whereas if it is described by the structure on the left it would be purely covalent. Studies on the actual charge distribution have, however, revealed that the actual state of the bond is somewhere in between these two extremes. Such situations are best depicted by writing a double-headed arrow separating the two extreme structures. In general the per cent ionic character of the C—M bond, depends largely upon the electronegativity of the metal, *i.e.*, the more electropositive the metal, the more ionic is the C—M bond. Thus, the ionic character of the C—M bond is in the order: C—K > C—Na > C—Li > C—Mg > C—Al > C—Zn > C—Cd > C—Hg. Further, as expected the reactivity of organometallic compounds increases with increase in the ionic character of the C—M bond. It has indeed been found that highly ionic compounds such as alkylsodiums and alkylpotassiums are so reactive that they inflame in air, and react explosively with water. Because of increased ionic character they are the strongest bases.

$$RNa + H_2O \longrightarrow R—H + NaOH$$

These compounds are generally named as alkyl or aryl derivatives of metals.

CH_3Li	C_6H_5Li	$(C_2H_5)_2Mg$
Methyllithium	Phenyllithium	Diethylmagnesium

Mixed organomagnesium halides are named as alkyl- or arylmagnesium halides.

C_2H_5MgBr	C_6H_5MgBr
Ethylmagnesium bromide	Phenylmagnesium bromide

11.1 ORGANOMAGNESIUM COMPOUNDS: THE GRIGNARD REAGENTS

The alkyl-or arylmagnesium halides, R (or Ar)—Mg—X, are also known as *Grignard reagents* after the name of **Victor Grignard** who introduced them as versatile synthetic reagents in organic chemistry. Victor Grignard earned the coveted *Nobel Prize* for this pioneering work.

11.1.1 Formation

A Grignard reagent is prepared by adding an alkyl or aryl halide to a flask containing a suspension of magnesium turnings in dry absolute ether. A visible reaction takes place at the surface of magnesium which slows down during the course of the reaction.

$$\text{R(Ar)—X} + \text{Mg} \xrightarrow{\text{Ether}} \text{R(Ar)—Mg—X}$$

A number of other solvents such as tetrahydrofuran (THF), tertiary amines and ethylene glycol dimethyl ether have also been used. The best solvent, is THF as it increases the reactivity of organic halides towards magnesium. It may also be mentioned that if the preparation of a particular Grignard reagent is difficult, addition of a drop or two of methyl iodide or a crystal of iodine will initiate the reaction. This technique is known as **entertainment technique**. The ease of the formation of Grignard reagents depends upon many factors including the nature of alkyl (or aryl) group and the halogen atom. For a given halogen increase in the number of carbon atoms makes the formation of the reagent rather difficult and for a particular alkyl group the ease of formation follows the order: I > Br > Cl. The alkyl (aryl) magnesium fluorides are practically unknown.

Precautions

While preparing Grignard reagents following guidelines should be strictly followed:

(*i*) *The apparatus should be flame dried and should be protected from moisture by using* $CaCl_2$ *guard tube.*

(*ii*) *The alkyl or aryl halide, magnesium metal and ether should be free from even traces of moisture.*

(*iii*) *Ether which is used as solvent, should not be contaminated with alcohol. To make it free from alcohol it must be first washed several times with water after which it should be dried over anhydrous* $CaCl_2$ *for 1–2 days and subsequently distilled over sodium.*

(*iv*) *The reaction should be carried out in the atmosphere of dry nitrogen (free from oxygen and carbon dioxide).*

The above precautions are absolutely essential. It is due to the fact that Grignard reagents are decomposed by moisture (water) and react with oxygen as well as carbon dioxide resulting in the formation of alcohols and carboxylic acids, respectively (*vide infra*).

The ethereal solution of Grignard reagents thus obtained is generally used as such for various reactions by adding to it the other reagents.

Mechanism

The mechanism of the formation of Grignard reagents is quite uncertain. According to Gomberg and Beckmann (1930) the reaction takes place by free radical mechanism. The reaction is initiated by $\dot{\text{M}}\text{gX}$ which is produced by interaction of Mg with MgX_2 also formed *in situ*.

$$\text{R—X} + \text{Mg} + \text{X—R} \longrightarrow \text{R—R} + \text{MgX}_2$$

$$\text{MgX}_2 + \text{Mg} \longrightarrow 2\dot{\text{M}}\text{gX}$$

$$R{-}X + \dot{M}gX \longrightarrow \dot{R} + MgX_2$$

$$\dot{R} + \dot{M}gX \longrightarrow RMgX$$

This mechanism also explains the role of iodine as initiator (*entertainment technique*) which may form magnesium iodide by reacting with magnesium.

11.1.2 Structure

The structure of Grignard reagents has been the subject of much controversy over the years. However, usually they are represented by the general formula R—Mg—X formed by the insertion of magnesium between the carbon and halogen. But this is an over-simplification of the true structure. Attempts to isolate these compounds have shown that they contain ether, commonly known as **ether of crystallization**.

The following three alternative formulations have been proposed, in which they may exist as monomers or dimers along with ether of crystallization.

I: $(R)(X)Mg \leftarrow (OEt_2)_2$

II: $Et_2O \rightarrow Mg(R)(\leftarrow Et_2O){-}X \rightarrow Mg(R)(X) \leftarrow Et_2O$

III: $R_2Mg(\mu{-}X)_2Mg(OEt_2)_2$

Structure I shows the simplest structure of the monomeric form which is formed by coordinating with two molecules of ether. The structures II and III represent the dimeric forms solvated in ether. In structure II the halogen atom of one molecule coordinates with the magnesium atom of the other, whereas in III, the compound is considered as existing in a 1:1 complex of dialkyl (or aryl) magnesium and magnesium halide. The halogen atoms of the magnesium halide coordinate with the magnesium atom of dialkyl (or diaryl) magnesium. In addition to these structures, it has been proposed that the composition of Grignard reagents in ether may be represented by the following equilibria.

$$\left[(R)(X)Mg \leftarrow X{-}Mg{-}R \right]_{\text{Dimer}} \rightleftharpoons \underset{\text{Monomer}}{2RMgX}$$

$$\text{Dimer} \rightleftharpoons R_2Mg + MgX_2 \rightleftharpoons 2RMgX$$

The crystalline Grignard reagents have been isolated by evaporation of their ethereal solution and their structures have been determined by X-ray diffraction technique. This study shows that these compounds exist as dietherates—$RMgX{\cdot}2Et_2O$. The four groups are tetrahedrally arranged around magnesium atom. Although less accurate but for the sake of convenience we shall be using the formulation of R—MgX or Ar—MgX in subsequent discussion.

11.1.3 Reactions of Grignard Reagents

Due to large difference in electronegativity of carbon (2.5) and magnesium (1.2), the C—Mg bond is highly polar as shown.

$$-\overset{|}{\underset{|}{C}}^{\delta^-} \leftarrow \overset{\delta^+}{MgX} \quad \text{or} \quad \overset{\delta^-}{R}-\overset{\delta^+}{MgX} \rightleftharpoons \overset{\ominus}{R} + \overset{\oplus}{MgX}$$

The greater electron density around the carbon atom makes the alkyl group in Grignard reagent to behave as a carbanion or a nucleophile and hence the main reactions of these reagents are either nucleophilic addition or nucleophilic substitution reactions as shown below:

(*i*) Nucleophilic Addition Reactions

Grignard reagents upon treatment with compounds having multiple bonds form addition compounds. Hydrolysis of such adducts gives rise to a variety of compounds. In all such reactions, the addition of the nucleophilic group, $\overset{\ominus}{R}$, occurs at the positive end of the polar molecule and $\overset{\ominus\oplus}{RMgX}$ gets bonded to the more electronegative atom of the multiple bond.

Thus,

$$\overset{\delta^-}{>C}=\overset{\delta^+}{O} + \overset{\delta^-}{R}-\overset{\delta^+}{MgX} \longrightarrow -\overset{R}{\underset{|}{\overset{|}{C}}}-\overset{\ominus\oplus}{OMgX} \xrightarrow{\overset{\oplus}{H}/H_2O} -\overset{R}{\underset{|}{\overset{|}{C}}}-OH + \overset{OH}{\overset{|}{MgX}}$$

Carbonyl compound; Grignard reagent; Addition compound; An alcohol

(*ii*) Nucleophilic Substitution Reactions

Due to carbanion character of alkyl group, Grignard reagents attack as nucleophiles on positively charged centres having an acidic hydrogen or halogen attached to them.

$$\overset{\delta^-}{R}-\overset{\delta^+}{MgX} + A-B \longrightarrow R-A + \overset{B}{\overset{|}{Mg}}-X$$

In carrying out a chemical reaction with Grignard reagent, it is customary to add the reactant solution to the ethereal solution of the reagent. After sometime the reaction mixture is decomposed with water containing a little mineral acid or ammonium chloride. The Grignard reagents are never isolated from the solution which are prepared just before use.

11.1.4 Synthetic Applications of Grignard Reagents

These reagents are so reactive that they react with most of the functional groups. Their reactivity has made them versatile reagents. Some chemical reactions highlighting their synthetic applications are listed below:

(*i*) Nucleophilic Addition to Carbonyl Compounds: Synthesis of Alcohols and Ketones

As mentioned earlier Grignard reagents, being nucleophilic in character, add to carbonyl group of aldehydes, ketones, carboxylate esters, anhydrides and amides. In these reactions the alkyl group, having carbanionic character, becomes bonded to the carbonyl carbon and the magnesium halide to the oxygen of the carbonyl group. The general picture of the reaction would be as under:

$$\text{>C=O} + \overset{\ominus\oplus}{\text{RMgX}} \longrightarrow \text{R—}\overset{\overset{\ominus\oplus}{\text{OMgX}}}{\text{C<}}$$

The complex so formed after the decomposition with water etc., produces alcohols. Since this reaction results in the formation of a carbon-carbon bond it is also known as **Grignard synthesis**, a term which includes all the reactions of Grignard reagents that result in the formation of a carbon-carbon bond. The reactions of these reagents with various carbonyl functionalities are given below:

(a) Reaction with formaldehyde: Synthesis of primary alcohols

Addition of a Grignard reagent to formaldehyde produces a primary alcohol in which the original carbon chain is extended by one.

$$\overset{\ominus\oplus}{\text{RMgX}} + \text{CH}_2\text{=O} \longrightarrow \text{R—CH}_2\text{—}\overset{\ominus\oplus}{\text{OMgX}} \xrightarrow{\text{H}_3\overset{\oplus}{\text{O}}} \underset{\text{A primary alcohol}}{\text{R—CH}_2\text{OH}}$$

For example,

$$\underset{n\text{-Propyl bromide}}{\text{CH}_3\text{CH}_2\text{CH}_2\text{Br}} \xrightarrow[\text{Ether}]{\text{Mg}} \text{CH}_3\text{CH}_2\overset{\ominus}{\text{CH}_2}\overset{\oplus}{\text{MgBr}} \xrightarrow[(ii)\ \text{H}_3\overset{\oplus}{\text{O}}]{(i)\ \text{CH}_2\text{O}} \underset{\text{1-Butanol}}{\text{CH}_3\text{CH}_2\text{CH}_2\text{CH}_2\text{OH}}$$

(b) Reaction with aldehydes other than formaldehyde: Synthesis of secondary alcohols

Reaction of a Grignard reagent with an aldehyde other than formaldehyde followed by treatment of the complex with acidulated water, produces a secondary alcohol.

$$\overset{\ominus\oplus}{\text{RMgX}} + \text{R}'\text{CH=O} \longrightarrow (\text{R})(\text{R}')\text{CH—}\overset{\ominus\oplus}{\text{OMgX}} \longrightarrow \underset{\text{A secondary alcohol}}{(\text{R})(\text{R}')\text{CH—OH}}$$

Chapter 11

For example,

$$CH_3CH_2MgBr \xrightarrow{CH_3CHO} CH_3CH_2-\underset{}{\overset{CH_3}{\overset{|}{C}H}}-\overset{\ominus\oplus}{O}MgBr \xrightarrow{H_3\overset{\oplus}{O}} CH_3CH_2-\overset{CH_3}{\overset{|}{C}H}-OH$$

2-Butanol

(c) *Reaction with ketones: Synthesis of tertiary alcohols*

Similarly, the reaction of ketones with Grignard reagents gives rise to tertiary alcohols.

$$\overset{\ominus\oplus}{RMgX} + \begin{matrix}R'\\ \\R'\end{matrix}\!\!>C=O \longrightarrow R-\overset{R'}{\underset{R'}{\overset{|}{\underset{|}{C}}}}-\overset{\ominus\oplus}{O\,MgX} \xrightarrow{H_3\overset{\oplus}{O}} R-\overset{R'}{\underset{R'}{\overset{|}{\underset{|}{C}}}}-OH$$

A tertiary alcohol

For example,

$$C_6H_5MgBr + \begin{matrix}C_6H_5\\ \\C_6H_5\end{matrix}\!\!>C=O \longrightarrow C_6H_5-\overset{C_6H_5}{\underset{C_6H_5}{\overset{|}{\underset{|}{C}}}}-\overset{\ominus\oplus}{O\,MgX} \xrightarrow{H_3\overset{\oplus}{O}} C_6H_5-\overset{C_6H_5}{\underset{C_6H_5}{\overset{|}{\underset{|}{C}}}}-OH$$

Triphenylcarbinol

(d) *Reaction with esters: Synthesis of secondary and tertiary alcohols*

The reaction of Grignard reagents with formate ester gives rise to a secondary alcohol when treated with 2 molar equivalents of the Grignard reagents. The reaction may be considered as taking place in two steps, first step leading to the formation of aldehyde by addition-elimination mechanism. In the subsequent step the aldehyde reacts with the second molecule of the Grignard reagent in the usual way to form a *secondary alcohol.*

$$\overset{\ominus\oplus}{RMgX} + \begin{matrix}H\\ \\EtO\end{matrix}\!\!>C=O \longrightarrow R-\overset{\overset{\ominus\oplus}{OMgX}}{\underset{OEt}{\overset{|}{\underset{|}{C}}}}-H \longrightarrow R-\overset{O}{\overset{\|}{C}}-H$$

$$R-\overset{O}{\overset{\|}{C}}-H \xrightarrow[(ii)\ H_3\overset{\oplus}{O}]{(i)\ RMgX} R_2CHOH$$

A secondary alcohol

It may, however, be mentioned that treatment of Grignard reagent with formate in 1:1 ratio results in the formation of aldehydes.

$$\overset{\ominus\oplus}{\text{RMgX}} + \text{H}-\underset{\text{OEt}}{\text{C}}=\text{O} \longrightarrow \text{R}-\underset{\text{OEt}}{\overset{\text{H}}{\text{C}}}-\overset{\ominus\oplus}{\text{OMgX}} \longrightarrow \text{R}-\overset{\text{H}}{\text{C}}=\text{O} + \text{Mg(OEt)X}$$

The reaction of Grignard reagent with any other carboxylate ester, however, leads to the formation of a tertiary alcohol containing at least two identical alkyl groups. The reaction takes place exactly as described in the case of formate esters.

$$\overset{\ominus\oplus}{\text{RMgX}} + \text{R}'(\text{R}''\text{O})\text{C}=\text{O} \longrightarrow \text{R}-\underset{\text{OR}''}{\overset{\overset{\ominus\oplus}{\text{OMgX}}}{\text{C}}}-\text{R}'$$

$$\longrightarrow \text{R}-\overset{\text{O}}{\overset{\|}{\text{C}}}-\text{R}' \;\text{(A ketone)} \xrightarrow[2.\ \overset{\oplus}{\text{H}_3\text{O}}]{1.\ \text{RMgX}} \text{R}-\underset{\text{R}'}{\overset{\text{R}}{\text{C}}}-\text{OH} \;\text{(A tertiary alcohol)}$$

A ketone

A tertiary alcohol

(*e*) *Reaction with acid chlorides: Synthesis of tertiary alcohols*

Grignard reagents react with acid chlorides, again in two stages, ultimately forming tertiary alcohols.

$$\overset{\ominus\oplus}{\text{RMgX}} + \text{R}'-\underset{\text{Cl}}{\text{C}}=\text{O} \longrightarrow \text{R}-\underset{\text{R}'}{\overset{\overset{\ominus\oplus}{\text{OMgX}}}{\text{C}}}-\text{Cl}$$

$$\longrightarrow \text{R}-\overset{\text{O}}{\overset{\|}{\text{C}}}-\text{R}' + \text{MgXCl} \xrightarrow[2.\ \overset{\oplus}{\text{H}_3\text{O}}]{1.\ \text{RMgX}} \text{R}-\underset{\text{R}'}{\overset{\text{R}}{\text{C}}}-\text{OH}$$

A tertiary alcohol

Chapter 11

(f) Reaction with acid anhydrides: Synthesis of tertiary alcohols

Similarly, Grignard reagents react with acid anhydrides forming tertiary alcohols.

$$\overset{\ominus\oplus}{RMgX} + \begin{matrix} R'-C(=O) \\ \quad O \\ R'-C(=O) \end{matrix} \longrightarrow R'-\underset{R}{\overset{\overset{\ominus\oplus}{OMgX}}{C}}-O-\overset{O}{\overset{\|}{C}}-R'$$

$$\underset{\text{A tertiary alcohol}}{R-\underset{R}{\overset{OH}{C}}-R'} \xleftarrow[2.\ \overset{\oplus}{H_3O}]{1.\ RMgX} R'-\overset{O}{\overset{\|}{C}}-R$$

(g) Reaction with amides: Synthesis of ketones

Grignard reagents upon treatment with amides form ketones.

$$\overset{\ominus\oplus}{RMgX} + \begin{matrix} R' \\ H_2N \end{matrix}\!\!>C=O \longrightarrow R-H + R'-\underset{\overset{\ominus}{N}HMgX^{\oplus}}{\overset{O}{\overset{\|}{C}}} \xrightarrow{\overset{\ominus\oplus}{RMgX}}$$

$$\begin{matrix} R \\ R' \end{matrix}\!\!>C<\!\!\begin{matrix} \overset{\ominus\oplus}{OMgX} \\ \overset{\ominus\ \oplus}{NHMgX} \end{matrix} \xrightarrow{\overset{\oplus}{H_3O}} R-\overset{O}{\overset{\|}{C}}-R'$$

(h) Reaction with carbonyl compounds giving abnormal products

A number of cases are known where the Grignard reagent fails to react with a particular functional group with which it is known to react or if it reacts at all it gives rise to 'abnormal' products. In general branching of the carbon chain near the carbonyl group prevents the nucleophilic addition by the reagent due to steric hindrance. Alternatively, if the Grignard reagent has a bulky alkyl or aryl group it fails to attack the electrophilic centre of the substrate molecule. For instance, even methylmagnesium bromide or iodide fail to react with *di-t*-butyl ketone; methyl isopropylketone reacts with methylmagnesium halide but not with *t*-butylmagnesium halides. Reactions are, however, known when suitably substituted ketones and reagents react together to form products, which are normally not obtained in a Grignard reaction. Two such reactions namely reduction and enolisation reactions are illustrated as follows:

(A) Reduction: When diisopropyl ketone is treated with isopropyl magnesium bromide, it gives rise to diisopropyl carbinol.

$$(H_3C)_2CH-\overset{O}{\overset{\|}{C}}-CH(CH_3)_2 \xrightarrow{(H_3C)_2\overset{\ominus}{C}H\overset{\oplus}{Mg}Br} (H_3C)_2CH-CHOH-CH(CH_3)_2 + CH_3-CH{=}CH_2$$

This reaction probably takes place by a hydride transfer from alkyl group of the reagent involving a six-membered cyclic transition state.

$$\underset{\text{Grignard reagent}}{H_3C-CH(MgBr)-CH_2-H} + \underset{\text{Diisopropyl ketone}}{(CH_3)_2HC-\overset{O}{\overset{\|}{C}}-CH(CH_3)_2} \longrightarrow [\text{cyclic transition state: } CH_3-CH\cdots Mg(Br)\cdots O=C(CH(CH_3)_2)_2,\ CH_2\cdots H\cdots C]$$

$$\longrightarrow \underset{\text{Propene}}{CH_3-CH{=}CH_2} + ((CH_3)_2HC)_2\overset{\ominus}{CHO}\overset{\oplus}{Mg}Br$$

Thus this reaction is analogous to **Meerwein-Pondorf-Verley reduction.**

Instances are also known where the reaction between an aldehyde and a Grignard reagent gives a ketone and a primary alcohol, probably by a hydride transfer.

Chapter 11

$$RMgX + R'-\overset{\overset{\displaystyle O}{\|}}{C}-H \longrightarrow R(R')(H)C-\overset{\ominus}{O}\overset{\oplus}{Mg}X$$

$$R'-\underset{\underset{\displaystyle O}{\|}}{C}-H \;\longrightarrow\; R'-CH_2-\overset{\ominus}{O}\overset{\oplus}{Mg}X + R(R')C{=}O \xrightarrow{H_3\overset{\oplus}{O}} R'-CH_2-OH$$

(B) Enolization: If the congested carbonyl substrate has an α-hydrogen, enolization of the ketone becomes a competitive process. The carbanion, $\overset{\ominus}{R}$, abstracts α-hydrogen forming a hydrocarbon R—H and enolate anion which can be isolated as $\overset{\oplus}{Mg}X$ salt of an enol. Hydrolysis with acidulated water gives unstable enol which ketonizes to the starting ketone.

$$>\!\underset{|}{C}(H)-C(=O)\cdots Mg(R)X \longrightarrow >\overset{\ominus}{C}-C=O\,\overset{\oplus}{Mg}X + RH \longrightarrow >C{=}C<\overset{\ominus}{O}\overset{\oplus}{Mg}X \xrightarrow{H_3\overset{\oplus}{O}} >C{=}C<OH \rightleftharpoons >\underset{|}{C}(H)-C(=O)<$$

(*ii*) Nucleophilic addition to nitriles: Synthesis of ketones

Like carbonyl group, there is electronic imbalance in the nitrile group—the carbon acquiring a slight positive polarity and nitrogen developing slight negative charge. Due to this polarity the Grignard reagent adds as a nucleophile forming a ketimine salt which upon hydrolysis gives the corresponding ketone.

$$\overset{\ominus\oplus}{RMgX} + R'-\overset{\delta^+}{C}\equiv\overset{\delta^-}{N} \longrightarrow \begin{matrix} R \\ R' \end{matrix}\!\!>C=\overset{\ominus\oplus}{NMgX} \xrightarrow{\overset{\oplus}{H_3O}} \left[\begin{matrix} R \\ R' \end{matrix}\!\!>C=NH\right] \xrightarrow{H_2O} \begin{matrix} R \\ R' \end{matrix}\!\!>C=O$$

***(iii)* Reaction with carbon dioxide and carbon disulphide: Synthesis of carboxylic acids and dithionic acids**

When a Grignard reagent is treated with solid carbon dioxide (dry ice) or carbon disulphide, nucleophilic addition products are obtained which on decomposition with acids yield monocarboxylic acid and a dithionic acid, respectively.

$$\overset{\ominus\oplus}{RMgX} + O=C=O \longrightarrow R-\overset{O}{\overset{\|}{C}}-\overset{\ominus\oplus}{OMgX} \xrightarrow{\overset{\oplus}{H_3O}} R-\overset{O}{\overset{\|}{C}}-OH$$

$$\overset{\ominus\oplus}{RMgX} + S=C=S \longrightarrow R-\overset{S}{\overset{\|}{C}}-\overset{\ominus\oplus}{SMgX} \xrightarrow{\overset{\oplus}{H_3O}} \underset{\text{Dithionic acid}}{R-\overset{S}{\overset{\|}{C}}-SH}$$

The acid is obtained in good yield if carbon dioxide gas is passed in an ethereal solution of Grignard reagent at 273 K. This procedure is particularly useful for preparing acids of the type R_3C—COOH, which usually cannot be prepared by nucleophilic displacement reactions.

***(iv)* Reaction with oxygen, sulphur and halogens: Synthesis of alcohols, thiols and organic halides**

The action of oxygen, sulphur and halogen on Grignard reagents results in the formation of alcohols, thiols and organic halides, respectively.

$$RMgX + O_2 \longrightarrow R-O-O-MgX \xrightarrow{RMgX} 2ROMgX \xrightarrow{\overset{\oplus}{H_3O}} 2ROH^-$$

$$8\,RMgX + S_8 \longrightarrow 8\,RSMgX \xrightarrow{\overset{\oplus}{H_3O}} 8\,RSH$$

$$RMgX + X'_2 \longrightarrow RX' + MgXX'$$

(X′ = other than X)

As such this is not a good synthetic method for the preparation of these compounds since all of them can easily be obtained by the usual nucleophilic displacement reactions. However, in such cases where both S_N1 and S_N2 reactions are extremely slow, this sequence becomes quite useful. For instance, neopentyl chloride can be easily converted into neopentyl iodide since, the usual nucleophilic displacement with iodide is a slow process.

$$H_3C-\underset{CH_3}{\overset{CH_3}{C}}-CH_2-Cl \xrightarrow{Mg/Et_2O} H_3C-\underset{CH_3}{\overset{CH_3}{C}}-CH_2MgCl \xrightarrow[-MgCl]{I_2} H_3C-\underset{CH_3}{\overset{CH_3}{C}}-CH_2I$$

The importance of the reaction of Grignard reagent with oxygen lies in the fact that hydroperoxides can be obtained if **inverse addition*** of the Grignard reagent is carried out at low temperature (203 K).

$$RMgX + O_2 \longrightarrow ROOMgX \xrightarrow{\overset{\oplus}{H_3O}} \underset{\text{Hydroperoxide}}{ROOH}$$

(*v*) Nucleophilic displacement: Synthesis of hydrocarbons

Grignard reagent, being source of carbanions, can participate in nucleophilic displacement reactions.

$$\overset{\ominus\ \oplus}{RMgX} + R'-X \longrightarrow R-R' + MgX_2$$

Such reactions are more pronounced with alkyl halides having high reactivity towards nucleophilic displacements. However, a few other derivatives of alcohols such as alkyl sulphates, alkyl sulphonates and α-chloroethers which have good leaving groups can also undergo nucleophilic displacements.

(*vi*) Reaction with epoxides: Synthesis of alcohols

When epoxides are treated with a Grignard reagent, they undergo ring opening to form alcohols having two carbon atoms more than the alkyl or aryl group. The reactions may be regarded to be analogous to the base catalyzed ring-opening of epoxide by an S_N2 reaction.

$$\overset{\ominus\ \oplus}{RMgX} + \underset{\diagdown O \diagup}{CH_2-CH_2} \longrightarrow R-CH_2-CH_2-\overset{\ominus\oplus}{OMgX} \xrightarrow{\overset{\oplus}{H_3O}} RCH_2CH_2OH$$

(*vii*) Reaction with compounds containing active hydrogen atoms: Formation of hydrocarbons

When the Grignard reagents are treated with compounds containing active hydrogen, a hydrocarbon is

*Normally the other reactant is added to an ethereal solution of the Grignard reagent. In the inverse addition, however, the solution of the Grignard reagent is added to ether through which oxygen is bubbled.

formed. By *active hydrogen* we mean acidic hydrogen and the Grignard reagents being strong carbon bases react violently even with water which is regarded as a very weak acid.

$$RMgX + H_2O \longrightarrow R—H + Mg(OH)X$$

Although it is an undesired side reaction, it can be useful in the conversion of an alkyl halide into a hydrocarbon.

$$R—X \xrightarrow[Et_2O]{Mg} RMgX \xrightarrow{H_2O} R—H + MgX(OH)$$

For instance, conversion of cyclobutyl bromide into cyclobutane.

$$\text{Cyclobutyl bromide (Br)} \xrightarrow[Et_2O]{Mg} \text{Cyclobutyl-MgBr} \xrightarrow{H_2O} \text{Cyclobutane} + MgX(OH)$$

This procedure of alkane synthesis is very useful in the preparation of deuterated hydrocarbons. Some examples are outlined below:

$$C_2H_5I \xrightarrow[Et_2O]{Mg} C_2H_5MgI \xrightarrow{D_2O} C_2H_5D$$

Monodeuteroethane

$$CH_3—CH(Br)—CH_3 \xrightarrow[Et_2O]{Mg} (H_3C)_2CHMgBr \xrightarrow{D_2O} (H_3C)_2CHD$$

2-Deuteropropane

In addition to water all acidic compounds including alcohols, primary and secondary amines, carboxylic acids and terminal alkynes react with the Grignard reagents forming the hydrocarbons.

$$RMgX + R'OH \longrightarrow R—H + R'OMgX$$

$$RMgX + R'NH_2 \longrightarrow R—H + R'NHMgX$$

$$RMgX + R_2'NH \longrightarrow R—H + R_2'NMgX$$

$$RMgX + R'COOH \longrightarrow R—H + R'COOMgX$$

$$RMgX + R'—C\equiv CH \longrightarrow R—H + R'—C\equiv CMgX$$

It may be interesting to note that the last reaction affords alkynylmagnesium halide which in turn can be a useful reagent for the synthesis of higher alkynes.

$$R'-C\equiv\overset{\ominus\oplus}{C}MgX + R''-X \xrightarrow{S_N2} R'-C\equiv C-R'' + MgX_2$$

In the reactions of Grignard reagents with active hydrogen compounds, the yields of hydrocarbons are quantitative. This forms the basis of *Zerewitinoff active hydrogen determination.* The procedure consists of treating weighed quantity of an unknown compound with methylmagnesium iodide. The volume of methane evolved is measured and the number of moles of active hydrogen in the sample are then calculated. If we also know the molecular weight of the compound, the number of active hydrogen atoms per molecule can be readily calculated.

(*viii*) Addition to carbon-carbon double bond of α, β-unsaturated compounds

We know that ordinarily the alkenes are not susceptible to attack by nucleophiles, say $\overset{\ominus}{R}$ in the case of Grignard reagent. The reason for this is that in contrast to carbonyl compounds there is very little or no electronic imbalance in the π-electron cloud and as such it would repel the approaching nucleophile. However, if the double bond is polarized by a neighbouring electron-withdrawing group, the nucleophilic addition to carbon-carbon double bond takes place as we have seen in the case of **Michael addition**. The addition of Grignard reagents to α, β-unsaturated esters, for instance, leads to the formation of a saturated ester, which probably arises from an initial 1, 4-addition.

$$\overset{\ominus\oplus}{RMgX} + >C=C<-C(OR')=O \longrightarrow R-\overset{|}{\underset{|}{C}}-\overset{|}{C}=\underset{OR'}{\underset{|}{C}}-\overset{\ominus\oplus}{O}MgX$$

$$\downarrow \overset{\oplus}{H_3O}$$

$$R-\overset{|}{\underset{|}{C}}-\overset{|}{\underset{H}{\underset{|}{C}}}-\underset{OR'}{\underset{|}{C}}=O \rightleftharpoons \left[R-\overset{|}{\underset{|}{C}}-\overset{|}{C}=\underset{OR'}{\underset{|}{C}}-OH\right]$$

Enol form of ester, unstable

Thus overall reaction is an addition to a carbon-carbon double bond. Such addition also take place with conjugated aldehydes and ketones. The reaction mixture generally contains products arising from overall addition to carbon-carbon as well as carbon-oxygen double bond formed through the initial 1,4-and 1,2-addition, respectively. Thus reaction of ethylmagnesium iodide with 3-pentene-2-one gives a mixture containing 3 parts of 4-methylhexanone (arising from initial 1, 4-addition) and 1 part of 2-ethylpent-3-en-2-ol (arising from initial 1,2-addition). In other words, *substitution at the carbonyl group favours* 1,2-*addition while substituion at the* β-*position of the double bond favours* 1,4-*addition.*

$$\underset{\text{3-Penten-2-one}}{CH_3-CH=CH-\overset{\overset{O}{\|}}{C}-CH_3} \xrightarrow[2.\ H_3\overset{\oplus}{O}]{1.\ C_2H_5MgI} \underset{\text{3 parts}}{CH_3-\underset{\underset{C_2H_5}{|}}{CH}-CH_2-\overset{\overset{O}{\|}}{C}-CH_3} + \underset{\text{1 part}}{CH_3-CH=CH-\overset{\overset{OH}{|}}{\underset{\underset{C_2H_5}{|}}{C}}-CH_3}$$

An interesting feature of the reaction is that the ratios of two products are sensitive to steric hindrance. Thus if another methyl group is introduced in the β-position of 3-pentene-2-one, as in 2-methyl-2-penten-4-one, the corresponding unsaturated alcohol is obtained in 100% yield. This can be due to the fact that the attack of the nucleophilic carbanion at the β-carbon is hindered because of the presence of two methyl groups.

$$\underset{\text{2-Methyl-2-penten-4-one}}{(H_3C)_2C=CH-\overset{\overset{O}{\|}}{C}-CH_3} \xrightarrow[2.\ H_3\overset{\oplus}{O}]{1.\ C_2H_5MgI} \underset{\text{3, 5-Dimethyl-4-hexen-3-ol}}{(H_3C)_2C=CH-\overset{\overset{OH}{|}}{\underset{\underset{C_2H_5}{|}}{C}}-CH_3}$$

Similar results are obtained when the bulk of the substituent around the carbonyl group is reduced. Thus crotonaldehyde also affords the corresponding unsaturated alcohol in 100% yield.

$$\underset{\text{Crotonaldehyde}}{H_3C-CH=CH-\overset{\overset{O}{\|}}{C}-H} \xrightarrow[2.\ H_3\overset{\oplus}{O}]{1.\ C_2H_5MgI} \underset{\text{4-Hexen-3-ol}}{CH_3-CH=CH-\overset{\overset{OH}{|}}{\underset{\underset{C_2H_5}{|}}{C}}-H}$$

11.1.5 Limitations in the uses of Grignard synthesis

1. These reagents are highly sensitive to moisture, oxygen and carbon dioxide. They must, therefore, be protected from them. Due to this problem the handling of Grignard reagents becomes difficult.
2. These reagents cannot be prepared from alkyl halides containing functional groups with which they are known to react. Such groups are —OH, —SH, $—NH_2$, —COOH, $—SO_2$,

—CHO, > C = O, —COOR, —CN, etc. Only alkyl halides containing a few other groups such as —R, —OR, —Ar, —X, etc. can be used to form Grignard reagents.

3. Presence of bulky groups near the functional group hinders the approach of the reagent. For example, $CH_3Mg Br$ and CH_3MgI do not react with di-*tert*-butyl ketone. Similarly, if the Grignard reagent contains bulky alkyl group, the reaction does not take place due to steric hindrance. For example, isopropyl methyl ketone reacts with CH_3MgBr but not with $(CH_3)_3CMgCl$.
4. Enolisable carbonyl compounds librate hydrogen when treated with Grignard reagents to give alkanes. For example

$$\underset{\substack{\text{Ethyl acetoacetate} \\ (\textit{keto form})}}{CH_3-\overset{O}{\overset{\|}{C}}-CH_2-\overset{O}{\overset{\|}{C}}-OC_2H_5} \rightleftharpoons \underset{(\textit{Enol form})}{CH_3-\overset{OH}{\overset{|}{C}}=CH-\overset{O}{\overset{\|}{C}}-OC_2H_5} \xrightarrow{CH_3MgI}$$

$$CH_3-\overset{OMgI}{\overset{|}{C}}=CH-\overset{O}{\overset{\|}{C}}-OC_2H_5 + \underset{\text{Methane}}{CH_4}$$

5. Nitro compounds cannot be used in Gringard synthesis since they oxidise these reagents.

It may be pointed out here that the above limitations have not hampered the use of these versatile reagents.

11.2 ORGANOZINC COMPOUNDS

Although organozinc compounds, R_2Zn, are not commonly used in organic synthesis, they are particularly useful when milder reagents are required. Historically they were the first organometallic compounds to be prepared in the laboratory.

11.2.1 Formation

Following methods have been used for the preparation of organozinc compounds:

(*i*) From alkyl halides: Treatment of freshly prepared zinc fillings with alkyl iodides results in the formation of alkylzinc iodides, which upon distillation in carbon dioxide atmosphere undergo disproportionation giving rise to dialkylzinc derivatives.

$$RI + Zn \longrightarrow RZnI$$

$$2RZnI \xrightarrow[\text{Distil.}]{CO_2} R_2Zn + ZnI_2$$

In a less hazardous procedure, high boiling ethers are used as solvents and zinc-copper alloy is employed in place of metallic zinc.

(*ii*) **From organoaluminium compounds:** Treatment of trialkylaluminium with zinc chloride results in the formation of dialkylzinc derivatives.

$$2R_3Al + ZnCl_2 \longrightarrow R_2Zn + 2R_2AlCl$$

The above method forms the basis of the laboratory preparation of these compounds.

11.2.2 Physical Properties

The zinc atom in these compounds is *sp*-hybridized resulting in the linear geometry. They can exist in higher coordination number in combination with electron-donor molecules. They are nonpolar, low boiling liquids or low melting solids, soluble in organic solvents. The lower zinc compounds are highly inflammable in air and are, therefore, prepared and used under carbon dioxide atmosphere.

11.2.3 Chemical Reactions

The chemical reactions of organozinc compounds are similar to those of the Grignard reagents. The only difference is that they are less reactive than the latter. For instance, they do not react with carbon dioxide and react only slowly with cyanides, ketones and esters. Some of the important chemical reactions are discussed as under:

(*i*) **Reaction with water:** Dialkylzinc compounds react readily with water forming hydrocarbons.

$$2R_2Zn + 2H_2O \longrightarrow 2R\text{—}H + R\text{—}Zn\begin{smallmatrix}OH\\ \\OH\end{smallmatrix}Zn\text{—}R$$

(*ii*) **Reaction with alkyl halides:** These compounds do not react with primary and secondary alkyl halides. However, hydrocarbons are formed by treating dialkylzinc compounds with tertiary alkyl halides.

$$R_2Zn + (H_3C)_3CCl \longrightarrow (CH_3)_3CR + RZnCl$$

(*iii*) **Reaction with acid chlorides:** They react readily with acid chlorides forming ketones.

$$R'\text{—}Zn\text{—}R' + R\text{—}\overset{O}{\overset{\|}{C}}\text{—}Cl \longrightarrow R\text{—}\underset{R'}{\overset{\overset{\ominus\oplus}{OZnR'}}{C}}\text{—}Cl \xrightarrow{-R'ZnCl} R\text{—}\overset{O}{\overset{\|}{C}}\text{—}R'$$

As the reaction between ketones and dialkylzinc compounds is extremely slow, the former can be isolated from the reaction mixture. It may be noted that the reactions of dialkylzinc compounds with esters, cyanides, etc. is extremely slow and hence they have practically no synthetic importance.

(*iv*) **Reformatsky reaction: Formation of β-hydroxyesters:** If an aldehyde or a ketone is added to a solution of an α-haloester in presence of zinc, a nucleophilic addition reaction, analogous to

Grignard reaction, takes place yielding a β-hydroxy ester. The reaction is known as *Reformatsky reaction* and has been developed as a general procedure for the preparation of β-hydroxy acids and their derivatives.

Detailed studies of the reaction have revealed that it takes place by the initial reaction of the α-haloester and zinc, leading to the formation of the corresponding organozinc halide. This adds on the carbonyl carbon of the aldehyde or ketone forming an alkoxyzinc halide *in situ*. The reaction mixture upon treatment with acidulated water forms the corresponding α-hydroxy ester.

$$BrCH_2COOC_2H_5 \xrightarrow{Zn} \overset{\ominus\oplus}{BrZn}\,[\overset{\ominus}{C}H_2{-}\underset{\underset{O}{\|}}{C}{-}OC_2H_5 \longleftrightarrow CH_2{=}\underset{\underset{O^{\ominus}}{|}}{C}{-}OC_2H_5]$$

Organozinc compound

$$C_6H_5\underset{\underset{H}{|}}{\overset{\overset{O}{\|}}{C}} + \underset{\underset{\overset{\oplus}{BrZn}}{}}{\overset{\ominus}{C}H_2COOC_2H_5} \longrightarrow C_6H_5\underset{\underset{H}{|}}{\overset{\overset{\ominus\oplus}{OZnBr}}{\overset{|}{C}}}{-}CH_2COOC_2H_5 \xrightarrow[\text{(Hydrolysis)}]{H_2O,\ H^{\oplus}} C_6H_5\underset{\underset{H}{|}}{\overset{\overset{OH}{|}}{C}}{-}CH_2COOC_2H_5$$

β-Hydroxy ester

The β-hydroxy esters thus produced can be easily dehydrated forming α, β-unsaturated esters.

$$R{-}\underset{\underset{R'}{|}}{\overset{\overset{OH}{|}}{C}}{-}CH_2{-}COOEt \xrightarrow{H_2SO_4} R{-}\underset{\underset{R'}{|}}{C}{=}CH{-}COOEt + H_2O$$

The Reformatsky reaction is also known to occur with nitriles to give β-keto esters.

$$\begin{matrix} R{-}C{\equiv}N \\ + \\ BrCH_2COOEt \end{matrix} \xrightarrow[Et_2O]{Zn} R{-}\underset{\underset{\ominus\,NZnBr\,\oplus}{\|}}{C}{-}CH_2COOEt \xrightarrow{H_3\overset{\oplus}{O}} R{-}\underset{\underset{O}{\|}}{C}{-}CH_2COOEt$$

This reaction provides a convenient route for the synthesis of a large number of organic compounds and is particularly versatile for preparing higher carboxylic acids from lower ones. The general outline of one such synthesis is given as under:

$$R-CH_2COOH \xrightarrow[\textbf{HVZ}\text{ reaction}]{Br_2/P} R-CHBr-COOH \xrightarrow[H^{\oplus}]{EtOH} R-CHBr-COOEt$$

$$\downarrow Zn,\ Et_2O;\ R'-CO-R'$$

$$R'-\underset{\underset{R'}{|}}{CH}-\underset{\underset{R}{|}}{CH}-COOEt \xleftarrow{H_2,\ Pt} R'-\overset{R'}{\overset{|}{C}}=\overset{R}{\overset{|}{C}}-COOEt \xleftarrow{H_3\overset{\oplus}{O}} R'-\overset{R'}{\overset{|}{\underset{OH}{\underset{|}{C}}}}-\overset{R}{\overset{|}{CH}}-COOEt$$

Iodomethylenezinc Iodide

An interesting organozinc compound is iodomethylenezinc iodide, ICH_2ZnI. It reacts with olefins to form cyclopropane derivatives.

$$\text{>C=C<} \xrightarrow{ICH_2ZnI} \text{cyclopropane} + ZnI_2$$

Initially it was believed that the reaction proceeds by the decomposition of the organozinc compound to a highly reactive bivalent carbon species—methylene ($:CH_2$)—by an α-elimination process.

$$I-CH_2-Zn-I \xrightarrow{\alpha\text{-elimination}} :CH_2 + ZnI_2$$

$$\text{>C=C<} + :CH_2 \longrightarrow \text{cyclopropane}$$

Recently it has been shown that the decomposition of the reagent and addition to double bond take place in a concerted manner involving a *cyclic transition state*. This reaction is known as *Simmons-Smith reaction* and iodomethylenezinc iodide is called *Simmons-Smith reagent*.

$$\text{alkene} + \text{I—CH}_2\text{—Zn—I} \longrightarrow [\text{transition state}] \longrightarrow \text{cyclopropane} + ZnI_2$$

Since, this reagent behaves like a carbene, it is also known as a ***carbenoid***. A number of carbenoids derived from other metals are also known.

11.3 ORGANOLITHIUM COMPOUNDS

Among the organometallic compounds derived from alkali metals, the organolithium compounds are by far the most important reagents for organic synthesis. They possess a polar-covalent carbon-lithium bond and are represented by the general formula R-Li where R may be an alkyl or aryl group. Organolithium compounds are more reactive than Grignard reagents and undergo some synthetically useful reactions.

11.3.1 Formation

Following methods are generally employed for the preparation of lithiumalkyls:

(*i*) From alky or aryl halides: When an alkyl or aryl halide is treated with lithium metal at low temperature in a hydrocarbon solvent and in an atmosphere of nitrogen or argon, an alkyllithium compound is obtained.

$$\text{R—X} + 2\text{Li} \longrightarrow \text{R—Li} + \text{LiX}$$

$$\text{Ar—X} + 2\text{Li} \longrightarrow \text{Ar—Li} + \text{LiX}$$

(*ii*) By halogen-metal exchange reaction: A number of organolithium derivatives have been obtained by treating organic halides with lithiumalkyls.

$$\text{R—X} + \text{R}'\text{—Li} \longrightarrow \text{R}'\text{—X} + \text{R—Li}$$

Aryllithium derivatives are best prepared by this method which consists in treating the aryl halide with an alkyllithium. Thus phenyllithium is conveniently prepared by treating bromobenzene with *n*-butyllithium

$$C_6H_5Br + nC_4H_9Li \longrightarrow C_6H_5Li + nC_4H_9Br$$

(*iii*) Direct metallation: Hydrocarbons containing relatively more acidic hydrogen atoms on treatment with an organolithium compound produce the corresponding lithium compound.

$$\text{Cyclopentadiene (H, H)} + C_6H_5Li \longrightarrow \text{Cyclopentadienyl (H, Li)} + C_6H_6$$

When a solution of organolithium compound is needed hydrocarbons are generally used as solvents (ethers are attacked by these reagents).

11.3.2 Physical Properties

Organolithium compounds are either liquids or low-melting solids which are readily soluble in non-polar hydrocarbon solvents.

11.3.3 Chemical Reactions

Because of the strong bond polarization and small size of lithium, alkyllithium compound exist as polymeric aggregates of two or more molecules in solution as well as in gas phase.

(*i*) **Reaction with acidic substances:** Like Grignard reagents the strongly basic alkyllithiums react with acidic compounds, such as carboxylic acids, alcohols, water, primary and secondary amines, liberating the hydrocarbon in quantitative yield.

$$\overset{\delta^-}{R}—\overset{\delta^+}{Li} + \overset{\delta^+}{H}—\overset{\delta^-}{A} \longrightarrow R—H + \overset{\oplus\ominus}{LiA}$$

(*ii*) **Reaction with carbonyl compounds: Formation of alcohols:** Reactions of alkyl (or aryl) lithiums with carbonyl compounds are similar to those of the Grignard reagents, *i.e.*, with formaldehyde they yield primary alcohols, with other aldehydes they form secondary alcohols and with ketones they produce tertiary alcohols.

$$R—Li + CH_2{=}O \longrightarrow R—CH_2—OLi \xrightarrow{H_3\overset{\oplus}{O}} \underset{\text{A primary alcohol}}{R—CH_2OH}$$

$$R—Li + R'CH{=}O \longrightarrow \begin{matrix} R \\ R' \end{matrix}\!\!>\!CH—\overset{\ominus\oplus}{OLi} \xrightarrow{H_3\overset{\oplus}{O}} \underset{\text{A secondary alcohol}}{\begin{matrix} R \\ R' \end{matrix}\!\!>\!CH—OH}$$

$$R—Li + \begin{matrix} R' \\ R' \end{matrix}\!\!>\!C{=}O \longrightarrow \begin{matrix} R' \\ R \end{matrix}\!\!>\!\underset{R'}{C}—\overset{\ominus\oplus}{OLi} \xrightarrow{H_3\overset{\oplus}{O}} \underset{\text{A tertiary alcohol}}{\begin{matrix} R' \\ R \end{matrix}\!\!>\!\underset{R'}{C}—OH}$$

The alkyllithiums have an advantage over **Grignard reagents.** While the latter fail to give tertiary alcohols when treated with highly hindered ketones, the former react with such ketones giving corresponding tertiary alcohols.

$$(CH_3)_3C-\overset{O}{\overset{\|}{C}}-C(CH_3)_3 + (CH_3)_3CLi \xrightarrow{195\ K} [(CH_3)_3C]_3C-\overset{\ominus\oplus}{OLi} \xrightarrow{\overset{\oplus}{H_3O}} [(CH_3)_3C]_3C-OH$$

Di-*t*-butyl ketone — *t*-Butyllithium — Tri-*t*-butylcarbinol

Thus, these compounds are less susceptible to steric crowding than Grignard reagents.

***(iii)* Reaction with carbon dioxide: Formation of ketones:** We have seen earlier that when a Grignard reagent is treated with carbon dioxide the reaction stops after one alkyl or aryl group has added to the carbonyl group of carbon dioxide, giving rise to a carboxylate anion. This anion being resonance stabilized is not attacked by the second molecule of Grignard reagent. However, with organolithium compounds which are more reactive, second molecule also attacks the carboxylate anion producing a ketone.

$$R-Li + O=C=O \longrightarrow R-\overset{O}{\overset{\|}{C}}-\overset{\ominus\oplus}{OLi} \xrightarrow{R-Li} R-\underset{R}{\overset{\overset{\ominus\oplus}{OLi}}{C}}-\overset{\ominus\oplus}{OLi} \xrightarrow{\overset{\oplus}{H_3O}} R-\overset{O}{\overset{\|}{C}}-R$$

Treatment of a carboxylic acid also gives a ketone as the main product.

$$\text{trans-}H_5C_6\text{-cyclopropyl-}COOH + 2CH_3Li \longrightarrow \text{trans-}H_5C_6\text{-cyclopropyl-}\overset{O}{\overset{\|}{C}}-CH_3$$

trans-2-Phenylcyclopropanecarboxylic acid — *trans*-2-Phenylcyclopropyl methyl ketone

***(iv)* Reaction with alkenes:** Unlike Grignard reagents, organolithium compounds add to the olefinic double bond to give a lithiumalkyl which again adds to the double bond of second alkene forming the corresponding lithiumalkyl. This process can continue to a certain extent depending upon the relative amount of alkene in the reaction mixture, thus giving rise to alkanes of varying chain length.

$$Li—R + CH_2{=}CH_2 \longrightarrow R—CH_2—CH_2—Li \xrightarrow{CH_2=CH_2} R—(CH_2—CH_2)_2Li \xrightarrow{CH_2=CH_2} R—(CH_2CH_2)_3Li$$

and so on

(*v*) Nucleophilic displacement: Alkyllithium compounds can be used in substitution reactions of alkyl halide. Usually such reactions do not succeed with Grignard reagents. The mechanism of the reaction is quite similar to heterolytic (S_N2) mechanism of Würtz coupling reaction. For instance,

$$R—Li + R'—X \longrightarrow R—R' + LiX$$

Phenyllithium on treatment with 1-chloronaphthalene gives a product containing both α-phenylnaphthalene and β-phenylnaphthalene.

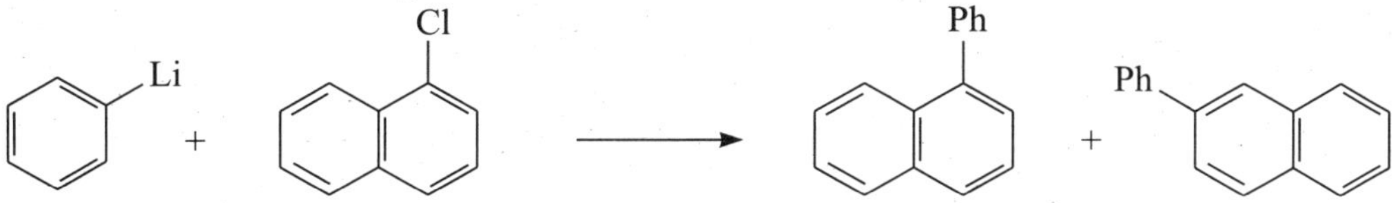

The formation of a mixture of isomeric phenylnaphthalenes indicates that the reaction does not take place *via* the usual mechanistic path of aromatic substitution reaction. It rather takes place *via* an elimination-addition mechanism involving naphthyne (*c.f.* Chapter 10).

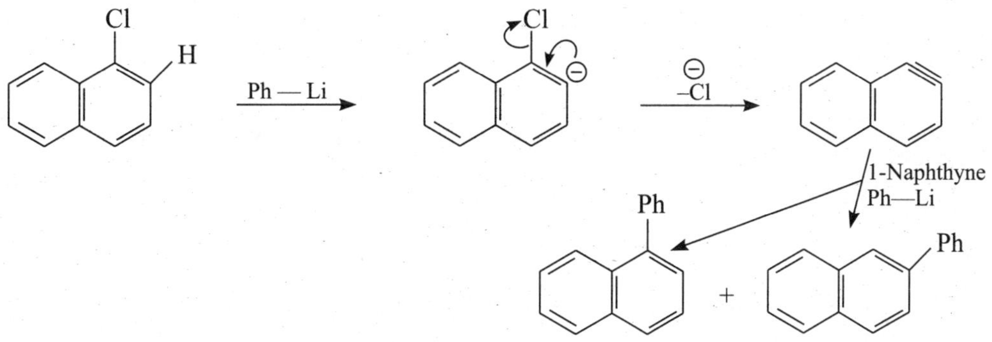

Here phenyllithium plays a dual role. It acts as a strong base to remove the *o*-hydrogen generating the corresponding carbanion, and it also acts as a nucleophile in addition to naphthyne. Cases are, however, known where nucleophilic displacement by addition-elimination mechanism takes place on aromatic systems having electron-withdrawing groups on the ring. Thus, for instance, formation of 2-phenylpyridine by treating pyridine with phenyllithium can be rationalized through this mechanism.

$$\text{Pyridine} \xrightarrow{C_6H_5—Li} \text{N-Li-2-}C_6H_5\text{-1,2-dihydropyridine (H)} \xrightarrow{H_2O} \text{2-}C_6H_5\text{-pyridine}$$

(vi) **Metal-halogen exchange: Electrophilic displacement:** The formation of an organometallic derivative by the treatment of an organic halide with an organometallic compound is known as a metal-halogen exchange reaction.

$$R—M + R'—X \longrightarrow R'—M + RX$$

This is an example of electrophilic substitution. Thus treatment of vinyl bromide with *n*-butyllithium gives rise to vinyllithium.

$$nC_4H_9Li + CH_2{=}CH—Br \longrightarrow nC_4H_9Br + CH_2{=}CH—Li$$

(vii) **Reaction with α, β-unsaturated carbonyl compounds:** In contrast to Grignard reagents, they give rise to products arising predominantly from an initial 1,2-additon on carbonyl function forming mainly the unsaturated alcohol.

$$R—CH{=}CH—COEt \xrightarrow[\text{Hexane}]{R—Li} R—CH{=}CH—C(OH)(R)—CH_2—CH_3$$

11.4 ORGANOLEAD COMPOUNDS

The lead tetraalkyls are the most common organometallic derivatives of the group IV elements. These compounds are largely covalent in character and have tetrahedral shapes as the metal atom uses sp^3-hybrid orbitals for bonding with four substituents. Tetraethyllead (TEL) is very common industrial compound and is used as a good antiknock agent in gasoline in the concentration of 1 to 3 ml. per gallon.

It has been prepared as follows:

(i) **Reaction of ethyl chloride with lead alloy**

$$4C_2H_5Cl + 4PbNa \longrightarrow (C_2H_5)_4Pb + 4NaCl + 3Pb$$

(ii) **Metal-metal exchange reaction:** Lead chloride on treatment with ethylmagnesium bromide forms TEL.

$$4C_2H_5MgBr + 2PbCl_2 \longrightarrow (C_2H_5)_4Pb + Pb + 4MgBrCl$$

Tetraethyllead is a colourless liquid having strong smell. It is covalent compound as evident from the fact that it is insoluble in water but soluble in organic solvents.

11.5 ORGANOCADMIUM COMPOUNDS

Organocadmium compounds are quite similar to organozinc compounds. These compounds are best prepared from Grignard reagents by the action of cadmium chloride on them.

$$2RMgX + CdCl_2 \longrightarrow R_2Cd + 2MgClX$$

These reagents have linear geometry, as the metal uses *sp*-hybridized orbitals for bonding. They are volatile liquids. In chemical reactions they are similar to dialkylzincs but are less reactive. When treated with acid chlorides they from the corresponding ketones.

$$R_2Cd + 2R'-\overset{\overset{\displaystyle O}{\|}}{C}-Cl \longrightarrow 2R-\overset{\overset{\displaystyle O}{\|}}{C}-R' + CdCl_2$$

It may be mentioned here that the cadmium dialkyls do not react with carbonyl compounds. Hence they are useful for the selective synthesis of ketones.

A BRIEF REVIEW

- Organic compounds in which metal atom is linked directly to a carbon atom are called *organometallic compounds*. For example, Grignard reagents (RMgX), C_6H_5Li, $(CH_3)_2Cd$, $(C_2H_5)_2Zn$, etc.
- **Grignard reagents** are prepared by treating an alkyl/aryl halide with dry magnesium turnings in dry ether or THF in an inert atmosphere of N_2 or Ar. The Grignard reagents of less reactive aryl and vinyl halides are often prepared in THF.
- **Alkyl or aryllithium** compounds are prepared by reactions of a suitable alkyl/aryl halide in hydrocarbon solvents such as hexane, benzene, etc. but not in ethers because these reagents react with them.
- Grignard reagents because of the large electronegativity difference between carbon and magnesium (2.5 – 1.2 = 1.3) are highly polar and thus, the alkyl group in Grignard reagents can act as carbanion or a nucleophile. In other words, Grignard reagents undergo *nucleophilic substitution and nucleophilic addition reactions.*
- Grignard reagents and organolithium compounds react with compounds containing active hydrogen such as H_2O, ROH, H_2S, RSH, NH_3, RNH_2, R_2NH and terminal alkynes ($HC\equiv CH$, $RC\equiv CH$) etc. to form alkanes/arenes corresponding to the alkyl or the aryl group of the Grignard reagent.
- Grignard reagents and organolithium compounds react with reactive halids such as *t*-butyl chloride, benzyl chloride to form alkanes.
- Alkyl/aryllithium are *even more polar* (electronegativity difference: 2.5 – 0.9 = 1.6) hence are *more reactive than Grignard reagents.*
- Grignard reagents and organolithium reagents react with reactive alkenyl halides such as allyl halides to form higher alkenes.

- Termınal alkynes react with Grignard reagents/organolithium compounds to from alkynyl Grignard reagents which upon subsequent treatment with alkyl halides give higher alkynes.
- Grignard reagents and organolithium compounds react with aldehydes and ketones to form alcohols. **With HCHO**, 1° alcohols are produced, **with other aldehydes** 2° alcohols are produced but **with ketones** 3° alcohols are produced.
  - (*i*) **Primary alcohols** can also be prepared by bubbling dry O_2 through the solution of an alkyl Grignard reagent or organolithium compound followed by acid hydrolysis of the resulting addition product. Similar treatment with aryl Grignard reagents gives **phenols.**
  - (*ii*) Treatment of alkyl/aryl Grignard reagent or organolithium compound with ethylene oxide gives 1° *alcohols containing two carbon atoms more than the corresponding alkyl/aryl halide.*
  - (*iii*) **Secondary alcohols** can also be obtained by addition of two moles of a Grignard reagent or organolithium compound to *formic ester while with other esters* **3° alcohols** are produced.
- Grignard reagents and organolithium comounds react with *HCN or ethyl orthoformate* to form **aldehydes**. However, with *alkyl/aryl nitriles or ortho esters of any acid other than formic acid*, **ketones** are produced.
- Alkyl Grignard reagents react with S to form an addition product which upon hydrolysis gives thioalcohols.
- **Grignard reagents** react with dry ice to form **carboxylic acids** but **organolithium compounds** being more nucleophilic react further to form **ketones**.
- Grignard reagents react with α,β-unsaturated aldehydes and ketones to form both 1,2-and 1,4-addition products. The ratio of the two products formed, however, depends upon steric hindrance. *Whereas substitution at the carbonyl group favours,* 1,2-*addition, substitution at the* β-*position of the double bond favours* 1, 4-*addition.* The proportion of 1,4-addition product can, however, be increased by carrying out the reaction in presence of CuCl or $Cu(OAc)_2$. This is called **conjugate addition**.

  In contrast, organolithium reagents being more nucleophilic add to α, β-unsaturated compounds to form 1, 2-addition products.
- Grignard reagents do not add to sterically hindered ketones. *If, however, the Grignard reagent has a branching at the* α-*position*, sometimes *reduction of the keto group by the Grignard reagents occurs.* For example, addition of isopropyl magnesium bromide to *di*-isopropyl ketones, gives *di*-isopropyl carbinol and propylene but not the expected sterically hindered 3° carbinol. However *organolithium compounds being more nucleophilic add to sterically hindered ketones* to form 3° **alcohols**.
- **Dialkyl organozinc compounds** are obtained by treating Zn—Cu couple with alkyl iodides followed by distillation of the resulting alkylzinc iodide in an atmosphere of CO_2.These can also be prepared by reacting trialkylaluminium with zinc chloride.

- Organozinc compounds being less polar (electronegativity difference, 2.5 – 1.6 = 0.9) than Grignard reagents and organolithium compounds are less reactive as compared to these reagents. Even then they react with H_2O producing hydrocarbons corresponding to the alkyl or the aryl group. They also react with more reactive *tert*-alkyl halides to form alkenes containing a quaternary carbon atom.
- Being less reactive dialkylzinc compounds react with acid chlorides to form ketones in good yield. Under these conditions, Grignard reagents and organolithium compounds give 3° alcohols.
- α-Haloesters such as ethyl bromoacetate react with zinc to form first *organozinc compounds* which add to the carbonyl group of aldehydes and ketones in a manner similar to the Grignard reagents to from β-*hydroxy esters*. This reaction is called **Reformatsky reaction** and is widely used in synthesis of β-hyroxy esters/acids.
- **Iodomethylenezinc iodide** is an important organozinc compound. It is obtained by reacting Zn—Cu couple with CH_2I_2. This reagent which is a source of carbenoid (carbene like species) adds to olefins to give *cyclopropanes*. This reaction is called **Simmons-Smith reaction**.

SOLVED PROBLEMS

Q. 1. Give the product obtained by the reaction of Mg/ether with

(*i*) $BrCH_2CH_2Br$ (*ii*) $BrCH_2CH_2CH_2Br$ and (*iii*) $BrCH_2CH_2CH_2CH_2Br$

Ans. (*i*) Initially formed Grignard reagent eliminates $MgBr_2$ to form ethylene.

$$Br—CH_2—CH_2—Br + Mg \xrightarrow{\text{Ether}} [Br—CH_2—CH_2—MgBr] \longrightarrow CH_2{=}CH_2 + MgBr_2$$

(*ii*) With 1,3-dibromopropane, similar reaction occurs to form cyclopropane.

$$BrCH_2CH_2CH_2Br \xrightarrow{\text{Mg/ether}} \left[[BrMg—CH_2—CH_2—CH_2—Br] \right] \longrightarrow \underset{\text{Cyclopropane}}{CH_2—CH_2—CH_2 \text{ (ring)}} + MgBr_2$$

(*iii*) Since four-membered rings are most difficult to form, only a Grignard reagent is produced from 1,4-dibromobutane.

$$\underset{\text{1,4-Dibromobutane}}{Br—CH_2CH_2CH_2CH_2—Br} + 2Mg \xrightarrow{\text{Dry ether}} \underset{\text{Grignard reagent}}{BrMg—CH_2CH_2CH_2CH_2—MgBr_2}$$

Q. 2. The solution obtained by treating CH_3—CH = CH—CH_2Cl with magnesium was treated with acetone. After usual work up, it was found that the product contained a mixture of two alkanols. Identify the alkanols and suggest a suitable mechanism for their formation.

Ans.

$$CH_3CH{=}CHCH_2—Cl + Mg \longrightarrow [CH_3—CH{=}CH—\dot{C}H_2 \longleftrightarrow CH_3—\dot{C}H—CH{=}CH_2] + \dot{M}gCl$$

$$\longrightarrow CH_3—CH{=}CH—CH_2MgCl + CH_3—\underset{\underset{MgCl}{|}}{CH}—CH{=}CH_2 \xrightarrow[\text{(ii) } H^+/H_2O]{\text{(i) } CH_3COCH_3}$$

$$CH_3—CH{=}CH—CH_2—\overset{\overset{OH}{|}}{\underset{\underset{CH_3}{|}}{C}}—CH_3 \quad + \quad CH_3—\overset{\overset{CH_3}{|}}{\underset{\underset{HO}{|}}{C}}—\underset{\underset{CH_3}{|}}{CH}—CH{=}CH_2$$

2-Methylhex-4-en-2-ol 2, 3-Dimethylhex-4-en-2-ol

Q. 3. Suggest the reactants for preparing following alcohols from CH_3MgI:

(*i*) CH_3CH_2OH (*ii*) $CH_3CH_2CH_2OH$

Ans. (*i*) HCHO (*ii*) ethylene oxide.

Q. 4. Commencing with *tert*-butyl chloride, how will you synthesize $(CH_3)_3CD$?

Ans.

$$CH_3—\overset{\overset{CH_3}{|}}{\underset{\underset{CH_3}{|}}{C}}—Cl \xrightarrow{\text{Mg/ether}} CH_3—\overset{\overset{CH_3}{|}}{\underset{\underset{CH_3}{|}}{C}}—MgCl \xrightarrow[-Mg(OD)Cl]{D_2O} CH_3—\overset{\overset{CH_3}{|}}{\underset{\underset{CH_3}{|}}{C}}—D$$

tert-Butyl chloride 2-Deutero-2-methylpropane

Q. 5. Write the product formed when methylmagnesium iodide is reacted with (*i*) CO_2 (*ii*) SO_2 and the addition product thus formed is hydrolysed.

Ans.

$$CH_3MgBr \xrightarrow[(ii)\ H^+/H_2O]{(i)\ CO_2} CH_3COOH$$

$$CH_3MgBr \xrightarrow[(ii)\ H^+/H_2O]{(i)\ SO_2} CH_3—SO—OH$$

Methanesulphinic acid

Q. 6. Outline a reaction sequence depicting the use of Grignard reagent to convert RX into RCH_2CH_2OH.

Ans. $$R—X \xrightarrow{\text{Mg/ether}} R—MgX \xrightarrow[\text{(ii) } H^+/H_2O]{\text{(i) Ethylene oxide}} RCH_2CH_2OH$$

Q. 7. What happens when CH_3MgBr is added to crotonaldehyde? Write the equation for the reaction.

Ans. $$\underset{\text{Crotonaldehyde}}{CH_3—CH{=}CH—\overset{O}{\overset{\|}{C}}H} \xrightarrow[\text{(ii) } H^+/H_2O]{\text{(i) } CH_3MgBr} \underset{\text{Pent-3-ene-2-ol}}{CH_3—CH{=}CH—CH—\overset{OH}{\overset{|}{C}}H—CH_3}$$

Q. 8. What happens when phenylmagnesium bromide is treated with ethylene oxide followed by acid hydrolysis?

Ans. $$C_6H_5MgBr + H_2C\overset{O}{—}CH_2 \longrightarrow C_6H_5CH_2CH_2OMgBr \xrightarrow{\overset{\oplus}{H_3O}} C_6H_5CH_2CH_2OH$$

2-Phenyl ethanol

Q. 9. Suggest reagents for the following conversion:

Ans. $$C_6H_{11}—Br \longrightarrow C_6H_{11}—D$$

$$C_6H_{11}—Br + Mg \xrightarrow{\text{ether}} C_6H_{11}—MgBr \xrightarrow{D_2O} C_6H_{11}—D$$

UNSOLVED PROBLEMS

1. What are Grignard reagents? Describe the method of preparing ethylmagnesium bromide. Sketch a mechanistic scheme depicting addition of a Grignard reagent to a carbonyl group.
2. Suggest a plausible mechanism for the formation of a Grignard reagent from an alkyl halide and magnesium.
3. Discuss the structure of alkyl magnesium halide.
4. What happens when

(*i*) Di-*tert*-butyl ketone is made to react with CH_3MgBr ?

(*ii*) Methyl isopropyl ketone is made to react with *tert*-butyl magnesium halide ?

5. Discuss the limitations in use of Grignard synthesis.

6. Predict the products of the following reactions:

(*i*) $C_6H_5MgBr \xrightarrow[(ii)\ H_3\overset{\oplus}{O}]{(i)\ CH_2O}$

(*ii*) $C_3H_7MgBr \xrightarrow{SnCl_4}$

(*iii*) $CH_3MgBr \xrightarrow[(ii)\ H^+/H_2O]{(i)\ HC(OC_2H_5)_3}$

(*iv*) $CH_3MgI \xrightarrow[(ii)\ H^+/H_2O]{(i)\ CH_3CN}$

(*v*) $C_6H_5MgBr \xrightarrow{PCl_3}$

(*vi*) $CH_3MgBr \xrightarrow{C_6H_5OH}$

(*vii*) $CH_3MgBr \xrightarrow[(ii)\ H^+/H_2O]{(i)\ CO_2}$

(*viii*) $CH_3MgCl \xrightarrow{PbCl_2}$

7. What happens when ethylmagnesium bromide and isopropylmagnesium bromide are treated separately with CO_2 and the adduct formed is decomposed with dil. HCl. Can the compounds obtained above be obtained from ethyl bromide and isopropyl bromide without using CO_2? If so, explain how? [**Hint**. By hydrolysis of cyanides]

8. Using a suitable Grignard reagent, sketch the following transformations:

(*i*) 1-Bromocyclohexane ⟶ 1-Cyclohexyl-1-ethanol.

(*ii*) $C_6H_5C \equiv CH \longrightarrow C_6H_5—C \equiv C—CH_2OH$

(*iii*) $CH_3CH_2CH_2Br \longrightarrow CH_3CH_2CH_2CH_2OH$

(*iv*) $CH_3C \equiv CH \longrightarrow (CH_3)_3COH$

(*v*) $C_6H_5CH_3 \longrightarrow C_6H_5CH_2—C(OH)(CH_3)_2$

(*vi*) $CH_3CH = CH_2 \longrightarrow DCH_2CH = CH_2$

(*vii*) $(C_6H_5)_2CH_2 \longrightarrow (C_6H_5)_2CD_2$

9. Addition of Grignard reagents to carbon-carbon double bonds may result in the formation of compounds of the type $CH_2 = CHX$ where X is an electron-withdrawing group such as carbonyl, cyano or nitro group. How do you account for this observation?

10. Discuss briefly 1,2-and 1,4-addition of Grignard reagents to α, β-unsaturated ketones. How does steric hindrance affect the course of these addition reactions ?

11. Explain why treatment of di-isopropyl ketone with isopropylmagnesium bromide gives di-isopropylcarbinol instead of the expected tri-isopropylcarbinol ?

12. Arrange the following in order of increasing ease of formation of Grignard reagents and assign reason for the same:

(*i*) C_2H_5Cl (*ii*) CH_3Cl (*iii*) $CH_3CH_2CH_2Cl$

13. Give equations for the preparation of the following form CH_3MgBr:

(*i*) CH_4 (*ii*) CH_3CHO (*iii*) CH_3COCH_3

(*iv*) CH_3COOH (*v*) $(CH_3)_2CHOH$ (*vi*) $CH_3OCH_2CH_3$

14. How are the following compounds prepared from suitable Grignard reagents ?

(*i*) Methyl-*n*-propyl ether (*ii*) Acetaldehyde (*iii*) Acetone

(*iv*) Diethylcarbinol (*v*) Tetraethyllead (*vi*) Ethyl thioalcohol.

15. How will you prepare the following using suitable Grignard reagents? Give chemical equations ?

(*i*) 2-Butanone (*ii*) Ethyldimethylcarbinol (*iii*) Ethanoic acid.

16. Give one method of preparation of $(C_2H_5)_2Zn$. How can it be used to prepare 2-butanone? What would have happened if $(C_2H_5)_2Zn$ is replaced by C_2H_5MgI in the above reaction?

17. Write the organozinc compound formed when zinc reacts with ethyl α-bromopropionate.

18. How will you prepare methyllithium in the laboratory? How can it be used to synthesise the following:

(*i*) Acetophenone (*ii*) *tert*-Butyl alcohol (*iii*) Acetaldehyde?

19. Explain why?

(*i*) Methylmagnesium bromide reacts with CO_2 to form acetic acid while methyllithium bromide under similar conditions acetone.

(*ii*) Grignard reagents react with α, β-unsaturated compounds to give 1,4-addition products but organolithium compounds give 1,2-addition products.

(*iii*) Alkyllithiums add to sterically hindered ketones but Grignard reagents do not.

20. A carbanion is generated on the middle carbon of 1,4-pentadiene by reaction with an organolithium reagent. This upon treatment with ethylene oxide produced a mixture of two compounds. Write the structures of these products.

21. Explain with the help of equations what happens when:

(*i*) n-Butyllithium is treated with bromobenzene ?

(*ii*) An alkyllithium is treated with an olefin ?

(*iii*) Methyllithium is treated with carbon dioxide ?

22. Write short notes on:

(*i*) Reformatsky reaction (*ii*) Simmons-Smith reaction.

23. Offer suitable explanation why treatment of 1-chloronaphthalene with phenyllithium yields a mixture of both α- and β-chloronaphthalenes.

Index

C

D

T

U

V

W

Z